THE GALILEO MISSION

THE GALILEO MISSION

Edited by

C.T. RUSSELL
Institute of Geophysics and Planetary Physics,
University of California at Los Angeles.

Reprinted from Space Science Reviews, Volume 60, Nos. 1–4, 1992

KLUWER ACADEMIC PUBLISHERS
DORDRECHT / BOSTON / LONDON

Library of Congress Cataloging-in-Publication Data

The Galileo mission / edited by C.T. Russell.
p. cm.
ISBN 0-7923-1719-X (hb : acid-free paper)
1. Galileo Project. I. Russell, C. T.
QB661.G37 1992
629.43'545--dc20 92-9943

ISBN 0-7923-1719-X

Published by Kluwer Academic Publishers,
P.O. Box 17, 3300 AA Dordrecht, The Netherlands.

Kluwer Academic Publishers incorporates
the publishing programmes of
D. Reidel, Martinus Nijhoff, Dr W. Junk and MTP Press.

Sold and distributed in the U.S.A. and Canada
by Kluwer Academic Publishers,
101 Philip Drive, Norwell, MA 02061, U.S.A.

In all other countries, sold and distributed
by Kluwer Academic Publishers Group,
P.O. Box 322, 3300 AH Dordrecht, The Netherlands.

Printed on acid-free paper

Printed in Belgium

TABLE OF CONTENTS

FOREWORD

Jupiter, as one of the brightest objects of the night sky, has always held a fascination for mankind. That fascination intensified with the invention of the telescope. The discovery by Galileo of a small 'planetary' system in orbit about Jupiter revolutionized the world's view of its place in the Universe. No longer did the Universe revolve around the Earth. The Earth became but one of many worlds. Telescopic observations of the Jovian system also made a major contribution to modern physics. The timing of eclipses of the Galilean moons led to the first measurement of the velocity of light. The telescope also gives atmospheric scientists an opportunity to study a 4 century old enigma, the Great Red Spot. Is it a giant storm that has been raging for over 4 centuries or is it related to some structure in the interior of Jupiter?

The fascination deepened once again with the invention of another type of telescope, the radio telescope. In 1955 B. F. Burke and K. L. Franklin discovered that Jupiter emitted intense radio waves. This led almost immediately to the conclusion that Jupiter had a strong magnetic field and an intense radiation belt. Jupiter soon became the target of our fledgling planetary program, with visits by Pioneer 10 in 1973, Pioneer 11 in 1974, and Voyager 1 and 2 in 1979. These four space Probes provided brief glimpses into the magnetosphere and the atmosphere of the planet and returned pictures of the Galilean satellites, worlds quite alien to our own.

The momentum had been building through these years for the next step in the exploration of the Jovian system, an orbiter. When this Jovian orbiter was approved by NASA, it was only fitting that it be named for the first explorer of the Jovian system, Galileo. The 1980's was a decade of planning, designing, building, and rebuilding as the Galileo mission got off to a slow start because of the ever changing availability of launchers in NASA's fleet during these years. Eventually in October 1989, Galileo was launched and sent on its way via Venus and the Earth (twice) on its voyage to Jupiter with arrival in December 1995.

The articles that follow in this volume represent our attempt to document the Galileo mission and make it accessible to the broader scientific community. The volume begins with a mission overview by the project scientist, Torrence Johnson, and his colleagues which details the science objectives of the mission and describes how these objectives will be addressed. This article is followed by one by L. A. D'Amario and colleagues on the design of the spacecraft trajectory, a key element of the mission planning process that allows the often conflicting objectives of the mission to be met. Following these articles are descriptions of investigations by the various investigator groups associated with the mission. Most of these involve dedicated hardware but some, such as the article by Interdisciplinary Scientist J. Pollack and colleagues and those by the Radio Science and the Gravitation and Celestial Mechanics teams, discuss the use of spacecraft telemetry signals to further the scientific objectives of the missions.

We have divided the mission into three sections. The Probe articles are presented first, as appropriate for these investigations that will provide much of the initial science return

Space Science Reviews **60:** 1–2, 1992.

in the mission as the Probe descends through the atmosphere on the day of orbit insertion. Then we present the 'magnetospheric' instruments which provide local measurements of the Jovian system. Finally we present the remote sensing instruments and the radio science investigations.

It is hoped that these articles will provide interested planetary and space scientists some insight into what Galileo is expected to achieve and how the requisite measurements will be made. The compilation of this volume is due to the efforts of many individuals, especially the referees and the authors, who worked together to develop readable and complete descriptions of the investigations. We also wish to remember Clayne Yeates of the Galileo Science Office who assisted in the very early phases of this project but whose untimely death did not allow him to see its completion.

C. T. RUSSELL

SPACE SCIENCE REVIEWS VOLUME ON GALILEO MISSION OVERVIEW

T. V. JOHNSON, C. M. YEATES

Jet Propulsion Laboratory, California Institute of Technology, Pasadena, CA 91109, U.S.A.

and

R. YOUNG

NASA Ames Research Center, Moffett Field, CA 94035, U.S.A.

Abstract. The Galileo Mission is an extremely complex undertaking. This paper provides a brief historical overview, a discussion of broad scientific objectives, and a description of the spacecraft and trajectory characteristics.

1. Introduction

The Galileo Mission evolved from studies in the early- and mid-seventies designed to develop outer solar system mission concepts to follow the Pioneer and Voyager flybys. As the nearest and largest of the giant planets of the outer solar system, Jupiter was regarded as the logical first target of the next stage of exploratory missions. The mission concept studies included a number of different options, ranging from entry Probes with relatively simple carrier spacecraft to Voyager-like orbiting spacecraft. The mission finally recommended to NASA by a working group chaired by J. Van Allen combined the ideas of an entry Probe and a sophisticated Orbiter to study the planet, magnetosphere and satellites in a mission initially dubed Jupiter Orbiter Probe (JOP). The JOP mission was approved by Congress for start in October of 1977 and was shortly thereafter renamed Project Galileo. Galileo will provide several unique additions to the planetary exploration program: the first *in-situ* sampling of the atmosphere of one of the outer planets; the first extended study of an outer planet system; and, because of Galileo's unusual trajectory to Jupiter, the first close up reconnaissance of an asteroid.

Originally scheduled for launch in January 1982, Galileo was delayed several times by development and scheduling problems in the Space Transportation System (Shuttle); and associated upper stage launch vehicles. Following the tragic Challenger accident in 1986 and the subsequent cancellation of the Centaur upper stage vehicle for use with the Shuttle, Galileo was rescheduled for an October or November 1989 launch using a solid fuel Inertial Upper Stage (IUS) rocket for an upper stage. This required the use of a gravity assist trajectory to obtain the launch energy required to reach Jupiter (called the Venus–Earth–Earth–Gravity–Assist or VEEGA trajectory). On October 18, 1989 STS-34 Atlantis, with the Galileo spacecraft and IUS in its cargo bay, was successfully launched from the Kennedy Space Center with a crew of five astronauts commanded by Mission Commander D. Williams. Following deployment from Atlantis'

Space Science Reviews **60:** 3–21, 1992.

cargo bay, Galileo was placed on an interplanetary trajectory by the IUS, heading initially toward a flyby of Venus. The Venus flyby was accomplished on February 9, 1990 and the first Earth flyby on December 8, 1990. Figure 1 shows the VEEGA trajectory, and Table I gives the major mission milestones.

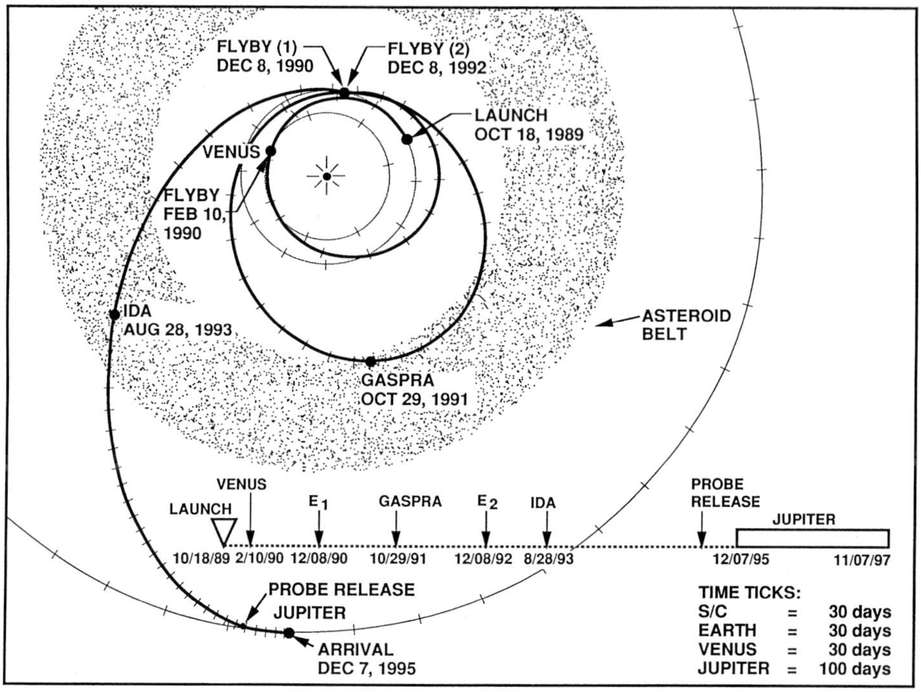

Fig. 1. Galileo Venus–Earth–Earth Gravity Assist (VEEGA) trajectory: the path of the Galileo spacecraft is shown along with key events and dates. In the lower portion of the figure a simplified time line is provided. The spacecraft (s/c) position along the trajectory is indicated by ticks every 30 days.

TABLE I

Major mission events

Year	Months	Event
1989	October	Launch
1990	February	Venus encounter
	December	Earth encounter
1991	October	Gaspra encounter
1992	December	Earth encounter
1993	August	Ida encounter
1995	August	Probe release
	December	Io encounter Probe/relay Jupiter orbit insertion
1996	July	First Ganymede encounter
	October	Second Ganymede encounter
1997	October	Completion of 10 targeted encounter tour and tail petal

Galileo is an international effort. Under an agreement with NASA, Germany provided the retro propulsion module (RPM), consisting of a 400 N main engine for large maneuvers (including orbit insertion) and a system of small 10 N thrusters for attitude control. Germany also provided two major instruments, significant subsystems for several other instruments and has selected investigators on several teams, including the radio and imaging system facility teams. In addition, Germany is providing some tracking support during the cruise phase of the mission. Investigators from several other countries are also included on the various teams, including France, the U.K., Canada, and Sweden.

2. Objectives

Galileo is not a 'Jupiter mission' in the traditional sense of a focused space mission to a given target. It is rather an integrated project addressing multidisciplinary objectives concerning the entire Jovian system. The scientific objectives for the Galileo Mission derive from this desire to understand the Jovian system as a whole. The system is important both for the insights it gives us about conditions in the early solar nebula, four and a half billion years ago, and for the understanding it helps provide concerning the bewildering array of processes and phenomena which have affected the evolution of the planets and which control their environments and futures.

The Jovian system is particularly interesting for many reasons. First, Jupiter itself is the largest planet in the solar system (it contains about 75% of the total nonsolar matter in the solar system). The planet is essentially a transitional object between terrestrial planets and stars, being composed primarily of hydrogen and helium in approximately solar proportions, with an internal heat source from ancient accretionary heating and gravitational collapse. Its composition may be little altered from that of the original solar nebula from which it formed. Not only are composition and internal heat effects of prime scientific interest, but the dynamic meteorology of Jupiter's atmosphere exhibits many intriguing phenomena not present in atmospheres of the terrestrial planets. Some key unresolved questions with regard to the planet which will be addressed by Galileo include; what is the precise ratio of hydrogen to helium and what does it imply for the planet's evolution; how does the internal heat affect the outer regions of the planet; are there water and ammonia hydrosulfide cloud decks below the ammonia cloud deck observed in all the images; what is the nature and origin of the banded east–west wind system observed for centuries and how deep does it extend into the planet; what drives motions in the great red spot?

The second principal reason for interest in the Jovian system is the regular satellite and ring system. From the time of its discovery by Galileo, the Jupiter system has been regarded as a 'mini-solar system', with many features in common with the larger system, including a clear change in the composition of the satellites with distance from the primary. During the early accretionary phases of Jupiter's evolution, Jupiter is calculated to have emitted about 10 million times more energy than it presently does, thereby leading to the observed compositional gradient in the satellites. It is known that current interactions among the satellites and the planet are important and affect observed features of the satellites, such as causing volcanoes on Io, the innermost Galilean satellite. Further, each of the planet-sized Galilean satellites merits serious study in its own right, since each is a unique and fascinating object. Io is the most volcanically active body in the solar system, Europa and Ganymede are apparently tectonically active bodies possessing 100 km or thicker water ice crusts, and Callisto shows little tectonic activity but preserves a record of ancient bombardment by large impacting bodies in its frozen ice crust. Galileo will enormously increase our knowledge of the Galilean satellites by providing extremely high resolution imaging and spectral mapping during a series of very close encounters over the nearly 2-year lifetime of the Orbiter mission, addressing questions such as: what are the important tectonic processes on Europa and Ganymede, how thick are their ice crusts and is there a liquid water mantle on Europa, how do the volcanoes of Io evolve and what is their composition? In addition to the satellites, a detailed mapping by Galileo of the ring system discovered by the Voyager flybys should provide clues as to its origin and age, two presently unknown fundamental quantities. Are there processes currently active which deplete and replenish the ring material?

The magnetosphere of Jupiter is the largest of the planetary magnetospheres. Its great expanse could envelope the Sun and much of its corona. Could the magnetosphere be seen from Earth, it would appear larger than the Earth's moon in the sky. Its size, the rapid rotation of the planet and the volcanoes of Io conspire to produce the richest array of magnetospheric phenomena in the heliosphere. The volcanic gases in the tenuous atmosphere of Io are bombarded by the intense particle fluxes of the radiation belts of Jupiter. The ions released in this sputtering process encircle Jupiter in a doughnut shaped Io torus which in turn supplies mass to the entire magnetosphere. The centrifugal force associated with this spinning plasma stretches the magnetosphere into a disk quite unlike that of any other planet.

Galileo will for the first time provide extended observations of this dynamic magnetospheric system. It will probe the configuration of the Jovian magnetodisk and its tilted warped current sheet. It will map the spatial and energy distributions of the trapped charged particles in Jupiter's radiation belts. It will measure the density, composition, and configuration of the Io torus. It will observe the magnetic pulsations, plasma waves and radio waves which fill the Jovian magnetosphere and alter the motion of its particle populations. It will probe the workings of the Jovian magnetotail, a structure that is thought to extend as far as the orbit of Saturn.

The initial flyby reconnaissances by the Pioneer and Voyager spacecraft discovered many important features of the system and provided our first close-up look at Jupiter and its sateliites. Galileo will be remote also, except for the Probe. However, the best Voyager images of the satellites, for instance, were only as good as pre-space age telescopic images of the Earth's moon. In addition, these missions spent a very limited time in the near vicinity of Jupiter (only a few days within the magnetosphere for each spacecraft). Thus, they provided only a brief 'snapshot' of the many dynamic processes operating in the system, such as Io's changing volcanic activity, atmospheric dynamics, and magnetospheric processes. The Galileo Mission is designed to take the next major step in the exploration of this system, combining *in-situ* atmospheric Probe measurements with an Orbiter which will provide a long-term magnetospheric survey, remote sensing of the atmosphere, and extremely close passes of the major satellites.

3. Scientific Objectives

The specific objectives of the Galileo Mission at Jupiter are given in Table II. Other papers in this issue detail the various instruments and their measurement goals and objectives. Here we will provide only a brief discussion of the rationale for the major objectives listed in Table II.

TABLE II

Science objectives of the Galileo Mission

Atmosphere

- Determine the chemical composition
- Determine the structure to a depth of at least 10 bars
- Determine the nature of the cloud particles and location and structure of cloud layers
- Determine the radiative heat balance
- Investigate the circulation and dynamics
- Investigate the upper atmosphere and ionosphere

Satellites

- Characterize the morphology, geology, and physical state of the surfaces
- Investigate the surface mineralogy and surface distribution of minerals
- Determine the gravitational and magnetic fields and dynamic properties
- Study the atmospheres, ionospheres, and extended gas clouds
- Study the magnetospheric interactions of the satellites

Magnetosphere

- Characterize the energy spectra, composition, and angular distribution of energetic particles throughout the magnetosphere to 150 R_J
- Characterize the vector magnetic fields throughout the magnetosphere to 150 R_J
- Characterize the plasma energy spectra, composition, and angular distribution throughout the magnetosphere, including plasma wave phenomena, to 150 R_J
- Investigate satellite-magnetosphere interactions

3.1. Atmosphere

The atmospheric objectives are focused on making major advances in our understanding of Jupiter over what we have been able to discern from ground-based and previous spacecraft studies. The chemical composition is obviously extremely important, both from the point of view of the history of the planet and current processes. While the general composition of the Jovian atmosphere has been known for some time (hydrogen, helium, and small amounts of methane and ammonia), the precise abundances even of major constituents such as helium are uncertain by significant amounts and many species, such as noble gases and most important isotope ratios, cannot be detected or measured accurately using only remote techniques. Other constituents, such as water, can be detected, but the applicability of the measurements to global average abundances is hotly debated. Among the key issues concerning composition are abundance of heavier elements, such as oxygen and carbon, compared to 'solar' abundances, the abundances of noble gases, which provide clues to the original solar nebular values and to the degree of modification and evolution of Jupiter's atmosphere since formation. The atmospheric entry Probe is the single most important experiment set for answering many of these questions, providing as it does direct sampling of the atmospheric gases as a function of depth. However, it will provide data for only one place and time in the atmosphere, and interpretation of Probe data will rely heavily on remote measurements from the Orbiter instruments to provide the context and relation to previous remote sensing data.

Atmospheric structure (that is, temperature and pressure as a function of altitude) is a vital piece of information for the study of any atmosphere, being a critical constraint on models of composition and dynamics. Structure can be inferred from a variety of remote measurements, notably radio occultations and infrared sounding measurements. Again, the Galileo entry Probe will provide definitive, *in-situ* measurements for the Probe entry location, without the assumptions and ambiguities involved in interpretation of remote observations. The Probe measurements will thus provide ground-truth for this critical data type, allowing occultation and other remote observations at other sites to be more accurately interpreted and extending our knowledge of the structure to deeper levels than obtainable with other techniques.

Ever since telescopic observations of Jupiter began in the seventeenth century there has been great interest in the nature of the clouds which cover the visible disk of the planet. The known atmospheric constituents provide important clues to the candidate condensable species; it is generally believed for instance that the high white clouds are made of ammonia ice particles, which should be the only cloud forming material at those temperatures and pressures. The darker clouds with yellowish and brownish hues (the Great Red Spot is not really red but has somewhat more absorption at violet and blue wavelengths than the other dark regions) are at lower levels in the atmosphere. It has been suggested that some of these clouds are composed of compounds such as ammonia hydrous sulfide (NH_4SH), while water should be capable of making a cloud deck below this (and hidden by the overlying clouds) if present in sufficient quantities. Most of the proposed cloud-forming compounds should be white in their pure forms, so the nature of the coloring material in the clouds is also of great interest. Among the suggestions for the colored impurities in recent literature are elemental sulfur or sulfur compounds, phosphorus, and small quantities of organic material produced by energetic processes such as lightning in the Jovian atmosphere. The entry Probe will provide key data for resolving many of our questions about the clouds, specifically the exact composition as a function of temperature, pressure and depth from the mass spectrometer and the precise location of cloud layers from the nephelometer. Other measurements such as the amount of sunlight and upwelling infrared radiation at a given depth (from the net flux radiometer) and the characterization of lightning activity by the LRD experiment will also be important in understanding the clouds.

Jupiter is the first planet that was discovered to have a major source of internal heat (all planets including the Earth have some degree of energy supplied by the decay of radioactive elements, but this is generally orders of magnitude smaller than the energy received from the Sun), emitting at long infrared wavelengths more than twice as much heat as it receives from the Sun. The source of this heat is believed to be 'fossil' energy from the initial formation of Jupiter four and a half billion years ago, emanating from the deep interior of the planet. It is a critical factor in determining the energy drives for the interior and perhaps the atmospheric dynamics manifested in cloud motions at upper levels. An accurate measure of the total energy budget of the planet is critical to many theoretical models explaining atmospheric processes and also provides constraints for formation theories. The measurements required include precise observations of the ingoing and outgoing radiation over all angles and relevant wavelengths. Galileo Orbiter measurements of most of these quantities will be by far the most accurate yet made and will extend the measurement of outgoing radiation to longer wavelengths than previously achieved while Probe measurements will provide detailed energy balance data at the entry location.

Jupiter's atmosphere is in constant motion, a veritable laboratory for fluid dynamics on a grand scale, with some weather systems such as the Great Red Spot being several times larger than the Earth. Understanding the fundamental drives for Jovian atmospheric dynamics and circulation pushes current theories of atmospheric motion into regions of parameter space beyond those available on the terrestrial planets, with gravity, spin rate, scale, and energy input all having values very different than encountered in terrestrial settings. Weather systems on Jupiter have been studied for centuries with telescopes and can be observed with increasingly better resolution with modern systems such as the Hubble Space Telescope. The Voyagers carried out two 'observatory phase' sequences lasting for several months each, where the resolution in returned images was better than ground-based pictures. Galileo will add to the study of atmospheric dynamics by conducting high resolution observations over a two year period, using improved techniques for obtaining time lapse 'movies' of weather patterns and the power of coordinated visual, near-infrared and far-infrared observations.

The upper reaches of the Jovian atmosphere are the interface between the planet and its external environment, where gases escape the planet's gravity, a complex ionosphere is formed by ultraviolet and charged particle irradiation, airglow and auroral phenomena occur, and minute ring particles and other interplanetary debris rain into the atmosphere. The Probe will provide some data on the structure of this region in one area during its entry phase, but global characterization of the upper atmosphere and ionosphere will require the coordinated study of radio occultations, ultraviolet, infrared, and visual images.

3.2. Satellites

The large satellites of Jupiter are major elements of Galileo's exploration of the system. As individual objects, they each have unique characteristics, ranging from Io's sulfur-rich volcanism to large scale ice tectonics on Ganymede and Callisto. As a collection of outer planet satellites they provide clues to satellite forming processes at Jupiter, and, by comparison with other satellite systems, to the nature of circum-planetary nebular material throughout the outer solar system. Finally, the satellite system has strong interactions with the magnetospheric environment, supplying major amounts of material to the charged particle populations and, in turn, being themselves bombarded and modified by particle irradiation of all energies.

Galileo will utilize multiple close (well within one satellite radius typically) flybys and greatly expanded instrumentation (over previous missions) to achieve its satellite objectives. The close encounter ranges provide obvious improvements in remote sensing resolution (images will typically have 100 m to 1 km resolution) and in *in-situ* observations of gravity fields and magnetospheric interactions. The expanded range of Galileo instrumentation includes completely new types of experiment, such the dust detector and the near infrared mapping spectrometer (NIMS), and also extended spectral, energy and angular ranges for many other experiments (such as the near infrared capability of the imaging system's CCD sensor).

One of the primary objectives for satellite studies is the characterization of their surfaces, primarily through high resolution imaging but including cooperative data from a number of remote sensing techniques. The goals of such studies are to understand their geologic histories and the nature of the processes which have been or are active on these bodies. Among the specific problems which will be addressed are: (1) the nature of volcanic activity on Io – what changes have occurred since Voyager, how much extrusive flow activity is going on compared with phase change geyser-like eruptions, whether there are patterns of volcanic activity which provide clues to the thickness of the crust and the nature of the partially molten mantle, (2) the relative age of the smooth surface of Europa and whether there are active resurfacing processes ocurring in the present epoch, (3) the thickness of the ice outer layer of Europa and whether there is evidence for liquid water under it, (4) the tectonic histories of Ganymede and Callisto – why are they so different in appearance today and are there general processes which have shaped their surfaces which can be extrapolated to 'cryo-volcanic' icy volcanism on other outer planet satellites.

In addition to morphological studies, Galileo will address a range of compositional questions. Spectral maps from NIMS and multispectral images from the camera system will be used to identify the surface mineralogies of the satellites and the nature and extent of water ice, silicate and carbonaceous mixtures on the icy surfaces. The compounds present on Io's surface will provide clues to the degree of differentiation of the source regions for the satellite's volcanism, and the identification and distribution of the dark material mixed with the ices on the other satellites will allow study of the roles of exogenic modification and endogenic process on these satellites.

The interiors of the satellites can be studied indirectly by making precise measurements of their gravity fields and searching for evidence of intrinsic magnetic fields. Both of these objectives can be best addressed by the multiple close flybys of the satellites. Observations of the higher order moments of the satellites' gravity may allow the degree of differentiation and the size of possible cores to determined for at least some of the satellites, with important consequences for understanding their histories. Likewise the presence or absence of intrinsic magnetic fields provides important information on the composition and internal state of the satellites.

There are a variety of objectives connected with the satellites' tenuous atmospheres and their interactions with the magnetosphere. Io has the only positively identified atmosphere and ionosphere, but the other satellites, with water ice exposed on their surfaces must have some, possibly extremely tenuous, amount of water, hydrogen, and oxygen in their 'atmospheres' due to sublimation and sputtering processes. These regions are also key to the interactions with the magnetosphere. Atmospheric and magnetospheric satellite processes will be studied using both remote observations (imaging, infrared, and most importantly ultraviolet) and *in-situ* measurements by a variety of charged particle and dust experiments.

3.3. Magnetosphere

Magnetospheric objectives concentrate on understanding the structure, composition, and dynamics of the huge Jovian magnetospheric environment. A major goal is to characterize much better than Voyager the composition of the energetic and plasma particle populations. In addition to being a key to understanding the energetics powering the processes in the magnetosphere, such information is also obviously valuable in studying the interactions with the satellites (and potentially the rings and upper atmosphere of Jupiter). By conducting observations from within the magnetosphere for an extended time while exploring various regions, Galileo experimenters hope to understand more fully the dynamic character of the system and the many still poorly understood processes of particle acceleration, wave phenomena and diffusion within the magnetosphere. One goal of such studies will be a global view of the processes producing major changes in the magnetosphere and the relative roles played by solar wind and satellite interactions.

The satellite interactions themselves are one of the highest priority objectives of the magnetospheric experimenters. The satellites can be both a source and sink of particles to the Jovian magnetosphere. Radiation particles which hit the surface of the satellites are absorbed but in the process they may liberate several low energy atoms and ions. These particles can become ionized (if neutral) and accelerated to high energies by magnetospheric processes to replace ultimately the lost radiation belt flux. To understand the energetics and the physical processes in this chain of acceleration and loss is one of the exciting challenges of the Galileo project.

4. Galileo Spacecraft

The Galileo spacecraft design reflects the very demanding objectives of the Galileo mission. It builds upon experience gained in the Pioneer, Mariner, and Voyager programs. Primary requirements include: (1) reliable, long-lived operation; (2) high rate (134 kbs) telecommunications from 4 to 6 AU distances; (3) radiation hard parts and designs; (4) autonomous computer-controlled operation and onboard fault protection; (5) accommodation of both remote sensing and fields and particles experiments; (6) non-solar, radioisotope thermoelectric generator power source; (7) a thermal design to accommodate operation at 0.7 and 6 AU; and (8) a large propulsion system for orbit insertion and trajectory trim maneuvers. In addition to these requirements for the Jupiter phase of the Galileo Mission, the VEEGA trajectory also placed some stringent requirements on telecommunications and attitude control of the spacecraft while inside 1 AU.

The spacecraft consists of an Orbiter and an atmospheric entry Probe, which is carried by the Orbiter and deployed 150 days prior to Jupiter encounter (Figure 2). The Orbiter uses a dual spin attitude stabilization system which allows easy accommodation of both remote sensing experiments (generally requiring stable, high accuracy pointing) and space physics instruments (generally desiring repetitive, broad angular coverage). In this design, the main body of the Orbiter, consisting of the electronics bays, propellant system, RTG and science booms and main antenna, spins at about three RPM, providing basic stabilization for the spacecraft. Immediately aft of the main body is the 'despun' section of the spacecraft, which uses an electric motor to drive this section counter to the rotation of the main section, stabilizing it in inertial space. Attached to the despun section are the scan platform and the Probe radio relay antenna and receivers and supporting engineering electronics. The scan platform uses another actuator aligned orthogonally to the spin axis to provide precision pointing for the remote sensing instruments in any direction, unobscured by the spacecraft body or booms. The Probe relay antenna also has some articulation to allow accurate pointing at the Probe entry site during Probe entry (see description of mission below).

During the early phases of the VEEGA mission, telecommunications is via one of two low-gain antennas, one located near the feed of the main antenna and the other mounted on an RTG boom.

After deployment of the 4.8-m diameter high-gain antenna following the first earth encounter, most downlink communications are carried out at X-band frequencies at data rates up to 134 kbs, while uplink communications are performed at S-band frequencies. A tape recorder with $\sim 10^9$ bit capacity provides the capability to buffer about 120 frames of imaging data or $\sim$ 36 hours of low rate science.

Key features of the Orbiter are:

Telecommunications System

Key elements include the furlable 4.8-m circular parabolic HGA used for receiving and transmitting both S- and X-band signals, redundant transponders, redundant dual-power S- and X-band power amplifiers, and two S-band low-gain antennas. A late addition to the telecommunications equipment includes the X to S down-converter receiver to implement an X-band uplink receive capability. In addition to the RF equipment, redundant digital command detector and modulation units are used to detect and route ground commands to the central computer and modulate spacecraft data onto a subcarrier for downlink transmission.

The telecommunications channels provide two 'classes' of data quality to the scientific

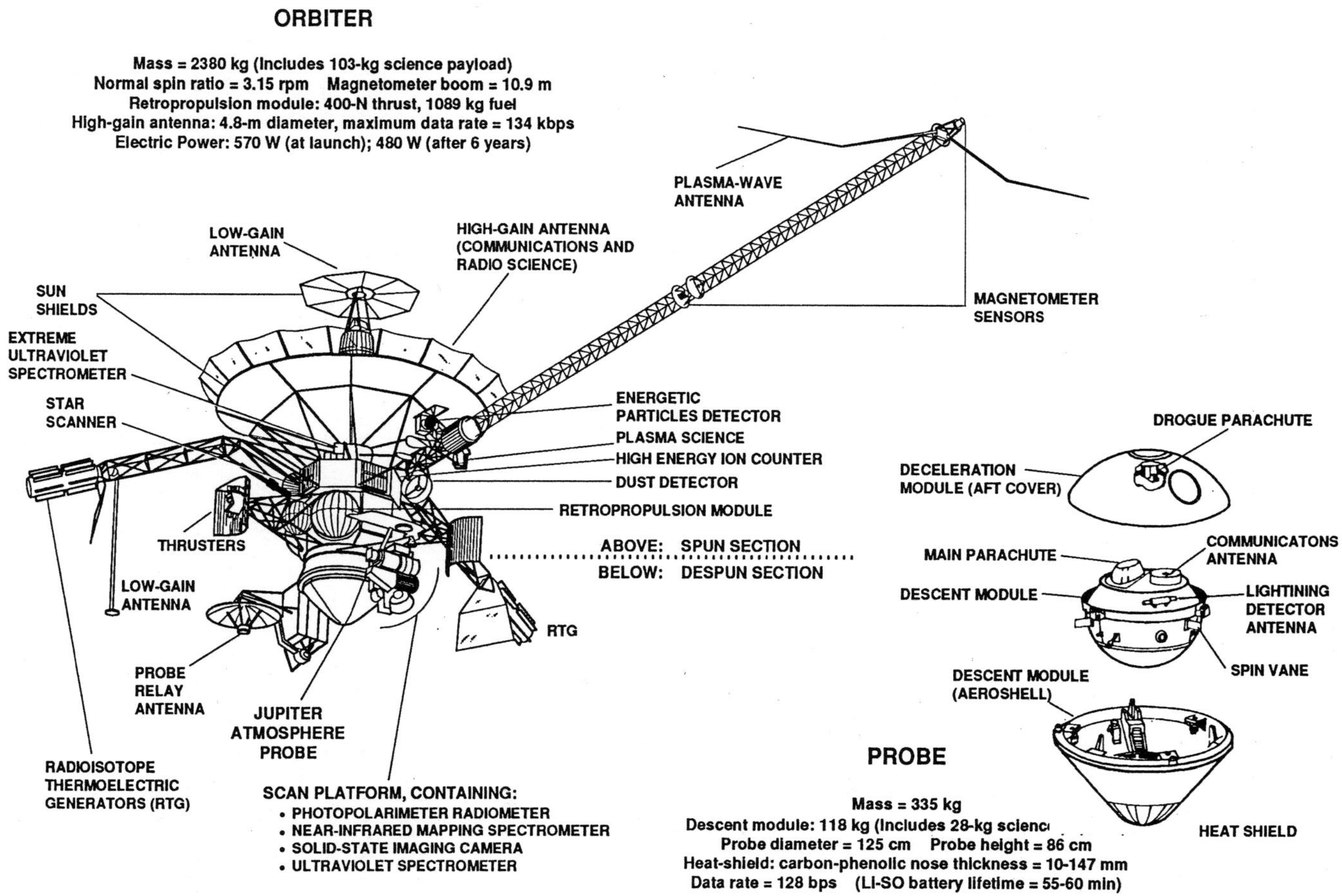

Fig. 2. Galileo spacecraft: views of the Galileo spacecraft and an expanded view of the Jupiter Probe are shown.

user – high quality (bit error rates less than 5×10^{-5}) for non-imaging science and a lower quality (bit error rates less than 5×10^{-3}) for imaging. This is achieved by operating the downlink channel at a signal to noise ratio equivalent to a bit error rate of about 5×10^{-3}, thus maximizing data quantity returned for imaging. Non-imaging science is coded onboard the Orbiter using a Golay error-detecting, error-correcting code (located in the CDS) to allow reconstruction of the data received at Earth to the quality needed for those science users.

Propulsion System

The Galileo RPM system located on the Orbiter spun side and supplied by Germany, is based on earlier bi-propellant Symponie designs. The system provides all directed impulse for attitude control, trajectory correction, and Jupiter orbit insertion. The propulsion functions consist of spin rate control, fine turning to point the HGA to Earth, and orienting the spacecraft for propulsive or science maneuvers. The RPM includes four propellant tanks (two fuel tanks containing monomethylhydrazine and two oxidizer tanks containing nitrogen tetroxide), two helium pressure tanks, twelve 10-N thrusters (six each mounted on separate cantilevered booms), one 400-N engine, and necessary isolation and control elements. The Orbiter mission and science requirements require a propulsive capability of nearly 1600 m s^{-1} for Probe delivery, orbit insertion, satellite tour and other attitude control functions. The thrusters are mechanized on two separate branches providing redundancy for spin control, HGA pointing, and trajectory correction. Only three uses of the 400-N engine are planned; all are subsequent to Probe separation.

Control of propellant to the 10-N thrusters and the 400-N engine is accomplished by opening and closing fuel and oxidizer solenoid latch valves via electrical signals from the attitude control system propulsion drive electronics. The propulsion drive electronics also provide the control signals for opening and closing the thruster and 400-N engine valves.

Command and Data System

This system consists of the data storage element which is a four-track digital tape recorder (capable of data recording and playback) and the CDS. Galileo is the first planetary spacecraft to use distributed computer control of its subsystems. Instead of the traditional central flight data computer, Galileo provides a Command Data System (CDS), which controls spacecraft activities through a data bus architecture. Each subsystem, including virtually all of the science experiments, is run by a local microprocessor with an interface to the CDS bus, providing great flexibility in the programming of individual instrument and subsystem functions through flight software.

The primary responsibility for command, control, and data handling is performed by the actively redundant command and data subsystem (CDS). The major functions of this computer system include processing and issuing real-time commands from Earth and executing sequences of stored commands either as part of a normal pre-planned flight activity or in response to the actuation of various fault recovery routines, and

TABLE III

Galileo's scientific payload

Experiment	Mass (kg)	Range	Objectives
Probe			
Atmospheric Structure Instrument (ASI)	4	Temp.: 0–540 K Pres.: 0–28 bars	Determine temperature, pressure, density, and molecular weight as a function of altitude
Neutral Mass Spectrometer (NMS)	11	Covers 1–150 amu	Determine chemical composition of atmosphere
Helium Abundance Detector (HAD)	1	Accuracy: 0.1%	Determine relative abundance of helium
Nephelometer (NEP)	5	0.2–20-μm particles, as few as 3 cm^3	Detect clouds and infer states of particles (liquid versus solid)
Net-flux Radiometer (NFR)	3	6 infrared filters from 0.3 to 100 μm	Determine ambient thermal and solar energy as a function of altitude
Lighting and Energetic Particles (LRD/EPI)	2	Fisheye lens sensors; 1 Hz–100 kHz	Verify the existence of lightning and and measure energetic particles in inner magnetosphere
Orbiter			
Solid-State Imaging (SSI)	28	1500 mm, f/8.5 800 × 800 CCD, 8 filters, 0.47° field-of-view	Map Galilean satellites at roughly 1-km resolution, and monitor atmospheric circulation over 20 months while in orbit around planet
Near-Infrared Mapping Spectrometer (NIMS)	18	0.7–5.2-m range, 0.03-μm resolution 0.5m rad IFOV	Observe Jupiter and its satellites in the infrared to study satellite surface composition, Jovian atmospheric composition and temperature
Ultraviolet Spectrometer (UVS)	4	1150–4300 Å	Measure gases and aerosols in Jovian atmosphere
Extreme Ultraviolet Spectrometer (EUV)	13	54 to 128 nm	Investigate S, O ion emissions of the Io torus, and atomic and molecular H auroral and airglow emissions of Jupiter
Photopolarimeter-Radiometer (PPR)	5	Discrete visible and near-infrared bands, radiometry to > 42 μm	Determine distribution and character of atmospheric particles; compare flux of thermal radiation to incoming solar levels
Magnetometer (MAG)	7	32–16384 γ	Monitor magnetic field for strength and changes
Energetic Particles Detector (EPD)	10	Ions: 0.020–55 MeV Electrons: 0.015–11 MeV	Measure high-energy electrons, protons, and heavy ions in and around Jovian magnetosphere and study processes affecting these populations
Plasma Detector (PLS)	13	1 eV to 50 keV in 64 bands	Assess composition, energy, and three-dimensional distribution of low-energy electrons and ions
Plasma Wave (PWS)	7	E: 5 Hz to 5.6 MHz B: 5 Hz to 160 kHz Wideband 1 kHz 10 kHz 80 kHz	Detect electromagnetic waves and analyze wave-particle interactions
Dust Detector (DDS)	4	10^{-16} g–10^{-6} g, 2–50 km s^{-1}	Measure particles' mass, velocity, and charge
Radio Science (RS): Celestial Mechanics	–	*S*- and *X*-band signals	Determine mass of Jupiter and its satellites (uses radio system and high-gain antenna)
Radio Science (RS): Propagation	–	*S*- and *X*-band signals	Measure atmospheric structure and objects' radii (uses radio system and high-gain antenna)
Heavy Ion Counter (HIC)	8	Ions from carbon to nickel range -6 to > 200 MeV $nucl^{-1}$	Monitor the fluxes and composition of energetic heavy ions in the inner Jovian magnetosphere, and high energy solar particles in the outer magnetosphere, characterize the ionizing radiation

controlling and selecting data modes and collecting and formatting science and engineering data for downlink transmission. The CDS employs a new architecture using multiple microprocessors and a high-speed data bus for both internal and user communication, thus taking advantage of the state of the art (as of the mid-seventies) in data system engineering.

The majority of CDS electronics are located on the Orbiter spun side in proximity to the data storage, science, and telecommunication equipment. CDS despun elements are limited to those necessary to support the Probe and relay radio hardware equipment, the scan platform mounted remote science instruments, the launch vehicle, and sequence operations. Six 1802 microprocessors, memory units, and the data bus comprise the 'heart' of the CDS. Four of the microprocessors (two high-level modules and two low-level modules) and four memory units contain a total of 192 K words of random access memory and are located on the spun section along with supporting electronics. The low-level modules of the remaining two microprocessors, each with 32 K RAM, are located on the despun section. The data bus comprises three dedicated busses.

Attitude and Articulation Control Subsystem

The attitude and articulation control subsystem (AACS) is responsible for maintaining spun section spin rate; orienting the spin vector; controlling propulsion isolation valves, heaters, 10-N thruster firing, and 400-N engine firing; and controlling the science platform containing the remote sensing instruments on the despun section.

The central element of the AACS is the attitude control electronics (ACE) that control the AACS configuration; monitor its health; perform executive, telemetry, command, and processing functions; provide spun position data to other subsystems; and provide AACS fault recovery. The 'heart' of the ACE is a high-speed 2900 ATAC-16 processor, and memory containing 32 K words times 16-bit RAM and 1 K words times 16-bit ROM. Other major functional elements of AACS include:

– A radiation-hardened star scanner employing photomultiplier tubes for star field identification for in-flight attitude determination.

– Linear actuators for raising or lowering the RTG booms to reduce wobble and maintain stability.

– Acquisition sensors for Sun pointing and spin rate data.

– Spin detector accelerometer for providing spin rate data.

– Sun gate sensor for providing off-Sun point fault detection.

– Propulsion drive electronics to control the RPM latch valve, thrusters, and 400-N engine valves.

– Spin bearing assembly to provide the mechanical and electrical interfacing between the spun and despun sections of the Orbiter as well as to provide despun orientation in clock degree-of-freedom motion.

– Gyros mounted on the despun scan platform to control platform articulation and stabilization.

– Accelerometers mounted on the despun side diametrically opposite to each other and aligned parallel to the Orbiter spin axis to measure velocity changes during propulsive burns.

– Scan actuator subassembly to provide the scan platform cone actuation and positioning in formation. This subassembly contains a flex capsule consisting of 200 printed wire tracks for electrical interfacing between the scan platform instruments and the despun electronics.

Probe

The Probe is composed of two major segments, the deceleration module and the descent module. The deceleration module includes the fore and aft heat shields, the structure that supports the heat shields, and the thermal control hardware for mission phases up through entry. The descent module is the package that descends through the Jovian atmosphere by parachute while the prime science data are gathered. It contains the science instruments and the Probe subsystems required to support the instruments and transmit the data back to the overflying Orbiter, which serves as a relay to Earth.

Deceleration Module. The general shape was selected to provide a minimum weight Probe, primarily because this allows more forward placement of the pressure vessel than do more blunt vehicle shapes, leading to minimum base diameter and minimum ballast required for aerodynamic stability.

Thermal protection during entry is provided by a carbon phenolic forebody heat shield and a phenolic nylon afterbody heat shield. Although these materials have been used extensively for Earth re-entry vehicles, on the Galileo Mission they will be subjected to environments never before experienced in flight. Entry velocity relative to the atmosphere is 48 km s^{-1}, far higher than any atmospheric entry attempted to date. For both the forebody and afterbody, radiative transfer is the primary energy transport mechanism. An additional consideration included in the final heat shield design is mechanical erosion (spallation) of heat shield material. The shield is subjected to a hot atmospheric shock layer (14000 K). The heat transfer at the nose of the vehicle at peak heating exceeds 42 kW cm^{-2}. The approximate mass of the forebody heat shield is 145 kg, of which about 90 kg is expected to be lost by ablation during entry.

The main parachute is used to separate the descent and deceleration modules and to control the rate of descent through the atmosphere. The parachute diameter (projected) is 2.5 m, a value selected to guarantee separation of the Probe in worst-case conditions, that is, the aeroshell structure without the forebody heat shield. Nominal conditions at parachute deployment are Mach 0.9 and a dynamic pressure of 6000 N m^{-2}. A swivel is used to decouple Probe and parachute spin. The pilot parachute is injected into the wake by a mortar fired at a velocity of 30 m s^{-1}. After the pilot chute is established, the separation nuts are fired to release the aft cover. The aft cover, pulled through the wake by the pilot chute, in turn pulls out and strips the bag from the main parachute. The deployment process, from initiation through full opening of the main chute, is designed to take less than 2 s.

Descent Module Most of the equipment is mounted on a shelf; the aft side of this shelf is used primarily for the science instruments. The Probe radio frequency communication subsystem and the pyrotechnic control unit are located in the aft compartment; the main parachute, the lightning and the radio emissions detector (LRD) antenna, and energetic particles instrument (EPI) sensor are located on top of this compartment. Note that in contrast to the Pioneer Venus design, which included a sealed pressure vessel, the weight is minimized by venting the Galileo descent module and by protecting individual units as necessary with hermetically sealed housings. The housings are designed for survival to 20 bars and were tested to 16 bars.

Electronics. To eliminate single-point, catastrophic mission failures, the Probe electrical and electronic subsystems, including the radio frequency link with the Orbiter and the receiver on the Orbiter, are redundant designs. Two parallel and simultaneous data streams go from the instruments to the Orbiter.

Communications Subsystem. The communications subsystem provides two *L*-band microwave channels, one at 1387.0 MHz and one at 1387.1 MHz, each circularly polarized, one with right-hand and the other with left-hand polarization. Each channel consists of an exciter, a power amplifier which puts out 23 W of radio frequency power, and either an ultrastable oscillator or a standard, thermally-controlled oscillator. The channel with the ultrastable oscillator will be used for a Doppler wind measurement.

Power Subsystem. The power subsystem consists of two electronics units, a lithium/sulfur dioxide ($Li-SO_2$) battery and a set of thermal batteries. The main power supply for the Probe mission is the $Li-SO_2$ battery. This battery is made up of three modules of 13 D-size, high-rate cells. The cells are connected in series with a bypass diode across each cell so that the failure of one cell will not cause the loss of an entire module. Total battery capacity at manufacture is expected to be about 21 A hr^{-1} (about 730 W hr^{-1}). Probe mission requirements are for about 16.3 A hr^{-1}, leaving a capacity margin at the end of the mission (defined as 48 min after entry) of 1.7 A hr^{-1}. The primary source of energy for pyrotechnic events is a set of four thermal batteries. Pyrotechnic pulses for actuation of these batteries and for neutral mass spectrometer functions later in the descent are provided by a tap on the main battery at the eight cell.

Command and Data Handling Subsystem. The command and data handling subsystem consists of the data and command processor, the pyrotechnic control unit, and four acceleration switches (two redundant parts). The data and command processor is a micro-controller-based unit which provides and controls all system command, telemetry, data storage, and timing functions.

During the period from separation of the Probe and Orbiter to the end of the mission, the data and command processor operates in several different modes. For the coast phase, only the coast timer operates. Its purpose is to 'wake up' the Probe about 6 hours before entry. During the pre-entry period, one string is used to gather and store LRD and EPI data. Both strings operate starting about 20 min before entry through the end of the mission. The entire sequence from separation to the end of the mission, with the exception of the magnitude of the time for the coast phase, is contained in permanent memory. Entry and descent sequence functions are tied directly to the atmosphere by the acceleration switches. The switches provide an adaptive parachute deployment which removes uncertainties in deployment conditions associated with uncertainties in entry angle and in the characteristics of the atmosphere. The acceleration switches also provide backup of the descent sequence in the event or early or late time-out of the coast timer.

Mass. The total mass of the Probe is about 331 kg, that of the Probe adapters is 6.8 kg, and that of the RRH equipments is 23.2 kg. Almost half of the total Probe mass is dedicated to the heat shield, and the electronic units are considerably heavier than one would expect for standard spacecraft hardware because they include hermetically sealed housing designed to withstand 20 bars.

Relay Radio Hardware. The relay radio hardware (RRH) receivers, ultrastable oscillators, and antenna are mounted on the despun section of the Orbiter to receive the Probe signal.

The RRH receivers are a key element of the data link between the Probe and the Orbiter during the Probe's descent through the Jovian atmosphere. Each receiver is designed to acquire, coherently track, and process the Probe's data along with RRH radio science and engineering data. The receivers also format and buffer the data for transmission to the Orbiter command and data system. The acquisition, tracking, and data-handling functions are performed by the microprocessor-controlled section of the receiver. The main functions in the acquisition and tracking process are coarse frequency acquisition through the implementation of a sequential probability radio test; frequency synchronization, using a frequency-lock loop; symbol synchronization, using a symbol-lock loop; ambiguity resolution; and phase synchronization and signal tracking, using a phase-lock loop. Uncertainties and signal perturbations, including uncertainty in Probe signal frequency and spin rate, and non-steady motion of the Probe caused by wind shear, wind turbulence, and nonsteady aerodynamics, make the design and test of these functions complex and difficult. The unit must acquire the Probe signal within 50 s, with an acquisition probability of 0.995 and a false-lock probability of less than 5×10^{-5}. Minimum strength at acquisition is specified as 31 dB Hz^{-1}. The unit is required to track a signal strength as low as 26 dB Hz^{-1}. In addition to providing Probe data, the receiver measures received Probe signal strength and Doppler rates for radio science use.

The RRH antenna is a 1.1-m parabolic reflector with a dual feed for receiving the two signals from the Probe. It is mounted on a deployable mechanism for storage at launch and for pointing during the Probe mission. This mechanism is provided by the Orbiter.

5. Mission Design

While the Probe's success is keyed to its ability to penetrate the Jovian atmosphere, the Orbiter's success depends on its unique trajectory, which provides for unprecedented new measurements. Once captured by Jupiter's gravity, the Orbiter would repeat its initial 200-day orbit if nothing were done; this would allow several Voyager-like passes through the system before the spacecraft 'died' from radiation effects or actually dropped at its low point into the atmosphere due to gravitational perturbations of the Sun. For Galileo to be utilized more effectively during its limited lifetime, the orbital period must be shortened and the spacecraft targeted to make very close flybys of the Galilean satellites.

If this had to be accomplished by rocket propulsion, the mission would be impossible – too much propellant (and weight) would be required to do the necessary maneuvers in Jupiter's strong gravitational field. Fortunately, mission designers have found a way to fly a very demanding, complicated mission using little propellant. They will manage this trick by employing the gravity-assist technique that successfully redirected the Voyagers and other spacecraft as they swung by various planets along their routes.

In the case of Galileo, a celestial 10 cushion billiard shot will be set up, using the gravity of the massive Galilean moons to modify the Orbiter's course during each pass. This simultaneously sends the craft on toward the next encounter and provides extremely close approaches to the satellites for scientific measurements. As a result, the entire 'satellite tour' can be flown so that rockets need supply only about 100 m s^{-1} of velocity change – 60 times less than what would be needed without the satellites' help!

The Galileo mission is the first mission to use an orbiting spacecraft to perform an intensive scientific investigation of Jupiter, its environment, and its Galilean satellites – Io, Europa, Ganymede, and Callisto. The Orbiter will perform a series of 10 highly elliptical revolutions about Jupiter. On each of these revolutions, a close, targeted flyby of one of the Galilean satellites is planned. This series of targeted encounters is termed a 'tour'.

The influence of the gravity of any of the Galilean satellites may be used to shape and control the Orbiter's trajectory. Consequently, repeated close flybys with Europa, Ganymede, and Callisto are used to target the Orbiter to the next satellite encounter. In this manner, tightly limited Orbiter ΔV propellant need only be used for fine tuning navigational purposes.

The design of a successful tour is constrained by other factors in addition to the laws of orbital mechanics. Specifically, budgetary and hardware constraints significantly influence the tour, as well as the need to design encounters that permit the most useful scientific observations to be made.

At each satellite encounter, different aimpoints exist that allow the Orbiter to return to the same satellite or to encounter a different satellite. Consequently, an abundance of aimpoints at each satellite encounter can result in many possible tours, each of which will satisfy many of the science objectives in different ways. While it is possible to design a tour to satisfy any single science requirement, the tendency for various science desires to be mutually incompatible rules out the possibility that any single tour can satisfy equally all the science requirements. Therefore, strategies must be developed to maximize the satellite tour science. Tour design strategies and their application to a representative tour are discussed in detail in following articles. Subsequent articles describe each instrument on the Probe and Orbiter.

Acknowledgements

This work was carried out at Caltech/Jet Propulsion Laboratory under a contract from NASA. Dr C. Russell, an Interdisciplinary Scientist on Galileo, acted as editor of this set of papers and made significant improvements to the magnetospheric portion of this paper with his comments and suggestions.

Note Added in Proof

On October 29, 1991 Galileo completed one of its first scientific milestones when it flew by the asteroid Gaspra at a range of 1600 km. Preliminary data, including four images returned in early November 1991, indicate that this first ever spacecraft encounter with an asteroid was highly successful.

Bibliography

Due to the extremely broad nature of the science and engineering background discussed in this paper, a detailed reference list would be far too massive to be included easily. A short list of recent review articles and books as well as the Voyager Jupiter reports is provided to guide the reader to the relevant literature on these subjects.

General:
Yeates *et al.*: 1985, Galileo: *Exploration of Jupiter's System*, NASA SP-479.
Miner, E. D., 1990: 1990, Voyager 2's Encounter with the Gas Giants', *Phys. Today* **000**, 000.
Spacecraft and Mission:
Landano, M. R. and Jones C. P.: 1983, *The Galileo Spacecraft System Design*, AIAA 83-0097, AIAA 21st Aerospace Sciences Meeting, Reno, NV.
O'Neil, W.: 1990, *Project Galileo*, AIAA 90-3854, AIAA Space Programs and Technologies Conference, Huntsville, AL.
Jupiter System Science:
Science **204**, 945, 1979.
Nature **280**, 725, 1979.
Geophys. Res. Letters **7**, 1, 1980.
Science **206**, 925, 1979.
J. Geophys. Res. **85**, 8123, 1981.
Satellites of Jupiter, Morrison and Matthews (eds.), University of Arizona Press, Tucson, AZ, 1982.
Planetary Rings, Greenberg and Brahic (eds.), University of Arizona Press, Tucson, AZ, 1984.
Satellites, Burns and Matthews (eds.), University of Arizona Press, Tucson, AZ, 1986.
The New Solar System, 3rd edition, Beatty and Chaikin (eds.), Sky Publishing, Cambridge, MA and Cambridge University Press, Cambridge, 1990.

GALILEO TRAJECTORY DESIGN

LOUIS A. D'AMARIO, LARRY E. BRIGHT, and ARON A. WOLF

Jet Propulsion Laboratory, California Institute of Technology, Pasadena, CA 91109, U.S.A.

Abstract. The Galileo spacecraft was launched by the Space Shuttle Atlantis on October 18, 1989. A two-stage Inertial Upper Stage propelled Galileo out of Earth parking orbit to begin its 6-year interplanetary transfer to Jupiter. Galileo has already received two gravity assists: from Venus on February 10, 1990 and from Earth on December 8, 1990. After a second gravity-assist flyby of Earth on December 8, 1992, Galileo will have achieved the energy necessary to reach Jupiter. Galileo's interplanetary trajectory includes a close flyby of asteroid 951-Gaspra on October 29, 1991, and, depending on propellant availability and other factors, there may be a second asteroid flyby of 243-Ida on August 28, 1993. Upon arrival at Jupiter on December 7, 1995, the Galileo Orbiter will relay data back to Earth from an atmospheric Probe which is released five months earlier. For about 75 min, data is transmitted to the Orbiter from the Probe as it descends on a parachute to a pressure depth of 20–30 bars in the Jovian atmosphere. Shortly after the end of Probe relay, the Orbiter ignites its rocket motor to insert into orbit about Jupiter. The orbital phase of the mission, referred to as the satellite tour, lasts nearly two years, during which time Galileo will complete 10 orbits about Jupiter. On each of these orbits, there will be a close encounter with one of the three outermost Galilean satellites (Europa, Ganymede, and Callisto). The gravity assist from each satellite is designed to target the spacecraft to the next encounter with minimal expenditure of propellant. The nominal mission is scheduled to end in October 1997 when the Orbiter enters Jupiter's magnetotail.

List of Acronyms

ASI	Atmospheric Structure Instrument
EPI	Energetic Particles Instrument
HGA	High Gain Antenna
IUS	Inertial Upper Stage
JOI	Jupiter Orbit Insertion
JPL	Jet Propulsion Laboratory
LRD	Lightning and Radio Emissions Detector
NASA	National Aeronautics and Space Administration
NEP	Nephelometer
NIMS	Near-Infrared Mapping Spectrometer
ODM	Orbit Deflection Maneuver
OTM	Orbit Trim Maneuver
PJR	Perijove Raise Maneuver
PM	Propellant Margin
PDT	Pacific Daylight Time
PST	Pacific Standard Time
RPM	Retropropulsion Module
RRA	Radio Relay Antenna
SSI	Solid State Imaging
TCM	Trajectory Correction Maneuver
UTC	Universal Time Coordinated
UVS	Ultraviolet Spectrometer
VEEGA	Venus–Earth–Earth Gravity Assist

Space Science Reviews **60:** 23–78, 1992.

1. Introduction

The Galileo spacecraft was successfully launched by the Space Shuttle Atlantis on October 18, 1989. After a 6-year-long interplanetary cruise, including flybys of Venus (once), Earth (twice), and at least one asteroid, Galileo will arrive at Jupiter on December 7, 1995.

Galileo has three major and equally important scientific objectives at Jupiter: (1) to investigate the chemical composition and physical state of the Jovian atmosphere, (2) to investigate the chemical composition and physical state of the Jovian satellites, and (3) to investigate the structure and physical dynamics of the Jovian magnetosphere (O'Neil, 1990; Yeates *et al.*, 1992). To accomplish these objectives, the Galileo spacecraft includes an Orbiter and an atmospheric Probe (O'Neil, 1990; Yeates *et al.*, 1992). The Probe will obtain the first *in-situ* measurements of the Jovian atmosphere. The Orbiter will relay the Probe data to Earth during the Probe's descent through the Jovian atmosphere and then conduct an intensive study of the Jupiter system, including the Galilean satellites, the Jovian atmosphere, and the Jovian magnetosphere.

The 22-month orbital phase of the mission is referred to as the satellite tour. During this time, the Orbiter will complete 10 orbits about Jupiter. On each of these orbits, there is a close encounter with one of the Galilean satellites Europa, Ganymede, or Callisto, and on several of these orbits, there is a second, more distant, encounter with a different satellite. (The actual tour to be flown at Jupiter will not be selected until the summer of 1992.) The nominal mission is scheduled to end in October 1997 when the Orbiter enters Jupiter's magnetotail.

In the design of the Galileo satellite tour, extensive use is made of gravity-assist trajectory techniques. The encounter conditions for each satellite flyby in the tour are chosen so that the gravitational perturbation from the close satellite flyby modifies the orbit about Jupiter in the precise manner needed to target the spacecraft to an encounter with another (or the same) satellite on the next orbit. The large orbit changes achieved by the satellite flybys in a typical Galileo satellite tour translate to a total spacecraft velocity change (ΔV) of 6000–8000 m s^{-1}. This amount of ΔV is between four and five times the total ΔV capability of the spacecraft. Thus, gravity-assist techniques are instrumental in making the Galileo satellite tour feasible.

Galileo's interplanetary trajectory also utilizes gravity-assist flybys. Galileo is flying a Venus–Earth–Earth Gravity Assist (VEEGA) trajectory to Jupiter. As the name implies, the VEEGA trajectory uses one gravity-assist flyby of Venus followed by two gravity-assist flybys of Earth to provide the energy necessary to reach Jupiter. The VEEGA trajectory was selected for Galileo because the two-stage Inertial Upper Stage (IUS), which propelled Galileo out of Earth orbit, could not provide the necessary energy for any of the more direct Earth–Jupiter transfer options (Byrnes *et al.*, 1987; D'Aamario *et al.*, 1987). In addition to lowering injection energy requirements, the Galileo VEEGA trajectory requires only a small amount of ΔV for trajectory shaping between launch from Earth and arrival at Jupiter.

The remainder of this paper addresses the interplanetary trajectory, the Probe mission, satellite tour design, and mission performance. The dynamics of gravity-assist flybys are discussed in greater detail in the Appendix.

2. Interplanetary Trajectory

2.1. Mission constraints

The two major mission constraints that determined the selection of the 1989 VEEGA trajectory as the Earth–Jupiter transfer mode for Galileo can be expressed in terms of injection energy* and interplanetary ΔV requirements. First, the injection energy required for the interplanetary trajectory could not exceed the capability of the two-stage IUS. Second, the interplanetary ΔV needed for trajectory shaping maneuvers between Earth launch and Jupiter arrival could not be so large that overall propellant requirements for the mission would exceed the propellant capacity of the onboard propulsion system.

At injection, the total mass of the Galileo spacecraft was 2717 kg. For this mass value, the maximum C_3 available from the two-stage IUS is about 17 km^2 s^{-2} assuming a 160-nm circular parking orbit. The minimum C_3 for the 1989 VEEGA opportunity is about 13 km^2 s^{-2}. Thus, the 1989 VEEGA trajectory requires a launch energy that is well within the capability of the IUS. For a December 7, 1995 Jupiter arrival date, the Galileo launch period (the number of consecutive days on which launch can occur) extended for 41 days from October 12, 1989 through November 21, 1989.

The total amount of usable propellant contained in the onboard propulsion system at injection for attitude and trajectory control amounted to 925 kg. After subtracting propellant required for attitude control, trajectory correction maneuvers (TCMs) for navigation, and ΔV for Jupiter orbit insertion (JOI) and the satellite tour, the amount of propellant available for interplanetary ΔV is very small, corresponding to about 100 m s^{-1}. This upper limit on interplanetary ΔV represented a very stringent constraint that ruled out many interplanetary transfer options. Without modifications to the propellant capacity of the Galileo spacecraft, the only interplanetary transfer option that did not cause the total propellant capacity to be exceeded was the VEEGA trajectory.

For the 1989 VEEGA opportunity to Jupiter, there existed a wide range of Earth launch dates and Jupiter arrival dates that satisfied both the injection energy and propellant constraints (D'Amario *et al.*, 1987).

2.2. Interplanetary trajectory design constraints

Given selection of the 1989 VEEGA transfer for Galileo, there were several constraints that significantly affected the design of the interplanetary trajectory. These constraints are associated with asteroid flyby opportunities, injection biasing, and the Earth flyby navigation strategy. Each of these constraints is explained below.

* Injection energy is measured in terms of the quantity C_3, which is the square of the asymptotic speed with respect to Earth on the departure hyperbola.

2.2.1. *Asteroid Flyby Opportunities*

It is current NASA policy that all interplanetary missions that traverse the asteroid belt include a close asteroid flyby if at all possible. This requirement is, in and of itself, a major factor in the interplanetary trajectory design process. The search for suitable asteroid flyby candidates represented a major challenge for VEEGA trajectory design (Johannesen *et al.*, 1989). Since most asteroids lie between about 2.0 AU and 3.5 AU from the Sun, the VEEGA trajectory passes through the asteroid belt twice: between the two Earth flybys and again after the second Earth flyby on the final transfer to Jupiter. Closest approaches for over 4000 asteroids were checked for a large number of trajectories covering the entire range of allowable Earth launch and Jupiter arrival dates in order to determine suitable candidates for close flybys. For each candidate, the trajectories were then reoptimized to include an arbitrarily close flyby of the asteroid in order to evaluate the propellant cost of achieving the flyby. One factor in the selection of December 7, 1995 as the Galileo arrival date at Jupiter was the availability of the Gaspra/Ida dual-asteroid flyby option.

2.2.2. *Injection Biasing*

The trajectory design process allows for a number of TCMs during the course of the mission. Some of these maneuvers are deterministic in nature; these ΔV's are required for an optimal (minimum propellant) trajectory. Other maneuvers are statistical in nature and thus are zero in the absence of errors. Due to orbit determination and planetary ephemeris uncertainties and maneuver execution errors, both kinds of maneuvers are slightly different during the course of an actual flight. In particular, the statistical maneuvers are non-zero due to these error sources. In addition, the deterministic maneuvers also change in magnitude and direction from their prelaunch design values.

The first TCM after launch, called TCM-1, would normally have been a statistical maneuver to correct for the injection errors of the IUS. It was necessary, however, to bias TCM-1 to include a deterministic sunline component. This TCM, and all other TCMs that occur at distances less than 1 AU from the Sun, must be performed with the spacecraft pointed at the Sun. This pointing requirement is caused by thermal constraints on the spacecraft. Since adding ΔV in the anti-Sun direction is much less efficient in terms of propellant consumption, a bias was introduced effectively to eliminate any chance of having to perform a ΔV with an anti-sunline component. A TCM-1 bias of 17 m s^{-1} in the sunline direction was incorporated as a constraint in the trajectory optimization process to reduce the probability of having an anti-sunline component to less than 1%.

2.2.3. *Earth Flyby Navigation Constraints*

A major feature of the VEEGA trajectory is the inclusion of two close flybys of Earth to provide the necessary increase in heliocentric energy to reach Jupiter. It is a Galileo project requirement, however, that no action be taken that would have a greater than 10^{-6} *a priori* probability of resulting in an Earth-impacting trajectory (Maize, 1989; Mitchell, 1988). To meet this Earth navigation requirement for the first Earth flyby, it was necessary to introduce constraints on both the direction and magnitude of the post-Venus deterministic maneuver in order to bias the post-Venus trajectory away from Earth. For launch dates earlier than October 15, this strategy resulted in a small ΔV penalty.

The situation for the second Earth flyby is even more complicated. During the traverse of the asteroid belt, there is some probability of a micrometeoroid impact disabling the spacecraft. (Although the existence of an asteroidal micrometeoroid component is questionable, it has been accounted for in the Earth navigation strategy.) Since such an impact may lead to a ΔV caused by outgassing of propellant, there is a requirement that a ΔV of 3 m s^{-1} or less in any direction not be capable of placing the spacecraft on an Earth-impacting trajectory. Satisfaction of this requirement is accomplished by introducing constraints on both the maneuver between the first Earth flyby and Gaspra and the maneuver between Gaspra and the second Earth flyby in order to bias the trajectory away from Earth. This latter maneuver must also be delayed beyond its optimal time until the spacecraft has moved inside of the inner edge of the asteroid belt on its way to the second Earth encounter. These trajectory modifications also resulted in a small ΔV cost.

2.3. Interplanetary cruise

2.3.1. *Launch Pause*

The Galileo launch period opened on October 12, 1989, but two delays occurred that postponed the launch 6 days until October 18. The first delay, to October 17, was caused by the need to replace a faulty Space Shuttle main engine controller. The second delay, which prevented launch on October 17, was caused primarily by unacceptable weather at the Shuttle Landing Facility at Kennedy Space Center. (This landing site is used if an abort occurs during the first few minutes of flight.) In addition, there was considerable concern over the weather at the primary trans-Atlantic abort site in Ben Guerir, Morocco.

Launch of Galileo finally occurred on October 18, 1989 with lift-off of the Space Shuttle Atlantis at 16:54 UTC (09:54 PST). At 6 hours 21 min after launch, the Galileo spacecraft, attached to a two-stage IUS, was deployed from Atlantis. One hour later, at 00:15 UTC (17:15 PST) the IUS ignited the first of its two solid rocket motors to propel Galileo out of Earth orbit. The primary effect of the IUS burns and the Earth departure was to decrease the heliocentric velocity of the spacecraft by 3.1 km s^{-1} so that it would fall inward toward the Sun and encounter Venus.

2.3.2. *Earth to Venus*

The Galileo 1989 VEEGA trajectory is illustrated in Figures 1 and 2. Figure 1 shows a close-up of the inner-solar-system portion of the trajectory, and Figure 2 shows the complete trajectory to Jupiter. Also shown on Figure 1 are the locations of the major deterministic TCMs.

The first trajectory correction maneuver (TCM-1) was initiated on November 9, 1989,

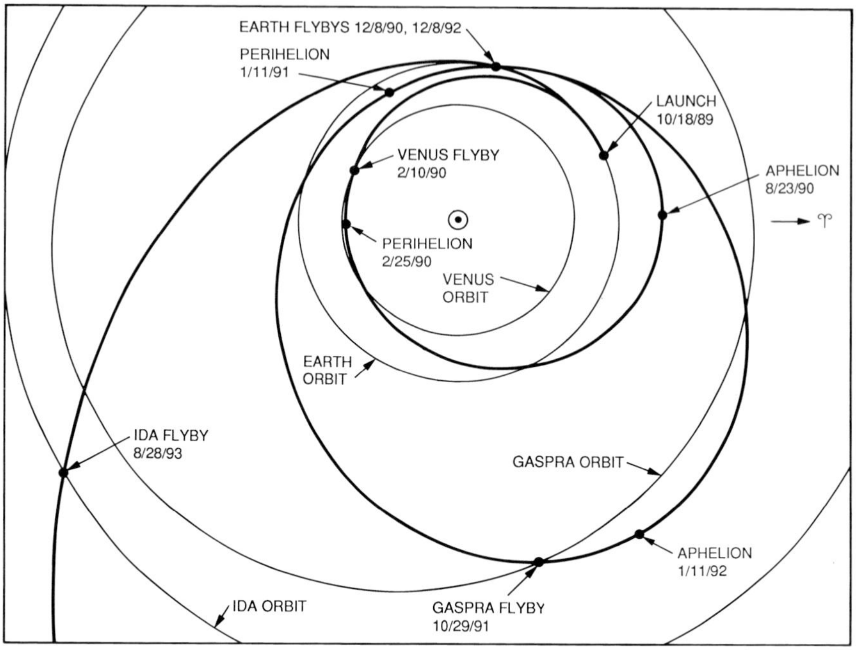

Fig. 1. Galileo 1989 VEEGA interplanetary trajectory: launch through Ida flyby.

21 days after departure from Earth. This three-portion (i.e., three-day) maneuver, totaling 15.7 m s^{-1} of ΔV, was completed on November 11. TCM-1 removed the deterministic sunline bias mentioned above and corrected IUS injection errors. The IUS injection burn was extremely accurate, as evidenced by the fact that the part of TCM-1 that corrected IUS injection errors was only 1.7 m s^{-1}. TCM-2 was performed on December 22, 1989. This small maneuver (0.7 m s^{-1}) removed the maneuver-execution errors of TCM-1 and targeted Galileo to the desired aimpoint at Venus. The third scheduled maneuver on the Earth–Venus trajectory leg (TCM-3) was not required because the aimpoint at Venus achieved by TCM-2 was virtually perfect.

The Venus encounter occurred on February 10, 1990 at 05 : 59 UTC (February 9 at 21 : 59 PST). The Venus flyby trajectory is shown in Figure 3. The spacecraft approached Venus from the anti-Sun direction, with closest approach occurring approximately over the terminator. At closest approach the altitude was 16 123 km, and the Venus-relative velocity was 8.2 km s^{-1}.

Also shown in Figure 3 is the gravity-assist 'velocity-vector diagram' for the Venus flyby. A velocity-vector diagram is a representation of a gravity-assist flyby which illustrates how spacecraft velocity is changed by the gravitational perturbation of the flyby. Velocity-vector diagrams and the dynamics of gravity-assist flybys are explained in the Appendix. The velocity-vector diagram for the Venus flyby in Figure 3 indicates that the gravity assist from Venus added 2.3 km s^{-1} to the spacecraft's heliocentric

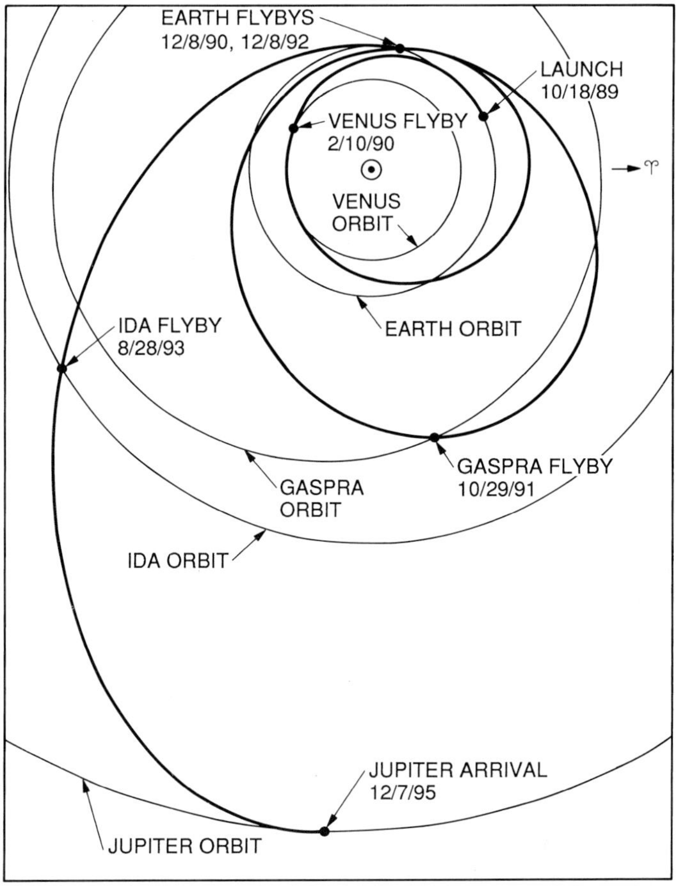

Fig. 2. Galileo 1989 VEEGA interplanetary trajectory: launch through Jupiter arrival.

velocity. The Venus flyby also redirected Galileo's trajectory so that it would return to the vicinity of Earth for the first Earth gravity-assist flyby. After the Venus flyby, the period of the orbit about the Sun was 1 year.

2.3.3. *Venus to Earth 1*

On February 25, 1990 (two weeks after the Venus encounter), the spacecraft passed through perihelion at a distance from the Sun of 0.70 AU. At this point on the trajectory, the thermal stress on the spacecraft reached its most severe level. During the entire time period between launch and the first Earth encounter, Galileo is maintained in a nearly Sun-pointed orientation so that its various sunshields protect sensitive spacecraft components from excessive heating.

On the Venus–Earth trajectory leg, a large deterministic deep-space maneuver (TCM-4) was performed to target Galileo to the first Earth encounter. For mission operations reasons, this maneuver was divided into two parts: TCM-4A and TCM-4B.

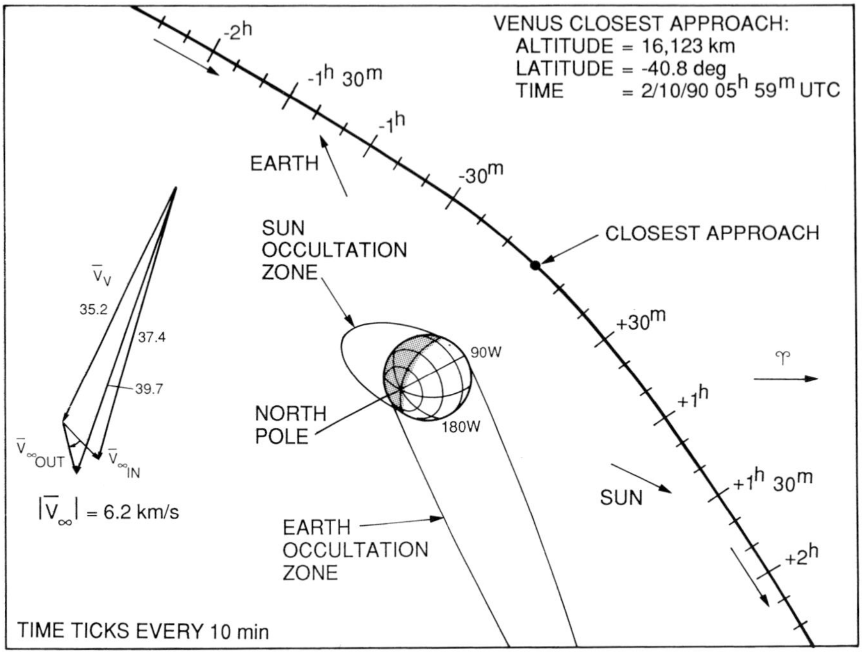

Fig. 3. Trajectory pole view of Venus flyby on February 10, 1990.

TCM-4A, a four-portion maneuver, was performed on April 9–12, 1990 and imparted a ΔV of 24.2 m s^{-1}. TCM-4B, a two-portion maneuver, was performed on May 11 and 12, 1990 and imparted a ΔV of 11.0 m s^{-1}. On July 17, 1990, the first of four small maneuvers (about 1 m per second or less) was performed to adjust the aimpoint at the first Earth encounter. This maneuver, TCM-5, also served to correct the maneuver execution errors associated with TCM-4B. The next significant trajectory event occurred on August 23, 1990 when Galileo passed through aphelion at a distance from the Sun of 1.28 AU. The remaining three Earth 1 approach TCMs (TCM-6, TCM-7, and TCM-8) were performed on October 9, November 13, and November 28, 1990.

The first Earth encounter occurred on December 8, 1990 at 20:35 UTC (12:35 PST). The traverse of the Earth/Moon system is shown in Figure 4, and the Earth flyby trajectory is shown in Figure 5. The spacecraft approached Earth from the anti-Sun direction and passed through Earth's shadow before closest approach. The occultation lasted for 1 hour and 22 min, ending 21 min before closest approach. At closest approach, the spacecraft was approximately over the terminator. The closest approach altitude was 960 km, and the Earth-relative velocity was 13.7 km s^{-1}. The ground track of the flyby is shown in Figure 6. The sub-spacecraft point at closest approach was over the Atlantic ocean about 1700 km (1050 mi) due east of the southern tip of Florida.

In order for Galileo to reach Jupiter from Earth, it must have a heliocentric velocity of about 39 km s^{-1}. Earth's velocity in its orbit about the Sun is 30 km s^{-1}; therefore

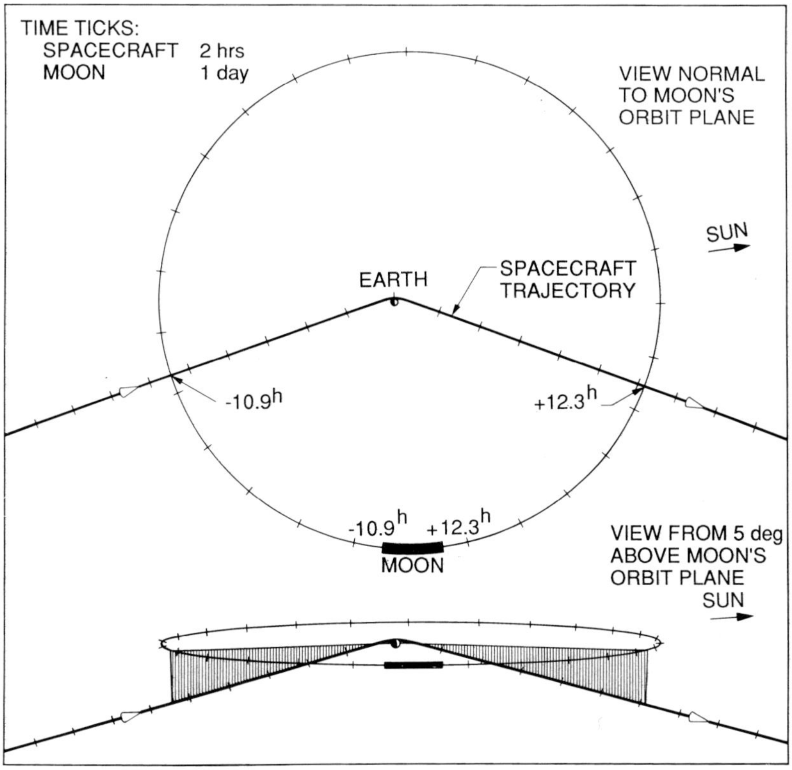

Fig. 4. First traverse of Earth/Moon system (December 8, 1990): views normal to and 5° above lunar orbit plane.

Galileo must acquire about 9 km s^{-1} of Earth-relative speed (V_∞) in a direction approximately parallel to Earth's velocity vector. From the velocity-vector diagram in Figure 5, it is apparent that, as a result of the Venus gravity assist, the spacecraft has been returned to Earth with an incoming V_∞ vector of the desired magnitude. This incoming V_∞ vector, however, is not parallel to Earth's velocity vector, but rather almost normal to it. The two Earth gravity-assist flybys must rotate this 9 km s^{-1} velocity to be approximately parallel to Earth's velocity vector, so that Galileo has the 39 km s^{-1} of heliocentric velocity needed to reach Jupiter. The first Earth gravity-assist flyby rotates the V_∞ vector by 48°, increasing the heliocentric velocity by 5.2 km s^{-1}, from 30.1 km s^{-1} to 35.3 km s^{-1}. This velocity change increases the orbit period from 1 year to 2 years. After completion of this 2-year orbit, Galileo will return to this same point in space for the second Earth encounter, the Earth having completed two full orbits about the Sun.

2.3.4. *Earth 1 to Earth 2*

At the time this paper was written, Galileo had just completed its first Earth gravity-assist flyby. On January 11, 1991 (approximately one month after the first Earth

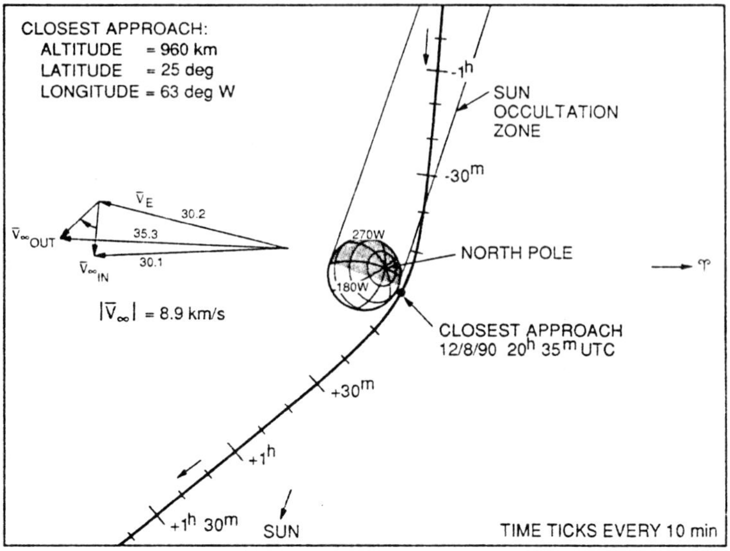

Fig. 5. Trajectory pole view of Earth 1 flyby on December 8, 1990.

encounter), the spacecraft will pass through its second perihelion at a distance from the Sun of 0.90 AU. On May 10, 1991, Galileo will be at solar opposition.

There are three TCMs required between the first Earth flyby and the Gaspra encounter. TCM-9A (5.3 m s^{-1}) was performed on December 19, 1991 (11 days after the first Earth encounter). TCM-9A partially corrected the interplanetary trajectory for the dispersions introduced by the first Earth encounter. The remainder of the correction will be accomplished on March 20, 1991, when TCM-9B (2.5 m s^{-1}) will be performed. The last maneuver required before the Gaspra encounter is TCM-10 (3.8 m s^{-1}), scheduled for July 2, 1991.

On October 29, 1991, slightly before aphelion of the 2-year Earth–Earth transfer, Galileo will encounter 951-Gaspra, a 16-km diameter type S asteroid. Because the mass of Gaspra is extremely low, its gravity has no appreciable effect on the spacecraft trajectory; therefore, the flyby conditions have been selected solely to optimize science return. Galileo will be targeted to a closest approach point almost directly over the dark side of Gaspra at a distance of 1600 km and an ecliptic latitude of 3.9° (north)*. This aimpoint provides optimal solar phase angles for the final approach observations of Gaspra. (Imaging at, or very near, closest approach is not possible due to the excessively high angular rate of Gaspra.) The spacecraft velocity relative to Gaspra during the flyby is 8.0 km s^{-1}.

Aphelion on the Earth–Earth trajectory leg will occur on January 11, 1992 at a distance from the Sun of 2.27 AU. Galileo will be at solar conjunction on January 22, 1992.

About four months before the second Earth encounter, in August 1992, a relatively

* Some interest has been expressed in whether Galileo can be targeted to pass through Gaspra's solar wind 'wake', represented by a cylinder that points generally in the anti-sunline direction. A wake passage would be nearly impossible because of the small size of the asteroid wake relative to expected trajectory dispersions.

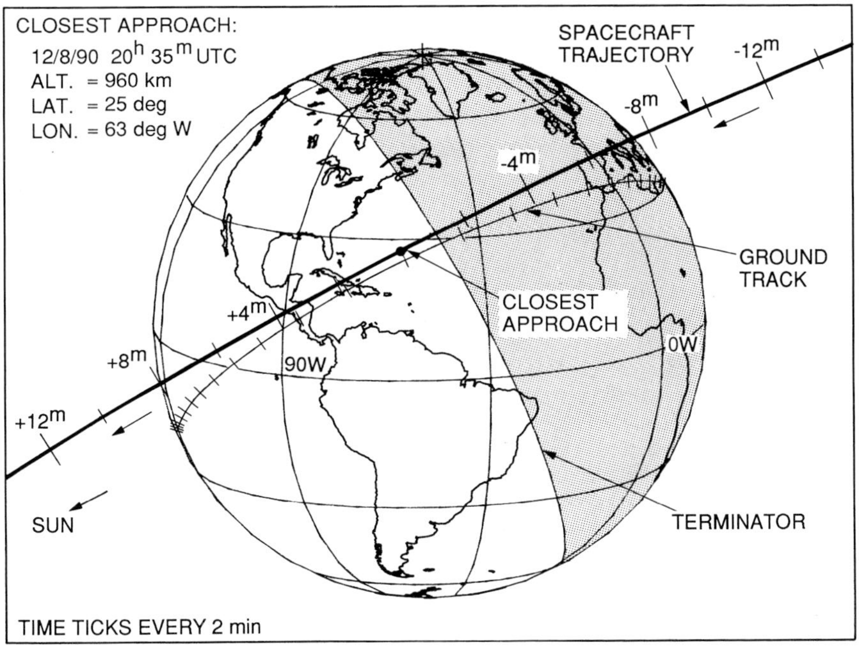

Fig. 6. Close-up view of spacecraft trajectory and ground track at Earth 1 flyby (December 8, 1990).

large deep-space maneuver (21 m s^{-1}) will be performed to target Galileo to the second Earth encounter. This maneuver is designated TCM-14.

The second Earth encounter will occur on December 8, 1992 at about 15 : 35 UTC (07 : 35 PST), two years after the first Earth encounter*. The traverse of the Earth/Moon system is shown in Figure 7. About 11 hours before closest approach to Earth, the spacecraft passes nearly over the north pole of the Moon at an altitude of about 111 000 km. The Earth flyby trajectory for the second Earth encounter is shown in Figure 8. As for the Venus and first Earth encounters, the spacecraft approach is generally from the anti-Sun direction, but the closest approach point is further from the terminator and closer to the subsolar point. At closest approach, the altitude is 300 km, and the Earth-relative velocity is 14.1 km s^{-1}. This is the lowest altitude of the three planetary flybys. The ground track of the flyby is shown in Figure 9. The closest approach point is over the South Atlantic ocean near the southern tip of Africa.

The velocity-vector diagram for the second Earth gravity-assist flyby is shown on Figure 8. The gravity assist rotates the V_∞ vector by 51° to a direction approximately parallel to Earth's velocity vector. As a result, the heliocentric velocity increases by 3.7 km s^{-1}, from 35.3 km s^{-1} to 39.0 km s^{-1}. At this point, Galileo has acquired the heliocentric velocity necessary to reach Jupiter. The orbit period is now 5.6 years.

* The information presented here about the second Earth encounter assumes that the Ida flyby option is exercised.

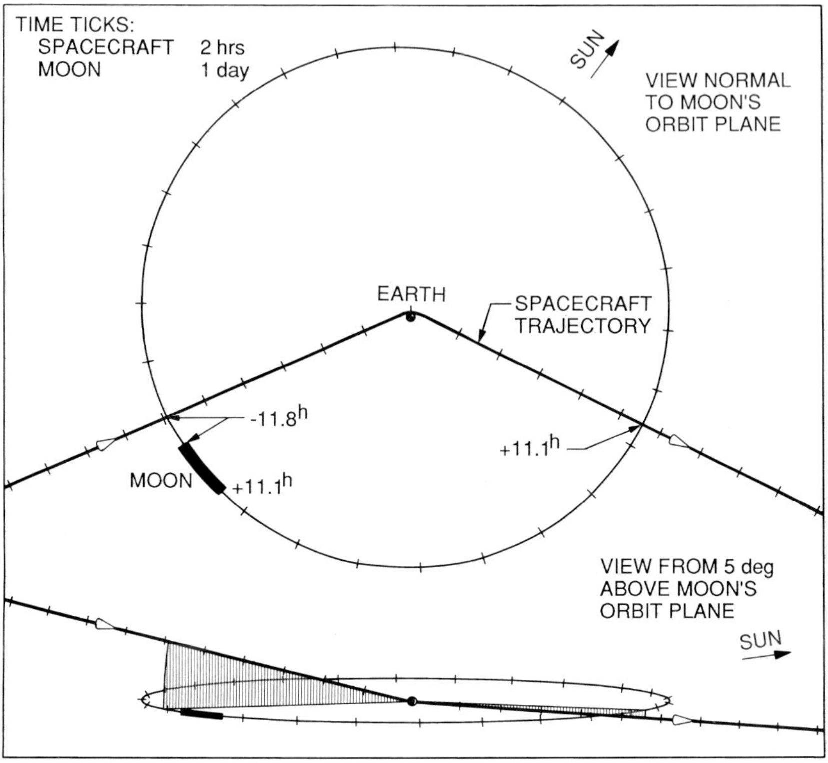

Fig. 7. Second traverse of Earth/Moon system (December 8, 1992): views normal to and 5° above lunar orbit plane.

The reason that the Galileo VEEGA trajectory requires two Earth gravity-assist flybys can be explained with the help of the hyperbolic flyby equation (see Appendix) that relates approach speed (V_∞), closest approach distance (r_p), the GM of the body (μ), and the bend angle (θ). In order for the first Earth flyby to increase the heliocentric velocity to 39 km s^{-1}, the V_∞ vector would have to be rotated through 99°. The hyperbolic flyby equation shows that, in order to rotate the V_∞ vector through 99°, it would be necessary to fly by Earth at a distance of 1600 km, which is below Earth's surface. The second Earth gravity-assist flyby is required to keep both Earth flybys above the surface.

2.3.5. *Earth 2 to Jupiter*

On December 13, 1992, just five days after the second Earth encounter, the spacecraft will pass through its third, and final, perihelion at a distance from the Sun of 0.98 AU. Galileo will be at solar opposition on February 28, 1993 (March 1 if the Ida option is not exercised).

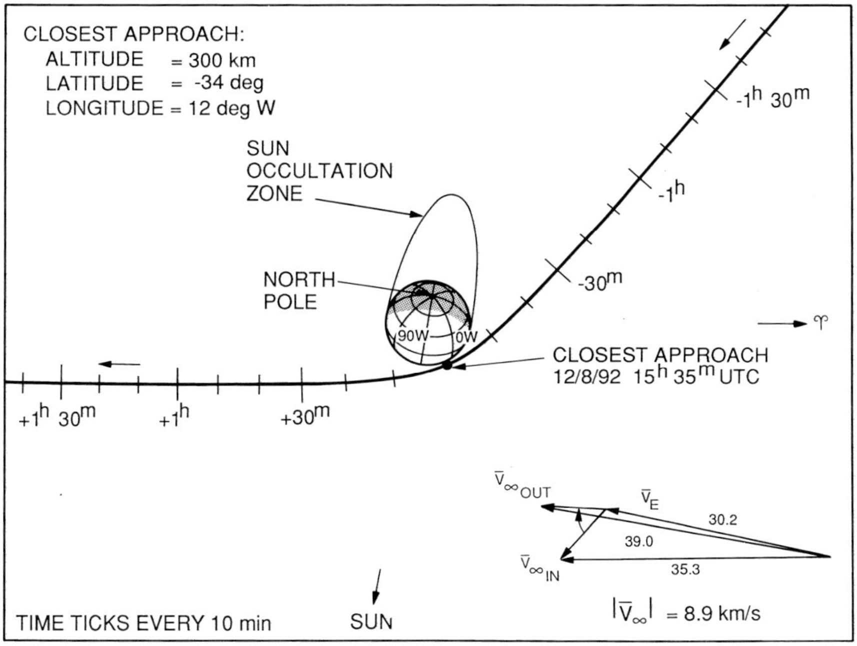

Fig. 8. Trajectory pole view of Earth 2 flyby on December 8, 1992.

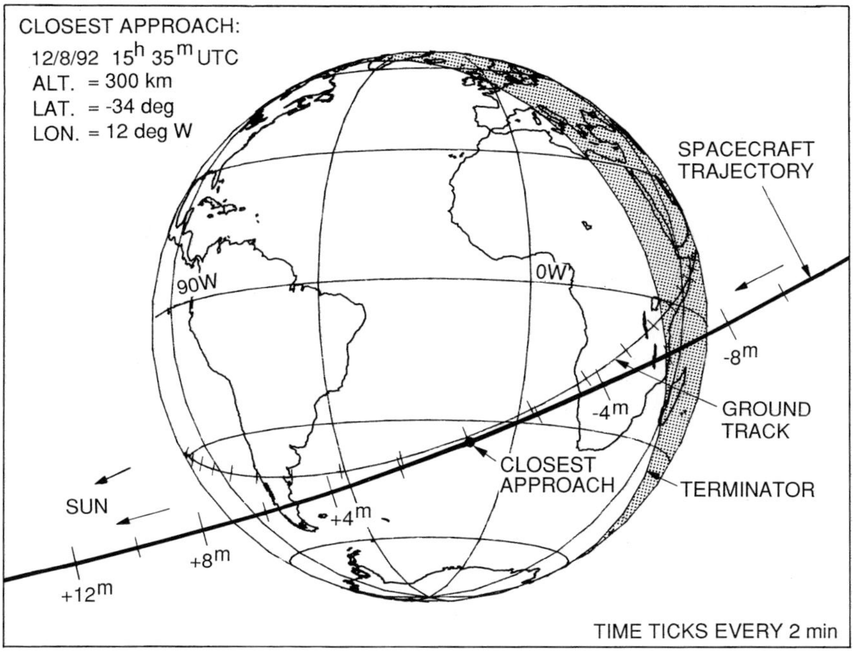

Fig. 9. Close-up of spacecraft trajectory and ground track at Earth 2 flyby (December 8, 1992).

Depending on propellant consumption and other factors, Galileo may be targeted for a second close asteroid flyby of 243-Ida, a 32-km diameter type S asteroid. If the Ida flyby is included, it will occur on August 28, 1993, about nine months after the second Earth encounter. The spacecraft velocity relative to Ida during the flyby is 12.4 km s^{-1}. The effects of the Ida decision on propellant consumption and the mission at Jupiter are discussed below in the section on mission performance.

If the Ida flyby option is exercised, a large deep-space maneuver (38 m s^{-1}) will be performed shortly after the Ida encounter to target the spacecraft to Jupiter. This maneuver is designated TCM-22.

Between the Ida flyby and arrival at Jupiter, there are a number of solar conjunction and opposition events for Galileo: solar conjunction on November 3, 1993, solar opposition on May 18, 1994 (May 19 if the Ida flyby option is not exercised), solar conjunction again on December 1, 1994, and finally solar opposition on June 9, 1995.

The sequence of events for the interplanetary trajectory is summarized in Table I. Although the trip time to Jupiter is 6 years, the planetary and asteroid encounters on the interplanetary trajectory are distributed in such a way that there are no large gaps in time between major mission events. This progression can be seen clearly in the timeline shown in Figure 10.

2.4. JUPITER APPROACH/INITIAL ORBIT

On July 10, 1995, about five months prior to arrival at Jupiter, the Galileo atmospheric Probe will be released from the Orbiter. Since the Probe has no onboard propulsion

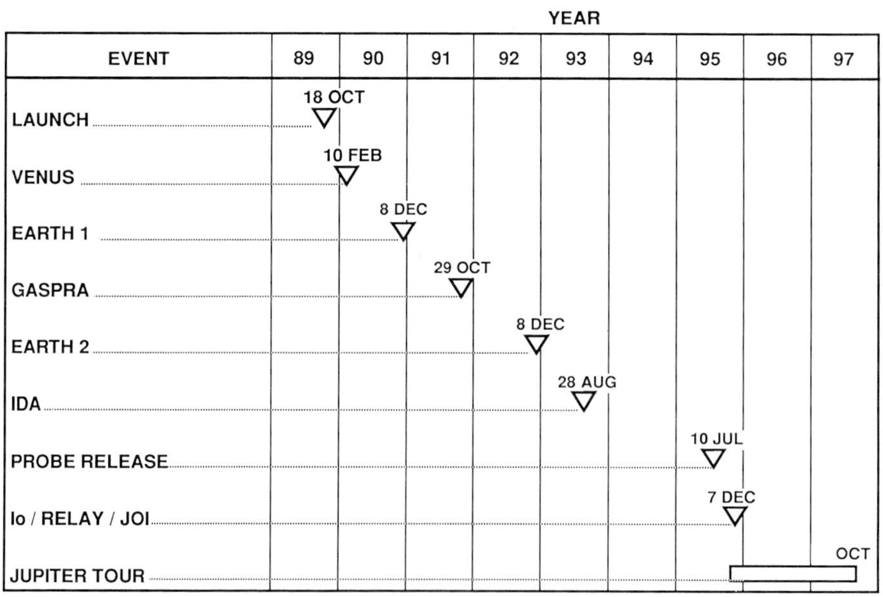

Fig. 10. Timeline of major Galileo events between launch and end of mission.

TABLE I

Galileo 1989 VEEGA interplanetary trajectory sequence of events

Event	Date	Flyby altitude, km	Sun–S/C distance, AU
Earth launch	18 Oct., 1989		
Venus flyby	10 Feb., 1990	16 123	
Perihelion	25 Feb., 1990		0.70
Aphelion	23 Aug., 1990		1.28
Earth flyby No. 1	8 Dec., 1990	960	
Perhelion	11 Jan., 1991		0.90
Gaspra flyby	29 Oct., 1991	1 600 [a]	
Aphelion	11 Jan., 1992		2.27
Earth flyby No. 2	8 Dec., 1992	300	
Perihelion	13 Dec., 1992		0.98
Ida flyby	28 Aug., 1993	TBD	
Probe release	10 July, 1995		
ODM	17 July, 1995		
Io flyby	7 Dec., 1995 [b]	1 000	
Probe entry	7 Dec., 1995 [b]		
JOI	8 Dec., 1995 [b]		

[a] Distance from center of asteroid.

[b] Event times:

Io flyby	7 Dec. 17:46 UTC	7 Dec. 09:46 PST
Probe entry	7 Dec. 22:04 UTC	7 Dec. 14:04 PST
Probe relay end	7 Dec. 23:22 UTC	7 Dec. 15:22 PST
JOI start	8 Dec. 00:27 UTC	7 Dec. 16:27 PST
JOI end	8 Dec. 01:16 UTC	7 Dec. 17:16 PST

system for trajectory corrections, the combined Orbiter/Probe spacecraft must be placed on a ballistic trajectory targeted such that, once released, the Probe will achieve the desired atmospheric entry aimpoint. Seven days after the Probe is released, the Orbiter performs an Orbit Deflection Maneuver (ODM). This 59 m s^{-1} maneuver establishes the Orbiter on a trajectory that will overfly the Probe during its descent such that the Probe relay link and the Jupiter Orbit Insertion (JOI) maneuver can be accomplished. Without the ODM, the Orbiter would follow the Probe into Jupiter's atmosphere.

The Orbiter and Probe trajectories at arrival at Jupiter are shown in Figure 11. About 4 hours before closest approach to Jupiter (4.4 hours before the start of the Probe relay), the Orbiter will pass 1000 km above the surface of the innermost Galilean satellite Io. The Io flyby trajectory is shown in Figure 12. This close Io flyby shows the Orbiter, thus reducing the orbit insertion ΔV by about 175 m s^{-1} and, in addition, provides the only opportunity during the mission for close-up observations of Io. As an added bonus which accrues from the selection of December 7, 1995 as the Jupiter arrival date, the Orbiter will have a relatively close (32 000 km altitude) flyby of Europa 4.6 hours before

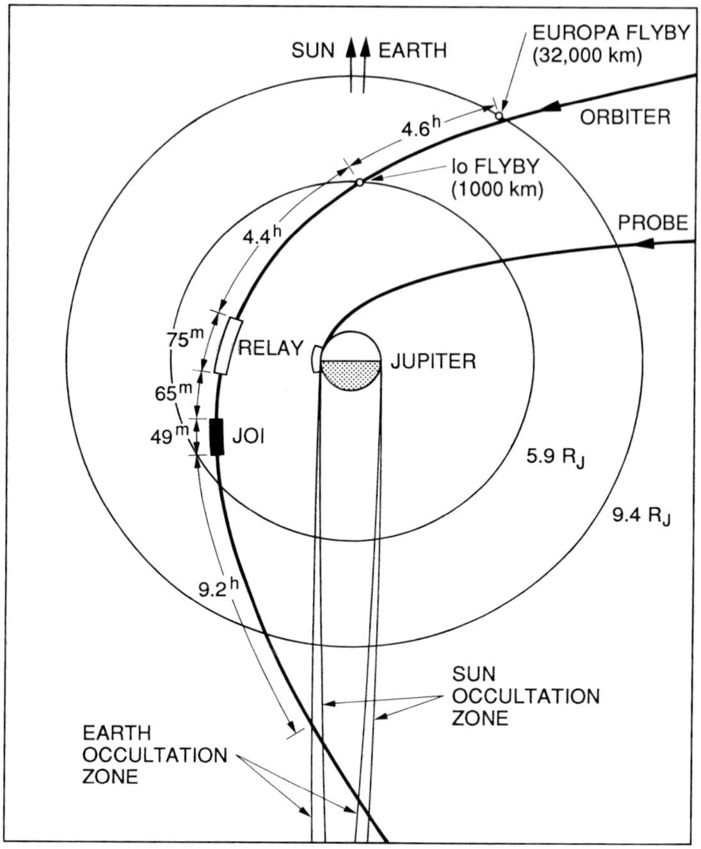

Fig. 11. Jupiter arrival (December 7, 1995): Io flyby, Probe relay, and Jupiter orbit insertion.

the Io flyby; the Europa flyby trajectory is shown in Figure 13. At 21:54 UTC on December 7, 1995, the Orbiter will make its closest approach to Jupiter at a distance of 4 R_J* (286 000 km), which is 3 R_J (214 500 km) above the Jovian cloud tops.

Eleven minutes after the Orbiter's closest approach to Jupiter, the Probe will enter the Jovian atmosphere. The Orbiter will relay Probe data back to Earth for up to 75 min starting 3 min after Probe entry. At 2.6 hours after Jupiter closest approach (65 min after the end of Probe relay), the JOI maneuver begins. This maneuver, which has a duration of 49 min, will impart a retro-velocity increment of 630 m s^{-1} in order to establish a 230-day orbit about Jupiter.

As the Orbiter heads out away from Jupiter after JOI, the Earth and Sun will both be occulted by Jupiter as seen from the spacecraft (see Figure 11). Entry into Earth occultation occurs 12.6 hours after Jupiter closest approach (9.2 hours after the end of

* 1 R_J = 1 Jupiter radius (equatorial) = 71 492 km.

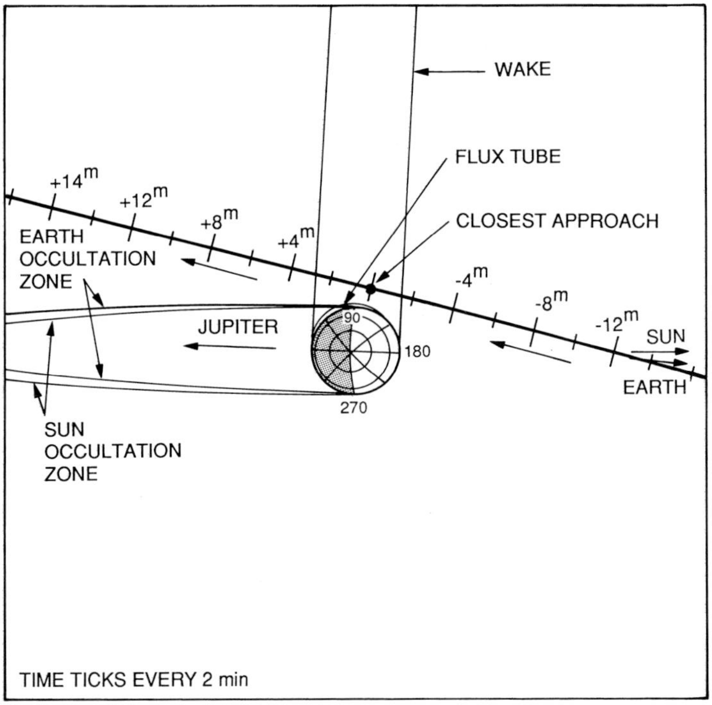

Fig. 12. Trajectory pole view of Io flyby 4 hours before Jupiter closest approach (December 7, 1995).

JOI). Entry into solar occultation occurs about 1 hour later. The durations of the Earth and Sun occultations are 3.5 and 3.3 hours, respectively.

Near apojove of the initial orbit, a Perijove Raise Maneuver (PJR) is performed; this maneuver occurs about four months after JOI (nominally on March 31, 1996). PJR increases the speed of the Orbiter by 370 m s^{-1} in order to raise the perijove distance to about 11.5 R_J and target the Orbiter to the first encounter of the satellite tour. It is necessary to raise the perijove distance at the start of the satellite tour because of the intense radiation environment at Jupiter. After the first 4 R_J perijove passage, the spacecraft has already received about one-third of the total permissible radiation dosage.

The period of the initial orbit and the perijove raise distance depend strongly on the satellite tour, which will not be selected until the summer of 1992. The values quoted above for orbit period, PJR date and ΔV magnitude, and perijove raise distance are nominal design values that are used for benchmark mission performance calculations. The representative satellite tour described in Section 4 requires an initial orbit period and a perijove raise distance that are slightly different from the nominal values.

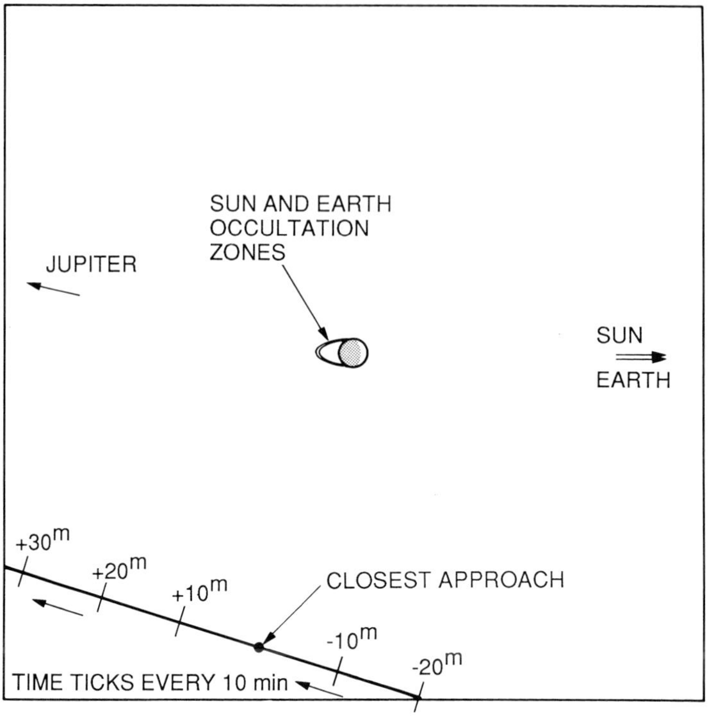

Fig. 13. Trajectory pole view of Europa flyby 9 hours before Jupiter closest approach (December 7, 1995).

3. Probe Mission

This section addresses the major elements of the design of the Probe mission with emphasis on the relationship between the Orbiter and Probe missions. The Probe constraints that affect the selection of the Jupiter arrival date and the choice of the Probe release time are treated first. These constraints are referred to as the Orbiter aspect angle constraint and the solar conjunction constraint. Next, those requirements that affect the Probe mission more or less independently of the Orbiter's mission are dealt with. The relevant parameters are the Probe's entry speed and flight-path angle and the ring plane crossing radius. These three parameters are then related to the process of Probe entry site selection. Next there follows a discussion of the Orbiter trajectory requirements that are relevant to the Probe mission and how these are integrated with the Probe requirements. The section concludes with a synopsis of the Probe atmospheric descent profile and sequence of events.

3.1. Probe mission design constraints

The following are the principal constraints governing the design of the Probe mission. The expected lifetime of the Probe battery precludes releasing the Probe from the Orbiter much earlier than 150 days before arrival at Jupiter; nor may Probe release occur during a period of approximately 35 days around solar conjunction. For safety reasons, the Probe must be at least 3.8 R_J from the center of Jupiter when it crosses the equatorial plane, which contains the ring discovered by Voyager. In order to survive the enormous thermal and mechanical stresses of the entry event, the Probe must enter the Jovian atmosphere within a narrow entry angle corridor of approximately $-7.2°$ to $-10.0°$ and with a speed relative to the atmosphere that is not greater than 47.8 km s^{-1}. Because of possible interference from rotating spacecraft booms, the relay antenna on the Orbiter must not be pointed more than 92° from the Orbiter's spin axis. These and other constraints affect a number of crucial mission design choices, including arrival date, the time at which to release the Probe from the Orbiter, and the Probe entry site.

3.1.1. *Arrival Date Selection*

Two Probe mission constraints bear directly on the choice of arrival date: the Orbiter aspect-angle constraint and the requirement to avoid solar conjunction near Probe release.

The Orbiter aspect angle is the angle between the Orbiter $+Z$ axis and the line of sight to the Probe during the relay period. If the Orbiter aspect angle is large enough, there is a danger of serious degradation to the relay link from the 'boom interference' effect described in the following paragraph.

The Probe data relay requires that the Orbiter simultaneously point its High Gain Antenna (HGA) at Earth and the Radio Relay Antenna (RRA) at the Probe. Because the HGA is boresighted along the Orbiter $-Z$ axis, the Orbiter must be oriented with the $-Z$ axis aimed at Earth. Consequently, for certain arrival dates, the RRA must look towards the magnetometer boom and the radioisotope thermoelectric generator booms, which are mounted on the spun side of the Orbiter, in order to point at the Probe. The presence of these booms in the RRA's field of view can cause a significant degradation of the signal from the Probe. For this reason, the Project applied an Orbiter aspect-angle criterion in the arrival date screening process, rejecting those dates for which the Orbiter aspect angle would exceed 92° at 41 min past entry.

In order to deliver the Probe accurately to its targeted entry conditions and subsequently to reconstruct the achieved conditions, navigation requires a substantial amount of high-quality tracking data in the time period near Probe release. If the S/X-band radio signal from the Orbiter to Earth passes too near the Sun, a significant degradation in the quality of radiometric data results. To protect these data, a constraint was adopted that prohibits placing the Probe release event within a period of approximately 35 days around solar conjunction.

3.1.2. *Probe Release Time*

In addition to the above constraint on the date of Probe release, there are important limitations on the timing of release in relation to arrival at Jupiter. First, there is a propellant penalty associated with performing Probe relase later (i.e., nearer to Jupiter arrival). The propellant cost of the ODM, which occurs seven days after Probe release, increases monotonically as it is performed later. For each 10 days of delay, about

5 m s^{-1} of additional ΔV is required. Secondly, the limited lifetime of the batteries that power the Probe after release from the Orbiter requires that release occur not much earlier than 150 days before Jupiter arrival. Finally, the accuracy of delivery of the Probe to its targeted entry conditions improves if the release is performed later. The tradeoff among these competing factors has resulted in setting the time for probe release at 150 days before Jupiter arrival.

3.1.3. *Probe Trajectory Design Constraints*

The major constrained parameters governing the design of the Probe's trajectory are: entry speed, entry flight-path angle, and the distance from Jupiter at which the Probe crosses the equator or ring plane.

The single most important factor determining the speed of the Probe at entry is the mass of the target body. Jupiter's great mass implies a typical inertial entry speed of about 60 km s^{-1}. Because such a high entry speed represents a severe challenge to Probe designers, it is of considerable importance to reduce this speed through judicious mission design. Direct reduction by means of interplanetary trajectory shaping or propulsive maneuvers near the planet would be prohibitively costly in terms of propellant consumption. A substantial indirect reduction *is* realizable, however, by an appropriate choice of the Probe entry site.

It is, of course, the Probe's speed relative to the atmosphere that must be minimized rather than the inertial speed. If the plane of the Probe's trajectory is such that the Probe is moving due east at the entry point, its speed relative to the atmosphere is simply its inertial speed minus the planet's rotational speed at the entry latitude. Since the maximum rotational speed of Jupiter (at the equator) is about 13 km s^{-1}, the minimum atmosphere-relative entry speed is approximately 47 km s^{-1}. Thus, the ideal entry latitude for minimizing entry speed is at the equator with the Probe moving due east. The Galileo Probe was designed for an entry near these ideal conditions with some margin to allow for entry a few degrees from the equator. The adopted requirement is that the entry speed must be less than 47.8 km s^{-1}.

Another critical entry parameter is the entry flight-path angle, which measures the steepness of the Probe's entry into the atmosphere. The flight-path angle is the angle between the local horizontal plane and the velocity of the Probe relative to the Jovian atmosphere. (Negative values indicate that the velocity vector is below the horizontal plane, which is necessarily the case at the entry point.) If the entry is too steep (magnitude of flight-path angle too large), the Probe can be destroyed by high levels of mechanical loading and heating. If entry is too shallow (magnitude of flight-path angle too small), the Probe bounces off or 'skips out' of the atmosphere like a stone skipping across a lake. The range of values between the skip-out flight-path angle and the angle at which physical damage to the Probe can occur is called the *entry flight-path angle corridor*. The boundaries of this corridor depend on the density and scale height* of the planet's atmosphere and on the ballistic coefficient and entry speed of the Probe. For

* Scale height is the altitude range over which the atmospheric density changes by a factor of e ($e \simeq 2.718$).

the Galileo Probe, the corridor lies between approximately -7.2 and $-10.0°$. This small range is further narrowed by consideration of navigation and other uncertainties affecting the delivery of the Probe to its target entry conditions. The Probe is targeted to an entry flight-path angle of $-8.6°$, the middle of the range given above.

It is known from Voyager observations that a ring of particulate material orbits Jupiter in its equatorial plane. The outer boundary of material visible to Voyager's camera is at about 1.8 R_J from the center of Jupiter. Since the ring material represents a potentially serious impact hazard to the Probe, the Galileo Project has adopted a requirement that the Probe may not cross the ring plane at a distance of less than 3.8 R_J from the center of Jupiter. This value allows for the possibility that there is additional material at greater distances from Jupiter than that seen by Voyager. In addition, dynamical analysis based on material observed demonstrates that no significant particle hazard exists at 3.8 R_J.

3.2. Probe Entry Site Selection

Figure 14 shows the great circle ground tracks of four possible Probe trajectories having the required entry flight-path angle of $-8.6°$. The dashed portion of each ground track represents a hypothetical continuation of the trajectory (beyond the entry point) that ignores aerodynamic effects on the Probe's motion. The ground tracks intersect at a point determined by the V_∞ vector of the Jupiter approach trajectory.

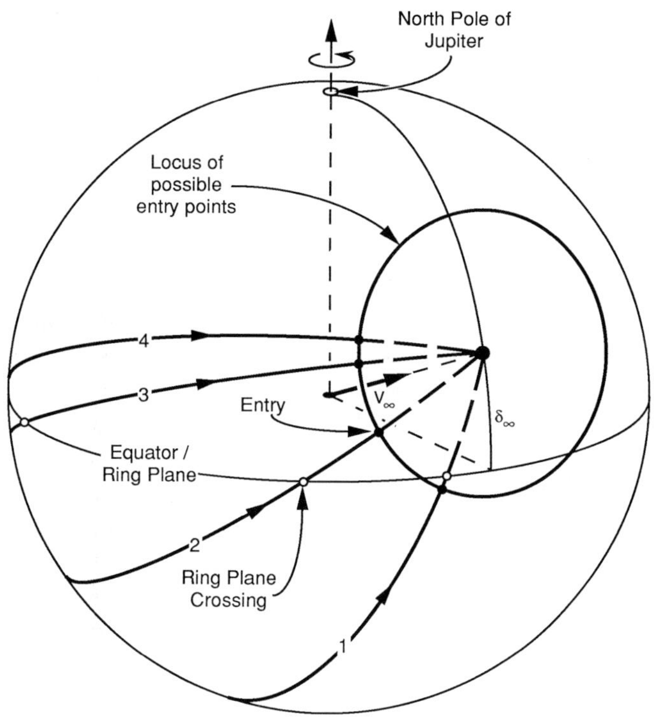

Fig. 14. Probe targeting geometry: effects of entry latitude on ring-plane crossing radius and entry speed.

The figure illustrates the set of all possible entry points having a fixed entry flight-path angle of $-8.6°$; this locus is approximated by a small circle on the planet's surface*. The size and location of this circle are determined by the magnitude and direction, respectively, of the V_∞ vector. Thus, the entry flight-path angle and the approach velocity together determine the range of entry latitudes that are accessible to the Probe.

Trajectory 1 is distinguished by a high inclination and an entry point in the southern hemisphere. On this trajectory, there is no hazard of impact with ring material because there is no ring-plane crossing (i.e., entry occurs before reaching the ring plane). The major disadvantage of Trajectory 1 is an excessive entry speed, a consequence of the fact that the Probe's inertial velocity is not closely aligned with the planet's rotational velocity. For this reason, entry in the hemisphere opposite to that of the V_∞ vector is ruled out.

On Trajectory 2 the Probe enters at a low latitude in the northern hemisphere, very shortly after crossing the equator. The Probe flies through the ring-plane relatively close to Jupiter and thus penetrates the visible part of the ring.

The plane of Trajectory 3 in the figure is chosen so that the entry latitude, φ_e, is equal to the declination of the approach V_∞ vector, δ_∞. In terms of angular distance along the trajectory, the ring plane crossing occurs relatively far from the entry point and so at a correspondingly large distance from Jupiter, although not large enough. This trajectory crosses the ring plane at about 2.0 R_J, which is well below the 3.8 R_J constraint. Note that, since the Probe's velocity at entry is very nearly parallel to the rotational velocity of the planet, the entry speed relative to the Jovian atmosphere is the lowest among all the trajectories shown.

Trajectories 1 through 3 illustrate the principle that the distance of the ring-plane crossing from Jupiter increases monotonically with entry latitude. It can be shown that the ring-plane crossing is at a distance at or above 3.8 R_J provided that φ_e is not less than 1.24 times δ_∞. Trajectory 4 represents a trajectory with an entry latitude chosen according to this rule. It is thus the only trajectory shown in the figure that meets all requirements.

Applying the entry latitude formula to the actual Galileo approach trajectory ($\delta_\infty = 5.3°$) gives the entry latitude of 6.6° (north) to which the Probe is targeted. The error in the achieved Probe entry latitude is expected to be $\pm 0.05°$ (99%). An error of this magnitude has no appreciable effect on the ring-plane crossing radius.

3.3. Probe/Orbiter trajectory compatibility

Independent of any considerations of the Probe mission or relay, the basic features of a desirable Orbiter trajectory are: (1) a close flyby of Io prior to the initial perijove, (2) a low-inclination trajectory, and (3) as low an initial perijove distance as possible.

The Io flyby is a required feature of the Orbiter trajectory because of the significant gravity-assist effect it has on the trajectory, which reduces the propellant expenditure required at the JOI maneuver.

* This discussion assumes that the Probe entry time is fixed, an assumption which is justified below in 'Probe/Orbiter Trajectory Compatibility'.

A low-inclination (near-equatorial) trajectory is desirable because the Orbiter's primary objective after completion of the Probe relay is a tour of the Galilean satellites. Since the satellites orbit Jupiter essentially in the equatorial plane, a near-equatorial trajectory both simplifies the design of the complex series of flybys comprising the tour and reduces propellant requirements by minimizing the need for costly plane-change maneuvers.

A low initial perijove is desirable because it further reduces the ΔV required for JOI. However, a perijove lower than 4 R_J has been ruled out because of hazards to the spececraft electronics from the intense radiation environment near Jupiter. The trajectory is consequently designed to have a perijove radius of 4 R_J.

Given that its perijove radius must be 4 R_J, there is a unique location (i.e., time) in each Io orbit at which a flyby can occur. The Orbiter's arrival time is therefore fixed by the requirement to fly by Io. This requirement also effectively fixes the Probe entry time, since the Probe's atmospheric descent must occur while the Orbiter is overhead in order to relay the Probe data to Earth.

The objectives of both Probe and Orbiter missions favor low-inclination trajectories. The Probe benefits by reduction of its entry speed while the Orbiter benefits by staying near the plane of the Galilean satellites. The overall mission benefits because it is easier to perform a data relay through the Orbiter during the Probe's descent into the Jovian atmosphere. Both missions also benefit from a low Orbiter perijove radius; the Probe benefits in terms of better link performance, and the Orbiter benefits from a decrease in propellant consumption.

3.4. Probe Mission Profile

Activation of the Probe science instruments begins at about six hours from entry (E), corresponding to an altitude* of 7.5 R_J. Three lightning and energetic particle samples are taken between $E - 3$ hr and $E - 2$ hr (i.e., between about 4 and 2 R_J altitude). At about $E - 17$ min, the Nephelometer and Atmospheric Structure Instruments are calibrated. Data taken before activation of the Probe transmitter are stored for playback to the Orbiter during the relay period.

Ten minutes after the Orbiter's perijove, the Probe enters the Jovian atmosphere. Probe entry occurs, by definition, when the Probe reaches an altitude of 450 km. Inertial entry speed is 59.9 km s^{-1}. Because the entry direction is nearly aligned with the planet's rotational velocity, atmosphere-relative speed is reduced to 47.4 km s^{-1}. Entry is at 6.6° north latitude. The nominal entry flight-path angle is $-8.6°$. In order to maximize stability and assure symmetrical distribution of the heat load, the Probe's attitude is controlled at release from the Orbiter such that the entry angle of attack is near zero.

In about 2 min of deceleration, an ablative heat shield of carbon phenolic material dissipates the enormous kinetic energy of entry, reducing the Probe velocity to Mach 1 (800 m s^{-1}). Maximum dynamic pressure is about 5.0×10^5 N m^{-2} (1.0×10^4 lb ft^{-2})

* All altitudes are relative to the 1-bar pressure level in the Jovian atmosphere.

and the maximum deceleration level is about 225 g*. At a Mach 1 altitude of about 49 km, the Probe deploys a parachute and jettisons the heat shield. The top-mounted medium-gain Probe antenna now points generally toward the Orbiter which is passing overhead. The Probe transmitter is activated and an uplink to the Orbiter is established by about 3 min after entry. Sensors are deployed into the Jovian atmosphere and data-taking begins on a pre-programmed schedule. The Probe's data system interleaves data recorded during the Jupiter approach and atmospheric entry and deceleration phases with real-time data. The Probe data are simultaneously recorded by the Orbiter and relayed to Earth in real time.

A possible water-cloud layer is traversed nominally between about $E + 22$ min and $E + 24$ min. The latest time of passage through the water-cloud layer is estimated to be at $E + 32$ min. At this time, the Orbiter performs the first of four adjustments of the pointing direction of its receiving antenna in order to maintain near-peak antenna gain. Subsequent pointing updates occur at 10 min intervals. Day-to-night terminator crossing occurs at about $E + 42$ min.

The Probe is designed to survive to at least the 10-bar pressure level (equivalent to an altitude of -92 km), which is reached nominally at $E + 38$ min, but no later than $E + 41$ min. It is likely that the Probe will continue to function at much deeper levels. The Probe and the relay link eventually succumb to a combination of high pressure, high temperature, Probe battery depletion, and degraded link performance margins. The Orbiter is configured to relay Probe data to Earth until 78 min after entry, giving a maximum relay duration of 75 min. This corresponds to an altitude of about -164 km and a pressure of 30 bars.

Figure 15 (not to scale) illustrates, in a planar approximation, the dynamic behavior of the link geometry as a function of the time past entry. Note that at signal acquisition, the signal enters the RRA to the left of its axis, while 30 min later it enters to the right of the axis. Therefore, at some point between $E + 2$ min and $E + 30$ min, the signal crosses the RRA axis. Near the end of Probe relay, the signal would arrive far off the RRA axis if the RRA were to remain inertially fixed. The antenna position shown at $E + 60$ min represents a repointing of the antenna to improve relay link performance in the later portion of the mission. (The dashed figure of the RRA indicates the unadjusted orientation.) It may also be inferred that environmental losses will increase during the mission as the Probe falls deeper into the atmosphere and penetrates the clouds, resulting in attenuation of the radio signal as the path length increases.

A representative sequence of events for the Probe mission is shown in Table II. Figure 16 illustrates the Probe descent profile in terms of pressure, temperature, and altitude as a function of time.

4. Satellite Tour

After insertion into orbit about Jupiter, the Galileo Orbiter will complete a series of 10 highly elliptical orbits about Jupiter during its baseline mission. Each orbit will contain

* One 'g' is the acceleration of gravity at Earth's surface and is equal to 9.806 m s^{-2}.

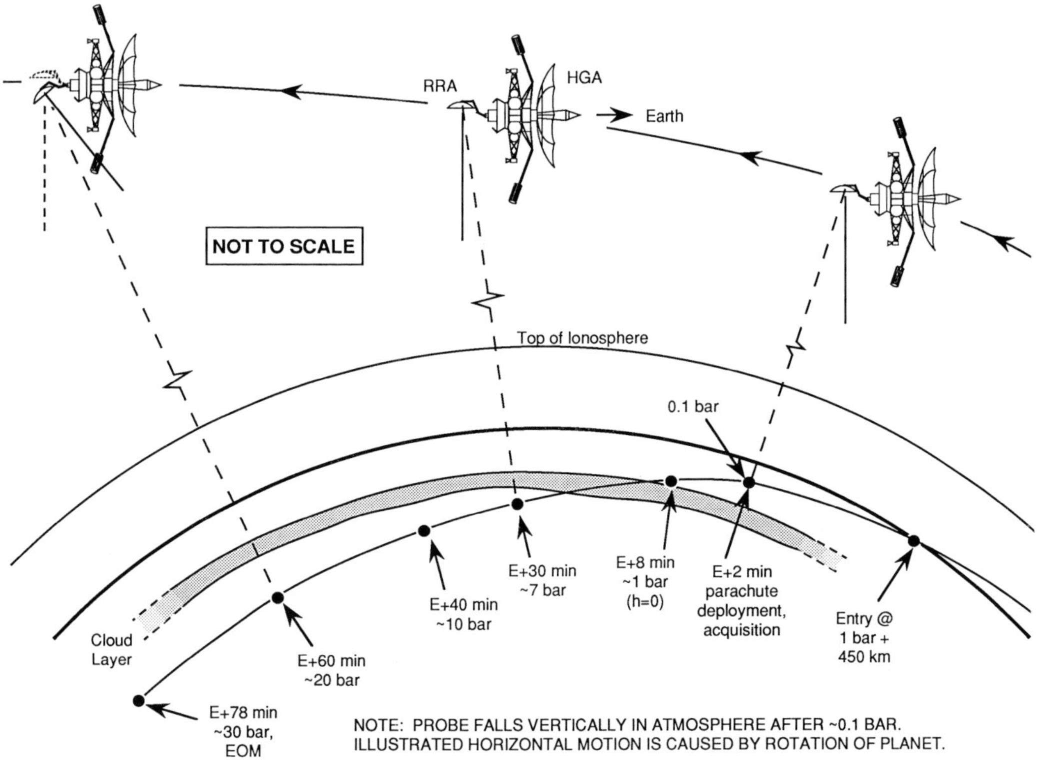

Fig. 15. Relay link geometry as function of time past entry.

a close 'targeted' flyby of one of the Galilean satellites. A targeted flyby is one where the Orbiter's trajectory has been designed to pass through a specified aimpoint at closest approach in order to use the satellite's gravitational influence to produce a desired change in the trajectory. This series of 10 orbits containing 10 targeted satellite encounters is termed a 'satellite tour'. The 10 orbits are to be completed within 23 months, and this constitutes the baseline operational mission. It is expected that an extended mission will proceed thereafter until all spacecraft expendables are exhausted. The Orbiter will, of course, continue to orbit Jupiter indefinitely.

Each targeted flyby is used to direct the Orbiter to the next flyby. At each satellite encounter, different aimpoints exist that allow the Orbiter to return to the same satellite, or to target to a different satellite. The abundance of aimpoints at each satellite encounter makes possible a large number of unique tours, each of which may satisfy many of the science objectives in a different manner. While it is relatively easy to design a tour to satisfy any individual science requirement, it is difficult to design a single tour to satisfy all the science requirements, because the trajectories needed to satisfy individual science requirements tend to be dissimilar. Therefore, strategies must be developed which maximize total science return during the tour. Tour design strategies are discussed below, and their application in a sample tour is shown.

TABLE II

Probe mission sequence of events

Time from entry	Probe altitude	Atmospheric pressure, bars	Event
– 165 days			Final Probe entry targeting maneuver
– 150 days			Probe release
– 143 days			Orbit deflection maneuver
– 140 days			Establish initial RRA actuator position
– 17 days			Update RRA pointing based on latest navigation data
– 6 hr	7.5 R_J		Initiate Probe pre-entry sequence
– 4.5 hr	5.8 R_J		Io encounter (stator movement allowed for Io science)
– 3 hr	4.1 R_J		First LRD/EPI sample; fix stator for remainder of Probe mission
– 17 min	0.5 R_J		NEP and ASI calibrations
0 s	450 km	5×10^{-8}	Probe entry
57 s	100 km	0.007	Maximum dynamic pressure (5×10^5 N m^{-2})
58 s	97 km	0.008	Maximum deceleration (225 g's)
112 s	49 km	0.08	Deploy main parachute
114 s	48 km	0.09	Jettison aeroshell; active radio transmitter
120 s	46 km	0.10	Nominally at 0.1 bar pressure
135 s	42 km	0.12	Begin relay receiver acquisition sequence
3.1 min	34 km	0.20	Latest time of acquisition (99.5% probability)
8.5 min	0 km	1.0	Nominally at 1 bar pressure
21 min	– 48 km	4.0	Orbiter overflies lngitude of Probe; minimum Probe aspect angle (2.5°)
22 min	– 51 km	4.3	Enter water-cloud layer
24 min	– 57 km	5.0	Exit water-cloud layer (nominal)
32 min	– 78 km	7.7	Exit water-cloud layer (latest), begin periodic repointing of RRA
38 min	– 92 km	10.0	Nominally at 10 bars pressure
42 min	– 100 km	11.7	Cross day-to-night terminator
78 min	– 164 km	30.0	End of Probe relay

RRA = Radio Relay Antenna
LRD = Lightning and Radio Emissions Detector
EPI = Energetic Particles Instrument
NEP = Nephelometer
ASI = Atmospheric Structure Instrument

4.1. Satellite tour design constraints

The design of a successful satellite tour is constrained by other factors in addition to the laws of orbital mechanics. After completion of RJR near apojove of the initial orbit about Jupiter, only a limited amount of propellant is available for spacecraft trajectory and attitude control during the satellite tour. Propellant is used in the tour to provide small Orbiter velocity changes both for deterministic maneuvers that are required for

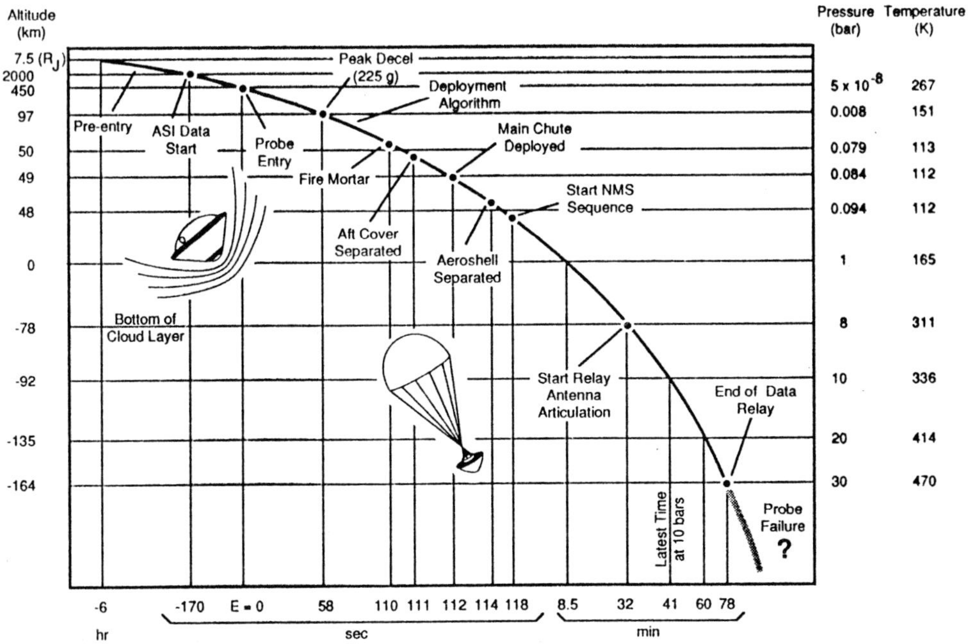

Fig. 16. Probe descent profile: pressure, temperature, and altitude as a function of time from entry.

the optimal, minimum-propellant trajectory and for statistical maneuvers that are required to correct for trajectory dispersions. In addition, propellant is used to repoint the Orbiter's HGA periodically to maintain communications with Earth and occasionally for special science-gathering opportunities. Plans for 10 orbits are based on predictions of propellant requirements during the tour and other parts of the mission, and assume a ΔV allocation of about 150 m s^{-1} (statistical plus deterministic). Any propellant remaining at the end of 10 orbits could be used to conduct an extended mission. After all propellant is exhausted, it will no longer be possible to provide the small trajectory adjustments needed to target the Orbiter to satellite flybys or to control its attitude to maintain the HGA pointed towards Earth. Effective communication from the Orbiter will then be lost.

Energetic electrons and ions (both protons and heavy ions) trapped in Jupiter's radiation belts can cause interference and damage in electronic parts in the Galileo Orbiter. The measure of radiation dosage adopted is the dose in kilorads (krad) that would penetrate shielding equivalent to 2.2 g cm^{-2} of aluminum (spherical shell analysis) in the Jovian radiation environment. To ensure that such effects do not seriously degrade spacecraft performance, the maximum acceptable radiation dose for the baseline mission has been set at 150 krad. The amount of radiation shielding incorporated in the spacecraft is such that no part should fail during the baseline mission as long as the design of the Orbiter's trajectory would cause no more than 150 krad to accumulate inside the 2.2 g cm^{-2} aluminum shell. It is necessary to raise the Orbiter's perijove above the more intense regions of Jupiter's radiation belts in order not to exceed the 150 krad limit. Incorporating significantly more radiation shielding in the Orbiter would have resulted in prohibitively large increases in the Orbiter's mass.

Since Io orbits deep within the radiation belts, only one Io flyby is permited during the mission. A close Io flyby is used on Jupiter approach to slow the Orbiter; this flyby decreases JOI by about 175 m s^{-1} and significantly reduces the amount of propellant needed to accomplish insertion into orbit about Jupiter. During the initial perijove passage at 4 R_J (after the Io flyby), the Orbiter absorbs 40–50 krad, approximately one-third of the dose allowed during the entire mission. PJR is performed near the first apojove in order to raise perijove high enough so that the 150-krad constraint is not exceeded during the remainder of the tour. In general, perijove must be kept near or beyond Europa's orbit during the remainder of the tour in order not to exceed the 150-krad constraint.

The requirement to navigate the Orbiter accurately necessitates that no more than one targeted satellite encounter occur on any one orbit. This requirement is due to the differences expected between the predicted and the actual post-flyby orbit which accrue due to pre-flyby maneuver execution errors, imperfect knowledge of the pre-flyby orbit and the orbits and masses of the satellites, and other factors. Furthermore, based on mission operations considerations, it has been adopted as a constraint on tour design that the minimum time between targeted encounters is 35 days. Navigational constraints also impose a minimum satellite flyby altitude of 500 km for the first targeted flyby, and 200 km for subsequent flybys.

4.2. Satellite Tour Gravity-Assist Considerations

The basis of the gravity-assist concept as applied to satellite tours is that, when a spacecraft is traveling in a system involving satellites orbiting a planet (Jupiter, in the case of the Galileo tour), the spacecraft's velocity with respect to the planet and, consequently, the characteristics of the planet-centered orbit can be changed significantly by a close satellite flyby. The gravity-assist concept is explained in more detail in the Appendix. As illustrated in Figure A-1, during a gravity-assist flyby the spacecraft's satellite-centered asymptotic velocity vector (V_∞) changes only in direction, but this change causes the spacecraft's velocity with respect to the planet to change in both magnitude and direction.

Satellite flybys can be used to change orbital energy with respect to the planet; an energy change is equivalent to a change in orbital period. For tours like that of Galileo in which the plane of the Orbiter's trajectory lies near the planes of the satellite orbits, an orbital period change is accomplished by flying approximately over the satellite's equator. Increasing the orbital period with respect to the central body is accomplished by flying behind a satellite's trailing edge; decreasing the orbital period involves flying ahead of its leading edge. (See Figure A-1.)

Satellite flybys can also be used to change the orbit inclination without changing the orbital period. For tours in which the Orbiter's trajectory lies near the planes of the satellite orbits, this is accomplished by flying approximately over one of the satellite's poles. Flying over a satellite at latitudes between the equator and a pole produces a change in both orbital period and inclination.

Flybys that change the orbital period also rotate the line of apsides and change the perijove distance. For a given orbital period change, a flyby that occurs far from the Orbiter's perijove rotates the line of apsides and changes perijove distance more than a flyby occurring close to perijove. A flyby occurring exactly at perijove changes orbital period with minimal changes to the line of apsides and perijove distance. The direction in which the line of apsides is rotated depends on whether the orbital period is increased or decreased and whether the satellite flyby occurs before perijove ('inbound') or after perijove ('outbound'). This is shown in Table III.

TABLE III

Orbit rotation rules

Flyby location	Energy (period) increasing flyby	Energy (period) reducing flyby
Inbound (pre-perijove)	Clockwise	Counterclockwise
Outbound (post-perijove)	Counterclockwise	Clockwise

Notes:
Clockwise rotation is in the direction from the initial orbit orientation (near the dawn terminator of Jupiter) toward the anti-Sun direction, or, equivalently, in the direction opposite to the orbital motion of the spacecraft.

Each of the Galilean satellites provides unique gravitational assist capabilities due to differences in their masses and orbit radii. Table IV provides a summary of important physical and orbital data for each of the Galilean satellites. These satellites are in

TABLE IV

Galilean satellite physical and orbital data

Satellite	*GM*, $km^3\,s^{-2}$	Radius, km	Orbital radius		Eccentricity	Inclination, deg	Period, days
			10^3 km	R_J			
Io	5934	1815	422.2	5.91	0.003	0.037	1.769
Europa	3196	1569	671.4	9.40	0.009	0.418	3.551
Ganymede	9885	2631	1071.1	15.00	0.001	0.254	7.155
Callisto	7172	2400	1883.4	26.38	0.008	0.163	16.689

Notes:
GM = Newton's gravitational constant times satellite mass.
1 R_J = 1 Jupiter radius = 71 492 km.

near-equatorial orbits, and the inner three satellites' orbital periods are locked in resonance. Europa's period is approximately twice Io's period, and Ganymede's period is approximately twice Europa's period. It should be noted that Jupiter has satellites other than the Galilean satellites; there are 12 additional known satellites, all of which are much smaller than the Galilean satellites. Because these satellites have very low masses, their gravitational influence on the Orbiter's trajectory is negligible. These smaller satellites are, therefore, not used for targeted flybys during the satellite tour. In any case, the radii and inclinations of their orbits would make close targeted flybys impractical.

Ganymede, the most massive satellite, is effective for changing orbital period and inclination. It is particularly effective at accomplishing large reductions in orbital period without loweirng perijove into the regions of strong radiation. Consequently, most candidate tour designs have more encounters with Ganymede than with any other satellite.

Callisto, the outermost Galilean satellite, is less massive than Ganymede and orbits at a greater distance from Jupiter. Therefore, Callisto is less effective than Ganymede in changing orbital period. However, because Callisto orbits further from Jupiter than Ganymede, a given period change at Callisto results in a greater change in perijove distance and a greater apsidal rotation than at Ganymede. Although not as massive as Ganymede, it can be more effective in changing inclination because the Orbiter's velocity is lower at the greater distance from Jupiter at which Callisto is encountered.

Europa, the least massive of the Galilean satellites, is least effective in changing orbital period and inclination. Since the Orbiter's perijove cannot be much closer to Jupiter than the distance at which Europa orbits due to radiation considerations, targeted flybys with Europa must occur close to perijove. Therefore, the amount of change in perijove distance and rotation of the line of apsides caused by energy changes at Europa is small. Observations of Europa are a high-priority scientific objective, which makes it necessary to include as many Europa flybys as possible in the tour.

4.3. Satellite tour science requirements

The scientific investigations to be performed by the Orbiter can be divided into three areas: investigations of Jupiter's atmosphere, investigations of the fields and particles in Jupiter's magnetosphere, and investigations of the Galilean satellites (Io, Europa, Ganymede, and Callisto).

The angle measured clockwise at Jupiter from the Sun direction to the apojove point, referred to as the 'orbit orientation', is an important science consideration for reasons that will become apparent. Arrival conditions at Jupiter fix the orientation of the initial orbit at about 126°. Due to the motion of Jupiter around the Sun, the orbit orientation increases with time at a rate of about 2.5° per month. During the interval of approximately 18 months from PJR to the end of the satellite tour, the total amount of 'drift' in orbit orientation is 46°. This means that the orbit orientation at the end of the tour would be 171° (that is, within 9° of the anti-Sun direction) if changes in orbit orientation caused only by orbit drift are considered. Period-changing targeted flybys (which, as we recall, rotate the line of apsides) can be used to add to this drift in orbit orientation (to *rotate* the orbit) or to counteract it (to *counterrotate* the orbit).

Figure 17, referred to as a 'petal plot' because of the resemblance of the orbits to petals of a flower, illustrates the Jupiter-centered orbits in the sample tour presented here. The

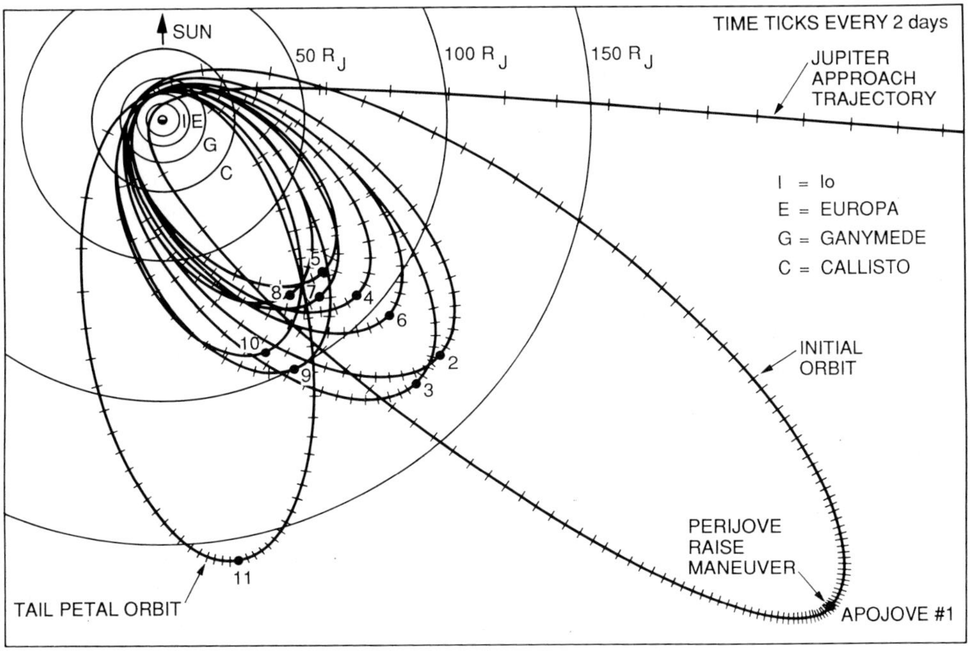

Fig. 17. Satellite tour petal plot (representative tour).

coordinate system used in this figure is one in which the direction to the Sun is fixed. The rotation of the orbit from the initial orientation to the anti-Sun direction during the tour, which is due to both the satellite encounters and the motion of Jupiter about the Sun, is apparent in this figure.

4.3.1. *Observations of Jupiter's Atmosphere*

Both the Solid State Imaging (SSI) camera and the Near Infrared Mapping Spectrometer (NIMS) are used to observe Jupiter's atmosphere. Cloud features and other atmospheric dynamics are observed in the visual spectrum by the SSI camera. These SSI observations can only be made of the sunlit portions of Jupiter and are best done at great distances in order to view the planet with Galileo's narrow angle camera. Over Jupiter's dark side, the utility of the SSI camera is limited, although it can be used some atmospheric phenomena such as lightning. Unfortunately, the spacecraft spends a great deal of time over Jupiter's dark side. Figure 17 shows that the time spent by the Orbiter over sunlit portions of Jupiter decreases as the orbit rotates toward the anti-Sun direction. The NIMS instrument is used to image atmospheric features over a wide spectral range. NIMS imaging is conducted over both sunlit and dark portions of Jupiter; however, the small amount of time available for light-side observations makes the time over the sunlit side a valuable commodity for NIMS as well as SSI.

In order to maximize the amount of time available to observe sunlit portions of Jupiter, the orientation of the spacecraft's orbit must be kept close to the initial orientation for as long as possible. Consequently, targeted flybys near the beginning of the tour are designed to counteract the clockwise orbit drift.

4.3.2. *Magnetospheric Science*

The two most important magnetospheric science requirements are: (1) to pass through the Jovian magnetotail (specifically, within $10°$ of the anti-Sun direction) at a distance of at least 150 R_J and (2) to pass through the 'wake' and 'flux tube' regions surrounding each satellite. The magnetotail streams out in a shape roughly resembling a windsock in the direction opposite the Sun from Jupiter. Wakes are created as charged particles trapped in Jupiter's magnetic field sweep by the satellites. Jupiter's magnetic field rotates at the same rate that Jupiter revolves on its axis, and this rate is much faster than the rates at which the satellites orbit the planet. Therefore, the wakes stream out in front of each satellite. Flux-tube regions connect the satellites, rotating through Jupiter's magnetic field, with the Jovian ionosphere along magnetic field lines. Flux tubes are located approximately over the satellites' poles.

If the first few flybys in the tour are used to keep the orbit close to the initial orientation to maximize the time available for observations of Jupiter's atmosphere, the drift in orbit orientation due to Jupiter's motion about the Sun is not sufficient to reach the anti-Sun direction, where the magnetotail lies, by the end of the tour. Therefore, some subsequent flybys must be employed to help rotate the orbit toward the anti-Sun direction in order to achieve magnetotail passage at the end of the tour. The conflicting needs of atmospheric observations and magnetotail passage make it desirable to use as many flybys as possible at the start of the tour to counterrotate and to use all subsequent flybys to rotate toward the magnetotail, achieving the magnetotail passage on the last orbit. The last satellite flyby must increase the orbital period to approximately 90 or more days in order to achieve a distance of 150 R_J at the apojove of the 'tail petal' orbit (the orbit on which magnetotail passage is achieved).

Wake passes are achieved with flybys near a satellite's equator and in front of the satellite. Thus wake flybys reduce the orbital period. Flux-tube passes are achieved by flying nearly over a satellite's pole. Thus, flux-tube flybys change inclination, but do not appreciably change the orbital period. The requirement to include in the tour flux-tube passes, which do not appreciably change orbital period, complicates the orbit-orientation strategy further. The tour cannot consist solely of counterrotating flybys at the beginning followed by rotating flybys at the end; there must be a few flybys designed to produce flux-tube passes, which provide little or no apsidal rotation (i.e., which neither rotate nor counterrotate the orbit).

4.3.3. *Satellite Imaging*

4.3.3.1. *Satellite Imaging Priorities.* The goal given the highest priority in satellite imaging is to maximize the number of SSI images at resolutions* of 80 m or better. The SSI camera is capable of obtaining this resolution at altitudes of about 4000 km or less when passing over the sunlit side of a satellite. Unfortunately, due to competing constraints and requirements on the trajectory (some of which have been enumerated above), closest approach cannot always be situated over the sunlit side of the satellite. In fact, in previously designed tours, about half the closest approaches in the tour occur over the dark side. Also, the SSI has a narrow field of view, and the Orbiter passes so quickly over the satellite surface near closest approach that there is not enough time to mosaic all the available surface area. Therefore, it is possible to image only a small portion of each satellite at high resolution.

The next highest priority for SSI imaging is placed on images with resolutions better than 250 m with low-Sun conditions. Low-Sun conditions with respect to a surface region or feature are defined as follows: (1) the Sun-surface-zenith angle (or incidence angle) must be greater than 45° in order to create shadows, and (2) the Sun-surface-spacecraft angle (or phase angle) must be greater than 30° to allow shadows to be seen. Shadows aid in determining the dimensions of surface features. Low-Sun coverage is obtained when the Orbiter passes nearly over the satellite's terminator. To obtain resolutions of better than 250 m, the Orbiter must pass within about 12 500 km of the surface.

The third priority SSI goal is to obtain coverage under both high- and low-Sun conditions at lower resolutions (1 km or better). These resolutions are obtained at distances of about 50 000 km or less, which are achieved along the approach and departure asymptotes of target flybys.

Satellites are also imaged with the NIMS instrument which measures the spectrum of sunlight reflected from the surface in order to determine satellite composition. High-Sun coverage is needed by NIMS in order to minimize local temperature variations from shadows. The highest priority NIMS objective is to maximize the amount of coverage obtained at resolutions** of 25 km or better under high-Sun conditions. Because of the lower resolving capability of the NIMS instrument, the distance at which NIMS coverage is obtained at 25-km resolution is the same at which SSI coverage is obtained at 1-km resolution (per line pair). In the interest of brevity, only SSI resolutions are used in the following discussion; however, 1-km SSI resolution may be taken to be synonymous with 25-km NIMS resolution.

4.3.3.2. *Satellite Longitudinal Coverage.* Each of the Galilean satellites is in synchronous rotation with Jupiter; that is, the same satellite hemisphere always points

* The resolution of the SSI camera is expressed per line pair, where a line pair corresponds to two rows of camera picture elements; the resolution per line pair for the SSI camera is 2×10^{-5} radians.

** The resolution of the NIMS instrument is expressed in terms of a single NIMS picture element and is equal to 5×10^{-4} radians.

toward the planet. (According to IAU convention, the prime meridian of each satellite is located at the approximate center of the hemisphere facing Jupiter.) Because the same hemisphere always faces Jupiter, changes to the shape, size, and orientation of the spacecraft's orbit do not appreciably change the 'real estate' viewed along the approach or departure asymptotes to a flyby. Consequently, the approach to an outbound flyby (that is, one occurring after perijove) allows viewing of the region near 0° longitude, which faces Jupiter, while the approach to an inbound (pre-perijove) flyby allows viewing of the region near 180° longitude, which faces away from Jupiter. This is illustrated in Figure 18. Therefore, coverage of both hemispheres of a satellite requires at least two separate encounters, one inbound and one outbound.

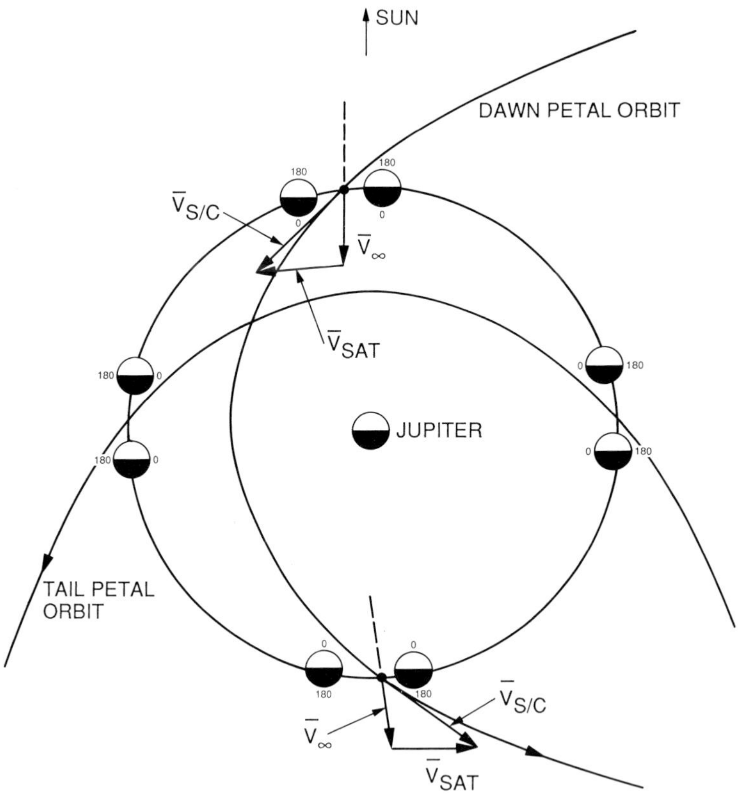

Fig. 18. Flyby geometries and satellite longitudes during Galileo satellite tour.

The areas near 90° and 270° longitude can never be viewed from the asymptotes. This gives rise to gaps in coverage at 1-km resolution near the 90° and 270° loingitude regions, which unfortunately coincide with regions not observed by the Voyager spacecraft. Both the Voyager and Galileo trajectories are posigrade, with perijoves far beneath the orbits of Ganymede and Callisto.

The areas viewed at closest approach to a satellite are nearly 90° in longitude away from the centers of the areas viewed along the asymptotes. Therefore, the areas viewed at high resolution near closest approach lie near 90° and 270° longitude regions; however, the total area viewed at high resolutions is too small to fill the coverage gaps.

It is possible to fill the coverage gaps by designing satellite flybys with closest approach distances between about 20 000 km and 50 000 km. At such flybys, the closest approach (which always occurs over a region near 90° or 270° longitude) is far enough away to mosaic the entire satellite at 1-km resolution. The gravitational effect of these flybys on the Orbiter's trajectory is small, due to the much greater distance at which the flyby occurs. These distant flybys are called 'nontargeted' flybys, because their flyby aimpoints are not tightly controlled, as is the case for targeted flybys. It is permissible for a single orbit to contain one nontargeted flyby in addition to the targeted flyby.

It is, of course, desirable to include as many nontargeted flybys as possible in the tour. The availability of nontargeted flyby opportunities depends on the orbit-orientation, perijove distance, and targeted encounter time. Therefore, the tour must be designed to arrive at the combinations of orientation and perijove distance necessary to achieve nontargeted encounters in conjunction with targeted encounters at various points in the tour. This, of course, must be accomplished within the constraints of the orbit orientation profile discussed above (which itself is a compromise between the conflicting requirements to maximize time over Jupiter's sunlit side and to reach the magnetotail at the end of the tour) and the perijove profile required to manage the Orbiter's exposure to radiation.

Because of the resonance between the orbital periods of Europa and Ganymede, opportunities to encounter both of these satellites (on the same orbit) can occur on succeeding orbits at certain points in the tour. Callisto's orbital period is not resonant with the other satellites; therefore, satellite pairings involving Callisto are less frequently available. Tours designed for the 1986 Galileo launch opportunity featured as many as six nontargeted flybys. The sample tour included here, which was developed as part of a preliminary analysis of tour opportunities for the 1989 launch opportunity, contains three nontargeted flybys, one each with Europa, Ganymede, and Callisto. In addition, a nontargeted flyby with Europa occurs on approach to Jupiter before the Io flyby.

4.3.3.3. *Satellite Lightning Conditions.* Although the 'real estate' viewed along the asymptotes does not change with orbit orientation, the lighting conditions under which this 'real estate' is viewed do change. Figure 18 shows that for orbits whose apojove lies nearly over Jupiter's dawn terminator ('dawn petal' orbits), the Orbiter sees a nearly full moon from the approach asymptote, passes near the terminator region at closest approach, and on departure sees an almost completely darkened satellite. Good high-Sun coverage at lower resolutions is obtained on the approach asymptote to these flybys, and good low-Sun coverage at high resolution is obtained near closest approach. The orbit orientation is closest to the dawn petal early in the tour.

For orientations near the tail petal, which occur late in the tour, the Orbiter sees a half-moon satellite on the approach and depature departure asymptotes, and either a full moon or a nearly completely darkened satellite at closest approach, depending on whether the Orbiter passes in front of or behind the satellite. Low-Sun coverage at 1-km resolution is obtained along the approach and departure asymptotes to these flybys, and high-Sun coverage at high resolution is obtained near closest approach, if the Orbiter passes over the sunlit side of the satellite. Of course, if the Orbiter passes over the dark side, no visible imaging (SSI) coverage will be obtained at closest approach. However, if the closest approach altitude is low enough, the Orbiter may attain a low enough altitude while still over the sunlit portion before crossing the terminator to allow some high-resolution coverage to be obtained.

4.3.3.4. *Satellite Imaging Summary.* In summary, high-Sun coverage of the 0° and 180° longitude regions at 1-km resolution is obtained on the approach asymptotes to target flybys *early* in the tour. Low-Sun coverage of these regions at 1-km resolution is obtained on the approach asymptotes to targeted flybys *late* in the tour. With regard to high-resolution coverage (250 m or less), the total area imaged is small, because the Orbiter does not spend considerable time at low altitudes. The regions near 90° and 270° longitude can be covered at 1-km resolution only during nontargeted flybys. Nontargeted flybys occurring *late* in the tour can completely fill these regions. Non-targeted flybys occurring *early* in the tour can only partially fill these regions, because the Orbiter sees only a partially sunlit view of the satellite near closest approach. The three nontargeted flybys included in the sample tour fall near the end of the tour. The satellite coverage obtained from these flybys is close to the best that can be achieved with nontargeted flybys.

4.3.4. *Radio Science*

Radio science experiments support all three major science areas: atmospheric science, magnetospheric science, and satellite science.

When the spacecraft passes behind Jupiter as viewed from Earth, radio signals from the Orbiter are not cut off. Instead, they are refracted by the thick Jovian atmosphere on their way to Earth. Polarized radio signals are also influenced by the magnetic field of Jupiter, through a phenomenon called the Faraday effect. Because a great deal of information on the atmosphere and magnetic field can be learned by analysis of the refracted polarized signals, passes behind Jupiter are desirable in the tour. Such passes are called occultations of Earth by Jupiter as viewed from the Orbiter. Occultations of the Sun by Jupiter as viewed from the Orbiter (i.e., passages through Jupiter's shadow) are also desirable, because they offer an opportunity to observe lightning and other atmospheric phenomena best seen in the dark.

Fortuitously, for tours associated with the 1989 launch opportunity, Jupiter's equator is viewed nearly edge-on from Earth. Occultations of Earth by Jupiter as viewed from the Orbiter are, therefore, achieved on many orbits without the use of flybys to change orbital inclination.

Occultations of Earth by the Galilean satellites as viewed from the Orbiter are also desirable. These occultations provide information on the satellite's radius and offer the opportunity to look for tenuous atmospheres and ionospheres. Occultations by Io, where the radio signal may be distorted due to outgassing from volcanic plumes, are particularly important. Since radiation considerations prohibit more than one encounter with Io, and since no occultation occurs during the single Io flyby on approach to Jupiter (see Figure 12), any occultation of Earth by Io must occur at a large distance. Therefore, orbital inclination must be tightly controlled to be assured of passing behind a small satellite at a large distance. It is necessary to use the targeted flyby immediately preceding a distant satellite occultation to change inclination to the precise value needed to achieve the occultation.

4.4. Representative Satellite Tour

A representative satellite tour is presented here to illustrate the application of the tour design strategies discussed above. This sample tour was designed as part of preliminary efforts to categorize characteristics of tours associated with the 1989 Galileo launch opportunity. This tour satisfies many of the science objectives; however, further investigation is required before a final tour is selected. The final tour will be selected in the summer of 1992 from a group of tours that will be designed as part of future efforts.

This sample tour begins at PJR. In designing PJR, it is necessary to take into account the fact that, since the period of the insertion orbit is 222 days, the first several flybys must reduce orbital period (and, therefore, depress perijove into regions of stronger radiation) in order to comply with the constraint that the duration of the tour not exceed 22 months. For this sample tour, PJR adds 361 m s^{-1} to the Orbiter's speed in order to raise perijove to 10.8 R_J and target the Orbiter to the first encounter of the tour.

The initial perijove distance affects the timing of the first Europa encounter in the tour. Since Europa orbits at a radius of 9.4 R_J, the first Europa flyby cannot take place until the perijove is reduced below that value. In this sample tour, and in most previously designed tours, the first encounter with Europa occurs on the fourth orbit.

Table V shows a summary of pertinent information about each satellite encounter in the tour. In the table, and in the following discussion, a flyby is described with the satellite name and the orbit number (e.g., Ganymede 1); an 'N' is added for nontargeted flybys (e.g., Europa 5N).

The individual satellite flybys are described briefly below. Figures 19–31 are trajectory pole views of each of the flybys, illustrating passages through regions of scientific interest such as wakes, flux tubes, and Sun and Earth occultation zones. Because the views are from the trajectory pole, the viewer is situated over a satellite pole for an equatorial flyby, in which case wake regions are visible, and over the satellite equator for a polar flyby, in which case flux tube regions are visible.

4.4.1. *Ganymede 1 Encounter*

The first encounter of the satellite tour must reduce orbital period significantly, so that the baseline tour can be completed within the 23 months following arrival at Jupiter. Since Ganymede is the most effective satellite for changing orbital period, the first encounter of the tour is with Ganymede (as is the case with most past tour designs). The Ganymede 1 flyby reduces the orbital period from 222 days to 71 days. Since the

TABLE V

Representative satellite tour

Encounter	Date	Satellite	Inbound/ Outbound	Altitude, km	Latitude, deg	Objectives
1	18 July, 1996	Ganymede	In	616	29	Wake, UVS, gravity, reduce period
2	28 Sept., 1996	Ganymede	In	555	87	Flux tube, gravity, reduce inclination
3	7 Dec., 1996	Callisto	In	455	10	Wake, UVS, counterrotate orbit, Jupiter occultation
4	23 Jan., 1997	Europa	In	1741	26	Wake, UVS, counterrotate orbit, Jupiter occultation
5N[a]	27 Feb., 1997	Europa	In	35058	1	Coverage (114° W longitude, $\varphi = 65°$)
5	28 Feb., 1997	Ganymede	Out	2088	10	Counterrotate orbit, high-resolution coverage, Jupiter occultation
6	22 Apr., 1997	Europa	Out	499	33	Rotate orbit, Jupiter and satellite occultations, wake
7	31 May, 1997	Europa	Out	354	−65	Flux tube, Jupiter occultation
8	4 July, 1997	Ganymede	In	1213	−48	Rotate orbit, high-resolution coverage
8N[a]	6 July, 1997	Callisto	Out	37157	−25	Coverage (284° W longitude, $\varphi = 56°$)
9N[a]	23 Aug., 1997	Ganymede	In	25002	−17	Coverage (93° W longitude, $\varphi = 37°$)
9	26 Aug., 1997	Callisto	Out	6001	−60	Rotate orbit, Jupiter and (distant) Io occultations
10	5 Oct., 1997	Ganymede	In	609	−1	Rotate orbit, Jupiter, and satellite occultations
Tail petal:						
Entry	9 Nov., 1997[b]					150 R_J from Jupiter, $\varphi = 177°$, 0.3° inclination, 23 months after JOI
Apojove	22 Nov., 1997					157 R_J from Jupiter, $\varphi = 170°$, 0.3° inclination, 23.5 months after JOI

[a] Nontargeted encounter.

[b] Tail petal entry is 1 month beyond nominal end-of-tour date.

φ = phase angle.

Total deterministic ΔV: 56 m s^{-1}.

Total radiation: 127 krad.

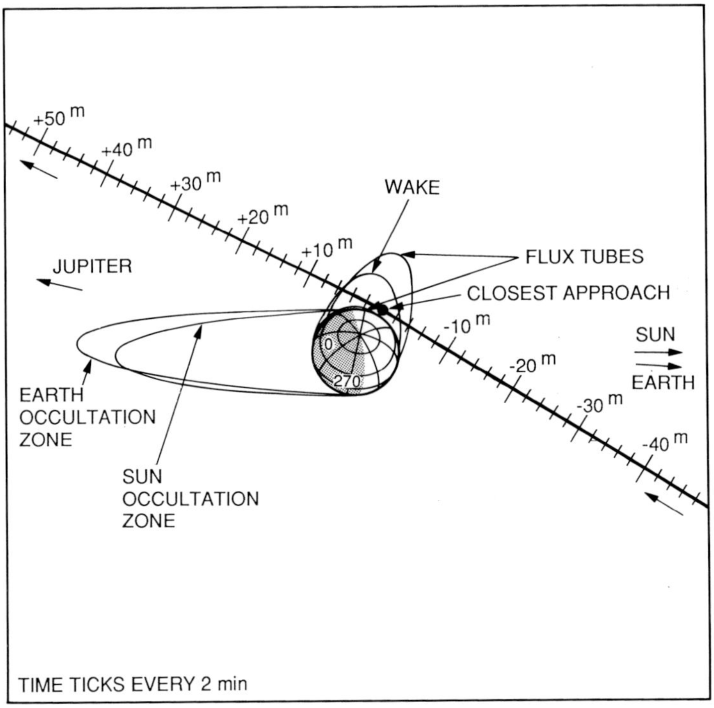

Fig. 19. Trajectory pole view of Ganymede 1 encounter.

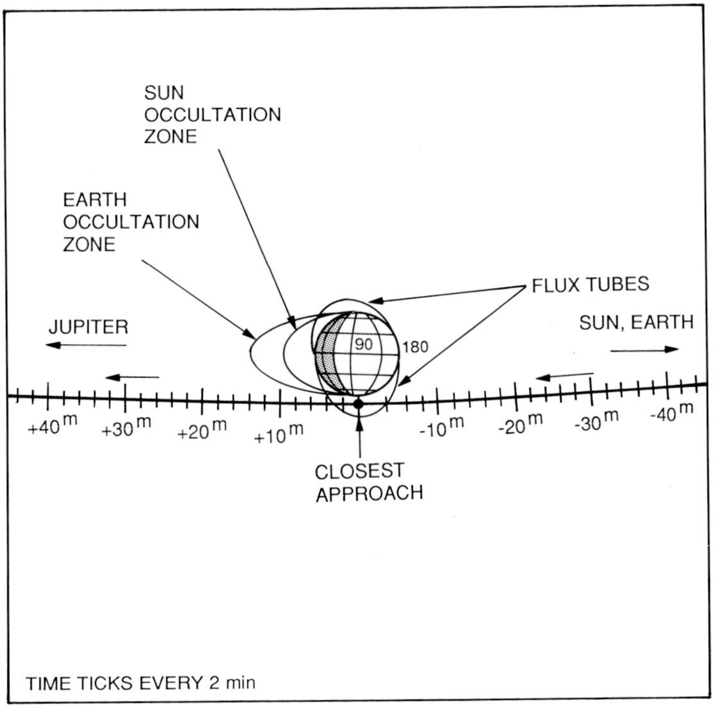

Fig. 20. Trajectory pole view of Ganymede 2 encounter.

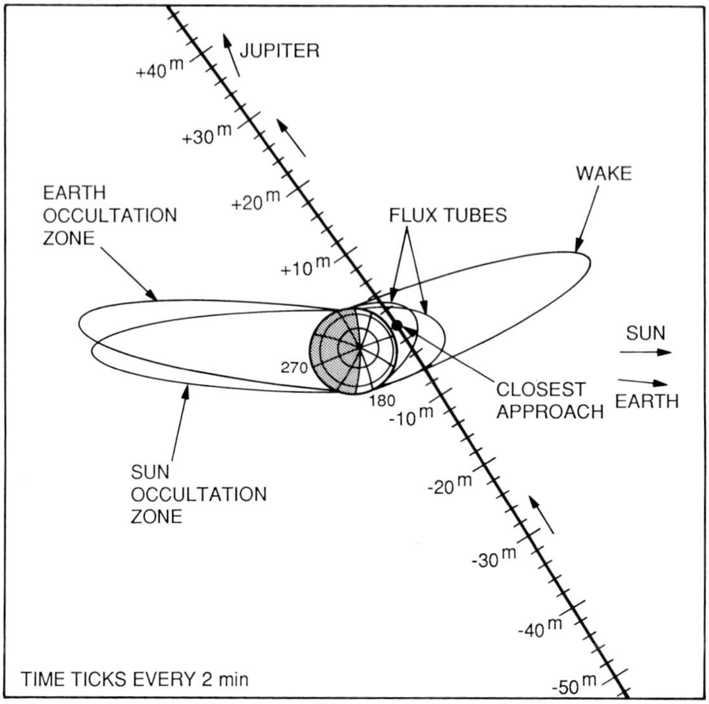

Fig. 21. Trajectory pole view of Callisto 3 encounter.

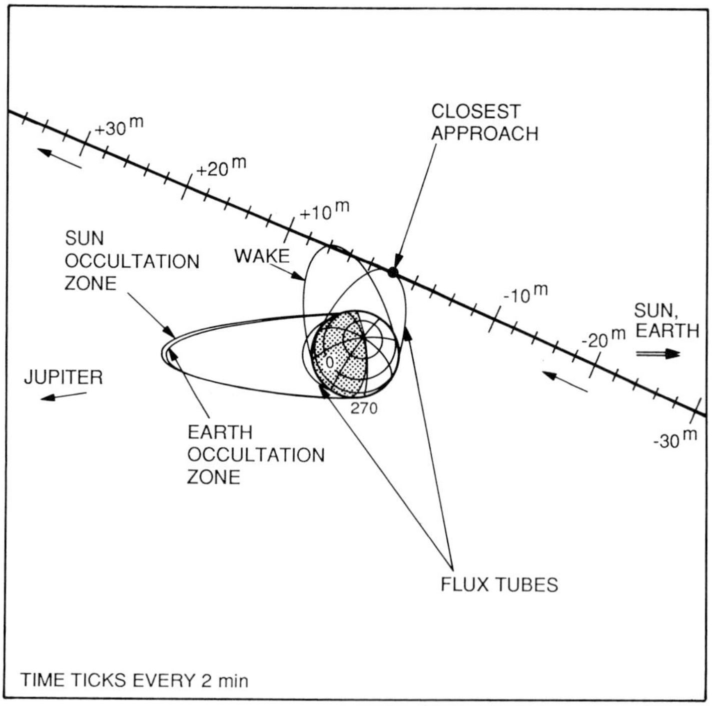

Fig. 22. Trajectory pole view of Europa 4 encounter.

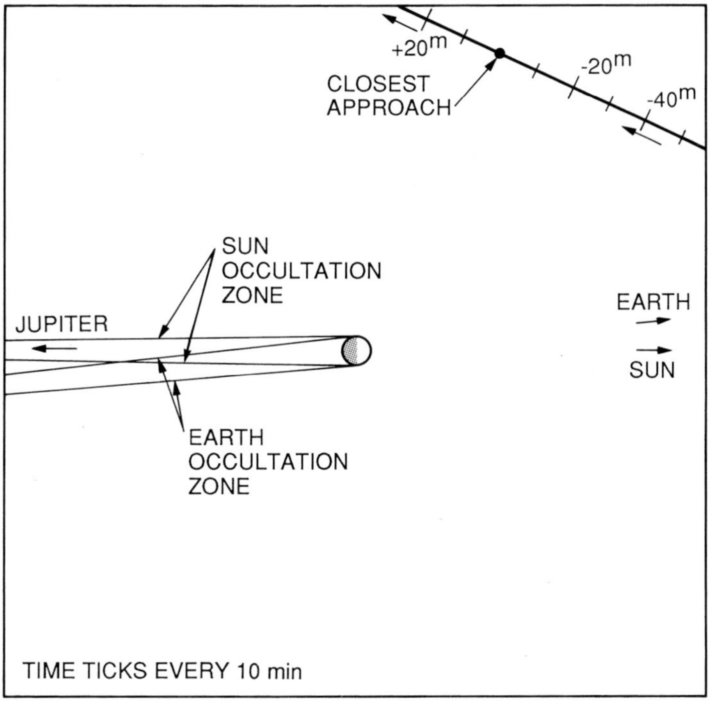

Fig. 23. Trajectory pole view of Europa 5N nontargeted encounter.

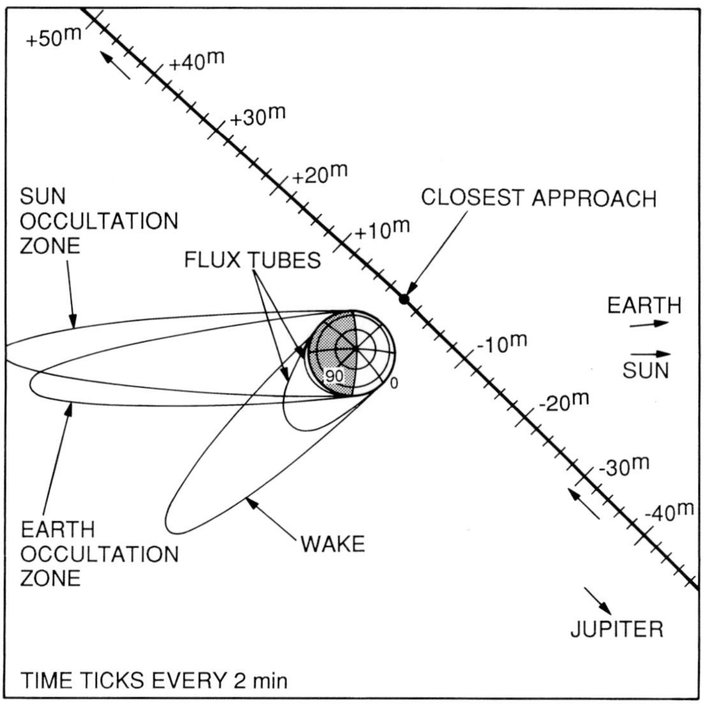

Fig. 24. Trajectory pole view of Ganymede 5 encounter.

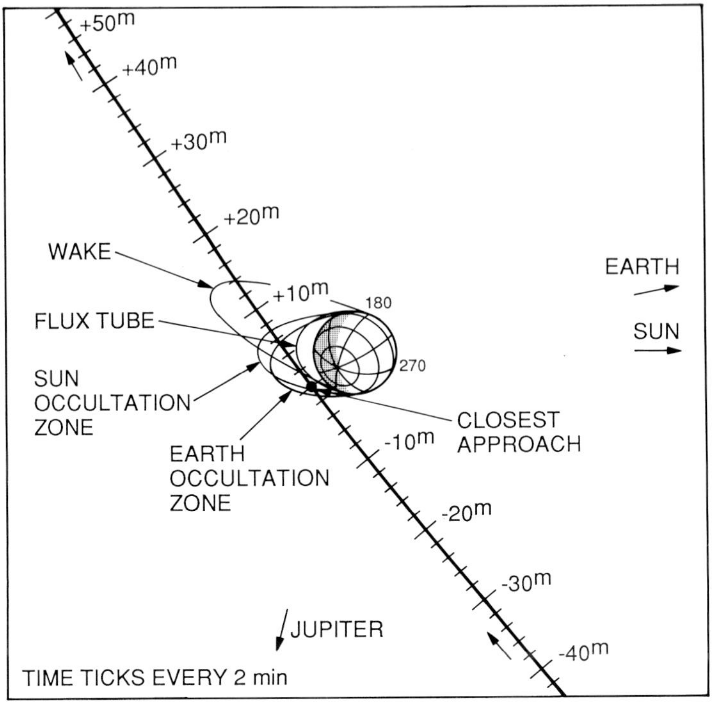

Fig. 25. Trajectory pole view of Europa 6 encounter.

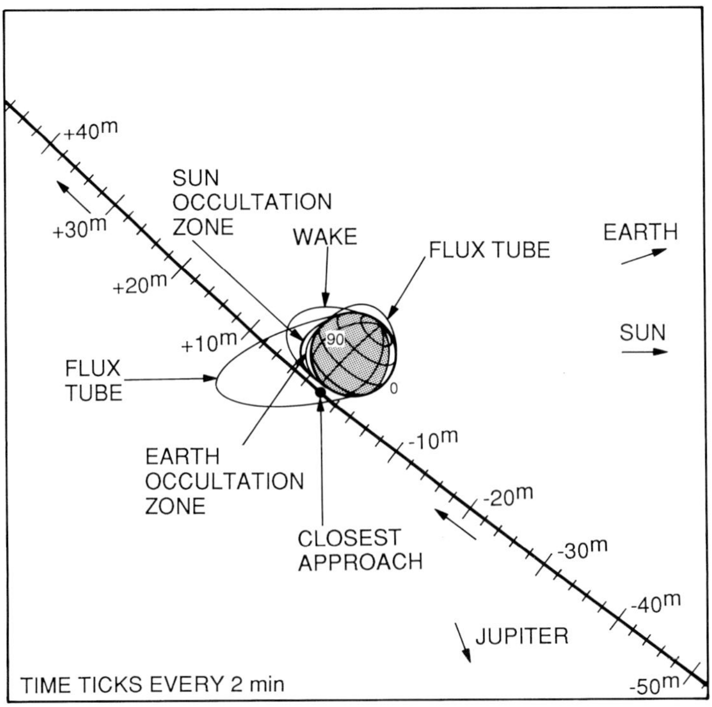

Fig. 26. Trajectory pole view of Europa 7 encounter.

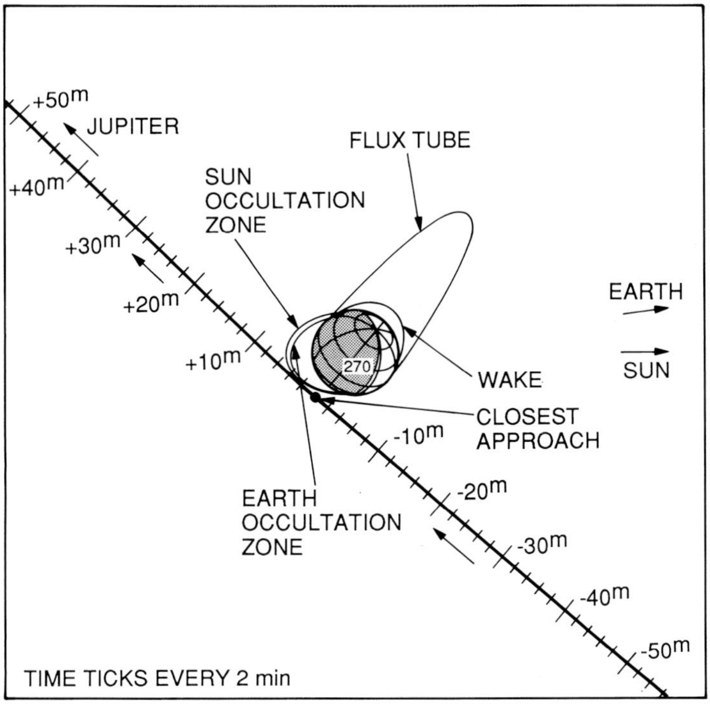

Fig. 27. Trajectory pole view of Ganymede 8 encounter.

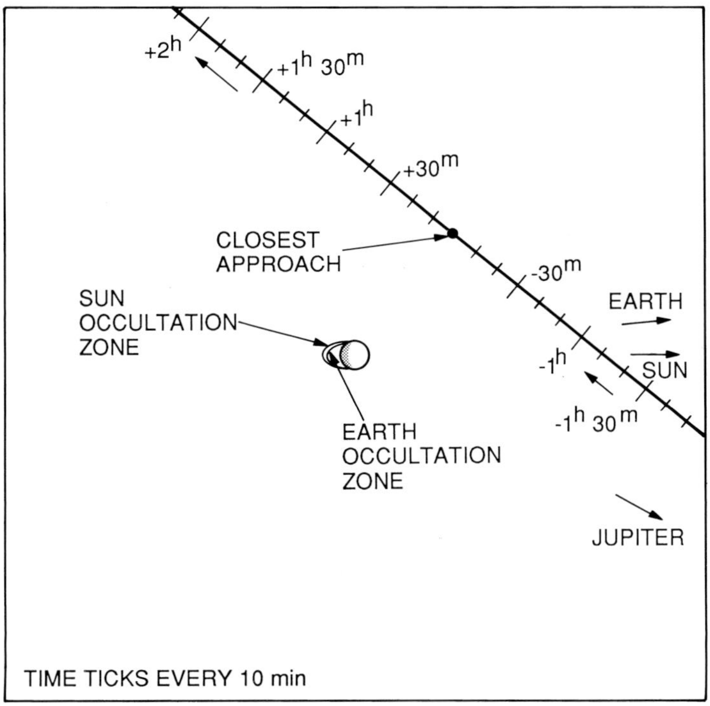

Fig. 28. Trajectory pole view of Callisto 8N nontargeted encounter.

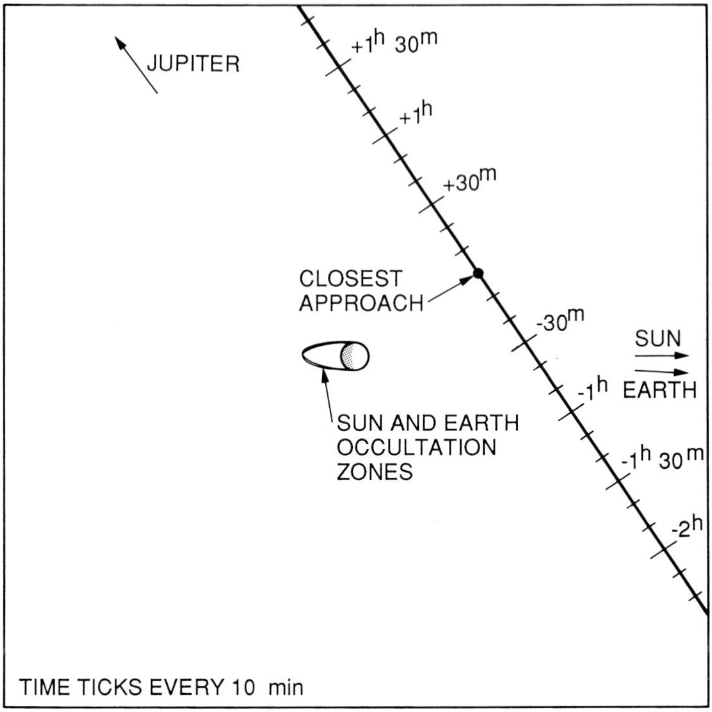

Fig. 29. Trajectory pole view of Ganymede 9N nontargeted encounter.

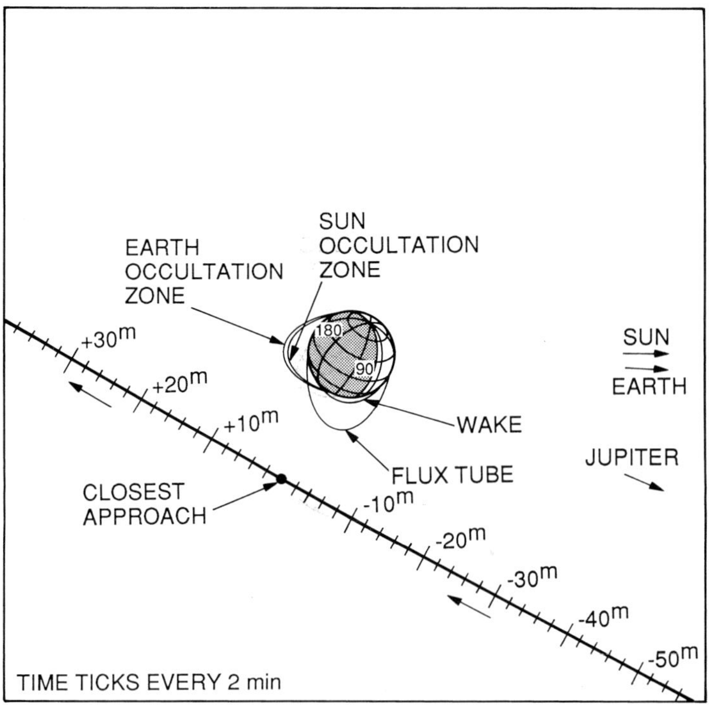

Fig. 30. Trajectory pole view of Callisto 9 encounter.

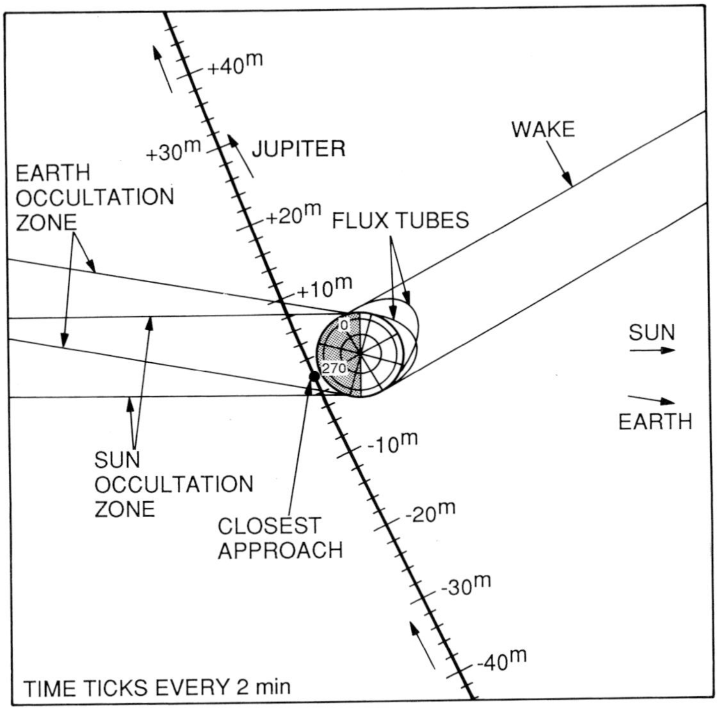

Fig. 31. Trajectory pole view of Ganymede 10 encounter.

Orbiter must fly in front of Ganymede to accomplish this orbital period reduction, a close wake pass is also achieved. Since it is both period-reducing and inbound, this is a counterrotating flyby (rotating the line of apsides away from the anti-Sun direction), which is favorable for Jupiter atmospheric observations. The closest approach point is on the sunlit side of the satellite. High-resolution imaging of the surface is obtained under both high-Sun and low-Sun conditions. This flyby also provides high-Sun coverage at low resolutions along the approach asymptote. At closest approach, bright-limb observations are possible, enabling the Ultraviolet Spectrometer (UVS) experiment to be performed.

Earth is not included by Ganymede during the flyby, which permits unobstructed tracking of the Orbiter. Tracking data from this pass, which occurs at a latitude of 29°, together with tracking data from a subsequent low-altitude pass near one of Ganymede's poles, will be used to construct a model of Ganymede's gravitational field.

A pass closer to Ganymede's equator would meet the needs of the gravity-modeling experiment better; however, the latitude of the closest-approach point cannot be reduced. The reason for this has to do with the need to reduce orbital inclination. The inclination of the initial orbit is about 5.6°, primarily as a consequence of the Jupiter approach declination associated with the interplanetary trajectory. Until inclination is reduced, only one satellite (the one chosen for the first flyby) can be encountered. The most desirable strategy (and the one used in tours designed for previous Galileo launch opportunities) is to have the first two encounters with Ganymede: the first a near-equatorial flyby to reduce period, and the second a near-polar flyby to reduce inclination. However, the 5.6° initial inclination is high enough that the second flyby alone cannot reduce inclination by the required amount. Therefore, the first Ganymede flyby must be used to reduce inclination slightly, even though the primary objective at this flyby is to reduce orbital period. As a result, closest approach is at 29° latitude, rather than nearer Ganymede's equator.

4.4.2. *Ganymede 2 Encounter*

The relatively high inclination of the initial orbit (5.6°), coupled with the choice of Ganymede for the first flyby, forces a return to Ganymede for the second flyby, with the encounter occurring at the same place in Ganymede's orbit. The primary goal at this flyby is to accomplish the remaining reduction in inclination needed to encounter the other Galilean satellites. The required reduction in inclination is close to the maximum amount Ganymede can achieve with one flyby; consequently, this flyby must be a low-altitude flyby nearly over Ganymede's north pole. This provides the opportunity to complete the Ganymede gravity-modeling experiment described above, as well as to accomplish a passage through Ganymede's flux tube. The close polar view allows high-resolution, low-Sun imaging, a high-priority science objective. The viewing conditions on the approach and departure asymptotes are similar to those of the Ganymede 1 flyby, covering much of the same area. Therefore, little additional low-resolution coverage is obtained.

4.4.3. *Callisto 3 Encounter*

Gallisto is used to further reduce the orbital period from 71 days to 45 days, resultng in a low-altitude pass through Callisto's wake. Since the encounter is also inbound, it rotates the line of apsides away from the anti-Sun direction, providing more time on Jupiter's sunlit side for atmospheric observations. The Orbiter passes over Callisto's sunlit side, allowing high-resolution imaging of the satellite under both high-Sun and low-Sun conditions. Viewing conditions are also favorable for low-resolution, high-Sun coverage along the approach asymptote. At closest approach, the bright limb is observable by the UVS. This flyby reduces the Orbiter's perijove below the orbit of Europa, so that Europa can be encountered on the next orbit. The orbital inclination after this flyby is such that an occultation of Earth by Jupiter occurs.

4.4.4. *Europa 4 Encounter*

Europa is used to reduce the orbital period still further, from 45 days to 36 days. This results in a wake passage; however, because of the combination of the altitude (1741 km) and the latitude (26°), the time spent in the wake region is short. Because this period-reducing flyby is also inbound, it counterrotates the orbit, which is favorable for viewing Jupiter's sunlit side. On the approach asymptote, almost a full moon is visible, providing an opportunity to obtain good low-resolution NIMS and SSI high-

Sun coverage. Closest approach occurs over Europa's sunlit side, providing an opportunity to image a portion of Europa at high resolution. This flyby also provides conditions favorable for observing Europa's bright limb with the UVS. This flyby produces an occultation of Earth by Jupiter on the post-flyby orbit, and puts the Orbiter on a transfer orbit to the next targeted flyby (with Ganymede) in such a way that a nontargeted encounter with Europa is achieved before arriving at the next targeted flyby.

4.4.5. *Europa 5N Encounter*

This is the first nontargeted flyby in the tour. At closest approach, which occurs at a longitude of 114°, about two-thirds of the Europa disc is sunlit, providing an opportunity to obtain coverage at low resolutions of some regions under high-Sun conditions, and of others under low-Sun conditions. This flyby also produces an unusual distant passage through Europa's wake.

4.4.6. *Ganymede 5 Encounter*

This encounter is outbound, which means that along the approach asymptote, the Orbiter sees the side of Ganymede not visible during the approaches to Ganymede 1 and 2. The satellite appears to be almost fully sunlit on the approach asymptote. The closest-approach point is situated well away from the terminator over Ganymede's sunlit side. Because the Orbiter spends a comparatively long time over sunlit portions of Ganymede at low altitudes, this flyby yields more high-resolution coverage than any other single flyby in the tour. Orbital period is increased to 55 days. Since this outbound flyby is period-increasing, the orbit is again counterrotated, which allows more time for viewing Jupiter's sunlit side. This flyby produces another occultation of Earth by Jupiter.

4.4.7. *Europa 6 Encounter*

This outbound flyby is period-reducing, causing the line of apsides to rotate towards the magnetotail. Because all the previous flybys have been used to counterrotate the orbit to maximize the time available for observations of Jupiter's sunlit side, this flyby and all succeeding flybys must be used to rotate the orbit towards the anti-Sun direction in order to achieve magnetotail passage at the end of the tour. An occultation of both Earth and the Sun by Europa occurs near closest approach. On the approach asymptote, about two-thirds of the Europa disc is sunlit; on the departure asymptote, about one-third of the disc is sunlit. Closest approach occurs over the dark portion of Europa. Some high-resolution, low-Sun coverage is obtained before the Orbiter crosses the terminator as it nears the closest approach point.

The Europa 6 flyby sets up conditions that facilitate a pass through Europa's flux tube at the Europa 7 flyby. A Europa flux-tube pass is an important scientific objective. In order to pass through the flux-tube region, the Orbiter must fly nearly over one of Europa's npoles, which results in a significant change in inclination without a large change in the orbital period. The Europa 6 flyby puts the Orbiter into a 360° transfer orbit, targeted to return to Europa at the same place in its orbit. Choosing to return to Europa at the same place in its orbit allows using Europa 6 to increase inclination without being restricted by the need to keep the orbital plane near Jupiter's equator in order to be able to encounter other satellites. Inclination must be removed at Europa 7 in order to return the Orbiter to near Jupiter's equatorial.

4.4.8. *Europa 7 Encounter*

The Europa 7 encounter accomplishes a relatively small reduction in orbital period and a relatively large change in inclination. The need for the inclination change is a result of the increase in inclination caused by the Europa 6 flyby, with the assistance of a small propulsive maneuver at apojove. This results in a passage through Europa's flux tube, as planned. The small amount of reduction in orbital period results in a slight rotation of the orbit towards the magnetotail since the flyby is outbound. No significant additional satellite imaging is obtained along the asymptotes, because the conditions are much the same as those of Europa 6. However, high-resolution coverage is obtained near the south pole, which is not the same region covered at high resolution during the Europa 6 flyby. An occultation of Earth as viewed from the Orbiter is obtained on the post-flyby orbit.

4.4.9. *Ganymede 8 Encounter*

This period-increasing, inbound flyby helps rotate the orbit further toward the magnetotail while providing additional high-resolution, low-Sun coverage before closest approach (which occurs over the dark portion of the satellite). This flyby also establishes the spacecraft on a trajectory on which it is possible to achieve a nontargeted flyby with Callisto on this orbit and a nontargeted flyby with Ganymede on the next orbit. About two-thirds of the Ganymede disc appears sunlit on approach, and about one-third of the disc is sunlit on departure.

4.4.10. *Callisto 8N Encounter*

This nontargeted flyby provides the opportunity to obtain coverage at 1-km resolution in the coverage gap surrounding the 270° longitude region.

4.4.11. *Ganymede 9N Encounter*

This nontargeted flyby offers coverage at 1-km resolution of a mostly sunlit Ganymede in the coverage gap surrounding the 90° longitude region.

4.4.12. *Callisto 9 Encounter*

At his outbound flyby, the orbital period is reduced slightly, rotating the orbit toward the magnetotail. The reduction in orbital period must be kept small because the tenth (and last) flyby must increase the orbital period to at least 90 days in order to achieve a distance of 150 R_J at the apojove of the tail-petal orbit. This flyby also sets the inclination at the precise value needed to obtain a distant occultation of Earth by Io, as viewed from the Orbiter, during the post-flyby orbit. This occultation takes place at a distance of 4.5 million km from Io. About two-thirds of the Callisto disc is sunlit on approach and about one-third of the disc is sunlit on departure. Closest approach occurs over the dark portion. The closest approach altitude is too high to allow any coverage at 80-m resolution before crossing the terminator; however, some coverage is obtained at resolutions better than 250 m.

4.4.13. *Ganymede 10 Encounter*

This inbound flyby increases orbital period to 95 days, sufficient to achieve passage through the magnetotail at a distance of 157 R_J (greater than the minimum required 150 R_J) within 10° of the anti-Sun direction. Again, about two-thirds of the Ganymede disc appears sunlit on approach, and about one-third of the disc appears sunlit on departure. A small amount of low-Sun coverage is obtained at resolutions of 250 m or better before closest approach, which occurs over the dark portion of the satellite.

5. Mission Performance

5.1. Propellant margin definition

Galileo's onboard propulsion system is called the Retropropulsion Module (RPM). Propellant is expended for TCMs, attitude control, RPM line-clearing maneuvers, and science turns. Propellant is supplied from the RPM to 12 10-N thrusters that provide maneuvering capability and attitude control, and to a 400-N engine that is used after Probe release to perform ODM, JOI, and PJR.

TCMs are categorized as either deterministic or statistical, and occur during both the interplanetary and satellite tour phases of the mission. Each interplanetary trajectory leg has a deterministic TCM to shape the flight path; ODM, JOI, and PJR also fall into the category of deterministic TCMs. TCMs to correct for navigation dispersions during both the interplanetary cruise and satellite tour phases of the mission and the Orbit Trim Maneuver (OTM) that occurs five days after JOI are considered statistical maneuvers. By convention, small TCMs required for biasing the aimpoints at the Earth flybys to satisfy the Earth navigation strategy are included as part of interplanetary statistical TCMs.

Mission performance is evaluated in terms of the quantity 'propellant margin' (PM) which is defined as the amount of usable propellant remaining in the onboard propulsion system after completion of a 10-encounter (10-orbit) satellite tour with 90% probability. Positive PM implies that there is unexpended propellant at the end of the 10-encounter satellite tour, while negative PM implies that the usable propellant supply is exhausted before completing the entire tour.

5.2. Propellant margin ground rules

The key mission ground rules utilized for the PM calculation are as follows:

(1) Calculations are based on 10 satellite encounters at Jupiter with 90% probability.

(2) ODM is performed 143 days before Probe entry.

(3) The altitude of the pre-JOI Io flyby is 1000 km.

(4) The Orbiter initial perijove radius is 4.0 R_J.
(5) The Probe relay link duration is 75 min.
(6) The JOI maneuver is in the earthline direction and is initiated 143 min after Probe entry. (The start of JOI has been delayed 20 min to allow for a pulsed spin-up.)
(7) OTM is performed five days after JOI.
(8) The initial orbit period after JOI is 230 days.
(9) PJR is targeted to an 11.5 R_J perijove.
(10) Science turns during the satellite tour are allocated 20 kg of propellant.

The above values for initial orbit and perijove raise distance are benchmark performance values. The final values will be determined at the time the satellite tour is designed.

5.3. Propellant Margin Status

The current PM status is shown in Table VI*. Two cases are shown: (1) the baseline mission which includes flybys of asteroids Gaspra and Ida and (2) a Gaspra-only

TABLE VI

Propellant margin status

Mission	Propellant margin, kg
Gaspra + Ida	−8
Gaspra	26

mission. PM is negative 8 kg for the Gaspra + Ida mission and positive 26 kg for the Gaspra mission. The elimination of the Ida flyby causes a 34-kg increase in PM.

According to the definition of PM, a zero value corresponds a 90% probability of completing 10 encounters. Nonzero values of PM can therefore be related either to some number of encounters other than 10 at 90% probability or to 10 encounters at some probability other than 90%. Figure 32 shows how the number of encounters achievable at various probability levels varies with PM. The PM values for Gaspra + Ida and Gaspra are indicated on the figure. The data in Figure 32 can be summarized as shown in Table VII. For the Gaspra mission (PM = 26 kg), there is greater than 99% probability of completing 10 (or fewer) encounters. For the Gaspra + Ida mission (PM = −8 kg), there is still an 80% probability of completing the full 10 encounter tour.

Figure 33 shows the amount of propellant remaining at 90% probability as a function of number of encounters. (By definition, the amount of propellant remaining at 90% probability for *10 encounters* is equal to PM.) The data in Figure 34 are also summarized in Table VII. For the Gaspra + Ida mission, propellant is exhausted (at 90% probabilty)

* The PM status reported here is somewhat conservative. It is based on a post-launch PM analysis that used the completion of TCM-2 (December 2, 1989) as a reference epoch for PM calculations. Because the trajectory delivery errors at the first Earth encounter were well below the 90% probability level, there has been an appreciable savings (about 9 kg for Gaspra + Ida and 6 kg for Gaspra) in PM that is not reflected in the data in this section. The next comprehensive PM analysis is scheduled to be carried out prior to the Gaspra encounter.

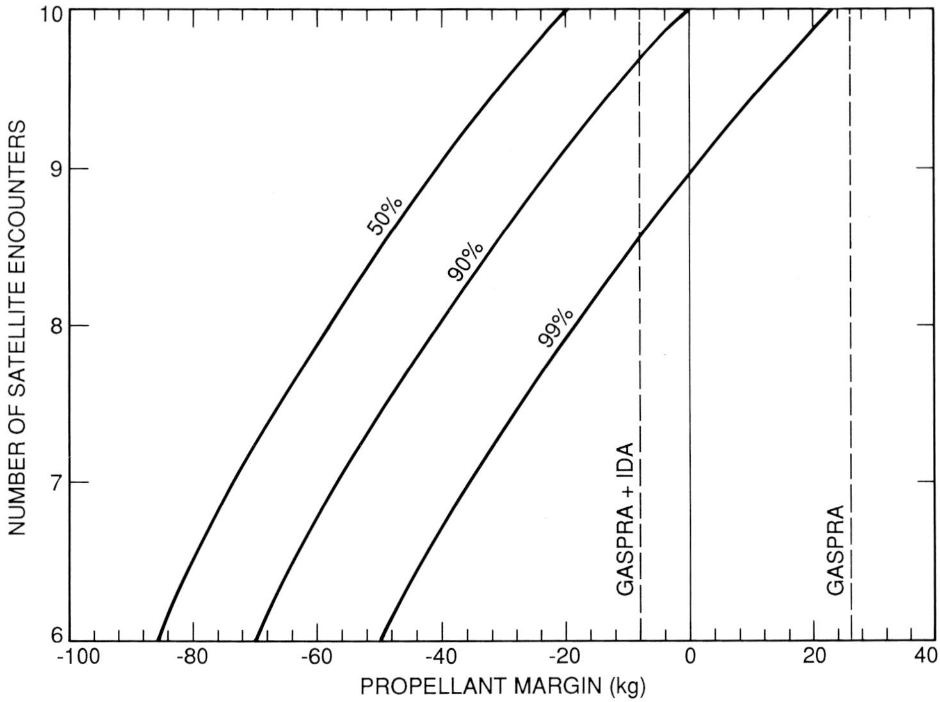

Fig. 32. Number of satellite encounters achievable at 50%, 90%, and 99% probabilities as function of propellant margin for Gaspra + Ida and Gaspra missions.

TABLE VII

Propellant remaining at 90% probability and probability of completion as function of number of encounters

Number of encounters	Probability of completion		90% propellant remaining, kg	
	Gaspra + Ida	Gaspra	Gaspra + Ida	Gaspra
10	80%	>99%	−8	26
9	97%	>99%	15	49
8	>99%	>99%	33	67

between the ninth and tenth encounters. After completion of nine encounters, the amount of propellant remaining (at 90% probability) is 15 kg.

Table VIII illustrates the PM calculation for the Gaspra + Ida mission. The initial spacecraft mass used for this PM calculation is the mass at completion of TCM-2. All of the interplanetary TCMs are combined into two components for the PM calculation, one for deterministic maneuvers and one for navigation (statistical) maneuvers. Propellant allocated for attitude-control maneuvers and RPM line-clearing allocations is also separated into interplanetary and satellite tour components. For any propulsive maneuver, the propellant consumption is computed from the rocket equation.

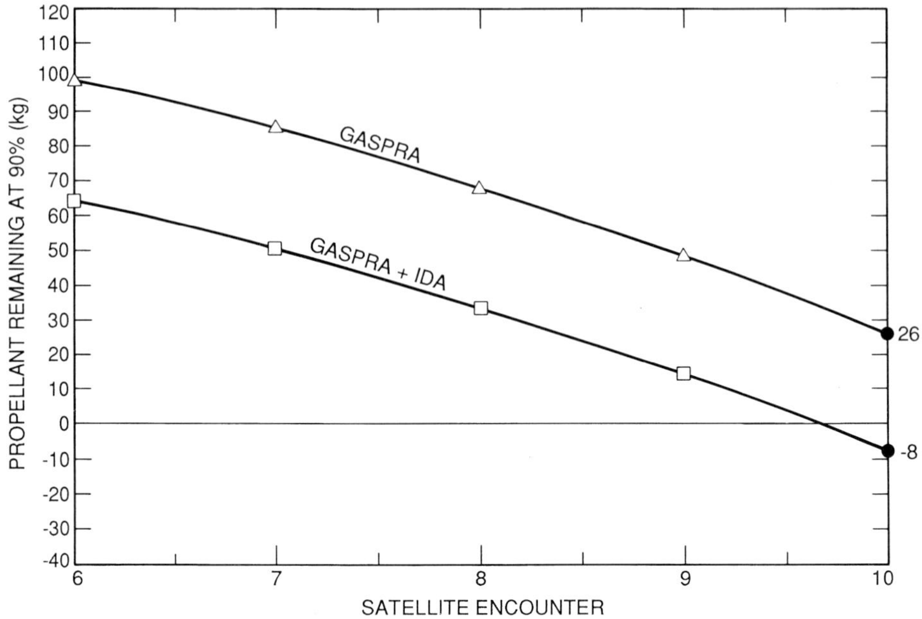

Fig. 33. Propellant remaining at 90% probability as function of satellite encounter for Gaspra + Ida and Gaspra missions.

5.4. IDA DECISION PLAN

The nominal design for Galileo's interplanetary trajectory currently includes a close flyby of Ida. As pointed out above, however, PM can be increased significantly by eliminating the Ida flyby. A Galileo Project plan has been developed for evaluating the mission tradeoffs with regard to the Ida flyby. Current Galileo Project policy states that implementation of a decision to eliminate the Ida flyby will not occur until after the Gaspra flyby. The additional PM gain from eliminating the Ida flyby earlier is small. For example, if Ida were eliminated as early as before the Earth 1 encounter, the additional gain would have been only 3 kg. By delaying the decision, the effects on propellant consumption of the delivery accuracy at the Earth 1 encounter and the targeting constraints at the Gaspra flyby will be known. In addition, a post-Gaspra decision allows Gaspra encounter science results to be factored into the process. Currently, the decision on whether to proceed with the Ida flyby is schedules to be made by July 1, 1992, about eight months after the Gaspra flyby. If the Ida flyby is eliminated, action to alter the trajectory will be taken at TCM-14, which is scheduled for August 1992.

6. Summary

The Galileo spacecraft began its 6-year-long journey to Jupiter on October 18, 1989. A new Venus–Earth–Earth Gravity Assist (VEEGA) trajectory mode, involving two phas-

TABLE VIII

Gaspra + Ida propellant margin calculation

Mission event	Isp, s	ΔV, m s^{-1}	M_0, kg	M_f, kg	ΔM, kg
Interplanetary statistical ΔV	274	45	2542	2500	42
Interplanetary deterministic ΔV	274	102	2500	2407	93
Probe release			2407	2068	339
ODM	309	59	2068	2028	40
Interplanetary RPM line flushing			2028	2024	5
Interplanetary attitude control			2024	1984	40
JOI	309	628	1984	1612	371
OTM	274	9	1612	1607	5
PJR	309	370	1607	1422	185
Satellite tour ΔV[a]	274	149	1422	1345	77
Science turns			1345	1325	20
Satellite tour RPM line flushing			1325	1324	2
Satellite tour attitude control			1324	1290	34
End-of-mission mass		1290 kg			
Orbiter dry mass		– 1298 kg			
Propellant margin		– 8 kg			

[a] Includes deterministic ΔV.
M_0 = spacecraft mass before event.
M_f = spacecraft mass after event.
$\Delta M = M_0 - M_f$.
Note: M_f and ΔM may not add to M_0 because of rounding.

ing orbits about the Sun and gravity-assist flybys with Venus (February 10, 1990) and Earth (December 8, 1990 and December 8, 1992), is being used for the transfer to Jupiter. The aggregate ΔV imparted to Galileo by the three planetary gravity-assist flybys is 18.3 km s^{-1}. In addition, Galileo will be targeted for a close flyby of asteroid 951-Gaspra on October 29, 1991, and a second asteroid flyby with 243-Ida on August 28, 1993 may also be included. Upon arrival at Jupiter on December 7, 1995, the Galileo Probe will make the first *in-situ* measurements of the Jovian atmosphere, and the Galileo Orbiter will use repeated gravity-assist flybys of the three outermost Galilean satellites to conduct a 22-month satellite tour during which Jupiter, its magnetosphere, and the Galilean satellites will be intensively studied.

Acknowledgements

The research described in this paper was carried out by the Jet Propulsion Laboratory, California Institute of Technology, under a contract with the National Aeronautics and Space Administration.

Certain sections of this paper contain revised versions of material originally included in an earlier Galileo mission description paper (D'Amario *et al.*, 1989). Dennis Byrnes and Jennie Johannesen contributed those portions of the earlier paper that were incorporated, in revised form, in this paper. Nancy Ericson and Daniel Scheeres are acknowledged for preparing some of the trajectory plots and numerical data included in this paper. Bill O'Neil and Dennis Byrnes provided valuable reviews of the paper, and their comments and suggested changes have significantly improved the final product. Editorial assistance was provided by Marilyn Morgan.

Appendix. Gravity-Assist Dynamics

When a spacecraft in orbit about a primary body has a close flyby of a secondary body (also in orbit about the primary body), the gravitational attraction of the secondary body on the spacecraft will alter its orbit about the primary. For interplanetary trajectories, the primary body is the Sun, and the secondary body is a planet. For planetary orbiters, the primary body is the planet, and the secondary body is one of the planet's satellites. Gravity-assist flybys are used primarily to change the energy and/or inclination of the spacecraft orbit about the primary body.

Galileo trajectory design makes extensive use of the gravity-assist effect. In the Galileo VEEGA interplanetary trajectory, planetary flybys of Venus (once) and Earth (twice) are used to add energy (i.e., velocity) to the spacecraft so that the injection energy at Earth departure can be reduced significantly from that required for a direct Earth–Jupiter transfer. On approach to Jupiter, a close (1000 km altitude) flyby of Io is used to slow the spacecraft to reduce the ΔV required to enter orbit at Jupiter. Close flybys of the Galilean satellites are used to increase and decrease both the energy (period) and the inclination of the orbit about Jupiter in order to accomplish the satellite tour.

The manner in which a gravity-assist flyby of a secondary body can change spacecraft velocity with respect to the primary body can be illustrated by what is commonly referred to as a 'velocity-vector diagram'. Figure A-1 shows two types of gravity-assist flybys, 'energy increasing' and 'energy decreasing', and their associated velocity-vector diagrams. (In what follows it is assumed that, on the time scale of the orbit about the primary, the effects of the hyperbolic flyby of the secondary occur instantaneously.)

First consider the energy-increasing case. The spacecraft trajectory with respect to the secondary is a hyperbola; the spacecraft approaches and departs along the asymptotes of this hyperbola with a constant speed, V_∞. The spacecraft achieves its maximum speed with respect to the secondary at closest approach. The angle θ between the incoming and outgoing V_∞ vectors is referred to as the 'bend angle' of the flyby. The sense of rotation of the V_∞ vector is clockwise, because the spacecraft passes behind the secondary at the flyby. The velocity vector of the secondary with respect to the primary at the time of the flyby is indicated by V_S. Adding the incoming and outgoing V_∞ vectors to the velocity vector of the secondary yields the velocity vectors of the spacecraft with respect to the primary before and after the flyby.

The effect of the hyperbolic flyby in a secondary-centered reference frame is simply to rotate the V_∞ vector through an angle equal to the bend angle θ; there is no net energy

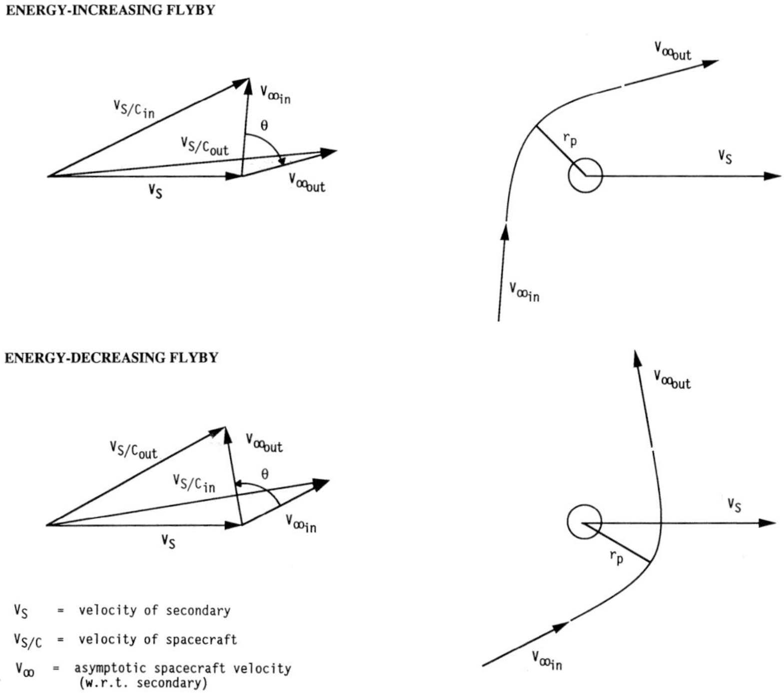

Fig. A-1. Velocity vector diagrams for 'energy-increasing' and 'energy-decreasing' gravity-assist flybys.

change for the spacecraft trajectory with respect to the secondary as a result of the flyby. However, the rotation of the V_∞ vector has the effect of increasing the magnitude of the velocity vector in a primary-centered reference frame.

For the energy-decreasing case, the spacecraft passes in front of the secondary and, therefore, the rotation of the V_∞ vector is in a clockwise direction, which is opposite to that for the energy-increasing case. As a result, the magnitude of the velocity vector with respect to the primary decreases.

The physical basis of the gravity-assist effect can be seen quite easily. Again, first consider the energy increasing case. Note that the direction of motion of the secondary is horizontally to the right, and the direction of the spacecraft's motion *with respect to the primary* is in the same general direction. Since the spacecraft passes directly behind the secondary, the gravitational acceleration of the secondary also points to the right. The acceleration from the secondary is in the same direction as the spacecraft's motion and therefore increases the velocity of the spacecraft. On the other hand, for the energy-decreasing case, the spacecraft is targeted to pass in front of the secondary. The acceleration of the secondary is in a direction opposed to the spacecraft's motion, and thus the spacecraft's velocity decreases. For gravity-assist flybys, it is generally true that in order to increase energy, the spacecraft must pass behind the secondary, and, conversely, a frontside pass will decrease energy.

An important relationship between the radius of closest approach r_p, approach speed V_∞, and the bend angle θ (a measure of the effectiveness of the gravity assist) is given by

$$\sin \theta/2 = 1/(1 + r_p V_\infty^2/GM) .$$

This equation demonstrates a basic fact of gravity-assist theory: namely, that either a lower approach speed (smaller V_∞) or a closer flyby (smaller r_p) results in a greater rotation of the V_∞ vector and, hence, a larger change in the primary-centered velocity. Disregarding the issues of atmospheric heating and navigation accuracy, the effectiveness of a gravity-assist flyby is limited by the physical radius of the flyby body.

References

Byrnes, D. V., D'Amario, L. A., and Diehl, R. E.: 1987, *Galileo Options After Challenger*, AAS Paper 87–420, presented at the AAS/AIAA Astrodynamics Specialist Conference, Kalispell, Montana, August 1987.

D'Amario, L. A., Byrnes, D. V., Johannesen, J. R., and Nolan, B. G.: 1987, *Galileo 1989 VEEGA Trajectory Design*, AAS Paper 87–421, presented at the AAS/AIAA Astrodynamics Specialist Conference, Kalispell, Montana, August 1987.

D'Amario, L. A., Bright, L. E., Byrnes, D. V., Johannesen, J. R., and Ludwinski, J. M.: 1989, *Galileo 1989 VEEGA Mission Description*, AAS Paper 89–431, presented at the AAS/AIAA Astrodynamics Specialist Conference, Stowe, Vermont, August 1989.

Johannesen, J. R., Nolan, B. G., Byrnes, D. V., and D'Amario, L. A.: 1987, *Asteroid/Comet Encounter Opportunities for the Galileo VEEGA Mission*, ASS Paper 87–422, presented at the AAS/AIAA Astrodynamics Specialist Conference, Kalispell, Montana, August 1987.

Maize, E. H.: 1989, *Earth Flyby Delivery Strategies for the Galileo Mission*, AAS Paper 89–427, presented at the AAS/AIAA Astrodynamics Specialist Conference, Stowe, Vermont, August 1989.

Mitchell, R. T.: 1988, *Galileo Earth Avoidance Study Report*, JPL Internal Document D-5580, Revision A (Galileo Project Document 1625–396, Revision A), 4 November, 1988.

O'Neil, W. J.: 1990, *Project Galileo*, paper presented at the AIAA Space Programs and Technologies Conference, Huntsville, Alabama, September 25–27, 1990.

Yeates, C. M., Johnson, T. V., and Young, R.: 1992, 'Galileo Mission Overview', *Space Sci. Rev.* **60**, 3 (this issue).

ENERGETIC PARTICLES INVESTIGATION (EPI)

H. M. FISCHER[1], J. D. MIHALOV[2], L. J. LANZEROTTI[3], G. WIBBERENZ[1], K. RINNERT[4], F. O. GLIEM[5], and J. BACH[5,6]

Abstract. The Energetic Particles Investigation (EPI) instrument operates during the pre-entry phase of the Galileo Probe. The major science objective is to study the energetic particle population in the innermost regions of the Jovian magnetosphere – within 4 radii of the cloud tops – and into the upper atmosphere. To achieve these objectives the EPI instrument will make omnidirectional measurements of four different particle species – electrons, protons, alpha-particles, and heavy ions ($Z > 2$). Intensity profiles with a spatial resolution of about 0.02 Jupiter radii will be recorded. Three different energy range channels are allocated to both electrons and protons to provide a rough estimate of the spectral index of the energy spectra. In addition to the omnidirectional measurements, sectored data will be obtained for certain energy range electrons, protons, and alpha-particles to determine directional anisotropies and particle pitch angle distributions. The detector assembly is a two-element telescope using totally depleted, circular silicon surface-barrier detectors surrounded by a cylindrical tungsten shielding with a wall thickness of 4.86 g cm^{-2}. The telescope axis is oriented normal to the spherical surface of the Probe's rear heat shield which is needed for heat protection of the scientific payload during the Probe's entry into the Jovian atmosphere. The material thickness of the heat shield determines the lower energy threshold of the particle species investigated during the Probe's pre-entry phase. The EPI instrument is combined with the Lightning and Radio Emission Detector (LRD) such that the EPI sensor is connected to the LRD/EPI electronic box. In this way, both instruments together only have one interface of the Probe's power, command, and data unit.

1. Introduction

The principal scientific objective of the Energetic Particles Investigation (EPI) is to measure the energetic particle population of the innermost region of the Jovian magnetosphere within 5 Jupiter radii (R_J). Both the Pioneer and the Voyager missions provided measurements in a portion of this inner magnetospheric region (Armstrong *et al.*, 1981; Schardt and Goertz, 1983). The most prominent findings include a continuing increase of the particle fluxes with decreasing radial distance to the planet down to the closest distance measured ($\approx 1.6\ R_J$ for Pioneer 11). However, the increase in fluxes is not simply monotonic, but rather shows a complex shell-like structure, which depends on energy and particle species (Fillius, 1976; Simpson and McKibben, 1976). These observations have been interpreted as indicating the absorption of the radially inward-diffusing particles by macroscopic objects orbiting the planet. There is also strong indication in the previous measurements for pitch angle scattering due to wave-particle interactions (Gurnett and Scarf, 1983; McDonald and Trainor, 1976; Simpson and McKibben, 1976; Van Allen, 1976). Closer to the planet than previous missions have

[1] Institut für Reine und Angewandte Kernphysik, Universität Kiel, Kiel, Germany.
[2] NASA Ames Research Center, Moffett Field, CA 94035, U.S.A.
[3] AT&T Bell Laboratories, Murray Hill, NJ 07974, U.S.A.
[4] Max-Planck-Institute für Aeronomie, Katlenburg-Lindau, Germany.
[5] Institute für Datenverarbeitanslagen, Universität Braunschweig, Germany.
[6] Donier Systems, Immenstadt, Germany.

Space Science Reviews **60:** 79–90, 1992.

gone ($\approx 1.6\ R_J$) other loss effects, such as radiation loss in the case of electrons and atmospheric absorption, will become more and more dominant. The measurements of EPI will be extended as close as possible to the planet's atmosphere ($\approx 1\ R_J$) in order to complete the knowledge about the energetic particle population at the inner edge of the Jovian magnetosphere and to make it possible to study the related magnetospheric transport processes. The operational period of the EPI is terminated by the beginning of the entry phase when the power supply for the sensor head is switched off.

The Voyager 1 mission detected, by imaging observations, a ring of dust particles around Jupiter near $1.8\ R_J$ as well as the small satellites Adrastea and Metis at the outer edge of this ring (Jewitt *et al.*, 1979; Owen *et al.*, 1979; Smith *et al.*, 1979). The existence of absorbing material in this region was already suggested from particle observations made by Pioneer 11 (Schardt and Goertz, 1983). Because of the absence of measurements, little is known about particle loss, source processes, and fluxes inside the closest approach of Pioneer 11. Particle losses due to the existence of Jupiter's ring and the satellites mentioned above are likely to be significant and to be dependent upon energy and pitch angle. Particle losses analogous to terrestrial conditions, such as Coulomb energy degradation in the atmosphere, may or may not be important, depending upon the density of free and bound electrons in Jupiter's upper ionosphere. A candidate for internal particle sources at low altitudes is cosmic-ray produced albedo neutron decay. This especially interesting region will be fully covered by EPI measurements.

In summary, electron and ion data acquired by the EPI instrument within 1.1–2 R_J will be used to:

- test the validity of radial diffusion as a transport and acceleration mechanism in the deep inner magnetosphere;
- infer the nature of field perturbations responsible for radial diffusion;
- infer the size distribution and radial structure of Jupiter's ring;
- identify possible additional inner magnetosphere source and loss mechanisms.

Moreover, comparisons of Galileo Probe and Pioneer 11 measurements in the region $\approx 5\ R_J$ to $\approx 1.6\ R_J$ will provide information on any temporal variations of inner-zone particle fluxes.

2. EPI Instrument

Sensor. Fundamental to the design of the detector configuration was the requirement to maintain expected counting rates below about 3 million counts per second under the most intense anticipated flux conditions. Another design constraint was the lower threshold for the particle species to be measured. This is defined by the material thickness of the Probe's aft heat shield. The heat shield consists of different layers of aluminum, silicon adhesive, kapton, mylar, dacron, and phenolic nylon. For perpendicularly-penetrating particles it has a resulting thickness of 1.34 cm, equivalent to about 1.87 g cm^{-2} 'shielding' (determined from a piece of heat shield material used in the calibrations; see last section).

The detector assembly is a two-element telescope using totally-depleted, circular

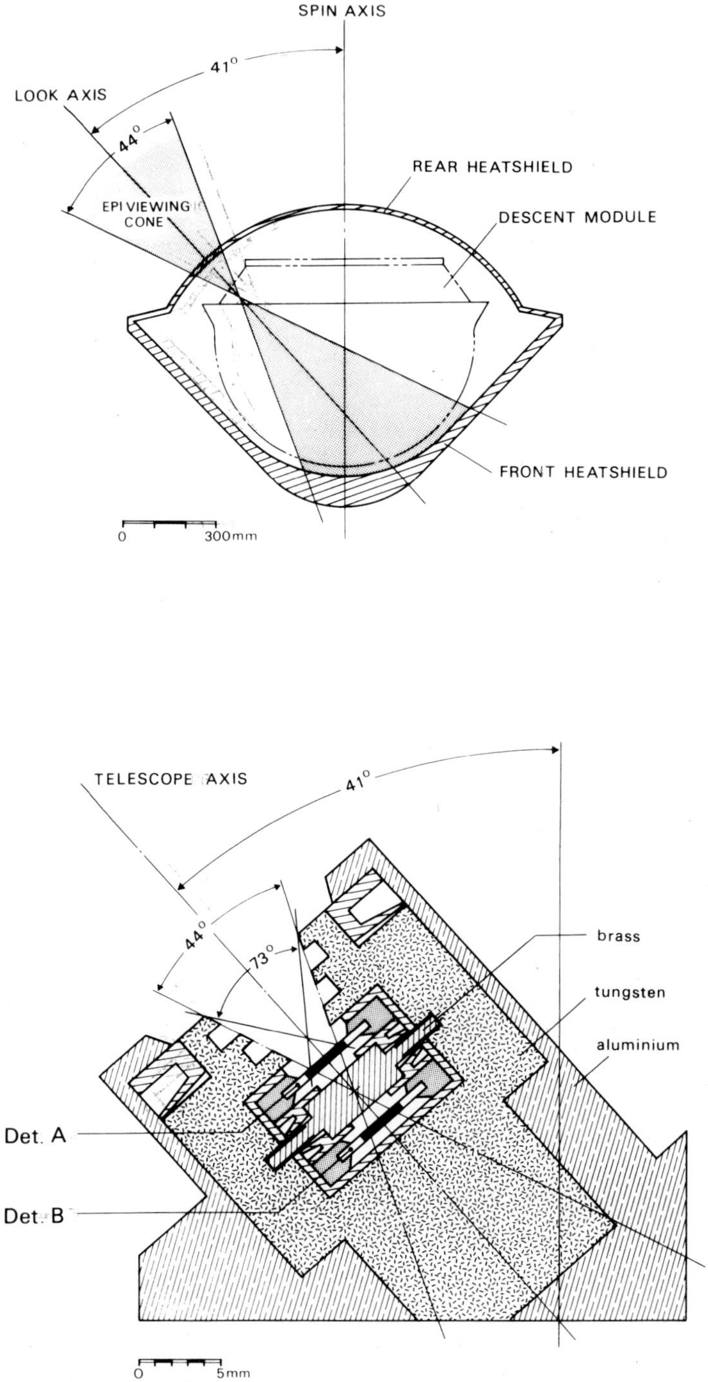

Fig. 1. Orientation of the EPI look axis with respect to the Probe spin axis, and detector and shielding configuration of the EPI sensor head.

silicon surface barrier detectors. Both detectors have a radius of 1.4 mm with sensitive areas of 6.2 mm^2 and thicknesses of 0.5 mm. A 3 mm thick (equivalent to 2.55 g cm^{-2}) brass absorber is inserted between the two detectors in order to expand the energy range of particles to be investigated. The total length of the telescope (upper surface front detector to lower surface back detector) is 6 mm. The detector assembly is surrounded by tungsten shielding with a cylindrical wall thickness of 2.7 mm (equivalent to 4.86 g cm^{-2}). The entire Probe and its contents, estimated to be equivalent to at least 80 g cm^{-2}, forms the shielding at the rear of the telescope. A reduced background rate is thus expected from the rear.

The tungsten shielding cylinder is elongated beyond the front detector surface and forms an open aperture cone of length 6.7 mm with an opening diameter of 9 mm. The aperture opening angle of 44° and the detector area define an effective opening angle of 73° and a geometrical factor of 0.045 cm^2 sr, for single particle counting events in the front detector. The corresponding values for coincidence events are 44° and 0.01 cm^2 sr, respectively.

The axis of the telescope is mounted at an inclination angle of 41° with respect to the Probe's spin axis (Figure 1). The orientation of the Probe's spin axis in space and the local magnetic field direction define the range in which pitch angle distributions can be determined by the instrument.

Telescope and analog electronics, the detector bias supply converter, and three housekeeping channels for monitoring leakage currents in the two detectors and the ambient temperature are contained in the sensor box (Figure 2). The scaling, data

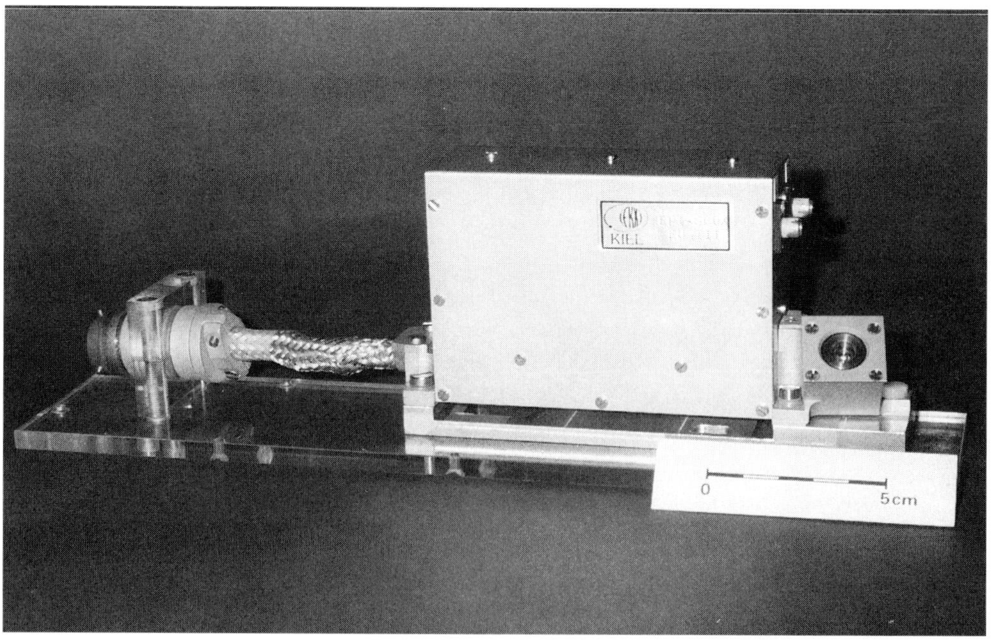

Fig. 2. Photograph of the EPI sensor box.

processing, and data formatting are executed with the Lightning and Radio Emission Detector (LRD) data in the central electronics box (see Lanzerotti *et al.*, 1992). A summary of the energy channels for the different particle species is given in Table I.

Electronics. The limits on the available Probe power prohibited the design and development of extremely fast circuitry for the EPI. As a compromise between the demands of power consumption and high-frequency response, a pulse rate of 3 million

TABLE I

EPI energy channels. The channels are valid for each particle type but they are associated with the given 'particle species' and the 'energy range' related to this species.

Channel No.	Particle species	Detector	Energy range (MeV $nucl^{-1}$)
E1	electron	A	≥ 3.2
E2	electron	AB	≥ 8
E3	electron	B	>8
P1	proton	A	42–131
P2	proton	AB	62–131
P3	proton	AB	62–92
HE	alpha-particle	AB	62–136
HVY	heavy particle	AB	^{12}C: 110–910 ^{32}S: >210

counts per second can be handled by the EPI analog electronics. An overview of the EPI analog electronics is given in Figure 3.

The detectors must be maintained in operational condition during the ≈ 6 years cruise phase from Earth to Jupiter. There is no power available from the Probe's main battery supply, which will only be activated shortly before the pre-entry part of the cruise phase. Therefore, in order to supply the detectors with bias voltage, a compact battery package with a capacity of about 250 mA-h is contained in a special fixture on the front side of the electronic box. The battery consists of Li/CrO_x cells having an open circuit voltage of approximately 3.6 V. The assembly is configured in two stacks with ten cells each. The overall voltage output is ≈ 75 V, which has been monitored in different test assemblies for more than 4 years. The batteries have been carefully tested and selected from an especially fabricated series with respect to the operational requirements under vacuum conditions.

3. Modes of Operation

A limited data rate is available for the Energetic Particles Investigation during the pre-entry phase. The general mission design approach was to make (for comparison purposes) some measurements in regions of the Jovian magnetosphere where particle observations from previous missions already exist and to concentrate the EPI investi-

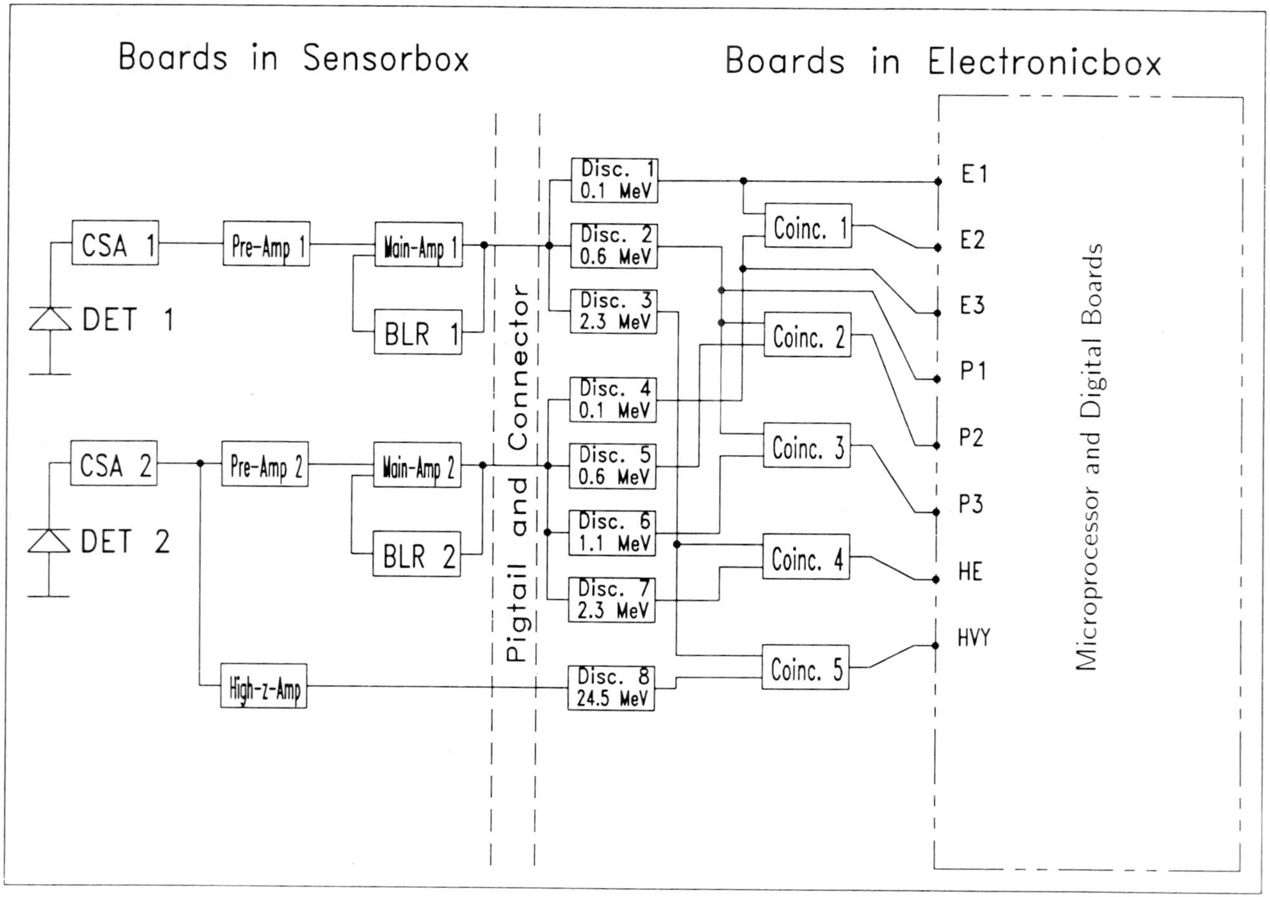

Fig. 3. Block diagram of the analog electronics of the EPI sensor system. As indicated, the electronics are contained partly within the sensor box shown in the photograph in Figure 2 and in the electronics box, respectively. (CSA – charge sensitive pre-amplifier, BLR – baseline restorer, Disc. – discriminator, Coinc. – coincidence circuit.)

gation on the innermost edge of the radiation zone, below $\approx 2\,R_J$. Therefore, the measurements must be distributed with respect to detected particle species and with respect to spatial locations in an appropriate manner, in order to obtain as much information as possible.

The data coverage of the investigation consists of three data samples made near the equatorial region at 5, 4, and 3 R_J, and in a continuous series of measurements between 2 and 1.1 R_J. Limited radio-frequency data will also be obtained at these locations by the Lightning and Radio Emission Detector on the Probe (see Lanzerotti *et al.*, 1992). During the arrival at Jupiter it will be of particular advantage for data analysis and interpretation if particle measurements are made simultaneously on the Orbiter and on the Probe in the region 5 to 4 R_J. The measurements from $\approx 4\,R_J$ down to the innermost edge of the trapping region will be continued by the Probe instrument.

The distribution of the EPI samples, the related position of the Orbiter, and the trajectories of both spacecraft at Jupiter arrival are presented in Figure 4. The instru-

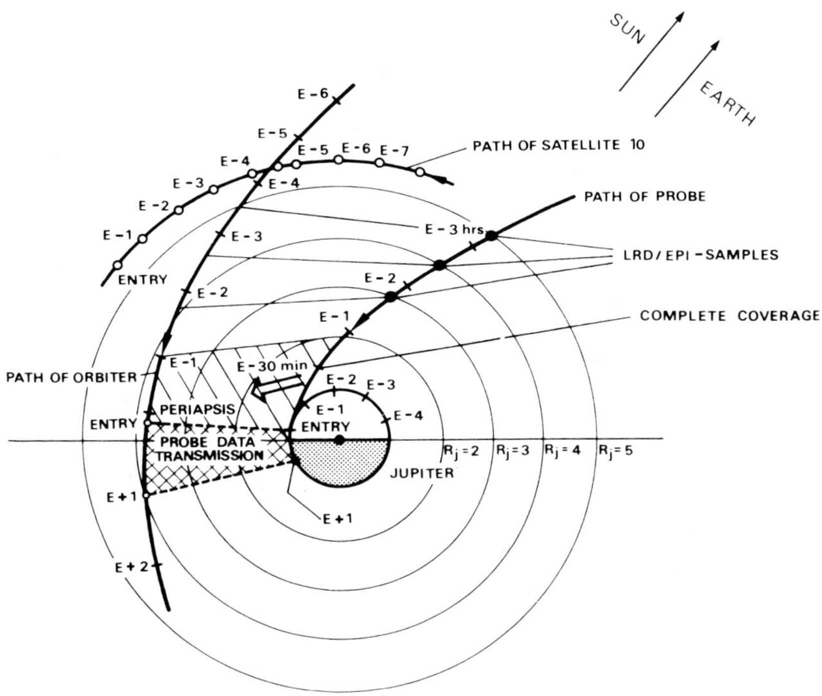

Fig. 4. Orbiter and Probe trajectories at Jupiter arrival, showing locations of the EPI data acquisition and corresponding locations of the Orbiter and its science instruments.

ment will make omnidirectional measurements of the particle species listed in Table I. Time-intensity profiles of these species, with a spatial resolution of about 0.02 R_J during the period of continuous measurement between 2 R_J and 1 R_J, will be recorded. Three different energy range channels are allocated to both electrons and protons to provide a rough estimate of the spectral index of the energy of each species in the energy range measured.

In addition to the omnidirectional measurements, angular sectored data will be obtained for electrons, protons, and alpha particles in certain energy ranges in order to determine directional anisotropies and particle pitch angle distributions. Because of the

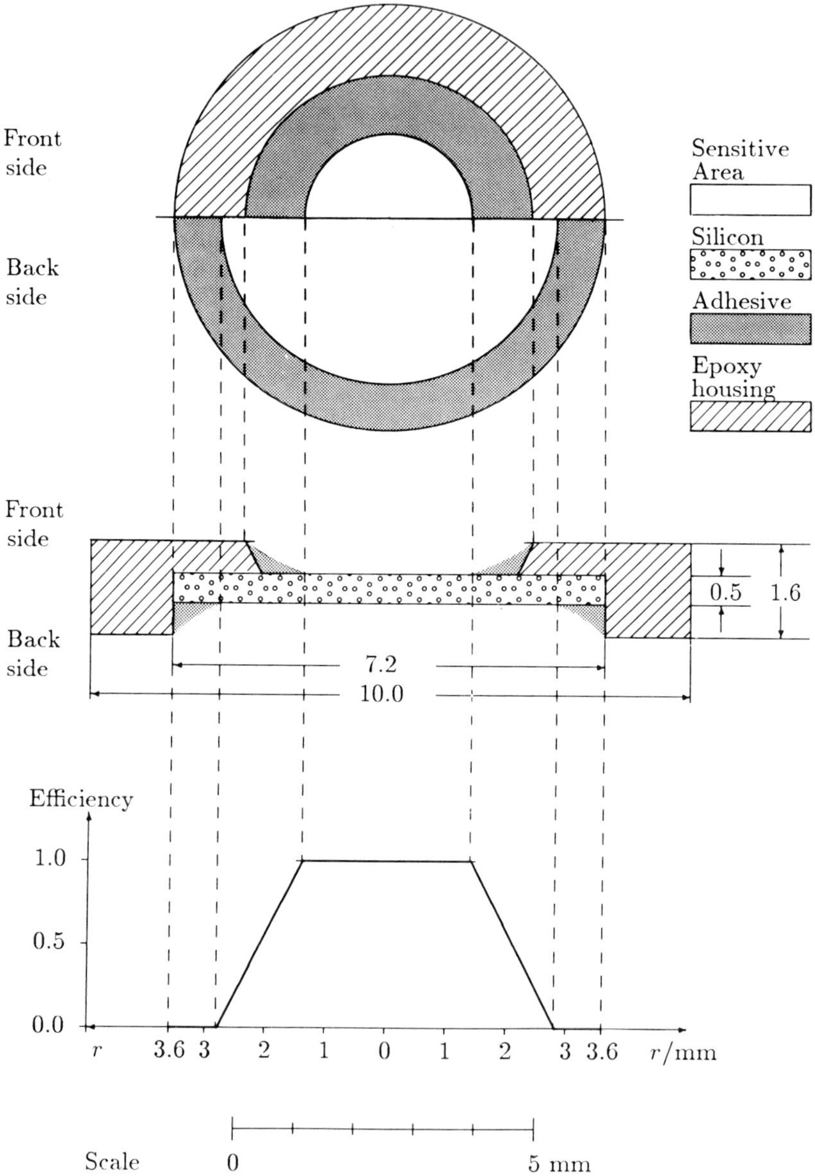

Fig. 5. Size, sensitive area, and efficiency of the EPI detectors.

expected low statistics, the heavy particle measurements are accumulated over longer time periods. The LRD-determined main magnetic field direction will be used for these sectored measurements (see Lanzerotti *et al.*, 1992).

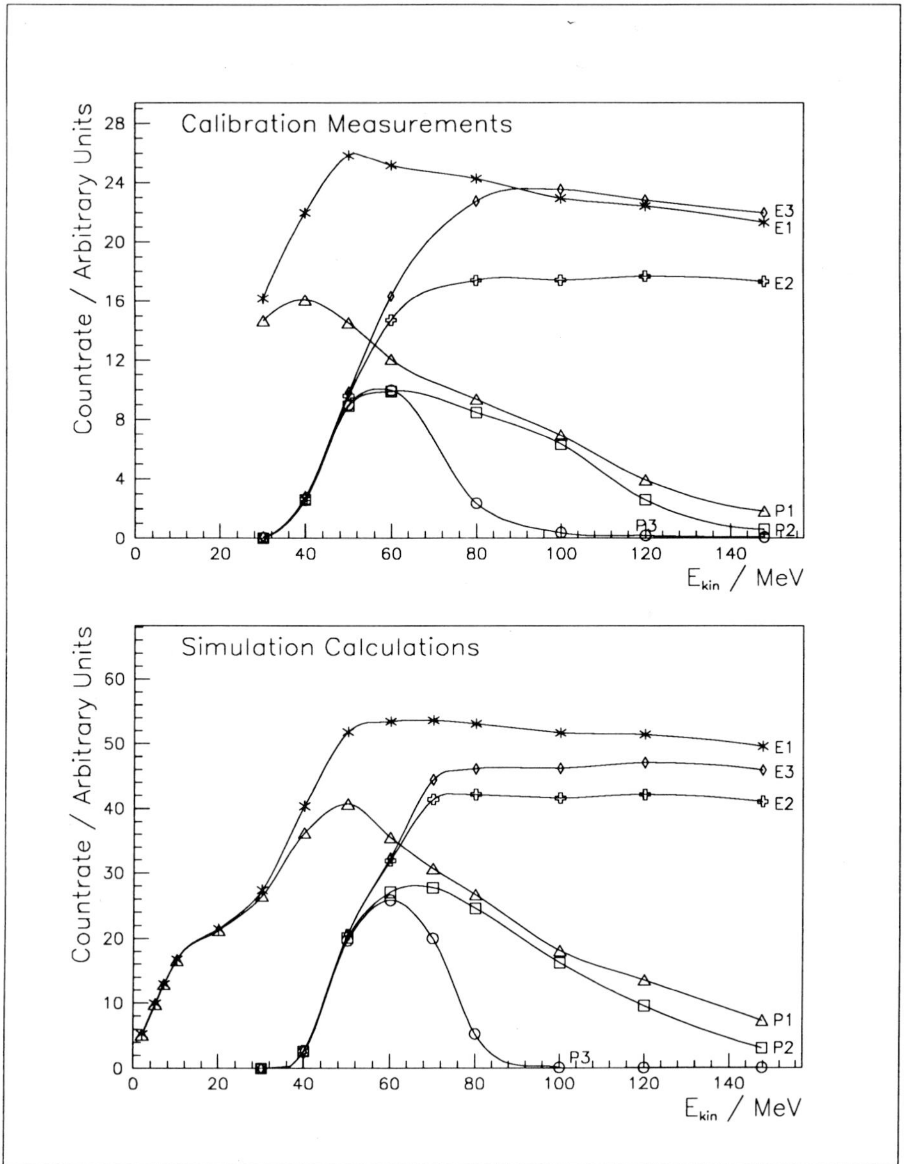

Fig. 6. Energy dependence of the count rates in selected EPI energy channels (cf. Table I). The calibration measurements and the simulation calculations both were performed without the heat shield and for protons only.

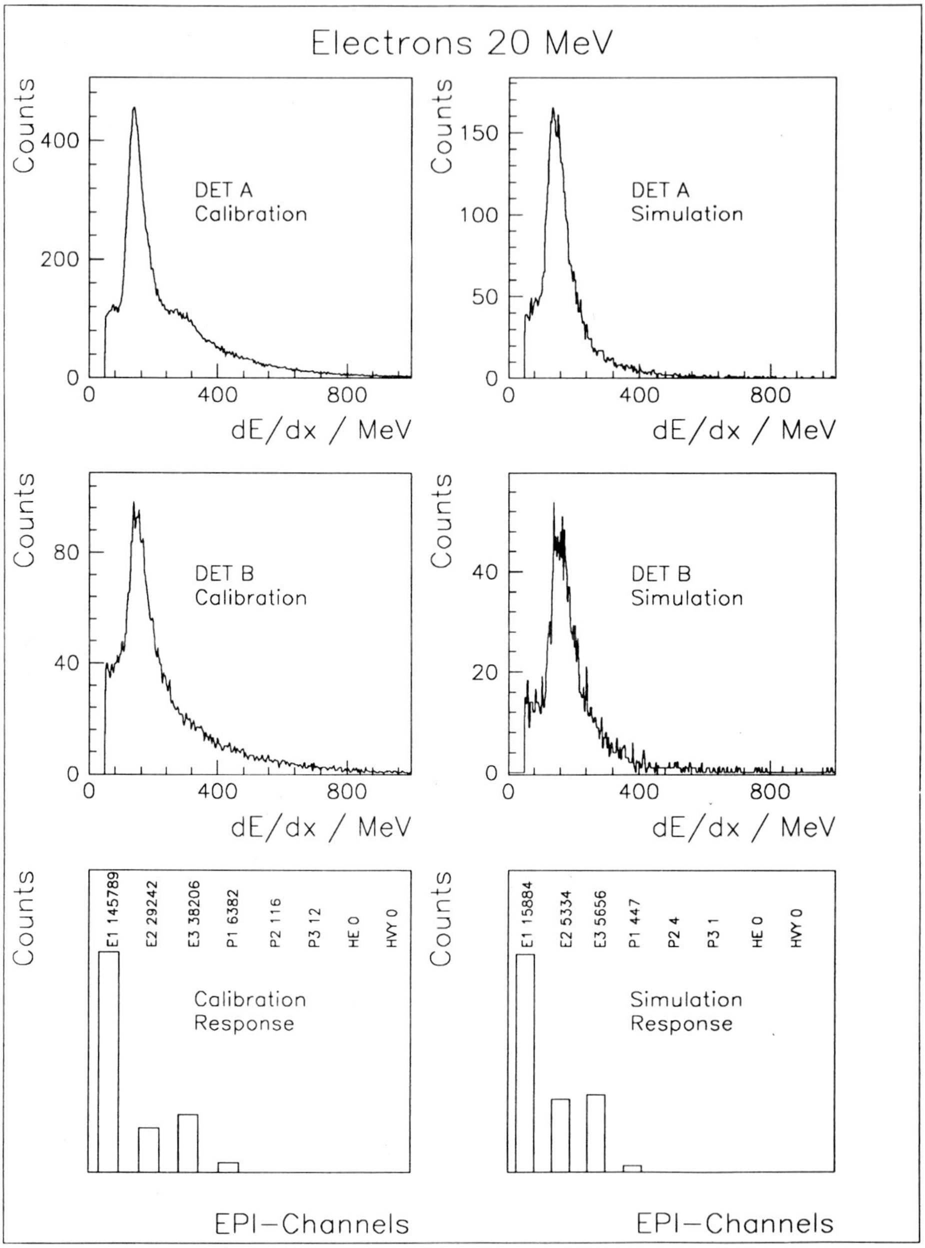

Fig. 7. Comparison of EPI calibration measurements with Monte-Carlo simulations for 20 MeV electrons including the heat shield material. Presented are the energy loss distributions in detectors A and B from calibration and simulation, respectively. The response of the EPI counting channels is demonstrated for comparison.

4. Tests and Calibrations

There has been a long series of detector calibration measurements with radioactive sources over a period of 10 years to determine the long-term stability of the detectors under normal and thermal-vacuum conditions. For the higher energies, the instrument was calibrated with protons at the Harvard Cyclotron (30–150 MeV), with electrons at the Lawrence Livermore accelerator (20–30 MeV), and for heavier particles (He, O, Si, Ar) at the Berkeley Bevatron (70–340 MeV $nucl^{-1}$). The calibration measurements were performed in part with a layer of original heat shield material in front of the aperture in order to determine the influence of the absorber on the instrument's response functions.

Subsequent to the calibration measurements, a model of the detector assembly was developed for the investigation of more physical details by means of a Monte-Carlo simulation method (Hoop, 1989; Pehlke, 1988). In a first attempt, the calibration measurements without the heat shield were simulated in order to compare the experimental and calculated data. By this procedure the influence of the detector edges on the counting efficiency could be determined. This is of special importance in case of such tiny detectors having a sensitive area diameter as small as 2.8 mm (which has to be considered in comparison to the diameter of the original silicon disc (7.2 mm) mounted in the detector housing). In the case of an ideal detector, the efficiency for particle detection should steeply drop from 100% to zero at the edges of the active area.

A model of the real efficiency decrease in the edge regions of the detectors is presented in Figure 5. This model was a necessary assumption in order to explain in detail the distribution of the energy losses of particles which were obtained from the accelerator measurements. Figure 6 shows the energy dependence of the count rates in selected EPI energy channels (cf. Table I) resulting from calibration measurements, and simulation calculations performed without the heat shield and for protons only. Protons in the energy range below 30 MeV could not be provided by the accelerator. In this range, data were only obtained by means of simulation. A comparison of the calibration and simulation data yields a reasonable correspondence. Therefore, further simulations incorporating the heat shield were done. Figure 7 shows a comparison between the 20 MeV electron calibration results and the simulation data, both with the heat shield. In the lower part of this figure the response of the EPI counting channels is added.

The final aim of the simulation method is to determine a response function of the instrument operating behind the heat shield in a particle population with a spectral and spatial distribution as might be expected in the innermost part of the Jovian radiation belt. These simulation capabilities, together with the calibrations, will be invaluable for interpreting the data to be acquired in the harsh inner Jovian magnetosphere.

Acknowledgements

Several of the authors (H. M. F., L. J. L., and J. D. M.) would like to thank A. Koehler and the staff of the Harvard Cyclotron Lab as well as I. Proctor and the staff of the

LLNL rf Linac for their generous support during the calibration measurements with protons and electrons, respectively. The same thanks are due to the staff of the LBL Bevatron for the calibration procedures with heavy particles. The results from the Monte-Carlo simulation were obtained by a CERN computer program entitled GEANT. Thanks are extended to E. Böhm (IFKKI) for his efforts in promoting the simulation program and to K.-D. Hoop and E. Pehlke (both from IFKKI) for their contribution of computing results and valuable discussions.

This work has been supported by the Bundesministerium für Forschung und Technologie (BMFT), Germany, contract 01QJ018 to the Christian-Albrechts-Universität of Kiel.

References

Armstrong, T. P., Paonessa, M. T., Brandon, S. T., Krimigis, S. M., and Lanzerotti, L. J.: 1981, 'Low-Energy Charged Particle Observations in the 5–20 R_J Region of the Jovian Magnetosphere', *J. Geophys. Res.* **86**, 8343.

Fillius, W.: 1976, in T. Gehrels (ed.), *The Trapped Radiation Belts of Jupiter, in Jupiter*, The University of Arizona Press, Tucson.

Gurnett, D. A. and Scarf, F. L.: 1983, in A. J. Dressler (ed.), 'Plasma Waves in the Jovian Magnetosphere', *Physics of the Jovian Magnetosphere*, Cambridge University Press, Cambridge.

Hoop, K.-D.: 1989, *Rechnersimulation der Protonen-Eichmessungen für das Kieler Experiment an Bord der Raumsonde GALILEO-Probe, Staatsexamensarbit*, Institut für Reine und Angewandte Kernphysik der Christian-Albrechts-Universität, Kiel, IFKKI Gelbe Reihe 90/1.

Jewitt, D. C., Danielson, E. D., and Synnott, S. P.: 1979, 'Discovery of a New Jupiter Satellite', *Science* **206**, 4421.

Lanzerotti, L. J., Rinnert, K., Dehmel, G., Gliem, F. O., Krider, E. P., Uman, M. A., Umlauft, G., and Bach, J.: 1992, 'The Lightening and Radio Emission Detector (LCD) Instrument', *Space Sci. Rev.* **60**, 91 (this issue).

McDonald, F. B. and Trainor, J. H.: 1976, in T. Gehrels (ed.), 'Observation of Energetic Jovian Electrons and Protons', *Jupiter*, The University of Arizona Press, Tucson.

Owen, T., Danielson, E. D., Cook, A. F., Hansen, C., Hall, V. L., and Duxbury, T. C.: 1979, 'Jupiter's Rings', *Nature* **281**, 442.

Pehlke, E.: 1988, *Auswertung von Protonen-Eichmessungen für das Kieler Experiment an Bord der Raumsonde GALILEO, Diplomarbeit*, Institut für Reine und Angewandte Kernphysik der Christian-Albrechts-Universität, Kiel.

Scarf, F. L.: 1976, in T. Gehrels (ed.), 'Plasma Physics and Wave-Particle Interaction at Jupiter', *Jupiter*, The University of Arizona Press, Tucson.

Schardt, A. W. and Goertz, C. K.: 1983, in A. J. Dressler (ed.), *'High Energy Particles', Physics of the Jovian Magnetosphere*, Cambridge University Press, Cambridge.

Simpson, J. A. and McKibben, R. B.: 1976, in T. Gehrels (ed.), 'Dynamics of the Jovian Magnetosphere and Energetic Particle Radiation', *Jupiter*, The University of Arizona Press, Tucson.

Smith, B. A. *et al.*: 1979, 'The Jupiter System through the Eyes of Voyager 1', *Science* **204**, 4396.

Van Allen, J. A.: 1976, in T. Gehrels (ed.), 'High-Energy Particles in the Jovian Magnetosphere', *Jupiter*, The University of Arizona Press, Tucson.

THE LIGHTNING AND RADIO EMISSION DETECTOR (LRD) INSTRUMENT

L. J. LANZEROTTI[1, 6], K. RINNERT[2], G. DEHMEL[3], F. O. GLIEM[4], E. P. KRIDER[5], M. A. UMAN[6], G. UMLAUFT[2], and J. BACH[4, 7]

Abstract. The Lightning and Radio Emission Detector (LRD) instrument will be carried by the Galileo Probe into Jupiter's atmosphere. The LRD will verify the existence of lightning in the atmosphere and will determine the details of many of its basic characteristics. The instrument, operated in its magnetospheric mode at distances of about 5, 4, 3, and 2 planetary radii from Jupiter's center, will also measure the radio frequency (RF) noise spectrum in Jupiter's magnetosphere. The LRD instrument is composed of a ferrite-core radio frequency antenna ($\sim$100 Hz to $\sim$100 kHz) and two photodiodes mounted behind individual fisheye lenses. The output of the RF antenna is analyzed both separately and in coincidence with the optical signals from the photodiodes. The RF antenna provides data both in the frequency domain (with three narrow-band channels, primarily for deducing the physical properties of distant lightning) and in the time domain with a priority scheme (primarily for determining from individual RF waveforms the physical properties of closeby-lightning).

1. Introduction

Prior to the Voyager 1 flyby of Jupiter, the presence of lightning in that planet's atmosphere had been speculatively suggested as an energy source both for the non-thermal radio emissions from the planet (which are easily detectable at Earth) and for the production of certain nonequilibrium chemical species. The suggestion regarding nonthermal radio emissions was soon effectively dismissed by the work of several researchers, in particular Zheleznyakov (1970). However, the second suggestion, related to possible non-equilibrium chemical processes, may well have some validity (Sagan *et al.*, 1967; Woeller and Ponnamperuma, 1969; Chadha *et al.*, 1971; Bar-Nun, 1975; Bar-Nun *et al.*, 1984; Bar-Nun and Podolak, 1985; Prinn and Owen, 1976). In spite of the uncertainty prior to the Voyager 1 encounter regarding the existence of lightning on Jupiter, NASA believed the question of Jovian lightning to be of sufficient importance that a joint U.S.–German lightning experiment was selected for flight on the Galileo Probe (Lanzerotti *et al.*, 1983).

The Voyager 1 mission produced two night-side photographs showing strong optical evidence for lightning in Jupiter's atmosphere (Cook *et al.*, 1979; Borucki *et al.*, 1982). Data from Voyager's plasma wave experiment (PWS) show the existence of whistler waves, most probably generated by lightning discharges, propagating in Jupiter's mag-

[1] AT&T Bell Laboratories, Murray Hill, NJ 07974, U.S.A.
[2] Max-Planck-Institut für Aeronomie, D-3411 Katlenburg-Lindau, Germany.
[3] Institut für Nachrichtentechnik, Univ. Braunschweig, Braunschweig, Germany.
[4] Institut für Datenverarbeitungsanlagen, Univ. Braunschweig, Braunschweig, Germany.
[5] University of Arizona, Tucson, AZ 85721, U.S.A.
[6] Department of Electrical Engineering, University of Florida, Gainesville, FL 32611, U.S.A.
[7] Dornier Systems, Friedrichshafen, Immenstadt, Germany.

Space Science Reviews **60**: 91–109, 1992.

netosphere (Gurnett *et al.*, 1979; Kurth *et al.*, 1985). However, these observations cannot provide information on several key parameters related to possible Jovian lightning (e.g., frequency of occurrence, intensity, distribution, global distribution, source height distribution, cloud charging, and discharging). Hence, measurements of lightning *in situ* in the Jovian atmosphere are crucial to understanding some of the electrical properties, as well as the chemistry and dynamics, of this giant planet's atmosphere.

Because of the possible existence of lightning on Venus (e.g., Ksanfomality, 1979; Taylor *et al.*, 1979) and Saturn (e.g., Kaiser *et al.*, 1983; Burns *et al.*, 1983; Zarka, 1985), as well as on Jupiter, there has emerged a new field of research that could be called planetary lightning. Several reviews of planetary lightning have already been written (Levin *et al.*, 1983; Williams *et al.*, 1983; Rinnert, 1982, 1985; Lanzerotti *et al.*, 1988a, 1989a) and a book chapter devoted to the subject is contained in a recent monograph on the lightning discharge (Uman, 1987). At least one thesis already exists on Jovian lightning (Williams, 1986).

The LRD instrument has been designed to take into account large uncertainties in the nature of possible Jovian lightning. For example, since Jupiter has no well-defined surface close to the cloud system, there will be no cloud-to-ground discharges, which are the best understood type of lightning on Earth. Lightning in general, and cloud discharges specifically, are very complex physical phenomena and can generate a large variety of RF pulse types and trains: unipolar pulses, bipolar pulses, asymmetric pulses, groups and bursts of pulses. The LRD instrument is designed as a compact and versatile instrument which allows a characterization of these signals with maximum sensitivity and maximum dynamic range. During the design phase, prototype instruments have been intensively tested with Earth lightning during several measuring campaigns. The final instrument characteristics have been set with acceptable margins for the unknown conditions to Jupiter.

Modelling of the propagation of RF signals in the frequency range of the LRD instrument in Jupiter's atmosphere shows that direct propagation of signals will occur to distances of order 10^4 km (Rinnert *et al.*, 1979). Hence, it is likely that Jovian atmospheric discharges with the energy of a typical cloud-to-ground discharge on Earth (order 10^8 J) will be detected at 10^4 km or more distance within the atmosphere with the LRD instrument. As noted below, the LRD instrument also includes a 'superbolt' channel, in order to count extremely large events.

Hence, in light of all the above, the flown lightning detector instrument must be designed to be as sensitive as possible, limited only by spacecraft noise. The instrument must also cover as large a dynamic range as possible.

2. Cloud Structure and Lightning

Most Earth lightning is produced by the charge generated and separated in cumulo nimbus clouds that develop in unstable air containing water vapor (see, for example, the reviews in Uman, 1987). The instability is most often produced by solar heating at the ground or by the interaction of air masses driven together by winds and convection.

As the moist air rises and cools, condensation occurs and ultimately precipitation forms. Charging of particles in the air can occur by various processes, including a thermoelectric effect and collisions between water droplets, supercooled water, and ice in various forms. On Earth, it is thought that the positive charges are predominant on the smaller particles and are raised preferentially by vertical convective motion. In Earth's atmosphere, charges can be separated to distances of the order of kilometers. Most lightning discharges in the Earth's atmosphere occur as intracloud lightning and neutralize tens of Coulombs of charge. Cloud-to-ground lightning generally transfers negative charge from the lower portion of the cloud to the Earth's surface.

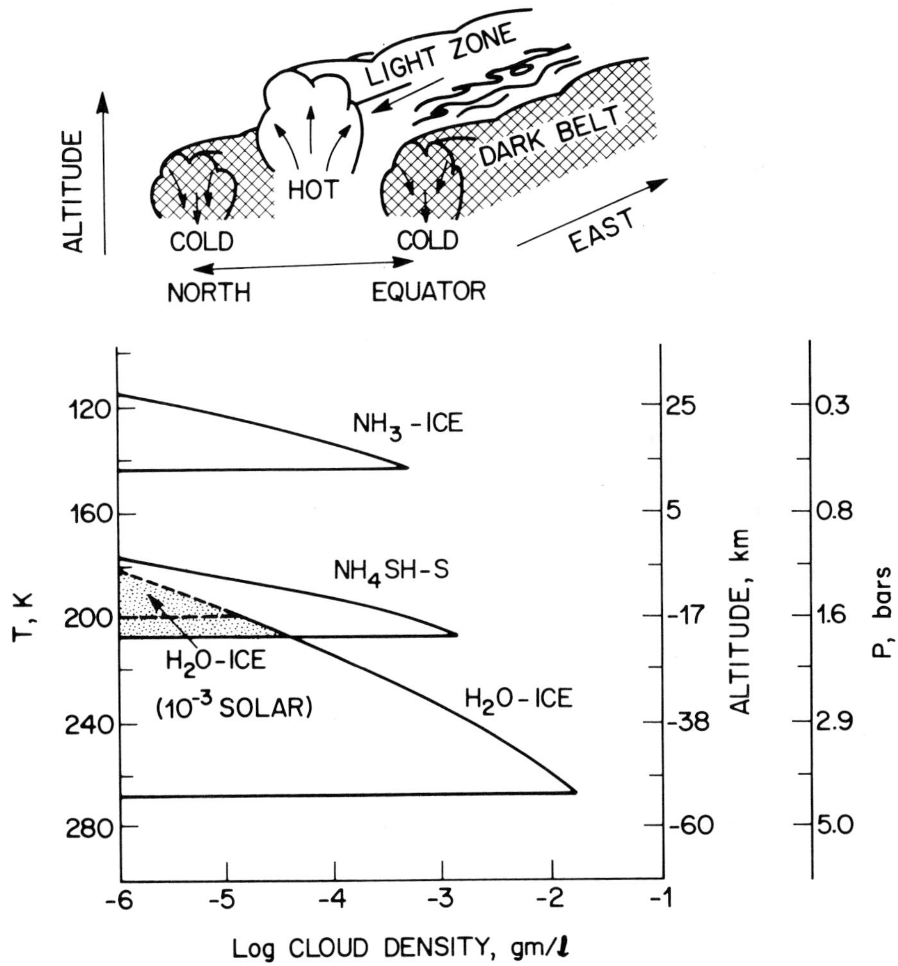

Fig. 1. Temperature, altitude, and pressure in the Jovian cloud system as given by the model of Weidenschilling and Lewis (1973). The water ice cloud for a depleted water abundance (10^{-3} solar abundance) at Jupiter is also shown (adapted from Atreya, 1986). At the top is a cartoon schematic of the global atmospheric convective motion.

A sketch of the temperature-density-altitude-pressure profile of the Jovian atmosphere is shown in Figure 1, based upon the cloud model of Weidenschilling and Lewis (1973). The height region for a water ice cloud for an atmospheric mixing ration 10^{-3} that of the solar abundance is indicated by the hatched area (Atreya, 1986). At the top of Figure 1, the global convection motions in the Jovian system are indicated schematically.

The cloud model in the lower portion of Figure 1 assumes stationary atmospheric conditions, solar composition, and chemistry in thermal equilibrium (i.e., no sources such as lightning to change energy from any existent electromagnetic fields to sound, heat, RF, etc.). More recent cloud models, deduced primarily from spacecraft optical measurements, contain either no, or only a thin,NH_4SH cloud, and add aerosol layers to account for the haze seen above the ammonia clouds (Sato and Hansen, 1979; West and Tomasko, 1980; Marten *et al.*, 1981; Orten *et al.*, 1982; Owen and Terrile, 1982). Further, Bjoraker *et al.* (1986) deduced from Voyager infrared interferometer spectrometer (IRIS) measurements at 5 μm that water is depleted by a factor of ~50 relative to solar abundances in the 2–6 bar atmosphere region.

With the type of atmosphere illustrated in Figure 1, a cartoon comparison can be made of Earth's and Jupiter's cloud structures, as is done in Figure 2. The dominant

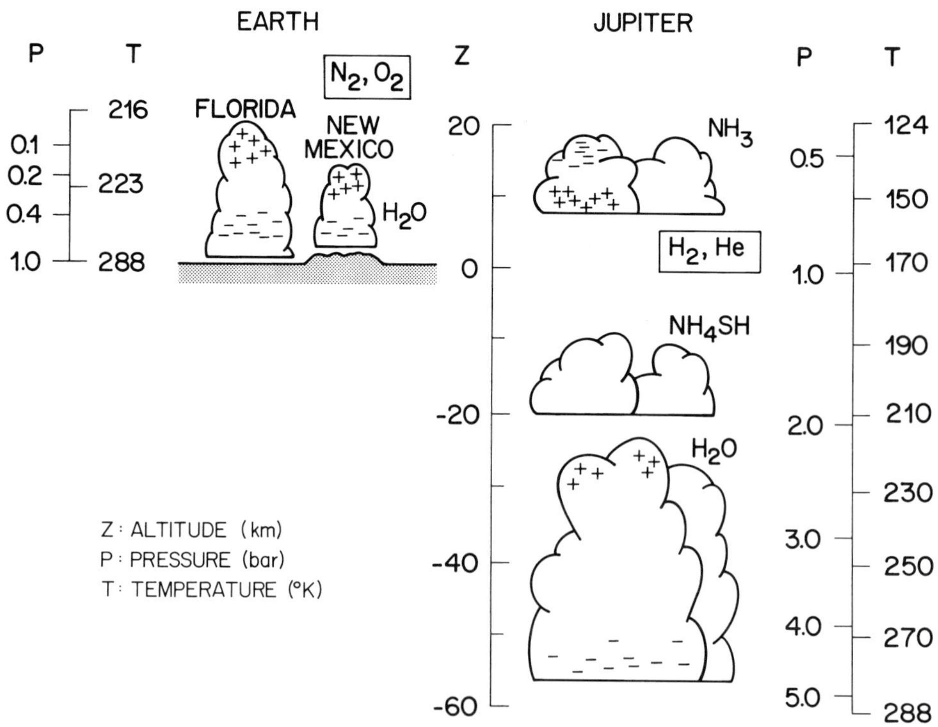

Fig. 2. Cartoon representation of the possible locations of separated charges in the atmospheres of Earth and Jupiter, showing comparisons of the cloud locations, altitudes, and temperatures.

atmospheric chemical species in each planet's atmosphere are boxed. On Earth, the topography of the land surface can at times also determine cloud location for the separated charges. Obviously, such an orographical influence should not exist at Jupiter, and the lightning should primarily be of the intra-cloud, cloud-to-cloud, and cloud-to-'air' types.

Especially important for establishing the conditions for localized, intense, upward convective motions in the Jovian atmosphere has been the recent modeling work of Stoker (1986) to understand the vertical structure of the equatorial plumes. She finds that, for appropriate parameters for cumulus cloud formation, moist convection can produce the equatorial plumes. For a solar water abundance in the atmosphere, peak vertical convective velocities of $\sim 150 \text{ m s}^{-1}$ are found.

The model results of Stoker (1986) suggest that the water abundance is closer to solar than to the depleted water conditions suggested by Bjoraker *et al.* (1986). Lunine and Hunten (1987) use the idea of moist convective plumes and the model plume of Stoker (1986) to conclude that such plumes can explain the apparent depletion of the IRIS-detected water vapor on Jupiter without invoking a global depletion of water from solar abundance. It is interesting to note that neither Stoker (1986) nor Lunine and Hunten (1987) made mention of the possibility of lightning production in convective regions such as these plumes, even though atmospheric conditions in the models, such as aerosols and upward convective velocities, are present to produce electrical charge separations and discharges (e.g., Williams *et al.*, 1983; Levin *et al.*, 1983).

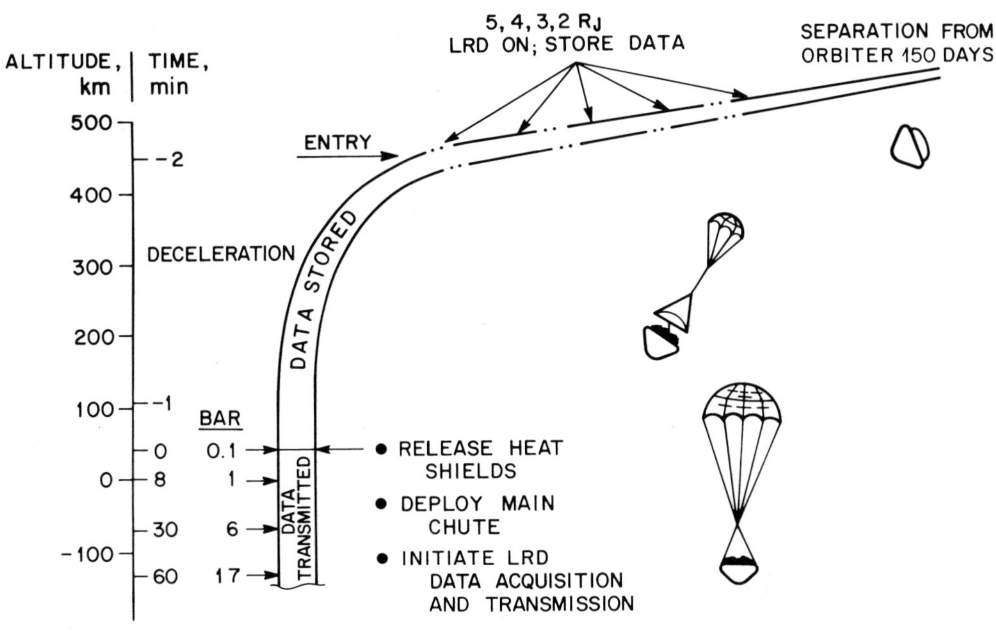

Fig. 3. Schematic illustration of Probe descent into Jupiter's atmosphere and the locations in the magnetosphere and atmosphere of LRD data acquisition.

3. Overview of LRD Objectives

As illustrated schematically in Figure 3, the Galileo Probe will be released from the Orbiter spacecraft 150 days prior to atmospheric entry. Radio frequency measurements are made in a reduced mode of operation at altitudes of ~ 5, 4, 3, 2 planetary radii from the center of Jupiter. These data are stored in the Probe memory and then read out during the atmospheric descent phase of the mission. During the atmospheric descent, the full complement of LRD data are acquired until the loss of the Probe signal by the over-flying Orbiter and/or the demise of the Probe due to atmospheric pressure and heat.

The science objectives of the LRD instrument are shown in Figure 4 for both magnetosphere and the atmosphere measurements. Measurements will not be made in the Jovian ionosphere (schematically shown by the horizontal dashed line). The RF data obtained in the magnetosphere will be analyzed also jointly with the Probe Energetic Particle Instrument (EPI) data to gain understanding of magnetospheric particle dynamics. In the magnetosphere, statistics on the characteristics of individual waveforms measured during a sampling interval will be accumulated at the four different altitudes. In addition, noise levels at three different spectral frequencies (3, 5, 90 kHz) will be determined during the measurement intervals.

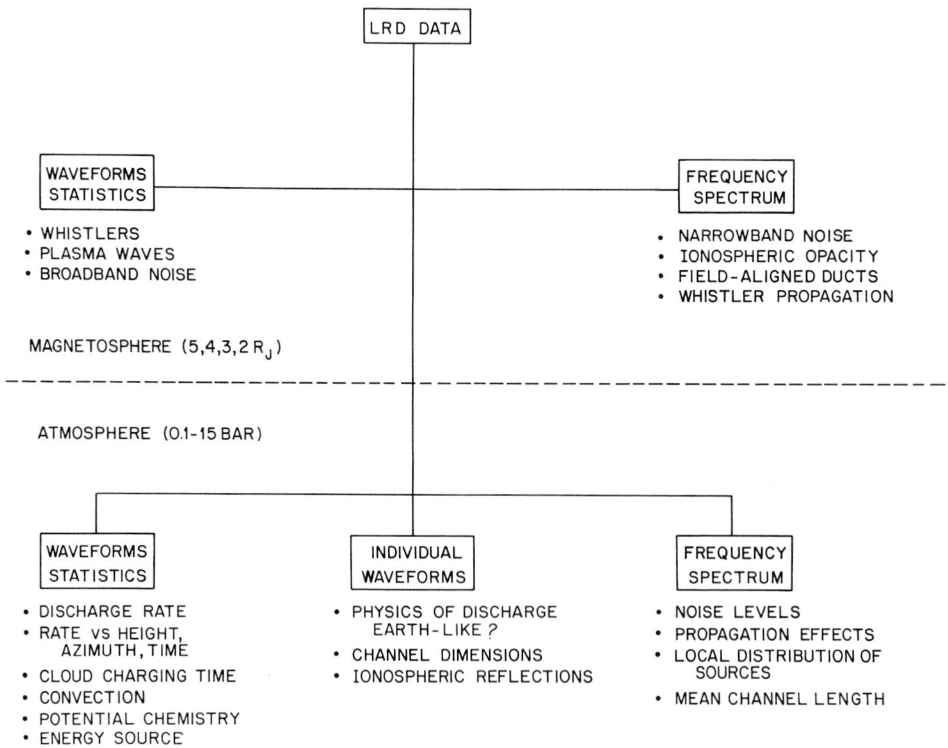

Fig. 4. Science objectives of the LRD instrument.

In the atmosphere mode, in addition to statistics on the waveforms and the spectral noise levels at the three narrow-banded frequencies, individual waveforms will be detected, saved, and transmitted to Earth. Such waveforms will provide powerful additional diagnostic capabilities for Jovian RF signals.

The LRD instrument, as noted above, has been designed to be as sensitive as possible, limited only by the spacecraft noise, and to be as versatile as possible, limited only by the imposed limitations on power, bit rate, and reliability considerations. It is within these constraints that the scientific objectives (Figure 4) will attempt to be achieved. Extensive measurements with Earth lightning have been made (see below) and these will be continued in order to gain the maximum understanding of the operational characteristics of the instrument, and therefore the maximum science from the Probe descent through Jupiter's atmosphere.

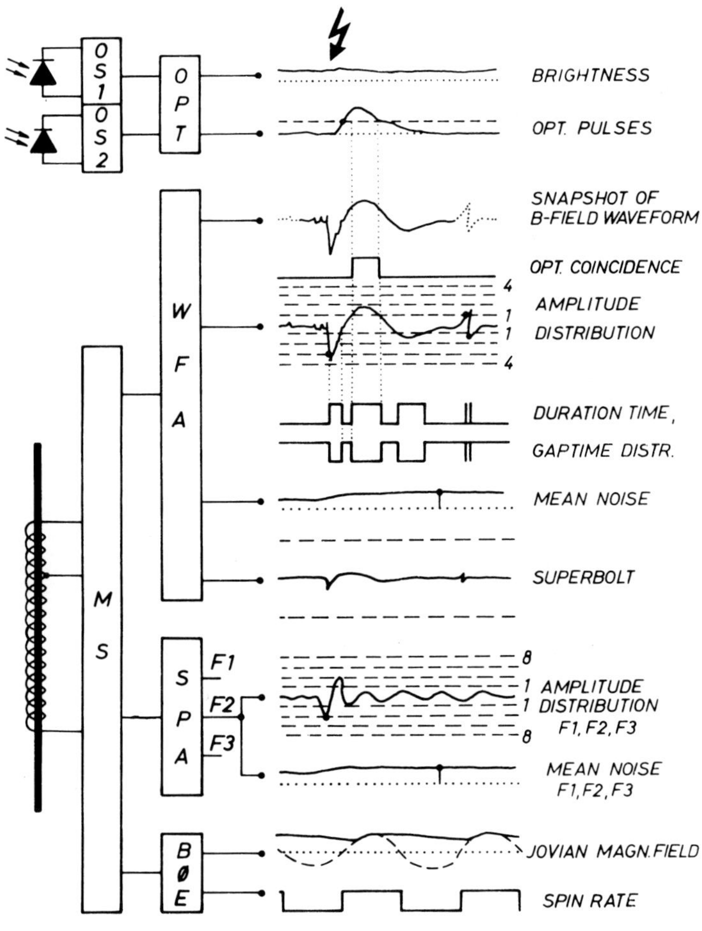

Fig. 5. Diagram of the sensors and the several different data outputs of the LRD instrument.

In both the magnetosphere and atmosphere modes the component of the Jovian magnetic field perpendicular to the Probe spin axis will be determined. These data will be used for analyses of EPI data and for determining the spatial distribution of the sources of some of the detected lightning signals. Further, these data will give engineering data on the Probe spin rate.

The output data from the LRD instrument is schematically shown with the two sensor types in Figure 5. The single ferrite core antenna ('magnetic sensor', MS) and two optical sensors (OS1; OS2) are shown. The optical sensors provide the optical (OPT) signals. The MS provides the input signal for the three principal RF data channels: the waveform analyzer (WFA) for statistics and waveform snapshot, the spectrum analyzer (SPA) with the three narrow-banded frequencies F1, F2, F3, and the magnetic field/spin rate determination B0E.

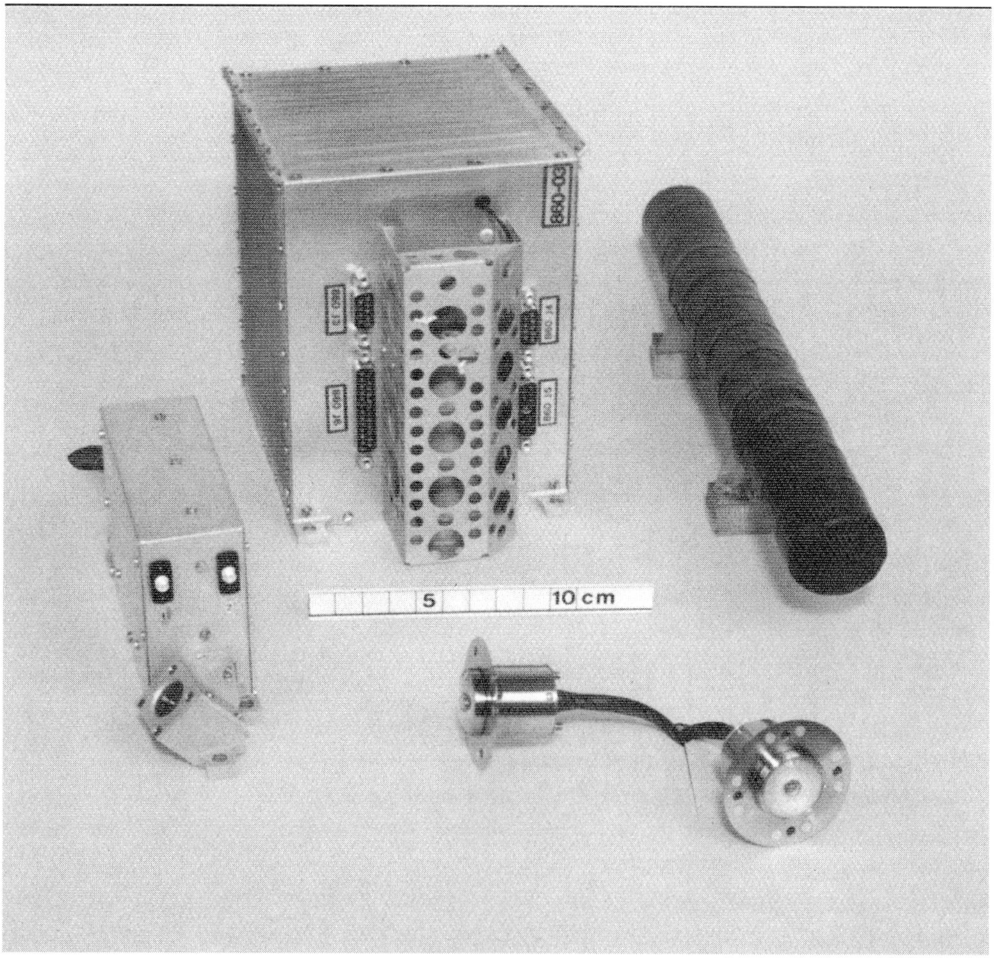

Fig. 6. Picture of LRD instrumentation.

4. LRD Instrument

Overview. A picture of the electronics box, assembled antenna, optical sensors, and the EPI detector housing is shown in Figure 6. The parameters of the instrument (including the EPI) are given in Table I.

TABLE I

LRD and EPI instrument parameters

Weight:	2.5 kg	
Size:	Electronics plus four sensors	
	RF antenna:	32 cm × 3.3 cm dia.
	Electronics box:	13 cm × 11.5 cm × 15 cm
	Optical sensors:	4 cm × 2.3 cm dia. each, two pieces
	EPI sensor box:	4 cm × 15.2 cm × 7.4 cm
Power:	EPI bias:	1 W for 1 hour
	Pre-entry:	3 W for 84 min (nominal)
	Entry:	3 W for duration
Data rate:	10^4 bits stored; 544 bits/64 s at entry; stored and real time data transmitted at 8 bps. During descent, a measuring period lasts 256 s and provides one data frame of 256 bytes.	

Radio wave sensor. The elements of the RF sensor are pictured in Figure 7. The RF signals in the Jovian atmosphere in the frequency range ~100 Hz to ~100 kHz are detected using the 32 cm long ferrite core antenna (Dehmel, 1989). An electric antenna is not used because of expected electrostatic noise during Probe descent through the Jovian atmosphere. The MS is mounted in a plane perpendicular to the spin axis of the Probe. As the Probe rotates with a nominal 10 rpm rate, a relatively high voltage is induced due to the strong Jupiter magnetic field (~4 G). This signal has to be suppressed and, in effect defines the lower band limit. The upper band limit was set to 100 kHz according to the sampling frequency of the signal (see below) and in order to suppress the detection of any high-frequency noise generated in the Jovian magnetosphere which can penetrate the Jovian ionosphere at frequencies higher than its maximum critical frequency.

The antenna is mounted outside the Probe body and is thus subject to large ambient temperature changes: from about 0 °C during cruise, to −150 °C after heat shield release, to +150 °C during descent. The strong ferrite core support structure is filled with insulating ceramic material to smooth this temperature variation. The antenna is wrapped in a conducting fabric as a shield for high-frequency stray fields from the nearby telemetry antenna. The antenna also contains a low-noise preamplifier with a dynamic range of 80 dB.

The LRD instrument performs its magnetospheric measurements when the Probe is encapsuled in the Probe heat shield. Hence, the after heat shield cover was designed with a special RF window over the antenna location. All of the thermal and structural

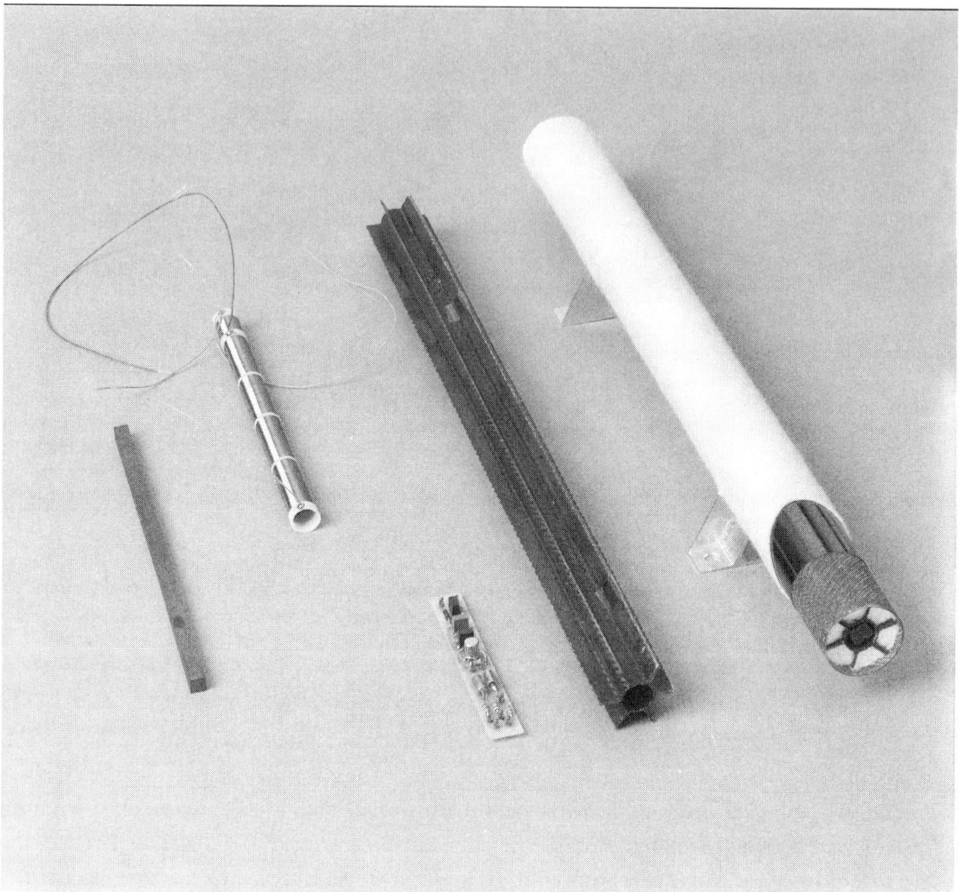

Fig. 7. Picture of the elements of the LRD ferrite core antenna. *Left to right*: Piece of ferrite rod, coil, electronics, support structure, mounted antenna filled and enveloped with thermal isolation. (Not shown: outer-most cover of microwave absorbing material.) The dimensions of the assembled antenna shown on right are 32 cm long by 3.3 cm diameter.

loads, as well as the electromagnetic responses with and without the heat shield, including operation with the Probe telemetry, have been extensively tested during the development and fabrication program.

Optical sensors. For an unambiguous verification of lightning as the source of measured magnetic pulses, a coincidence with light pulses is important. For this purpose, two identical sensors are mounted to look in opposite directions about the perpendicular to the Probe spin axis. Each sensor consists of a photodiode with an amplifier behind a fisheye lens. Both sensors are sensitive to the visual spectrum and together cover nearly a full 4π field of view. Special care has been taken to thermally decouple the lenses (exposed to the ambient temperature) and the electronics (which are coupled to the Probe structure). The signals of both sensors are added and provide measurements of the overall atmospheric brightness and of the light pulses (Figure 4).

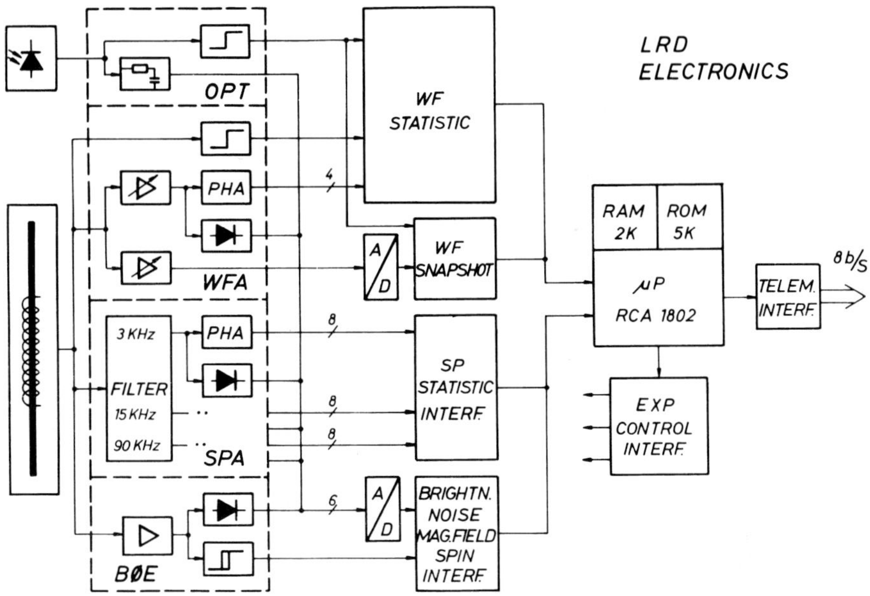

Fig. 8. Overview block diagram of the LRD electronics. The numbers on the lines indicate the actual number of connections.

Electronics. Figure 8 gives an overview of the analog and digital electronics and the signal paths of the sensor outputs. Four major sub-elements of the electronics are indicated by the dashed boxes and are described below.

B0E: The voltage induced due to the rotation of the antenna within the Jovian magnetic field provides a measure of the magnitude of the magnetic field component perpendicular to the Probe spin axis and of the Probe spin period. This signal is also used to sector RF measurements with respect to the ambient magnetic field direction.

SPA: The SPectral Analyzer consists of three narrow-band frequency channels centered at 3, 15, and 90 kHz. In each of the three channels the rectified and averaged signal is output as a measure of the narrow-band noise level, and a pulse height analyzer measures the amplitude distribution with 8 amplitude levels spaced by 10 dB. One of the narrow-band channels is sectored using the spin phase information of B0E. During the Pre-Entry (magnetosphere) mode of Operation (see below) the 3 kHz-channel is subdivided into two channels: parallel and perpendicular to the magnetic field. During the Descent mode of operation the 15 kHz channels is subdivided into 4 channels. This is the most 'sensitive' portion of the instrument. For example, an 'average' Earth lightning at a distance ~20 km could trigger a threshold ~60 dB above the present Probe noise level (see Rinnert *et al.* (1979) for discussions of propagation conditions in Jupiter's atmosphere).

WFA: The Wave Form Analyzer uses the full antenna bandwidth. Since the output signal of the magnetic antenna is proportional to the time derivative of the magnetic wave field, the antenna output first is integrated to regain the correct waveform. This signal is then fed to a gain change amplifier and is continuously digitized every 4 μs to provide the input for the waveform snapshot (see below). The integrated signal is amplified, rectified and averaged to provide the wideband signal noise levels and to obtain statistical characteristics of the waveforms. There are four threshold levels spaced by 10 dB following a gain change amplifier (0, 10, 20, 30 dB). Signals exceeding the lower threshold are treated as events and are characterized by the highest threshold which is exceeded and by the duration between consecutive crossings of the lower threshold, which corresponds to either a 'pulse duration time' or a 'pulse gap time'. As noted, there are 4 amplitude levels; the duration times and the gap times are sorted into 8 time bins for each distribution:

Duration time distribution	Gap time distribution
<8 μs	<8 μs
8–24 μs	8–40 μs
24–40 μs	40–168 μs
40–72 μs	168–680 μs
72–138 μs	0.68–8.7 ms
138–266 μs	8.7–72.2 ms
266–522 μs	72.2–1100 ms
>522 μs	>1.1 s

The gain change amplifier adjusts the pulse height analyzer to the appropriate noise level as it is controlled by the number of events in order to guarantee that pulses are characterized which sufficiently exceed the background noise level. A bipolar pulse, for example, is characterized by two amplitudes, two duration times (one for each half period) and one gaptime between the threshold crossings of the two half periods. Another channel is devoted to extremely large events, so-called 'super bolts' (see Figure 5). The threshold of this channel is set 20 dB above the highest WFA threshold.

OPT: The OPTical channel adds the output signals of both optical detectors. An averaged d.c. signal from the combined detectors provides a measure of the 'brightness' of the Probe's atmospheric environment. The pulses from the combined signals correspond to flashes. There is no information on the optical spectrum, the pulse amplitude or other pulse characteristics.

Not shown in Figure 2 is the In Flight Testgenerator (IFT). This device inserts 128 well-defined electrical pulses into the electronics in place of the MS signals. These pulses are processed the same way as the antenna signals and give test data on the operation of the instrument.

WF SNAPSHOT: The Wave Form Snapshot channel provides a window of 1 ms out of the continuously sampled signal (sample period 4 μs). A block diagram of the WF Snapshot electronics is shown in Figure 9. The peak amplitude of the measured pulse defines the trigger point. The 1 ms signal interval contains 64 pre-trigger samples covering 256 μs before the pulse peak, 64 post-trigger samples covering 256 μs after the pulse peak, and 61 samples (every second sample) covering 488 μs. The time resolution of the first half of the time interval is therefore 4 μs and the second half is 8 μs. Although

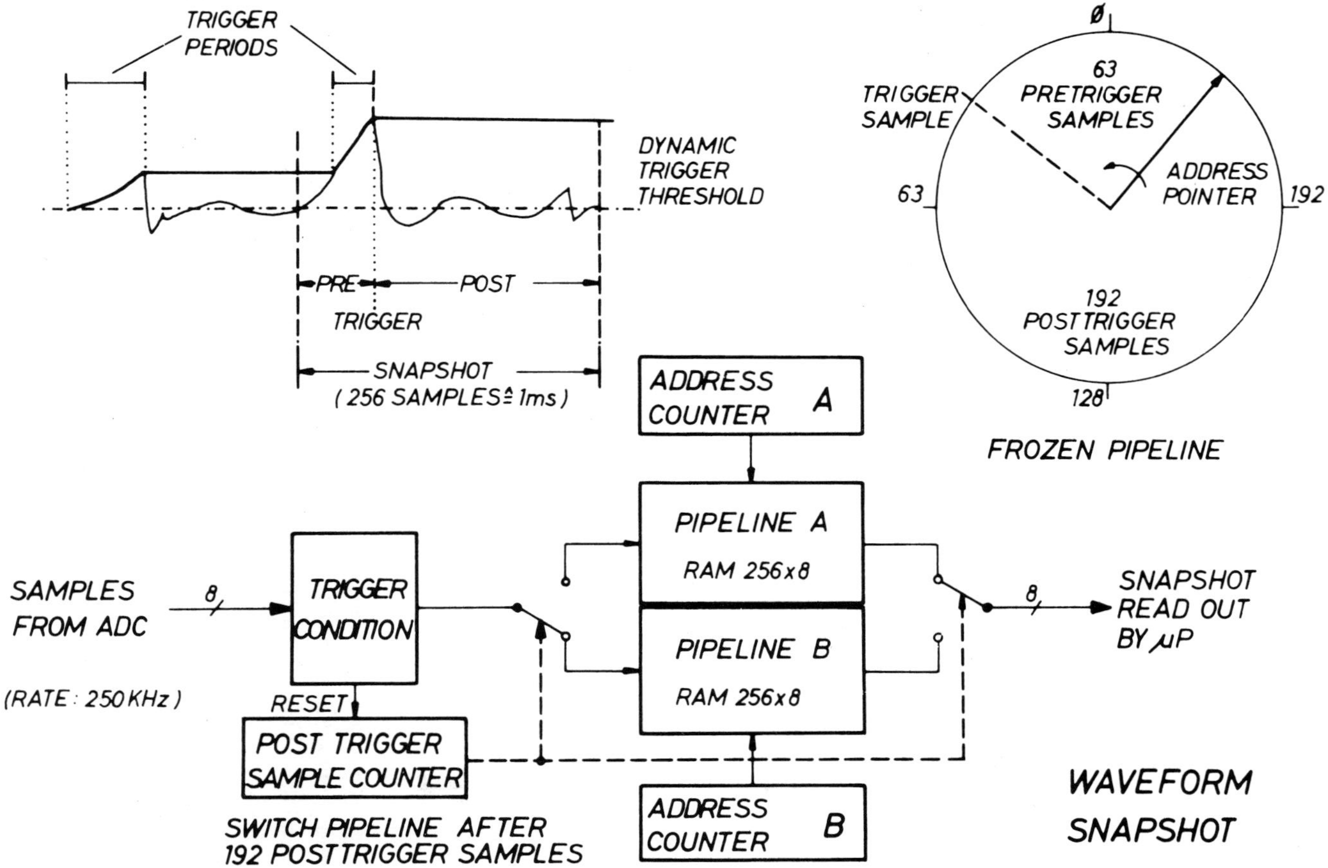

Fig. 9. Block diagram of the waveform 'snapshot' analyzer of the LRD instrument.

the antenna signal is continuously sampled, only one one-millisecond interval is saved for transmission during each measuring period of 256 s. This is because of the limited data rate available for the instrument. Two switched pipeline memories are used for this purpose. The signal interval that is transmitted during each measuring interval (see below) is defined by a rotating priority scheme (largest amplitude waveform with optical coincidence; largest amplitude waveform; and random).

The instrument is controlled by an 1802 microprocessor which determines the modes of operation (including the EPI measurements), generates the statistical distributions, determines coincidences between optical and magnetic pulses and performs the data formatting (most of the science data are logarithmically compressed). One standard data frame consists of 256 bytes in total and contains the information gathered during four major frame periods of 64 s each. The first one quarter (64 bytes) of the LRD science data frame contains all of the LRD statistics information (spectral and wäveform statistics), number of events, average levels, and gain information. The other three quarters (192 bytes) of the data frame contain the waveform samples as signed numbers (− 128 to + 127) and waveform status information. The second and third quarters of the LRD science data frame contain the values of the samples which are made each 4 μs, while the fourth quarter contains the values of those samples which are each 8 μs. In addition, the data frame contains engineering data such as voltages, currents, and temperatures.

5. Modes of Operation

Magnetosphere mode. The LRD instrument will operate in the pre-entry phase at distances from the planet's center of about 5, 4, 3, and 2 R_J. The instrument is switched on by the Probe timer at these locations. In this 'magnetosphere mode' the EPI is also in operation. As the Probe is still encapsulated within the heat shield, the MS is less sensitive and the optical sensors are covered. The outputs of the LRD instrument are as noted in the previous section, but without the waveform snapshots. The magnetosphere mode data set at each of the four locations consists of a 64 byte data frame with statistics and the 3 kHz spectral channel subdivided into parallel and perpendicular (to the magnetic field) channels. The data are stored in the Probe memory for transmission during the atmospheric descent phase of the mission.

Atmosphere mode. When the LRD instrument is switched on at descent the instrument begins with a test cycle (IFT) and the first data set contains the test pulse data. After that, the instrument runs continuously until the end of the mission and outputs a complete data set every four major frame periods, 256 s. These data sets contain spectral data (the 15 kHz channel being sectored), waveform statistics data, a 1 ms time interval with a selected waveform, optical data and miscellaneous data such as magnetic field component, spin period, and engineering data. The number of complete data sets achieved during descent depends upon the length of time that the Probe survives and/or the length of time that the Probe relay signal is successfully acquired by the over-flying Orbiter. For example, if the total atmosphere data time is ~ 48 min, then 10 data sets would be sent back (the 11th would be acquired but there would be no time for transmittal). The 10 data sets would contain one test data set and 9 science data sets.

6. Tests and Calibrations

Because of severe constraints as to weight and power for Probe subsystems, the LRD instrument is very compact. Further, extensive on-board compression of the data is necessary because of the limited available data rate.

All sensors and instrument characteristics, of course, have been extensively tested and calibrated. For example, radiation tests were carried out on the sensor electronics and pressure tests were made of the vented electronics box. These latter tests caused a stiffening piece to be added to the microprocessor chip. The calibrations could be verified over long periods because of the delays of the launch of the Galileo spacecraft. A further verification of the instrument parameters is provided by the on-board implemented test generator (ITG), as noted in Section 4.

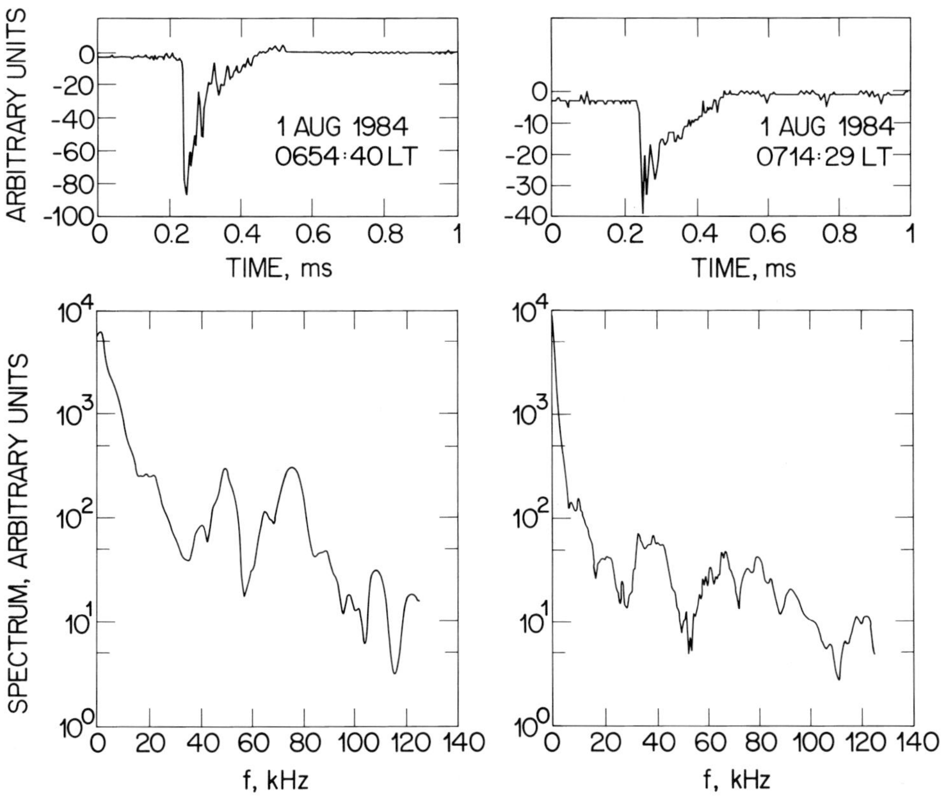

Fig. 10. *Upper panels*: Two cloud-to-ground discharges measured with the Galileo Probe lightning experiment engineering unit during a lightning storm near Lindau in August 1984. *Lower panels*: The power spectrum of each discharge.

Most important for the definition, design verification, and test of the instrument have been the extensive campaigns devoted to measurements of Earth lightning. Some of the results of these campaigns and their relevance to both characterizing the instrument and to providing new data on Earth lightning have been reported in the scientific literature over the years (Rinnert *et al.*, 1984, 1985, 1988, 1989; Lanzerotti *et al.*, 1988b; 1989b).

On Earth, cloud-to-ground discharges are quite different in their RF characteristics

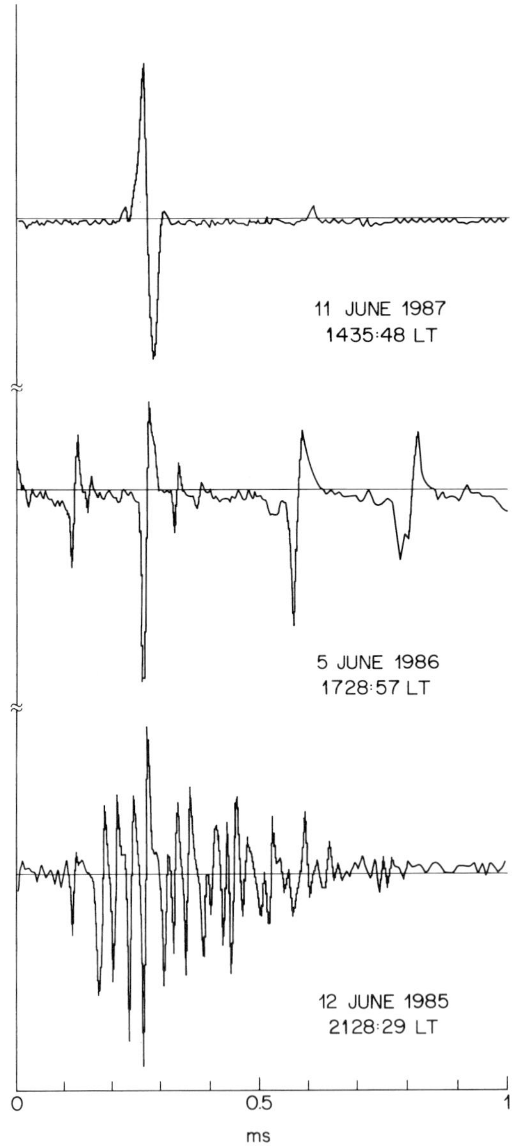

Fig. 11. Examples of types of intra-cloud discharge magnetic field waveforms measured in Lindau during three different lightning storms.

than are intra-cloud discharges. These differences are being carefully studied by our team using the LRD engineering unit (which has identical electrical characteristics to the instrument flying on the Probe). These studies are being made in order to eventually be able to evaluate the atmospheric breakdown processes, the electrical currents, and the possible chemical processes associated with Jovian lightning.

As an example, shown in the upper portions of both sides of Figure 10 are the magnetic field wave forms of two cloud-to-ground return strokes measured in Lindau with the Galileo LRD engineering instrument unit during a nearby lightning storm in 1984. The power spectrum of the RF signals for each of these discharges was calculated for us by D. J. Thomson (AT&T Bell Laboratories) after first mathematically removing the impulse response of each signal. The power spectra were then calculated using four prolate spheroidal data windows in the time domain and a fast Fourier transform algorithm (see Lanzerotti *et al.*, 1989b). Each RF power spectrum clearly falls with increasing frequency and each spectrum has definite structure that corresponds to the 'secondary peaks' evident in the time domain traces following the initial impulse.

The RF signals of intra-cloud Earth lightning discharges have been much less studied than those of cloud-to-ground discharges (see, e.g., Uman, 1987). On average, ground discharges are much more powerful than cloud discharges. The ratios of the amplitudes of return strokes to cloud discharges appear to range from about 20 to 1 at ~3 kHz, to ~10 to 1 at 10 kHz, to ~1 to 1 at 100 kHz (Brook and Ogawa, 1977). Shown in Figure 11 are examples of different types of magnetic field waveforms produced by cloud discharges and measured using the engineering unit (Lanzerotti *et al.*, 1989b). From top

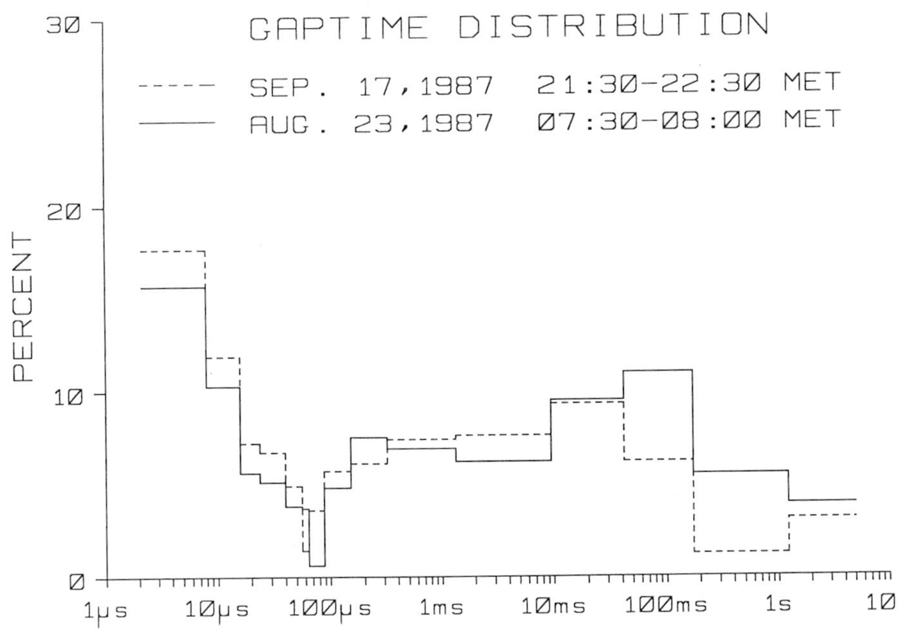

Fig. 12. The gap time distribution provided by the LRD instrument during two lightning storm intervals.

to bottom in the figure, the cloud discharges observed in 1 ms include a single bipolar pulse, a series of quasi-periodic bipolar pulses filling the entire measuring internal, and a sequence of (possibly approximately periodic) bipolar pulses apparently modulated by an overall signal envelope.

LRD analysis of the intervals between pulses (gap times) obtained during two storm periods is shown in Figure 12 (Rinnert *et al.*, 1989). The large numbers of gap times less than 8 μs are probably due to the gaps between successive portions of rectified bipolar pulses. There is a significant minimum in gap times between 56 and 88 μs, and then a maximum for gap times between ~ 10 and ~ 170 ms. Additional studies have been made on Earth lightning with the LRD engineering unit to study gap time and other distributions as a function of whether the originating RF signals were predominantly intra-cloud or cloud-to-ground (Rinnert *et al.*, 1989). The field campaigns with the LRD instrument have yielded interesting new scientific results in the discipline of lightning research. Such studies will be continued in the years to come until the Probe descends through the Jovian atmosphere in December 1995.

Acknowledgements

This work has been supported in part by NASA contract NAS2-997 to the University of Florida and in part by the Bundesministerium für Forschung und Technologie, Germany by contract 01QJ0976. We thank the NASA Ames Galileo Project Office personnel over the many years of the program, especially Al Wilhelmi, Benny Chin, Ed Tischler, and Charles Sobec.

References

Atreya, S. K.: 1986, *Atmospheres and Ionospheres of the Outer Planets and their Satellites*, Springer-Verlag, Heidelberg.

Bar-Nun, A.: 1975, *Icarus* **24**, 86.

Bar-Nun, A. and Podolak, M.: 1985, *Icarus* **64**, 122.

Bar-Nun, A., Noy, N., and Podolak, M.: 1984, *Icarus* **59**, 162.

Bjoraker, G. A., Larson, H. P., and Kunde, V. G.: 1986, *Astrophys. J.* **311**, 1058.

Borucki, W. J., Bar-Nun, A., Scarf, F. L., Cook, A. F., and Hunt, G. E.: 1982, *Icarus* **52**, 492.

Brook, M. and Ogawa, T.: 1977, in R. H. Golde (ed.), *Lightning*, Academic Press, New York, p. 191.

Burns, J. M., Showalter, M. R., Cuzzi, J. N., and Durison, R. H.: 1983, *Icarus* **54**, 280.

Chadha, M., Flores, J., Lawless, J., and Ponnamperuma, C.: 1971, *Icarus* **15**, 39.

Cook, A. F., Duxbury, T. C., and Hunt, G. E.: 1979, *Nature* **280**, 794.

Dehmel, G.: 1989, in R. Ball and K. J. Overshott (eds.), *Sensors, 5*, VCH Publ., New York, p. 205.

Gurnett, D. A., Shaw, R. R., Anderson, R. R., Kurth, W. S., and Scarf, F. L.: 1979, *Geophys. Res. Letters* **6**, 511.

Kaiser, M. L., Connerney, J. E. P., and Desch, M. D.: 1983, *Nature* **303**, 50.

Ksanfomality, L. V.: 1979, *Kosm. Issled.* **17**, 747.

Kurth, W. S., Strayer, B. D., Gurnett, D. A., and Scarf, F. L.: 1985, *Icarus* **61**, 497.

Lanzerotti, L. J., Rinnert, K., Krider, E. P., Uman, M. A., Dehmel, G., Gliem, F. O., and Axford, W. I.: 1983, in L. H. Ruhnke and J. Latham (eds.), *Proc. Atmospheric Electr.*, Deepak Publ., Hampton.

Lanzerotti, L. J., Rinnert, K., Krider, E. P., and Uman, M. A.: 1988a, *Proc. VIII Int. Conf. Atmospheric Electr.*, Uppsala, p. 749.

Lanzerotti, L. J., Thomson, D. J., Maclennan, C. G., Rinnert, K., Krider, E. P., and Uman, M. A.: 1988b, *Proc. VIII Int. Conf. Atmospheric Electr.*, Uppsala, p. 480.

Lanzerotti, L. J., Rinnert, K., Krider, E. P., and Uman, M. A.: 1989a, in M. J. S. Belton, R. A. West, and J. Rahe (eds.), *Time-Variable Phenomena in the Jovian System*, NASA SP-494, p. 374.

Lanzerotti, L. J., Thomson, P. J., Maclennan, C. G., Rinnert, K., Krider, E. P., and Uman, M. A.: 1989b, *J. Geophys. Res.* **94**, 13221.

Levin, Z., Borucki, W. J., and Toon, D. B.: 1983, *Icarus* **56**, 80.

Lunine, J. I. and Hunten, D. M.: 1987, *Icarus* **69**, 566.

Marten, A., Rovan, D., Baluteau, J. P., Chedin, A., and Scott, N.: 1981, *Icarus* **46**, 233.

Orten, G. S., Appleby, J. F., and Martonchick, J. V.: 1982, *Icarus* **52**, 94.

Owen, T. and Terrile, R. J.: 1982, *J. Geophys. Res.* **87**, 11193.

Prinn, R. G. and Owen, T.: 1976, in T. Gehrels (ed.), *Jupiter*, University of Arizona, Tucson, p. 319.

Rinnert, K.: 1982, in H. Volland (ed.), *Handbook of Atmospherics*, CRC Press, Boca Raton, FL, p. 99.

Rinnert, K.: 1985, *J. Geophys. Res.* **90**, 6225.

Rinnert, K., Lanzerotti, L. J., Krider, E. P., Uman, M. A., Dehmel, F., Gliem, G. O., and Axford, W. I.: 1979, *J. Geophys. Res.* **84**, 5181.

Rinnert, K., Lanzerotti, L. J., Dehmel, G., Gliem, F. O., Krider, E. P., and Uman, M.: 1984, *Proc. VII Int. Conf. Atmospheric Electr.*, p. 481.

Rinnert, K., Lanzerotti, L. J., Dehmel, G., Gliem, F. O., Krider, E. P., and Uman, M. A.: 1985, *J. Geophys. Res.* **90**, 6239.

Rinnert, K., Lauerdale, R., Lanzerotti, L. J., Krider, E. P., and Uman, M. A.: 1899, *Proc. VIII Int. Conf. Atmospheric Electr.*, Uppsala, p. 882.

Rinnert, K., Lauerdale, R., Lanzerotti, L. J., Krider, E. P., and Uman, M. A.: 1989, *J. Geophys. Res.* **94**, 13229.

Sagan, C. E., Lipincott, E. R., Dayhoff, M. D., and Eck, R. V.: 1967, *Nature* **213**, 273.

Sato, M. and Hanse, J. E.: 1979, *J. Atmospheric Sci.* **36**, 1133.

Stoker, C. R.: 1986, *Icarus* **67**, 106.

Taylor, W. W. L., Scarf, F. L., Russell, C. T., and Brace, L. H.: 1979, *Nature* **279**, 614.

Uman, M. A.: 1987, *The Lightning Discharge*, Academic Press, New York.

Weidenschilling, S. J. and Lewis, J. S.: 1973, *Icarus* **20**, 465.

West, R. A. and Tomasko, M. G.: 1980, *Icarus* **41**, 278.

Williams, M. A.: 1986, 'An Analysis of the Voyager Images of Jovian Lightning', Ph.D. Thesis, University of Arizona, Tucson.

Williams, M. A., Krider, E. P., and Hunten, D. M.: 1983, *Rev. Geophys.* **21**, 892.

Woeller, F. and Ponnamperuma, C.: 1969, *Icarus* **10**, 386.

Zarka, P.: 1985, *Astron. Astrophys.* **146**, L15.

Zheleznyakov, V. V.: 1970, *Radio Emission of the Sun and Planets*, Pergamon Press, Oxford.

GALILEO PROBE MASS SPECTROMETER EXPERIMENT

H. B. NIEMANN, D. N. HARPOLD

NASA-Goddard Space Flight Center, Greenbelt, MD 20771, U.S.A.

S. K. ATREYA, G. R. CARIGNAN

University of Michigan, U.S.A.

D. M. HUNTEN

University of Arizona, Tucson, U.S.A.

and

T. C. OWEN

University of Hawaii, U.S.A.

Abstract. The Galileo Probe Mass Spectrometer (GPMS) is a Probe instrument designed to measure the chemical and isotopic composition including vertical variations of the constituents in the atmosphere of Jupiter. The measurement will be performed by *in situ* sampling of the ambient atmosphere in the pressure range from approximately 150 mbar to 20 bar. In addition batch sampling will be performed for noble gas composition measurement and isotopic ratio determination and for sensitivity enhancement of non-reactive trace gases.

The instrument consists of a gas sampling system which is connected to a quadrupole mass analyzer for molecular weight analysis. In addition two sample enrichment cells and one noble gas analysis cell are part of the sampling system. The mass range of the quadrupole analyzer is from 2 amu to 150 amu. The maximum dynamic range is 10^8. The detector threshold ranges from 10 ppmv for H_2O to 1 ppbv for Kr and Xe. It is dependent on instrument background and ambient gas composition because of spectral interference. The threshold values are lowered through sample enrichment by a factor of 100 to 500 for stable hydrocarbons and by a factor of 10 for noble gases. The gas sampling system and the mass analyzer are sealed and evacuated until the measurement sequence is initiated after the Probe enters into the atmosphere of Jupiter. The instrument weighs 13.2 kg and the average power consumption is 13 W.

The instrument follows a sampling sequence of 8192 steps and a sampling rate of two steps per second. The measurement period lasts appropriately 60 min through the nominal pressure and altitude range.

Table of Contents

Space Science Reviews **60:** 111–142, 1992.

1. Goals and Capabilities

1.1. Introduction

The Voyager observations, complemented by ground-based measurements, have contributed immensely to our knowledge of the atmosphere of Jupiter. Despite this success, however, the inventory of known constituents in the atmosphere is severely limited, the gross composition still includes large statistical uncertainties in the mixing ratios of various species, and the noble gas abundances (with the exception of helium) are completely unknown. Isotopic ratios, which might harbor the secret of the origin, and to some extent, evolution of the atmosphere of Jupiter, are hardly measured at all – the only exceptions being certain isotopes of carbon and hydrogen.

It is already apparent from a few isolated measurements that the atmosphere does not fully mimic the solar ratios of the elements. For example, C/H, O/H, and N/H (from CH_4, H_2O, and NH_3 measurements, respectively) all appear to be enhanced relative to solar by factors of 2–3, whereas He/H and Ge/H apparently have sub-solar values (Noll *et al.*, 1988; Gautier and Owen, 1989; Carlson *et al.*, 1991). Both NH_3 and H_2O condense and form clouds in the region of the atmosphere probed by Earth-based observations, so their absolute abundances are model-dependent. The situation becomes still worse for sulfur, which must be present but has not yet been detected, presumably because of its incorporation into NH_4SH clouds (Weidenschilling and Lewis, 1973; Larson *et al.*, 1984; Atreya and Romain, 1985). The current upper limit is 10^{-3} of the solar value.

Altitude-dependent mixing ratios are also exhibited by the disequilibrium species: those formed through photolysis and charged particle bombardment of various constituents in the upper atmosphere, others that are created by lightning discharges within the clouds, and a third set consisting of molecules formed under thermoequilibrium conditions at the high temperatures existing at great depths in the atmosphere, some of which are brought up to altitudes where we can observe them by strong vertical convection.

With *in situ* measurements, the Galileo Probe Mass Spectrometer (GPMS) is in a position to fill the gaps in our knowledge of the composition of the atmosphere of Jupiter. Furthermore, sample enrichment provides an opportunity to address fundamental questions of the origin and evolution of the Jovian atmosphere by allowing measurements of trace gases and removing interference (by overlapping ions) from key isotopes. Repeated sampling of the atmosphere at different altitudes will permit the determination of the altitude dependence of constituent abundances, thus enabling correlations with the locations of cloud layers and an identification of the source(s) of energy driving various chemical reactions.

The primary objective of the Probe mass spectrometer is thus to determine abundances and isotope ratios of 'major' (i.e., mole fraction $\geq 10^{-7}$) constituents as a function of altitude. The best precision is obviously achieved for the most abundant gases; it is estimated that the mixing ratio of He and Ne will be measured to an accuracy of 2% and molecules difficult to measure with mass spectrometers such as H_2O, NH_3, and H_2S to a factor of two or better. To search for trace constituents at concentrations on the order of 1 ppbv, a sample enrichment system is added to the basic mass spectrometer, and for noble gases a purification cell is included. We discuss the expected results in the following sections.

The instrument was developed at the Goddard Space Flight Center in Greenbelt, Maryland, in collaboration with the University of Michigan Space Physics Research Laboratory in Ann Arbor, Mich., which developed the Electronics Subsystem, and the General Electric Company which developed and fabricated the hybrid electronics circuits. The Principal Investigator of the Galileo Probe Mass Spectrometer is H. B. Niemann of NASA-Goddard Space Flight Center, and the Co-Investigators are S. K. Atreya, G. R. Carignan, and T. M. Donahue of the University of Michigan, R. E. Hartle of NASA-Goddard Space Flight Center, D. M. Hunten of the University of Arizona, T. C. Owen of the University of Hawaii, and N. W. Spencer of the University of Maryland.

1.2. Science goals and GPMS capabilities

As already mentioned in the Introduction, the mixing ratios of even the principal minor constituents in the Jovian atmosphere appear to be non-solar. This is consistent with the current paradigm for outer planet formation, which postulates the initial growth of a core by accretion of icy planetesimals to a mass of 10–15 M_E, accompanied in its later stages with the gravitational attraction of surrounding nebular gases. Thus one anticipates atmospheres consisting of a mixture of two components: gases released from the planetesimals during accretion and/or through dissolution in the forming gaseous envelope, and the gases acquired directly from the solar nebula. This should lead to a general enrichment of all heavy elements except neon, which is not trapped in ices except at temperatures below 20 K (Bar-Nun *et al.*, 1987). In fact, carbon, as manifested in methane, becomes progressively more enriched in the series Jupiter, Saturn, Uranus, Neptune, which is also a progression in the relative sizes of the planetary cores. This conforms to the predictions of the model. However, the elements arsenic (as AsH_3) and phosphorus (as PH_3) are more enriched on Saturn (compared to Jupiter) than are carbon and nitrogen (Noll *et al.*, 1989, 1990). The dichotomy even exists in Jupiter itself (Table I). There must be some clues here about the differing internal structures of these two bodies, or differences in a way the elements were incorporated into the planets. To understand these differences, we need to know the abundances of more elements, in order to identify the source of the fractionation. For example, does it depend on the ability of a given element to form a gas that would be trapped or frozen into ices of an

TABLE I

Gross atmospheric composition

A. Dominant isotope

Gas	Solar mixing ratio	Observations
H_2	0.84	0.90
He	0.16	0.10
H_2O	1.7×10^{-3}	$\sim 2 \times$ solar
CH_4	7.2×10^{-4}	1.7×10^{-3}
Ne	2.5×10^{-4}	
NH_3	2.2×10^{-4}	$\sim 4 \times 10^{-4}$
SiH_4	7.2×10^{-5}	$< 2.5 \times 10^{-9}$
H_2S	3.7×10^{-5}	$< 3.3 \times 10^{-9}$
^{36}Ar	7.2×10^{-6}	
PH_3	7.5×10^{-7}	$7 \pm 1 \times 10^{-7}$
HCl	3.8×10^{-7}	
HF	6.0×10^{-8}	

B. Isotope ratios

Ratio	Solar mixing ratio	Observations
D/H	3.4×10^{-5}(?)	$2 \pm 1 \times 10^{-5}$
$^3He/^4He$	1.4×10^{-4}(?)	
$^{18}O/^{16}O$	2.0×10^{-3}	
$^{13}C/^{12}C$	1.1×10^{-2}	$1.1 + 0.3 \times 10^{-2}$
$^{15}N/^{14}N$	3.6×10^{-3}(?)	
$^{22}Ne/^{20}Ne$	7.3×10^{-2}(?)	
$^{34}S/^{32}S$	4.4×10^{-2}	
$^{38}Ar/^{36}Ar$	1.9×10^{-1}	
$^{21}Ne/^{20}Ne$	2.4×10^{-3}(?)	
$^{37}Cl/^{35}Cl$	3.2×10^{-1}	
$^{33}S/^{32}S$	7.9×10^{-3}	

accreting planetesimal? The abundances of the noble gases could prove to be very enlightening as a discriminant between internal and external processes. The apparent depletion of helium on Jupiter (Table I) is probably a mild example of the much more severe situation on Saturn, where helium is dissolving in the metallic hydrogen surrounding the core of the planet. The precise values of He/H to come from the HAD experiment on Galileo (von Zahn and Hunten, 1992) will allow this interpretation to be sharpened, as will the GPMS determinations of other noble gas abundances.

In spite of these deviations from solar abundances, the absence of nuclear reactions on Jupiter and the vertical convection in its atmosphere suggest that primordial values of isotopic ratios will be preserved, provided that we measure them in dominant carriers. Thus we have an opportunity to determine the starting value for D/H in the hydrogen gas in the solar nebula. The most recent estimate of this cosmologically important ratio from studies of $^3He/^4He$ in the solar wind and meteorites leads to

$D/H = 3.4 \pm 1 \times 10^{-5}$ (Anders and Grevesse, 1989). Measurements of CH_3D/CH_4 in Jupiter's atmosphere yield $D/H = 2 \pm 1 \times 10^{-5}$ (Gautier and Owen, 1989). Although the error bars overlap, there is a suggestion that these values may be higher than the local interstellar ratio ($D/H = 1.5 \pm 1 \times 10^{-5}$) and such a difference is recognized as a natural consequence of the nuclear burning of deuterium in stars, e.g., Geiss and Reeves (1981). A good determination of D/H in the primordial solar nebula can thus be used (in principle!) to constrain models of Big-Bang nucleosynthesis and galactic evolution, particularly if abundances of other key elements can be determined in that same reservoir (Gautier and Owen, 1983). Jupiter offers us this opportunity, and the GPMS is designed to exploit it.

Yet another problem is posed by an observed secular increase of $^{15}N/^{14}N$ in lunar soils (Geiss and Bochsler, 1982; Kerridge, 1989). The two leading interpretations of this effect are an as yet undefined nuclear process in the convection zone of the Sun (Kerridge, 1989) and the existence of an as yet unidentified reservoir of 'light' nitrogen in condensed matter in the solar system (Geiss and Bochsler, 1982). Both camps agree that a precise determination of $^{15}N/^{14}N$ on Jupiter would provide an important constraint, since it will give the starting value for the nitrogen isotope ratio in the solar nebula. This value will also be important as an anchor for the interpretation of the $^{15}N/^{14}N$ enrichment found in the Martian atmosphere (Nier and McElroy, 1977).

Finally, there has been great interest in disequilibrium chemistry on Jupiter. What are the agents responsible for the pastel colors in Jupiter's clouds? Why is sulfur so deficient in the regions of the atmosphere that we can investigate by remote sensing? What is the relative importance of solar UV compared with lightning discharges in driving various chemical reactions? Do any of these reactions have any relevance to prebiotic chemistry on the primitive Earth? All of these questions have been discussed at length in the literature, but we have no good answers to any of them. We need more information, the kind we can only get by sampling the atmosphere directly, repeatedly, over a range of altitudes. This is just what the GPMS will do.

For the purposes of illustrating the capabilities of this instrument, we use a fictitious 'solar composition' atmosphere as a reference. The following tables thus reflect the solar ratios of elements (Anders and Grevesse, 1989) and the gases in which they are primarily incorporated in the Jovian atmosphere. Where available, these values are compared with actual observations. The reader is referred to Chapter 1 of Atreya (1986), Gautier and Owen (1989), Noll *et al.* (1989, 1990), and Carlson *et al.* (1991), for additional discussion.

1.3. Direct investigations

The projected dynamic range of the Probe mass spectrometer is 10^8, hence in principle, gases with mixing ratios greater than a few tens of ppbv can be detected. In practice, the precision of the measurement deteriorates rapidly as the limit of instrumental capabilities is approached. Nevertheless, the experiment can improve on some of the large uncertainties inherent in remote sensing for mixing ratios of most species above 5×10^{-8}.

Inspection of Table I indicates that there are prospects of a rich harvest: at the very least eight elemental abundances and eight isotopic ratios; nine if sulfur is detected, as it probably will be. Silicon may not be found even at the deepest level reached by the Probe, but the search is an important one (see below). The over-solar abundances of heavy elements (except neon) predicted by the nucleation-accretion model for giant planet formation can be quantitatively evaluated and there is an opportunity to obtain several primordial isotope ratios. Important aspects of Jovian chemistry and cloud physics will be investigated by recording the abundances of NH_3, H_2S, and H_2O at various altitudes. At the present time, the mixing ratios of these condensible gases and the locations of the clouds they are expected to produce (ammonia cirrus, ammonium hydrosulfide, ammonium hydroxide, water) are very poorly defined.

In particular, the abundance of water in the upper troposphere is so low that some authors have suggested it implies a global depletion of oxygen on Jupiter (Bjoraker *et al.*, 1986). This conclusion is contradicted by models for the chemistry of the deep atmosphere that are constrained by observations of CO (Noll *et al.*, 1988), requiring near solar abundances of water. Lunine and Hunten (1987) have suggested that moist vertical convection on Jupiter is limited to narrow plumes, resulting in the very low relative humidity observed in the upper atmosphere. The global abundance has been re-assessed by Carlson *et al.* (1991), in the most comprehensive model of the Voyager IRIS observations presented to date. These authors find the H_2O abundance on Jupiter to be at least 1.5 times the solar value with a factor of two enrichment providing the best fit to the observations. The Carlson *et al.* model has an NH_4SH cloud layer at 1.9 bar and the H_2O cloud base is at 2.3 bar. This location for the ammonium hydrosulfide cloud agrees well with the model developed by Larson *et al.* (1984) and Atreya and Romain (1985) to explain their low upper limit on H_2S (Table I). We have adopted the Carlson *et al.* best-fit value for the H_2O mixing ratio in Table I, although it is strongly model dependent. The measurements with the GPMS down to at least 10 bar will reveal the sulfur and oxygen mixing ratios by direct measurement for the first time. With careful interpretation, it may be possible to explore the role played by sulfur (if it has one!) in generating observable chromophores in the Jovian clouds.

1.4. The Use of Enrichment Cells

1.4.1. *Noble Gases*

Removal of hydrogen and other chemically-active species by the noble gas cell gives the following mixing ratios for the noble gases:

TABLE II
Noble gas abundances

He	1.0
Ne	1.3×10^{-3}
Ar	3.7×10^{-5}
Kr	1.7×10^{-8}
Xe	1.7×10^{-9}

With a dynamic range of 10^8, it may just be possible to reach krypton and xenon. This will depend on the abundances in Jupiter's atmosphere. It should be recognized that cosmic abundances of these important constituents are poorly determined, since they have never all been measured in a single, undisturbed reservoir. Jupiter offers the first opportunity to do this, although the analysis will have to include possible solubility effects in the interior of Jupiter and the differences in the ways these gases can be incorporated in the ices of planetesimals (Lunine and Stevenson, 1985; Bar-Nun *et al.*, 1987). The scaling to hydrogen will come from the unpurified sample where mixing ratios of He, Ne, and Ar to H_2 will be measured directly.

Removal of hydrogen increases the relative abundances of *all* species by approximately a factor of 10. Thus the isotopic ratios of the constituents discussed in the previous section can be measured with greater precision in this mode. Removal of hydrogen will greatly diminish interference with the $^3He/^4He$ determination, which is of fundamental importance in comparing the deduction of D/H in the primordial solar nebula from meteoritic helium measurements with the direct measurement that will be made by the GPMS.

1.5. Other Possibilities

1.5.1. *Photochemistry, Lightning, and Internal Heat*

For future reference, the expected abundances of a number of other classes of compounds are listed in Tables III–V, even though it cannot be asserted that they are

TABLE III

Photochemically produced species

Gas		'Maximum' mixing ratio
C_2H_2	(observed)	3×10^{-6}
C_2H_6	(observed)	3×10^{-5}
C_2H_4	(observed)	$7 \pm 3 \times 10^{-9}$ in north polar region
C_3H_8	(observed)	$<6 \times 10^{-7}$ in north polar region
C_3H_4	(observed)	$2.5^{+2}_{-1} \times 10^{-9}$ in north polar region
		$<7 \times 10^{-10}$ at mid latitudes
C_4H_2	(observed)	$<3 \times 10^{-10}$
C_6H_6	(observed)	$2^{+2}_{-1} \times 10^{-9}$ in north polar region
		$<2 \times 10^{-10}$ at mid latitudes
HCN		2×10^{-9}
N_2	(predicted)	2×10^{-9}–10^{-6}
N_2H_4	(predicted)	3×10^{-10}
CH_3NH_2	(predicted)	$\sim 10^{-9}$
P_2, P_4	(predicted)	?

detectable. With a potential enrichment for some of these species by a factor 100 to 500, several of them may in fact be brought into the measurable range. Abundance determinations will then depend on measurements relative to species that can be detected directly (without enrichment) as calibrated by the laboratory model of the instrument.

TABLE IV

Chemically active trace constituents

Gas	Mixing ratio	Observations
CO	1×10^{-9}	$1.3 \pm 0.3 \times 10^{-9}$
GeH_4	8.5×10^{-9}	$0.7 \pm 0.2 \times 10^{-9}$
H_2Se	4.5×10^{-9}	
HBr	8.5×10^{-10}	
B_2H_6	7.5×10^{-10}	
AsH_3	4.7×10^{-10}	$2.2 \pm 1.1 \times 10^{-10}$
H_2Te	3.4×10^{-10}	
SnH_4	2.7×10^{-10}	
HI	6.7×10^{-11}	
SbH_3	2.2×10^{-11}	

TABLE V

Species from electric discharges.
Concentrations unknown

Species
CH_3S
$(C_2H_5S)_2$
CH_3CN
HCN

The abundances of the gases produced photochemically in the planet's upper atmosphere that will be measured by the GPMS are difficult to evaluate from present information for two reasons: (a) both the analysis of observations and theoretical calculations are strongly dependent on model atmospheres and (b) the amounts of these gases actually sampled by the Probe mass spectrometer will be a strong function of altitude. The figures given are attempts to define maximum values; the actual amounts may be lower.

The discovery of lightning on Jupiter by the Voyager spacecraft has emphasized the possibility of the production of non-equilibrium species in the Jovian troposphere. This means that one should search for examples of the compounds that have been produced in laboratory experiments simulating an electric discharge in a Jovian mixture of gases. Of particular interest are HCN, C_2H_2, CH_3CN and a variety of organic compounds including some containing sulfur.

At the base of the proposed Probe trajectory, at a pressure of 15–20 bar, there is a possibility that trace compounds will be sampled that are only at equilibrium at much lower, hotter levels of the atmosphere. Traces of these constituents might survive to the deepest level where sampling is possible in much the same way CO, GeH_4, PH_3, and AsH_3 are brought to still higher levels of the atmosphere where they have been observed spectroscopically. An obvious and important candidate in this category is silane-SiH_4- whose enriched solar abundance mixing ratio of $\sim 10^{-4}$ should only be reached at temperatures above 1200 K.

2. Instrument Description

2.1. Introduction

Within the constraints of volume, power, and bit rate, a number of compromises must be made in the design of an atmospheric composition experiment. An important advantage of a mass spectrometer is its impartiality: within its mass and sensitivity range it detects everything admitted to it, and it is therefore ideal for an exploratory mission like the Galileo Probe. The instrument chosen is a quadrupole mass spectrometer supplemented by rather complex 'plumbing' with sample inlets and gas-handling, pressure-reducing, and pumping systems. To enhance the range of measurements the basic sample inlet is supplemented by three selective subsystems: a noble-gas purification cell and two sample enrichment cells for the more complex compounds. Advantage is taken of non-evaporable chemical getter technology for pumping hydrogen and the reactive minor constituents and thus avoid the increased complexity and reliability risk associated with moving parts in mechanical pumps such as turbomolecular pumps.

The approach to the electronics support system is also straightforward and conventional. Extensive use was made of hybrid circuit packaging because of weight and size limitation. The complexity of microprocessor control is not desirable given the devastating radiation environment that must be traversed before entry into the atmosphere of Jupiter. It is also not necessary since all control functions can be handled with conventional electronics and the measurement sequence followed during descent is stored reliably in read only memory. A summary of the instrument specification is given in Table 6.

2.2. Sample inlet and enrichment systems

The inlet system shown schematically in Figure 1 consists of two fully self-contained units which operate in time sequence as the Probe descends through the atmosphere. Both units contain an ambient atmosphere flow system whose gas inlets are placed near the apex of the Probe, and whose exit ports are placed at the minimum pressure point inside the Probe. Ambient atmospheric pressure at the exit is assumed as a worst case condition. The pressure difference (approximately 6 mbar) between the stagnation point and the ambient pressure point causes a flow past the pressure reducing leaks. Inlet and outlet ports are sealed by metal-ceramic devices and kept under vacuum prior to entry. They will be opened in sequence after entry by redundant pyrotechnic actuators.

A small fraction of the gas flowing through the high pressure flow system from Inlet 1 to Outlet 1 or Inlet 2 to Outlet 2 is conducted through the pressure reducing leaks L1 or L2 into the ionization region of the ion source. The leaks, which are arrays of glass capillaries (typically seven capillaries per leak) with inside diameters ranging from 1.5 μm to 6 μm have conductances chosen so that the pressure in the ion source does not exceed 10^{-4} mbar. The capillaries were fabricated in a proprietary process by the Galileo Electro Optics Corp. of Sturbridge, Massachusetts. The materials used for the inlet system plumbing are primarily nickel and inconel. The surfaces in contact with the sampled gases were passivated by a silinizing process.

TABLE VI

Instrument parameter summary

Parameter	Characteristic
Inlet system	2 glass capillary array direct leaks 1 scrubber (noble gas analysis) 2 enrichment cells
Ambient pressure range	0.15 to 20 bar nominal
Mass analyzer	Quadrupole; field radius 5.0 mm field length 150 mm Operating frequency; 2.83 MHz, 2 to 19 amu 1.13 MHz, 20 to 150 amu
Ion source	Electron impact/dual filament/variable energy (75 eV, 25 eV, 15 eV)
Ion detector	Secondary electron multiplier/pulse counter
Mass range	2 to 150 amu
Scan format	Stepping, Integer or 0.125 amu, 0.5 s per mass step. Nominal 75 s, scan period (2 to 150 amu)
Dynamic range	10^8
Mass analyzer resolution	Unit mass, flat top peaks, nominal 10^{-8} crosstalk for adjacent masses from 2 to 60 AMU; larger crosstalk for higher masses and for ions with non-thermal energies
Pressure range	10^{-13} to 10^{-4} mbar (ion source)
Data rate	32 bps
Electronics	Read Only Memory (ROM) Controlled descent sequence 8192 program steps (16 bits/step). Ground command override capability during checkout and cruise
Temperature range	-20 °C to 50 °C operating
Viewing requirements	Sample inlet port near stagnation point; Sample outlet port near minimum pressure point
Deployment mechanism	(1) Metal ceramic breakoff caps, pyrotechnically actuated (2) Valves, solenoid operated
Pumps	Non-evaporable getters and sputter ion pump
Power	(1) Instrument: 13 W avg (2) Pumps and heaters: 12 W
Total energy for 60 min descent time	120 kJ
Weight	13.2 kg (29.11 lbs)

The sample enrichment systems are an integral part of the ambient atmosphere flow systems. Atmospheric gas, after passing by the capillary leaks L1 or L2, will be conducted via valve V1 or V6 to the enrichment cells which contain getters G1 through G4 that, when activated, chemically bind reactive gases during the measurement sequence and thus allow a pure noble gas analysis. The enrichment cells C1 and C2 are packed with adsorbing compounds chosen to adsorb complex hydrocarbons. The material found through testing to be most suitable for this experiment is CARBOSIEVE, 80–100 mesh, a porous carbon structure of fixed pore size used commercially as a stationary phase in packed gas chromatographic columns (available from Supelco, Inc.

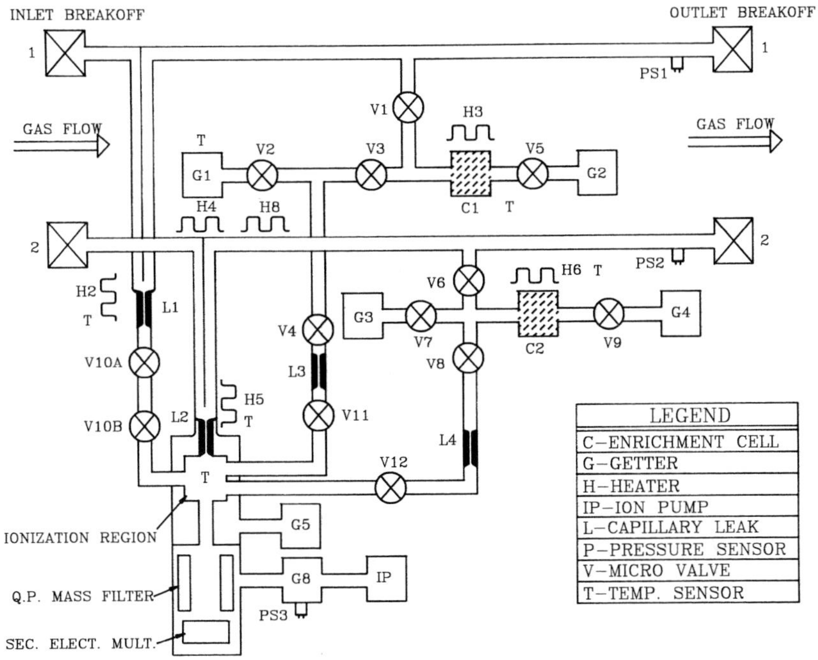

Fig. 1. Schematic of the gas inlet system and connection to the mass spectrometer sensor. Two parallel gas inlet/outlet systems are employed to provide gas samples to the direct leaks, L1 and L2, and to the two sample enrichment systems centered around C1 and C2.

in Bellefonte, PA). Gases adsorbed by the enrichment cells will be released by a programmed heating cycle during the Probe descent. During these cycles the cells are isolated from the flow system and the getters by valves V1 through V9. The valves are solenoid operated microvalves manufactured by Aker Industries (formerly Energy Research & Generation, Inc.) of Oakland, California. At specified times the gas trapped in the cells is conducted through separate capillary leaks L3 and L4 to the ionization region. The enrichment factor depends on the chemical and physical properties of the sample gas. In general it increases with the molecular weight of the species. This is qualitatively illustrated in Figure 2 where mass spectra obtained with analog laboratory recorders are shown for noble gases and hydrocarbon gas mixtures before and after enrichment. The substantial xenon enrichment further improves the concentration measurement of xenon since the expected mixing ratio is marginally measurable by direct analysis.

After use the sample enrichment leaks L3 and L4 and the first direct leak L1 will be isolated from the ion source by solenoid operated valves V10, V11, and V12 to prevent the ion source pressure from exceeding its optimum value, and also to permit repeated observation of system background pressure after the initial sampling and enrichment sequences are completed. The second independent inlet system will be opened to the atmosphere after the first system has been isolated from the ion source. Sample

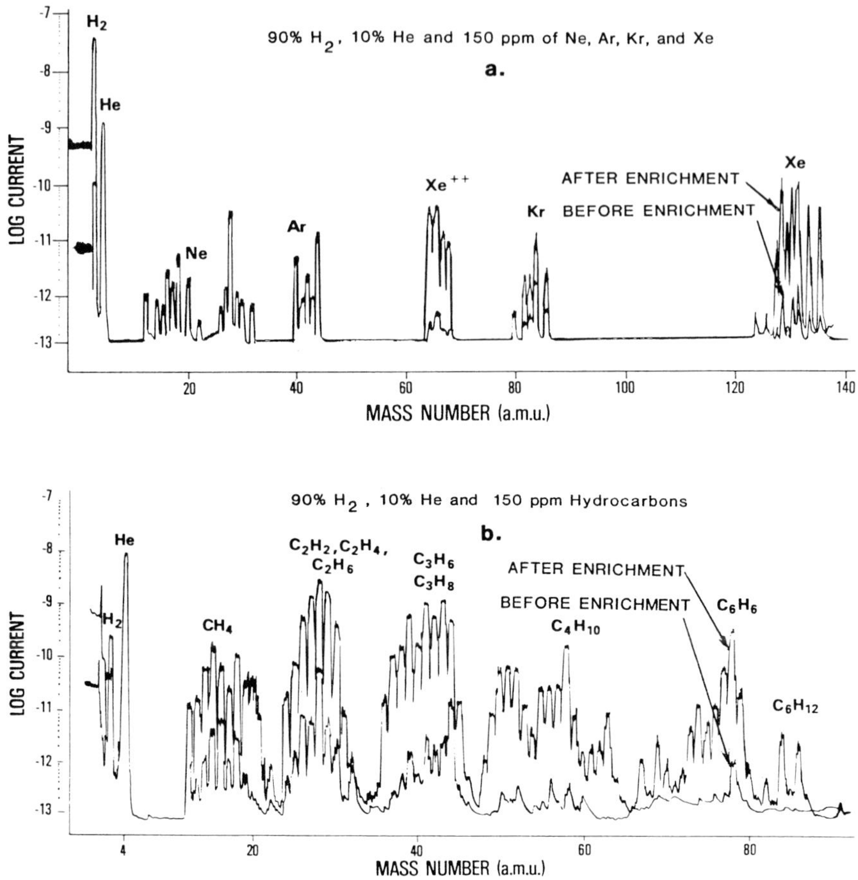

Fig. 2. Mass spectra showing enrichment obtained when gas processing is used to remove the major constituent H_2 from a 90% H_2 and 10% He mixture containing constituents each of 150 ppmv. (a) Rare gas enrichment. Note the substantial enrichment of Xe obtained. Expected Xe ratios are marginably measurable by direct analysis. (b) Hydrocarbon enrichment. There is a substantial enrichment of the C_3–C_4 hydrocarbons with a somewhat less enrichment for methane. For illustration the spectra were recorded analog with laboratory recording equipment.

distortion caused by gas-surface interactions is minimized by directing the high pressure flow against the capillary leak and by locating the leaks in the ion source so that the gas leaves the capillaries on the ion source side directly through the ionizing electron beam. The gas emitted from the capillaries can thus be ionized and analyzed without having experienced previous surface collisions with the ion source walls. Limits of the pumping speeds imposed by weight and power constraints in a flight system do not permit instant removal of the gases from the ion source after they initially pass the ionization region. A component of randomized gas therefore contributes to the measurement. The ratio of the direct beaming to the randomized component strongly depends on the system geometry. A ratio of approximately 5:1 was achieved with the ion source design of the flight instrument. It is most critically affected by the distance between the electron beam and the capillary exit and the cross section of the electron beam.

A cross-sectional view of the actual ion source is shown in Figure 3. The gas flow path for the second inlet system near leak L2 is marked by arrows. The gas flow path was

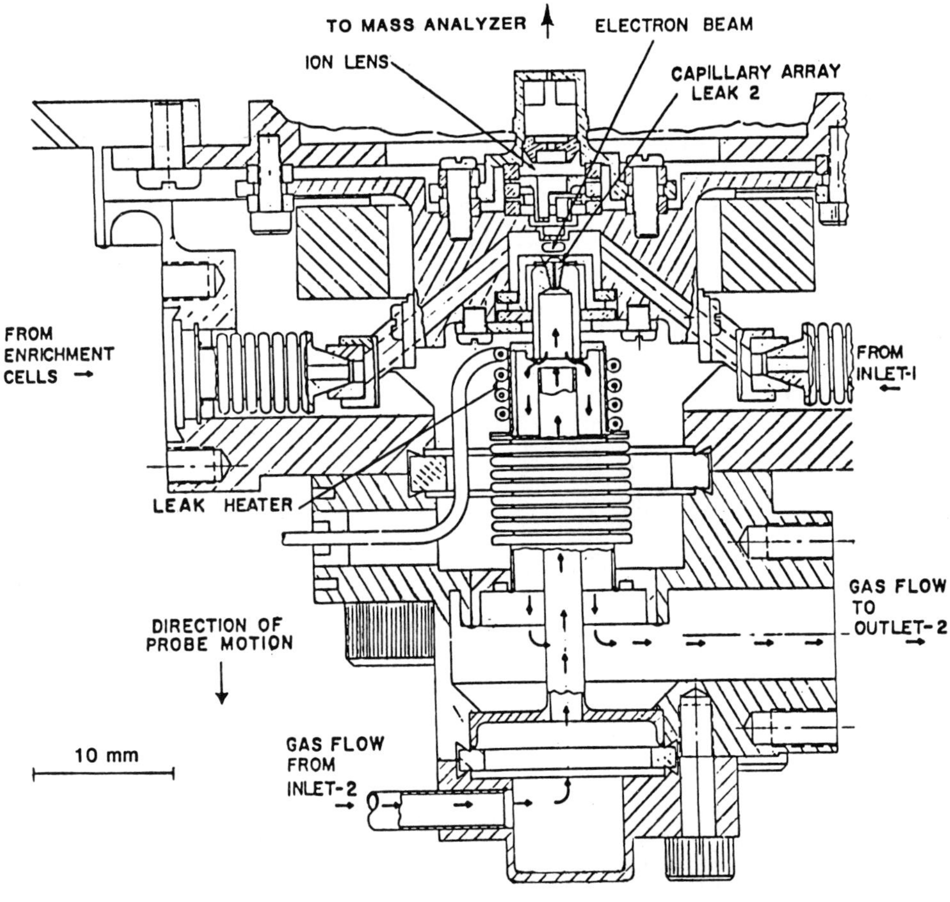

Fig. 3. Cross section of the Galileo Probe Mass Spectrometer ion source showing direct beaming capillary leak and the various gas flow connections.

designed to minimize clogging of the capillary leak array by condensible gases, e.g., water droplets should they exist. It is believed that the cloud particles on Jupiter are condensates in vapor pressure equilibrium with their surroundings. They are unlikely to clog any part of the inlet system, which is considerably warmer than the ambient atmosphere and will quickly evaporate them. Large droplets, should they overcome the gravitational potential while moving through the inlet lines, will be trapped in the cup, shown for example at the bottom of Figure 3. Involatile smog particles, such as the concentrated sulfuric acid on Venus are unlikely to be encountered. A droplet trap arrangement for inlet 1 is similar to that shown for inlet 2. Also shown in Figure 3 are the connecting vacuum flanges through which the sample gases coming from the first leak and the leaks that supply the enriched and purified samples enter the ion source.

The measurement sequence, illustrated in Figure 4(b), shows the time dependence of the total ion source pressure during the Probe descent. For reference the ambient pressure and temperature are shown in Figure 4(a) as a function of time. During the cruise phase of the mission all microvalves are open. Following the sequence start event,

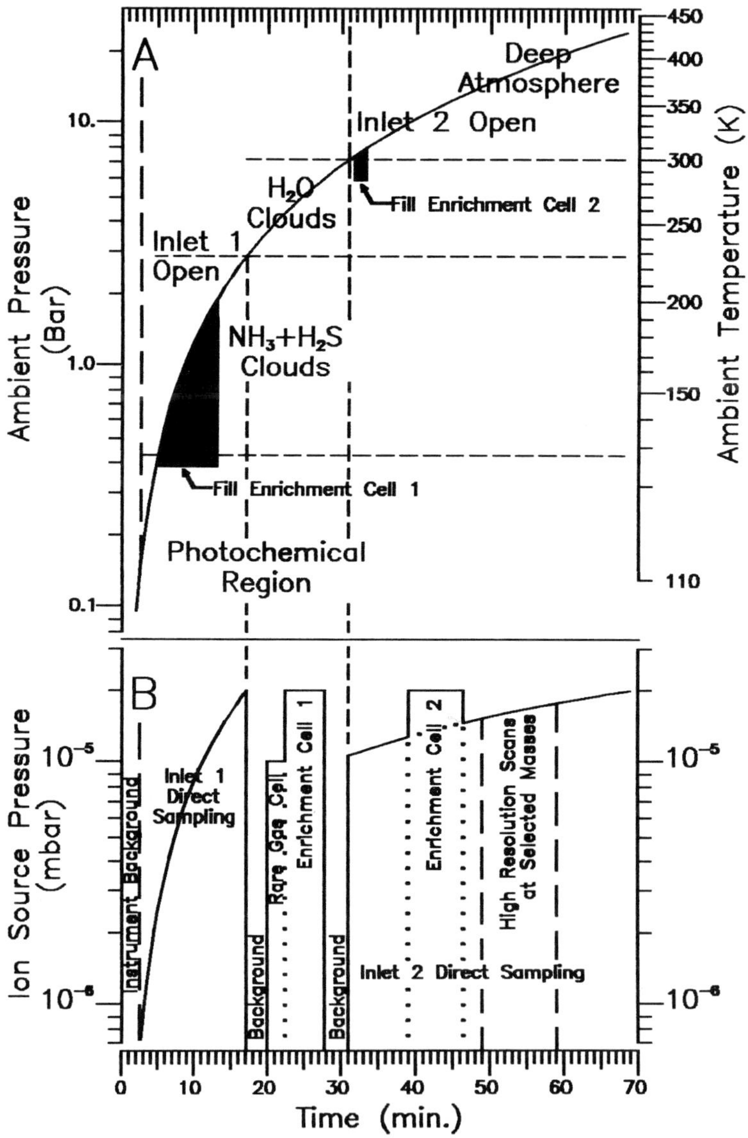

Fig. 4. (A) Ambient Pressure-Time Profile. (B) Typical Measurement Sequence; Ion Source Pressure-Time Profile.

microvalves V1, V3, V4, V6, V7, and V8 are closed which is the initial condition for sampling. The first inlet breakoff is opened when the Probe reaches a pressure level of approximately 150 mb. Approximately one second later the first outlet breakoff is opened, causing the ambient atmosphere to flow past the capillary leak L1 through which a small fraction of the atmosphere is conducted into the ion source. While the direct analysis is made, valve V1 is opened for 8 min to allow ambient gas to flow through enrichment cell C1 until getter G2 is saturated with hydrogen and the getter housing volume is filled with gas. Next valves V1 and V5 are closed, valves V2 and V3 are opened and the remaining reactive gas, mostly hydrogen, is absorbed by getter G1. Remaining in the enclosed volume are the noble gases and some residual hydrogen whose partial pressure is reduced by approximately five orders of magnitude determined by the equilibrium vapor pressure of the hydrogen dissolved in getter G1. In the third step valves V2 and V3 are closed and valve V4 is opened introducing a fraction of the gas into the volume enclosed by valves V2, V3, and V4 through leak L3 into the ion source. Prior to opening valve V4, valves V10A and V10B are closed and the ion source and analyzer section of the mass spectrometer are evacuated to background via getters G5 and G8 and the sputter ion pump IP. Mass spectra are obtained during the pump down time to record the remaining background gases. The noble gases subsequently introduced into the ion source through leak L3 are analyzed for 2.5 min. During this time enrichment cell C1 is heated for 5 min to 200 °C, the gases are desorbed, and, after opening valve V3, they are added to the noble gases flowing through L3 into the ion source. After 5.0 min of analysis of these gases during which the cell temperature is maintained at 200 °C, valve V11 is closed and the mass spectrometer is pumped once more to background pressure levels while the partial pressure of the background gases are measured. Inlet system 1 is now completely and irreversibly isolated from the mass spectrometer and the sequence described is repeated with inlet system 2 for which a separate noble gas measurement cycle is not provided. Leak L2 cannot be isolated from the ion source in order to allow direct gas beaming into the ionization region. Hence, the contents of enrichment cell C2 will be superimposed on the direct measurement as illustrated in Figure 4(b).

2.3. Ion source

The importance of minimizing gas-surface interactions in the high vacuum side requires that the ion source be very compact and an integral part of the sample inlet system. Electron impact ionization is used in a miniature, dual filament ion source. The second filament provides redundancy and is turned on automatically should the first filament break or burn out. A collimated electron beam is directed through the ionization region past the capillary leak L2 shown schematically in Figure 1 and the physical location in Figure 3. The electron beam energy is varied to permit species identification and discrimination by observing spectra of fragmentation patterns at several different electron energies. The other leaks L1, L3, and L4 are connected to the ionization region via short tubes as indicated schematically in Figure 1. Chemical reactions of gases on the surfaces of the hot filament are minimized by isolation through narrow slits and by separate pumping of the filament region. Ions are focused into the mass analyzer for filtering by a three-element ion lens.

2.4. Quadrupole mass analyzer

The quadrupole mass filter was first described by Paul and Steinwedel (1953), and Paul *et al.* (1958). The reader is also referred to the book by Dawson (1976) for detailed discussions of quadrupole mass spectrometers and applications.

The quadrupole analyzer filters the ion beam produced by the ion source, transmitting ions of a chosen charge to mass ratio only. The transmitted ions are focused on a secondary electron multiplier detector. The radius of the quadrupole field is 5 mm and the field length is 150 mm. Mass selection is achieved by application of radio frequency and static potentials of varying magnitude to diagonal rod pairs. The selected mass value is determined by the relation $m = 0.55V/f^2$ (amu) where V is the amplitude of the applied radio-frequency voltage in V and f is the frequency in MHz. To allow voltage scanning over a sufficient amplitude range two separate radio frequencies were used, 2.83 MHz for 2 to 19 amu and 1.13 MHz for 20 to 150 amu. The quadrupole mass filter has been chosen in preference to a magnetic sector spectrometer because of its light weight, versatility, high resolution, and simplicity. The only dimensionally critical element in the system is the precision rod assembly whose rigid and compact design has been proven by application in previous flight experiments and vibration and temperature testing to be extremely stable.

2.5. Pumping system

The pumping system establishes a flow of sample gas though the ion source at a particular pressure when a sampling device is opened, and, after analysis and closure, removes the sample from the ion source region. Non-evaporable getters and a sputter ion pump are used because of their simple adaptation to space flight systems. They depend only on electrical power for operation, without moving parts. Their application to the Galileo Probe Mass Spectrometer requires care because hydrogen and helium are the major gases in the atmosphere. While hydrogen is absorbed at a very high rate by getter materials, helium absorption is practically zero. Sputter ion pumps also pump hydrogen with high efficiency but hydrocarbons are synthesized in the pump by reactions of hydrogen ions with carbon trapped in the pump surfaces. The effective pumping speed for helium is usually small because of the low ionization cross section of helium and the requirement that helium be buried physically in the pump elements since it does not become chemically bound or go into solution like hydrogen. This requires sputtering of comparatively large amounts of cathode material which tends to release larger quantities of gases previously entrapped in the pump surfaces. Thus, to eliminate the synthesis of hydrocarbons in the sputter pump a cascaded pump system is used. A high capacity baffled getter pump is operated in cascade with a sputter ion pump. The gas flow from the mass spectrometer into the getter pump is conductance limited to maintain a constant pumping speed during the measurement phase. Hydrogen and all other reactive gases are adsorbed by the getter before they reach the sputter pump. Gases emitted by the pump pass the getter first before they can enter the mass spectrometer and their contribution to the ion source gas is therefore significantly reduced. Small sputter pump instabilities, not fully avoidable with the present state of technology, are buffered by the preceding getter chamber. The getter material used is sintered zirconium graphite available from SAES Getter of Milan, Italy as type ST171. It is fabricated in a proprietary process. After activation at approximately 900 °C for 45 min while the instrument is connected to the laboratory vacuum pumping system, the getters remain active at room temperature until they become saturated at the end of the Probe descent. Getters G5 and G8 are dimensioned approximately five times larger than required for the gas load expected during the descent in order to maintain constant pumping speed. The cathode materials of the sputter ion pump are tantalum and titanium and the electrode geometry is optimized to enhance pumping of helium by ion implantation and burial. Pumping speed is conductance limited at the pump flange to $2\ \mathrm{l\ s^{-1}}$. The magnetic field of the sputter ion pump is 0.2 Tesla over an effective area of 35 $\mathrm{cm^2}$. The yoke is designed to minimize the stray field. Magnetic shielding was also provided to the ion source housing to cancel the stray field of the pump because of its location directly above the ion source. The sample inlet system and vacuum pump assembly are shown in Figure 5.

2.6. Electronic system

In adapting previous concepts for the electronics to the Galileo requirements, simplicity has been a major consideration. Power commands and timing events are accepted and processed by the logic system. The data of the primary measurement are stored in an output register for interrogation by the spacecraft telemetry system. A block diagram of the electronic system is shown in Figure 6.

The instrument is under control of the programmer which is an array of Read-Only-Memory (ROM) devices which produces an 8192-word, 16-bit look-up table and an output register to hold the 16-bit word in current usage. Each one half second the ROM is incremented and the instrument is configured for the next measurement. Thus during each of 8192 one half second intervals each of six instrument variables, e.g., mass number, ionization energy, inlet system configuration, etc., can be configured to any allowable state. The application of power causes the instrument to begin executing a programmed 256 step 'test' sequence. The instrument remains locked in this 'test' mode until the SEQUENCE START command is received from the Probe.

As mentioned above mass number selection in a quadrupole is a function of amplitude and frequency of the RF signal applied to the rods. Two frequencies are being used to cover the mass range of 2 to 150 amu (Table VI). In each frequency range mass number is proportional to the RF amplitude. The actual mass scanning range is slightly larger than the nominal values quoted above for occasional high-resolution peak scanning past the center of the mass peaks to verify tuning and mass resolution. The mass peak width is kept approximately constant over the mass range 2 to 150 amu by proper choice of the dc and RF voltages.

The ion source requires an electrode supply of well-regulated voltages and a feedback

Fig. 5. Gas sample inlet system for the Galileo Probe Mass Spectrometer.

controlled emission regulator for the ionizing electron beam gun. For the Galileo application three different ionization energies are programmer selected by changing appropriate ion source potentials. The ion source, as previously noted, is provided with two redundant filaments which are powered by redundant emission regulators. This implementation simplifies the design and, of course, increases reliability.

A high voltage supply of nominal 3 kV operates the secondary electron multiplier ion

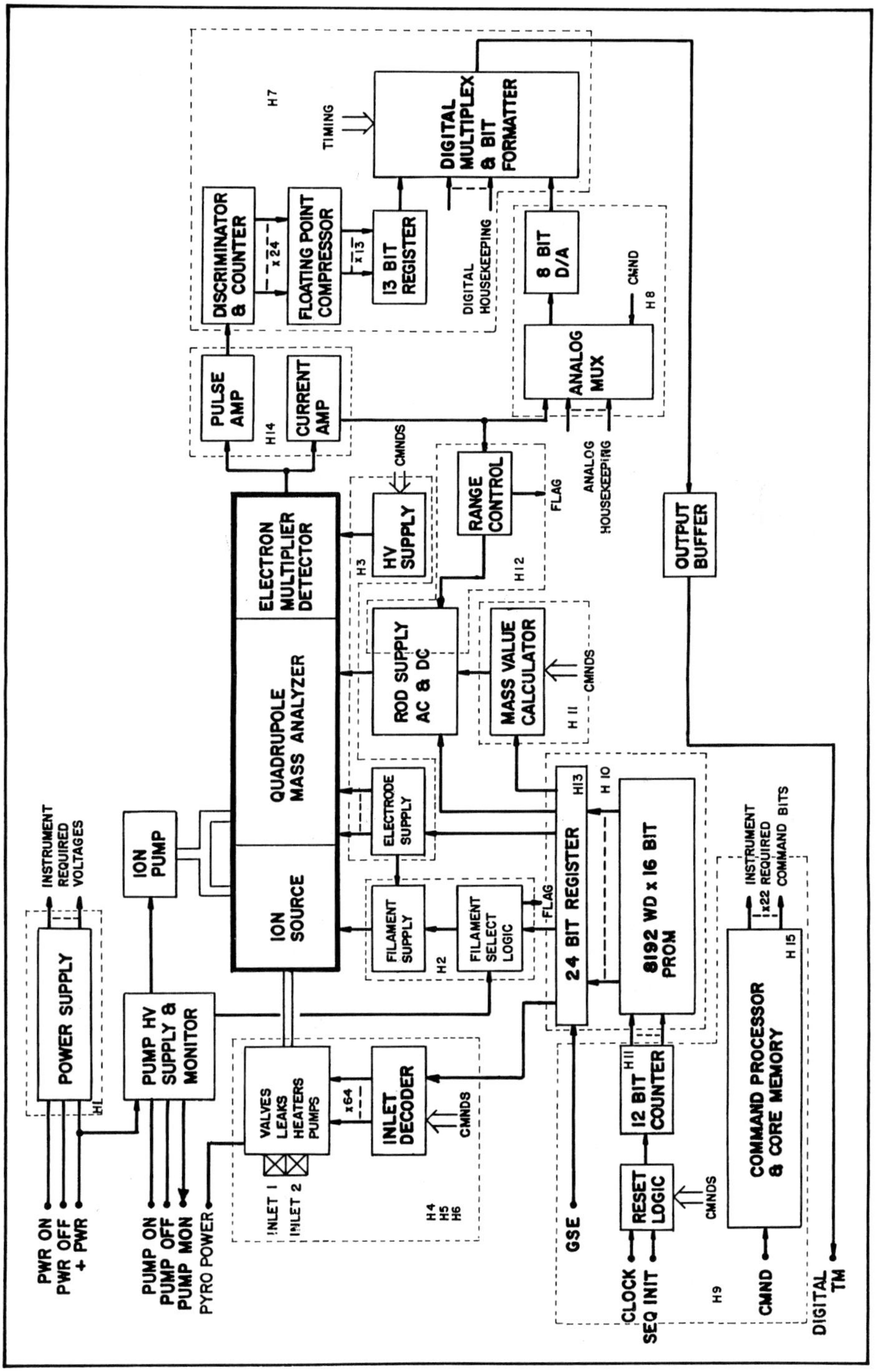

Fig. 6. Electronics System Block Diagram of the Galileo Probe Mass Spectrometer.

detector. Command capability to optimize the secondary electron multiplier gain through the selection of one of four values is provided. The ion arrival rate at the detector during each one half second during the descent constitutes the primary measurement. Pulse counting will range from a rate of about 5×10^7 cts s^{-1} down to rates as low as one count per half second integration. At ion arrival rates exceeding the upper count limit the instrument will be desensitized automatically. A logarithmic (base 2) compressor is used with a 9-bit mantissa and 4-bit exponent. This provides a full scale of 3×10^7 per range and a resolution of one part in 512.

The electronics system was constructed using multi-layer printed circuit board technology. Weight and size constraints for the Galileo Probe instruments required that approximately 90% of the electronics components were packaged in multi-layer hybrid form. To meet structural requirements the circuit boards were mounted on a cross web structure enclosing the quadrupole analyzer and ion detector section. A photograph of the electronics assembly is shown in Figure 7. The average power consumption of the electronics system is 13 W. Additional pump and heater power of 12 W maximum is required during the descent.

Since the Probe is not pressurized the instrument is enclosed by a pressurized envelope made of titanium to save weight. Pressurization (1.2 bar N_2) prevents corona and condensation during the descent, and defers collapse. The weight of the instrument is 13.2 kg (29.11 lbs.). The fully asembled instrument is shown in Figure 8. Also shown in the figure are the inlet and outlet breakoff devices of the sample inlet system.

2.7. Reliability and Redundancy

Much consideration is always given to reliability in the design of scientific space flight instruments. Of particular concern for the Galileo mission is the requirement for long instrument life which has increased from the original 3 years to more than 10 years. This increase was due to various programmatic changes in the mission and schedule delays caused by the tragic Challenger accident. Although there are no special long-life verification test results available for the instrument, an examination of the materials and components used in the instrument let us conclude that we expect reliable operation of the instrument over the required time period.

Of particular concern is the vacuum integrity of the mass spectrometer system. To maintain a sufficiently clean ion source environment for trace gas detection, it is necessary to keep the sample inlets and the analyzer section under high vacuum and sealed until the measurement sequence begins. Getter and sputter ion pump loading due to small vacuum leaks from the pressurized housing or from surface outgassing can seriously degrade or even destroy the capacity of the instrument to perform during entry. However, leak and outgassing tests performed after pinch off from the vacuum stand and vibration tests indicated leak rates low enough to exceed the required lifetime by more than a factor of three. Post launch and first earth encounter check out of the instrument also verified that the vacuum integrity was maintained during and after the launch and after vehicle separation maneuvers were completed. The background gas spectra obtained during the first Earth return of the spacecraft showed peaks of He, Ar,

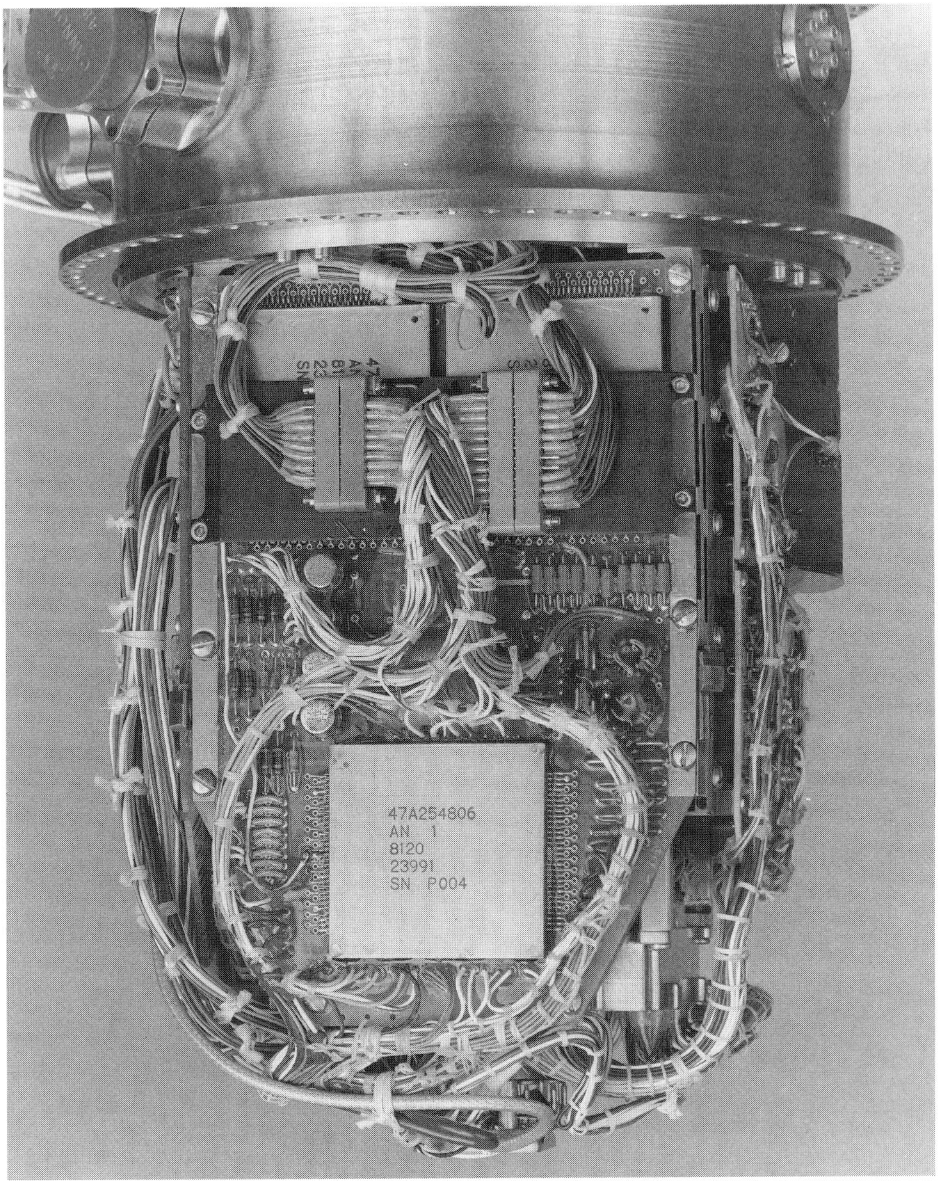

Fig. 7. Electronics System Assembly of the Galileo Probe Mass Spectrometer.

CH_4, H_2O, CO, and CO_2. Helium, argon and methane background pressures seen at the instant of turn on were 1×10^{-9} mbar, 2×10^{-9} mbar, and 2×10^{-12} mbar, respectively. This background was removed within several seconds by the sputter ion pump leaving a residual pressure of 1×10^{-12} mbar for CH_4 and CO_2, 3×10^{-12} mbar for H_2O and less than 2×10^{-13} mbar for He, Ar, and CO.

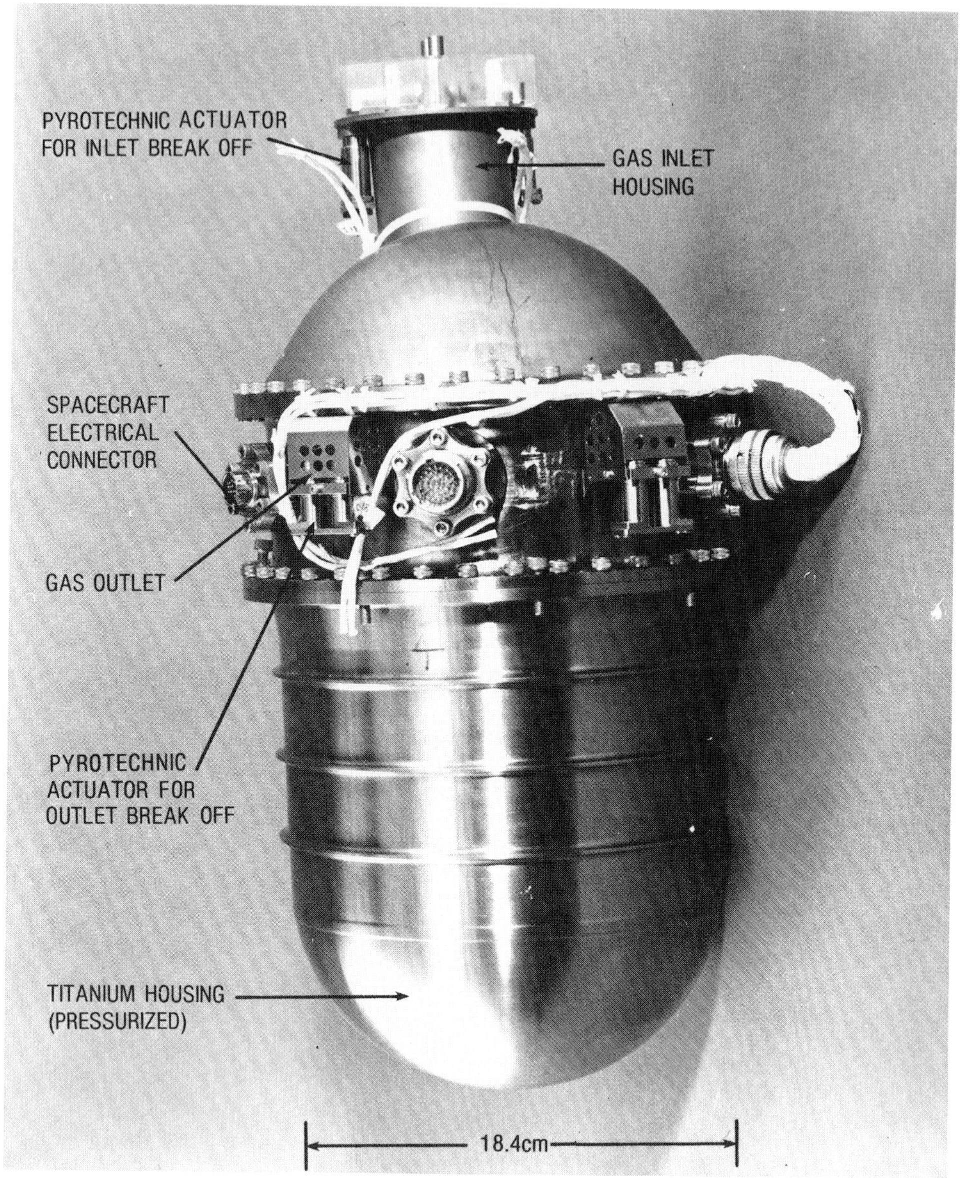

Fig. 8. Fully assembled Galileo Probe Mass Spectrometer.

For the remaining cruise time the instrument will be turned on once a year to operate the sputter ion pump for removal of noble gases, e.g., Ar, resulting from possible sudden release from gas pockets which may exist in the metal surfaces and welds. It is also necessary to exercise the micro valves, the only devices in the instrument containing moving parts, to prevent sticking under ultra high vacuum conditions. Since sticking in the closed position after long storage is the most likely failure mode of the valves, all valves are kept in the open position during the cruise phase of the mission.

Component redundancy is implemented in spacecraft support systems, but is usually not permitted in scientific instruments because of weight, size, power, and cost constraints. Hence, many single point failure possibilities exist in the instrument. Redundancy is provided, however, for the filaments because of the limited lifetime; for the pyrotechnic devices because of possible misfire; and for the isolation of inlet leak 1 by using two micro valves in series V10A and V10B because of catastrophic pressure increase in the ion source which would occur if the flow through leak 1 is not stopped at the designated closing time. The life expectancy of the electronic components in space is usually very high except for those which are radiation sensitive, e.g., microprocessors which are not part of our design.

The sampling sequence for the nominal one hour descent time was programmed on read only memory (ROM). Programs could be changed or corrected on the ground by substituting ROM units in the instrument but it remains fixed in flight. This approach avoided risks of radiation damage when using programmable onboard memory for which radiation hardened parts were not available.

3. Laboratory Calibration of the Instrument

Accurate instrument calibration is required for quantitative composition measurements in the atmosphere. The transfer characteristics of mass spectrometers depend on the species to be measured. Factors which affect signal levels include the ionization cross section, the effective pumping speed of the vacuum pumps, and the conversion efficiency and pulse height distribution of the secondary electron multiplier.

The instrument was calibrated on a dynamic flow system where the time, pressure, and temperature profile to be encountered during the Probe descent was simulated. Gas mixtures containing known mixing ratios of gases expected in the atmosphere of Jupiter were introduced into the high pressure flow system connected to the sample inlet system of the flight instrument. Ultra-pure, commercially available gas mixtures were used. To establish mixing ratios representative of the atmosphere of Jupiter those mixtures were further diluted with pure gases (H_2 and He). The final mixing ratios of the minor constituents was checked or verified by a laboratory gas chromatograph mass spectrometer system as well as a stand-alone gas chromatograph and mass spectrometer. The design of the sample inlet system allowed complete and separate instrument calibration of all components used in flight except for the replacement of getter materials and the sputter ion pump. This was required after calibration to assure that maximum pump capacities are available for the experiment. All pumps on the flight instrument are designed to operate in a conductance limited mode so that small variations in pumping speed of the getter materials have a negligible effect on the instrument transfer function.

The calibration system consists of two parts, the high-pressure gas flow and sample mixing system, and the ultra-high vacuum pumping stand. A schematic diagram of the high pressure flow system is shown in Figure 9. Calibration gas mixtures are circulated

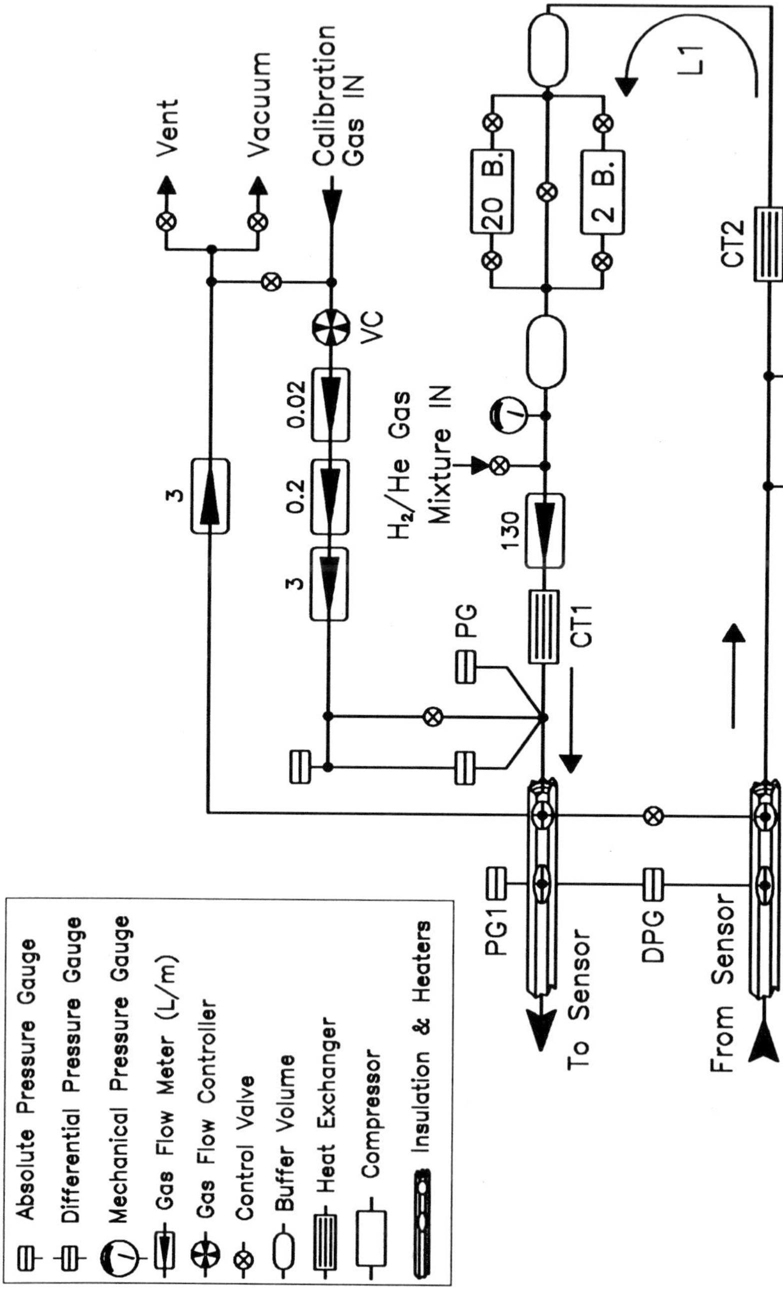

Fig. 9. Schematic diagram of the high-pressure gas circulation and mixing system of the Galileo Probe mass spectrometer calibration facility.

through loop L1 in the direction shown by the arrows. Two gas compressors are available for alternate use for gas compression and circulation through the circulating loop. They operate over the pressure range from 10 mbar to 2 bar and 1.0 to 20 bar. The flight instrument inlets and outlets were connected directly with vacuum flanges to the appropriate terminals as indicated in the schematic on the left side of Figure 9. All lines are thermally insulated and the gas temperature was controlled by heaters attached to the flow lines and heat exchangers CT1 and CT2 to heat or cool the circulating gas mixture as required. Absolute system pressure was monitored with pressure gauge PG1 and the differential pressure across the inlet was monitored with gauge DPG. The flow was adjusted so that the differential pressure was always equal to the Probe stagnation pressure (6 mbar) expected in flight. Trace gas mixtures were added in known quantities through the calibration gas line and valve VC. The exact quantities of added calibration gas was determined by measuring the gas flow with the flow meters and the pressure at the injection port with the pressure gauge PG.

The ultra-high vacuum pumping system was connected to the flight instrument through pumpout ports for evacuation, bake out at $\sim 300\ ^{\circ}C$ and getter pump activation. This is needed before calibration can begin and between calibration runs for getter reactivation and final instrument preparation for flight. Oil-free turbomolecular pumps and a closed circuit liquid helium pump provided the pumping capacity and ultra-high vacuum capability needed for thorough and rapid evacuation of the flight instrument system. Typical background pressures in the instrument, before calibration gases were introduced, were between 5×10^{-11} mbar and 2×10^{-10} mbar where the major component was H_2 followed by H_2O, CO, and CO_2.

Detailed calibration was carried out with the gas mixtures containing primarily H_2 and He 10% plus trace gases each at the 100 ppmv mixing ratio. A list of the calibration gases used either directly or in various combinations is given in Table VII. To establish

TABLE VII

Calibration gas mixtures

	Mixture	Inlet
1.	H_2 + He + Ne + Ar + Kr + Xe	1, 2
2.	$H_2 + PH_3$	1, 2
3.	H_2 + HCN	1, 2
4.	$H_2 + H_2S$	1, 2
5.	H_2 + He + $AsH_3 + SiH_4 + PH_3$	1, 2
6.	H_2 + He + $CH_4 + C_2H_6 + C_3H_8$	2
7.	$H_2 + C_2H_2 + C_2H_4 + C_3H_6$	2
8.	He + H_2O	2
9.	H_2 + He + NH_3	2
10.	H_2 + He + $CH_4 + C_2H_2 + C_2H_4 + C_2H_6 + C_3H_6$ + + $C_3H_8 + C_4H_{10} + C_6H_6 + C_6H_{12}$	2

Calibration gas mixture used to calibrate the GPMS. The mixture consisted of either H_2 or H_2 90% and He 10% and trace gases as indicated each at the 100 to 150 ppmv mixing ratios.

detailed relative sensitivities of the trace constituents with respect to H_2 and He and to evaluate the fractionation pattern at the three energy levels of the ionizing electron beams, the gas mixtures were introduced into the sample inlet system without further dilution, i.e., with trace gas mixing ratios each of order of 100 ppmv but at several fixed pressure levels between 100 mbar and 20 bar. Finally the descent profile, i.e., pressure and temperature vs time was simulated in the high-pressure flow system exposing the flight instrument to flight-like operating conditions. Predicted pressure-time and temperature-time profiles shown in Figure 4(a) were approximated as closely as possible for that test. All calibration data were recorded and processed on automated data processing equipment using software generated during the instrument development phase.

Some concern has been expressed that the presence of hydrogen in the ion source might impede the measurement of isotope ratios. Tests of the noble gas mixture with and without added H_2 at high ion source pressure, where hydride formation should be most pronounced, showed that the effect in flight will be too small to be observed. In addition, the purification cell efficiently removes the hydrogen and further alleviates any possible concern for the measurement of noble gas isotope ratios. This system shows great promise for the measurement of helium-3. An example of this is given in the graphic illustration shown in Figure 10 for H_2 and He. The purification permits the ^{3}He isotope to be measured well above background. Tests also verified that the carbon isotope ratio measurement using the methane spectrum is not affected by hydride formation in the ion source as shown in Table VIII.

TABLE VIII

CH_4–hydrides

	Source pressure	Fragmentation peaks		
		15	16	17
Pure	$\approx 2.0 \times 10^{-6}$ mb	0.767	1.0	1.01×10^{-2}
$H_2 + CH_4$	$\approx 4.0 \times 10^{-6}$ mb	0.762	1.0	1.02×10^{-2}

Normalized peak heights of methane at 15, 16, and 17 amu of pure CH_4 at 2.0×10^{-6} mbar ion source pressure and equal parts H_2 and CH_4 mixture at 4×10^{-6} mbar total ion source pressure indicating no measurable hydride formation in this pressure range.

Since it was not possible to calibrate the flight instrument with all the gases which might be found in the atmosphere of Jupiter and since the unambiguous interpretation of mass spectra obtained in flight can be difficult, a laboratory reference unit has been built. The engineering unit instrument was refurbished to be identical to the flight instrument. Post encounter simulations of flight results are planned using the laboratory unit and the calibration system to further aid in the data interpretation by reproducing experimentally as close as possible the data obtained in flight using various test gas mixtures.

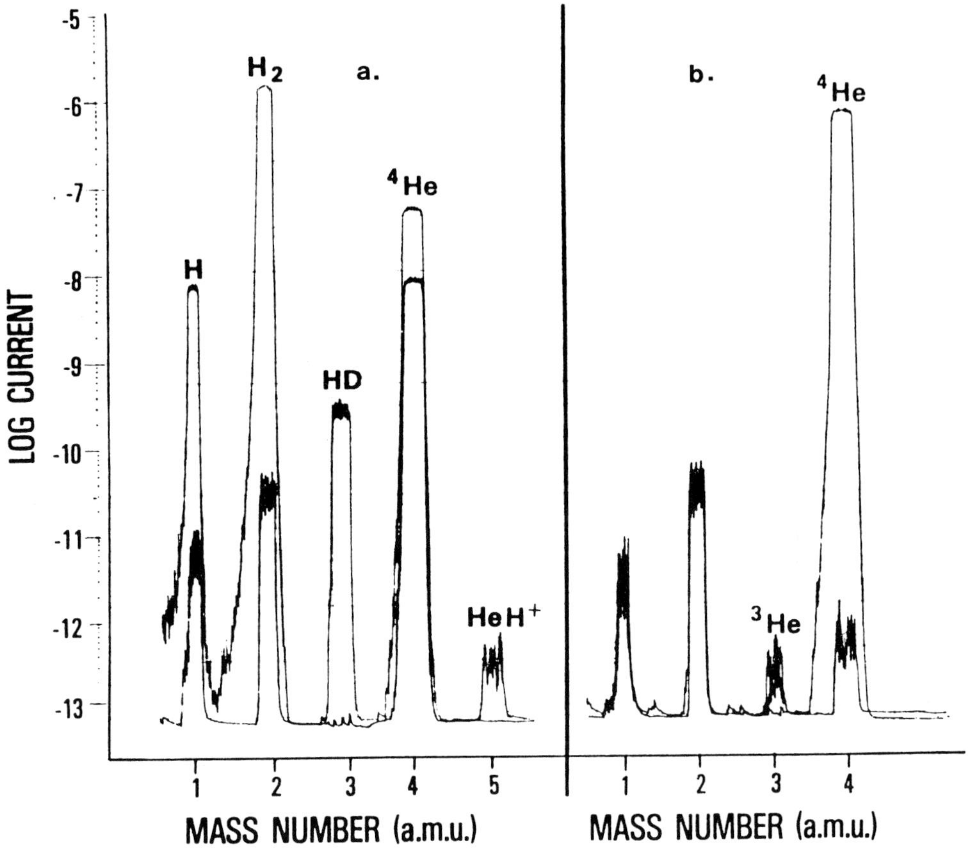

Fig. 10. Effect of purification cell on He isotope measurement. (a) Top spectrum shows a 90% hydrogen–10% helium mixture through direct leak 2 at 3 bar inlet pressure. The bottom spectrum is the same mixture after the hydrogen has been removed by the getter to the background level. The helium decrease in the bottom spectrum is from the volume expansion in the getter volume. (b) Top spectrum is a pure helium sample through direct leak 2 at an inlet pressure of 3 bar. The bottom spectrum shows the background gases. The helium 3 level in this example is approximately 10^{-7} below the helium 4 level. For illustration the spectra were recorded analog with laboratory equipment introducing also extra system noise at low level signals.

4. Expected Results and Measurement Accuracies

The objective of this experiment, as stated above, is the measurement of the chemical composition of the atmosphere of Jupiter, i.e., the mixing ratio of the major and minor constituents during the Probe descent. The basic accuracy of the measurements and the detection limits of trace gases are determined by the sensitivity of the instrument, the maximum operating pressure of the ion source and the dynamic range of the mass spectrometer. The ion source sensitivity varies with species depending on the ionization cross sections, for example, the sensitivity is 1.6×10^{13} counts per second mbar for molecular nitrogen, 2.2×10^{12} counts per second mbar for helium and 8.9×10^{12} counts per second mbar for hydrogen. The background count rate of the secondary electron multiplier ion detector is less than 0.1 counts per second. An ion source pressure of 2×10^{-5} mbar will be reached at the end of the direct leak 1 and leak 2 measurement phases and during the enrichment cell analysis periods. The exact ion source pressure cannot, of course, be predicted since assumptions had to be made about the descent pressure-time profiles. A maximum ion source pressure of 1×10^{-4} mbar is acceptable.

In the spectral regions where the maximum dynamic range of the instrument can be used, the accuracy will be limited by pulse counting statistics at low mixing ratios. At large mixing ratios where counting statistics become negligible, the measurement accuracy will be $\pm 2\%$ for peak height ratio measurements of adjacent mass peaks and $\pm 10\%$ for peak height ratios with wide mass separation. This is because of the error accumulation resulting from calibration uncertainties, small temperature drifts in the supply electronics and change in the gain of the secondary electron multiplier.

In the analysis of gas mixtures, ambiguities in species identification and additional accuracy limitations are expected for species with overlapping mass spectra. Some ambiguities will be resolved by using different energies for the ionizing electron beam. This will be effective for species with strongly differing ionization potentials and energy dependent ionization and dissociation cross sections. Space flight constraints of low weight, power, and volume put restrictions on the size of vacuum pumps and thus limiting the admissible sample size and the achievable background pressure. The available power for the RF generator for the quadrupole puts limits on the achievable mass resolution. The short descent time also restricts the chance for extensive signal averaging. This is particularly important for traces of reactive gases which may initially be absorbed or decomposed on the surfaces of the sample inlet system before they reach the ion source. Reactions of hydrogen with loosely bound carbon or oxygen on the vacuum surfaces in the ion source regions cause the formation of methane and water vapor. This effect was significantly suppressed by surface processing, e.g., carbon depletion and formation of stable surface oxides. Methane and water vapor buildup in a hydrogen atmosphere are, however, still limiting the measurement accuracy for these gases.

Early methane concentration measurements will therefore be most accurate, e.g., $\pm 10\%$ at the expected mixing ratio of 10^{-4}. Measurements made near the end of the mission are expected to be uncertain by a factor of two. Water vapor measurements in the mid pressure range (≈ 7 bar) are expected to be also uncertain by a factor of two if the mixing ratio is $\approx 10^{-5}$. Better accuracies ($\pm 50\%$) are expected if the mixing ratios are close to solar values of $\approx 10^{-3}$. Other gases causing chemical noise are carbon monoxide, carbon dioxide, and ethane. Carbon monoxide and carbon dioxide form a constant background of several hundred counts per second which does not appear to change significantly when hydrogen enters the ion source. Ethane, on the other hand, does not occur in the background spectrum prior to exposure to hydrogen but it appears in small quantities of 1 ppmv levels when hydrogen enters the inlet. This background level can conceivably change in flight depending on its origin. If ethane is formed in a similar method as methane the background is expected to be the same as during the laboratory tests. Should ethane slowly accumulate on the vacuum surfaces and be desorbed when hydrogen is introduced, the ethane background could become larger or smaller depending on which process, outgassing or pumping, is dominating over the long time period of the cruise phase of the mission. The background level of ethane will ultimately determine the detection threshold of gases with similar fragmentation patterns such as silane, ethylene, and acetylene.

No background gases or chemical noise is expected above mass 44 at high ion source pressures and hence constituent measurement accuracies in this region are expected to be determined by the statistical uncertainties associated with the respective sampling times at low concentrations or by the uncertainties caused by temperature drift, etc., to $\pm 10\%$ as stated above at large concentration levels. Examples of gases which will be detected in this higher, chemically cleaner mass range are krypton, xenon, arsine, and the higher molecular weight hydrocarbons.

Reactive trace gases of particular interest which are affected by surface adsorption or reactions are ammonia (NH_3), hydrogen sulfide (H_2S), and phosphine (PH_3). All three gases are transmitted without delay or apparent surface loss through inlet 2 because of the direct gas beaming feature of the ion source. Ammonia, the most reactive gas of the set, will not be transmitted to the ion source from leak 1 at a mixing ratio of $\approx 10^{-5}$ in any reasonable amount of time for a measurement. Evaporating cloud droplets are expected to be detectable, however, although quantitative calibration was not possible because of lack of time and facilities to produce droplets in known quantities during the calibration and evaluation period. Spectral interference with methane at mass 17($^{13}CH_4$) will limit the detection threshold of ammonia through leak 2 to a mixing ratio of approximately 10^{-5} (equal signal from ammonia and methane). Hydrogen sulfide is measurable through inlet 1 after a few seconds delay depending on the mixing ratio. At the predicted mixing ratio of $\approx 10^{-5}$ we expect a $\pm 50\%$ accuracy of the measurement. Transmission of phosphine was observed without any noticeable delays and the measurement accuracy will be primarily affected by spectral interference at the parent peak from hydrogen sulfide. Phosphine can be measured, however, by observing the mass 31 peak which occurs with an intensity of 0.25 of the parent peak and is free from spectral interference except from ethane ($^{13}C^{12}CH_6$). Assuming the predicted composition and carbon isotope ratio of ethane, the phosphine mixing ratio will be determined at least to a factor of two at the 10^{-7} mixing ratio. Enrichment factors of 100 are expected for ethane and propane and of 500 for higher order hydrocarbons. Krypton and xenon will be enriched approximately 10 times and 100 times, respectively. The actual accuracies can only be established after the data are received and analyzed, because the spectral interference depends on the actual data and background gas uncertainties resulting from the long storage and cruise time.

5. Sampling Sequence Summary

A brief description of the sampling sequence which will be followed by the instrument during the entry phase is given below.

The instrument contains an unalterably programmed 8192 step sampling sequence designed to optimize the data return at a sampling rate of two steps per second. The SEQUENCE START command is sent from the Probe Command and Data System to the instrument after the Probe parachute has been deployed and the heat shield has been jettisoned. All reference times quoted are with respect to the SEQUENCE START command. The ambient pressure values were taken from the engineering pressures vs. descent time model provided by the Galileo Project. After an automatic self-configuration which includes setting valves and heaters in the proper positions for the operations that follow, the instrument measurement sequence begins at t = 1.97 min, $P \approx 0.1$ bar by monitoring the 'residual' gas content (background).

Photochemical Sampling

Inlet 1 and Outlet 1 open at t = 2.66 min, $P \approx 0.15$ bar to expose the first gas handling system to the ambient atmosphere. The atmosphere will be sampled in seven sequences where data corresponding to masses 2–4, 12–68, 72–78, 80–84, 122, and 124 amu are transmitted.

NH_3 Cloud Sampling

In the region where ammonia (NH_3) cloud are predicted, t = 6.70 min, $P \approx 0.7$ bar, the mass range from 2 to 150 is scanned seven times. Approximately every 30 s (ΔP = 0.1 bar), the scan is interrupted to repeat the measurements at masses 14–18, 32–36, 64, and 66 amu.

Filling of Enrichment Cell 1 and Rare Gas Cell

During direct sampling in the photochemical region, the enrichment cell C1 is exposed to the ambient atmosphere; t = 4.71 min, $P \approx 0.4$ bar until t = 13.05 min, $P \approx 1.9$ bar. After isolation of the cell from the ambient atmosphere, gases not trapped by the sorbent material of the cell are expanded into the rare gas cell; t = 13.12 min to t = 16.10 min.

Background Gas Observation

At time t = 17.06 min the instrument inlet system is closed to allow observation of background gases accumulated during the measurement phase. The baseline condition is monitored by executing two decremental integer mass sweeps over the range of 150 to 2 amu and four measurements in the 0.125 amu mass incremental mode at masses 28, 44, 16, and 4 amu.

Rare Gas Sampling

The rare gas cell content is analyzed from t = 19.98 min to t = 22.40 min in six sequences. The first and last sequences include mass sweeps over the range 2–28 amu plus those masses corresponding to the isotopes of Kr and Xe. The other sequences monitor those masses corresponding to the isotopes of He, Ne, Ar, Kr, and Xe. To separate Ne from Ar^{++} several scans are made at an ionization energy of 25 eV.

Rare Gas and First Enrichment Cell Sampling

At t = 22.40 min the enrichment cell C1 processed sample from the enrichment cell is added to the rare gas cell sample for analysis. Four sequences are performed: 3 sweeps of the mass range 2–150 amu, and 1 covering masses 2–87 amu and 123–137 amu. An ionization energy of 15 eV is used during one full mass scan.

Background Gas Observation

After the enrichment cell analysis the cell is sealed off and the analyzer is pumped down to background pressure. At t = 27.78 min the instrument program executes three sequences where the background gas content is recorded once more after the enrichment cell analysis. Two mass scans cover the mass range 150–2 amu while the third mass scan covers masses 68–2 amu.

Direct Atmospheric Sampling

The second gas inlet system is opened to the ambient atmosphere starting at t = 30.99 min, $P \approx$ 7.0 bar and remains open until the end of the mission. Thirty-one sampling sequences are performed, twenty covering the mass range 2–150 amu, two scan the 2–68 amu region while the other nine monitor selected masses. Selected scans are also programmed for ionization energies of either 15 eV or 25 eV.

Filling of Enrichment Cell 2 and Analysis

While sampling the atmosphere directly, enrichment cell C2 will be filled from time t = 31.68 min to time t = 33.27 min corresponding to a predicted pressure range from $P \approx$ 7.3 to 7.8 bar. Following the isolation of the cell from the gas handling system, sample processing begins. The sample analysis will start at t = 38.99 min and end at t = 46.40 min. The predicted pressure levels are $P \approx$ 10.0 to 13.0 bar, respectively. During this time, five scans covering the mass range from 2–150 amu are performed. The ionization energies chosen for these scans is 75, 15, 75, 25, and 75 eV.

High-Mass Resolution Scan

A high-resolution mass scan for selected mass ranges (2–46, 47–90, 121–140) amu is performed at t = 48.97 min, $P \approx$ 14.0 bar. During this scan, a mass step size of 0.125 amu is used.

End of Mission Sequence

At the completion of the High-Mass Resolution Scan at t = 59 min, $P \approx$ 19.0 bar, the instrument will begin repeating a 2–150 amu scan every 75 s until the end of mission.

Acknowledgements

The dedication and efforts of many people made this complex and difficult experiment possible. The authors and the Experimenter Team wish to acknowledge particularly the contributions at Goddard Space Flight Center of J. Cooley for the excellent work as Instrument Manager and S. Way for his excellent support for the electronics system development and the hybrid fabrication.

Others at Goddard deserving thanks are R. Abell, R. Arvey, H. Benton, A. Doan, J. Fedor, J. Haberman, R. Lott, P. Mahaffy, H. Mende, H. Powers, J. Richards, T. Tyler, and J. Westberg of the instrument development team, J. Milligan, leader of the systems review team, and G. Peters for support in quality assurance.

At the University of Michigan we thank the electronics system development team of Bruce Block, J. Maurer, J. Caldwell and W. Pinkus and J. Eder for his support of the quality assurance program.

At the General Electric Co. we thank V. Geracy, L. Milton and their team members for the hybrid circuit fabrication.

At NASA Ames Research Center we thank the Galileo Probe Project Office for their support and guidance in this program. In particular we are indebted to A. Wilhelmi for his considerable effort as technical coordinator between the Project and the Experimenter Team. Last but not least the very capable management provided by J. Sperans, E. Tischler, and B. Chin is gratefully acknowledged.

References

Anders, E. and Grevesse, N.: 1989, *Geochim Cosmochim. Acta* **53**, 197.

Atreya, S. K.: 1986, *Atmospheres and Ionospheres of the Outer Planets and Their Satellites*, Chapter 1, Springer-Verlag, New York.

Atreya, S. K. and Romain, P. N.: 1985, in G. E. Hunt (ed.), 'Photochemistry and Clouds of Jupiter, Saturn and Uranus', *Planetary Meteorology*, Cambridge University Press, Cambridge, p. 17.

Bar-Nun, A., Dror, J., Kochavi, E., and Laufer, D.: 1988, *Phys. Rev.* **B35**, 2427.

Bjoraker, G. L., Larson, H. P., and Kunde, V. G.: 1986, *Astrophys. J.* **311**, 1058.

Carlson, B. E., Lacis, A. A., and Rossow, W. B.: 1991, *Astrophys. J.* (in press).

Dawson, P. (ed.): 1976, *Quadrupole Mass Spectrometry and Its Applications*, Elsevier Scientific Publ. Co., New York.

Gautier, D. and Owen, T.: 1983, *Nature* **302**, 215.

Gautier, D. and Owen, T.: 1989, in S. K. Atreya, J. B. Pollack, and M. S. Mathews (eds.), *Origin and Evolution of Planetary and Satellite Atmospheres*, University of Arizona Press, Tucson, p. 487.

Geiss, J. and Bochsler, P.: 1982, *Geochim. Cosmochim. Acta* **46**, 529.

Geiss, J. and Reeves, H.: 1981, *Astron. Astrophys.* **93**, 189.

Kerridge, J. F.: 1989, *Science* **245**, 480.

Larson, H. P., Davis, D. S., Hoffman, R., and Bjoraker, G. L.: 1984, *Icarus* **60**, 621.

Lunine, J. I. and Hunten, D. M.: 1987, *Icarus* **69**, 566.

Lunine, J. I. and Stevenson, D. J.: 1985, *Astrophys. J. Suppl.* **58**, 493.

Nier, A. O. C. and McElroy, M. B.: 1977, *J. Geophys. Res.* **82**, 4341.

Noll, K. S., Knacke, R. F., Geballe, T. R., and Tokunaga, A. T.: 1988, *Icarus* **75**, 409.

Noll, K. S., Gaballe, T. R., and Knacke, R. F.: 1989, *Astrophys. J.* **338**, L71.

Noll, K. S., Larson, H. P., and Geballe, T. R.: 1990, *Icarus* **83**, 494.

Paul, W. and Steinwedel, H.: 1953, *Naturforschung* **A8**, 448.

Paul, W., Reinhard, H. P., and von Zahn, U.: 1958, *Phys.* **152**, 143.

von Zahn, U. and Hunten, D. M.: 1992, *Space Sci. Rev.* **60**, 263 (this issue).

Widenschilling, S. J. and Lewis, J. S.: 1973, *Icarus* **20**, 465.

RETRIEVAL OF A WIND PROFILE FROM THE GALILEO PROBE TELEMETRY SIGNAL

JAMES B. POLLACK

NASA Ames Research Center, Moffett Field, CA 94035, U.S.A.

DAVID H. ATKINSON

University of Idaho, Moscow, ID 83843, U.S.A.

ALVIN SEIFF

San Jose State University Foundation, San Jose, CA 95106, U.S.A.

and

JOHN D. ANDERSON

Jet Propulsion Laboratory, Pasadena, CA 91109, U.S.A.

Abstract. Ultrastable oscillators onboard the Galileo Probe and Orbiter will permit very accurate determinations of the frequency of the Probe's telemetry signal as the Probe descends from a pressure level of several hundred mb to a level of about 20 bars. Analysis of the time-varying frequency can provide, in principle, a unique and important definition of the vertical profile of the zonal wind speed in the Jovian atmosphere. In this paper, we develop a protocol for retrieving the zonal wind profile from the Doppler shift of the measured frequency; assess the impact of a wide range of error sources on the accuracy of the retrieved wind profile; and perform a number of simulations to illustrate our technique and to assess the likely accuracy of the retrieval.

Because of unavoidably large uncertainties in the absolute frequencies of the oscillators, we use time-differenced frequencies in our analysis. Nevertheless, it is possible to recover absolute wind speeds as well as wind shears, since the Orbiter/Probe geometry changes significantly during the Probe relay link. We begin with the full relativistic Doppler shift equation. Through the use of power series expansions and a basis function representation of the wind profiles, we reduce the basic equation to a set of linear equations that can be solved with standard linear least-squares techniques.

There are a very large number of instrumental and environmental factors that can introduce errors into the measured signal or to the recovery of zonal winds from the data. We provide estimates of the magnitudes of all these error sources and consider the degree to which they may be reduced by *a posteriori* information as well as the results of calibration tests. The most important error source is the *a posteriori* uncertainty in the Probe's entry longitude. The accuracy of the retrieved winds is also limited by errors in the Probe's descent velocity, as obtained from atmospheric parameters measured by several Probe experiments, and in the *a posteriori* knowledge of secular drifts in the oscillators' frequencies during the relay link, due, for example, to aging and radiation damage.

Our simulations indicate that zonal winds may be retrieved from the Doppler data to an accuracy of several m s^{-1}. Therefore, it may be possible to discriminate among alternative models for the basic drive of the zonal winds, since they differ significantly in the implied zonal wind profile.

1. Introduction

The Galileo spacecraft consists of a Probe that will enter Jupiter's atmosphere and an Orbiter that will make multiple close flybys of the Galilean satellites of Jupiter. The Probe contains six science instruments that will conduct the first *in-situ* measurements

Space Science Reviews **60:** 143–178, 1992.

of the structure and composition of the Jovian atmosphere. The data from these measurements will be sent to a receiver on the Orbiter over approximately a 60-min interval that occurs soon after a close flyby of Io and shortly before commencement of orbit insertion maneuvers. The presence of ultra-stable oscillators on both the Probe and Orbiter will provide very accurate measurements of the frequency of the telemetry signal. As a result, it may be possible to derive useful information on the vertical profile of the horizontal winds. Such a profile could provide a valuable constraint on alternative models for the basic drive of the Jovian winds, as described shortly.

In this paper, we describe an in-depth analysis of the methodology and practicality of retrieving a wind profile from the Probe's Doppler signal that the authors of this paper have been carrying out over the last 10 years. Although we will present a thorough discussion of this problem here, a much more detailed presentation can be found in the Ph.D. dissertation of one of us (Atkinson, 1989; henceforth Paper I). In the remainder of this section, we summarize past attempts to derive data on winds of planetary atmospheres other than that of the Earth, indicate the aspects of the Galileo Doppler data that makes it challenging to retrieve wind formation, and describe the organization of this paper.

Estimates of wind speeds in planetary atmospheres have been obtained in a number of ways. These include tracking cloud features seen on images taken from space and the ground, measuring the Doppler shift and positional shift of entry Probes and balloons, and applying the thermal wind equation to measured temperature fields. All three techniques have been applied with considerable success to Venus' atmosphere (e.g., Counselman *et al.*, 1980; Kerzhanovich *et al.*, 1972; Marov *et al.*, 1973; Preston *et al.*, 1986; Limaye, 1985; Rossow *et al.*, 1980; Sagdeev *et al.*, 1986; Schubert *et al.*, 1980; Taylor *et al.*, 1980). To date, only the first and third of these techniques have been used to study winds in the Jovian atmosphere (e.g., Ingersoll *et al.*, 1981; Conrath *et al.*, 1981; Gierasch, 1976).

The three approaches to deriving winds are complementary. Feature tracking can provide a good definition of the global distribution of wind speeds at one or at most several vertical layers in the atmosphere (namely, at the level of the first optically thick cloud layer). Measurements of the motion of entry Probes yield a detailed vertical profile of the horizontal wind speeds (and, in restricted circumstances vertical wind speeds) at isolated locations. Finally, winds obtained from measurements of three-dimensional temperature distributions can provide useful estimates of the vertical shear in the east–west (zonal) wind speed (and, in certain circumstances, the north–south or meridional wind speed) over a range of altitudes, especially ones above the cloud tops. These latter estimates are subject to the joint caveats of certain dynamical assumptions (e.g., geostrophic balance, which breaks down near the equator, or cyclostrophic balance, which breaks down near the poles) and the need to have independent estimates of wind speed at some reference altitude (e.g., the cloud tops) (Seiff *et al.*, 1980). Wind speeds obtained from Probe tracking are valuable both in their own right (see below) and for providing 'ground-truth' for remote sensing techniques. With regards to the latter, Probe-derived winds help to define the average altitude to which the feature-tracked winds pertain and help to test the accuracy of winds found from measured temperature fields.

Because of Venus' proximity to the Earth, it has proved possible to obtain both accurate Doppler shift and angular displacement measurements of the time-varying positions of entry Probes and balloons. As a result, it was possible to determine the three-dimensional velocity field of the winds (especially, the usually much larger horizontal wind speeds) (e.g., Counselman *et al.*, 1980; Preston *et al.*, 1986). Unfortunately in the case of the Galileo Probe, it is possible only to measure the Doppler shift of the Probe's telemetry, but not its angular displacement, as a result of the much greater distance of Jupiter from the Earth and the Probe's antenna being pointed at the Orbiter (which puts the Earth in the distant lobes of the Probe's telemetry antenna). Nevertheless, the Doppler data can yield some very important and unique information on the variation of the zonal wind speed with altitude over a pressure range from about 100 mb to about 20 bars. The probe is expected to traverse the atmosphere at approximately 6.5° north, which lies well within Jupiter's equatorial jet.

In principle, all three wind speed components as well as the Probe's descent velocity and the Orbiter's motion can contribute to the observed Doppler shift. However, the Probe's descent velocity will be determined very accurately from analysis of temperature and pressure measurements made by the atmospheric structure experiment and the Orbiter's motion can be extremely accurately determined from other data. Among the three wind components, the zonal speed is expected to be the dominant contributor to the measured frequency shift in all, but the most pathological cases for the following reasons: (1) The projection of the Probe/Orbiter line of sight onto the local horizontal plane of Jupiter lies almost in an east–west direction, except at times very close to overflight. (2) At the level of the visible cloud layer, zonal wind speeds are typically more than an order of magnitude larger than the meridional wind speed, including at the latitude of the Galileo Probe (Ingersoll *et al.*, 1981). (3) It is easy to show from the continuity equation that the vertical wind speeds associated with the large-scale zonal wind speeds near the cloud tops are more than three orders of magnitude smaller than the horizontal wind speeds (Holton, 1979). However, vertical velocities associated with thermal convection driven by the internal heat source can have somewhat larger values in the deeper portions of the atmosphere that will be examined by the Galileo Probe experiments (Gierasch, 1976). Thus, it would be possible to retrieve the vertical profile of the zonal wind speed from the time-varying frequency of the Probe's telemetry signal.

Differences in the amount of sunlight absorbed at different latitudes provides the basic drive for the winds in the atmospheres of the terrestrial planets. However, a much wider suite of potential drives exists for the Jovian winds. They may be driven principally by differential solar heating, latent heat released by condensing water vapor, heat being advected upwards from the deep, hot interior, or heat generated from ortho/para transitions of molecular hydrogen (Stone, 1967; Gierasch, 1976; Ingersoll and Pollard, 1982; Conrath and Gierasch, 1984). Below, we briefly review the nature of these different drives and present schematic profiles of zonal winds that might result from each of them.

The magnitude of Jupiter's internal heat source is nearly comparable to the amount of sunlight that the planet absorbs (Hanel *et al.*, 1981). The thermal convective heat flux associated with this internal heat source almost totally compensates for the global-scale variation in absorbed sunlight from equator to pole so that the amount of thermal radiation emitted to space is almost constant on this length scale (Ingersoll and Porco, 1978). Consequently, no significant global-scale temperature gradient results from differential absorption of sunlight and hence no significant global-scale circulation pattern can result from this drive, contrary to some pre-Pioneer models of Jovian dynamics (e.g., Stone, 1967). Nevertheless, this compensating effect of the interior heat flux does not occur on smaller length scales: the difference in the thermal radiation emitted by adjacent belts and zones is somewhat larger than the non-trivial difference in the amount of sunlight they absorb (Pirraglia, 1984). Consequently, differential solar heating could be an important drive for wind systems having length scales on the order of the bands.

Significant absorption of sunlight occurs near the level of the uppermost, optically thick cloud layer, a putative ammonia ice cloud deck, whose bottom lies near the 700 mb level. This cloud layer typically has an optical depth of several and a non-unit single-scattering albedo in the visible (West *et al.*, 1986). Additional absorption of sunlight takes place in the lower stratosphere (1–100 mb) due to gases (especially methane) and aerosols and at greater depths in the troposphere due to gases and other cloud particles. According to the calculations of Hunten *et al.* (1980), almost all the solar energy absorbed by Jupiter's atmosphere is deposited at altitudes above the 2 bar pressure level. Therefore, we approximate the winds resulting from differential solar heating as having a maximum slightly below the ammonia cloud deck and tending towards zero near the 2 bar level (cf. Figure 1; P. Gierasch, private communication).

On the length scale of the Jovian bands, the cloud-top winds are approximately in geostrophic balance, i.e., the basic balance in the meridional momentum equation is between the Coriolis force and the pressure-gradient force (e.g., Gierasch, 1976). Consequently, the thermal wind equation can be used to place a constraint on the magnitude of the temperature difference between belts and zones that is needed to produce the observed zonal winds. This temperature difference scales inversely with the depth of the source region for the winds. For winds being driven by latent heat that is released by water vapor condensing near the 7 bar level of the atmosphere, the required temperature difference is comparable to the difference expected from wet and dry adiabats for solar abundances of water (Gierasch, 1976). Thus, this type of model envisions that water preferentially condenses in regions of large-scale rising motion. In this sense, the latent heat drive strongly reinforces any small differential rising and sinking motions. Note that although almost all the water condensation occurs within a fraction of a scale height of the water cloud base, the temperature difference between wet and dry adiabats persists throughout the convectively unstable region lying above the water cloud base. We represent the winds resulting from the latent heat drive as beginning at the base of the water clouds and reaching their maximum near the top of the convection zone (cf. Figure 1; P. Gierasch, private communication).

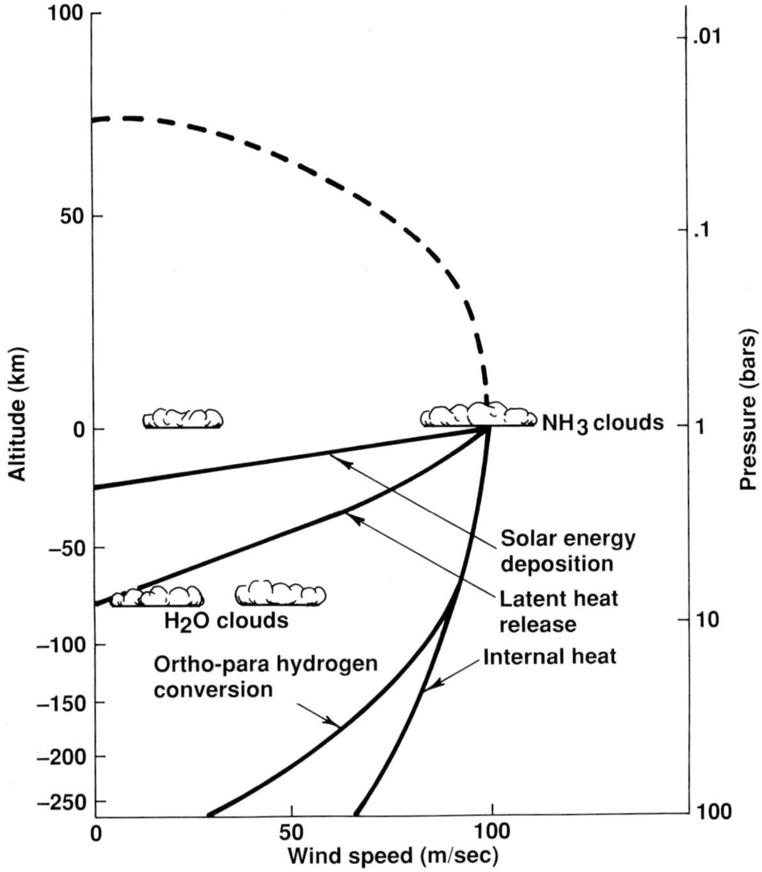

Fig. 1. Vertical profiles of the zonal wind speed predicted by alternative models of the basic drive for the Jovian atmospheric circulation.

Temperature differences on constant pressure surfaces in the convective region of the Jovian troposphere can also be produced by horizontal variations in the relative abundance of molecular hydrogen in its ortho and para states (Gierasch, 1983). Since the specific heat at constant pressure of these two states of hydrogen differ, variations in the ortho to para ratio lead to variations in the adiabatic lapse rate and hence temperature. As the interconversion of ortho and para hydrogen is spin forbidden, it is expected to take place slowly in the colder, upper region of the Jovian troposphere (Massie and Hunten, 1982). In fact, there is evidence from the Voyager IRIS data that significant horizontal variations in the ortho to para ratio exist in the uppermost troposphere (Conrath and Gierasch, 1984). However, analysis of ground-based spectra implies that only equilibrium hydrogen is present at deeper levels of the Jovian atmosphere (Cunningham *et al.*, 1988).

The wind profile in Figure 1 associated with horizontal variations in the ortho/para ratio was obtained by assuming that extreme differences in this ratio occur in neighboring bands, that temperature gradients vanish at great depths in the troposphere and at stratospheric altitudes, and that the circulation extends fairly deep within the troposphere (Gierasch, 1983). Thus, like all the other profiles in Figure 1, the one shown for an ortho/para drive is simply meant to be one plausible example of this class of dynamical model.

In all the above models, the drive for the circulation is located at relatively shallow depths in a very deep molecular hydrogen envelope for the planet. An alternative viewpoint is to consider a much deeper circulation pattern that persists throughout the molecular hydrogen envelope and that is driven by the internal heat source (Ingersoll and Pollard, 1982). Thereotical and laboratory studies show that columnar convection cells, aligned with the spin axis, are the preferred mode of convective instability in rapidly and uniformly rotating, viscous, conducting fluids (e.g., Busse, 1976). A somewhat more generalized state of motion involves cylindrical cells that interact with a basic cylindrical, differential rotation to produce zonal winds near the top of the adiabatic portion of the fluid (Ingersoll and Pollard, 1982). We crudely represent the zonal wind profile for this type of model as having a modest wind shear throughout the region measured by the Galileo Probe and as having a non-zero velocity at the base of this region (cf. Figure 1; P. Gierasch, private communication).

According to Figure 1, the 'representative' zonal wind profiles associated with the four dynamical models differ substantially from one another. Thus, one might hope to distinguish among them by comparing their predictions with zonal wind profiles derived from the Galileo Probe's telemetry frequency. However, there is a significant variance possible in the wind profiles corresponding to each of the dynamical models. Conceivably, data taken by a variety of Galileo Probe and Orbiter experiments will lead to well-constrained predictions for these various models in the post-Galileo era. For the purposes of this paper, we merely wish to show that winds can be recovered with sufficient accuracy from the Galileo Doppler wind experiment to provide useful constraints on the underlying drives for the winds. We also recognize that winds associated with eddy motions may contribute to the derived wind profile, although eddy wind speeds in the Jovian atmosphere are typically almost an order of magnitude smaller than the zonally average zonal wind speed near the cloud tops (Ingersoll *et al.*, 1981).

Given the experience in deriving wind speeds from Doppler data for probes into Venus' atmosphere and given the well-known and relatively simple form of the Doppler equation, it might be thought that its application to the Galileo Probe would be quite straightforward. However, this is not the case for several reasons. First, the geometry of the Probe and Orbiter positions changes significantly during the communication period. In particular, the angle between the line of sight of the probe and orbiter and the local vertical (probe aspect angle), which determines the contribution of horizontal wind speeds to the Doppler shift, varies from essentially $0°$ at overflight to values that approach $10°$ near the beginning and end of the communication period. As shown in the next section, such variations both complicate the analysis of the Doppler data and open up opportunities (e.g., it is possible to recover absolute wind speeds without knowing well the oscillators' zero-point frequency).

A second reason that the analysis of the Galileo Probe Doppler data is less than straightforward is the need to do careful study of a wide range of possible error sources that make deriving winds for Jupiter's atmosphere more problematic than was the case for Venus' atmosphere. Except very close to the surface of Venus, the zonal wind speeds are much larger than the speed of the planet's solid body rotation. In contrast, Jupiter's rotational velocity near its equator has a value of approximately 10 km s^{-1}, which is about two orders of magnitude larger than the measured zonal wind speed near the ammonia cloud tops. Consequently, small inaccuracies in the *a posteriori* knowledge of the Probe's location might seriously compromise the ability to derive accurate relative wind speeds from the Doppler shift data. Other aspects of the spacecraft geometry and Jovian environment also make the retrieval of accurate wind speeds more challenging than was the case for Venus. These include the small values of the Probe aspect angle and the strong Jovian magnetosphere (the high flux of high-energy particles can cause drifts in the Orbiter and Probe oscillators). Despite this highly unfavorable situation for performing Doppler wind measurements, we will show that accurate wind retrievals are possible.

In the next section of this paper, we describe the mathematical basis of our approach that begins with the relativistic Doppler shift equation and reduces it down to the solution of a set of linearized equations. In the subsequent section, we investigate the effects of a large number of possible error sources that limit the accuracy in the recovered winds and provide estimates of their magnitudes. Next, we present full-up simulations of our approach by attempting to recover the four classes of wind profiles given in Figure 1 in the presence of realistic errors in the data and our knowledge of critical parameters. Finally, we summarize the conclusions of this paper by assessing the accuracy with which a wind profile can be retrieved from the Galileo Probe's telemetry frequency, key sources of error, and the ability of the retrieved profiles to discriminate among the major candidate drives of the circulation.

2. Approach

Beginning at about the 100 mb level in the atmosphere, when the Galileo Probe's descent velocity has become subsonic and its parachute systems have been deployed, the Probe's radio system will send telemetry signals to the Orbiter's radio system. This communication will last approximately 60 min, at which point the probe will have descended to about the 20 bar level in the Jovian atmosphere. The Probe's telemetry will be sent by two independent channels to help ensure the successful receipt of the results from the Probe experiments. One of these channels is controlled by an ultra-stable oscillator (USO) on the Probe. An identical USO on he Orbiter will provide a highly accurate measurement of the frequency of the carrier, f, whose nominal value is 1387 MHz. The received frequency will be measured every $\frac{2}{3}$ s during the Probe/Orbiter relay link. We wish to use the Doppler shift in the measured frequency due to the zonal winds to retrieve an estimate of their vertical profile.

To understand the essence of the Doppler wind problem for the Galileo Probe, we suppose for the moment that the contributions to the Doppler shift from all sources other than the atmospheric winds can be removed exactly, we consider only the effects of the zonal component of the winds for reasons given in the Introduction, and we use the classical form of the Doppler shift equation for planar geometry

$$f_o = f_p + f^{mis} + f_p \frac{U}{c} \sin\psi \cos\alpha\,, \tag{1}$$

where f_o is the received frequency at the Orbiter, f_p is the transmitted frequency at the Probe, f^{mis} is the Doppler shift due to all factors other than the zonal wind speed, U is the zonal wind speed, c is the speed of light, ψ is the angle between the local vertical and the line of sight between the Probe and Orbiter, and α is the angle between the east–west direction and the projection of the line of sight on the local horizontal plane. The product $\sin\psi \cos\alpha$ is simply the fraction of U that lies along the line of sight.

Unfortunately, there are large uncertainties in the zero-point frequency of the Probe and Orbiter's oscillators. While the USOs are quite stable over the period of the communication link, they are expected to undergo a significant drift over the 6 to 7 year cruise interval from launch to arrival at Jupiter. Since there will be no 2 way links between the ground and the USOs following launch, there is no procedure for accurately determining the actual long-term drift. Ground-based tests that have been conducted on the USOs suggest that they may drift by about 500 Hz during the cruise period. A drift of this magnitude is equivalent to the Doppler shift produced by a 1000 m s^{-1} zonal wind for the geometry of the Galileo mission!

The solution to the problem posed by the large uncertainty in the USOs' zero-point frequency is to use differences in the measured frequencies:

$$\Delta f_o - \Delta f^{mis} = f_p \frac{\Delta[U]}{c} \sin\psi \cos\alpha + f_p \frac{U}{c} \Delta(\sin\psi \cos\alpha)\,, \tag{2}$$

where Δx means the difference in the value of parameter x at any time t during the relay link and some reference time, t_r. The left-hand side of the above equation contains a combination of the measured frequencies, now having a high accuracy, and Doppler shifts due to a variety of well-measured or known velocities. According to the above equation, the zonal velocity affects the measured frequency difference in two ways: through changes in zonal wind speed with depth in the atmosphere and through the changing geometry of the line of sight, as influenced by the time-integrated effect of the wind on the Probe's position.

Figure 2 shows the variation of ψ and α as a function of time after entry for the nominal Probe mission, with Orbiter overflight of the Probe occurring 21 min after Probe entry. The geometrical factors change significantly over the course of the relay link, making the magnitude of the second term on the right-hand side of the above equation comparable to or greater than that of the first term during much of the relay link. As a result, it is possible in principle to determine absolute wind speeds as well as wind shears from measured frequency differences.

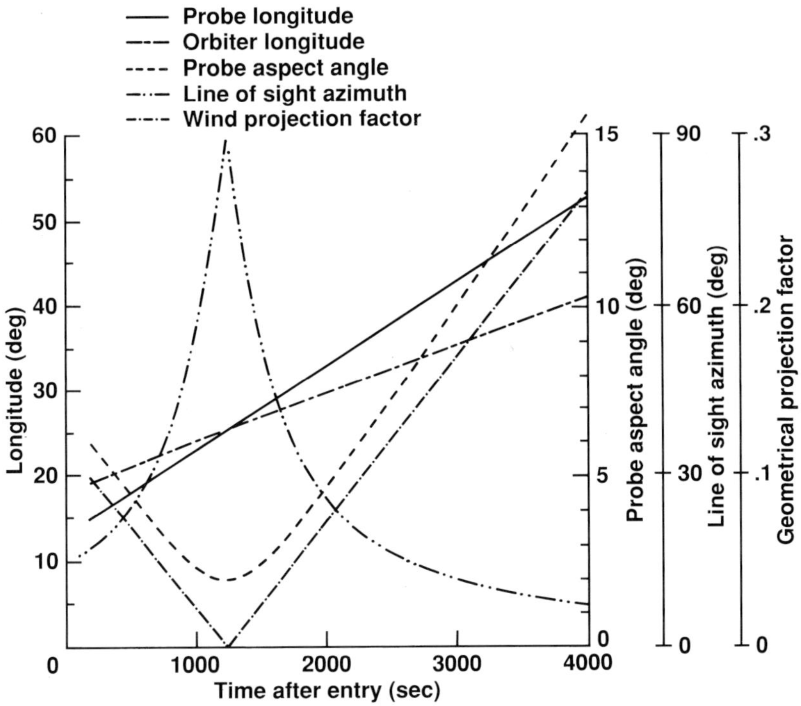

Fig. 2. Probe and Orbiter longitudes, Probe aspect angle (ψ), azimuth angle (α), and projection factor (fraction of the zonal wind projected onto the line of sight) as function of time from the Galileo Probe's entry into the Jovian atmosphere.

However, the above equation cannot be readily inverted to yield wind speeds directly. This last point is brought out even more clearly when it is realized that the winds significantly alter the longitudinal location of the probe (in a time-integrated fashion) and therefore significantly affect the values of the angular quantities appearing in these equations, as verified by our detailed calculations. For these reasons, we will perform a number of transformations on the Doppler shift equation, beginning with the more rigorous relativistic counterpart of the above equation, to reduce the problem to a set of linearized equations that can be solved with least-squares techniques.

In more detail, the derivation of the final linearized set of equations given below follows the following path: (1) The full relativistic Doppler equation is expanded to second-order in velocity. (2) We express the resulting equation as an equality between the measured Doppler shift and accurately known quantities on one side of the equation and the velocity along the line of sight on the other side of the equation. (3) This velocity is expanded about a state of zero horizontal wind speed. This expansion is useful since the horizontal wind speeds are much less than the planet's rotational velocity. (4) The horizontal wind speed's dependence on atmospheric pressure is expressed as a Legendre polynomial expansion. This expansion is needed since the Doppler shift depends, in part, on the time integral of the wind speed, which affects the instantaneous location of the Probe (cf. Equation (2)). (5) The problem at this point has been reduced to a linear least-squares problem involving the coefficients of the Legendre polynomials. The singular value decomposition method is used to invert the relevant equations, reduce them to a form required by standard least-squares techniques, and insure good numerical stability.

Since the magnitude of the relative velocity between the Probe and Orbiter, V_{tot}, is much less than the speed-of-light, c, we can use a Taylor series expansion on the full relativistic Doppler shift equation and retain terms only up to order $1/c^2$ (Kerzhanovich *et al.*, 1969; Paper I). The resulting equation is given by

$$f_o - f_p = -\frac{f_p}{c}\left[V_{\|} + \frac{1}{c}\left(\frac{V_{\text{tot}}^2}{2} - V_{\|}^2\right) - cG_r\right], \tag{3}$$

where $V_{\|}$ is the projection of V_{tot} along the line of sight between the Probe and Orbiter and G_r is the general relativistic red shift in the weak field limit (Paper I). The first term on the right-hand side of this equation is the familiar classical Doppler shift term. Our sign convention in this equation and the ones below is that a positive value of $V_{\|}$ implies an increasing distance between the Probe and Orbiter and, hence, a red shift to the measured frequency.

The gravitational red shift has a value of approximately 20.5 Hz for the nominal distances of the Probe and Orbiter from the center of Jupiter. However, it changes by only a few tenths of a Hz during the relay link and even this relatively trivial effect can be removed to very high accuracy. The second term on the right-hand side of the equation has a magnitude of several tens of Hz and varies by about this much over the relay link. Since a zonal wind speed of 1 m s^{-1} produces a Doppler shift that averages about 0.5 Hz over the relay link, these higher-order terms are important and must be taken into account in the ultimate analysis. Fortunately, these higher-order terms are dominated by the projections of the Orbiter's motion and the planet's rotational velocity along the line of sight and, hence, these terms can be evaluated with high accuracy from the start of the inversion procedure.

We rewrite the relativistic Doppler shift equation with 'knowns' on the left-hand side of the equation and the unknowns on the right-hand side:

$$f'_{\text{Dop}} = -f_p\,\frac{V_{\|}}{c}, \tag{4}$$

$$f'_{\text{Dop}} = f_o - f_p(1 + G_r) - \frac{1}{c^2}\left[\frac{V_{\text{tot}}^2}{2} - V_{\|}^2\right]. \tag{5}$$

We next cast the above equation in terms of the difference in the Doppler shift at an arbitrary time, t, during the relay link and some reference time, t_r:

$$\Delta f'_{\text{Dop}} = -f_p\,\frac{\Delta V_{\|}}{c}. \tag{6}$$

Here, as earlier, Δx means the difference in the value of x at time t and its value at time t_r.

It is useful at this point to introduce a spherical-coordinate system whose origin lies at the center of Jupiter, as illustrated in Figure 3. The positions of the Probe and Orbiter

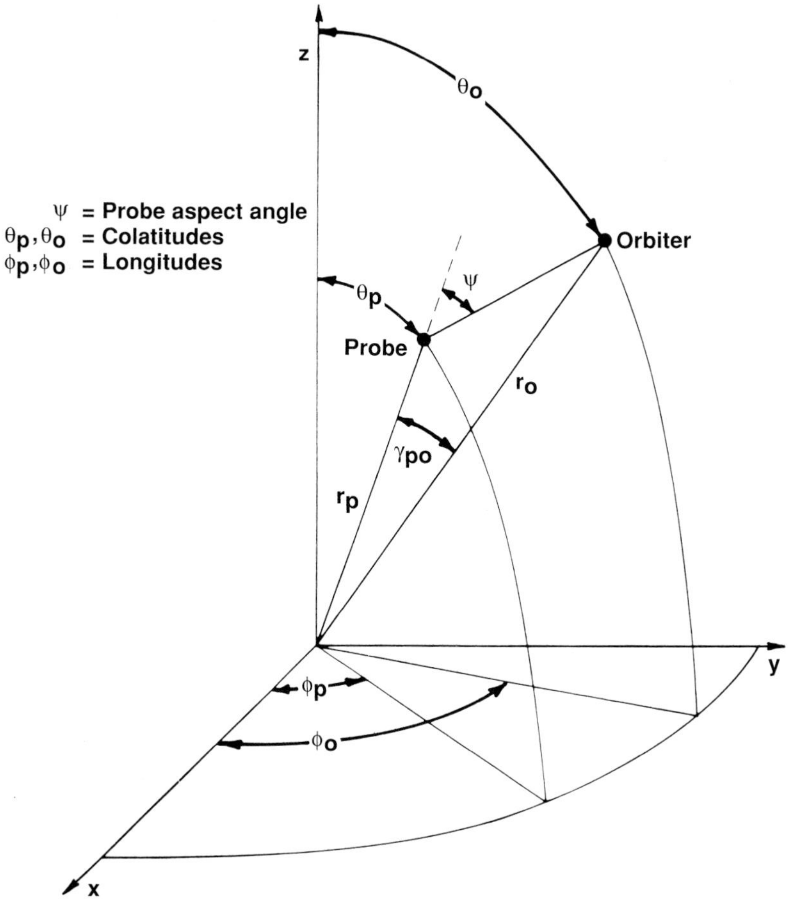

Fig. 3. The Probe/Orbiter geometry in a Jovian centered coordinate system. Various basic angles used in the text are defined in this figure.

are specified by a radial distance r from the origin and colatitudes θ and longitudes ϕ. Auxiliary angles in this coordinate system include γ, the angle between the direction to the Probe and Orbiter, as measured from the origin, ψ, the Probe aspect angle, and α, the line-of-sight azimuth. The mathematical relationships between the auxiliary angles γ, ψ, and α and the primary spherical coordinates and between $V_{\|}$ and these primary coordinates are given in Paper I.

It is useful to write $V_{\|}$ in terms of a Taylor series expansion of the spherical coordinates and their time derivatives both to obtain a linear set of equations for the zonal wind velocities and to provide a mathematical basis for performing the error analysis discussed in the next section. For our present purposes, we will assume that all the relative motions contributing to $V_{\parallel}$ are either known with negligible uncertainties or are unimportant, except for the zonal wind speed. In this case, only the coordinate ϕ_p and its time derivative enter into the second term on the right-hand side of the Taylor-series expansion. We obtain the following linearized equations for the zonal wind speed, U, in terms of the measured Doppler shift and other, known velocities:

$$\Delta f_{\text{Dop}} = -\frac{f_p}{c}\,\Delta\left[\frac{\mathrm{d}V_{\parallel}}{\mathrm{d}\phi_p}\,\delta\phi_p + \frac{\mathrm{d}V_{\parallel}}{\mathrm{d}\dot{\phi}_p}\,\delta\dot{\phi}_p\right], \tag{7}$$

$$\Delta f_{\text{Dop}} = \Delta f'_{\text{Dop}} + \frac{f_p}{c}\,\Delta V_{\parallel}^{0}, \tag{8}$$

$$\delta\phi_p(t) = \delta\phi_p(t_r) + \int_{t_r}^{t} \frac{U(t')}{r_p(t')\sin(\theta_p(t'))}\,\mathrm{d}t', \tag{9}$$

$$\delta\dot{\phi}_p(t) = \frac{U(t)}{r_p(t)\sin(\theta_p(\mathrm{t}))}, \tag{10}$$

where δx_i is the difference between the true of x_i and the value determined from the best available information (e.g., Probe descent velocity, Orbiter trajectory, etc.) and subscript p refers to the Probe.

As a final step in linearizing the equation relating measured and known quantities to the unknown zonal wind speed, we express the zonal wind speed in terms of an Nth order expansion in Legendre polynomials:

$$U(t) = \sum_{l=0}^{N} a_l P_l(z(t)), \tag{11}$$

$$z(t) = 2\,\frac{\log(pr(t)) - \log(pr(t_r))}{\log(pr(t_f)) - \log(pr(t_r))} - 1, \tag{12}$$

where P_l is a Legendre polynomial of order l, pr is the atmospheric pressure at the Probe's location, and subscripts r and f denote the times of the first (or reference) and last Doppler shift data. The unknowns now are the constants a_l.

Finally, we combine the above two sets of equations to obtain a matrix equation for the unknowns, a_l, which we now write as a vector $\mathbf{a}$ to simplify the notation:

$$\mathbf{Ca} = \mathbf{b}, \tag{13}$$

$$\mathbf{b} = \Delta f_{\text{Dop}}, \tag{14}$$

$$\mathbf{C} = -\frac{f_p}{c}\left[\frac{\mathrm{d}V_{\parallel}}{\mathrm{d}\phi_p}(t_m)\beta_1 + \frac{\mathrm{d}V_{\parallel}}{\mathrm{d}\dot{\phi}_p}(t_m)\beta_2\right], \tag{15}$$

$$\beta_1 = \sum_{m'=1}^{m} \frac{\delta t}{r_p(t_{m'}) \sin(\theta_p(t_{m'}))} P_l(z(t_{m'})), \tag{16}$$

$$\beta_2 = \frac{P_l(z(t_m))}{r_p(t_m) \sin(\theta_p(t_m))} - \frac{P_l(z(t_r))}{r_p(t_r) \sin(\theta_p(t_r))}; \tag{17}$$

vector **a** has L elements ($= N + 1$), corresponding to the number of terms used in the Legendre polynomial representation of the wind speed; vector **b** has M elements, corresponding to the number of discrete time points, t_m, at which the measurements are made (or a subset that have been appropriately time averaged over small time segments); matrix **C** is of dimension M by L, as are matrices β_1 and β_2. In the above equations, δt is the time between successive samples. There are many more time points, M, than unknowns, L.

In principle, the above matrix equation could be solved by simply multiplying the equation by $\mathbf{C}^{-1}$ on the left side of both sides of the equation, where $\mathbf{C}^{-1}$ is the inverse of matrix **C**. In practice, such a procedure is susceptible to numerical problems that arise when dealing with matrices that may be nearly singular. For this reason, we express **C** in terms of the product of three matrices, **U**, **W**, and **V**, in accord with the method of singular value decomposition (Golub and Van Loan, 1983). The resultant equation for **a** is then (Paper I):

$$\mathbf{a} = [\mathbf{V}\mathbf{W}^{-1}\mathbf{U}^T]\mathbf{b}, \tag{18}$$

where superscripts -1 and T means the inverse and transpose of a matrix, respectively. Given a set of values of b_m ($m = 1$ to M), obtained from Doppler shift measurements and the removal of well-measured values of velocity components other than the zonal wind speed, one can obtain a best estimate of the zonal wind speed coefficients a_l ($l = 0$ to N) by using standard least-squares techniques for solving the above matrix equation.

3. Error Analysis

There are a very large number of factors that can influence the measured Doppler shift. In this section, we determine the uncertainties in the knowledge of all factors aside from the zonal wind profile and their impact on the measured Doppler shift to ascertain the feasibility and limitations on recovering the zonal wind profile in Jupiter's atmosphere from the frequency of the Probe's telemetry signal. In the next section of this paper, we will provide quantitative estimates of the errors in the zonal wind speed profile due to the cumulative effects of the uncertainties in the *a posteriori* knowledge of all the relevant factors that affect the Doppler shift.

It is useful to place the factors of interest into several broad categories. These categories include the trajectories of the Probe and Orbiter during the relay link, as specified in terms of the six basic variables and their time derivatives; factors that influence the frequency of the USOs; propagation effects that influence the phase path of the Probe's signal; and a variety of miscellaneous factors that include such quantities as the effects of special and general relativity. In the discussion below, we emphasize those factors that can potentially generate the largest errors in the recovered winds.

As formulated in the previous section, the primary variable used to determine the zonal wind speed profile is the *difference* in the frequency of the Probe's signal at some arbitrary time t during the relay link and that measured at some fixed reference time, t_r, that we take to be the first such measurement. These frequency determinations are made by a USO on the Galileo Orbiter. Thus, factors that cause a time-independent alteration of the measured frequency ('constant' errors) will have no effect on the recovered winds. Rather systematic distortions to the recovered winds can be introduced by factors that cause a time-varying error in the measured frequency. Here, we distinguish between time-varying errors having a consistent trend during the link ('systematic' errors) and ones having a random or periodic nature on time scales short compared to the duration of the relay link ('random' errors). We emphasize that it is only the uncertainty in the values of the various factors subsequent to the relay link that are of interest: we intend to remove the influence of all factors from the measured frequency difference to the degree that their values are known and their impact can be appropriately modeled.

The influence of the zonal winds on the measured frequencies provides a convenient standard for assessing the importance of various error sources. Over the course of the relay link, the internal heat wind model of Figure 1 produces approximately a 100 Hz shift in the measured frequency due to the wind's effect on $\dot{\phi}_p$ and about 100 Hz in the measured frequency due to its effect on ϕ_p. Comparable shifts apply to the other wind models of Figure 1. Thus, factors that produce time-varying errors in frequency of 100 Hz or more, 10–100, 1–10, and < 1 Hz would have lethal, serious, non-trivial, and very minor impacts, respectively, on the recovered winds.

Finally, the measured frequency differences are linked to the cumulative effects of winds at all earlier times as a result of the changing geometry of the observations (cf. Equations (7)–(10)). As a result of this relationship and our use of a least-squares solution algorithm, it is possible to recover winds near the beginning of the relay link with comparable accuracy to those towards the end, despite our use of the initial frequency measurement as our reference frequency. This claim will be demonstrated when we show the frequency residuals of simulated cases.

3.1. Trajectory uncertainties

Several sets of observations will provide an accurate determination of variables that define the positions and velocities of the Probe and Orbiter, aside from the horizontal winds. These include Earth-based tracking of the Orbiter's position prior to, during, and following the relay link; reconstructions of the Probe's trajectory following its release from the Orbiter and, hence, its entry location in Jupiter's atmosphere that are based on careful tracking of the Orbiter's location around the time of Probe release; determination of the Probe's descent velocity from data obtained by several Probe experiments, notably the Atmospheric Structure Instrument (ASI); and current knowledge of the planet's solid body rotation rate and its figure. As we will shortly see, errors in the descent velocity and Probe entry location, especially its entry longitude, constitute by far the most important error sources among the ones listed above and these are among the largest error sources of all the factors. We, therefore, begin our discussion by focussing on these two error sources.

The descent velocity, $\dot{r}_p$, will be derived from frequent measurements of pressure, P, and temperature, T, made by the ASI experiment by using the following equation:

$$\dot{r}_p = -\frac{RT}{g\mu P}\frac{\mathrm{d}P}{\mathrm{d}t}, \tag{19}$$

where R is the universal gas constant, g is the acceleration of gravity, and μ is the mean molecular weight of the atmospheric gases (Seiff *et al.*, 1980). This equation is based on the equation of hydrostatic equilibrium and the perfect gas law, relationships that should be obeyed very closely in the region of the Jovian atmosphere of interest. To evaluate the uncertainty in $\dot{r}_p$, we consider errors arising from each of the four variables in Equation (19).

Very accurate values of g (to better than 1 part in 10^4) may be obtained from Jupiter's mass, its radius at the 100 mb pressure level as a function of latitude, and the Probe's position relative to the 100 mb pressure level. The first two of these variables are currently known to a very high accuracy from past spacecraft missions to Jupiter, while the third variable can be inferred from the data of ASI (from $\dot{r}_p$ in a slightly iterative fashion). The mean molecular weight, μ, is expected to be measured to a relative accuracy of 2 parts in 10^3 by the Helium Abundance Experiment of the Galileo Probe. This uncertainty makes a non-trivial contribution to the overall uncertainty in the values of $\dot{r}_p$ derived from the above equation.

It is useful to partitition the errors in P and T into scale factor, zero point or offset, and random errors (Seiff *et al.*, 1989). An extensive set of calibration tests of ASI indicate that the scale factor, offset, and random errors in the temperature measurements will be approximately 0.002, 0.15, and 0.087 K, respectively. Scale factor errors in pressure do not affect $\dot{r}_p$. Offset and random errors in pressure vary among three intervals of pressure values spanned by the separate sensors of the ASI, but typically correspond to fractional errors of 0.002 and 0.001, respectively.

Collectively, the errors in pressure, temperature, and mean molecular weight produce scale, offset, and random fractional errors in the inferred descent velocity of about 0.002, 0.0035, and 0.0016, respectively. When these separate components are combined in a root sum square fashion (RSS), a typical total fractional error in the descent velocity of about 0.0043 results. Thus, when the descent velocity is 50 m s^{-1}, as occurs at about the 5 bar pressure level, the corresponding uncertainty in this value would be about 0.2 m s^{-1}. Such an uncertainty may not seem to be very significant – indeed it represents a real triumph in measurement accuracy! However, it is non-trivial since almost all of the descent velocity lies along the line of sight between the Probe and Orbiter and only about 0.1 of the zonal wind speed projects along the line of sight. Thus, errors in descent velocity can cause an order of magnitude larger error in the recovered zonal wind speed.

Table I summarizes the effects of the errors in the inferred descent velocity, $\dot{r}_p$. by giving the resulting constant, systematic, and random errors in the measured frequency. *A priori* errors refer to ones expected in the absence of relevant data (e.g., measurements of the descent velocity by the Probe's ASI experiment) or the use of ground-based calibration data, while *a posteriori* errors refer to ones remaining after such information is employed. Thus, the magnitude of the *a priori* error provides an indication of the need to partially correct it, if possible, while the difference between the *a priori* and *a posteriori* errors show the degree to which the former can be eliminated. Table I also gives the corresponding errors due to a large number of other factors, most of which will be discussed in the remainder of this section.

Another important error results from the uncertainty in the entry location of the Probe. Its trajectory can be reconstructed accurately from a knowledge of its relative velocity to the Orbiter at the time of separation, the Orbiter's trajectory, and the laws of celestial mechanics. (Note that the Probe does not have any active propulsion system.) The Probe release dynamics can be calculated by tracking the Orbiter prior to, during, and after Probe release, since the Probe's relative momentum at release is equal and opposite to the change in momentum of the Orbiter at release.

Table II provides our current estimates of the 1σ uncertainties in the entry parameters for the Probe that result from uncertainties in the Orbiter's trajectory. Of these errors, the most important one is the uncertainty in the entry longitude of the Probe, since it produces a non-trivial error in the angle between the zonal direction and the line of sight. Figure 4(a) shows the change in relay link frequency near the start of the link as a function of the uncertainty in the entry longitude, $\delta\phi_p$. For the uncertainty in entry longitude given in Table II, there is a resulting uncertainty of almost 100 Hz in relay link frequency. The uncertainty in the entry longitude arises almost entirely from uncertainties in the time interval between Probe release from the Orbiter and entry into the Jovian atmosphere. $\delta\phi_p$ strongly affects the analysis of the relay link frequency because of the error it causes in the projection of the planet's rotational velocity (which is shared by the Probe) along the line of sight. Fortunately, as illustrated in Figure 4(b), the impact of the entry longitude on the relay link frequency difference, the primary quantity used in our analysis, is considerably smaller than its impact on the absolute frequency. As summarized in Table I, there still results a systematic error of about 8 Hz in the frequency difference, making the error in Probe entry longitude perhaps the single most important source of error in the recovered winds.

The Orbiter's trajectory during the relay link will be known to a very high degree of accuracy as a result of its being continually tracked by a series of techniques. The Orbiter's location and velocity will be determined from Earth-based Doppler tracking, ranging (signal propagation time), and angular position measurements made with a Differential Very Long Baseline Interferometer. Reconstruction of the Orbiter's trajectory will include allowance for stochastic accelerations due to spacecraft venting, gas leaks, and attitude reorientation maneuvers.

Table III summarizes the estimated 1σ errors in the position and velocity of the Orbiter during the Probe relay link and their associated frequency errors. As can be seen,

TABLE I

Summary of Galileo error sources[a]

Error	Magnitude	Constant (Hz)	Random (Hz)	Systematic (Hz)	
				A priori	*A posteriori*
		Probe trajectory errors			
θ_p	0.006 deg	0.1		0.0	0.0
$\dot{\theta}_p$	10.0 m s^{-1}	1.1			0.14[b]
$\delta\theta_p$[c]	0.02 deg	1.53		NA[d]	NA
ϕ_p	0.231 deg	121.4		7.6	
	0.153 deg	81.6			4.6[e]
$\dot{\phi}_{p,\,\mathrm{wind}}$	100–60 m s^{-1}[f]	47.5		108.0	
$\delta\phi_p$	0.2 deg[g]	0.0		109	
$\dot{\phi}_{p,\,r_p}$[h]	4 km	0.0			0.3
r_p	4 km	0.2			0.4
$\dot{r}_p$	0.2 m s^{-1}[i]	2.4			1.8
		Orbiter trajectory errors[j]			
θ_0	6×10^{-5} deg	7×10^{-4}			0.0
$\dot{\theta}_0$	5.5×10^{-8} deg s^{-1}	7.7×10^{-3}			0.0
ϕ_0	4.7×10^{-5} deg	0.025			0.0
$\dot{\phi}_0$	2.2×10^{-9} deg s^{-1}	1.25×10^{-3}			0.0
r_0	0.094 km	2.5×10^{-4}			0.0
$\dot{r}_0$	7.9×10^{-3} m s^{-1}	0.036			0.0

Error	Magnitude[l] $\Delta f/f$	Constant (Hz)	Random (Hz)	Systematic (Hz)	
				A priori	*A posteriori*
		Probe oscillator effects[k]			
Bus voltage	$\pm 2.3 \times 10^{-11}$	0	unknown	± 0.03	0.016
Gravitational accel.	8×10^{-12}	0	unknown	0.011	0.011
Spin accel.	4×10^{-11}	0	unknown	0.055	0.03
Temperature	1.2×10^{-10}	0	unknown	0.17	0.08
Pressure	10^{-10}	0	unknown	0.14	0.03
Magnetic field	3.6×10^{-11}	0	unknown	0.05	0.025
Short-term aging	4.3×10^{-10}	0	unknown	0.596	0.14
Allan variance[m]	2.0×10^{-12}	0	unknown	0.0	0.003
Radiation settling[n]		0	unknown	0.7	0.35

Error	Magnitude	Constant (Hz)	Random (Hz)	Systematic (Hz)	
				A priori	*A posteriori*
		Orbiter oscillator effects[o]			
Short-term aging	2×10^{-11}	0	unknown	0.027	0.027
Allan variance	2×10^{-12} s^{-1}	0	unknown	0.0	0.003
Radiation settling		0	unknown	-0.28	-0.28

Table I (continued)

Error	Magnitude	Constant (Hz)	Random (Hz)	Systematic (Hz) *A priori*	Systematic (Hz) *A posteriori*
		Probe spin[p]			
Spin axis offset		0	unknown	± 0.112	unknown
Polarization		0	unknown	± 0.083	unknown
		Frequency measurement error[q]			
Measurement error	0.18124	0	0.052[r]	0	0
		Gravitational redshift			

Error	Magnitude	Constant (Hz)	Random[s] (Hz)	Systematic (Hz) *A priori*	Systematic (Hz) *A posteriori*
Gravitational redshift		-20.5	2.3×10^{-4}	-0.20	0.0082
		Atmospheric effects			

Error	Magnitude	Constant (Hz)	Random[u] (Hz)	Systematic (Hz) *A priori*	Systematic (Hz) *A posteriori*
Neutral atmosphere[u]		0	0.04	0.30	0.043[v]
Ionosphere[w]		2×10^{-6}	2×10^{-7}	0	0

[a] At a probe aspect angle of 5°, a 1 Hz error in frequency will lead to a 2.48 m s^{-1} error in recovered zonal wind velocity.
[b] Assuming a constant 10 m s^{-1} meridional wind profile.
[c] Due to integrated effect of meridional wind on Probe latitude after 30 min.
[d] Not applicable.
[e] Due primarily to time of entry uncertainty of 14.95 s.
[f] Internal heat wind profile.
[g] Due to integrated effect of internal heat wind profile on Probe longitude after 30 min.
[h] Due to uncertainty in Probe radial location.
[i] Assuming a fractional uncertainty in Probe descent velocity of about 0.005 (0.5%).
[j] From Kenyon, 1986.
[k] From Garriga, 1981; and Garriga, 1984.
[l] 30-min drift rate.
[m] Variance of the difference between consecutive 1 s USO frequency averages.
[n] For nominal Probe entry radiation conditions. See text.
[o] From Gussner, 1980.
[p] Assuming Probe spin rate of 5 rpm.
[q] Caused by digitization of Probe signal during frequency measurement process on the Orbiter.
[r] Averaged over 1 s.
[s] Due to uncertainties in Probe and Orbiter radial locations, and uncertainty in GM_J.
[t] Random error based upon assumed 10% error in integrated electron density.
[u] Based on Orton III atmosphere and refractive properties of H_2, He, NH_3, and CH_4 mixtures at Jovian temperatures and pressures. See Atkinson and Spilker, 1989.
[v] Random error based upon assumed 10% uncertainty in NH_3 mixing ratio as measured by the Probe Neutral Mass Spectrometer, and 0.1% error in hydrogen and helium mixing ratios as measured by the Probe Helium Abundance Detector. See Hunten *et al.*, 1986.
[w] Assuming model ionosphere of constant electron density equal to 5×10^4 cm^{-3} over a 4000 km layer.

TABLE II

Standard deviation in Probe entry parameters

Entry parameter	1-σ uncertainty
Latitude	0.0064 deg
Longitude	0.153 deg
Time of entry	14.95 s
Atmosphere relative velocity	1.212 m s^{-1}
Atmosphere relative entry angle	0.0205 deg
Heading angle	0.0203 deg

TABLE III

Standard deviation in Orbiter trajectory parameters[a]

Parameter	1-σ uncertainty	1-σ frequency uncertainty (Hz)
r_0	94.32 m	0.000246
ϕ_0	4.712×10^{-5} deg	0.0249
θ_0	6.011×10^{-5} deg	0.000722
$\dot{r}_0$	7.902×10^{-6} km s^{-1}	0.0365
$\dot{\phi}_0$	2.207×10^{-9} deg s^{-1}	0.00125
$\dot{\theta}_0$	5.534×10^{-8} deg s^{-1}	0.0077

[a] At time of Probe entry.

these errors are extremely small. Furthermore, these values are primarily constant errors and so the systematic errors associated with the uncertainties in the Orbiter trajectory would be much smaller (cf. Table I).

Uncertainties in the planet's solid body rotational velocity introduce errors in the absolute zonal wind speeds, i.e., the zero point of the wind speeds. They arise from two sources: the uncertainty in the planet's solid body rotation rate and the uncertainty in the Probe's distance from the planet's center. We assume that the planet's interior rotates at the same rate as the magnetic field. The fractional uncertainty of 10^{-6} in the period of system III (Smith and West, 1982) produces an uncertainty in the planet's rotational velocity of about 1 cm s^{-1}, which is negligible compared to many other error sources.

A larger uncertainty in the value of the rotational velocity results from the uncertainty in the radial distance of the Probe from the center of Jupiter. By combining occultation profiles obtained by the Voyager and Pioneer spacecraft, Lindal *et al.* (1981) have determined the gravitational equipotential surface of Jupiter at the 100 mb level. Based on the goodness of fit of the individual measurements to the proposed zenoid, the uncertainty in the radius of the 100 mb pressure level at the Probe's location will be $\pm$ 4 km. This uncertainty produces a 0.6 m s^{-1} uncertainty in the planet's rotational velocity at this level.

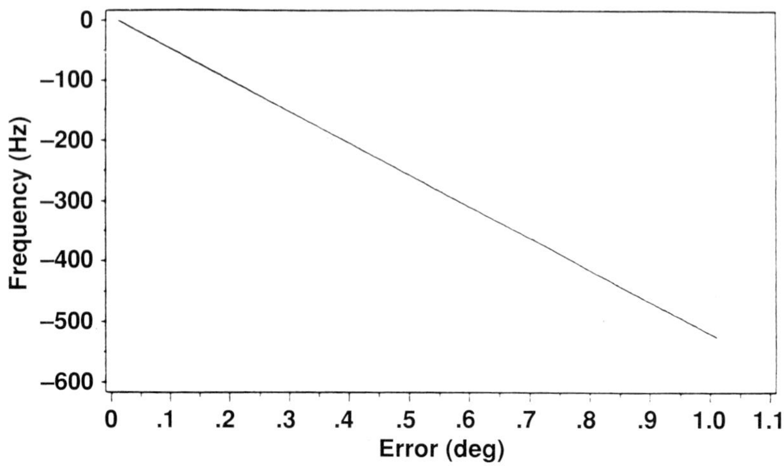

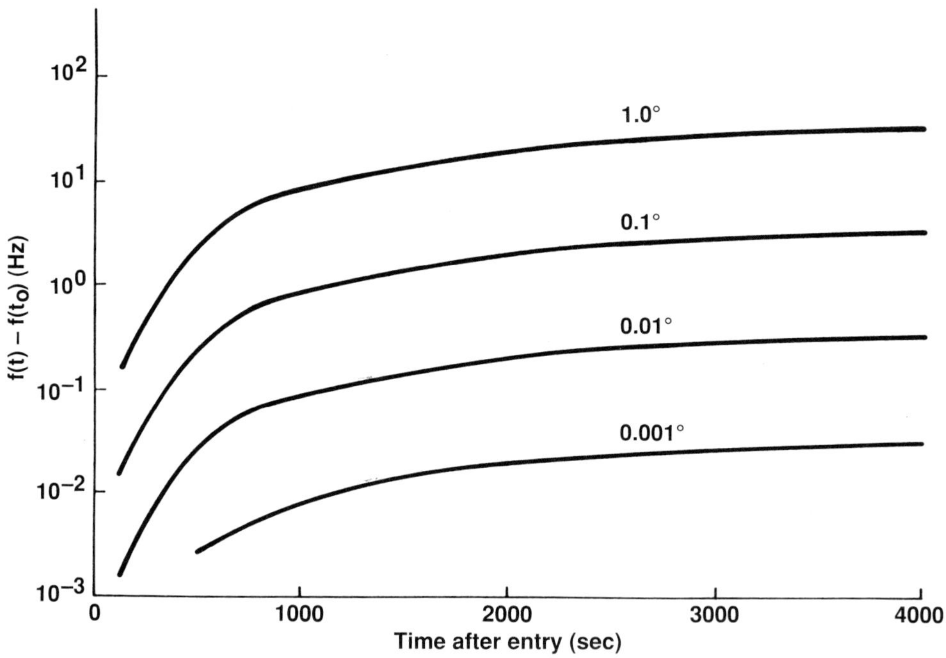

Fig. 4. (a) Change in relay link frequency at 200 s after entry as a function of the offset in the Probe's entry longitude. (b) Change in the differential relay link frequency (instantaneous value – value at the reference time) as a function of time from entry for alternative values of the offset in the Probe's entry longitude.

In addition to this constant error, there will also be a small systematic error in the rotational velocity due to errors in deriving the altitude separation between the 100 mb level and some deeper level, based on the ASI data. By the end of the relay link, the error in this altitude separation may reach about 0.05 km. Such an error produces a systematic error in velocity of about 0.9 cm s^{-1}, which is negligibly small compared to other error sources.

Finally, two small changes in the measured frequency arise from the Probe's spin. First, the deliberate misalignment of the Probe's antenna and its spin axis results in a periodically varying velocity of the antenna along the line of sight and, hence, a periodic modulation of the measured frequency that depends on the spin rate, distance of the antenna from the central axis, and its offset angle. As shown in Table I, a frequency modulation (peak to peak) of about 0.2 Hz results for a nominal spin rate of 5 rpm. The magnitude of this modulation is proportional to the spin rate and has a period equal to the spin rate. Conceivably by appropriately filtering the measured frequency, we may be able to recover the spin rate as a function of time during descent (it will vary slowly). Since the Probe's antenna transmits circularly polarized signals, the transmitted frequency is slightly altered by the antenna's rotation (Paper I). The observed frequency is shifted by an amount equal to the spin rate (0.08 Hz for the nominal rate).

3.2. Oscillator errors

Much effort has been expended in constructing extremely stable oscillators for use on the Probe and Orbiter, in shielding them, and in calibrating their behavior, especially for frequency shifts induced by a variety of environmental factors. As a result, frequency drifts of the USOs on the Probe and Orbiter should not seriously compromise the retrieval of winds from the Doppler data (indeed, this experiment influenced the design of the USOs), but these drifts can still have some effect on the accuracy of the retrieved winds. Here, we first summarize pertinent characteristics of the USOs and then enumerate the drifts they can undergo due to a variety of factors, with emphasis on the largest drifts.

The heart of the USO is an 'SC' cut quartz crystal that is extremely stable in its resonance frequency and that is much less sensitive to *g*-loading, thermal disturbances, and radiation than the 'AT' cut used on the Voyager spacecraft. Furthermore, the Galileo crystals have been radiation hardened by exposure to a one Mrad dose of gamma rays from a cobalt-60 source to minimize their susceptibility to radiation-induced drifts. The housing for the crystal includes a double-insulated temperature control oven to maintain a highly stable operating temperature. The Orbiter USO is identical in terms of the crystal and its housing to the Probe USO.

Various environmental factors and the intrinsic properties of the USOs will cause their frequencies to change by small, but not totally trivial amounts over the course of the relay link. A series of calibration tests on the flight USOs have provided estimates of the magnitudes of these drifts as well as calibration curves that will enable them to be partially modeled, thereby reducing somewhat their impact on the wind recovery. As shown in Table I, the separate and the cumulative errors due to time variations of pressure, temperature, bus voltage, acceleration, and orientation of the magnetic field produce very small systematic errors in the Probe's carrier frequency.

Even in the absence of all known perturbations, the frequency of the oscillators will undergo some small drift in frequency. These effects are referred to as 'aging'. The calibration tests indicate that the USOs undergo secular frequency drifts of 0.6 Hz over a 30-min period, with an uncertainty of 0.14 Hz (Paper I). On much shorter time scales, the frequency experiences random variations of approximately 0.003 Hz ('Allen variance').

Jovian magnetospheric particles can alter the secular drift rate of the frequency of the Probe and Orbiter's USOs, thereby degrading the Doppler wind experiment. Figure 5(a) shows an estimate of the radiation dose rate (mostly bremsstrahlung X-rays) that the Probe's USO will experience as it passes through the inner magnetosphere on its way to the planet. This figure was derived from an empirical model of the Jovian magnetosphere, based on data from the Pioneer and Voyager spacecraft (e.g., Armstrong *et al.*, 1981; Galileo Project, 1982), the nominal Probe trajectory, and the shielding provided by the spacecraft and oscillator houding (P. Garcia, private communication). Since the radiation is highest in the equatorial plane of the magnetic field, which is tilted

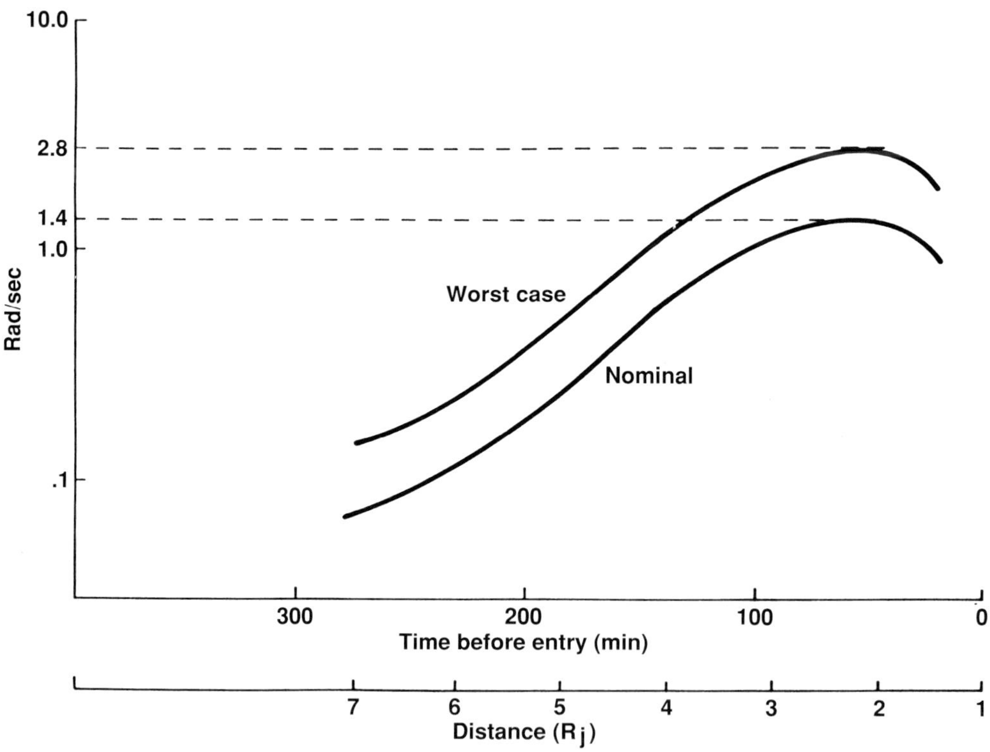

Fig. 5a.

Fig. 5. (a) Nominal and worst case dose rates that will be experienced by the Probe's ultrastable oscillator as a function of time before entry and distance from the center of Jupiter in units of the planet's equatorial radius. These rates include an allowance for shielding by the spacecraft and oscillator housing. (b) Model for the radiation-induced drift rate derived from the calibration test results. (c) Comparison of the predictions of the radiation model (solid curve) with a test result involving a dose rate of 1 rad s^{-1} applied for 1200 s. (d) Predicted radiation-induced drift rate for the Galileo Probe USO as a function of time after entry for the nominal and worst case radiation models of Figure 5(a).

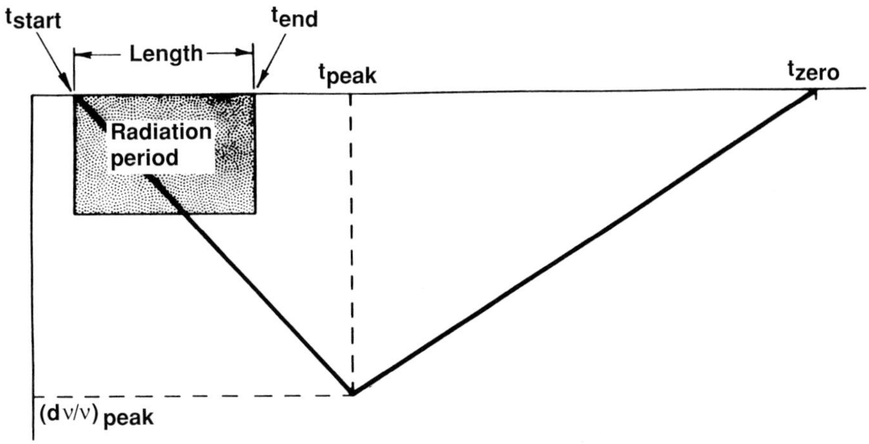

$$t_{end} = t_{start} + \text{length (sec)}$$
$$t_{peak} = t_{end} + 1070/\text{Radrate} - 300 \text{ (sec)}$$
$$t_{zero} = t_{peak} + 5000 \text{ (sec)}$$
$$(d\nu/\nu)_{peak} = -7.5 \times 10^{-10} \times \text{radrate (rad/sec)}$$

Fig. 5b.

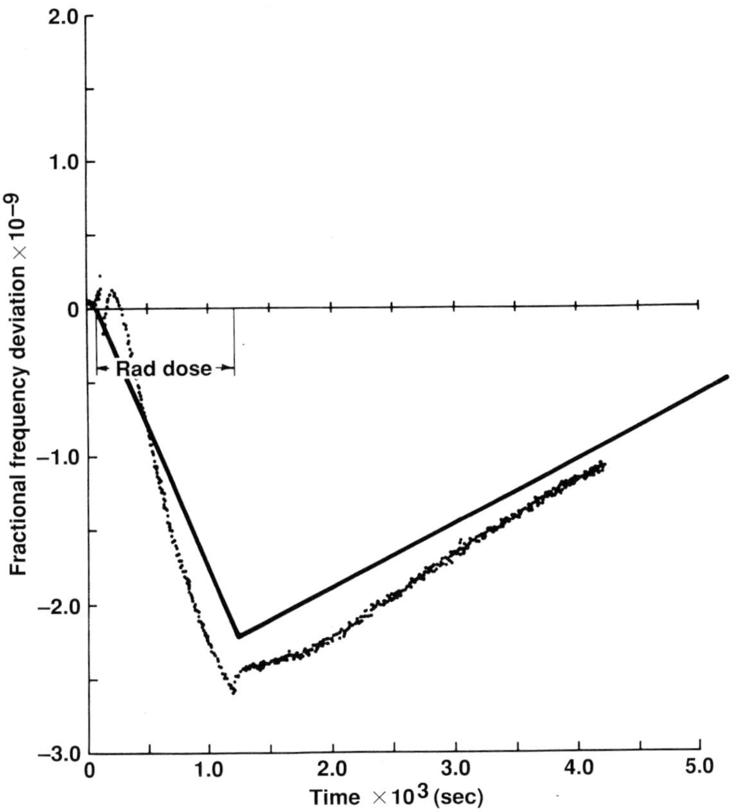

Fig. 5c.

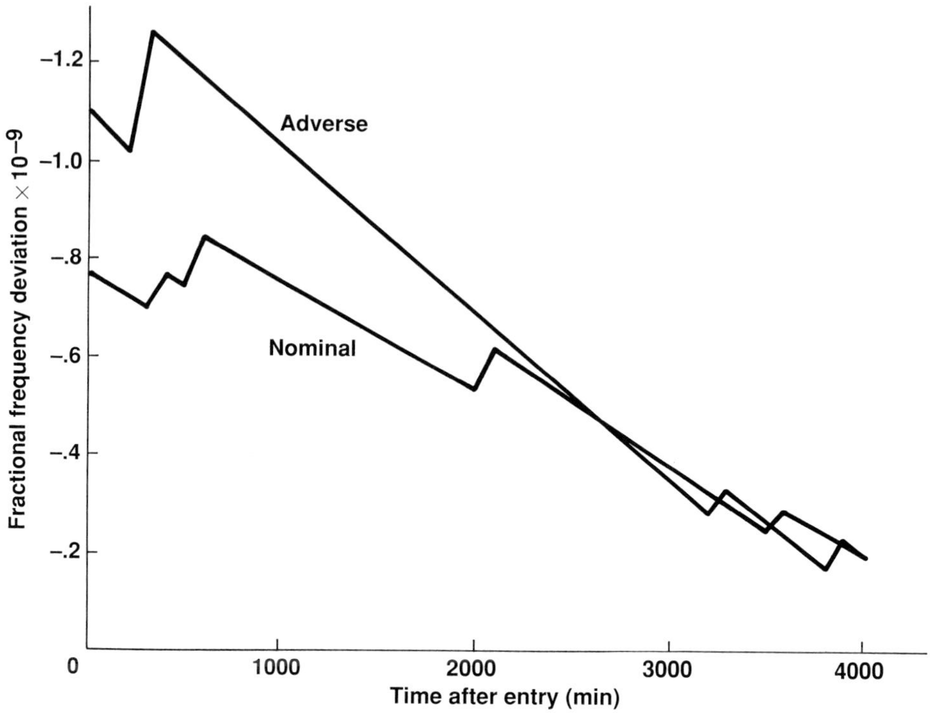

Fig. 5d.

by 11.5° to the rotational equatorial plane, the actual radiation seen by the probe will vary sinusoidally with time. The curve labelled 'nominal' in Figure 5(a) refers to conditions that have been averaged over system III longitude, while 'worst case' shows the dose rate for system III longitudes that place the Probe closest to the magnetic equator.

According to Figure 5(a), the Probe's USO experiences a maximum dose rate of about 1.4 rad s^{-1} at 2 R_J from the planet's center, which occurs about 50 min before entry, where R_J is the planet's equatorial radius. A somewhat smaller, but more time constant dose rate will be experienced by the Orbiter's USO, which will be located at about 4.0 R_J during the relay link.

In 1981, calibration tests were performed on the flight USOs to determine the effect of radiation on their frequency. Of greatest interest was the alteration to the secular drift rate of this frequency. The tests were performed following radiation hardening. They involved exposing the USOs to gamma rays from a cobalt-60 source. In the case of the Probe's USO, the dose rate and the total dosage were varied. In addition, one test involved the use of a time-varying dose rate. Less extensive tests were performed on the Orbiter's USO, in which the dose rate, but not the total dosage was varied. For most of these tests, the dose rates used (0.5 to 3 rad s^{-1}) were comparable to those expected in the Jovian environment.

We have developed an empirical model of the frequency drift induced by radiation by simulating the radiation calibration tests. Figure 5(b) illustrates this model and

Figure 5(c) shows its ability to reproduce one of the tests. In both cases, a constant dose rate was applied for a limited time interval. Clearly, the model approximately reproduces the data.

In actuality, especially for the Probe, the applied dose rates will vary significantly as the Probe approaches the planet (cf. Figure 5(a)). We can apply the model of Figure 5(b) to this situation by appropriately averaging, NOT summing, the response curves produced by a series of discrete and constant dose rates that approximate the actual time-varying behavior. Based on this approach and the dose rates of Figure 5(a), we estimated the radiation induced drifts that the Probe's oscillator will undergo during the relay link, as shown in Figure 5(d). The kinks in the predicted curves are artifacts of our discretizing the dose rate curves for use in our model. Averaged over the relay link, the nominal radiation-induced drift would produce a frequency change of about 0.4 Hz over a 30-min period. The corresponding curve for the 'worst case' dose profile yields a drift that is about a factor of 2 higher than the nominal case near the start of the link and fairly similar to the nominal case near the end of the link.

A somewhat smaller drift will be experienced by the USO on the Orbiter. This USO will experience a more constant radiation dose rate of about 1 rad s^{-1}. Tests on the Orbiter USO suggest that it will experience a radiation induced drift of about 0.07 Hz during the relay link.

3.3. Propagation Effects

The frequency of the Probe's signal, as measured at the Orbiter, can be affected by its passage through the Jovian atmosphere and ionosphere. First, the propagation time is altered by the signal's passage through a refracting medium. More precisely, the phase path length will be retarded by the neutral atmosphere and advanced by the ionosphere. Since the phase path length varies with time due to these effects, the measured frequency will be altered from the value that would be found in the absence of a refracting medium along the path. Second, the direction of the signal is slightly altered by refraction. This induces a small, time-varying, change to the Probe aspect angle for the ray that reaches the Orbiter, thereby altering the Doppler shift produced by the horizontal winds and the planet's rotational velocity. These effects, especially that due to the neutral atmosphere, can be quite large. But, fortunately, they can be allowed for with high precision in the *a posteriori* analysis, since the indices of refraction along the path length will be known quite well.

Based on the formalism developed in Paper I and Atkinson and Spilker (1990) and the characteristics of the Probe's descent profile during the relay link, we find that refraction by the neutral atmosphere can be expected to produce a systematic variation of less than a Hz in the measured frequency over the relay link (Table I). Furthermore, almost all of this variation can be removed *a posteriori* by using the highly accurate profiles of pressure and temperature, obtained by the ASI experiment, in conjunction with compositional information obtained from several Probe experiments.

In a similar vein, frequency drifts due to passage through the Jovian ionosphere can be estimated from the formalism of Paper I and Atkinson and Spilker (1990). Using electron density profiles derived from past spacecraft occultation experiments, we find that the frequency shifts are quite small, even *a priori* (cf. Table I). Even smaller shifts occur during the passage of the signal through the Jovian magnetosphere.

3.4. Miscellaneous effects

The measured frequency is also affected by the gravitational red shift of general relativity, the time dilation effect of special relativity, and other higher-order terms that appear when the exact form of the Doppler shift equation is used. Here, we simply provide estimates of the magnitudes of these effects to assess the importance of including them in the final analyses. All of these can be treated quite accurately and so need not generate significant errors in the recovered wind.

The gravitational red shift produces a large constant frequency shift and a very small systematic one, both of which can be reduced to negligible amounts by appropriately calculating them (Paper I; cf. Table I). Errors introduced by neglecting higher-order terms in the equation of special relativity (cf. Equation (3)) and by truncating the Taylor series expansion of the Doppler frequency residuals (cf. Equation (7)) result in systematic frequency errors on the order of 10^{-2} Hz (Paper I). Thus, they can be safely neglected.

4. Simulations

In this section, we use the linear least-squares algorithm developed in the approach section to recover wind profiles from simulated data. The simulated data were generated for each of the four wind profiles of Figure 1. In most cases, we included the more prominent errors of Table I to make the simulations realistic. These simulations are directed at addressing and answering several basic questions about our approach and the feasibility of recovering a wind profile from the Galileo Probe's telemetry: Are Legendre polynomials a useful set of basis functions for the zonal winds of Jupiter? How would we detect and analyze regions of sudden shears? What accuracy in the recovered wind profile is possible, given plausible error sources? What are the limitations on the vertical resolution of the recovered profile?

We carried out simulations for the nominal Probe and Orbiter trajectories for a time interval ranging from 200 to 300 s after Probe 'entry'. Probe entry occurs when it reaches an altitude of 450 km above the 1 bar pressure level. During the time interval of the simulations, the Probe traverses pressure levels ranging from 800 mb to 25 bars. In reality, the relay link could begin as early as 120 s after entry, when the Probe is at the 100 mb level. The conclusions drawn based on the chosen period of simulation are equally valid for other possible periods.

Figure 2 shows the variation of the Probe aspect angle, ψ, and azimuth angle, α, with time from entry. These angles define the relationship between the zonal direction and the line of sight and hence the projection of the zonal wind speed on the line of sight. At the start of the simulated relay link, ψ has a value of 6°; it reaches a minimum value of 1.6° at 21^{m}, the time that the Orbiter crosses the Probe's meridian; and it increases to a value of nearly 10° by the end of the link. Throughout almost all this time, α has a value close to 0°, since the Probe and Orbiter have nearly coplanar locations, but it goes through 90° at the time that the Orbiter crosses the Probe's meridian.

The first step in performing a simulation is to construct two frequency profiles. We refer to the first profile of frequency as a function of time as a 'true' profile, in the sense that it corresponds to what would really be measured and conveyed from the Orbiter to the Earth for analysis. It is generated by including the influence of the candidate wind profile on the measured frequency, by using the 'real' trajectories, and by including a variety of effects that would change the USOs' frequencies from their ground-based measured values, including oscillator aging, probe spin, atmospheric turbulence, measurement errors at the Orbiter, and the gravitational red shift. The second profile, the 'nominal' profile, is obtained by neglecting winds (they are not known *a priori*), using the ground-based frequency of the oscillators, and trajectory information that differs from the 'real' trajectories by 1σ errors based on Table I. Table IV summarizes the errors introduced into the two profiles that were used in the simulations discussed below.

TABLE IV

Doppler wind recovery simulation errors and uncertainties

Parameter	Magnitude	Reference
Probe USO zero-point frequency	500 Hz	RLIT report[a]
Probe spin	20 rpm	RLIT report
Probe buffetting	$0.36 \cos(24t)$[b]	RLIT report
Atmospheric turbulence	$0.08 \cos(5.6t)$[b]	RLIT report
Fractional Probe USO drift	1.0×10^{-9}/30 min	Paper I
Probe oscillator Allan variance	$\sigma = 2.0 \times 10^{-12}\ \mathrm{s}^{-1}$	Paper I
Gravitational redshift	– 20.53 to – 20.73 Hz	Paper I
Magnetometer boom interference	$0.05 \sin(7.9t)$[b]	RLIT report
Frequency measurement error	$\sigma = 0.0523$ Hz	Paper I
Probe longitude	0.231 deg	Paper I
Probe latitude	0.006 deg	Paper I
Probe radius	21.4 km	Paper I
Orbiter longitude	4.7×10^{-5} deg	Paper I
Orbiter latitude	6.0×10^{-5} deg	Paper I
Orbiter radius	0.094 km	Paper I
Probe descent velocity	$0.0043 V_d$[c]	Paper I

[a] Galileo Probe–Orbiter Relay Link Integration Team Report (Bright, 1984).
[b] Time after Probe entry (s).
[c] Probe descent velocity m s^{-1}.

Once the two frequency profiles have been generated, we subtracted the nominal profile from the true one to generate frequency residuals. Next, we subtracted from the residuals at all times t the residuals at the reference time, t_r, which we take as the initial time of the link. These operations produce the Doppler data, Δf_{Dop}. We apply the least-squares algorithm to this Doppler data to recover the zonal wind speed profile and compare the recovered profile with the 'true' profile to assess the accuracy of the recovery. This assessment is done by plotting the two wind profiles as well as evaluating their root mean square (r.m.s.) difference and their mean difference. The latter provides a measure of the systematic offset between the two profiles. In order to keep the computational time at a reasonable level, we generated data every 10 s during the relay link (280 samples altogether).

It might seem that the time of overflight offers a better choice for the reference time, t_r, since the zonal winds do not affect the measured frequency at overflight. Indeed, such a choice might neatly yield the Probe USO's zero-point frequency, if the meridional and vertical wind speeds are sufficiently small. Unfortunately, the 0.23° uncertainty in the Probe's descent longitude, when combined with the planet's rotational speed of 10 km s^{-1}, leads to a 26 m s^{-1} error in the Probe–Orbiter range rate. This offset is essentially constant throughout the relay link so that it does not provide a marker of the actual time of overflight. Thus, the uncertainty in the Probe's position prevents a useful determination of the USO's zero-point frequency from being made. But, the near constancy of the resulting error permits its impact to be substantially reduced by using frequency differences as the primary data to be analyzed.

We first assess the adequacy of a sum of Legendre polynomials to represent the wind profiles by determining the r.m.s. error in the recovered wind profile as a function of the order of the polynomial expansion for the four profiles of Figure 1. For all four wind profiles of Figure 1, an asymptotic error of a few m s^{-1} is closely approached by expansions of order 5 and larger (Paper I). Below, we perform further simulations using an expansion of order 5. We conclude that the Legendre polynomial expansion provides a good representation of the wind profiles of Figure 1, since only a small number of terms are needed and since the r.m.s. error is comparable to that expected from the error sources of Table IV.

We next assess the influence of random errors in the measured frequency on the accuracy of the recovered wind profile. For this purpose, we used covariance analysis. The mathematical procedure used to relate variances in the measured frequency to variances in the recovered wind speed is given in Paper I. Using the random errors given in Table I for the Allen variances of the Probe and Orbiter USOs and for the digitization error, we find that the total random error in the measured frequency is approximately 0.053 Hz. The corresponding random errors in the recovered winds at several times during the relay link are given in Table V for the latent heat wind profile of Figure 1. Similar results pertain to the other wind profiles.

The variances in the wind speed shown in Table V are less than 1 m s^{-1}. Thus, our least-squares formalism is stable for reasonable choices of random errors and the implied variances in wind speed will not significantly degrade its recovery. The application of this formalism to our standard model in which the entry longitude of the Probe was added as an unknown showed that it was not possible to also derive this key parameter from the Doppler data: very large variances in wind speed were obtained in this case.

We next present the results of full-up simulations that incorporate the ensemble of constant, random, and systematic errors given in Table IV. Figures 6(a), 7, 8, and 9

TABLE V

1-σ deviation in recovered winds

Time after entry (s)	σ_{v_h} [a, b] (m s^{-1})
200.0	0.088
1000.0	0.264
2000.0	0.384
3000.0	0.474

[a] 1σ-deviation in horizontal wind velocity at Probe location.
[b] Includes Probe and Orbiter USO Allan variances, and frequency measurement error.

show comparisons of the recovered wind profiles with the 'true' profiles for the four standard wind models. Table VI summarizes the r.m.s. and mean wind errors for these 4 simulations. Finally, Figure 6(b) shows the corresponding least-squares residuals as a function of pressure in the atmosphere for the first of these recoveries. Similar residuals pertain to the other cases. The frequency residuals are the instantaneous differences between the frequencies derived from the recovered winds and the 'true' values. (In both cases, the frequency at the start of the relay link has been subtracted out.) As anticipated earlier in the paper, it is possible to recover not only the shape of the wind profiles, but also their absolute values. The residual profiles do not show any large biases over limited regions of pressure, indicating that the least-squares algorithm is working well.

To evaluate the relative importance of the various error sources of Table IV in affecting the accuracy of the wind recovery and the possible impacts of sources not considered in this table, we have carried out a series of sensitivity studies. For this purpose, we used the latent heat wind profile as a base profile. As a standard of comparison, we first performed a wind recovery with all the errors set equal to 0. As shown in Table VI, the recovered winds are not in exact agreement with the true ones. This difference is attributable to a combination of the truncation of the polynomial expansion at order 5 and the finite number of data points used in our analysis.

We next conducted several numerical experiments in which alternately only errors due to frequency effects, Probe descent velocity, and Probe entry longitude were considered. These three types of errors are the most significant ones, according to Table I. As shown in Table VI, the error in entry longitude has the biggest impact on the accuracy of the wind recovery.

The large impact of an error in Probe longitude on the accuracy of the recovered winds prompted us to evaluate the errors in these winds that resulted when the errors in Probe entry longitude and latitude were systematically varied. According to the results shown in Table VI, both the r.m.s. error and the mean error increase significantly, but not catastrophically, as the errors in these two coordinates are separately increased.

So far, we have considered only the nominal wind profiles of Figure 1, all of which

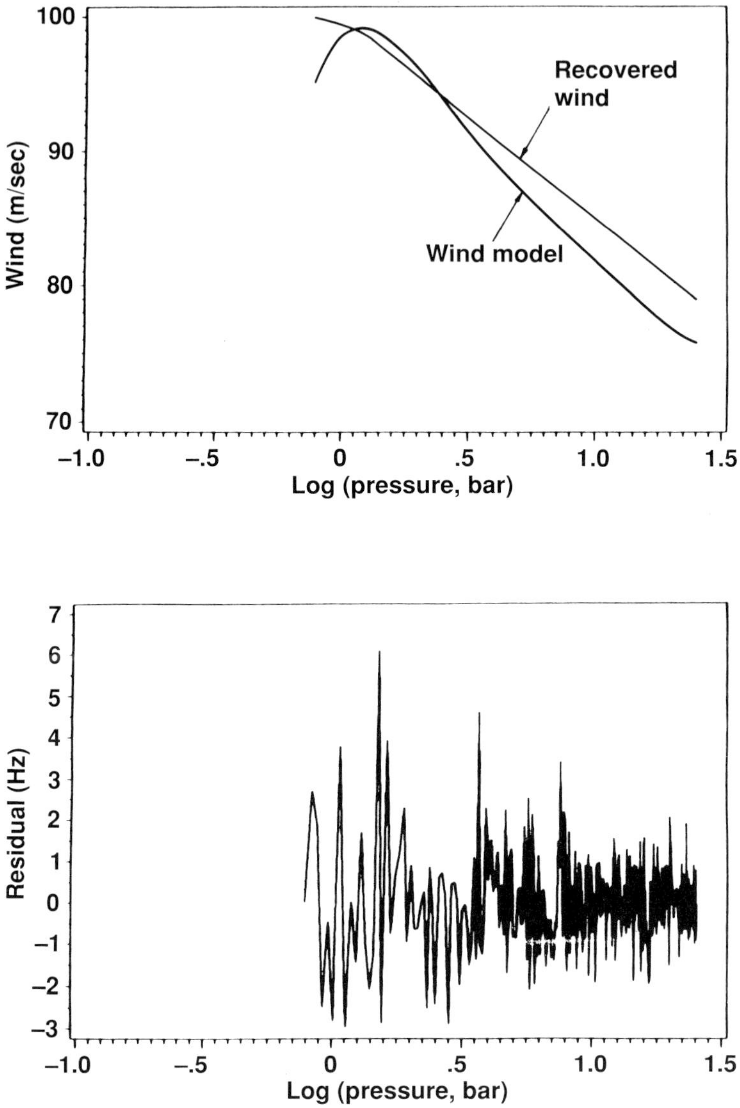

Fig. 6. (a) Comparison of the retrieved wind profile with the actual profile of the zonal wind speed of the internal heat wind model for simulations in which a nominal set of errors was used. (b) Residuals as a function of pressure for the retrieval of (a).

vary comparatively slowly with pressure. Conceivably, there could be localized regions of the Jovian atmosphere where much steeper shears exist. To simulate this possibility, we added a shear to the latent heat wind profile. The shear region was placed in the 2.3 to 4.4 bar location, which the Probe traverses in about 250 s. Thus, about 25 data points out of a total of 280 were affected by the shear. Figures 10(a) and 10(b) show the retrieval and residuals obtained for a case in which a shear of several tens of m s^{-1}

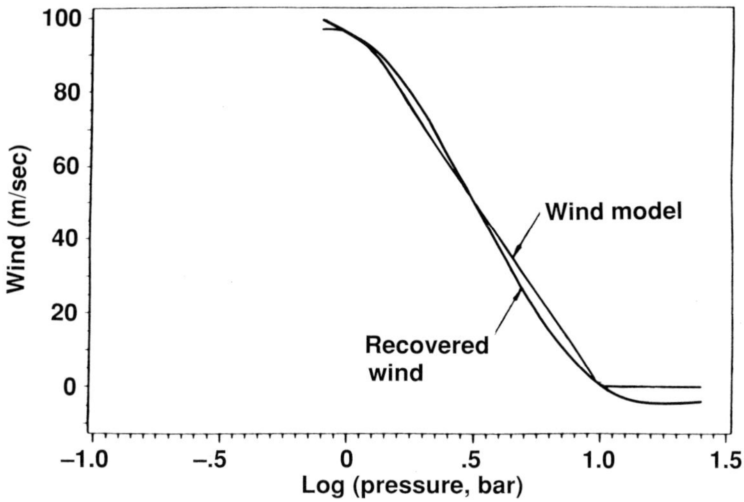

Fig. 7. Same as Figure 6(a) for the latent heat wind model.

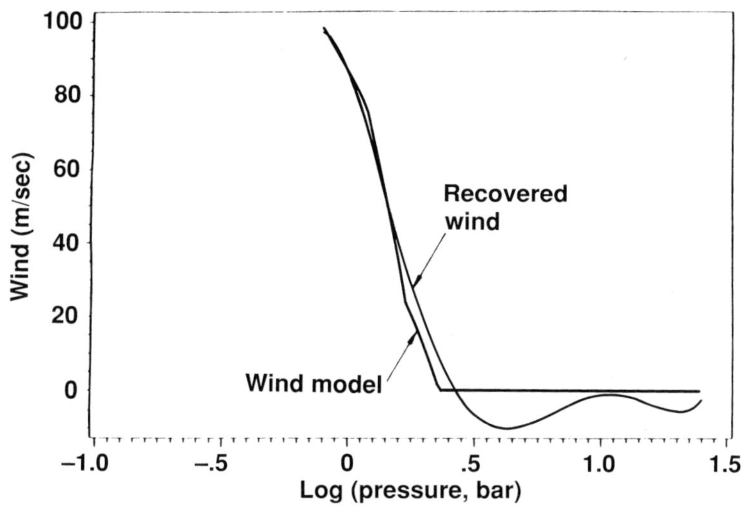

Fig. 8. Same as Figure 6(a) for the solar energy deposition wind model.

was imposed on the nominal wind profile (wind model of Figure 10(a)). The corresponding r.m.s. and mean errors are given in Table VI. Even in this extreme case a reasonably accurate recovery is possible, except of course in the shear zone itself. Furthermore, the residuals show an anomalous behavior in the region of the shear. In such a case, we will use a different type of analysis than the one used for the large-scale winds, one similar to the classical Doppler method. We are currently developing the protocol to recover both the shear and the large-scale wind profile in an iterative fashion.

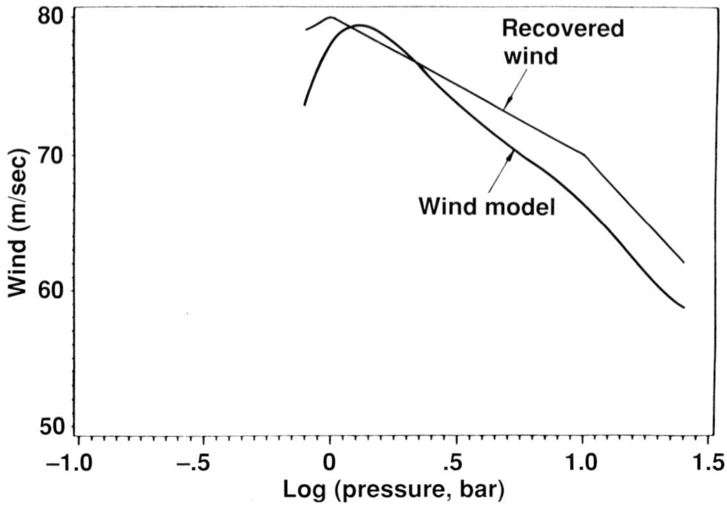

Fig. 9. Same as Figure 6(a) for the ortho/para hydrogen conversion wind model.

Finally, we considered the impact on the recovery of a time-independent meridional wind. A constant 10 m s^{-1} north–south wind had a negligible impact on the errors of the recovered wind, while even a 100 m s^{-1} north–south wind did not increase the r.m.s.

TABLE VI

Errors in the wind recoveries

Model	Input errors	r.m.s. errors (m s^{-1})	Mean errors (m s^{-1})
Latent heat	None	1.52	−0.03
Internal heat	Nominal [a]	2.92	−2.65
Latent heat	Nominal	3.53	−2.70
Solar energy	Nominal	4.86	−3.33
Ortho/para	Nominal	3.14	−2.91
Ortho/para	Only frequency	1.80	−0.99
Ortho/para	Only desc. vel.	1.72	0.40
Ortho/para	Only entry long.	2.41	−1.82
Ortho/para	Δ lat. = 0.05 deg	2.27	1.21
Ortho/para	Δ lat. = 0.10 deg	2.73	1.93
Ortho/para	Δ lat. = 0.20 deg	3.89	3.38
Ortho/para	Δ lat. = 0.50 deg	8.02	7.78
Ortho/para	Δ long. = 0.05 deg	1.85	0.09
Ortho/para	Δ long. = 0.10 deg	2.05	1.08
Ortho/para	Δ long. = 0.20 deg	3.53	1.55
Ortho/para	Δ long. = 0.50 deg	3.84	3.24
Latent plus shear	Nominal	6.46	−2.72
Latent plus meridional (10 m s^{-1})	Nominal	3.54	−2.63
Latent plus meridional (100 m s^{-1})	Nominal	4.42	−4.02

[a] Includes frequency errors, and uncertainties in Probe and Orbiter trajectories as listed in Table IV.

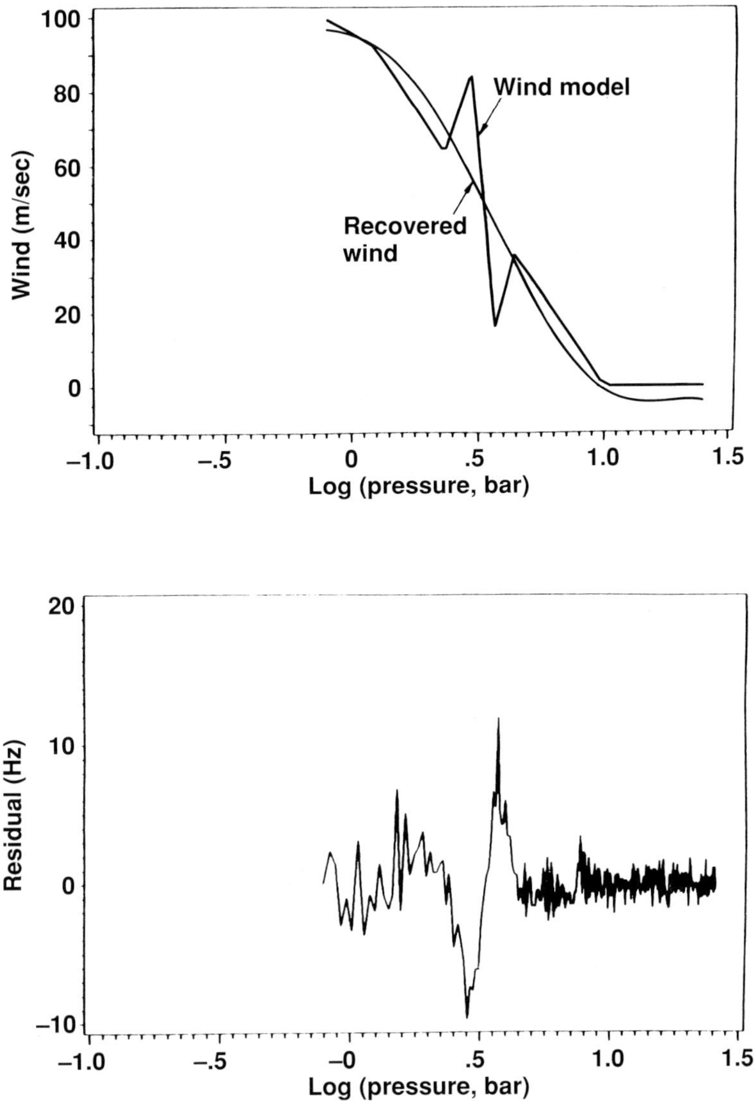

Fig. 10a–b. Same as Figures 6 (a) and 6(b) for the latent heat wind model with a strong shear added to it in a localized region of the atmosphere.

and mean errors by very much (Table VI). This insensitivity is due primarily to the small values of α during much of the descent sequence, which minimizes the projection of the meridional wind on the line of sight. In addition, the recovery is not too sensitive to the latitudinal location of the Probe.

The vertical resolution of the retrieved wind profile is fundamentally limited by the finite response time of the Probe to changes in the zonal wind speed. According to the derivations given in Seiff *et al.* (1980) and Paper I, the Probe's zonal wind speed adjusts to a change in the atmospheric value on a time-scale $\tau_h = w_p/g$, where w_p is the vertical descent speed of the Probe and g is the acceleration of gravity. During this characteristic time scale, the Probe moves vertically a distance, $l_{\min} = w_p \tau_h$.

Table VII gives estimates of the Probe's descent velocity, response time τ_h, and minimum vertical distance, $l_{\min}$, for a range of positions in the Jovian atmosphere. As indicated above, the relay link will begin when the Probe is at the several hundred mb level and end when the probe is at about the 25 bar level. According to Table VII, the

TABLE VII

Probe response time characteristics

Pressure (bar)	Density (kg m^{-3})	w_p [a] (m s^{-1})	τ_h [b] (s)	$l_{\min}$ [c] (m)
0.01	0.0019	770.55	32.72	25212.4
0.1	0.0244	213.66	9.20	1965.7
0.5	0.1020	104.50	4.50	470.2
1.0	0.1650	82.26	3.54	291.2
5.0	0.4990	47.25	2.03	95.9
10.0	0.8110	37.07	1.60	59.3
20.0	1.3190	29.06	1.25	36.3

[a] Probe terminal velocity.
[b] Time required for Probe to recover $1 - e^{-1}$ of step wind gust.
[c] Minimum detectable wind structure, equal to $w_p \times \tau_h$.

Probe adjusts to changes in the zonal wind speed on time scales ranging from almost 10 s at the start of the link to slightly more than 1 s at its end. The corresponding minimum distances range from on the order of 1 km to several tens of meters. Thus, the Probe responds quickly enough to changes in the zonal wind speed so that good vertical resolution is possible during the entire relay link. Near the end of the relay link the response time approaches the sampling time of $\frac{2}{3}$ s that separates successive measurements of the Probe's frequency made by the Orbiter's relay radio system.

5. Conclusions

We have developed a practical approach for estimating the zonal wind speed profile within Jupiter's upper troposphere from its influence on the frequency of the Galileo Probe's telemetry signal. This was achieved by linearizing the relativistic Doppler shift equation about a state of zero wind speed, expanding the zonal wind speed in a Legendre polynomial series, and reducing the resulting equations to a linear, least-squares problem. Since the instantaneous Doppler shift depends on both the instantaneous zonal wind speed and a time integral of this speed, it is possible to recover not only the vertical wind shear, but also the absolute wind velocity *without* an accurate knowledge of the USOs' zero-point frequencies.

There are a very large number of factors that can introduce errors into the wind profile retrieved from the Probe's Doppler-shifted frequency. Fortunately, the collective impact of these error sources appears to be sufficiently mild so that the uncertainty in the retrieved winds should be only a few m s^{-1}. The largest error source is the uncertainty in the Probe's entry location. Consequently, it is essential that the Orbiter's trajectory be carefully monitored before, during, and after Probe release to provide as accurate a determination of the Probe's trajectory subsequent to release and hence as accurate an estimate of its entry location as possible. Other potentially major error sources include uncertainties in the descent velocity and frequency errors, especially those produced by systematic drifts (e.g., radiation-induced drifts and aging). Finally, random variations in frequency can produce r.m.s. errors in the wind speeds of several tenths of a m s^{-1}, as determined by co-variance analysis.

If the error sources have magnitudes comparable to or smaller than the ones given here, the retrieved wind speed profiles can be used for a number of important purposes. These include placing a key constraint on the basic drive for the atmospheric circulation as well as anchoring winds derived from orbiter images to well-defined vertical positions in the atmosphere. With regard to the first of these objectives, it might seem a little naive to expect to learn something fundamental about the atmospheric circulation from a single vertical profile. Certainly, the retrieved profile will be influenced by eddy motions on a variety of scales and thus will not be an accurate representation of the time and longitudinal average profile at the entry latitude. However, experience with Venus indicates that the time and zonally-averaged zonal winds are the dominant component of zonal winds measured at a single entry site (Councelman *et al.*, 1980). Finally, as mentioned earlier, data taken by a variety of Galileo experiments should help to flesh out the various wind models of Figure 1, leading to more precise predictions that can be compared to and, hopefully, distinguished among by the wind profile retrieved by the Galileo Doppler wind experiment.

References

Armstrong, T. P., Paonessa, M. T., Brandon, S. T., Krimigis, S. M., and Lanzerotti, L. J.: 1981, *J. Geophys. Res.* **86**, 8343.

Atkinson, D. H.: 1989, 'Measurement of Planetary Wind Fields by Doppler Monitoring of an Atmospheric Entry Vehicle', Ph.D. Thesis, Washington State University, Pullman, WA.

Atkinson, D. H. and Spiker, T. R.: 1991, *Radio Sci.* (in press).

Born, M. and Wolf, E.: 1964, *Principles of Optics*, MacMillan Co., New York.

Budden, K. G.: 1961, *Radio Waves in the Ionosphere*, Cambridge University Press, Cambridge.

Busse, F. H.: 1976, *Icarus* **29**, 255.

Condran, L. D.: 1983, *Final Test Report for Galileo Probe System Test Drop No. 2*, General Electric Report 83 SDS 2148.

Conrath, B. J. and Gierasch, P. J.: 1984, *Icarus* **57**, 184.

Conrath, B. J., Flasar, F. M., Pirraglia, J. A., Gierasch, P. J., and Hunt, G. E.: 1981, *J. Geophys. Res.* **86**, 8769.

Councelman, C. C. III, Gourevitch, S. A., King, R. W., Loriot, G. B., and Ginsberg, E. S.: 1980, *J. Geophys. Res.* **85**, 8026.

Cunningham, C. C., Hunten, D. M., and Tomasko, M. G.: 1988, *Icarus* **75**, 324.

Galileo Project Document JP-590: 1982.

Gierasch, P. J.: 1976, *Icarus* **29**, 445.

Gierasch, P. J.: 1983, *Science* **219**, 847.
Golub, G. H. and Van Loan, C. F.: 1983, *Matrix Computations*, Johns Hopkins University Press, Baltimore, MD.
Hanel, R. A., Conrath, B. J., Herath, L. W., Kunde, V. G., and Pirraglia, J. A.: 1981, *J. Geophys. Res.* **86**, 8705.
Holton, J. R.: 1979, *An Introduction to Dynamic Meteorology*, Academic Press, New York.
Hunten, D. M., Tomasko, M., and Wallace, L.: 1980, *Icarus* **43**, 143.
Ingersoll, A. P. and Pollard, D.: 1982, *Icarus* **52**, 62.
Ingersoll, A. P. and Porco, C. C.: 1978, *Icarus* **35**, 27.
Ingersoll, A. P., Beebe, R. F., Mitchell, J. L., Garneau, G. W., Yagi, G. M., and Muller, J. P.: 1981, *J. Geophys. Res.* **86**, 8733.
Kerzhanovich, V. V., Gotlib, V. M., Chetyrkin, N. V., and Andreev, B. N.: 1969, *Kosmicheskie Issledovaniya* **7**, 592 (translated).
Kerzhanovich, V. V., Marov, M. Ya., and Rozhdestvensky, M. K.: 1972, *Icarus* **17**, 659.
Limaye, S. S.: 1985, *Adv. Space Res.* **5**, 51.
Lindal, G. F. *et al.*: 1981, *J. Geophys. Res.* **86**, 8721.
Marov, M. Ya. *et al.*: 1973, *Icarus* **20**, 407.
Massie, S. T. and Hunten, D. M.: 1982, *Icarus* **49**, 213.
Orton, G. S.: 1981, JPL Report 1625–125.
Pirraglia, J. A.: 1984, *Icarus* **59**, 169.
Preston, R. A. *et al.*: 1986, *Science* **231**, 1414.
Rossow, W. B., Del Genio, A. D., Limaye, S. S., Travis, L. D., and Stone, P. H.: 1980, *J. Geophys. Res.* **85**, 8107.
Sagdeev, R. Z. *et al.*: 1986, *Science* **231**, 1411.
Schubert, G. *et al.*: 1980, *J. Geophys. Res.* **85**, 8007.
Seiff, A. and Knight, T. C. D.: 1992, *Space Sci. Rev.* **60**, 203 (this issue).
Seiff, A. *et al.*: 1980, *J. Geophys. Res.* **85**, 7903.
Smith, R. E. and West, G. S.: 1982, *Space and Planetary Environmental Criteria Guidelines for Use in Spacecraft Development*, Vol. II, NASA Technical Memorandum 82478.
Stone, P. H.: 1967, *J. Atmospheric Sci.* **24**, 642.
Taylor, F. W. *et al.*: 1980, *J. Geophys. Res.* **85**, 7963.
West, R. A., Strobel, D. F., and Tomasko, M. G.: 1986, *Icarus* **65**, 161.

GALILEO PROBE NEPHELOMETER EXPERIMENT

B. RAGENT

San Jose State University Foundation, San Jose, CA 95192–0139, U.S.A.

C. A. PRIVETTE†

NASA Ames Research Center, Moffett Field, CA 94035, U.S.A.

and

P. AVRIN, J. G. WARING, C. E. CARLSTON, T. C. D. KNIGHT, and J. P. MARTIN

Martin Marietta Astronautics Group, Denver, CO 80201, U.S.A.

Abstract. The objective of the Nephelometer Experient aboard the Probe of the Galileo mission is to explore the vertical structure and microphysical properties of the clouds and hazes in the atmosphere of Jupiter along the descent trajectory of the Probe (nominally from 0.1 to >10 bars). The measurements, to be obtained at least every kilometer of the Probe descent, will provide the bases for inferences of mean particle sizes, particle number densities (and hence, opacities, mass densities, and columnar mass loading) and, for non-highly absorbing particles, for distinguishing between solid and liquid particles. These quantities, especially the location of the cloud bases, together with other quantities derived from this and other experiments aboard the Probe, will not only yield strong evidence for the composition of the particles, but, using thermochemical models, for species abundances as well. The measurements in the upper troposphere will provide 'ground truth' data for correlation with remote sensing instruments aboard the Galileo Orbiter vehicle. The instrument is carefully designed and calibrated to measure the light scattering properties of the particulate clouds and hazes at scattering angles of 5.8°, 16°, 40°, 70°, and 178°. The measurement sensitivity and accuracy is such that useful estimates of mean particle radii in the range from about 0.2 to 20 μ can be inferred. The instrument will detect the presence of typical cloud particles with radii of about 1.0 μ, or larger, at concentrations of less than 1 cm^3.

1. Introduction

The Galileo Jupiter mission atmospheric Probe (O'Neil and Mitchell, 1983; Givens *et al.*, 1983) includes a Nephelometer as part of the Probe instrument complement. The scientific objective of the Nephelometer experiment* is to explore the vertical structure and microphysical properties of the clouds and hazes in the Jovian atmosphere. The experiment will provide evidence for the existence or absence of particulates along the descent trajectory of the Probe (nominally from altitudes corresponding to ambient pressures of 0.1 bars to greater than 10 bars). It will yield data that will be used to infer the properties of these particulates, including mean size and particle number density, and to differentiate between liquid and solid particles (i.e., spherical and non-spherical).

† Deceased.

* The nephelometer experiment team includes Boris Ragent, San Jose State University Foundation, Principal Investigator, and Co-Investigators Philip Avrin, Martin Marietta Astronautics Group, Jacques E. Blamont, Service d'Aeronomie du C.N.R.S., David Colburn, San Jose State University Foundation, Gerald Grams, Georgia Institute of Technology, and James Pollack, NASA/Ames Research Center.

Space Science Reviews **60:** 179–201, 1992.

In conjunction with data from other Galileo Jupiter Probe and Orbiter experiments, the Nephelometer measurements will be used to confirm the consistency of predictions of the effective particle indices of refraction, including absorption. Approximate cloud opacities at a wavelength of 0.9 μ, a measure of the local cloud mass density along the descent trajectory, and the total columnar mass loading in each cloud will be obtained. All of the above results, together with the cloud base locations, that will be well documented by this experiment, will not only yield strong evidence for inferring the gross composition of the particles, but, by comparison with thermochemical models, will constrain the particles' species abundances. Finally, the data will provide *in situ* information about particles in the upper levels of the troposphere at the Probe entry site to provide 'ground truth' data for correlation with remote sensing experiments aboard the Galileo Jupter Orbiter vehicle.

The launch of the Galileo spacecraft took place on October 18, 1989, and the Probe will enter the Jovian atmosphere at a latitude of 6.5 N at about local Jovian sunset (nominally at 22:11 UT) on December 7, 1995. It is expected that, after deployment, Nephelometer measurements will be obtained, at a minimum, every kilometer for at least 100 km or more during the course of the Probe descent. In the following sections we summarize the presently held views of the clouds and hazes of Jupiter and what we may expect to learn from our experiment, describe the design of the instrument, and discuss its calibration, testing, and expected performance.

2. Background

Many of the aspects of the nature, structure, and variability of the clouds and hazes in the troposphere of Jupiter at pressures less than 5 bars are subjects of very active ongoing discussion in the literature. (See, for example, West *et al.*, 1986, who have recently provided an excellent review of the status of knowledge of the Jovian clouds and aerosols; Carlson *et al.*, 1987, 1988, 1990; Cunningham *et al.*, 1988; West, 1988; West *et al.*, 1989; Del Genio and McGrattan, 1990). The foundation for ideas of the aerosol and cloud structure comes from equilibrium thermochemical models that depend on the abundances of relevant species and the atmospheric structure functions (atmospheric temperature versus pressure curves). By assuming that pertinent species (N, O, and S) are present with solar abundances, the presence of three cloud layers in the upper troposphere are predicted by such models (cf. Lewis, 1969a, b; Weidenschilling and Lewis, 1973). NH_3 condenses at the altitude corresponding to a pressure near 0.7 bars, mixed NH_3 and H_2S and/or ammonia polysulfides are condensed at about 2 bars, and H_2O condenses at about 5 bars. The precise condensed particle compositions and pressure levels of these cloud bases are dependent on the actual local species abundances and temperature structure.

Model calculations based on microwave wavelength measurements have determined that NH_3 is undersaturated above the NH_3 cloud base predicted for solar abundance and at high levels (0.3–0.5 bars), depleted relative to solar abundance at 0.5 to 2 bars, and superabundant at higher pressures (de Pater, 1986). The depletions appear to be due to chemistry, cloud physics, and atmospheric dynamics. Taking these into account, some investigators currently believe that the NH_3 abundance above the clouds may be essentially solar (Carlson, 1990, private communication), so that the abundance at lower levels in the clouds is solar or larger, although there is some question about the very deep NH_3 mixing ratio. Most investigators agree that the current experimental data do not yield reliable information on sulfur abundances and that, therefore, they do not provide constraints on the NH_4SH cloud, even on the question of its existence. From the analysis of airborne telescope and Voyager IRIS spectra Bjoraker *et al.* (1986), inferred an H_2O abundance approximately 50 times less than solar abundance in the 2 to 6 bar region of Jupiter. Such an abundance would shift the predicted water cloud to a level much higher than that predicted for a solar abundance atmosphere, perhaps up to the 2 bar level. However, Carlson *et al.* (1990) have reanalyzed the same experimental data (including cloud scattering that had not been considered in the earlier analyses) and conclude instead that the H_2O abundance is solar or greater, which would place the cloud at a level close to the original level discussed above.

The equilibrium thermochemical models alone cannot yield realistic descriptions of the vertical or horizontal cloud structures, or predict details such as particle sizes and number densities. They cannot give the details of particle formation or destruction, or discuss the apparently large spatial and temporal variations. These characteristics are intimately related to the local and regional atmospheric dynamics, precipitation and the concentration of condensation nuclei, and involve the cloud microphysics. For example, suggestions have been made that molecular weight differentiation and latent heat release in condensible species may control the dynamics in the cloud regions (Barcilon and Gierasch, 1970; Gierasch, 1976; Ingersoll, 1976; Gierasch and Conrath, 1985, 1987), and that moist convection processes are important (Stoker, 1986; Lunine and Hunten, 1987; Del Genio and McGrattan, 1990). Due to the large range of length scales involved and the lack of fundamental information such as the concentration of condensation nuclei, the prediction from first principles of the detailed cloud properties appears to be well beyond the capability of existing dynamical models in the absence of *in situ* measurements (Gierasch, 1988). The comparison of remote sensing observations of scattered sunlight and thermal radiation with radiative transfer calculations provide several constraints on the cloud structures in Jupiter's troposphere. The complexity of the cloud structure, along with significant infrared opacity over most wavelengths, has, however, made the unique inversion of the remote sensing observations difficult. This uncertainty is evident in the continuing discussion in the literature among different investigators using the same and different data sets.

The following summary of the clouds of Jupiter in the upper troposphere is based on presently available data and analyses (West *et al.*, 1986; Tomasko, 1989, private conversation; Carlson, 1990, private conversation).

Haze is present in the stratosphere above, and perhaps persists down into the tropopause. This haze consists of meteoritic particles and/or energetic particle impact-generated or photochemically-produced aerosols with a mean radius of about 0.1 μ and a total optical depth of about 0.1 to 0.3 at visible wavelengths. There are more particles and greater optical depths at higher latitudes, presumably due to the more intense radiation belts and larger number of energetic particle impacts (Tomasko, 1989, private communication).

The pressure range from 200 to 700 mb contains the clouds and hazes that form the layer observed at visible wavelengths, although the precise level of the visible clouds within this range is a subject of some discussion (West *et al.*, 1986). Also, a comparison of longitudinal structures and zonal motions of features in infrared maps and visible images suggests that the visible features have tops throughout this pressure range (cf. Magalhaes *et al.*, 1990). Essentially no holes permitting direct viewing to deeper levels at visible wavelengths exist in this layer down to very small spatial scales (West *et al.*, 1986). Two particle populations are believed to exist in this pressure range. A diffuse ubiquitous layer of particles with mean radii of about 1–2 μ is present and may extend up to about 200–300 mb at low latitudes; this layer is required by the results of studies of scattered visible sunlight. The optical depth of this layer of particles at visible wavelengths is between 2 and 10, and particle densities (based on opacity arguments) are estimated to be of the order of 100–1000 cm^{-3}. The absence of identifiable spectral features of NH_3 ice particles implies that these aerosols may not be composed completely of NH_3 (Orton *et al.*, 1982; West *et al.*, 1989). However, the particles may consist of NH_3 with large occlusions, mixtures, or coatings, consisting, for example, largely of photochemical products of NH_3 such as N_2H_4, or of various proposed chromophores or other species.

A layer of larger particles (with radii of 3–100 μ that is horizontally patchy and confined to a fraction of the gas scale height above the NH_3 condensation level at approximately 700 mb is required by observations in the thermal infrared (cf. Gierasch *et al.*, 1986). The zonally-averaged latitudinal variation of the optical cloud depth of this cloud layer correlates well with the zonally averaged visible reflectivity ('the belt-zone structure of the planet') and the zonally-averaged NH_3 abundance at the 700 mb level (Gierasch *et al.*, 1986). The composition of the particles is presumably NH_3 and the optical depth at visible wavelengths is comparable to or less than that of the diffuse layer. Particle number densities are estimated to be of the order of 1–10 cm^{-3}.

The characterization of clouds at deeper levels is poorly constrained by remote sensing observations of cloud and aerosol structure. An inhomogeneous time-variable cloud is thought to be located in the vicinity of the altitude corresponding to a pressure of about 2 bars. Its composition is not well defined by observations. If the abundance of H_2S is solar, then this cloud contains NH_4SH and/or polysulfides particles. In addition, if the H_2O abundance is appreciably depleted from solar, then, as mentioned above, this cloud may contain appreciable amounts of H_2O. The cloud is patchy, may also contain chromophores, and also contributes to the appearance of the 5 μ images. The vertical extent of this cloud is ill-defined, and there are no estimates of particle sizes.

An H_2O or H_2O–NH_3 cloud may exist in the 4 to 6 bar region. If the mixing ratio of O is equal to or greater than corresponding solar mixing ratios, then this cloud may be quite massive, containing ice crystals of the order of a micron in size near its top and large droplets of the order of 100 μ or greater in size precipitating, perhaps, from the cloud base. This cloud, if it exists, is also spatially inhomogeneous, perhaps almost disappearing in the regions of the 5 μ features (West *et al.*, 1986). It should be emphasized that there is only weak experimental evidence for the existence of this cloud, evidence for which other explanations may be made without invoking its existence.

From the above discussion it is clear that there are many uncertainties about the clouds of Jupiter. By making *in-situ* measurements the Nephelometer experiment will help to resolve a number of the outstanding questions. The experiment is expected to yield information on the number of clouds, their location, especially of their bases, the amount of material in these clouds and its vertical distribution, the particle sizes and number densities (and hence opacities) in the clouds, and indications of the physical form of the particles. These results can then be used in models to attempt to obtain species abundances, to explain radiation balances, and to investigate atmospheric dynamics.

3. Instrument Design

3.1. General

The Nephelometer is designed to achieve the desired objectives by comparing simultaneous measurements of the light scattered at five angles from a well-defined volume of atmosphere in the vicinity of the Probe with theoretical models of light scattering from particulate matter. A similar approach was successfully used by Marov *et al.* (1980), for measurements made from the Venera Probes in the Venus atmosphere.

A cloud or haze is characterized by the way in which it scatters light. In particular, each unit of volume illuminated by a beam of light will scatter the light at a given angle, θ, in proportion to the product of the particle number density, n, and the probability of the particles in that volume to scatter light into a unit solid angle at that angle, the differential scattering cross section, $[d\sigma/d\Omega]_\theta$. The Nephelometer measured this quantity, at five angles. Measurements are then compared with calculations of the same quantities for model aerosols to obtain the best agreement with the experimental data. Results of such comparisons yield mean particle sizes, particle number densities, and indications of non-sphericity of the particles and/or absorption in the particles. The accuracy with which these quantities can be determined depends on the accuracy of the experimental data and, to a small extent, on the availability of subsidiary information, for example hints or particle composition from other experiments on the Probe. A description of one method of performing such comparisons to obtain the best fit to the data is given by Marov *et al.* (1980).

The instrument contains the following components: (1) pulsed solid state laser light sources, (2) solid state scattered light detectors, (3) collimating, defining, collecting optics, including a deployable axicon (axially-symmetric conical) mirror system, and spectral filters, (4) optical alignment, surface condensation and source output monitors, (5) other 'housekeeping' measurement systems to monitor instrument operation and performance, and (6) analog and digital electronics circuitry and power supplies. The mechanical structure, deployment system, and thermal design assure that the instrument will survive the severe launch, cruise phase, atmospheric entry, and descent environments.

A number of complicating factors must be considered in the design of the instrument. For example, the required high sensitivity to small scattered light signals and the relatively large background light levels (up to 10^6 times as large as the minimum signal levels), as well as the large dynamic range of expected signals (of the order of 10^5 to 10^6), necessitate very careful signal processing. An irradiating light beam collimated highly enough for the measurement of small angle scattering in the forward direction, yet powerful enough to provide sufficient scattered light for measurement of the relatively small scattering at wide angles is required. This requirement is further complicated by the need to reduce instrumentally scattered light, the severely limited space, and the need for reliable source operation after an extended multiple-year cruise phase. In addition, large zero-signal baseline effects may be caused by electrical signals induced by the operation of high power pulsed sources near very sensitive detector circuitry. There is a need to survive not only the severe launch, cruise phase, and atmospheric entry environments, but also the intense high-energy radiation in passing through the Jovian

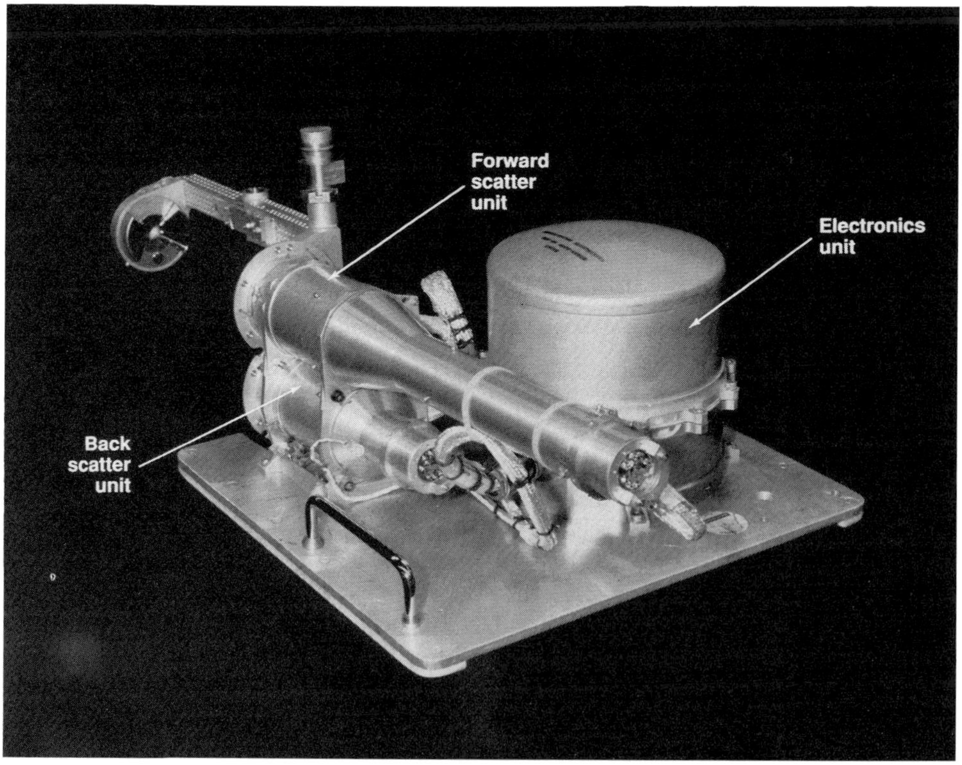

Fig. 1. Galileo Probe Nephelometer instrument. The instrument is positioned on a holding fixture. The forward scatter unit, shown with its mirror arm assembly in the deployed position, is directly mounted on the back scatter unit. For reference, the cylindrical electronics unit is 18.8 cm in diameter and 16.5 cm high.

radiation belts. The effects of this radiation on the reliability and stability of electronic components and circuitry need to be carefully considered in the instrument design. Finally, there are the requirements of relatively low allowable weight, space, power, and data rate.

Physically, the instrument is constructed in three parts, a vented sensor head containing the forward scatter unit, a vented sensor head containing the backward scatter configuration, and a pressure-tight electronics unit containing the bulk of the electronics. A photograph is shown in Figure 1. The sealed unit is capable of withstanding pressures of greater than 20 bars with negligible leakage. The vented sensor heads, containing components also capable of withstanding pressures greater than 20 bars, are connected to the electronics unit with cables terminating in pressure-tight connectors sealed into the wall of the electronics unit. Both units are mounted onto the aft side of the instrument shelf of the Probe. The faces of the sensor units are flush with the Probe skin, and the

TABLE I

Instrument characteristics. The dynamic range for all channels is approximately 10^6, and the mean source wavelength for both forward and backscatter sources is approximately 904 nm. The effective sampling volume decreases for strong signals as the number of sampled pulses is reduced.

Performance					
Scatter channels	5	16	40	70	180 (Bkwd)
Sensitivity, m^{-1} sr^{-1} $count^{-1}$	9.3×10^{-7}	5.1×10^{-7}	1.3×10^{-7}	1.5×10^{-7}	1.1×10^{-8}
Mean scattering angle, degrees	5.82	16.01	40.01	70.00	178.1
Angular resolution, FWHM, degrees	0.64	1.08	1.72	1.76	4.0
Effective sampling volume, l	1.25	0.63	0.65	0.40	16.4

Physical description	
Mechanical	
Weight, kg	
Sensor assembly	1.4
Electronics	3.0
Total	4.4
Dimensions, cm	
Sensor assembly	50.8 × 8.9 × 12.7
Electronics	18.8 dia × 16.5
Electrical	
Power, W	
Instrument	4.8
Heater	6.5
Total	11.3 average
Data rate	10 bps
Data storage on Probe	800 bits
Data output	a digital, 2 bilevel
Timing signals	minor frame
Commands	3 stored, 4 real time

instrument is oriented on the Probe so that sampled volumes extend out of the Probe essentially radially. A 'closeout' structure is used to seal the edges of the sensor faces to the Probe skin. A deployable arm containing the axicon mirror segments, as well as the pyrotechnic pin puller that activates the deployment mechanism, extends from the upper corner of the top of the sensor unit out through the Probe skin. This assembly allows forward scattering sample volumes to be situated in relatively undisturbed air, outboard of flow regimes near the skin of the Probe in which aerodynamic effects may severely modify the particle size distributions with respect to the true ambient free-stream distributions. Calculation of these effects for the present case have been performed using modified methods similar to those described by Chow (1979). The detector external windows and the axicon mirror assembly are electrically heated continuously during Probe descent to prevent condensation of atmospheric vapors. During transit to Jupiter and the period of high heating on entry into the Jovian atmosphere, the Probe is immersed in the heat shield with the axicon mirror arm stowed in its undeployed position. Targets are mounted on the inner surface of the heat shield, scattering fixed amounts of light from the forward and backward irradiating sources. This scattered light is measured by the instrument, permitting checks of calibration stability during the long test and cruise phases of the mission, and shortly before entry into the Jovian atmosphere. Initiation of the Nephelometer experiment begins after entry and deployment of the Probe parachute, removal of the Probe from the heat shield, and deployment of the axicon miror arm.

The instrument characteristics are summarized in Table I. A preliminary description of the instrument has been published earlier (Yeates *et al.*, 1985).

3.2. Optical design

A schematic of the optical system is shown in Figure 2, which illustrates the scattering configuration for each of the four forward scattering channels, Figure 2(a), and the backward scatter channel, Figure 2(b). For each channel the effective scattering volume is defined by the intersection of the source beam with the field of view of the detector. In the case of the forward scattering channels, light scattered in the forward direction impinges on a portion of an axially-symmetric conical mirror (axicon) and is reflected backward through a window onto the collecting lens of the detector assembly. The light is focussed at the plane of a field stop aperture through which it passes onto a spectral filter and into the detector. For the measurement of back-scattered light no mirror is necessary. The source beam and collecting lens optical axes are parallel and displaced from each other so that only light scattered from the source beam at very wide angles, close to 180°, is collected. A slightly displaced off-axis aperture is used. The scattering configuration and the effect of the aperture are shown schematically in Figure 2(b). Rays *a* are characteristic of light scattered at 180° (from an infinite distance along the irradiating beam), focussed onto and passing through the aperture onto the spectral filter and detector. Ray *c* represents an extremal ray, back-scattered at a minimal angle less than 180°, that, after passing through the lens, just grazes the edge of the aperture. Rays *b*, back-scattered at an angle between the minimal angle of ray *c* and 180°, are

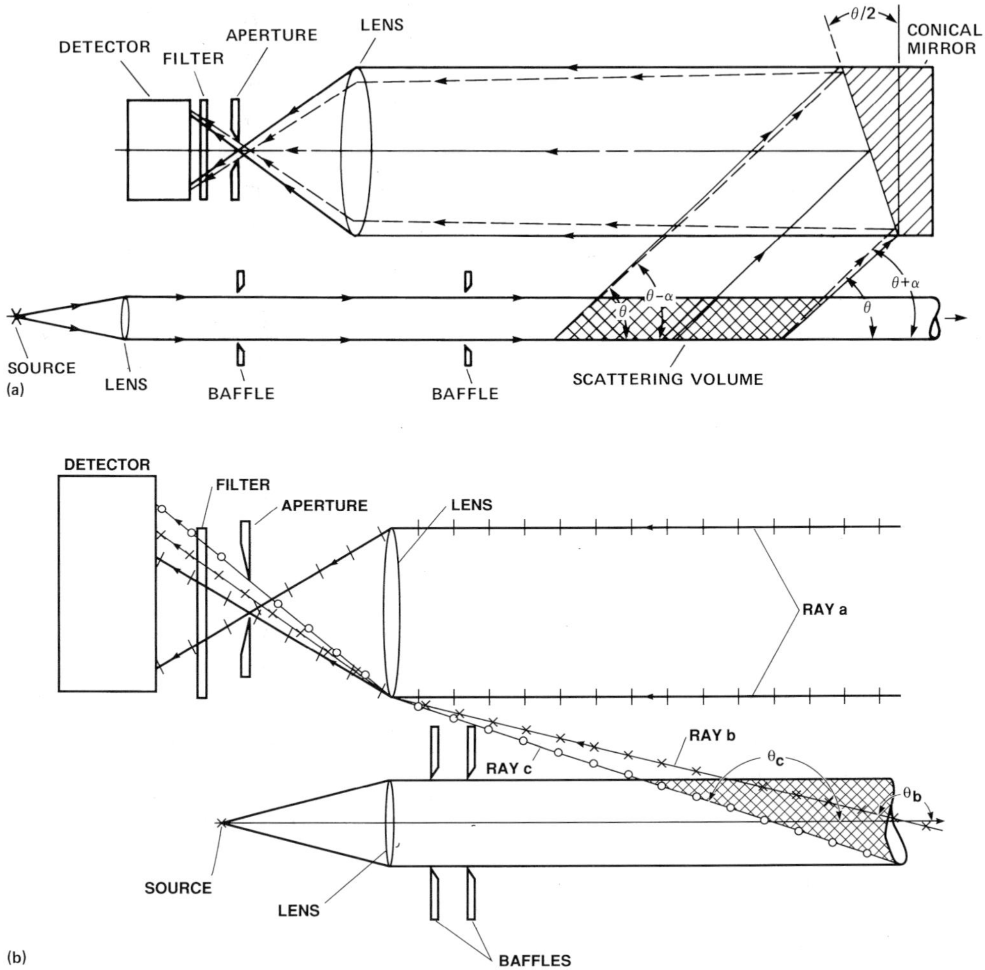

Fig. 2. Schematics of scattering configurations. (a) Forward scatter configuration. The mean scattering angle, θ, is defined by the conical mirror angle, $\theta/2$, and the angular acceptance angle, α, by the aperture that determines the angular limits at which scattered light rays may enter the detector. (b) Backward scatter configuration. Ray *a* corresponds to light back-scattered from the source beam at $\theta_a = 180°$. Ray *c* is an extremal ray, back-scattered from the source beam at an angle, θ_c, defined by the off-axis aperture, and ray *b* is for back-scattering at an intermediate angle θ_b. The mean scattering angle and the angular acceptance function are determined from ray tracing calculations, or empirically, from target scans of the scatter volume.

characteristic of all other rays scattered at intermediate angles passing through the aperture.

The configuration of the components, the aperture, lens characteristics, masking, and source beam dimensions determine the angular acceptance range, $\pm\,\alpha$, for light scattered at the angle θ (fixed by the mirror angle $\theta/2$, for forward scattered light). Ray-tracing calculations, verified by measurements performed by passing diffusely scattering targets through the sample volumes, have established that the angular acceptance functions for the four forward scattering channels are essentially symmetrical with mean values of 5.82, 16.01, 40.01, and 70.00°. Full widths at half maximum for the corresponding angular acceptance functions are 0.64, 1.08, 1.72, and 1.76°, respectively. The effective angular acceptance function of the back-scatter unit, determined by scanning the scattering volume with a diffusely reflecting target, is slightly asymmetric with a mean value of 178.1° and a full width at half maximum of 4.0° (cf. Ragent and Blamont, 1980).

A cutaway view of the vented optical sensor head and electronic unit is shown in Figure 3. The forward scatter assembly is mounted above the backward scatter assembly. A solid state laser light source was used in each assembly. This type of source was especially chosen to provide the small effective source size necessary to produce the low divergence light beam required for small angle forward scattering measurements. This source, a gallium arsenide laser injection diode (LID), Laser Diode Lab. Model LID-60, was selected because of its small active area (2.03×10^{-4} by 7.62×10^{-3} cm), its high optical power output (approximately 2 W peak when driven by a 200 ns wide, 9 A current pulse), and a favorable match between its spectral output (approximately 904 ± 5.0 nm) and the spectral response of the detectors. The output of the light source is collected by an anti-reflection coated lens corrected for spherical aberration. For the forward scattering unit the light output is further collimated by a 40 cm long baffled tube to reduce scattered light and to keep effective beam divergence to acceptable levels. The backscatter unit source-lens combination is recessed 10.2 cm into the backscatter housing so that no portion of the forward scatter mirror assembly or its mounting configuration is in the field of view of the backscatter unit.

The reflecting axicon mirror assembly, mounted on the end of the deployable arm, consists of four coaxial 90° segments of the frustums of gold-coated conical mirrors, each reflecting scattered light at a designated angle to one of the forward scatter channels. A hole through the center of the assembly allows unscattered light in the source beam to pass to the atmosphere outside of the collection volume. As mentioned above, the axicon mirrors are electrically heated during Probe descent to reduce the risk of condensation of atmospheric vapors onto their surfaces.

A grid-heated glass window on each of the forward and backward-scattering units helps to protect the detector assemblies from thermal effects, surface condensation, and electrostatic charging. These assemblies consist of light collecting lenses, lens masks, field stops, filters, and detectors. The lenses are light-weight, moderate resolution, acrylic Fresnel lenses, each cut to the same shape and size as the projection of the corresponding axicon element in the exterior plane of the detector module. Lens masks are used to further define the shape of the scattering volume. Field stops are placed at the paraxial image points of the collector lenses. In the case of the backscatter channel the field stop is slightly displaced from the lens axis so as to admit light scattered from angles of about 176° to 180°. The detectors are silicon p–i–n detectors (E. G. and G. Electrooptics Division Model 200), back-biased at 90 V, with a responsivity of about 0.64 A W^{-1} at 900 nm, and a rise time of 10 ns. Interference filters that reject light other

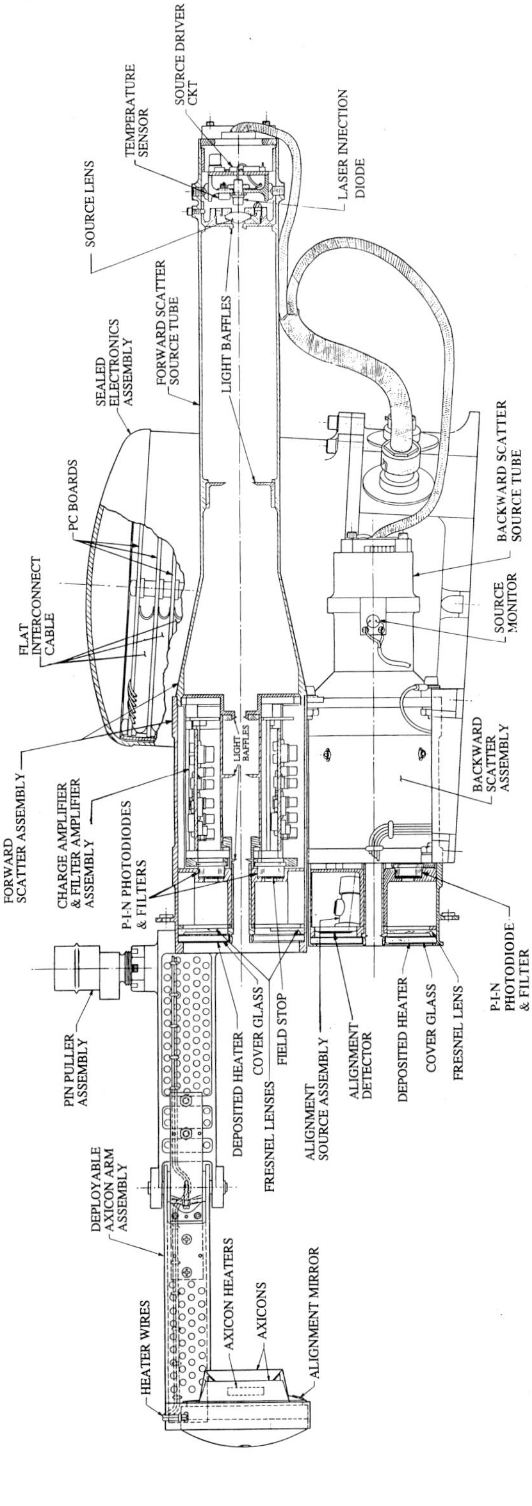

Fig. 3. Cutaway drawing of the vented sensor head and pressure-tight electronics unit.

than at wavelengths near that of the laser diode are used as cover glasses for the detectors. The bandpass curve for these filters is essentially flat with a full width at half maximum of about 66 nm, centered on 904 nm at a temperature of about 20° C. The width is sufficient to make allowance for non-perpendicular rays impinging on the filter, as well as wavelength shifts due to instrument temperature variations during the mission.

Several 'housekeeping' measurements documenting the instrument status are included in the design. These are a contamination monitor, an alignment monitor, light output monitors of the sources, three temperature monitors, and one voltage monitor. The contamination/alignment monitor uses a low power light-emitting-diode (LED) source (Texas Instrument, Inc. Model TIES 35) located in the back-scatter unit to produce a reference light beam. This beam is reflected by a flat mirror on the axicon assembly onto the contamination/alignment detector, a quadrature detector (E. G. and G. Electrooptics Division Model SGD 444-4), in the back-scatter unit. Any contamination on the mirror and on the window in front of the detector will reduce the total signal produced by summing the signals from the four segments of the detector, indicating contamination in the scattering channels. Alignment changes are manifested by motion of the reference beam on the four segments of the detector, changing the magnitude of the signals measured by each segment.

3.3. Electronic design

The Nephelometer electronic system acquires and processes all of the data from the forward- and backward-scatter channels and, also, from the 'housekeeping' sensors, formatting these data for transmission to the Probe telemetry system. It provides the drives for the laser injection diode (LID) and light emitting diode (LED) light sources, appropriate timing signals, and conditions the power received from the spacecraft 28-VDC power bus to provide regulated supplies to the instrument. An overall block diagram of the analog circuitry is shown in Figure 4 and a very simplified block diagram of the digital signal processing is shown in Figure 5.

3.3.1. *Scatter Data Collection and Processing Considerations*

The predicted Probe descent trajectory and range of ambient cloud parameters give rise to requirements that lead to a fairly complex data collection and processing system. The acceptable dynamic range of the measurements is to exceed 10^5. Scatter measurements are to be made at least once during each kilometer of Probe descent even though the descent velocity of the Probe varies greatly during descent. Because of the limitation on data transmission rate allocated to the Nephelometer, this means that the data collection rate exceeds the data transmission rate during the early part of the mission.

These considerations, as well as good design practice to maximize signal-to-noise ratios, lead to the following data collection and reduction scheme.

The LIDs are continuously pulsed at 2000 pps with the forward-scatter LID pulse leading the backscatter LID pulse by 250 μs. The data sampling sequence is controlled by the digital timing circuits. During a sampling period a measurement involves the detection, amplification, digitization, and digital integration of the detected scatter

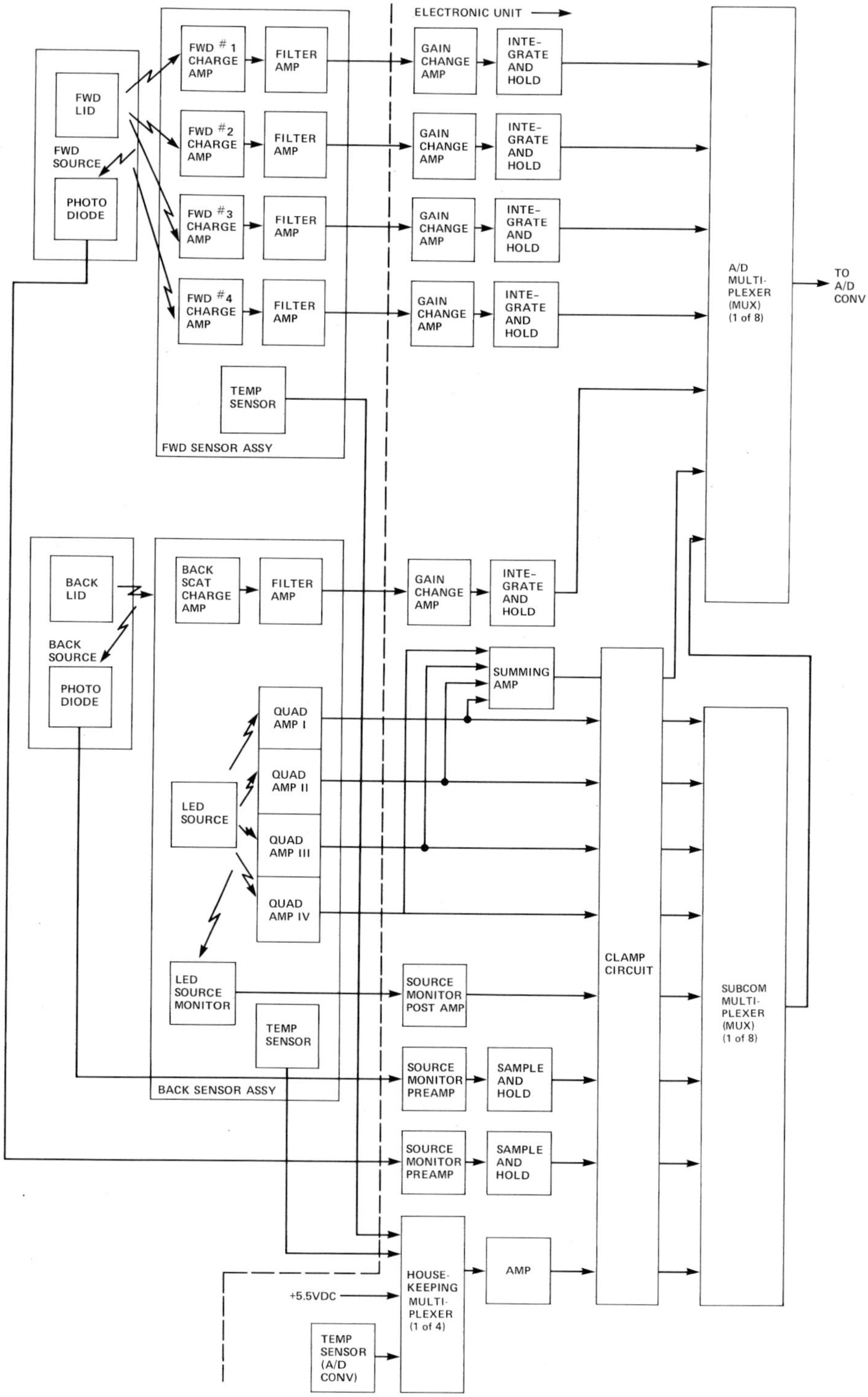

Fig. 4. Block diagram of the Nephelometer analog signal processing electronics.

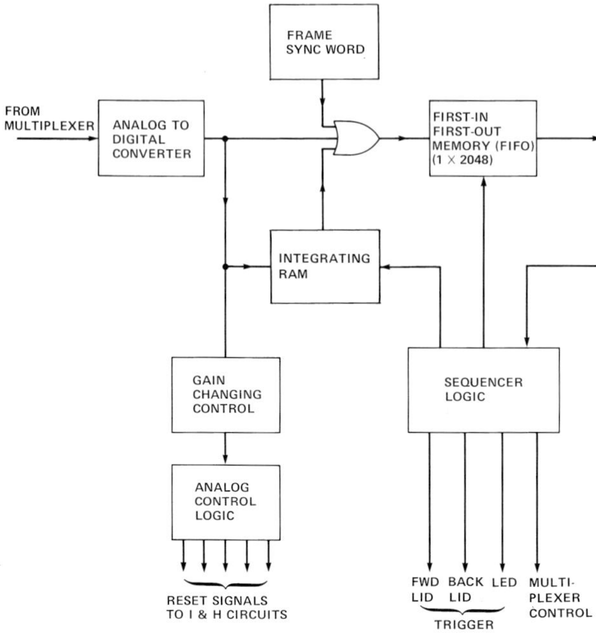

Fig. 5. Block diagram of the Nephelometer digital signal processing electronics.

signals in each scatter channel from 64 bursts, each burst consisting of 64 light pulses emitted at equal time intervals. The scatter channel detectors produce signals for each light pulse. However, for each burst of 64 pulses, only 1, 8, or 64 of the signals are digitally integrated, as controlled by the digital circuitry from a test of the output of the A/D converter. This scheme provides a digital gain change factor of 64. An additional gain change factor of 8 is provided by adjusting the gain of an analog amplifier in the signal processing circuitry for each scatter channel. The dynamic range of $> 10^5$ is achieved by using an 8-bit A/D converter along with the combined digital and analog gain change factors. The A/D converter used here allows for some possible negative baseline drift in the most sensitive ranges of each channel. The interval between the samples is varied so that at the beginning of the descent, data from 64 bursts are collected every 3 s (corresponding to predictions of less than 1 km of descent) for each of the scatter channels and this 3-s sampling period is repeated 10 times. The interval is then increased to 4 s for the next 10 records, then to 5 s for 10 records, etc., until the interval is 12 s for 10 records. At this time, 750 s after the start of data acquisition, the interval is reduced to a constant 8 s so that the data acquisition rate and data transmission rate are equal for the rest of the mission.

A Nephelometer data frame consists of the synchronization word, engineering data, and data from 10 scattering data records for a total of 800 bits. The data transmission rate to the spacecraft telemetry system is 128 bps, of which 10 bps is allocated to the Nephelometer. Since initially at deployment the data for the first sample are collected at 26.7 bps (corresponding to an 800 bit frame in 30 s), a buffer memory is provided in the form of a digital, first-in–first-out (FIFO), random-access-memory (RAM). The data stored in this FIFO RAM are transferred to it from an integrating RAM configuration in the form of a compressed 10-bit word using a sign bit, 3 bits for a power of 2 exponent, and 6 bits for the 6 active most significant bits (MSBs) in the 24-bit register of the integrating RAM.

The gain changing technique involves monitoring the A/D converter output word for each scatter channel conversion. If the four most significant bits (MSBs) of the word are all ones the channel gain is decreased by a factor of eight. If the 4 MSBs are all zeros, the gain is increased by a factor of 8. The digital gain at which the data are collected is taken into account in the integrating RAM by the proper shifting of the new data from the A/D converter output before it is added to the register contents. The gain (512, 64, 8, or 1) of the 16° forward scatter channel is also reported in the output data stream four times during each measurement interval, as a rough measure of the variability of the scatter signal (or cloud variability).

3.3.2. *Analog Electronics*

An overall block diagram for the analog circuitry is given in Figure 4. The high power LID light sources, driven by a train of 9 A, 200 ns pulses at 2000 pps from an SCR-controlled capacitor discharge, illuminate the scattering volume. The cloud-scattered light for each of the four forward and single backward-scatter channels is sensed by silicon p–i–n diodes operated in the photoconductive mode. Each photodiode signal is ac-coupled to its charge sensitive preamplifier to prevent the high background illumination signal from saturating the channel. The preamplifier, identical in each channel except for gain, is a wide bandwidth, low noise configuration with a very high dynamic range and slew rate. The output of each preamplifier is then fed to an active bandpass filter to optimize the signal to noise ratio and provide some gain before being multiplexed and sent from the sensor head to the electronics unit. An 8 : 1 gain change amplifier and a balanced integrate and hold circuit complete the analog processing of the signal prior to A/D conversion. Considerable care was taken in the design of all of these elements because of the low level, high-speed processing required, the constraints on power and size, and the need to survive and function properly in the high-energy radiation environment.

The alignment/contamination LED is driven by a train of 300 μs pulses at 1000 pps. The LED light reflected from the flat mirror mounted on the axicon base is detected by a quad photodiode each of whose diodes is operated in the photoconductive mode. The output signal from each section of the detector is amplified by a transimpedance amplifier and multiplexed to the A/D converter. For each fifth light pulse from the LED the four amplified photodiode signals are added in a summing amplifier, and the result multiplexed to the A/D converter.

The A/D converter is an 8-bit, bipolar, dual slope, integrating type with auto-zeroing. It contains integrated circuit operational amplifiers and comparators and CMOS central logic operating with a 2 MHz clock.

Source light output monitors for the LIDs are provided. Each monitor consists of a silicon p–i–n diode (E. G. and G. Electro-optics Division Model SGD 040B) operated in the photovoltaic mode, and is illuminated by a portion of its LID's light output. A charge sensitive preamplifier and a sample and hold circuit process the sampled signal from one light pulse for A/D conversion and inclusion in the output instrument condition data. A photodiode/preamplifier combination unit is used to monitor the output of the LED, and its output signal is also multiplexed to the A/D converter.

The 5.5-V reference and signals from three temperature sensors, one near each of the LIDs in the sensor head and one near the A/D converter in the electronics unit, are separately multiplexed in one of the four special multiplexers used for instrument condition signals, then amplified and multiplexed to the A/D converter.

3.3.3. *Digital electronics*

A simplified block diagram of the digital electronics is shown in Figure 5. The Nephelometer digital electronics provides all system timing, A/D converter control, data formatting and integration, and telemetry interface control. No microprocessor is used. Nephelometer operation requires three sequential operational modes, the gathering of scatter data 64 times and transfer to integrating memory, transfer of scatter data integrations into another memory awaiting transmission to telemetry, and preparation of preface information, comprising a sync word, frame count, and instrument status data. The oscillator-clock generator generates all of the system timing signals required for all of the subsystems, such as the power supply, and, especially, for commanding the system into one of the three modes of operation. The sequencer logic provides the LID and LED firing pulses, the integrate, sample, and reset commands, and commands to transfer data to the spacecraft telemetry system. The integrating RAM section performs the digital integration on 64 analog data measurements and provides signals to the analog gain changing circuitry. Transfer circuitry compresses the data in the integrating RAM and transfers it to the FIFO RAM for storage. During the preface mode of operation the frame and sync word generator and digitized instrument status data are transferred into the FIFO RAM storage.

3.4. Mechanical design

The Nephelometer instrument consists of three separate physical units, the forward scatter sensor unit on which the deployment mechanism is mounted, the backward scatter sensor unit, and the electronics enclosure. Figure 3 is a cutaway view of the Nephelometer with key components labeled. The two sensor units, vented to the ambient atmosphere to eliminate optical distortion or mechanical fracture, are fastened to each other and mounted to the Probe equipment mounting plate. The combined sensor units are slightly tilted on the mounting plate to allow for clearance with other equipment. Shims are used to align the optical systems except for a lateral adjustment of the laser diode sources. The electronics enclosure is sealed with one Earth atmosphere inside. A pressure sensor inside the sealed unit was used in laboratory tests in vacuum and under pressure to verify, within the test experimental limits, the integrity of the seal both for the long cruise phase and entry to at least the 20 bar level. The electronic boards inside are mounted in a manner to minimize the effects of being subjected to the 400 g entry deceleration pulse, as well as launch vibration. For example, the relatively heavy power supply is mounted near the bottom of the enclosure.

The deployment mechanism consists of a folded, hinged, spring-loaded arm on which the axicon mirror assembly is mounted, and a positive latching mechanism. The arm is folded through launch, cruise, and initial entry, and retained in position by a pin mounted in a pyrotechnically-initiated pin puller. At a pressure of approximately 0.07 bar the aeroshell is jettisoned, the pin puller fired, and the axicon assembly deployed under the action of a helical spring. The inboard section of the arm is a perforated rectangular tube and the outboard section a perforated channel to which the axicon assembly is mounted. The arm is downstream and off to the side of the centerline of the axicon assembly, to reduce possible flow field interference with the measurements, and is perforated to mitigate aerodynamic lift effects (that might affect Probe rotation) and sympathetic vibrations due to von Kármán vortex shedding. The end cap of the axicon assembly is convex and is vented to produce a pressure differential, helping to prevent particle condensation on mirror surfaces. The positive latching mechanism consists of the deployment spring, a counter-action leaf spring, and two overlapping spring plates, the ends of which are wedge-shaped and pass over each other, latching on the back ends of the wedges. The leaf spring forces the contact between these surfaces. The spring constants are chosen to assure deployment even at a zero Probe rotation rate, to absorb enough energy in latching to prevent arm distortion due to deformation during deployment at a possible 80 RPM Probe rotation rate, and to force close contact between the backs of the wedges. Materials are selected to minimize distortion by differential thermal expansion. Calculations, using a Monte-Carlo program, have indicated that, for the worst case (involving offsets in pitch angle creating opposite sensitivity shifts in the 5.8° and 16.0° channels), alignment offsets of less than 0.06° will produce sensitivity shifts of less than one digitization step. Acceptable performance has been verified from tests that have demonstrated alignment reproducibility to less than 0.05° on repeated deployment, and misalignment of less than 0.05° due to liquid nitrogen induced thermal shock.

Calculations of the predicted thermal behavior (temperatures, thermal contraction/expansion, stresses) of the Nephelometer during the course of the Probe descent in the Jovian atmosphere have been performed using the MITA (Martin Integrated Thermal Analysis) program along with a 118 node thermal model of the instrument. Input quantities are derived from sample entry trajectories, atmospheric structure models, transport and thermodynamic properties of the atmospheric gases, internal Probe temperatures and initial conditions, and free heat transfer coefficients. The results of these calculations for the final design of the Nephelometer are such that no serious thermal effects on instrument performance are predicted to occur during the course of the Probe mission.

4. Calibration, Tests, and Performance

4.1. General

In order to derive particle sizes, number densities and other properties, the measured relative magnitudes of the signals in each channel with respect to the other channels are compared for best fit with those obtained from Mie-scattering calculations. Model particle size distributions (normalized to one particle per cm^3) are used, with mean size and distribution width as parameters. Particle indices of refraction must, in general, be assumed or inferred from other sources for these calculations, although the calculated results are not strongly sensitive to the choice of index, especially in the forward direction. Conversely, if the particles are assumed to be only weakly absorbing and roughly spherical, the measured data, especially from the backward scatter channel, will provide constraints on allowable indices. Also, if the particles are not strongly absorbing, deviation of the measured scattering at wide angles from that calculated for spherical particle distributions is a strong hint that the particles may not be spherical. The calculated differential scattering cross section for each channel for the particle size distribution with these best fit parameters is then compared with the absolute value measured in each channel $n[\mathrm{d}\sigma/\mathrm{d}\Omega]_\theta$, yielding the particle number density, n. Inference of opacities and mass densities follows directly from the calculated total cross sections and reasonable assumptions about particle mass density.

4.2. Calibration

Two methods were used to calibrate the Nephelometer. The first is similar to the method described by Pritchard and Elliott (1960), as modified for application to the present case. This technique involves recording the response of each of the scattering channels to the scattered light produced by a diffusely scattering target positioned perpendicular to the source beam optical axis, as the target is stepped along the source beam until the sensitive volume for each channel has been traversed. For the forward-scattering channels a carefully documented diffusely transmitting screen mounted into the end of a set of telescoping tubes is used. The transmitting screen transmittance is carefully measured using a standard integrating sphere and the screen's angular response and polarization characteristics are documented with a specially constructed goniometer. Similar procedures are used to verify the characteristics of a large specially constructed Lambertian reflector that was used to calibrate the backward scattering channel. Calibrated neutral density attenuating filters are used in front of the collecting optics for the detectors in each channel to maintain the signals within the dynamic range of the instrument. The manner of relating the readings obtained using this scanning method to the calibration constants to be used in measuring actual aerosols is described below.

The Nephelometer instrument produces counts, C, in proportion to the product of particle differential scattering cross section, at angle θ, $[\mathrm{d}\sigma/\mathrm{d}\Omega]_\theta$ (with units of $m^2\ sr^{-1}$), and particle number density, n (with units of m^{-3}), with combined units for this product, $n[\mathrm{d}\sigma/\mathrm{d}\Omega]_\theta$, of $m^{-1}\ sr^{-1}$. The proportionality constant is the product of source intensity I_s, effective sampling volume V_{eff}, and detector/electronics/optics gain constant K. The instrument count output C can be written as follows:

$$C = (KI_s V_{\text{eff}})n[\mathrm{d}\sigma/\mathrm{d}\Omega]_\theta = (1/E)n[\mathrm{d}\sigma/\mathrm{d}\Omega]_\theta$$

and, the desired measured value,

$$n[\mathrm{d}\sigma/\mathrm{d}\Omega]_\theta = CE = C/(KI_s V_{\text{eff}}) .$$

In response to a diffuse calibration target normal to the source beam at position x, filling an effective area $A_{\text{eff}}(x)$, and having reflectivity (or transmission) at angle of $T\cos\theta$, the instrument count output will be given by

$$C(x) = t(x) = [KI_s A_{\text{eff}}(x)]\,(T\cos\theta)/\pi .$$

Because the normal calibration target is so bright, it is necessary to reduce the amount of scattered radiation reaching the detector with an attenuator of attenuation factor F. By moving this calibration target along the beam over all x at which response is obtained, and integrating the response over all x, we obtain

$$\int C(x)\,\mathrm{d}x = \int t(x)\,\mathrm{d}x = K(I_s/\pi F)\,(T\cos\theta)\int A_{\text{eff}}(x)\,\mathrm{d}x =$$
$$= K(I_s/\pi F)\,(T\cos\theta)\,V_{\text{eff}} .$$

Thus, the proportionality constant E, in units of $\mathrm{m}^{-1}\,\mathrm{sr}^{-1}\,\mathrm{count}^{-1}$, can then be evaluated from

$$E = (KI_s V_{\text{eff}})^{-1} = T\cos\theta\left\{(\pi F)\left[\int t(x)\,\mathrm{d}x\right]\right\}^{-1} .$$

In practice it is also necessary to make small corrections to account for the deviation of the reflection or transmission screens from true diffuse behavior, and polarization characteristics of the sources, screens, and detection system. The accuracy of this calibration procedure is a function of the accuracy of our knowledge of the reflection (or transmission) of the screen used to calibrate the Nephelometer and its simulation of diffuse reflection (or transmission), the accuracy of the measurement of the attenuation factor of the attenuator, the accuracy of the data taken at each target position, and the accuracy of the integration yielding the calibration factor. Estimates of the overall accuracy range from less than $\pm 5\%$ for the 5°, 15°, and 180° channels to less than $\pm 10\%$ for the 40° and 70° channels.

The second type of calibration method involves obtaining the response of the instrument to a well-documented 'standard' aerosol environment. These tests were performed in a large test chamber at Particle Measuring Systems, Inc. (PMS) of Boulder, Colorado. An aerosol with a very narrowly dispersed size distribution was produced by atomizing a suspension of spherical polystyrene or polyvinyl toluene particles into a large spherical chamber. The particle sizes were measured using standard electron microscope sizing techniques developed for aerosol research at Ames Research Center. The density of particles and the proportion of single particles to 'doublets', 'triplets', etc., in the actual aerosol was documented using standard particle sizing instrumentation manufactured and calibrated by PMS. Nephelometer responses were recorded for a variety of particle sizes, particle densities and particle composition. The calibration for each of the scatter channels was then determined, using Mie-scattering cross sections calculated for the PMS-documented aerosol distributions. In general, the results obtained were within 30 to 50% (often within 10%) of those measured using the first method. However, the variations in the results of repeated experiments in the particle chamber indicated that the results were less reliable than those of the target scanning technique. Closer investigation indicated a number of variables in the test conditions that were apparently difficult to control. For example, small persistent air currents in the test chamber were present, produced during aerosol injection, by thermal gradients, by the sampling of the PMS instrumentation, or by other causes. These currents introduced inhomogeneities and differences in the particle densities as measured by the test instrumentation and the Nephelometer. In addition, it proved to be difficult to produce an aerosol with a low enough content of aggregate particles, such that these larger particles did not appreciably affect the measured scattering cross sections. It was suspected that some of the particles might also have been electrically charged and that electrical effects, for example, at the chamber walls, may have produced differences between the aerosol sampled by the PMS instruments and the Nephelometer.

4.3. Tests

A number of tests have been performed to characterize and validate the erformance of the Nephelometer. Environmental tests have included vibration and acceleration tests, steady state and transient thermal tests, and vacuum and pressure tests (of the electronics unit). Particle flow experiments in a wind tunnel were performed on a simulated Probe configuration to validate theoretical particle flow calculations and the effectiveness of the design of the deployed arm in obviating distortions in the particle size distribution functions in the sample volumes (Ragent and Snyder, 1992). The arm-deployment mechanism was tested for its ability to repeatedly deploy the arm and mirror assembly accurately. Tests of the effects of possible thermal distortion or thermal shock on the scattering angle during deployment were also performed.

4.4. Performance

Instrument characteristics are summarized in Table I. The calibration constants obtained at a fixed instrument temperature of approximately 293 K for each of the five scatter channels are listed. As mentioned above, the absolute uncertainty of the calibration for each channel is estimated to be less than 10%. For actual measurements in the Jovian atmosphere we must include errors arising from possible telemetry effects, granulation uncertainties in the data extending from about + 1.7% to + 6.7% inherent in the data compression scheme used here, variations in the measured values due to noise in each channel, and particle sampling statistical fluctuations, which will be small for all reasonable signals (cf. Ragent and Blamont, 1980).

Multiplying the calibration constants by the adjusted number of counts attributable to scattering from particles in the channel sample volume yields the desired measured quantity $n[\mathrm{d}\sigma/\mathrm{d}\Omega]_\theta$ for each channel. The adjusted value is obtained by correcting the raw counts for temperature dependences of channel sensitivity and baseline values in each channel. Curves of these temperature corrections were obtained in the test programs mentioned above. Although these corrections vary nonlinearly with temperature and differ for each channel, a rough value of the magnitude of baseline shift with temperature is about 50 counts per 10° change from the nominal calibration temperature, and about a 10% change per 10° for the channel sensitivity change. Calculations of the thermal variation of the instrument during Probe descent predict that sensitive components will not vary by more than 10° and that, therefore temperature corrections of the adjusted data will be possible to within a few percent.

Noise values in each channel also vary with temperature and are greater in the more sensitive channels, the 40°, 70°, and 180° channels. The values vary from about ± 5 counts in the 5° channel to about ± 30 counts in the 180° channel at about 293 K, and increase by about a factor of 1.5 for a temperature increase of 10° above the nominal calibration temperature.

Our knowledge and uncertainties about the kinds and densities of particles we may expect to encounter duirng Probe descent in the Jovian atmosphere were summarized above. Although it is quite likely that we will encounter small haze or chromophore particles with sizes of the order of 0.1 μ, as well as large particles of NH_3 ice, H_2O ice and liquid, or other large particles, the only fairly well identified particles at the present time are the approximately 1.0–2.0 μ size particles (particle densities of the order of a few hundred per cm^{-3}) in the upper cloud. For this Jovian cloud, that may be roughly analogous to some optically thin Earth-type light clouds (cf., for example, Tables T.60 and T.105 of Deirmendjian, 1969), we may expect angular scattering cross sections varying from about 10^{-2} to $10^{-3}\ \mathrm{m}^{-1}\ \mathrm{sr}^{-1}$ at 6° from the forward direction to about 10^{-4} to $10^{-5}\ \mathrm{m}^{-1}\ \mathrm{sr}^{-1}$ in the backward direction, and a volume extinction coefficient of about $3 \times 10^{-3}\ \mathrm{m}^{-1}$. The instrument can be expected to read at least several thousand counts in all channels, except perhaps for the 70° channel, which would be expected to read somewhat less than 1000 counts. We may thus expect to begin to characterize this cloud at particle concentrations greater than about 5 to 10 cm^{-3}, and to detect it at considerably lower concentrations.

We have attempted to investigate the accuracy of recovery of cloud parameters from our measured data by the following procedure. We have assumed that our particles can be modeled by spherical particles using two parameters to characterize the particle size distribution, r_m, the mean particle radius, as weighted by the geometric cross section, and σ_m, the variance of the distribution, again weighted by the geometric cross section. We generated 'measured' values (labeled r_m^* and σ_m^*) by selecting values for r_m and σ_m, and used the Mie-scattering theory to find the scattering phase function $P(\theta_j)$ for a range of values of r_m and σ_m. From these calculated phase functions as a function of r_m and σ_m, we then computed the fractional deviation, FDV, defined as

$$\mathrm{FDV} = \{N^{-1}[w_j P^*(\theta_j) - w_j P(\theta_j)]^2\}^{1/2},$$

where $P^*(\theta_j)$ is the phase function at angle θ_j corresponding to r_m^* and σ_m^*, $N = 4$ for the current case, and w_j is a weighting function that depends on the accuracy of the measurements. For the case where the expected error in the measurement is independent of the value of the measured quantity ('constant error'), $w_j = (P^*_{max})^{-1}$, where P^*_{max} is the largest value of $P^*(\theta_j)$. Alternatively, if the measurement error is proportional to the value of the measured quantity ('constant relative error'), $w_j = [P^*(\theta_j)]^{-1}$. For our instrument, the errors lie between these two cases, but are closer to the latter.

Examination of the results of calculations of the FDV over the ranges of the parameters r_m and σ_m for various assumed forms of the particle size distribution function (e.g., log-normal and modified gamma function) and particle indices of refraction have indicated that, for the measurement errors discussed above, it should be possible to obtain useful estimates of the mean particle radius within less than a factor of 2 over the entire mean radius range of about 0.2 to 20 μ, and to within a factor of less than 1.4 over the range from about 0.3 to 6.0 μ. The loss of sensitivity for small values of r_m is due to the particles becoming Rayleigh scatterers, whereas for large r_m the diffraction peak lies entirely at angles smaller than the smallest angle of observation. The calculations have also shown that the sensitivity of the FDV to changes in σ_m is low, and consequently, only modest bounds may be placed on the width of the particle size distribution.

Acknowledgements

We are grateful for the dedicated efforts of many members of the staff of the Martin-Marietta Aerospace Division, Denver, CO, the Galileo Probe Project Office and the Electronic Instrument Development Branch of NASA Ames Research Center for the design, testing and calibration of the Galileo Probe Nephelometers. We wish to thank especially E. Tischler, J. Sperans, B. Chin, C. Sobeck, P. Melia, and M. Izadi of the Galileo Probe Project Office, G. J. Deboo and W. D. Gunter of the Electronic Instrument Development Branch of Ames Research Center, and W. Hirokawa, L. Thayne, C. Lord, W. Plaster, D. Lutze, J. Hayden, R. Burtzloff, M. Hinds, R. Callow, S. Rounds, W. Frazer, F. Bowers, L. Wilkin, G. Kyrias, and D. Morris of the Martin Marietta Aerospace Division. This work is supported under NASA Cooperative Research Agreement NCC 2–466 and NASA Contract NAS 2–10015.

References

Barcilon, A. and Gierasch, P.: 1970, *J. Atmospheric Sci.* **27**, 550.

Bjoraker, G. L., Larson, H. P., and Kunde, V. J.: 1986, *Astrophys. J.* **311**, 1058.

Carlson, B. E., Prather, M. J., and Rossow, W. B.: 1987, *Astrophys. J.* **322**, 559.

Carlson, B. E., Rosow, W. B., and Orton, G. S.: 1988, *J. Atmospheric Sci.* **45**, 2066.

Carlson, B. E., Lacis, A. A., and Rossow, W. B.: 1990, *Icarus* (in preparation).

Chow, C. Y.: 1979, *Introduction to Computational Fluid Mechanics*, John Wiley and Sons, New York, p. 51.

Cunningham, C. C., Hunten, D. M., and Tomasko, M. G.: 1988, *Icarus* **75**, 324.

Deirmendjian, D.: 1969, *Electromagnetic Scattering on Spherical Polydispersions*, American Elsevier Publ. Co., Inc., New York.

Del Genio, A. D. and McGrattan, K. B.: 1990, *Icarus* **84**, 29.

De Pater, I.: 1986, *Icarus* **68**, 344.
Gierasch, P. J.: 1976, *Icarus* **29**, 445.
Gierasch, P. J.: 1989, *Nature* **336**, 616.
Gierasch, P. J. and Conrath, B. J.: 1985, in G. Hunt (ed.), 'Energy Conversion Processes in the Outer Planets', *Recent Advances in Planetary Meteorology*, Cambridge University Press, Cambridge, pp. 121–146.
Gierasch, P. J. and Conrath, B. J.: 1987, *J. Geophys. Res.* **92**, 15019.
Gierasch, P. J., Conrath, B. J., and Magalhaes, J. A.: 1986, *Icarus* **67**, 456.
Givens, J. J., Nolte, L. J., and Pochettino, L. R.: 1983, *Galileo Atmospheric Entry Probe System: Design, Development and Test*, Proceedings of the AIAA 21st Aerospaces Science Meeting, January 10–13, 1983, Reno, Nevada.
Ingersoll, A. P.: 1976, *Icarus* **29**, 245.
Lewis, J. S.: 1969a, *Icarus* **10**, 365.
Lewis, J. S.: 1969b, *Icarus* **10**, 393.
Lunine, J. I. and Hunten, D. M.: 1987, *Icarus* **69**, 566.
Magalhaes, J. A., Weir, A., Conrath, B. J., Gierasch, P. J., and Leroy, S.: 1990, *Icarus* (in press).
Marov, M. Ya., Lystsev, V. E., Lebedev, V. N., Lukasevich, N. L., and Shari, V. P.: 1980, *Icarus* **44**, 608.
O'Neil, W. J. and Mitchell, R. T.: 1983, *Galileo Mission Overview*, Proceedings of the AIAA 21st Aerospaces Sciences Meeting, January 10–13, 1983, Reno, Nevada.
Orton, G. S., Appleby, J. F., and Matonchik, J. V.: 1982, *Icarus* **52**, 94.
Pritchard, B. S. and Elliott, B. J.: 1960, *J. Opt. Soc. Amer.* **50**, 191.
Ragent, B. and Blamont, J.: 1980, *J. Geophys. Res.* **85**, 8059.
Ragent, B. and Snyder, P. K.: 1992, in preparation.
Stoker, C. R.: 1986, *Icarus* **67**, 106.
Weidenschilling, S. J. and Lewis, J. W.: 1973, *Icarus* **61**, 311.
West, R. A.: 1988, *Icarus* **75**, 381.
West, R. A., Strobel, D. F., and Tomasko, M. G.: 1986, *Icarus* **65**, 161.
West, R. A., Orton, G. S., Draine, D. T., and Hubbell, E. A.: 1989, *Icarus* **80**, 220.
Yeates, C. M., Johnson, T. V., Colin, L., Fanale, F. P., Frank, L., and Hunten, D. M.: 1985, *Galileo: Exploration of Jupiter's System*, NASA SP-479, Scientific and Technical Information Branch, National Aeronautics and Space Administration, Washington, DC, p. 150.

THE GALILEO PROBE ATMOSPHERE STRUCTURE INSTRUMENT

ALVIN SEIFF

San Jose State University Foundation, Department of Meteorology, U.S.A.

and

T. C. D. KNIGHT

Martin Marietta Space Systems, Denver, Colorado, U.S.A.

Abstract. The Galileo Probe Atmosphere Structure Instrument will make *in-situ* measurements of the temperature and pressure profiles of the atmosphere of Jupiter, starting at about 10^{-10} bar level, when the Probe enters the upper atmosphere at a velocity of 48 km s^{-1}, and continuing through its parachute descent to the 16 bar level. The data should make possible a number of inferences relative to atmospheric and cloud physical processes, cloud location and internal state, and dynamics of the atmosphere. For example, atmospheric stability should be defined, from which the convective or stratified nature of the atmosphere at levels surveyed should be determined and characterized, as well as the presence of turbulence and/or gravity waves. Because this is a rare opportunity, sensors have been selected and evaluated with great care, making use of prior experience at Mars and Venus, but with an eye to special problems which could arise in the Jupiter environment. The temperature sensors are similar to those used on Pioneer Venus; pressure sensors are similar to those used in the Atmosphere Structure Experiment during descent of the Viking Landers (and by the Meteorology Experiment after landing on the surface); the accelerometers are a miniaturized version of the Viking accelerometers. The microprocessor controlled experiment electronics serve multiple functions, including the sequencing of experiment operation in three modes and performing some on-board data processing and data compression.

1. Introduction

When the Galileo Probe enters the atmosphere of Jupiter in December, 1995, an Atmosphere Structure Instrument (ASI) will make *in-situ* measurements of the thermal structure of the atmosphere, i.e., the variation of temperature, pressure, and density with altitude. Sensors of atmospheric temperature and pressure, and acceleration of the probe center of gravity will permit a number of key properties of the atmosphere ro be derived, as demonstrated by the Pioneer Venus probes (Seiff *et al.*, 1980a).

The Galileo experiment will operate over an altitude range of about 870 km in two measurement modes, one to accomodate conditions of high-speed entry at low ambient density; the other, the very different conditions of parachute descent. The entry mode begins at a nominal ambient density threshold of 10^{-11} kg m^{-3} (presently believed to be about 750 km altitude above the 1 bar level) where the Probe deceleration is expected to be about 15 μg. From measurements of Probe deceleration under the action of atmospheric drag, atmospheric densities will be derived. The density profile is integrated above a given altitude to define the pressure at that level, and the temperature profile is then obtained through the equation of state, given the variation of atmospheric mean

Space Science Reviews **60:** 203–232, 1992.

molecular weight with altitude. This mode of operation is continued until the Probe deploys its parachute, nomunally at the 100 mb level (47 km above the 1 bar level).

Then, during the subsequent parachute descent of the Probe to the nominal end-of-mission depth at 16 bars, the thermal structure of the atmosphere is defined by measurements of temperature, pressure, and acceleration in descent mode. Current estimates are that the mission could extend to below 20 bars, perhaps as deep as 25 bars.

At entry, the Probe is effectively a 48 km s^{-1} meteor, enveloped by a bow-shock wave and a thin shock layer of ionized, luminescent gases at extreme temperature ($\sim$ 15 000 K at peak). Under these conditions, measurements of the ambient atmosphere by means of conventional low-density sensors are clearly infeasible; it is futile to extend sensors into the shock layer or through the bow-shock wave, because they quickly burn away and, outside the probe bow wave, develop shock layers of their own. Measurements of atmospheric density by way of Probe decelerations, however, provides a direct means of sensing the atmosphere. This concept was tested and demonstrated in the Earth's atmosphere in 1973 (Seiff *et al.*) and has been applied with good results at Mars and Venus (Seiff and Kirk, 1977; Seiff *et al.*, 1980a).

After the parachute is deployed, the heat shield is jettisoned and temperature and pressure sensors are exposed to the ambient atmosphere. Acceleration measurements are continued, but at much less frequent intervals, and in the absence of large vertical winds, continue to define atmospheric density. From the three measured state variables, temperature, pressure, and density, the atmospheric mean molecular weight may be determined. Other instruments on the Probe, including the mass spectrometer and the helium abundance detector, will probably determine the molecular weight more accurately, however. In these circumstances, the accelerometer data may be used to define the magnitude of large vertical winds in waves or gusts.

Altitudes relative to any convenient reference level (e.g., the 1 bar level) are defined by the temperature and pressure data integrated in the equation of hydrostatic equilibrium. These data will be used to establish the altitudes of measurement for use by all Probe experiments. They also define the rate of descent, which is necesary to the analysis of Doppler wind experiment. In addition, atmospheric turbulence, radius to the center of the planet, and Probe angle of attack are sensed or derived.

The nominal end of mission, at 16 bars, is at an altitude of about $-$ 120 km below the 1 bar level. The vertical measurement range of about 870 km will be covered in a little over one hour – the initial 700 km in less than 4 min. The time in descent on the parachute at velocities from the order of 100 to 30 m s^{-1} could be up to 75 min.

The instrument electronics perform a number of essential functions. They receive and execute commands from the Probe systems and thereby set the experiment mode (calibrate, entry, or descent), control the measurement sequences, select sensor ranges, collect data from the three sensor sets, amplify signals, perform A/D conversions, do some on-board data processing, and condition instrument power. The microprocessor controlled electronics were designed and built by Martin Marietta Aerospace, in Denver, Colorado.

This is a fourth generation entry probe atmosphere structure instrument. Its predecessors were used on the four probes of the Pioneer Venus Mission (Seiff *et al.*, 1980b), on the Viking Mission to Mars (Seiff, 1976), and on the Planetary Atmosphere Experiments Test (PAET) entry probe into the Earth's atmosphere (Seiff *et al.*, 1973). Best features of earlier instruments were retained and some new capabilities have been added. In this paper, we discuss the design approach briefly, and describe the instrument, both the sensors and the electronics, but refer the reader to earlier publications for more thorough treatment of some topics.

Two photographs of the instrument are given in Figure 1. The first shows a close-up of the electronics housing with three pressure sensors mounted on its side; and the second, the complete instrument, with the pressure sensors thermally insulated for installation on the Probe. Instrument components will be discussed below.

2. Scientific Objectives and Expected Results

Some scientific objectives have been implied above. For convenience, they are more fully summarized here:

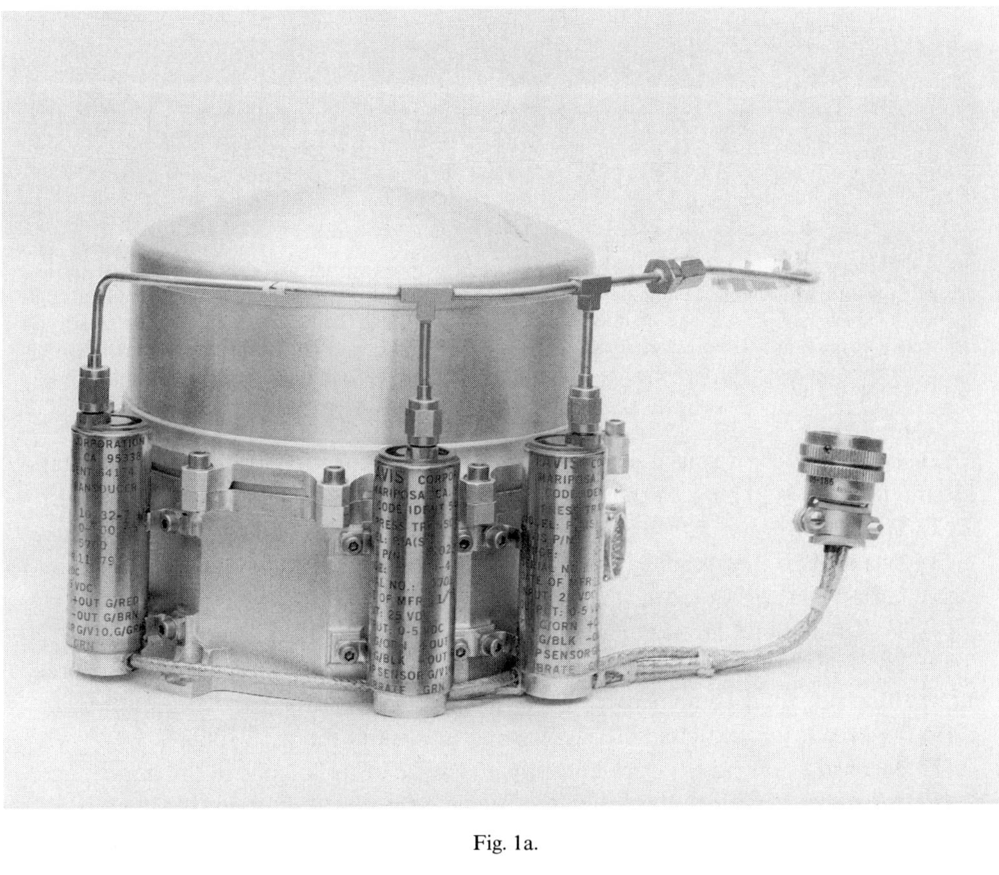

Fig. 1a.

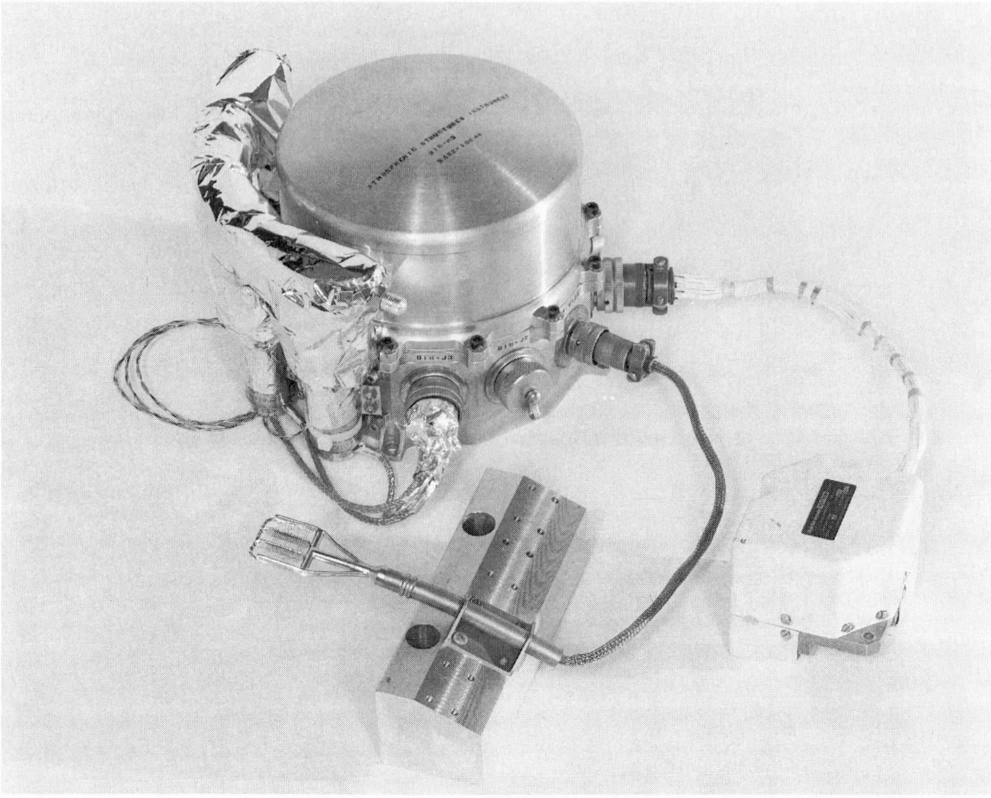

Fig. 1b.

Fig. 1. Photopraphs of the Galileo Probe Atmosphere Structure Instrument. (a) Electronics housing and pressure sensors, with the inlet manifold installed. Cables from the three pressure sensors join in a single connector which mates with one on the electronics housing. (b) The complete instrument, ready for installation on the Probe. The pressure sensors have been thermally insulated. The accelerometers are within the white housing. The temperature sensor is mounted on a block used in laboratory testing.

(1) To accurately define the state properties as a function of altitude below the 100 mb level to $\sim$20 bars. (They have never been measured below the 1 bar level.)

(2) To define the currently highly uncertain state properties of the upper atmosphere.

(3) To measure the stability of the atmosphere, and identify convective layers and stable layers, where they exist.

(4) To detect cloud levels from changes in lapse rate at their boundaries.

(5) To define state properties within the clouds, and thus provide supplementary information on cloud composition.

(6) To search for and characterize wave structures in the atmosphere.

(7) To search for and measure intensity and scale of turbulence in the atmosphere.

(8) To measure vertical flow velocities above a threshold of about 0.3 m s^{-1}.

(9) To establish an altitude scale for use in correlating all Probe experiment data.

(10) To define the probe vertical velocity, necessary to the analysis of the Doppler wind experiment.

One area of high interest to the experiment team is atmospheric dynamics. Key questions concern the nature of the Jovian circulation, and the convective processes (Ingersoll, 1981, 1990). Is the circulation very deep, extending well toward the planet center and driven by internal heat, or is it superficial, driven by energy of condensation of cloud species and/or solar heating, or is it a combination of these phenomena? Are gravity waves present and instrumental in maintaining the circulation? (They are known to be important to momentum transport in the atmospheres of Earth, Mars, and Venus.) Inferences from Atmosphere Structure Experiment data could provide insights into these questions.

The vertical lapse rate of temperature dT/dz is a key dynamical parameter which defines the static stability of the atmosphere against convective overturning. Adiabatic lapse rates imply convection; subadiabatic, stable stratification. Lapse rate will be determined as a function of altitude. Discontinuities in the lapse rate will provide information on condensation of condensible species (e.g., ammonia, water vapor, ammonium hydrosulfide) and locate cloud boundaries. Correlation of the measurements of temperature and pressure with the Nephelometer cloud observations will define altitudes and vertical separation of cloud layers, while state properties and lapse rates within the cloud layers will suggest their composition.

The magnitude of the vertical winds and their variability, inferred from measurements of descent velocity, dz/dt, could provide direct observations of vigorous convection, or of the presence and amplitude of gravity waves in stable layers. (Gravity waves can cause oscillatory variations in the descent velocity.) The descent velocities are also essential to the analysis of the Probe Doppler Wind Experiment, since they make it possible to resolve the line-of-sight velocity into vertical and horizontal components. Random variations in the instantaneous acceleration of the Probe center of gravity will indicate the presence, scale, and intensity of turbulence in the atmosphere. All of these measurements should provide a valuable 'ground truth' base for use by the remote sensing instruments on the Orbiter, which will then be better able to interpret their global observations. A discussion of the Galileo atmospheric science goals, and a brief description of the Atmosphere Structure Experiment has been given by Hunten *et al.* (1986).

These are ambitious and extensive goals, and to accomplish them, the experiment hardware must perform up to design expectations. To cope with the task of analyzing the data to satisfy the objectives, the following team of experimenters has been formed:

A. Seiff, San Jose State University Foundation, PI;
D. B. Kirk, University of Oregon;
R. E. Young, Ames Research Center, NASA;
G. Schubert, University of California, Los Angeles;
J. Mihalov, Ames Research Center, NASA;
T. C. D. Knight, Martin Marietta Aerospace;
R. C. Blanchard, Langley Research Center, NASA; and
S. C. Sommer, retired (formerly at Ames Research Center, NASA).

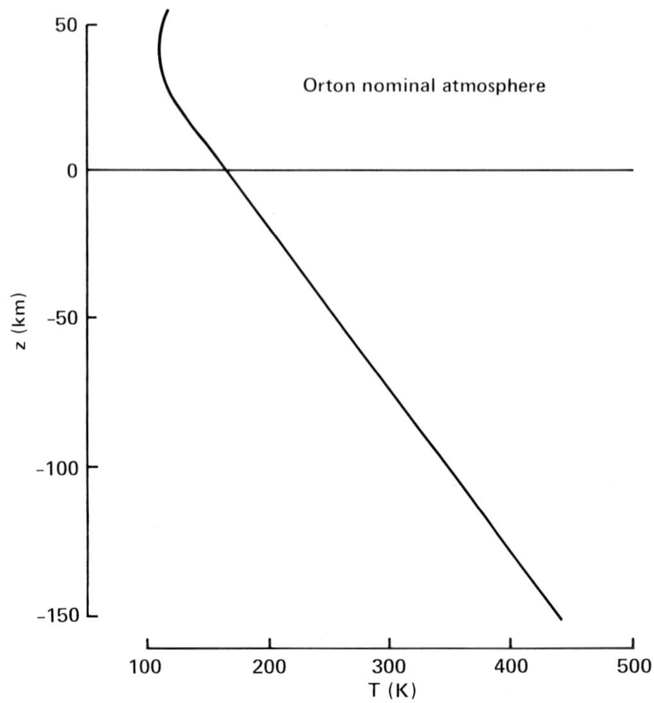
Orton nominal atmosphere
z (km)
50
0
-50
-100
-150
100
200
300
400
500
T (K)

Fig. 2a.

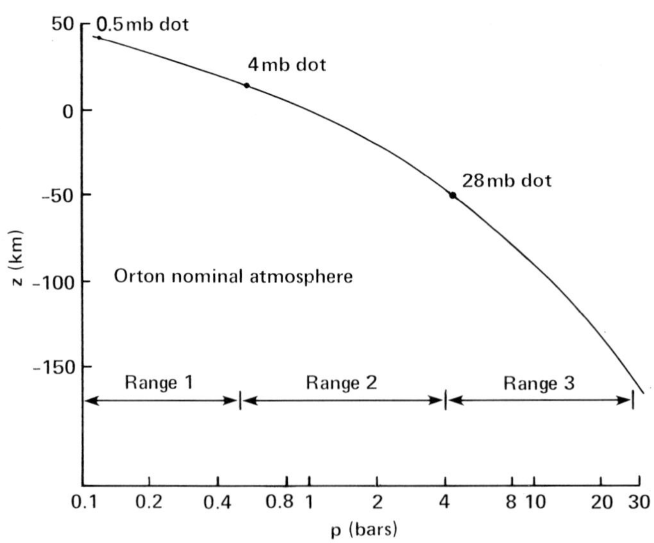
0.5 mb dot
4 mb dot
28 mb dot
Orton nominal atmosphere
Range 1
Range 2
Range 3
z (km)
50
0
-50
-100
-150
0.1
0.2
0.4
0.8
1
2
4
8
10
20
30
p (bars)

Fig. 2b.

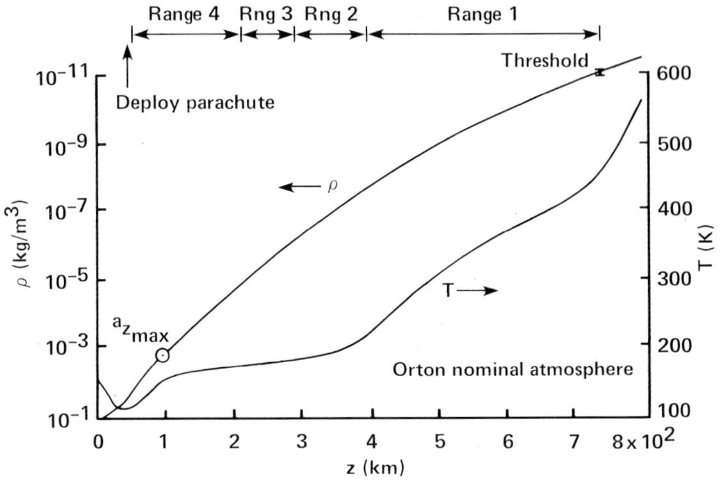

Fig. 2c.

Fig. 2. The measurement range and resolution of the Atmosphere Structure Instrument sensors, compared with Orton's nominal model of Jupiter's atmosphere: (a) Temperatures anticipated in the descent mode, from +50 to −150 km. The instrument has been designed to measure from 0 to 500 K, with overflow capacity to 550 K. Estimated accuracy ranges from ~0.1 K at the low end to ~1 K at 500 K (see text). (b) Pressures anticipated in descent mode. Sensor range boundaries and initial resolution on each range are indicated. See text for accuracy discussion. (c) Densities and temperatures expected in the middle and upper atmosphere. Acceleration sensor threshold and ranging boundaries are shown. See text for accuracy discussion.

2.1. Measurement ranges, resolution, and accuracy

An indication of measurement data we hope to obtain in descent mode is given in Figures 2(a) and 2(b), which show the instrument ranges compared with profiles of temperatures and pressure from a nominal atmosphere proposed by Orton (1981). (The deep atmosphere is modeled as adiabatic.) Pressure and temperature will be sampled at 2 s intervals in descent. Altitude resolution will be 0.2 to 0.06 km at descent velocities from 100 to 30 m s^{-1}. As can be seen in the figure, this will give a very dense array of points along the curves (50 to 160 points in 10 km), the number increasing with depth in descent as the Probe descent velocity decreases. Absolute uncertainty of the temperature measurements is expected to be ~1 K at the higher temperatures and <0.1 K near 100 K (hardly perceptible on the graph). The selected temperature range, 0 to 500 K, provides some margin beyond temperatures expected in descent. The measurement range extends to the 38 bar level in an adiabatic atmosphere.

Pressure sensors of three ranges, 0.5, 4, and 28 bars full scale, will be read to 10 bit resolution, giving least count values of 0.5, 4, and 28 mb, respectively. The size of the three dots in Figure 2(b), placed at locations of maximum uncertainty relative to pressure magnitude (low in the three ranges just after range change) indicate the resolution uncertainty. Absolute pressure errors could be as large as 3 counts, three times the dot sizes. Further discussion of accuracy is given below.

Figure 2(c) shows the density and temperature profiles from Orton's nominal model of the middle and upper atmosphere, as functions of altitude above the 1 bar level. The exospheric model temperature profile is guided by Voyager UV solar and stellar occultation data, which include a deduced temperature of 1100 ± 200 K at 1450 km altitude (Atreya *et al.*, 1981; Festou *et al.*, 1981). The model temperature variations induce the curvature shown in the density profile. (Conversely, from observed curvature, temperature variations may be deduced.) Notice the comparatively straight section of the density curve where the model atmosphere is nearly isothermal, between 100 and 350 km altitude.

The accelerometers to be used to define the atmosphere above 50 km have 4 ranges in the probe axial direction, with a dynamic range from 3 μg to 400 g. The altitude intervals of use for each range are indicated in Figure 2(c).

Altitude resolution is important to defining local curvature in the density profile and short vertical wavelength variations in the profile, such as gravity wave structures, thought likely to be present in the upper atmosphere. Altitude resolution is determined by sampling frequency. This was limited by the available Probe memory capacity, since the entry data are stored for transmission in descent. Samples will be taken at fixed time intervals of $\frac{5}{16}$ s, alternating between the primary z_1 sensor and the redundant or secondary sensor z_2. This leads to altitude resolution of ~2.0 km throughout ranges 1, 2, and 3, in which, because of the low ambient density, velocity and path angle do not change much from values at entry. At 100 km altitude (near peak deceleration), velocity has decreased, and altitude resolution is improved to ~1.0 km. Altitude resolution continues to improve to a value of ~0.15 km just prior to parachute deployment near 50 km altitude. The resolution above 200 km will permit waves of vertical wavelength ~10 km to be detected and defined. Wavelengths longer than 1 km vertically will be defined at 50 km altitude.

The measurement resolution of the axial accelerometers on the four ranges is nominally 3 μg, 0.1 mg, 3 mg, and 0.1 g, respectively. We take the range 1 threshold to be 15 μg (5 counts), which corresponds to 1.6×10^{-11} kg m^{-3} atmospheric density. In the model atmosphere, Figure 2(c), this places the threshold at 750 km altitude where density is defined within the ±10% resolution (indicated by the bar height at the threshold point). Measurement resolution imrpoves to ±1% at 615 km altitude, and, at full scale on ranges 1, 2, and 4, to 0.025% of reading (0.05% on range 3). After upranging, on each new range, initial resolution is 0.8%. Altitudes at which ranging will occur are indicated in Figure 2(c). Basic sensor accuracy is ~0.01%, so resolution, not accuracy, will always be the limitation.

Atmospheric temperature uncertainties from such a data set were shown for Pioneer Venus data (with less capable accelerometers and less frequent sampling relative to acceleration pulse duration) to be typically a few K, with maximum uncertainty ~10 K immediately after range change. However, this does not include the uncertainty in mean molecular weight of Jupiter's middle and upper atmosphere. (The mean molecular weight μ is required in the equation of state to derive temperature from density and pressure.) Errors in the molecular weight lead directly to the same fractional error in temperature. There is no present basis for estimating this uncertainty quantitatively, but we expect it will be $<5\%$ and possibly as small as 1% for the middle atmosphere, and $<10\%$ for the upper atmosphere, after the complete set of investigations to be performed by the Galileo Probe (including a neutral mass spectrometer and a helium abundance detector) and the Orbiter (including a UV spectrometer) has been analyzed.

3. Sensors

Sensors selected to make the above measurements are described here, along with expected accuracies and principal sources of error.

3.1. Temperature sensors

The temperature sensors for Galileo were modeled after those used on the four probes of the Pioneer Venus Mission. They were designed for fast response (i.e., good thermal coupling to the atmosphere and small thermal inertia), and to be insensitive to support conduction, thermal radiation, and self heating errors (Seiff, 1976; Seiff *et al.*, 1980b).

The sensors are deployed outside the probe boundary layer in a region where local flow velocity around the probe is high, to avoid thermal contamination and promote rapid heat transfer. Figure 3 shows their location, as well as the locations of the pressure sampling inlet, the accelerometers (near the Probe center of gravity) and of the electronics and pressure sensors within the Probe.

Figure 4 is a schematic of the temperature sensor, which is a dual element platinum resistance thermometer. Primary and secondary sensors, designated T_1 and T_2, are mounted on a single head. The primary sensor has very short response time and minimizes thermal errors. The secondary sensor has acceptable response and error rejection, and is better protected, less susceptible to damage in handling and in flight. Both were demonstrated by tests to survive the specified engineering environments. (Both survived the launch, entry, and descent environments on the Pioneer Venus Mission.) Both should measure temperatures accurately under conditions of descent in Jupiter's atmosphere.

The primary (T_1) sensor is a fine platinum wire directly exposed to the atmosphere. It is wound around an open frame of platinum–rhodium alloy, with atmospheric gases flowing through the frame (Figure 4). The wire is 0.1 mm in diameter by approximately 1.2 m long. It is insulated from the frame by a thin layer of glass, and is fixed in position by a second layer of glass along the outer posts. The secondary sensor (T_2) is a 6 cm length of 0.025 mm diameter wire configured as a 1 cm long raster over a thin glass film on the leading side of the outermost tube in the sensor frame (see enlarged inset in Figure 4). It is covered with a thin, protective coating of glass. To prevent resistance changes due to thermal strain, the glass used in the coatings was selected to match the thermal expansion coefficients of the frame and sensing elements. Any residual strain effects were included in the calibrations over the full use range of the sensors.

The sensor support stem is a thin-walled tube of low-conductivity stainless steel, which limits conduction between the sensor and the Probe shelf to which it is mounted.

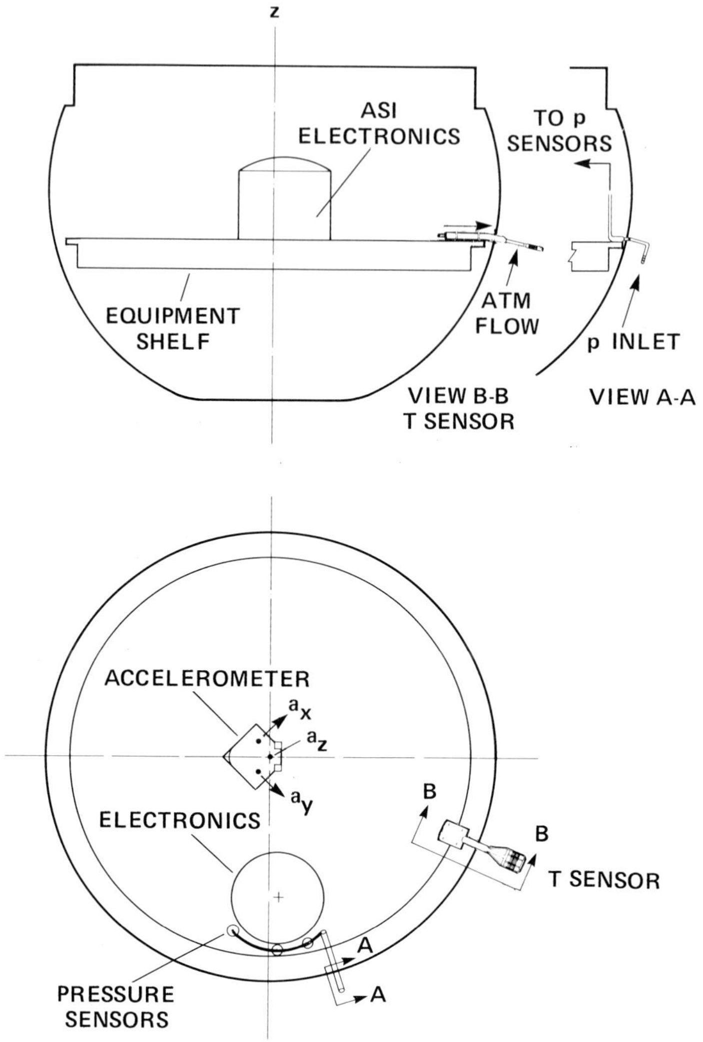

Fig. 3. Installation of the ASI on the Galileo Probe descent configuration. The temperature and pressure sensors sample flow outside the descent module boundary layer. The Probe center of gravity lies within the accelerometer package.

Pioneer Venus experience* led us to coat all external surfaces of the sensor with a 25 μm thick insulating polymer film (vapor deposited Parylene) in order to prevent shorting of the sensor by cloud materials. The polymer is, in turn, covered by a 1000 A film of gold, to ground the external surface to the Probe and prevent potential differences which could

* Partial shorting of the T_1 sensors occurred during passage through the clouds of Venus, due to accumulation of a conductive film of surfuric acid on the wire surfaces. The shorting cleared below the clouds, as the film was blown off by atmospheric flow (Seiff *et al.*, 1980a).

lead to static discharges. The low emissivity of the gold coating also limits radiative exchange with the environment.

A second modification to the Pioneer Venus design was the provision of an electrostatic discharge shield around the sensor. The shield, shown in Figure 4(b), is about 4.5 cm in diameter and is open at front and rear, so that the atmosphere flows through freely. Wires cross the openings at about 1 cm intervals, to enclose the sensor in a space at Probe potential. If, during descent, an electric charge builds up on the Probe, or if electrostatic fields in the atmosphere cause discharges to take place, the cage is intended to prevent currents from discharging to the sensor.

Tests of the shielded sensor were made at reduced pressures in a bell jar. Ten thousand volt discharges were created between the sensor assembly and a plate nearby. The discharges always took place to the shield, and sensor resistance measurements

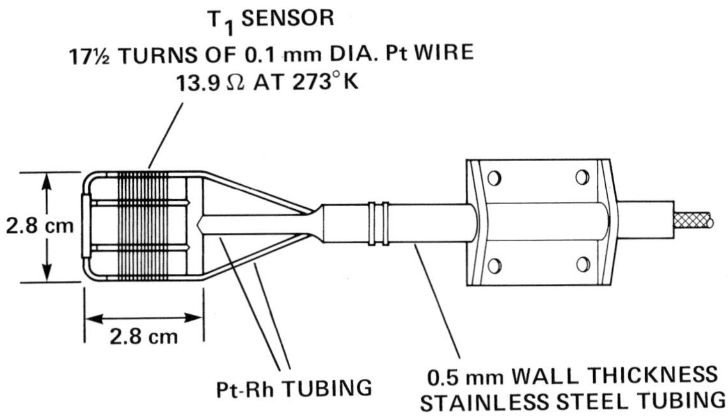

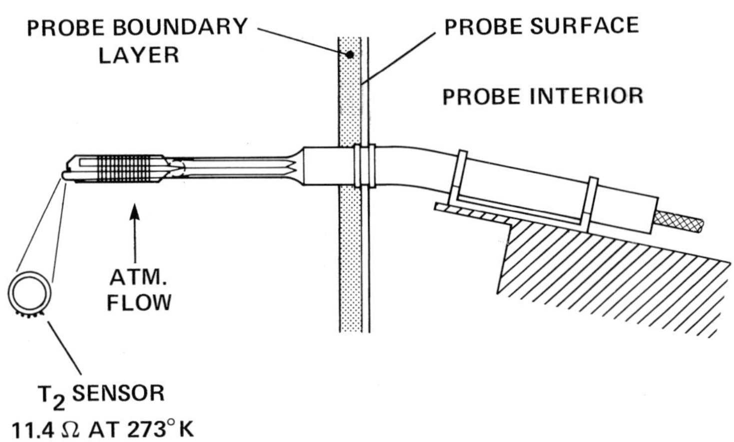

Fig. 4a.

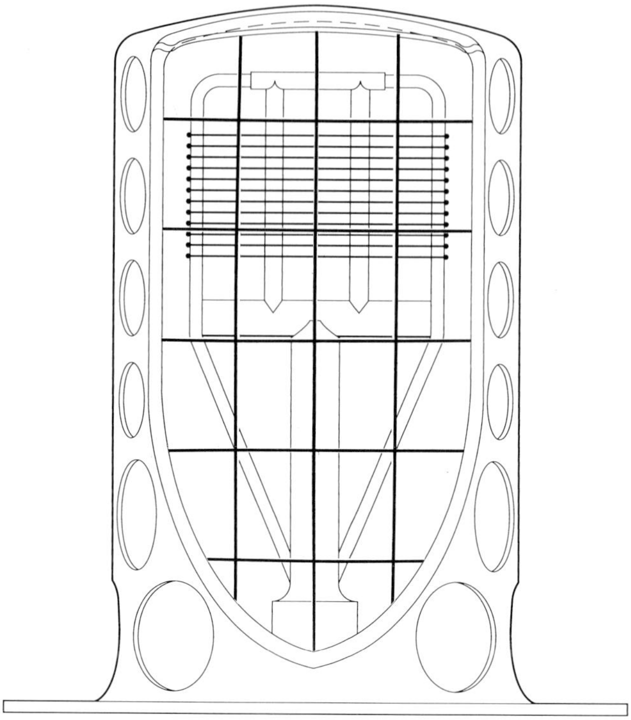

Fig. 4b.

Fig. 4. The temperature sensor. (a) Sketches showing the sensor configuration and mounting. The sensor is mounted to the Probe instrument shelf. Its stem passes through the Probe wall and boundary layer to deploy the sensor into external atmospheric flow. (b) View of the sensor enclosed in the electrostatic discharge shield.

taken simultaneously were found to be unaffected by the discharge. The shield could be important if the Probe encounters strong electrostatic fields or lightning discharges in the atmosphere of Jupiter.

The sensors were fabricated and calibrated by Rosemount, Inc. They were calibrated by immersion in baths of liquid nitrogen (77 K), liquid oxygen (89 K), ice water (273 K), boiling water (373 K), and against a secondary standard at 477 K. The sensors were enclosed in plastic bags and were fully immersed in the calibration bath and left until resistances were stable. The resistances at known temperatures were then used to calculate constants in the familiar Callendar–Van Dusen equation (see, e.g., Riddle *et al.*, 1973) for purposes of interpolation between calibration points above 273 K. (The resistance function $R(T)$ of platinum is the basis of the International Fractical Temperature Scale.) Below 273 K, data were referenced to the International Practical Temperature Scale (68), and correction curves for deviations from that scale were derived (Rosemount Report 27410, 1975).

Sensor calibrations were checked several times at Ames Research Center over a period of years. The initial Ames calibration was within a few hundredths K of the

Rosemount calibrations in liquid nitrogen. The repeatability of the Ames calibrations is shown in Figure 5 for two temperatures, 77.4 K (liquid nitrogen) and 273.15 K (ice bath). At 77 K, the T_1 calibration data group very closely around two lines differing by 0.15 K, which have the theoretical slope $\mathrm{d}R/\mathrm{d}T$ for platinum. The standard deviation of the points from their associated lines is 0.017 K. The left-hand line fits calibrations taken before November 1983, at which time the sensors were coated with parylene and

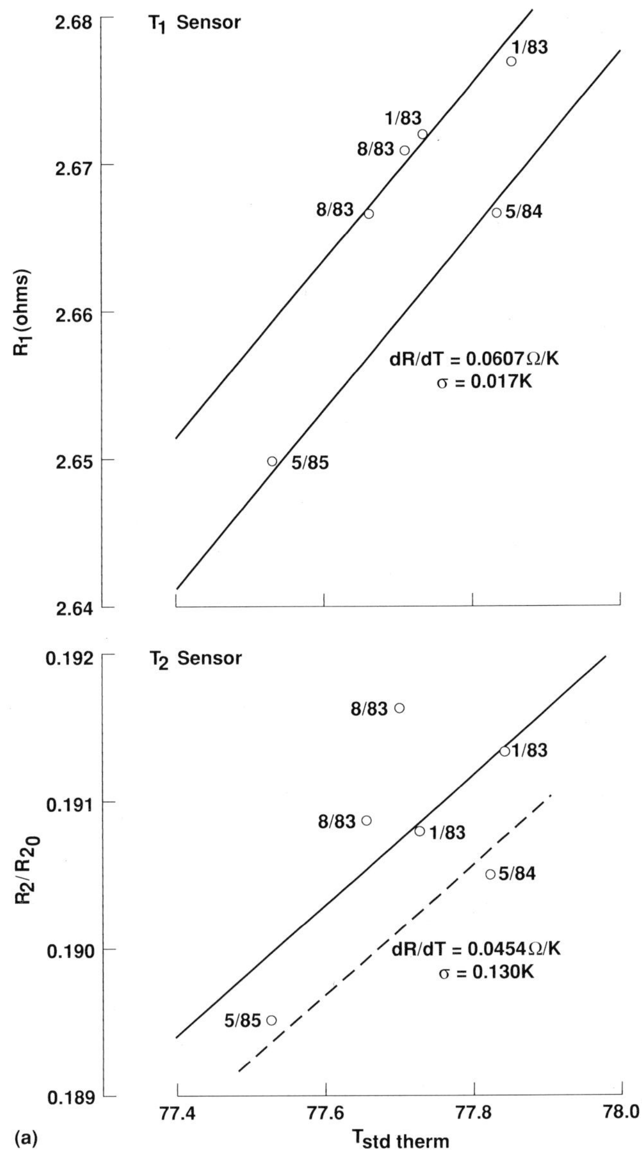

Fig. 5a.

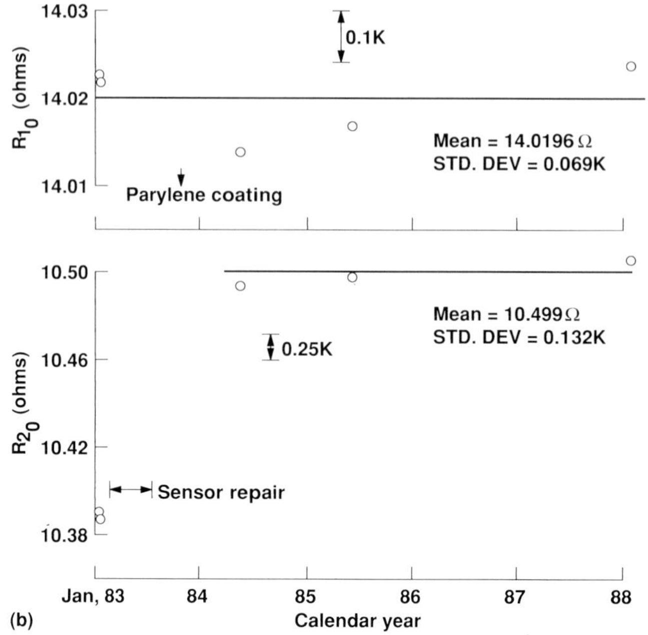

Fig. 5b.

Fig. 5. Temperature sensor calibration data taken over the period from 1983 to 1988. (a) In liquid nitrogen baths. (b) In water ice baths.

gold. The right-hand line fits two calibrations taken subsequently. It appears that the T_1 calibrations at 77 K were measurably affected (by 0.15 K) by application of the coatings, and the 1984, 85 calibrations will be used in reducing flight data.

In the lower figure, the full set of T_2 calibration data in liquid nitrogen shows a standard deviation of 0.13 K about the solid line. This sensor required a repair in March 1983, which increased its resistance at 0 °C by 0.12 Ω (1%). To take the effect of this change out of Figure 5(a), the resistance at 77 K has been ratioed to its value at 273.15 K. The data fall into 3 groups, one in early 1983, one after the repair (late 1983), and the third after coating application in November 1983. The dashed line of theoretical slope through the two data points taken in 1984 and 1985 (about 0.13 K to the right of the solid line) is considered the calibration of record.

Calibrations of sensor resistance in an ice bath over the period from 1983 to 1988 are shown in Figure 5(b) to an enlarged vertical scale. Mean values are shown by horizontal lines. Standard deviations about the mean are 0.069 K for T_1, and 0.13 K for T_2 over the 5-year period. A rising trend in resistance with time after May, 1984 is suggested by 3 data points. We can see no physical cause for this, and do not believe is real. However, we shall be alert to this possible trend in reducing the data at Jupiter. The bars are 0.1 K and 0.25 K high.

To functionally test the sensor over the full range to 500 K and to compare T_1 and

T_2 readings against one another more generally and against some calibrated thermocouples, a thermally insulated, closed circuit flow channel (a table-top wind tunnel) was built, in which gas (either air or helium) was circulated at atmospheric pressure. Electrical heaters on the backside of the circuit warmed the gas at a rate to match that expected in descent on Jupiter. The sensor projected through the channel wall into the flow about as far as it extends outside the descent module into the atmosphere, when mounted on the probe. The channel was made of ordinary 6-inch diameter, sheet metal duct pipe, insulated to reduce heat exchange with the ambient atmosphere. The two long legs of the channel were about 1.5 m long; the short legs, about 0.75 m long. A fan, mounted internally, circulated the flow at a velocity of 6 m s^{-1}.

It was found that there was considerable thermal inhomogeneity in this channel during the heating cycle, and the best data were taken when the environment was cooling from 500 K back to room temperature. Under these conditions, the following results are obtained: (1) T_1 and T_2 agreed typically within about 0.2 K at 450 K, their difference decreasing with temperature. (This was without correction for the lag of the T_2 sensor, which was greater in the test than it will be on Jupiter because of differences in gas density and flow velocity.) (2) The temperatures agreed with those given by the closest nearby thermocouples within 0.1 K typically. (3) Tests made with and without the electrostatic discharge shield showed that readings were not affected significantly by its presence.

Calibrations run end-to-end with the ASI sensing electronics, also over a period of years, were consistent with analog, sensor level calibrations within 1 count.

The above data suggest an accuracy of calibration of about 0.1 K T_1 and 0.2 K T_2 at 500 K, the absolute error tending to increase with temperature. However, in order not to promote extreme expectations, we presently suggest an uncertainty of 1 K at 500 K, and 0.1 K at 100 K when all other error sources are present. We will refine these uncertainties as warranted by the data returned. Relative accuracy over short time spans should be limited only by the resolution, 0.013 K.

Response time varies with ambient density and descent velocity, from 16 ms at deployment to 5 ms at 16 bars for the primary sensor, and from 300 ms to 80 ms for the secondary. The sensors will track the rise of atmospheric temperature during descent, T_1 to within 0.002 K, and T_2 within 0.038 K. A more severe requirement on response time is imposed by the goal of measuring turbulent temperature fluctuations. With the response achieved, turbulent fluctuations of 100 m scale and 1 K amplitude, for example, will be followed by the primary sensor within about 0.01 K at a descent velocity of 100 m s^{-1}.

The sensors will be sampled alternatively, at 2-s intervals. If either sensor fails, failure logic in the microprocessor provides that data from only the surviving sensor is to be taken at 2-s intervals. The failure logic simply looks for an open or shorted sensor, as indicated by overscale or zero readings which persist for one major frame (a 64-s period).

3.2. Pressure sensors

The pressure sensors in the Galileo Atmosphere Structure Instrument measure pressure from the deflection of a thin, stainless steel diaphragm, 1.5 cm in diameter, with a sealed vacuum reference on its backside. (The sensors were provided by the Tavis Corporation, and are commercially available.) A sensor of this type was used in the parachute phase of the Atmosphere Structure Experiment on each Viking Lander, and, after landing, was handed off to the Meteorology Team to measure daily and seasonal pressure variations on Mars. It did so successfully over a period of years.

The diaphragm deflection is measured by changes in reluctance of magnetic circuits which include the air gaps between the diaphragm and sensing plates on either side of the diaphragm. The sensor external dimensions are 2.5 cm diameter by 10 cm long. Each sensor weighs 129 g and requires 150 mW of power.

There are three sensors in the instrument, with full scale ranges of 500 mb, 4 bars, and 28 bars. (This geometric range scaling, by factors of 8 and 7, maintains roughly constant percentage reading accuracy over the descent.) The maximum pressure of 28 bars was selected to provide data in the event that the Probe survives to greater than the design depth on Jupiter. This sensor can also run overrange by about 10%. The three sensors can be seen mounted to the electronics housing and connected to the inlet manifold tubing in Figure 1(a). On the Probe, the manifold leads to a pitot tube inlet located along the side of the descent module, outside the Probe boundary layer, Figure 3. In Figure 1(b), the sensors and inlet are enclosed in a thermal insulation blanket (see below).

The signal level is 0 to 5 V full scale. With 10 bit A/D conversion, the resolution is 1/1024 of full scale on each range. Resolution uncertainty of a single reading thus varies from 0.5% of reading at first use to 0.1% at full scale.

Sampling is simultaneous with temperature sampling, at intervals of 2 s. At $p < 500$ mb, ranges 1 and 2 are sampled alternately. At $0.5 < p < 4$ bars, ranges 2 and 3 alternate. This sampling scheme provides redundancy and measurement accuracy checks, within the resolution of the higher range sensor. Above 4 bars, where only the 28 bar sensor is on range, it is sampled every 2 s.

Failure detection logic was provided for the two lower ranges. The sensor is considered to have failed if it gives zero output for one major frame (64 s), or if the reading is constant over one major frame. Failed sensors are checked continuously and reinstated if the above conditions disappear. Another failure protection implemented is that the lowest range sensor will not be read after 6 major frames into the descent, and the middle range will not be read after 28 major frames. These sensors should be well over range after these periods of descent, with deployment at 100 mb.

Accuracy and stability of the sensors were evaluated by repeated calibrations over a period of years while Galileo was awaiting launch. Typically, scale factors at a given temperature were found to be stable within 0.1% over these years. End-to-end calibrations, taken through flight electronics, were consistent with sensor level calibrations, within the digital resolution.

Sensor scale factors vary with temperature by 0.1% to 0.5% over the temperature range from -20 °C to $+50$ °C. The temperature dependence was repeatable within narrow limits over a period of years, and is part of the calibration data set. The sensor offsets vary with temperature to the extent of 20 to 50 mV (0.2% to 0.5% of the 5 V full scale signal) over the same temperature range. These variations are repeatable within 10 mV.

The signal at a given pressure is also sensitive to the rate of change of probe internal temperatures. It was to minimize this effect that the sensors were thermally insulated. The remaining magnitude of the effect is ~25 mV maximum in the nominal descent thermal environment, in which probe internal temperature changes at the rate ~1 °C min^{-1}. Corrections to the data will be made for these thermal offsets. To define the corrections, the Probe interior temperature history descent was simulated, and offsets induced were recorded. Tests were performed in the nominal and in a worst case environment with dT/dt up to 2.9 °C min^{-1}. The offsets were found to be correlated with the temperature difference between sensor internal thermometers and one mounted on the manifold coupling of the range 3 sensor. These four temperatures will be read as engineering data in flight. The correction is expected to be accurate within 5 mV (1 count or 0.1% of full scale signal) in the nominal descent environment.

Sensor offsets were typically set at 100 to 200 mV. They differ by about 10 mV from turn-on to turn-on. While this is undesirable, it is not expected to affect accuracy seriously, since offsets will be read just prior to entry into the atmosphere, and should be predictable through the ensuing hour of descent. At worst, it is estimated that offset uncertainty could introduce 1 or 2 counts of uncertainty (out of 1024, full scale) in the data.

The effect on sensor output of Jupiter's strong intrinsic magnetic field, which is the order of 10× that of Earth, was investigated and found to be significant. Pioneer 10 and 11 data (Smith *et al.*, 1976) indicate a magnetic field strength within Jupiter's equatorial region of approximately 4 G. The sensors were tested in magnetic fields up to 12 G. Sensor scale factors were unaffected by the applied field, but offsets were found to be sensitive to both axial and transverse fields by as much as 50 to 200 mV. This was not originally anticipated, but it is not surprising, considering that the diaphragm deflections are sensed magnetically. Offset shifts induced in these tests remained after the test field was turned off. To preserve the planned accuracy, the sensors were magnetically shielded in 'mu-metal' housings 0.5 mm thick, which extend over the sensing heads. (Shielding material was supplied and housings were fabricated by Advance Magnetics, Inc.) The shields reduce the field within by a factor of 20. With shields in place, sensor outputs were constant within 1 count (5 mV) in both transverse and axial 12 G fields, while the sensors were rotated through 360°.

3.3. Acceleration Sensors

The acceleration sensors on the Galileo Probe will play an important role by defining the structure of the upper and middle atmosphere (pressure levels 10^{-7} to 100 mb). Decelerations measured in the probe axis of symmetry (z-axis) provide the basic data, while the lateral (x and y) axis sensors define the Probe angle of attack history during entry and thus the magnitude and direction of the resultant aerodynamic force. In the lower atmosphere, accelerometers will establish vertical wind magnitude, and characterize turbulent motions in the atmosphere, in both axial and lateral directions.

To respond to these objectives, the sensors must have a broad dynamic range. In descent, measured axial accelerations will be close to Jupiter's gravitational acceleration, nominally $2.37g$ (g is Earth's gravitational acceleration, 9.806 m s^{-2}). During entry, the nominal deceleration peak is $260g$, with an upper limit near $400g$. The minimum detectable acceleration determines the experiment threshold, and for this purpose, a sensitivity of 3 μg $count^{-1}$ was selected, based on knowledge of capabilities of existing sensors. (Tests of the Galileo sensors indicate they are indeed capable of resolving <3 μg on the most sensitive range.) Thus, the dynamic range extends from 3 μg to $409g$.

The accelerometers were designed and manufactured by the Bell Aerospace Division of Textron. They were derived from Model XI guidance accelerometers, modified to meet specific experiment requirements. (Bell Model IX sensors were used in the Viking instrument.) A key modification was extension of maximum measuring range. External and sectional views of one sensor are shown in Figure 6. A photograph of the four-sensor flight unit is included in Figure 1(b).

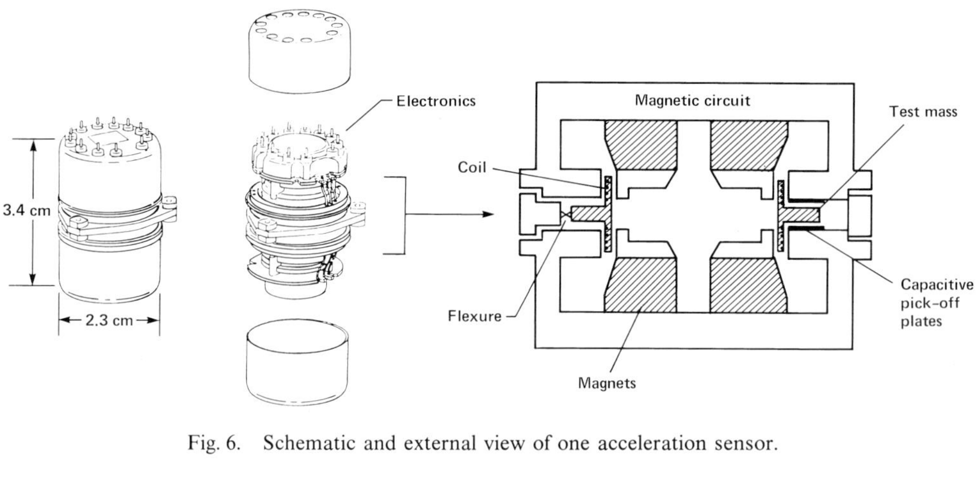

Fig. 6. Schematic and external view of one acceleration sensor.

The operating principle of the sensors may be understood from the sectional drawing in Figure 6. When the sensor experiences acceleration, its electronics detect movement of the flexure supported test mass from its null position (by capacitive sensing) and supply a servo-controlled current to the coil in the test mass to restore it to the null position. The coil field interacts with the intense field of the permanent magnets to provide necessary restoring force. The current in the coil is then the measure of the acceleration. The servo system constrains the test mass to its null position within 1 μrad g^{-1}. Thus, maximum deflection at $410g$ is 0.4 mrad, corresponding to test mass linear displacement $< 10\ \mu m$. This leads to exceptionally linear response to input accelerations over the full measurement range.

The sensor analog electronics generate a signal of 0 to 5 V, or, where both positive and negative accelerations must be sensed (x- and y-axes, and z-axes on Range 3), ± 2.4 V. Individual sensors fit within a 2.3 cm diameter $\times$ 3.4 cm long envelope (Figure 6) and weigh 48 g. The four sensors are rigidly mounted on a bracket, with the z-axis sensors on the probe axis (Figure 2), and the z_1 sensor very near the probe center of gravity during entry. The displacement of the lateral axis sensors from the Probe axis of symmetry is normal to the directions of their sensitive axes. Thus, the only rotational acceleration to which they respond is spin acceleration. The weight of the assembly is 540 g, much of the weight being in the bracket. Power required varies from 1.25 W at 0 g to 6.5 W at 410 g. A single cable brings conditioned power and accelerometer range change commands from the ASI electronics, and conducts analog signals from the 4 sensors to the ASI electronics (Figure 1(b)).

The two z-axis sensors operate on 4 ranges with full scale levels of $12mg$, $0.4g$, $\pm 6.4g$, and $410g$. The x- and y-axis sensors have 3 ranges with full scale levels of $\pm 12.5mg$, $\pm 8g$, and $\pm 12.8g$. The digital resolution is 12 bits in the z-axis, 8 bits in the two lateral axes. Thus, in the z-axis, 1 count ranges in value from 3 μg to $0.1g$.

Command capability was provided to choose how the two z-axis accelerometers will be used during Jupiter entry and descent, depending on how well each functions during interplanetary cruise. Either can be used alone, or both can be used with samples interdigitated. Either can also be selected as the 'primary sensor', the source of the z-axial turbulence data.

The data taken during the high-speed atmosphere entry have to be stored because of radio communication blackout induced by the plasma sheath around the Probe. The data are stored in the Probe memory and will be read out twice from designated words in the Probe data format during descent below the 0.1 bar level. The memory capacity provided was 12 kbits. This allocation, over the estimated maximum entry period of 240 s, allows a data collection rate of 50 bps, which had to be carefully budgeted. Thirty-six bps were allocated to axial decelerations, the basic data which define the density profile. A 12-bit sample is taken from each z-axis sensor every $\frac{5}{8}$ s. Samples from the two sensors are staggered, so that an axial acceleration is read every $\frac{5}{16}$ s. To conserve data, the resultant acceleration normal to the z-axis, termed a_n, is computed from a_x and a_y in the ASI microprocessor, and one sample/second is transmitted at 8-bit resolution. This is to be used to define the probe resultant angle of attack oscillations from observed oscillations in the lateral acceleration. The remaining 6 bps are used for engineering data, sensor ranges and temperatures.

In Descent Mode, the total experiment data rate is 18 bps. Of this, 6 bps are allocated to temperature, 6 bps to pressure, 4.5 bps to acceleration data, and 1.5 bps to engineering data. Within the acceleration data, 1.5 bps are used to define the total velocity change in the z-axial direction, as derived from each of the z_1 and z_2 sensors over staggered 16 sec intervals; 1.5 bps, to transmitting the mean, maximum, and minimum value of a_n in 16 s intervals; and 1.5 bps to transmit statistics on the axial and lateral turbulence levels every 96 s. The total velocity change in the z-direction is integrated by the microprocessor which samples the accelerations 32 times s^{-1}. Thus, it is evident that the use of data is very economical, and minimal. It is also evident that the data collection pattern is moderately complex, but is readily handled by programming the experiment control microprocessor. A more complete exposition of the sampling sequences and their implementation is given in the following section.

Turbulence data are compressed by counting excursions from the mean acceleration which exceed pre-specified levels, both in the z-axis and normal to the z-axis. The pre-specified deviation levels in a_z are $\pm 0.025\,g_J$, $\pm 0.05\,g_J$, $\pm 0.10\,g_J$, and $\pm 0.20\,g_J$ (where g_J is the acceleration of gravity on Jupiter, taken as 2.37 g_E). The pre-specified levels in a_n are 0.0019, 0.0056, 0.0113, and 0.0226 g_J. The number of excursions of the acceleration above each of these levels is counted in a 96-s sample period, and the number of counts is transmitted in 12, 12-bit words. (Positive and negative excursions in z are independently counted, while only the absolute magnitude of the excursions is read on n.) The interval mean acceleration in z is taken to be the average value in the 16-s period preceding the 96-s sample. In the two transverse axes (and hence in a_n), the Probe is assumed to be symmetric and the acceleration is assumed to be zero in the absence of turbulence.

The acceleration sensors were calibrated against the Earth's gravitational acceleration by rotation through 360° under temperature controlled conditions at Bell Aerospace in Buffalo, New York. The sensor level calibrations were done 3 times, before and after environmental testing (in December 1983 and February 1984), and on return of the instrument to Bell for a minor repair in mid-1985. The scale factors were defined to 5 significant figures, and they repeat within 0.3% on Range 1 (most sensitive) and 0.01% on Range 3 (6.5g full scale). The sensors are temperature compensated, but residual sensitivity was determined by calibration at four temperatures. End-to-end calibration checks have been repeated many times in the Earth's gravity field, the last being in July 1988, and they show no changes with time in the scale factors, within the digital resolution. The bias values have long term stabilities of about 150 μg's on all ranges, and the values in effect at the time of Jupiter encounter will be read at instrument turn-on prior to entry.

It was considered unsatisfactory to calibrate the scale factor on the 410g range with gravitational inputs limited to $\pm 1g$. As an alternative, precisely known stimulus currents were applied to the coil in the test mass to simulate input accelerations up to 410g. The sensor current scale factors (ma/g) determined to 5 significant figure accuracy on Range 3 at 1g input, were used to calculate acceleration values equivalent to the stimulus currents. It was found that the Range 4 scale factors so determined (g count^{-1}) checked those obtained from $\pm 1g$ calibrations within about 0.16%.

The sensor calibration on Range 4 was also checked by centrifuge tests with the complete ASI operating at inputs up to 210g. There was agreement within 0.1% between sensor acceleration data and input accelerations determined from centrifuge rotation rate. (This is the limit of accuracy to which input accelerations could be determined.) This test was a simulation of the full use cycle of the instrument during entry into Jupiter's atmosphere, and showed satisfactory range switching operation, and also tested the agreement between the z_1 and z_2 sensors.

4. Instrument Electronics Design

A critical factor in the electronics design was the fact that the Probe passes through the Jovian radiation belts prior to entering the atmosphere. As a consequence, the instrument has to operate wiithin specification after exposure to a significant radiation dose. Parts selection was severely constrained, radiation testing was performed on samples from every lot of parts used, and the potential effects of part parameter variation with radiation dose had to be considered along with other parametric variations in designing for stability of electronic circuit operation. In general, parts were selected that can be exposed to doses of 10^6 rad (Si) without unacceptable performance degradation.

Use of a microprocessor for control of this instrument allowed significant flexibility in instrument operation, data processing, and output data formatting. In addition, it permitted several failure sensing algorithms to be implemented, to sense gross sensor failures and substitute data from healthy sensors. The flexibility of operation made it necessary for the ground support equipment to be rather sophisticated in order to be able to exercise all possible operational branching paths within the instrument software.

The basic instrument electronics system design is shown in Figure 7. This design, completed during 1979, is centered around an RCA CMOS microprocessor system comprising the CDP 1802 central processing unit (CPU), 8192 8-bit words of read-only memory (ROM), and 256 8-bit words of read/write random access memory (RAM). The system is clocked at about 2.2 MHz, a low frequency used to retain an acceptable operating margin after radiation exposure degradation. The ROM contains the instrument's operating and control software and the RAM provides for temporary storage during data accumulation and manipulation and also provides the data output buffer.

Control signals from the Probe (comprising the operating mode commands, axial accelerometer selection commands, data frame markers, data transfer clocks, and the Probe clock) are used to control the operation of the instrument by generating system interrupts. Each system interrupt causes the microprocessor to suspend the sequence in process, interrogate the interrupt registers, execute the response, and return to the sequence in process, if appropriate. In the case of an operating mode command, the microprocessor branches to the correct mode, reinitializes, and proceeds in the new operating mode at the next Probe minor frame interrupt. For other interrupts, the microprocessor completes the interrupt action and then returns to where it was prior to the interrupt. The instrument is synchronized and slaved to the Probe timing as a result of this interrupt control. The control signals from the Probe and the functions performed are shown in Table I.

The design incorporates two 12-bit analog-to-digital (A/D) converters of the dual slope integrating type, each with an auto-zeroing loop. The A/D converter precision reference voltage is derived from a highly stable zener diode having internal temperature compensation. This same reference voltage is used to derive the precision stimulus

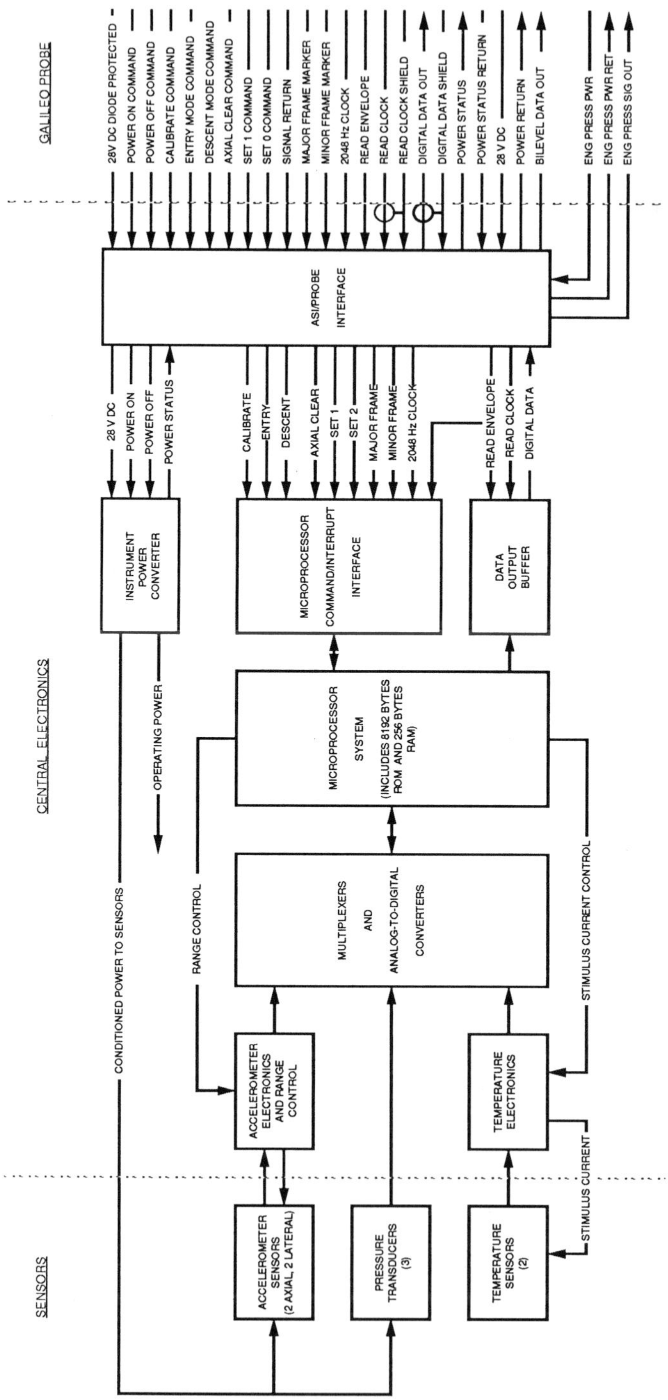

Fig. 7. Atmosphere Structure Instrument electronics block diagram.

TABLE 1 PROBE/ASI CONTROL SIGNALS AND FUNCTIONS.

CONTROL SIGNAL TYPE	CONTROL SIGNAL NAME	FUNCTION PERFORMED	
INSTRUMENT OPERATIONAL SEQUENCING CONTROL	CALIBRATE	INSTRUMENT PERFORMS A CALIBRATION SEQUENCE. DATA IS TRANSFERRED TO THE PROBE AND SOME DATA IS STORED IN THE ASI INSTRUMENT FOR USE DURING IN-FLIGHT DATA PROCESSING. THE SEQUENCE IS TO BE RUN IN-TRANSIT AND PRE-ENTRY AFTER PASSING THROUGH THE RADIATION BELTS.	
	ENTRY MODE	SEQUENCES THE SENSOR SAMPLING DURING HIGH SPEED ENTRY AND PRESENTS THE PROCESSED DATA FOR TRANSFER TO THE PROBE IN THE APPROPRIATE FORMAT.	
	DESCENT MODE	SEQUENCES THE SENSOR SAMPLING DURING DESCENT ON THE PARACHUTE AND PRESENTS THE PROCESSED DATA FOR TRANSFER TO THE PROBE IN THE APPROPRIATE FORMAT.	
AXIAL ACCELEROMETER SELECTION	AXIAL CLEAR	SETS AXIAL ACCELEROMETER BUFFER TO 00 STATE.	BUFFER STATE 00 = SELECT Z1 PRIME
	SET 0	SETS AXIAL ACCELEROMETER BUFFER BIT 0 TO THE 1 STATE	BUFFER STATE 01 = SELECT Z2 PRIME BUFFER STATE 10 = SUBSTITUTE Z1
	SET 1	SETS AXIAL ACCELEROMETER BUFFER BIT 1 TO THE 1 STATE	BUFFER STATE 11 = SUBSTITUTE Z2
SYNCHRONIZ-ATION WITH THE PROBE DATA SYSTEM	MAJOR FRAME	THE MAJOR FRAME MARKER IS USED TO COUNT THE TIME DELAYS ASSOCIATED WITH FAILURE DETECTION ALGORITHMS.	
	MINOR FRAME	THE MINOR FRAME MARKER (16 MINOR FRAMES PER MAJOR FRAME) IS USED TO PROVIDE ASI/PROBE SYNCHRONIZATION FOR SOME PARAMETERS AND TO REINITIATE OPERATION AT MODE CHANGES.	
	2048 Hz CLOCK	DATA SAMPLING WITHIN THE ASI IS SLAVED TO THE 2048 Hz PROBE CLOCK. THIS, TOGETHER WITH THE FRAME MARKERS, ASSURES THAT THE ASI DATA SAMPLING RATES AND DATA OUTPUTS ARE ALWAYS CORRECTLY SYNCHRONIZED WITH THE PROBE DATA SYSTEM.	
DATA TRANSFER TO THE PROBE	READ ENVELOPE	DATA TRANSFER TO THE PROBE OCCURS ON EIGHT READ CLOCK CYCLES WHILE THE READ ENVELOPE IS HIGH. THE NEXT WORD TO BE SENT FROM THE ASI TO THE PROBE IS WRITTEN TO THE ASI OUTPUT BUFFER WHEN THE READ ENVELOPE FALLS.	
	READ CLOCK	THE READ CLOCK TRANSFERS THE DATA FROM THE ASI OUTPUT BUFFER TO THE PROBE ONE BIT PER CLOCK CYCLE.	
POWER CONTROL	POWER ON	COMMANDS APPLICATION OF POWER TO THE ASI INSTRUMENT.	
	POWER OFF	COMMANDS REMOVAL OF POWER FROM THE ASI INSTRUMENT.	

currents and voltages for the sensors, including the constant current drive through the platinum resistance temperature sensors, the calibration currents supplied to the accelerometers, and the calibration voltages supplied to the pressure sensors. This approach results in an in-flight instrument calibration that is, to a first approximation, insensitive to the absolute stability of the reference source.

The accelerometers and pressure sensors provide 0 to + 5 V outputs to the central electronics. For the platinum resistance temperature sensors, however, the central electronics provide the stimulus currents and signal amplification. The simplified temperature measurement implementation is shown in Figure 8. A four wire configuration is used for each sensor, to eliminate measurement errors caused by voltage drop in the lead wires between the central electronics and the sensors. One pair of wires carries the stimulus current while the voltage measurement across each sensor is made through the second pair of wires, which look into the high impedance of the signal amplifier. The

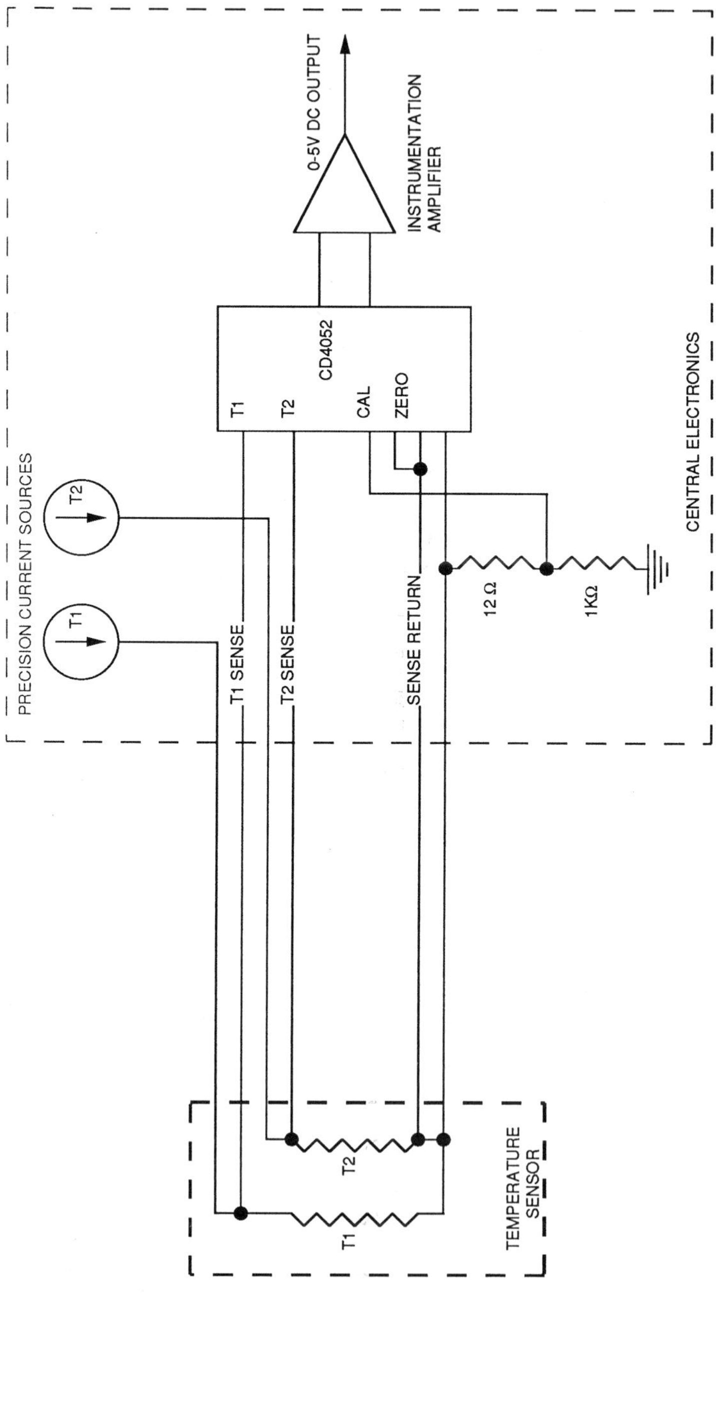

Fig. 8. Temperature measurement implementation.

configuration also provides a zero-offset determination and a scale factor calibration, by means of a stable, 12 Ω calibration resistor carried in the electronics. The analog outputs from the sensors, conditioned as necessary, are multiplexed to the A/D converters under control of the sequence programmed into the microprocessor.

Data sampling is slaved to the 2048 Hz probe clock, which is divided down to generate a data sampling interrupt at the rate of 128 samples per second. The microprocessor responds to the data sampling interrupt by setting appropriate system flags that are used later in processing the data. The data sampling interrupt itself initiates the 'start A/D conversion' in hardware, to avoid the possibility of small but variable delays in initiating it through the software interrupt process.

The basic sensor sampling sequence is a four data sampling interrupt cycle, as shown in Table II. Each of the accelerometer outputs is sampled 32 times per second, with the

TABLE 2 BASIC SAMPLING SEQUENCE CYCLE

DATA SAMPLING INTERRUPT NUMBER	SENSORS SAMPLED
1	SAMPLE OUTPUTS FROM AXIAL ACCELEROMETERS Z1 AND Z2. OUTPUTS SAMPLED EVERY CYCLE.
2	SAMPLE OUTPUTS FROM PRESSURE, TEMPERATURE, OR HOUSEKEEPING SENSORS AT APPROPRIATE TIMES TO SUPPORT DATA OUTPUT REQUIREMENTS.
3	SAMPLE OUTPUTS FROM LATERAL ACCELEROMETERS X AND Y . OUTPUTS SAMPLED EVERY CYCLE.
4	NO DATA SAMPLES ARE TAKEN

data samples for the pressure, temperature, and housekeeping sensors interspersed as necessary. The accelerometer sampling rate is the same in all operating modes and all of the data processing, including integration of acceleration to provide velocity change data, the computation of the resultant of the two lateral accelerometer samples of every measurement, and rescaling of data at accelerometer range changes, is performed digitally within the microprocessor system.

The logic for determining when to make accelerometer range changes, and sending the range change commands is also implemented within the microprocessor. Failure sensing techniques are applied to the temperature and pressure sensors to enable substitution of data in the event of a gross sensor failure, as discussed earlier.

The three instrument operating modes, calibration, entry, and descent may be functionally described as follows: The calibration mode is used at a time when the probe is still in free fall after passing through the radiation belts. It provides zero-offset and some scale factor calibrations of the sensors and will also calibrate any electronics changes. The entry mode sampling strategy emphasizes deceleration measurements, appropriate for the high speed ballistic entry portion of the mission (see above). In the descent mode, atmospheric temperature and pressure become the primary measurements.

Because of the limited data rate of the Probe communications link (128 bits per second), the data transfer and transmission rates allocated to the ASI are low, 18 bits per second in the descent mode (about $\frac{1}{7}$ of the Probe link capacity) and 50 bits per second in the entry mode, when data go into Probe memory for readout in the hour of descent. These rates critically affect the selection of the data parameters to be output, shown in Tables III and IV. The data output of 18 bits per second in the calibration mode is not critical because the quantity of data to be collected is limited, and there is no severe constraint on the collection interval, since these data are taken at a time before science data gathering begins, and are stored in the Probe memory. The parameters measured during the calibration sequence are shown in Table V. The data output for each calibrate mode parameter is the average of 16 separate measurements of that parameter. Data are taken relatively quickly, averaged, and stored in the instrument buffer memory while waiting for transfer to the probe. The zero-offset data for each sensor are also retained in instrument memory and multiplexed, as 16-bit accumulations, into the ASI descent mode data stream.

Some of the calibration data are required during on-board data processing, in both the entry and descent modes. Default values for these parameters are programmed into the microprocessor system to guard against a Probe failure in which a calibration sequence is not run.

The instrument software was written directly in the microprocessor machine code instruction set. This was necessary in order to optimize the processing and achieve the necessary processing throughout at the modest 2.2 MHz system clock frequency.

Packaging of the electronics used conventional techniques for this type of instrument. The parts were installed on custom designed multilayer (up to 9 layers) printed circuit boards. The packaging was designed and tested to survive a 400g deceleration along the flight path axis as well as normal launch environments. The other somewhat unusual requirement was to survive the high pressure of the Jovian atmosphere during descent. In order to avoid qualification problems with the large flatpacks being used, the approach taken was to seal the housing containing the electronics, initially at 1 atm of air pressure. The housing was subsequently tested over a 1-h period with external pressure gradually increasing to 20 bar to simulate descent on Jupiter. This test, performed three times over the years, showed small, but acceptable leakage. Based on these tests, the housing is expected to be evacuated in over 6 years in space, and to remain at pressures below about 0.35 atm during descent in Jupiter's atmosphere. There are no high voltages used in the electronics, and hence no danger of arcing at low internal pressures. The instrument was tested at vacuum to verify this.

TABLE 3. ENTRY MODE DATA OUTPUTS.

SENSOR	MEASUREMENT	WORD LENGTH	SAMPLING RATE	COMMENTS
PRIME AXIAL ACCELEROMETER	ACCELERATION	12 BITS	EVERY 0.625 SECONDS	THE Z1 AXIAL ACCELEROMETER IS SELECTED AS PRIME SENSOR AT INSTRUMENT TURN-ON. OTHER SELECTIONS CAN BE MADE BY COMMAND UPLINK AS SHOWN IN TABLE 1.
SECONDARY AXIAL ACCELEROMETER	ACCELERATION	12 BITS	EVERY 0.625 SECONDS	
LATERAL ACCELEROMETERS	ACCELERATION (LATERAL RESULTANT)	8 BITS	EVERY 1.0 SECONDS	THE RESULTANT LATERAL ACCELERATION IS COMPUTED FROM EACH PAIR (X,Y) OF LATERAL ACCELEROMETER MEASUREMENTS
PRESSURE (LOW RANGE)	OUTPUT OF SENSOR	12 BITS	SIX DATA SAMPLES AT MAXIMUM DECELERATION	THE G SENSITIVITY OF THE SENSOR WILL BE DETERMINED AND USED TO EVALUATE POTENTIAL SCALE FACTOR SHIFTS.
HOUSEKEEPING TEMPERATURES	TEMPERATURE OF EACH OF THE FOUR ACCELEROMETERS AND THE ELECTRONICS TEMPERATURE	8 BITS	EACH OF FIVE HOUSEKEEPING TEMPERATURES EVERY 40 SECONDS	ACCELEROMETER TEMPERATURES, AND LOW RANGE PRESSURE SENSOR TEMPERATURE AT MAXIMUM G, ARE REQUIRED TO CORRECT THE DATA FOR TEMPERATURE COEFFICIENTS.

TABLE 4. DESCENT MODE DATA OUTPUTS.

SENSOR	MEASUREMENT	WORD LENGTH	SAMPLING RATE	COMMENTS
PRESSURE	PRESSURE	10 + 2 BITS	EVERY 2 SECONDS	10 BITS FOR PRESSURE AMPLITUDE AND 2 BITS FOR SENSOR ID. THE LOGIC WITHIN THE ASI DETERMINES WHICH OF THE THREE SENSOR OUTPUTS IS WRITTEN TO THE DATA STREAM. ONE PRESSURE SENSOR CALIBRATION MEASUREMENT WILL BE WRITTEN INTO THE DATA STREAM EVERY 64 SECONDS IN PLACE OF PRESSURE DATA.
T1 SENSOR (FAST RESPONSE)	TEMPERATURE	12 BITS	EVERY 4 SECONDS	ONE SENSOR CALIBRATION MEASUREMENT WILL BE WRITTEN INTO THE DATA STREAM EVERY 64 SECONDS FOR EACH SENSOR IN PLACE OF THE APPROPRIATE SENSOR TEMPERATURE OUTPUT.
T2 SENSOR (SLOW RESPONSE)	TEMPERATURE	12 BITS	EVERY 4 SECONDS	
PRIME AXIAL ACCELEROMETER	VELOCITY CHANGE	12 BITS	EVERY 16 SECONDS	VELOCITY CHANGE IS OBTAINED BY DIGITAL INTEGRATION OF THE REGULAR (EVERY 32.25 MILLISECONDS) ACCELEROMETER MEASUREMENTS.
SECONDARY AXIAL ACCELEROMETER	VELOCITY CHANGE	12 BITS	EVERY 16 SECONDS	
LATERAL ACCELEROMETERS	MAXIMUM ACCELERATION	8 BITS	EVERY 16 SECONDS	LATERAL ACCELERATION IS COMPUTED FOR EVERY X AND Y MEASUREMENT PAIR (EVERY 32.25 MILLISECONDS). THIS RESULTANT ACCELERATION IS USED TO DETERMINE THE MAXIMUM, MINIMUM, AND MEAN DURING THE MEASUREMENT PERIOD.
	MINIMUM ACCELERATION	8 BITS	EVERY 16 SECONDS	
	AVERAGE ACCELERATION	8 BITS	EVERY 16 SECONDS	
PRIME AXIAL ACCELEROMETER	TURBULENCE LEVELS CROSSINGS	8 WORDS OF 12 BITS EACH	EVERY 96 SECONDS	THE NUMBER OF POSITIVE GOING TRANSITIONS BETWEEN EIGHT ACCELERATION VALUES ARE SUMMED FOR EACH TRANSITION LEVEL EACH MEASUREMENT PERIOD
LATERAL ACCELEROMETERS	TURBULENCE LEVELS CROSSINGS	4 WORDS OF 12 BITS EACH	EVERY 96 SECONDS	THE NUMBER OF POSITIVE GOING TRANSITIONS BETWEEN FOUR ACCELERATION VALUES ARE SUMMED FOR EACH TRANSITION LEVEL EACH MEASUREMENT PERIOD. THE LATERAL ACCELERATION RESULTANT IS USED FOR THIS MEASUREMENT.
VARIOUS	VARIOUS	12 BITS	EVERY 8 SECONDS	DATA TRANSMITTED WITHIN THIS ALLOCATION ARE: 1) PRE-ENTRY CALIBRATION DATA RETAINED IN ASI MEMORY FROM THE PRE-ENTRY CALIBRATION SEQUENCE. (16-BIT WORDS) 2) INSTRUMENT HOUSEKEEPING TEMPERATURE DATA TO PERMIT TEMPERATURE CORRECTIONS TO BE MADE TO THE SENSOR OUTPUTS. 3) TEMPERATURE SENSORS AMPLIFIER ZERO BIAS. 4) A/D CONVERTERS ZERO BIAS AND REFERENCE VOLTAGE. 5) ACCELEROMETER RANGE DATA.

TABLE 5. CALIBRATION MEASUREMENT SEQUENCE

A/D CONVERTER #1, ZERO OFFSET CAL
A/D CONVERTER #2, ZERO OFFSET CAL
LOW RANGE PRESSURE, 3.75 V CAL
MID RANGE PRESSURE, 3.75 V CAL
Z1 ACCELEROMETER, RANGE 2, I2 CAL
Z2 ACCELEROMETER, RANGE 3, I1 CAL
FAST RESPONSE TEMPERATURE CAL
HIGH RANGE PRESSURE, 1.25 V CAL
Z2 ACCELEROMETER, RANGE 4, I1 CAL
SLOW RESPONSE TEMPERATURE CAL
HIGH RANGE PRESSURE, 3.75 V CAL
Z1 ACCELEROMETER, RANGE 3, I1 CAL
Z2 ACCELEROMETER, RANGE 2, I2 CAL
SCIENCE TEMP AMP ZERO OFFSET CAL
Z1 ACCELEROMETER, RANGE 4, I1 CAL
TEMPERATURE OF LOW RANGE PRESSURE
X ACCELEROMETER, RANGE 2, I2 CAL
TEMPERATURE OF MID RANGE PRESSURE
Y ACCELEROMETER, RANGE 2, I2 CAL
TEMPERATURE OF HIGH RANGE PRESSURE
X ACCELEROMETER, RANGE 3, I1 CAL
TEMPERATURE OF Z1 ACCELEROMETER
Y ACCELEROMETER, RANGE 3, I1 CAL
TEMPERATURE OF Z2 ACCELEROMETER
TEMPERATURE OF X ACCELEROMETER
TEMPERATURE OF Y ACCELEROMETER
TEMPERATURE OF ELECTRONICS BOX
LOW RANGE PRESSURE, 1.25 V CAL
MID RANGE PRESSURE, 1.25 V CAL
A/D CONVERTER #1, 2.5 V CAL
A/D CONVERTER #2, 2.5 V CAL
LOW RANGE PRESSURE, ZERO OFFSET
MID RANGE PRESSURE, ZERO OFFSET
HIGH RANGE PRESSURE, ZERO OFFSET
Z1 ACCELEROMETER, RANGE 1, ZERO OFFSET
Z2 ACCELEROMETER, RANGE 1, ZERO OFFSET
X ACCELEROMETER, RANGE 1, ZERO OFFSET
Y ACCELEROMETER, RANGE 1, ZERO OFFSET
Z1 ACCELEROMETER, RANGE 2, ZERO OFFSET
Z2 ACCELEROMETER, RANGE 2, ZERO OFFSET
X ACCELEROMETER, RANGE 2, ZERO OFFSET
Y ACCELEROMETER, RANGE 2, ZERO OFFSET
Z1 ACCELEROMETER, RANGE 3, ZERO OFFSET
Z2 ACCELEROMETER, RANGE 3, ZERO OFFSET
X ACCELEROMETER, RANGE 3, ZERO OFFSET
Y ACCELEROMETER, RANGE 3, ZERO OFFSET
Z1 ACCELEROMETER, RANGE 4, ZERO OFFSET
Z2 ACCELEROMETER, RANGE 4, ZERO OFFSET

5. Concluding Remarks

In the design of the Atmosphere Structure Instrument for the Galileo Probe, the goal has been to obtain accurate and detailed information on the structure of the atmosphere from the exosphere to the maximum attainable depth in the lower atmosphere, 10^{-7} mb to 16 bar, nominally. A corollary goal is to define properties of the atmosphere important to atmospheric dynamics. Sensor selection was based on prior experience with conceptually similar instruments in the atmospheres of Mars and Venus, with modifications to avoid problems experienced in earlier missions and those anticipated in Jupiter's atmosphere. The electronics are microprocessor controlled and have added capabilities to those used previously. They permit versatile sampling sequences and make maximum use of the limited available data rate by means of some on-board data processing.

If the instrument performs up to design expectations, it should return valuable information about structure and dynamics of the largest planet in the Solar System.

References

Atreya, S. K., Donahue, T. M., and Festou, M. C.: 1981, 'Jupiter, Structure and Composition of the Upper Atmosphere', *Astrophys. J.* **247**, L43.

Festou, M. C., Atreya, S. K., Donahue, T. M., Sandel, B. R., Shemansky, D. E., and Broadfoot, A. L.: 1981, 'Composition and Thermal Profiles of the Jovian Upper Atmosphere Determined by the Voyager Ultraviolet Stellar Occultation Experiment', *J. Geophys. Res.* **86**, 5415.

Hunten, D. M., Colin, L., and Hansen, J. E.: 1986, 'Atmospheric Science on the Galileo Mission', *Space Sci. Rev.* **44**, 191.

Ingersoll, A.: 1981, 'Jupiter and Saturn', *Sci. Amer.* **245**, 90.

Ingersoll, A.: 1990, 'Atmospheric Dynamics of the Outer Planets', *Science* **248**, 308.

Orton, G. S.: 1981, *Atmospheric Structure in the Equatorial Region of Jupiter*, Galileo Project Document 1625–125.

Riddle, J. L., Furukawa, G. T., and Plumb, H. H.: 1973, *Platinum Resistance Thermometry*, NBS Monograph 126, U.S. Dept. of Commerce.

Seiff, A.: 1976, 'The Viking Atmosphere Structure Experiment – Techniques, Instruments, and Expected Accuracies', *Space Sci. Instr.* **2**, 381.

Seiff, A. and Kirk, D. B.: 1977, 'Structure of the Atmosphere of Mars in Summer at Mid-Latitudes', *J. Geophys. Res.* **82**, 4364.

Seiff, A., Reese, D. E., Sommer, S. C., Kirk, D. B., Whiting, E. E., and Niemann, H. B.: 1973, 'PAET, An Entry Probe Experiment in the Earth's Atmosphere', *Icarus* **18**, 525.

Seiff, A., Kirk, D. B., Young, R. E., Blanchard, R. C., Findlay, J. T., Kelly, G. M., and Sommer, S. C.: 1980a, 'Measurements of Thermal Structure and Thermal Contrasts in the Atmosphere of Venus and Related Dynamical Observations: Results from the Four Pioneer Venus Probes', *J. Geophys. Res.* **85**, 7903.

Seiff, A., Juergens, D. W., and Lepetich, J. E.: 1980b, 'Atmosphere Structure Instruments on the Four Pioneer Venus Entry Probes', *IEEE Trans. Geosci. Rem. Sens.* **GE–18**, 105.

Smith, E. J., Davis, L., Jr., and Jones, D. E.: 1976, in T. Gehrels (ed.), 'Jupiter's Magnetic Field and Magnetosphere', *Jupiter*, University of Arizona Press, Tucson, p. 788.

GALILEO NET FLUX RADIOMETER EXPERIMENT

L. A. SROMOVSKY, F. A. BEST, H. E. REVERCOMB

Space Science and Engineering Center, University of Wisconsin-Madison, WI 53706, U.S.A.

and

J. HAYDEN

Martin Marietta Astronautics Group, U.S.A.

Abstract. The Galileo Net Flux Radiometer (NFR) is a Probe instrument designed to measure the vertical profile of upward and net radiation fluxes in five spectral bands spanning the range from solar to far infrared wavelengths. These unique measurements within Jupiter's atmosphere, from which radiative heating and cooling profiles will be derived, will contribute to our understanding of Jovian atmospheric dynamics, to the detection of cloud layers and determination of their opacities, and to the estimation of water vapor abundance. The NFR uses an array of pyroelectric detectors and individual bandpass filters in a sealed detector package. The detector package and optics rotate as a unit to provide chopping between views of upward and downward radiation fluxes. This arrangement makes possible the measurement of small net fluxes in the presence of large ambient fluxes. A microprocessor-controlled electronics package handles instrument operation.

1. Introduction

The Galileo Probe Net Flux Radiometer (NFR) is designed to measure the net radiation flux and upward radiation flux within Jupiter's atmosphere as the Probe descends by parachute. The nominal measurement regime for the NFR extends from about 0.1 bar to at least 10 bars, which corresponds to an altitude range from 48 km above the 1-bar level to 91 km below it. These measurements will help to define sources and sinks of planetary radiation, regions of solar energy deposition, and provide constraints on atmospheric composition (primarily water vapor). The instrument was designed and built by Martin Marietta, later modified to incorporate design improvements developed at the University of Wisconsin-Madison where the instrument was calibrated. The NFR Principal Investigator is L.A. Sromovsky of the University of Wisconsin-Madison, and the Co-Investigators are J. Pollack of Ames Research Center, M. Tomasko of the University of Arizona, J. Hayden of Martin Marietta Astronautics Group, and H. Revercomb, also from the University of Wisconsin.

Following a brief description of the scientific objectives of the experiment, we describe the instrument hardware, how it functions, and how it is calibrated. Finally, we present a brief summary of expected errors and performance characteristics.

2. Scientific Objectives

2.1. Measuring driving forces for atmospheric circulation

Absorbed solar radiation and the flux of energy from the planet's interior provide the primary drive for atmospheric motions in Jupiter's atmosphere. Because solar radiation

Space Science Reviews **60:** 233–262, 1992.

absorption and planetary emission occur at different places and altitudes, net atmospheric heating and cooling result in buoyancy differences which force atmospheric motions. An understanding of Jovian circulation thus requires knowledge of the vertical profile of radiative heating and cooling and its horizontal distribution as well.

The NFR measurements contribute to this understanding by measuring the net radiation flux as a function of altitude. This net flux, the difference between upward and downward radiation powers crossing a horizontal surface per unit area, is directly related to the radiative heating or cooling of the local atmosphere: the radiative power per unit area absorbed by an atmospheric layer is equal to the difference in net fluxes at the boundaries of the layer. At any point in the atmosphere radiative power absorbed per unit volume is thus given by the vertical derivative of net flux ($\mathrm{d}F/\mathrm{d}z$), and the corresponding heating rate is just $\mathrm{d}T/\mathrm{d}t = (\mathrm{d}F/\mathrm{d}z)/(\rho C_p)$, where ρ is the local atmospheric density and C_p is the local atmospheric specific heat at constant pressure.

2.2. Objectives of Spectral Flux Measurements

Because the Galileo Probe will provide only one sample profile of Jovian atmospheric conditions at one location, it is especially important to understand why the measured radiative energy deposition occurs: we might then have some idea of how to apply the results to other atmospheric regions which were not sampled. The NFR experiment contributes to understanding horizontal variations by making spectral measurements which illuminate the mechanisms by which radiation interacts with the atmosphere. When these are interpreted with other Probe measurements and linked with Orbiter observations they provide a basis for using orbiter observations to extend radiative heating determinations to other locations on the planet (Hunten *et al.*, 1986).

Using six detector channels, the NFR measures net and upward flux densities in five spectral bands. The specific objectives of each channel are briefly described in the following paragraphs and also summarized in Table I.

TABLE I

Summary of NFR spectral channels and objectives

Channel	Bandpass	Objective
A	3–500 μm	Deposition/loss of thermal radiation
B	0.3–3.5 μm	Total deposition of solar radiation
C	3.5–5.8 μm	Water vapor and cloud structure
D	14–35 μm	Water vapor
E	0.6–3.5 μm	Solar deposition in methane absorption region
F	Blind	Witness to detector perturbations

Channel *A*: *Spectral range from 3 to 500 μm* (the broadband thermal channel). Net flux measurements define sources and sinks of planetary radiation. When used with calculations of gas opacity effects, these observations define the thermal opacity of particulates, which is much harder to calculate.

Channel *B*: *Spectral range from 0.3 to 3.5 μm* (the broadband solar channel). Net flux measurements determine the solar energy deposition profile; upflux measurements help discriminate between cloud particle absorption and extinction.

Channel *C*: *Spectral range from 3.5 to 5.8 μm* (the excess thermal channel). Net flux and upward flux measurements define the excess thermal energy emitted by Jupiter within this atmospheric window; this channel is very sensitive to both cloud opacity and the water vapor mixing ratio and will be used also to constrain other measurements of water.

Channel *D*: *Spectral range from 14 to 35 μm* (the water vapor channel). Net flux measurements define regions of combined hydrogen and water vapor absorption; water vapor will be deduced from deviations from known hydrogen effects. Where atmospheric opacity is high, net flux in this channel will be proportional to the product of local opacity (unknown) and the local temperature gradient (independently measured).

Channel *E*: *Spectral range from 0.6 to 3.5 μm* (the methane channel). Net flux and upward flux measurements in this band are strongly affected by cloud particles as are the broader band channel *B* measurements, but channel *E* is proportionately more strongly affected by methane band absorption; comparison with the broader band measurements of channel *B* will allow a 'methane-free' determination of cloud particle absorption effects.

Channel *F*: *A spectrally blind channel* (the witness channel). Measurements by this detector provide information needed to correct for extraneous detector perturbations.

It should be noted that none of the NFR measurements is really a direct measurement which is unencumbered by the need to assume some characteristics of the parameters to be measured. Rather than measuring net flux directly, the NFR samples the angular domain and weights non-uniformly the spectral variations in the radiation field. To interpret these measurements requires modelling of the radiation field as a function of composition, cloud, and thermal structure. The measurements then provide strong constraints on the models rather than provide direct determinations of model parameters.

3. Instrument Description

3.1. Physical Configuration

The NFR consists of two major sub-assemblies: the electronics module (EM), and optical head (OH), shown mated together in Figure 1. The electronics module is about 13 cm × 19.5 cm × 16 cm high while the optical head is about 8.5 cm × 8 cm high × 10.5 cm long. The total weight of the NFR is 3.134 kg, of which 0.672 kg is due to the optical head. The electronics module has four feet which are bolted to the Probe instrument shelf. The optical head, supported only by its attachment to the electronics module, extends out of the Probe through the Probe thermal blanket and aeroshell to allow the NFR sensors to view atmospheric radiation.

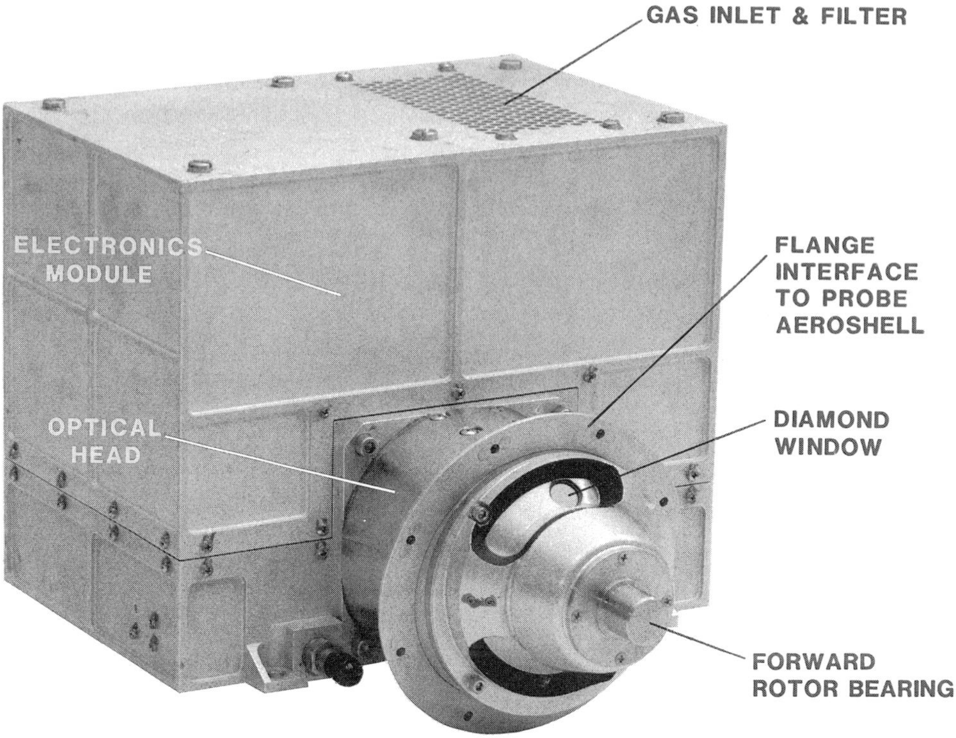

Fig. 1. Physical configuration of the NFR instrument. The total length of the optical head is approximately 10.5 cm. The part of the optical head beyond the flange is directly exposed to the ambient Jovian environment.

Optical Head

The NFR optical head, shown in cutaway view in Figure 2, contains optics, detectors, and preamplifiers which rotate as a unit as they chop between upward and downward atmospheric views, or between ambient and internal heated black-bodies for on-board calibration measurements. Flexible circuits carry power to the preamplifiers and analog signals from the pre-amplifiers to the electronics module for further processing. The rotating optics assembly is supported by a front and rear bearing and driven through a gear by a stepper motor. The front housing, which supports the front bearing, has two apertures through which atmospheric radiation can reach the diamond window. A curved band, or shroud, is attached to the rotating optics to restrict the flow of gas through the front housing apertures no matter what position the window is in.

Optics and Field of View

The optical system, shown in Figure 3, consist of a diamond window, a folding mirror, a toroidal mirror, a condensing cone, and spectral filters which are part of the detector package. An unfolded diagram of the optics is shown in Figure 3(b). The toroidal mirror images the diamond window at the entrance to the condensing cone. The detector and

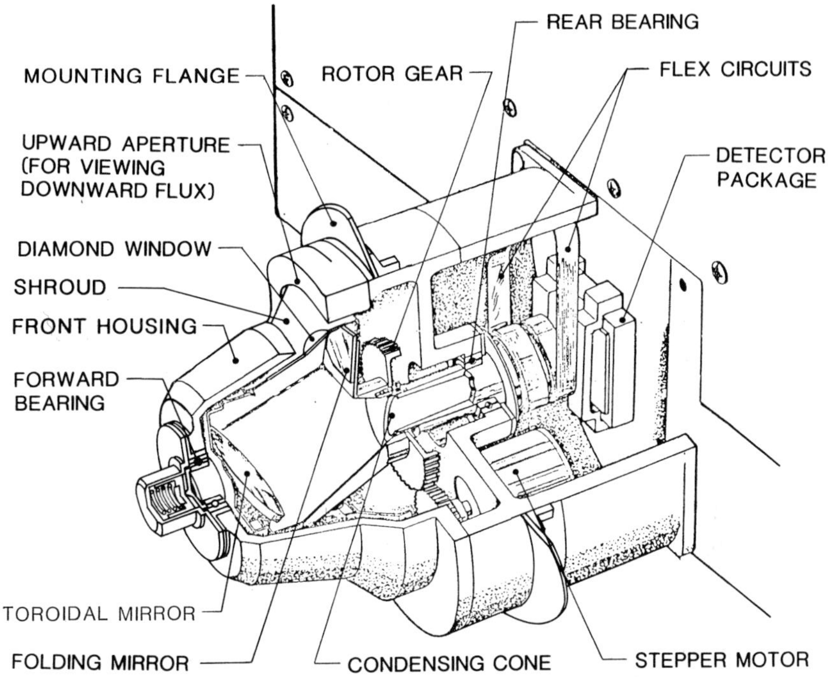

Fig. 2. NFR Optical Head. During net flux measurements the detector package and optics rotate between upward and downward viewing positions, which are respectively 45° below and above horizontal. Though it is not shown in this diagram, the front housing also has a bottom aperture to allow viewing upward flux (see Figure 1).

three mirror images are imaged approximately 0.76″ behind the diamond window. This arrangement allows the detector to be positioned in a thermally isolated location while still achieving a relatively large field of view of nominally 0.4 sr ($\pm 20°$ about the nominal viewing directions of 45° above and below horizontal).

Because of viewing and thermal constraints on the instrument, the NFR samples only part of upward and downward radiation fields, rather than integrating over the entire hemispheres. Thus, the NFR actually measures a weighted net spectral radiance rather than a net flux. The viewing directions of $\pm 45°$ from horizontal, instead of $\pm 90°$, actually provide a more representative sample because the largest contribution to the net flux occurs at $\pm 45°$ (when solid-angle weighting is included).

Detector Package

All NFR spectral channels use pyroelectric thermal detectors to convert absorbed radiation power to electrical signals (the operation of these detectors is described in a following section on analog signal characteristics). The NFR detector package contains an array of six lithium tantalate detectors mounted in close proximity on a single circuit board (Figure 4). This circuit board is bonded to a hermetically sealed hybrid pre-amplifier package to which the detectors are electrically connected (Figure 5). Spectral

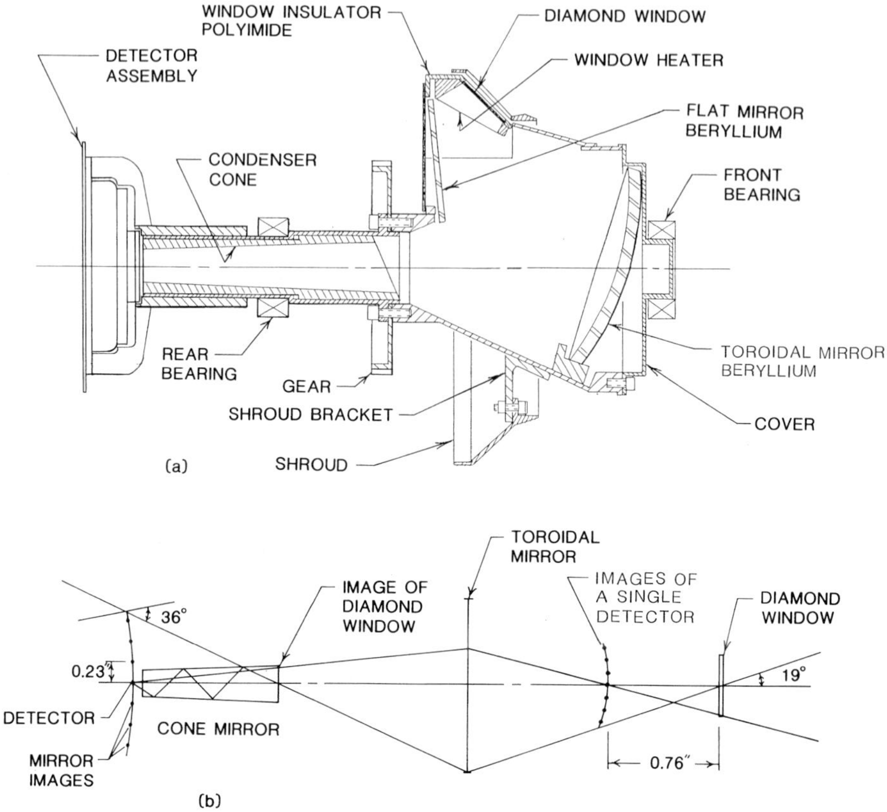

Fig. 3. Rotating optics layout (a) and unfolded ray-trace diagram (b). The condenser cone reflects three annular images of the detectors which the toroidal mirror brings to a focus 0.76″ behind the diamond window together with the prime image. The diameter of the base of the condenser cone is just large enough to prevent obscuration of the six-detector array. The diamond window is focussed at the entrance to the condenser cone.

filters are mounted in a filter frame which is pin-located and screw-mounted to the detector circuit board. The location pins (see Figure 4) are also used in detector fabrication to insure proper positioning of the detector elements relative to the filter frame openings.

The detectors were fabricated by Infrared Associates of Cranbury, New Jersey. Each pyroelectric detector element consists of a $LiTaO_3$ crystal approximately 1 mm × 2 mm × 25 μ (about 0.001″ thick). Lithium tantalate was chosen as the pyroelectric material because it has a Curie point well above the expected operating temperature and thus would not become depolarized. The NFR detector elements have unusually smooth edges because they are cut from a larger piece of crystal using ion-beam milling instead of a diamond saw. These smooth edges result in crystals which are exceptionally rugged. The bottom electrode of each element is opaque gold and is applied before the detector material is cut into individual elements. The top electrode

Fig. 4. Detector module photograph. The six pyroelectric elements in the center of the photograph have been overexposed to make them more visible (the paint coating the detectors is much blacker than it appears in this photo). Detector leads are soldered to eyelets which are connected to runs on the back side of the detector circuit board. The alignment pins are used during fabrication as well as for positioning the filter frame. Each detector element is approximately 1 mm × 2 mm.

is vacuum-deposited after the elements are mounted in the array and consists of a transparent layer of nichrome with small gold pads deposited for lead bonding. Platinum leads of 12.5 μ diameter are attached to top and bottom electrodes using electrically conductive epoxy. (The other ends of the leads are soldered to the circuit board.) The top electrode is sized to match the desired active area and is somewhat smaller than the crystal itself. Black paint (3M velvet) is applied to the top surface so as to cover the active area but not the entire detector. The typical paint thickness is 20–25 μ (slightly less than 0.001″). Because $LiTaO_3$ is a piezoelectric material, as well as a pyroelectric, the detector elements are mounted on 0.015″-high mesas made of a vibration dampening material called Visilox.

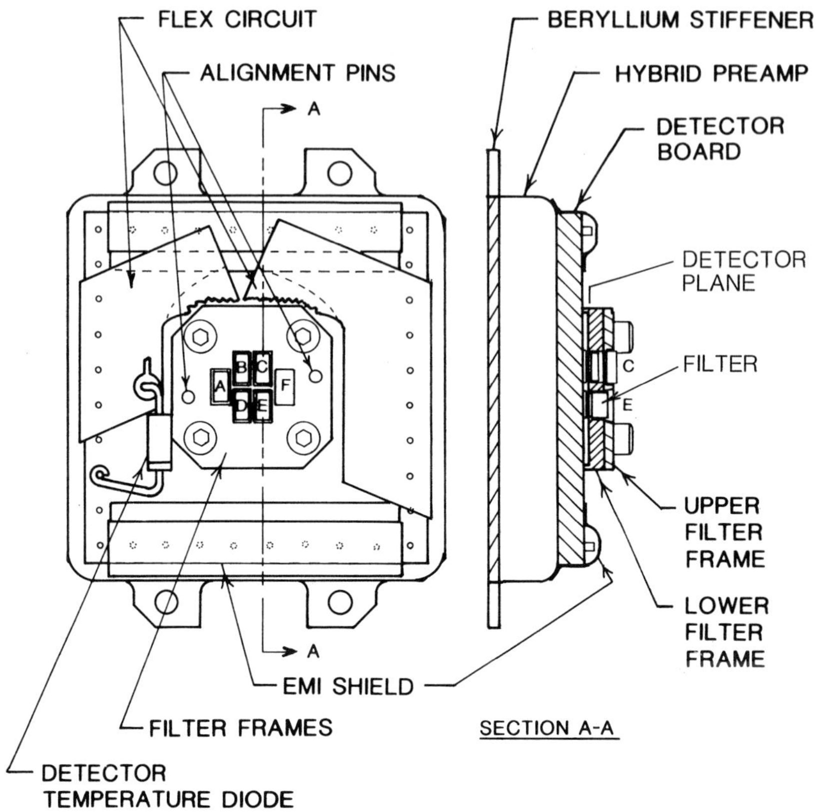

Fig. 5. NFR Detector/Hybrid Assembly. This drawing shows a 1″ × 1″ detector module bonded to the preamplifier can, which is in turn bonded to a beryllium stiffener. Upper and lower filter frames are shown mounted above the detector array. Note that a stack of two filter elements is used for channel *C*. The 0.033 cm^3 volume enclosed between the filter frame and the detectors (shown in the side view) is sealed and back-filled with xenon.

Detectors have a primary thermal time constant of approximately 110 ms, an electrical capacitance of about 40 pF, and a responsivity of approximately 1400 V/W. Noise associated with the detectors is generally dominated by Johnson noise of the detector load resistor.

The spectral filter materials and thicknesses are listed in Table II. (All channels experience additional spectral filtering by the 0.2 mm thick diamond window at the entrance to the rotating optics, the spectral reflectivity of the mirrors, and the spectral response of the detectors.) As indicated in Figure 5, there are two filter frames: an upper frame containing only a CaF_2 long-wave blocker for channel *C*, and a lower filter frame containing five spectral filters and one opaque blocker for the blind channel. All filters are epoxy bonded in place on shelves machined into the filter frame.

The detector package is hermetically sealed and backfilled with xenon gas at ambient pressure (about 800 mb). The seal alone eliminates an extreme sensitivity to external

TABLE II

NFR Spectral Filters. In the following table MLC denotes multi-layer interference coating. Filter thicknesses are approximate.

Channel	Bandpass	Upper filter	Lower filter
A	3–500 μm	none	MLC on 0.1 mm diamond
B	0.3–3.5 μm	none	Corning 056, 1.5 mm
C	3.5–5.8 μm	1 mm CaF_2	MLC on 1 mm Si
D	14–35 μm	none	MLC on 0.5 mm Ge
E	0.6–3.5 μm	none	Corning 2–64
F	Blind	none	Blocker

pressure modulations. The backfill of the very heavy xenon gas is used to buffer the small amount of hydrogen gas which will diffuse into the detector package during descent. The buffering effect maintains a low thermal conductivity inside the detector package and thereby eliminates significant thermal crosstalk which otherwise would occur via gas conduction between detector elements. Other benefits of the seal include making the detector responsivity independent of external pressure and protecting detector interior coatings and lead bonds from ambient atmospheric conditions. The seal is accomplished first by sealing the filters into the lower filter frame (then leak testing and patching if required). The second step is to epoxy both the lower and upper filter frame on the detector board with a hypodermic needle temporarily replacing one of the alignment pins. The needle provides access to the sealed volume so that leak testing and patching can be carried out until all leaks are sealed. The detector package (with hypodermic) is then backfilled with xenon and placed in a glove box filled with xenon. Within the glove box the hypodermic needle is removed and an alignment pin is epoxied in its place to complete the seal. After the epoxy is cured the entire package is leak tested using a radiometric technique to verify absence of a gross leak and then with a mass spectrometer scanning over the xenon peak to measure the fine leak. A leak rate of approximately 7×10^{-10} cc s^{-1} or less (of Xe at STP) was measured for the flight detector package, implying that about 100 mb of Xe will be present inside the detector package at entry, ensuring that the seal objectives will be met (see later discussion of expected performance).

Internal Sources

When the optical system is not viewing external radiation, it views one of two internal radiation sources: an ambient black-body source which is thermally coupled to the wall of the front housing, and a heated black-body source which is servo-controlled to a temperature of approximately 107 °C (the servo point is attained in air or vacuum, but generally is not attained in He or H_2 atmospheres where high gas conductivity limits the black-body temperature to a maximum differential above the ambient atmospheric temperature). By chopping the view between these two internal sources it is possible to carry out approximate internal calibrations during laboratory testing and as part of in-flight checkouts during the cruise phase of the mission. However, during the descent into the Jovian atmosphere, the hot black-body will operate much closer to the temperature of the ambient black-body and the accuracy of internal calibrations will not be very high.

Motor

A 90° step^{-1}, bipolar, 2-phase stepper motor and a 2 to 1 gear set are used to rotate the optics and detector system. Each step of the motor results in 45° of motion for the optics. Following the convention defined in Figure 6, the upward viewing position of the

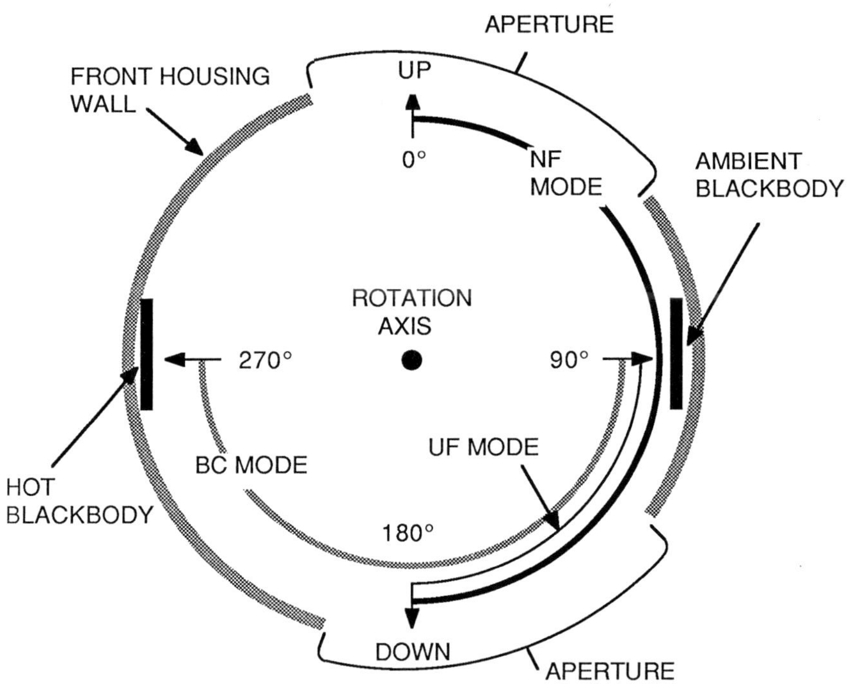

Fig. 6. NFR operating modes diagram, viewed from the rear of the optical head looking towards the nose. Cyclical NFR rotor motion is indicated by arcs for Net Flux (NF), Up Flux (UF), and Black-body Calibrate (BC) modes of operation. A hard stop prevents viewing near 315°. Radial arrows indicate viewing directions at the ends of each arc.

optics is defined to be at 0°, the ambient black-body position at 90°, the downward view at 180°, and the hot black-body at 270°. The motor has torque sufficient to move the rotor 180° in about 60 ms.

Window Heater

A 300 mW heater is used to maintain the external diamond window slightly warmer than the ambient atmosphere, thereby preventing condensation on the window.

Optical Head Ventilation Design

Because the rotating optics could not structurally support a high-pressure differential, the interior of the rotating optics is not sealed, and will admit ambient external gas during descent. Thermal modelling of the NFR shows that the detector package temperature will lag behind the external atmospheric temperature during all but the early part of descent. To inhibit possible condensation on interior optical surfaces, gas flow into the rotor is controlled by vents within the electronics module and at the base of the rotor. These vents take advantage of the dynamic pressure distribution around the Galileo descent Probe which provides a positive pressure differential between the inside of the Probe and the external atmosphere in the vicinity of the NFR apertures. The predicted gas flow is first into the Probe through a large vent at the back of the Probe. Some of this gas enters the electronics module through a Line 5A molecular sieve filter provided at the top of the electronics (see Figure 1). The filtered gas, also warmed by the electronics module, enters the rear of the optical head through a vent provided at the base of the electronics module. From this point the gas flows partly into the rotor through the rotor base vent (near the rear bearing) and partly through the motor and rotor bearings into the forward part of the optical head and then through the apertures into the external atmosphere. The differential pressure driving this flow is approximately 1 mb and at 1 atm the total flow rate through the optical head will be approximately 100 cc s^{-1}.

Method of Operation

The various modes of NFR operation are illustrated in Figure 6. To perform the net flux measurement the motor first positions the optics at 0°, then alternately cycles 4 steps down (45°, 90°, 135°, and 180°) and 4 steps up (135°, 90°, 45°, 0°). The optics is cycled down and up twice each second. During approximately 60 ms of each half cycle the optics is in transition, leaving approximately 190 ms for staring up or down. Eleven up-down cycles of the optics are integrated to make one net flux instrument cycle (IC) which is transmitted to the Probe data system at a rate of one IC every 6 s. Up flux and blackbody calibrate instrument cycles are performed in a similar fashion. To make an up flux measurement the rotor is first positioned at 90° to view the ambient blackbody then alternately cycled between down (180°) and the ambient blackbody (90°). For blackbody calibration, the optics is first positioned to view the ambient blackbody (90°) then cycled between the hot black-body (170°) and the ambient blackbody.

During the last powered pulse of each half of a chopping cycle the unpowered of the two motor coils is electrically shorted to activate eddy-current damping which provides greater stability of motion and more accurate pointing (especially in the Up Flux mode). A set of position sensors is used to verify correct rotor positioning within 5° of the nominal position. The LEDs used to stimulate the position sensors may degrade substantially during radiation exposure and therefore cannot be relied on to verify pointing during descent. However, LED degradation will not affect motor operation since the motor is driven with an open loop system which does not require position feedback.

Mode Sequences

The NFR operating modes are carried out in a sequence which repeats at 2-min intervals. Each complete sequence, called a data cycle (DC), consists of 20 six-second ICs. As defined in Table III, each DC uses one IC for the Blackbody Calibrate (BC) sample, one for the Up Flux (UF) sample, and one for Analog Zero (AZ), to be explained later. This leaves 17 ICs available for Net Flux (NF) measurements. Because of offset level shifts between different input sources, the net flux measurements following AZ and UF modes are 'short-cycled', meaning that only the last five of the eleven up-down flip cycles are included in the measurement. All AZ measurements, except the first, are also short-cycled because they immediately follow the large offset developed during the BC mode. Housekeeping data, listed at the right side of Table III, provides information on temperatures, voltages, and currents (the acronyms used here are defined in the table caption).

3.2. Electronics

A functional block diagram of the NFR electronics is shown in Figure 7. The design is centered around a CMOS microprocessor control system very similar to that used

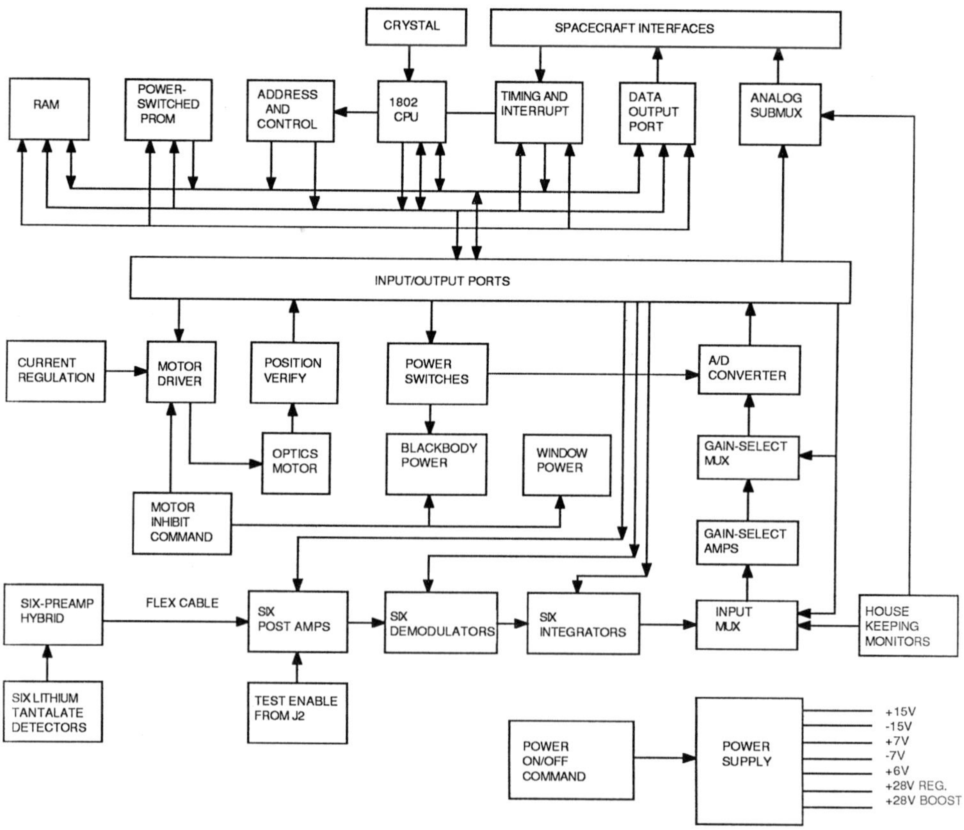

Fig. 7. Electronics block diagram. Instrument operation is controlled by software stored in PROMs.

TABLE III

NFR Data Cycle (DC) contents. Each row defines the data output for each instrument cycle (IC). AZ denotes analog zero, NF denotes net flux, UF denotes up flux, and BC denotes blackbody calibration. HB and BI denote the temperature and approximate current of the internal hot blackbody. A1 and A2 denote low range and high range thermistor measurements of the ambient blackbody. WT and WI denote window temperature and window heater current respectively. Detector temperature is denoted by DET and electronics temperature by ET. V1 and V2 denote voltage data for + 10 V ADC reference and + 7 V supply respectively. GSAC1, 8, 64 denote calibration data for gain-select amplifers, while GSAZ1, 8, 64 denote zero data for the gain select amplifiers.

IC	Time (s)	Mode	Flips used	Housekeeping
0	6	AZ	5	HB, A1
1	12	NF	5	A2, WT, WI
2	18	NF	11	V2, A2
3	24	NF	11	GSAC1, DET, A1, A2
4	30	NF	11	ET, WT, BI
5	36	NF	11	SW, SW
6	42	NF	11	GSAZ1, V1, WT
7	48	NF	11	DET, A1, A2
8	54	NF	11	HB, WT
9	60	UF	5	GSAC8, HB, WT, BI
10	66	NF	5	HB, WT
11	72	NF	11	DET, A1, A2
12	78	NF	11	V2, A1, WI
13	84	NF	11	GSAZ8, A2, WT
14	90	NF	11	ET, WT
15	96	NF	11	DET, A1, A2, BI
16	102	NF	11	GSAC64, V1, WT
17	108	NF	11	A1, A2, WI
18	114	NF	11	HB, WT
19	120	BC	11	GSAZ64, DET, HB, A2

on the Atmospheric Structure Instrument (ASI) for the Galileo Probe (Seiff *et al.*, 1990). Functional elements of the NFR electronics are described in the following paragraphs.

Digital Circuits

The microprocessor system consists of the 1802 CPU (Central Processing Unit), 256 words of RAM (Random Access Memory), 6144 8-bit words of PROM (Programmable Read-Only Memory), nine I/O (Input/Output) ports and a power-up reset circuit. The RAM is used to provide 256 bytes of temporary storage for data values during data accumulation and manipulation. The six 1-Kbyte PROMs contain the program necessary to operate the instrument. At any given time only the one of the six PROMS is turned on and for only 1 μs of the 8 μs machine cycle, providing a factor of 48 reduction of power consumption by the PROMs. Six 8-bit wide output ports are used to control the non-digital NFR subsystems. Three input ports are used to read data from the NFR subsystems.

The 2048 Hz spacecraft clock is divided down to a 4-Hz signal which is used to interrupt the microprocessor. The MINOR FRAME signal from the spacecraft is used to synchronize the 4-Hz timer and also to synchronize the software with spacecraft timing. The software does not begin cycling in its normal mode until the microprocessor detects a MINOR FRAME interrupt.

The PROM software controls the sequence, timing, and duration of motor control pulses. After each cycle of the optical rotor, position sensor photo transistors are read to determine if the optics are at the correct position. If any one of the 22 half cycles of one instrument cycle (IC) results in an incorrect optics position, the microprocessor notes this in the data stream by setting the motor position error bit to one for that IC.

Analog Circuits

There are six channels of analog processors, one for each detector. Each channel includes a detector signal pre-amplifier, a post-amplifier, a demodulator and an integrator. This six pre-amplifiers are housed in a hybrid package placed adjacent to the detectors on the rotating optics (see Figure 5). The rest of the analog circuits reside on two circuit boards within the electronics module.

Pre-Amplifiers

Each pre-amplifier is a DC differential amplifier with a gain of 6.67, using U423 dual JFET inputs. The effective input load resistance of $0.909 \times 10^{10}\ \Omega$ ($10^{10}\ \Omega$ in parallel with $10^{11}\ \Omega$) in combination with the typical detector capacitance of 40 pF leads to a detector electrical droop time constant of 0.36 s. With this droop a typical detector will generate an electrical offset of 0.055 V per $°\mathrm{C\ min}^{-1}$ of thermal ramp. During descent we expect a thermal ramp less than $7\ °\mathrm{C\ min}^{-1}$, so that the input offset will not exceed 0.38 V, comfortably less than the pre-amplifier saturation level of about 0.75 V. Because the load resistance so much less than the $10^{13}\ \Omega$ detector resistance, detector noise is dominated by the Johnson noise of the load resistors (modified by the detector shunt capacitance, of course).

Post-Amplifiers

The six parallel post-amplifiers each consists of three non-inverting amplifiers in series (except for channels *B* and *E* which have one inversion to correct for an inversion built into the detector package). Single pole RC filters are used to block the DC component from the hybrid and to tailor the frequency response of the circuit. The filter components are chosen to give a maximum response at 16 Hz. This may seem strange in view of our fundamental 2 Hz detector signal. However, this filter function acts somewhat like a differentiator, which, in combination with the following integrator, results in a very small sensitivity to the details of signal transitions and a high sensitivity only to the final values attained after each flip of the rotor. This effect minimizes asymmetry errors (described in a subsequent section).

FET (Field Effect Transistor) switches at the input of each post amplifier allow the inputs to be grounded through a $100\ \Omega$ resistor, providing a zero reading to be integrated as the Analog Zero (AZ) data. The AZ value is intended to be a measure of offset in the integration circuitry, but is also slightly affected by other factors which are still under investigation.

The gain of the post amplifiers is tailored to the dynamic range expected from each channel. To extend the dynamic range of channels *A*, *C*, and *D*, which receive much stronger signals from the internal heated black-body than they do from Jupiter's atmosphere, the third amplifiers of the circuits for those channels (and also for channel *F*) have two possible gains selectable with a FET switch. The gain is switched to 8 for analog zero (AZ), up flux (UF), and net flux (NF) measurements, and switched to unity for the black-body calibrate (BC) measurement. The solar channels, *B* and *E*, receive relatively weak signals from the on-board calibration source and thus do not need gain reduction capabilities.

Demodulator and Integrator

Each demodulator is a gain unity, reversible polarity amplifier, the polarity of which is controlled by two FET switches, synchronized to the 2-Hz NFR decommutation signal.

The integrator consists of an inverting amplifier with a 0.82 μF capacitor in the feedback loop and a 1 MΩ resistor connected between the output of the demodulator and the input to the integrating amplifier. A FET switch is placed in paralel with the capacitor to short out the charge after a measurement has been taken. Two other switches control input to the integrator. In one configuration the output of the demodulator is connected to the integrator (enabling integration); in the other configuration the demodulator output is disconnected and the integrator input is grounded (holding the integrated value for readout by the A/D converter).

Gain Select Amplifier (GSA)

All six integrator outputs and all housekeeping monitor outputs are routed by a 22-channel multiplexer (identified as INPUT MUX in Figure 7) to the gain selection circuits which properly scale those analog signals for input to the Analog to Digital Converter (ADC) described below). A 3-channel gain select multiplexer selects either the output of the 22-channel multiplexer or the output of one of two cascaded amplifiers, each with a gain of eight. The three multiplexed channels view the output of the 22-channel multiplexer at gains of 1, 8, and 64.

Analog to Digital Converter (ADC)

The ADC is a 12-bit, + 10 to − 10 V, successive approximation type converter, used over a + 5 to − 5 V 11-bit range only. The most significant bit (bit 11) indicates polarity of the signal, and the second most significant bit indicates a positive or negative overrange. Data reported to the Probe telemetry include the sign, two bits indicating the gain setting of the GSA, and nine bits of data from the ADC (bits 1 through 9 – bit zero is unused).

The microprocessor changes the gain of the GSA as required to obtain an on-scale reading for the ADC. The GSA is first set to its maximum gain of 64. If the processor detects that the ADC is in an overrange condition (input greater than + 5 or less than

– 5 V), it sets the GSA to a gain of 8. If this also leads to an overrange condition, the GSA is set to a gain of one.

Housekeeping Monitors

The housekeeping data that the NFR in the Probe telemetry stream is listed in Table III. The forward voltage drop change with temperature of a 1N4148 diode serves as the temperature sensor for the detector and electronics monitors (referenced in the table as DET and ET, respectively). The detector temperature sensing diode is mounted directly to the detector board on the rotating optics (see Figure 5). The hot black-body (HB) sensor is located on the back of a beryllium substrate on which the hot black-body heater resistor is painted. The ambient black-body low range (A1) and high range (A2) thermistors are located in the ambient black-body which is screw mounted and thermally coupled to the front housing wall. The window teperature (WT) sensor is located in the window housing structure on the rotating optics. The HB, A1, A2, and WT temperature sensors are Fenwal GB38SM43 thermistors.

Motor Driver

The motor driver is basically a pair of 28-V H-bridge circuits capable of driving up to 250 mA of reversible current through each of the two motor coils. The microprocessor controls the motor driver by writing a one into the appropriate latch bit to turn on one of two coils in one of two polarities. In addition to these four latch bits, there is one additional bit reserved for controlling eddy current damping by shorting one of the two coils. The exact time and duration of each motor coil pulse is controlled by the PROM software, and is turned, prior to burning PROMs, to obtain stable symmetric optical head rotation characteristics.

To obtain stable rotor motion characteristics under varying temperature conditions, the motor drive currents are stabilized by a current regulator circuit.

Optics Position Sensors

The rotor gear on the rotating optics has four slots cut into it so that four LED-photo transistor pairs mounted around the gear can determine if the rotor is at 0°, 90°, 180°, or 270° (± 5°). Only one photo-transistor will be turned on indicating the position of the rotor. If the correct transistor is not illuminated, this condition is reported in the data stream by setting the position error flag.

3.3. Analog signal characteristics

Radiation input to a given pyroelectric detector is absorbed by black paint covering the detector and subsequently heats the pyroelectric crystal, moving it towards a new thermal equilibrium temperature, which is ultimately attained when the crystal warms enough to conduct and radiate the absorbed power to its surroundings. The idealized DC pyroelectric detector generates charge across its capacitor plates which is proportional to the total temperature change since its last discharge. The charge which is generated slowly bleeds off through the load resistor and through the inherent leakage of the pyroelectric crystal material. For the NFR detectors the load resistors are about a thousand times more important than the crystal losses, producing a charge decay (or droop) time constant of 0.36 s.

As a result of chopping the external radiation field by flipping the rotating optics, the NFR detectors are exposed to a radiation input approximately given by the waveform in Figure 8(a). During the transition between up and down views there is a brief period

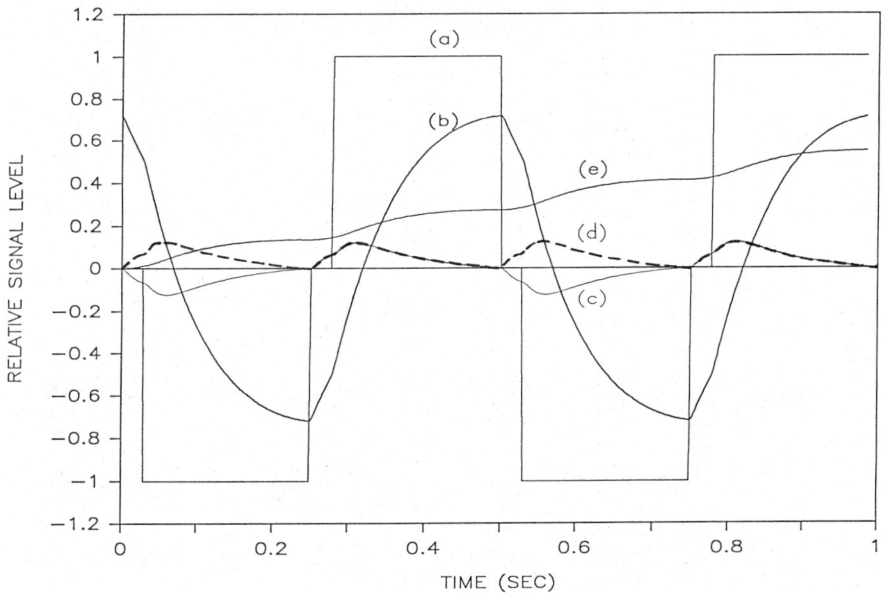

Fig. 8. Analog signal waveforms: (a) radiation input to detectors; (b) detector response; (c) after filtering by post amps; (d) after decommutation; (e) after integration. The plotted signals show the effects of filtering, decommutation, and integration, but not the rather large effects of amplification, which we excluded for clarity. These waveforms do not accurately simulate atmospheric input signals ((a) would have rounded corners during atmospheric measurements).

during which the detectors receive radiation from the wall of the front housing instead of the atmosphere; this contribution is clearly evident as the short flat spot in the irradiance profile. The approximate response of the detectors to the complete irradiance input waveform, shown in Figure 8(b), is a consequence of a thermal time constant of about 110 ms and an effective droop time constant of 1.1 s (the electrical droop time constant of 0.36 s appears to be partly offset from a long time constant component in the detector response). A rather drastic change in waveform is seen at the output of the post amplifier chain, shown in Figure 8(c), where the effects of the 16 Hz bandpass filter have resulted in a crude differentiation of the input signal. The decommutated and integrated waveforms are shown in Figures 8(d) and 8(e), respectively. What appear to be large distortions of the input signal should be thought of as signal conditioning to achieve maximum accuracy. The result of this signal processing is a very narrow bandpass filtering, with bandwidth of 0.09 Hz for 5.5 s of integration, which extracts the amplitude of the 2-Hz fundamental which is proportional to the net radiation flux. As noted earlier, the filtering acts like a crude differentiator which, in combination with the integrator, has the beneficial characteristic of being insensitive to the details of the transition between upward and downward signal levels, and thus insensitive to asymmetries in mechanical motions.

Significant asymmetry errors can arise when the emission from the wall (or ambient blackbody) is considerably different from that of the atmosphere, and when the wall pulses seen during up and down transitions are not cancelled perfectly by the decommutation system because the pulses themselves are either of different width during up and down transits, or occur with different delay times relative to the decommutation waveform, or both. The resultant error is proportional to the difference between the average atmospheric flux, F_a, and the flux emitted by the wall, F_w. The error is then conveniently characterized in terms of the asymmetry factor, given by X_a = net flux error$/(F_w - F_a)$. By using large front housing apertures to minimize the size of the wall pulse, and by adjusting the motor pulse train program to symmetrize the delay of up and down wall pulses, the NFR asymmetry factor has been reduced to approximately 10^{-3} to 5×10^{-4}. When thermal models are used to estimate the expected difference between F_w and F_a during descent, the resultant asymmetry error is almost everywhere negligible compared to other sources of error.

4. Instrument Calibration

4.1. Equations for Predicting Instrument Output

The object of the NFR calibration measurements is to define the instrument behavior sufficiently well so that, given a knowledge of the radiation field and the instrument environment, it is possible to predict the instrument output. The basic equations relating digital count output in a given channel to the external radiation field are essentially as follows:

$$C = Kf(T/T_0)\,(L_u - L_d)\,,$$

where K is an absolute calibration constant, $f(T/T_0)$ is a relative response function describing the instrument response dependence on detector temperature, and where $L_{u,\,d}$, the average radiances within the instrument FOV, are given by

$$L_{u,\,d} = \int_0^\infty s_\lambda\,\mathrm{d}\lambda \int_0^\pi \sin(\theta)\,\mathrm{d}\theta \int_0^{2\pi} \mathrm{d}\phi a_{u,\,d}(\theta, \phi) L(\theta, \phi, \lambda)\,,$$

where s_λ is the relative spectral response at wavelength λ; $a(\theta, \phi)$, relative angular response at θ (angle from vertical), ϕ (azimuth angle); $L(\theta, \phi, \lambda)$, spectral radiance at wavelength λ angles θ, ϕ, where s_λ and $a(\theta, \phi)$ are both normalized to have unit integrals. Because the NFR detector arrays are sealed in a rigid container, there is no dependence of response on external atmospheric pressure, and crosstalk between spectral channels is small enough to allow relatively accurate corrections.

A conversion factor is needed to translate the net average radiance into an estimate for the net flux for a typical atmosphere; but this is an estimate based on model results and is not part of the instrument calibration. However, it has been a habit of the NFR investigators to assume a working conversion factor in calculating sensitivity. The practice can be illustrated for a flat radiation field, as in the extended area blackbody calibration (EABB) configuration. In this case $L(\theta, \phi, \lambda)$ is given by the Planck radiance function $B_\lambda(T)$ and the integral collapses to the form

$$L_{u,d} = \int_0^\infty s_\lambda B_\lambda(T_{u,d})\,\mathrm{d}\lambda \int_0^\pi \sin(\theta)\,\mathrm{d}\theta \int_0^{2\pi} \mathrm{d}\phi a_{u,d}(\theta, \phi) = \int_0^\infty s_\lambda B_\lambda(T_{u,d})\,\mathrm{d}\lambda\,,$$

which reduces to the spectrally weighted average of the net Planck radiance. For a spectrally flat broadband channel, the average net radiation (in the EABB configuration) is just

$$L_n = (L_u - L_d) = \frac{\sigma}{\pi}(T_u'' - T_d^4) = \frac{1}{\pi}(F_u - F_d) = \frac{1}{\pi}F_{\text{net}}\,.$$

But in a relatively opaque atmosphere the net radiance varies as the cosine of the nadir angle and can be written as

$$L(\theta) \approx \frac{1}{2\pi}(F_u + F_d) + \frac{3}{4\pi}(F_u - F_d)\cos(\theta)\,,$$

and thus,

$$L_n = L_n = L(45°) - L(135°) \approx \frac{3\sqrt{2}}{4\pi}(F_u - F_d) = \frac{3\sqrt{2}}{4\pi}F_{\text{net}}\,.$$

The measured responsivity of an NFR channel is usually quoted in terms of the ratio Counts/(F_{net}(EABB)). Since $3\sqrt{2}/4\pi$ is only 6% larger than π^{-1}, the measurement sensitivity to the EABB flux difference is almost correct as a measure of sensitivity to atmospheric net flux. Appropriate conversion factors will be applied in deriving the best estimate of the descent profile of atmospheric flux.

4.2. Basic calibration measurements

To calculate the instrument output according to the above equations requires the following basic measurements:

(1) Relative angular response for each detector channel.
(2) Relative spectral response for each spectral band.
(3) Relative crosstalk coefficients between neighboring channels.
(4) Relative response versus temperature for each channel.

(5) Absolute responsivity measurements for each channel using a source which fills the NFR field of view.

Methods and sample results are described in the following paragraphs.

Relative Angular Response Measurements

Angular response was measured using a chopped stationary source of 4° angular diameter in conjunction with an articulating mount for the NFR optical head. The articulation was accomplished with crossed rotators scanned by computer control to map the solid angle domain with 2° resolution. The optical head was mounted so that the center of the diamond window remained fixed during the articulation. The rotating optics assembly was held motionless by disabling NFR motor power; a modulated signal was instead obtained by chopping the radiation source at a 2-Hz rate in synchronization with the NFR decommutation signal. This allowed the NFR data system to acquire the angular response data, while minimizing the wear and tear on the NFR gears. The angular scan was made twice, once using a short wavelength source consisting of an integrating sphere illuminated by a quartz-halogen lamp, and once using a long wavelength source which is just a small black-body operating at 128 °C. A sample response function is shown in Figure 9. A consequence of the NFR optical design (shown in Figure 3) is that the maximum response of each detector is in a slightly different location (adjacent detectors are separated by 5–6°).

Relative Spectral Response

Typical spectral response results are shown in Figures 10 and 11 (results from the final flight instrument calibration were not available at the time of this writing). The relative spectral response measurements were generally made using appropriate short or long wave sources, a monochromator to select wavelengths, and a reference detector to measure the monochromator output for comparison with the NFR measurements. Our primary reference detector is a deuterated triglycene sulfate (DTGS) pyroelectric detector with a thorium fluoride-coated KBr window. This detector was calibrated by D. Stierwalt at the Naval Oceans Systems Center (NOSC) in San Diego. Stierwalt's calibration is based on comparisons with a standard spectrally flat detector which uses a blackened conical receiver (Eisenman *et al.*, 1963; Eisenman and Bates, 1964). A conical detector was borrowed from the NOSC laboratory for additional comparisons at Wisconsin. Our comparisons and those at the NOSC laboratory were in significant disagreement concerning the long wavelength fall off in detector response. For example, at 30 μm our version of the detector response was 40% greater than the NOSC value. This discrepancy was investigated by making broadband filter measurements at Wisconsin, and performing a second calibration using a 2-Hz chopping frequency at NOSC (normally a 10-Hz chopping frequency is used at NOSC). Although more recent results are in much better agreement (the discrepancy is now less than 10% at 30 μm), this remains an area of study.

Our spectral response measurements used a quartz halogen source between 0.25 μm and 3 μm, and a 1000 K silicon carbide globar for wavelengths from 3 μm to about

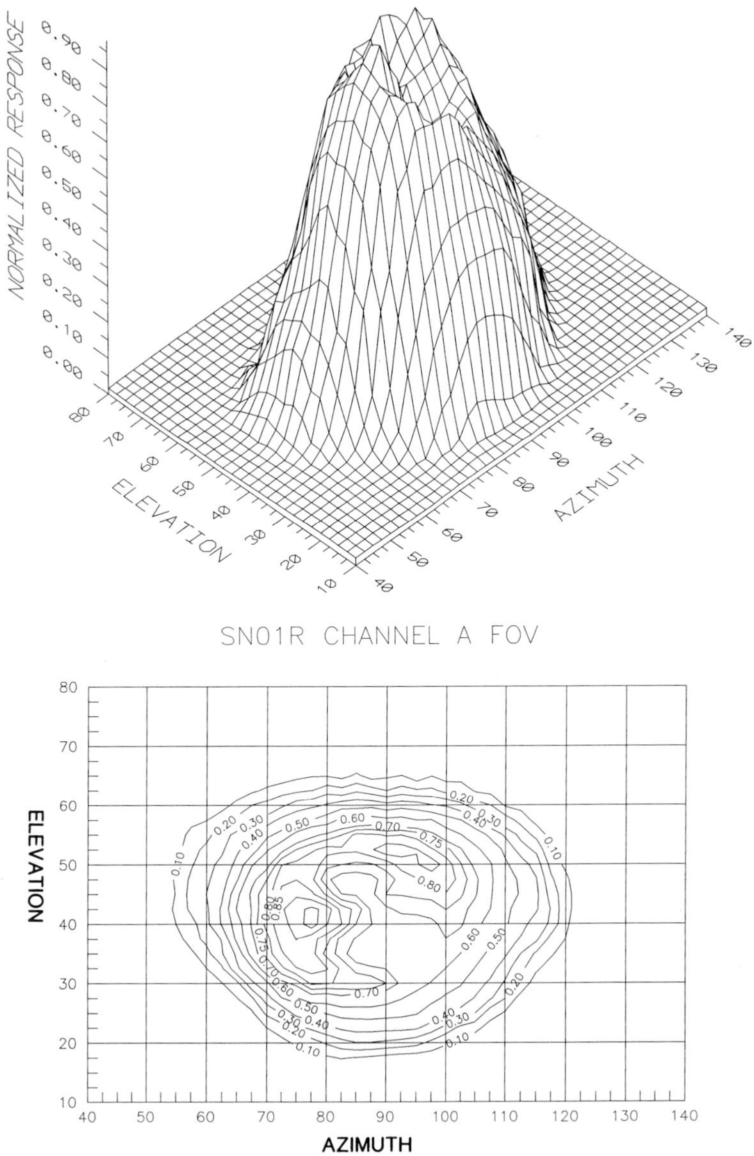

Fig. 9. Sample angular response functions. Here elevation refers to the angle above the horizon and azimuth refers to angle about the vertical (90° azimuth corresponds to the position of the NFR optical head rotation axis). During the early part of descent the Sun will have an elevation angle of 25°, and may intrude into the edge of the NFR field of view.

38 μm. Because of large variations in source output as a function of wavelength, it was necessary to use stray light filters at the shortest wavelengths, and neutral density filters at intermediate wavelengths, the latter to maintain on-scale detector signals. Multiple gratings and order-sorting filters were also required. Slit widths were chosen to obtain

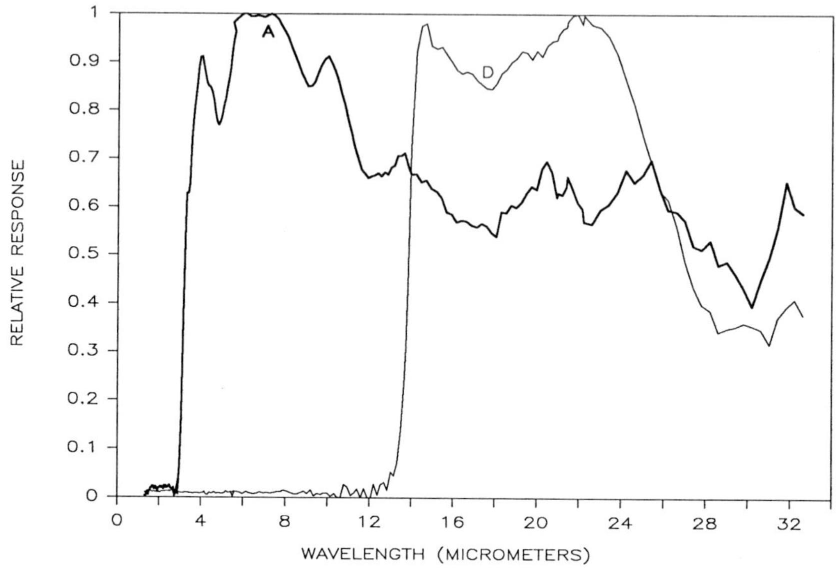

Fig. 10. Relative spectral response functions for NFR long wavelength channels *A* and *D*, normalized to have a maximum value of one. These were measured for the SN01R instrument prior to installation of new detectors. Analysis of final flight calibration results will yield very similar curves.

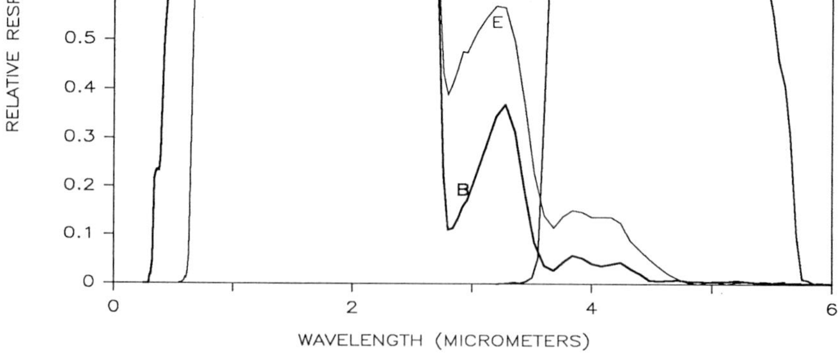

Fig. 11. Relative spectral response functions for NFR channels *B*, *E*, and *C*. Channels *B* and *E* use colored glass filters, while C uses an interference filter and a long wave CaF_2 blocker. Normalization and instrument used are the same as for Figure 10. Analysis of final flight calibration data and additional measurements using the spare NFR instrument and spare filters will allow us to define the spectral response much better at longer wavelengths.

high monochromator throughput with moderate spectral resolution (typically 4–5% of wavelength). Wavelength calibration was checked with a mercury vapor lamp.

Although all the radiation exiting the monochromator was focussed on the active area of the reference detector, the NFR detectors, because of optical limitations, could only sample a portion of the radiation. Because the radiation entering the NFR is not perfectly scrambled, sampling does lead to small spectral errors, the magnitude of which can be estimated by repeating scans with different NFR orientations, and by comparing overlapping scans with successive gratings.

Accurate measurements of system-level spectral response beyond approximately 35 μm were not practical with the monochromator system because of signal-to-noise limitations. To limit uncertainties in relative spectral response at longer wavelengths we also used broad-band and diamond dust cut-on filters (with known transmissions) in combination with a known black-body source. To refine the long wavelength spectral response characteristics these measurements of the flight instrument response will be supplemented by additional measurements of the spare instrument and spare filters and detectors.

Relative Crosstalk Coefficients

Crosstalk coupling coefficients were determined by exposing the NFR to spectrally independent sets of stimuli, then measuring the response of all channels. For any given stimulus vector there exists a direct detector response vector **R**, which is transformed by crosstalk matrix A_{ij} to a gain-corrected count response vector $\mathbf{C}'$, with elements

$$C_i' = \frac{C_i}{g_i} = \sum A_{ij} R_j \,,$$

where g_i is a gain conversion fator for channel i. When it is possible to stimulate directly just a single channel, then the coupling coefficients are trivially determined by the relationship

$$A_{ij} = \frac{C_i/g_i}{R_j} \approx \frac{C_i/g_i}{C_j/g_j} \,.$$

The approximation of R_j by C_j/g_j is warranted because the crosstalk coefficients are rather small (just a few tenths of 1% or less).

Relative Temperature Dependence

To measure responsivity as a function of temperature the NFR optical head was mounted on a thermoelectric cooler within a sealed chamber which was evacuated to maximize the performance of the temperature control system. While exposed to a constant radiation source the instrument temperature was changed from −20 °C to +50 °C. The test was carried out in two parts: first, extended area black-body sources were used and the NFR motor was used to chop between the two soures; and second, an external chopped quartz-halogen source was used and the NFR motor was turned off. A comparison of these two test results shows that the NFR motor operation does not vary with temperature sufficiently to have a significant effect on responsivity. We did find, however, that the NFR spectral response does vary somewhat with temperature, leading to different temperature dependencies in different channels. Figure 12

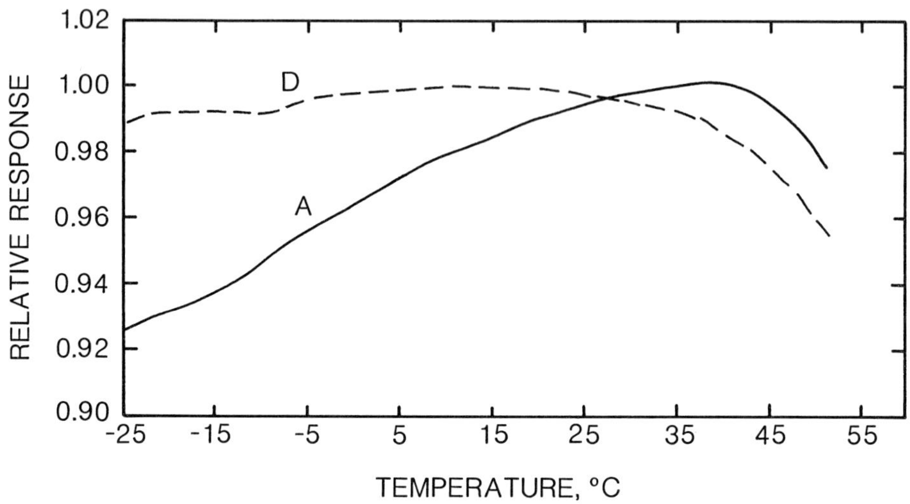

Fig. 12. Temperature dependence of responsivity for NFR channels *A* (3–500 μm) and *D* (14–35 μm). Results were obtained using Extended Area Black-Body (EABB) sources and NFR rotational chopping. The difference between these two curves is presumably a result of temperature dependence in the spectral characteristics of channel *D*. The response curve of channel *A* is dominated by the relative temperature dependence of detector responsivity. Both curves were normalized to have a maximum of one.

compares the response of channel *A* with that of channel *D*. Most channels, when illuminated mainly with in-band radiation, behave approximately like channel *A*. The responsivity decline at temperatures beyond 50 °C is sufficiently uncertain to warrant further investigation using the spare detector package.

Absolute Responsivity Measurements

The absolute calibrations used sources of well-known spectral content which uniformly fill the field of view of the NFR, and which are spectrally shaped to provide strong in-band stimulation.

Long Wave Channels. Long wavelength channels *A* (3.5–500 μm) and *D* (14–35 μm) were calibrated using two distinct black-body source configurations: one configuration uses a pair of extended area black-body (EABB) sources made by Advanced Kinetics, Inc. (Costa Mesa, CA), the second configuration used a pair of black-body cavities adapted from soruces used on a previous University of Wisconsin Pioneer Venus experiment (Sromovsky *et al.*, 1980). In both cases the pair of sources was arranged to fill the NFR field of view at 'up' and 'down' positions, and the NFR flipped between them using the rotating optical system. The EABB sources are 6-inch × 6-inch aluminum plates with a blackened honeycomb emitting surface and three independent thermoelectric servo systems trying to maintain a constant and spatially uniform temperature. Temperature measurements were made using platinum resistance thermometers. The emittance of the EABB sources is better than 0.995 by manufacturer calculation, but not independently measured. A concern with the EABB sources is the existence of temperature gradients within the aluminum radiating plate. The three independent temperature monitor readouts on each plate typically differ by 0.1–0.2 K. The platinum resistance sensors within the EABBs may also have experienced some minor shift in calibration due to repetitive thermal cycling. A comparison with recently calibrated precision thermistors indicates that such a shift is probably no more than 0.1 to 0.2 K. Both emissivity and temperature uncertainties are not expected to create calibration errors greater than 1%, although this estimate itself is sufficiently uncertain to warrant comparison with the independent set of cavity calibration sources.

The cavity sources are relatively small and thick-walled, leading to very uniform temperatures within the cavity. The cavity temperatures are monitored with YSI 44031 precision thermistors with a guaranteed interchangability of 0.1 K with respect to an absolute temperature scale. To minimize offset errors we compared a set of six thermistors, and selected four which were in closest agreement for use as cavity monitors (these differed by a maximum of 0.03–0.06 K). The cavity emittance, calculated according to formulas developed by Sydnor (1970), using paint properties measured by Stierwalt (1966), is estimated to be approximately 0.99, not counting the additional enhancement resulting from reflection of cavity radiation off the NFR back into the cavity. The disadvantage of the cavity sources relative to the EABB sources is that the former can only be heated above room temperature, while the latter can be cooled as well as heated.

The results of these calibrations for the SN02R2 instrument, summarized in Table IV, are very encouraging in that both source configurations lead to essentially the same calibration for both longwave channels (within 0.2% and 0.5% for A and D, respectively). This is better agreeement than the estimated uncertainty of about 1% (based on a temperature difference uncertainty of 0.2 K and an emissivity uncertainty of 0.5%). The responsivities in this table are calculated by dividing the digital count output by the net flux input, where the net flux input is 'in-band' net flux calculated by integrating the source emission spectrum over the relative spectral response of each channel, using a normalization of unity at the maximum responsivity. This able does not show EABB and cavity calibration results for channels B and E because the long wave sources are poorly suited for calibration of these short wave channels. (Most of the source energy is outside their spectral bandpasses making in-band flux calculations extremely dependent on details of the long wave cutoffs of these channels.)

Short Wave Channels. The calibration of the solar channels B (0.3–3.5 μm) and E (0.6–3.5 μm) is based on a spectral irradiance standard which illuminates a standard diffusing plate filling the NFR field of view. The irradiance standard is a 40 W quartz-

TABLE IV

Absolute calibration results. Columns labelled EABB and Cavity contain results obtained with long wavelength sources. The column labelled QH contains calibration results using a quartz-halogen standard lamp. The column labelled Globar lists calibration data obtained with a 1000 K globar source illuminating a gold integrating sphere and using a reference detector to measure the sphere output. Note the excellent agreement between EABB and cavity calibration results, and between these results for channel C and that obtained with a filtered globar source.

Channel	Responsivity in counts W^{-1} m^{-2}			
	EABB	Cavity	QH	Globar
A	241.7	242.2		
B			766.5	
C	1468.4	1475.0		1443.5
D	807.4	811.2		
E			1410.0	

halogen lamp calibrated by Optronics Laboratories, Inc. of Orlando Florida. The calibration extends from 0.25 to 3.5 μm, but is most accurate at wavelengths less than 2.5 μm. A water filter was placed between the lamp and the diffuser plate to block radiation at uncalibrated and at inaccurately calibrated wavelengths. The spectral transmission of the water filter was measured using the monochromator and a pyro-electric detector. The diffuser is a freshly prepared 6 × 10-inch × 5 cm thick Halon surface compressed to a density of 1 g cm^{-3} to obtain standard reflectivity properties (Weidner *et al.*, 1985). Over most of the visible spectrum, the surface reflectivity is approximately 0.99. Given the known source intensity, transmission of the water filter, distance to the diffusing screen, and reflection properties of the screen, it is possible to calculate the in-band flux in the short wave channels with an RSS uncertainty of about 2.5% (calculated as the RSS of the following contributions: 1.3% NBS uncertainty, 1% transfer uncertainty, 1.75% uncertainty in water filter transmission, 0.5% uncertainty in surface reflectivity). The calibration was performed twice: once with the radiation source chopped and the NFR rotor immobilized, and a second time with the source steady and the NFR chopping. The results of the absolute calibration of channels *B* and *E* are shown in Table IV.

Channel C Calibration. The excess thermal channel (*C*), with a spectral bandpass from 3.5 to 5.8 μm, was calibrated using both long wavelength broadband sources and also using a restricted wavelength source. The restricted source used a 1000 K globar filtered to restrict radiation to be within the channel *C* spectral band, then input to a gold diffusing sphere to provide the angular distribution needed to fill the NFR field of view. Because the output of the integrating sphere cannot be calculated with sufficient accuracy, it had to be measured. This was done using a reference detector calibrated at short wavelengths (using the same procedure as carried out to calibrate channels *B* and *E*) and adjusted for its relative spectral response. The uncertainty of this calibration is approximately 3.4%, estimated as the root sum square of the following uncertainties: 2.5% uncertainty in reference detector calibration at calibration wavelengths, 2% uncertainty in relative spectral response of the reference detector, and 1% variation in integrating sphere output due to NFR-reference detector reflectivity differences). The result of this calibration differs by only 1.9% from the average calibration using cavity and EABB broadband sources (see Table IV). This is a gratifying agreement between two completely independent calibration methods, one ultimately related to an NBS spectral irradiance standard, and the other dependent on temperature measurement and emissivity calculations.

5. Performance Expectations

5.1. Signal Characteristics

The Galileo Probe will enter the atmosphere near the evening terminator; the initial solar zenith angle for NFR measurements will be approximately 65° and will increase with time as Jupiter's rotation carried the Probe into sunset, which will occur near the 10-bar level. Depending on the Probe spin rate, the strong azimuthal variations in scattered solar radiation may be evident in channels *B* and *E* above the 1-bar level (Tomasko and Doose, 1988). Deeper in the atmosphere, the direct solar beam and azimuthal variations will both become insignificant. Gains on channels *B* and *E* are low enough to handle relatively large azimuthal variations, but not low enough to handle the direct solar beam (which will in the worst case only be near the edge of the NFR field of view (see Figure 9).

Predicted net flux profiles under expected descent conditions are shown in Figure 13. (These are only tentative estimates which will be considerably refined prior to Jupiter encounter.) Profiles for long wave channels *A*, *D*, and *C* (Pollack, 1986) are from calculations which do not include cloud opacity. Profiles for short wave channels *B* and *E* (Tomasko and Doose, 1988) do include effects of cloud opacity and haze, with boundaries as indicated in the figure. Expected measurement noise in individual net flux data samples is indicated by error bars on each profile. Errors due to calibration uncertainties are not shown. Detector noise (actually, this is mainly Johnson noise in the load resistors) is almost everywhere the dominant noise source, and quantization and asymmetry errors are almost everywhere insignificant. During thermal transients, there may also be noise contributions from detector arcing, as described later.

At the top of Figure 13 is a tabulation of the number of samples expected for each atmospheric layer. For example, within the NH_3 ice cloud between 300 and 700 mb we expect 29 net flux measurements, but less than two each of Up Flux, Black-body Calibrate, and Analog Zero measurements. The vertical resolution of net flux measurements is high enough to allow some noise reduction through averaging, especially at the higher pressures (where it is needed the most). Uncertainty in calibration, offset correction, and crosstalk couplings are, of course, systematic errors which cannot be reduced by vertical averaging.

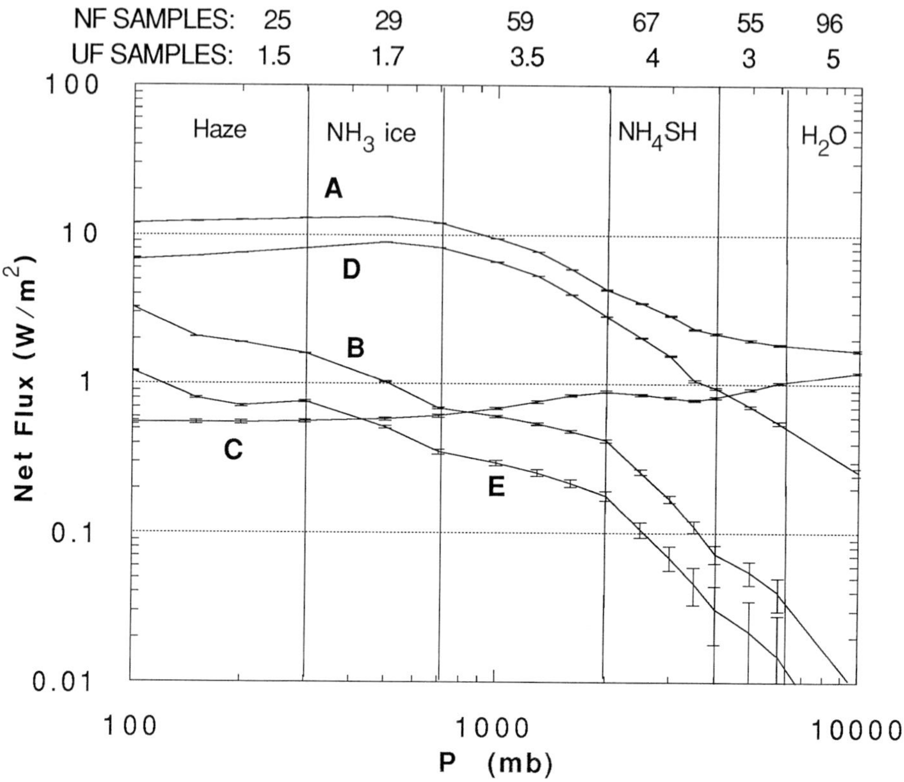

Fig. 13. Predicted profiles of net flux versus pressure during descent into Jupiter's atmosphere. Calculations for channels *A*, *D*, and *C* did not account for cloud opacity. Clouds were included in simulating measurements in the solar channels *B* and *E*. The cloud and haze boundaries used in these calculations are indicated by vertical lines and labelled by cloud composition. The number of samples of net flux (NF) and up flux (UF) data in each layer is listed across the top of the figure. Error bars indicate the expected measurement noise. Calibration errors are not included. The rapid decline of signal in channels *B* and *E* near 10 bars is a consequence of the Sun's setting.

5.2. Detector Arcing

NFR detector crystals develop potential differences proportional to the rate of change of detector temperature and to the bleed resistance. The electroded part of the crystal does not develop a substantial potential because the load resistors are very efficient at bleeding off the charge. The time constant for bleeding off charge from the unelectroded regions of the crystals is about 1000 times longer, so that thermal ramp-induced potentials can become large enough to cause sudden discharges. These discharges can become large enough so that a single event can invalidate a 5.5-s data sample. Fortunately, these events are likely to be relatively infrequent, occurring mainly during the second 5 min of descent, between the point of the maximum negative thermal ramp and the point of minimum detector temperature. This covers an approximate pressure range from 0.5 to 1.5 bars (see Figure 13). Only channels *C* and *E* exhibited arcing noise during laboratory testing of the flight instrument.

5.3. Expected performance of the detector package seal

The first objective of the seal, preventing pressure perturbations from reaching the detectors, appears to be met even with a relatively poor seal. This inference is based on comparing wind tunnel tests of an unsealed detector with those of a poorly sealed detector leaking at a rate of approximately 10^{-5} cc s^{-1} (xenon equivalent). A much better seal is required to prevent large crosstalk couplings and responsivity variation. With any more than 100 mb of He in the detector package crosstalk would be five times the vacuum level and responsivity would be degraded by 15%. (Hydrogen has approximately the same effect as He.) To make these effects negligible the seal should keep the thermal conductivity of the gas within the detector package no higher than that of air (about $\frac{1}{5}$ that of helium).

The upper bound of the detector leak rate measured by mass spectrometry shortly after sealing (in February 1989), is approximately 7×10^{-10} cc (of Xe at STP) s^{-1} with a pressure difference of approximately 840 mb. Given an initial volume of approximately 0.033 cc of gas (the internal volume of the detector package), and assuming that the leak is viscous in nature, we find that after 6 years of leaking the detector package will still retain about 100 mb of Xe. We measured the rate at which helium leaks into the package by watching the NFR response to a fixed radiation source as the helium diffuses into the detector package. After 23 hours of exposure to 1 bar of helium we saw a response drop of only 0.5%, implying a maximum He diffusion leak rate of about 10^{-7} cc s^{-1} (this rate takes several hours to become established). Because tests with a mockup detector seal show no significant difference between He and H_2 diffusion rates, we can conclude that no more than 10–20 mb of Jovian He and H_2 combined will leak into the detector during descent. This is well below the level which could affect detector response, especially in the presence of 100 mb of Xe, which buffers the conductivity of the lighter gases.

Acknowledgements

This difficult experiment would not have been possible without considerable effort and dedication by many individuals. Foremost among them is Robert W. Boese, the original P.I. from Ames Research Center, who conceived and nurtured this experiment until his untimely death in December 1985. Others from Ames Research Center deserving thanks are Jim Pollack for encouragement and technical support at several critical times, and Rick Twarowski and Benny Chin, who guided us through many complex difficulties in managing and funding instrument development.

At Martin Marietta we thank DeWane Shumaker (optical head fabrication), Tony Knight (management), Ruth Amundsen (engineering support and management), Tom Hopkins (quality assurance), Steve Shertz (engineering support), and Briggs Cunningham (program assistance).

At the University of Arizona we thank Martin Tomasko, Peter Smith, and Lyn Doose for excellent science impact evaluations and calibration support.

At the University of Wisconsin-Madison we wish to thank Patrick Fry (software, calibration, testing, data analysis), Donald Thielman (electronics design and testing), Jeffrey Vian (mechanical design, detector development support, and testing), Scott Ellington (electronics design), Jerry Sitzman (electronics design and testing), and Mike Dean (parts procurement).

We also thank the Galileo Project for providing support and guidance necessary to reach a satisfactory conclusion to this difficult effort. Bill O'Neil's efforts were particularly helpful.

The work carried out by the University of Wisconsin was supported by NASA Contract NAS2–12078.

References

Eisenman, W. L. and Bates, R. L.: 1964, 'Improved Black Radiation Detector', *J. Opt. Soc. Am.* **54**, 1.

Eisenman, W. L., Bates, R. L., and Merriam, J. D.: 1963, 'Black Radiation Detector', *J. Opt. Soc. Am.* **53**, 729.

Hunten, D. M., Colin, L., and Hansen, J. E.: 1986, 'Atmospheric Science on the Galileo Mission', *Space Sci. Rev.* **44**, 191.

Pollack, J.: 1986, personal communication.

Seiff, A. and Knight, T. C. D.: 1992, 'The Galileo Probe Atmosphere Structure Instrument', *Space Sci. Rev.* **60**, 203 (this issue).

Sromovsky, L. A., Revercomb, H. E., and Suomi, V. E.: 1980, 'Pioneer Venus Small Probes Net Flux Radiometer Experiment', *IEEE Trans. Geosci. Rem. Sens.* **GE-18**, 117.

Stierwalt, D. L.: 1966, 'Infrared Spectral Emittance of Optical Materials', *Appl. Optics* **5**, 1911.

Sydnor, C. L.: 1970, *A Numerical Study of Cavity Radiometer Emissivities*, NASA Technical Report 32–1463.

Tomasko, M. and Doose, L.: 1988, personal communication.

Weidner, V. R., Hsia, J. J., and Adams, B.: 1985, 'Laboratory Intercomparison Study of Pressed Polytetrafluoroethylene Powder Reflectance Standards', *Appl. Optics* **24**, 2225.

THE JUPITER HELIUM INTERFEROMETER EXPERIMENT ON THE GALILEO ENTRY PROBE

U. VON ZAHN
University of Bonn, Bonn, Germany

and

D. M. HUNTEN
University of Arizona, Tucson, AZ, U.S.A.

Abstract. We discuss the scientific objective, instrument design, and calibration of a miniaturized Jamin–Mascart interferometer which is to perform an accurate measurement of the refractive index of the Jovian atmosphere in the pressure range 2.5 to 10 bar. The instrument is to perform this measurement in December 1995 aboard the entry probe of the NASA Galileo spacecraft. From the data obtained the mole fraction of helium in the atmosphere of Jupiter is to be calculated with an estimated uncertainty of ± 0.0015. The instrument has a total mass of 1.4 kg and consumes 0.9 W of electrical power.

Table of Contents

1. Introduction

Hydrogen and helium are by far the most abundant elements in the Universe, the stars, the Sun, and the outer planets. If known, the mixing ratios of these elements in stellar and planetary bodies provide important constraints for theories about their origin and evolution. Yet, considerable uncertainty exists for these mixing ratios because they are difficult to measure.

The NASA Galileo mission to the planet Jupiter provides us with the first opportunity to measure the He/H_2 abundance ratio inside a heavenly body and with high accuracy. To this end, the Galileo entry probe carries a miniaturized interferometer which is to perform an accurate measurement of the refractive index of the Jovian atmosphere, after removal of NH_3, H_2O, and CH_4, in the pressure range from 2.5 to 10 bar. From these data the helium mole fraction (about 0.10) can be calculated with an estimated accuracy

Space Science Reviews **60:** 263–281, 1992.

of ± 0.0015. The instrument is called the Helium Abundance Detector, or HAD for short. It has a mass of 1.4 kg and an electrical power consumption of less than 1 W. Before it can perform its measurements within the Jovian atmosphere, the HAD instrument has to survive an in-space-storage period of more than 6 years, a radiation dose of 75 krad and a deceleration during entry into the Jovian atmosphere of approximately 300 g.

TABLE I

Definitions for abundance measures

'Mass fraction' Hydrogen mass fraction	$X \equiv \dfrac{N_H m_H + 2N_{H_2} m_H}{\sum_j N_j m_j}$
Helium mass fraction	$Y \equiv \dfrac{N_{He} m_{He}}{\sum_j N_j m_j}$
Mass fraction of all other elements	$Z \equiv 1 - X - Y$
with N_i the number density of particles of type i; m_i mass of a particle of type i	
'Mole fraction' (mixing ratio)	$q_i \equiv \dfrac{N_i}{\sum_j N_j}$
with	$\sum_j q_j = 1$
In particular at Jupiter	$q_{He} \approx \dfrac{N_{He}}{N_{H_2} + N_{He}}$
'Abundance ratio' of helium/hydrogen	$R_{He} \equiv N_{He}/N_{H_2}$

In astrophysics, element abundances are commonly quoted as mass fractions, whereas in planetary atmosphere studies these abundances are usually given as mole fractions ($\equiv$ mixing ratios). As we cannot avoid using both terminologies, we provide in Table I rigorous definitions of the more relevant abundance measures.

2. Scientific Objective

2.1. Background

Current theory holds that almost all of the hydrogen and helium now residing in the Universe formed billions of years ago in a cosmic fireball, called the Big Bang. It is of obvious scientific interest to find out what mass fractions these two elements had attained after the initial cool-down of the fireball and before additional, though limited, synthesis of helium in stars began. If known, the primordial mass fractions of H and He provide important constraints for theories trying to explain the formation and evolution of the Universe. The subject of the primordial helium abundance was treated extensively at an ESO workshop (Shaver *et al.*, 1983). Recent reviews of He observations and of the standard Big-Bang theory put the primordial helium mass fraction Y_P into the range $0.23 \leq Y_p \leq 0.24$ (e.g., Yang *et al.*, 1984; Pagel, 1989). The quoted range of uncertainty can be reduced by accepting the latest correction to the half-life of the neutron (Paul *et al.*, 1989).

After synthesis of the primordial helium in the Big Bang some additional helium has been produced by nuclear burning of hydrogen inside stars. Until the time of formation of the solar system 4.5 billion years ago, this process increased the helium mass ratio in the Universe by ΔY. Hence, one would assume the solar system to be formed from a pre-solar nebula having a helium mass ratio $Y_s = Y_p + \Delta Y$. The value of Y_s has turned out to be notoriously difficult to establish. Gough (1983) estimates $0.23 \leq Y_s \leq 0.27$. Stix (1989) writes: "... we see that presently the most accurate method to determine the solar abundance is the computation of solar models, and the adjustment of these models to the present Sun by a proper choice of the mass fraction Y_s of He. Virtually all standard solar models calculated so far yield Y_s values between 0.23 and 0.28."

In view of the astrophysical importance of the value of Y_s alternate approaches to just studying the Sun seem justified to arrive at a determination of Y_s. One alternative is to assume that the giant planets, in particular Jupiter and Saturn, have also been formed with the pre-solar helium mass fraction Y_s and that they have preserved it until today. Indeed, when the preparations for the HAD experiment began during 1976, many scientists assumed that the relative abundances of helium were equal in Jupiter, Saturn, and the Sun (see Lewis, 1974; Podolak and Cameron, 1975). At that time, the helium abundance in the atmosphere of Jupiter was, however, only poorly known. Orton and Ingersoll (1976) summarized the then available remote sensing measurements as indicating a Jovian He mole fraction $q_{\mathrm{He}} = 0.12 \pm 0.06$, which for a $H_2 + He$ atmosphere is equivalent to $Y_j = 0.214 \pm 0.11$. It does not appear to be very difficult to reduce the large uncertainty of either value significantly by a simple *in-situ* experiment aboard the Galileo entry probe.

So far we have implicitly made the assumption that the giant planets have a uniform composition throughout. Only then would a He mole fraction measured in the atmosphere of one of these planets resemble the bulk composition of that planet. As early as 1967 doubt was expressed about the validity of this assumption for different reasons (e.g., Smoluchowski, 1967; Salpeter, 1973). In addition, the processes, which were postulated to lead to an inhomogeneous composition of the giant planets, are of varying effectiveness in the different planets. Hence, the notion of similar He abundances in the atmospheres of Jupiter, Saturn, and the Sun also needed reconsideration. However, it was only after 1979 that those doubts could be verified experimentally through data obtained by the two Voyager spacecrafts. Today it appears established that (a) the atmospheres of Jupiter, Saturn, and Uranus each accommodate a different He mole fraction q_{He}, and that (b) the atmosphere of Jupiter contains somewhat less He than that of the Sun. This follows from the Voyager data concerning the helium abundances in the atmospheres of the giant planets (Table II).

TABLE II

Helium abundances in the giant planets and the Sun

	q_{He}	Y	Reference
Jupiter	0.101 ± 0.0206	0.18 ± 0.04	Conrath *et al.* (1984)
Saturn	0.032 ± 0.027	0.06 ± 0.05	Conrath *et al.* (1984)
Uranus	0.152 ± 0.033	0.262 ± 0.048	Conrath *et al.* (1987)
Sun		0.25 ± 0.02	Gough (1983)

The helium mass fractions Y for Uranus and the Sun, though uncertain, seem to agree; for Saturn, and perhaps Jupiter, helium may have been gravitationally separated in the form of 'raindrops' which may form at the high interior pressures (for a current review see Hubbard, 1989). Thus, it has become apparent that there is no simple relation between the current helium abundance in the atmospheres of the giant planets and that of the pre-solar nebula. Yet, the unexpectedly large differences in the He abundances of the giant planet atmospheres are themselves of eminent scientific interest and provide major inputs for our understanding of the origin and evolution of these atmospheres, as well as their energy balance.

We do not expect to observe any change of the He mole fraction over the (very limited) pressure range that we sample in the Jovian atmosphere. This is because the time scale for vertical mixing by turbulence is almost certainly much shorter than that for any process causing gravitational separation of He and H_2.

2.2. Scientific objective

The foremost scientific aim of the HAD experiment is to obtain an accurate measurement of the He abundance in the Jovian atmosphere. This datum will more accurately define the small difference between the helium mass fractions of Jupiter and the Sun and the large differences in the He mass fractions among the atmospheres of the giant planets. Beyond that, the helium mass fraction in the Jovian atmosphere represents an important lower boundary for the helium abundance in the pre-solar nebula. As such, it also impacts on theories about the origin of the solar system as a whole.

3. Instrument Description

3.1. Principle of operation

More than 99.5% of the Jovian atmosphere consists of hydrogen and helium. Hence, to a first approximation, we can consider this atmosphere to be a binary gas mixture, for which the mole fraction q_{He} of helium can be derived from the ratio of refractive indices

$$q_{He} = \frac{n_{H_2} - n_j}{n_{H_2} - n_{He}} , \tag{1}$$

with n_{He} the refractive index of helium; n_{H_2}, refractive index of hydrogen; n_j, refractive index of the $H_2 + He$ mixture (= Jovian gas).

The refractive index of gases is usually measured interferometrically and any two-beam interferometer can be used for this purpose. We have chosen a Jamin interferometer (see Jenkins and White, 1976), a schematic of which is shown in Figure 1. Let us assume that beam 1 passes through a sample gas cell CS filled with the Jovian gas and beam 2 passes through another gas cell CR filled with a reference gas of refractive index n_{ref}. Both cells are of length L. The optical path difference, OPD, between the beams then becomes

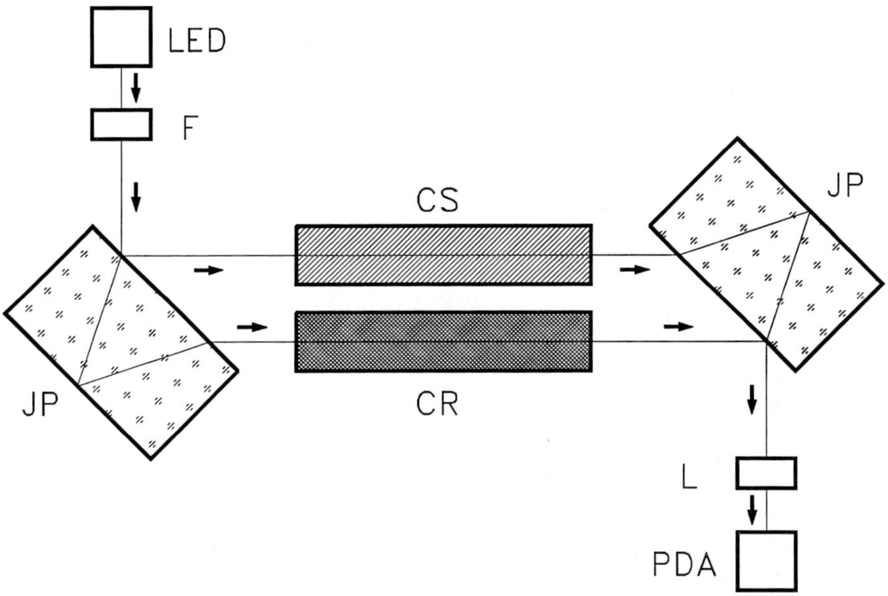

Fig. 1. Schematic of a Jamin interferometer. LED = light emitting diode; F = interference filter; JP = (two) Jamin plates; CS = sample gas cell; CR = reference gas cell; L = objective; PDA = photodiode array.

$$\mathrm{OPD} = (n_j - n_{ref})L$$

and the order m of the interference pattern at the detector is

$$m = \mathrm{OPD}/\lambda ,$$

hence,

$$n_j = n_{\text{ref}} + \frac{\lambda}{L}\, m\,. \tag{2}$$

In order to account for variations of the pressure p and gas temperature T in the gas cells we assume that refractive indices n vary with p and T as follows:

$$n(p, T) - 1 = \{n(p_0, T_0) - 1\}\, \frac{pT_0}{p_0 T}\,, \tag{3}$$

with p_0, T_0 are the standard pressure and temperature (= STP); $n(p_0, T_0) = n_0$, refractive index at STP.

This yields

$$(n_{j,0} - 1)\, \frac{p_j T_0}{p_0 T_j} = (n_{\text{ref},0} - 1)\, \frac{p_{\text{ref}} T_0}{p_0 T_{\text{ref}}} + \frac{\lambda}{L}\, m\,, \tag{4}$$

with p_j, T_j are the pressure and temperature in the Jovian gas cell; p_{ref}, T_{ref}, pressure and temperature in the reference gas cell.

Usually, the absolute value of m is not known. We can, however, measure the number of interference fringes Δm while going from an initial state (i) to an end state (e) in pressures and temperatures. Let us assume that we start our measurement with evacuated gas cells ($p_{j,i} = p_{\text{ref},i} = 0$). We then obtain for the end of the measurement

$$(n_{j,0} - 1)\, \frac{p_{j,e} T_0}{p_p T_{j,e}} = (n_{\text{ref},0} - 1)\, \frac{p_{\text{ref},e} T_0}{p_0 T_{\text{ref},e}} + \frac{\lambda}{L}\, \Delta m\,, \tag{5}$$

with $p_{j,e}$, $T_{j,e}$ are the pressure and temperature in the Jovian gas cell at the end of the measurement; $p_{\text{ref},e}$, $T_{\text{ref},e}$, pressure and temperature in the reference gas cell at the end of the measurement; $\Delta m = m_e - m_i$.

We now introduce $\Delta p_e = p_{j,e} - p_{\text{ref},e}$, the pressure difference between the two gas cells at the end of the measurement; $\Delta T_e = T_{j,e} - T_{\text{ref},e}$, the difference in temperature of the gases in two gas cells at the end of the measurement which allows us to develop (5) in combination with (1) into our final equation for determining the helium mole fraction q_{He}

$$q_{\text{He}} = \frac{n_{\text{H}_2,0} - n_{\text{ref},0}}{\text{n}_{\text{H}_2,0} - n_{\text{He},0}} + \frac{1}{n_{\text{H}_2,0} - n_{\text{He},0}}\, \frac{p_0 T_{j,e} \lambda}{p_{j,e} T_0 L}\, \Delta m + $$
$$+ \frac{n_{\text{ref},0} - 1}{n_{\text{H}_2,0} - n_{\text{He},0}} \left\{ \frac{\Delta p_e}{p_{j,e}} - \frac{\Delta T_e}{T_{j,e}} \right\}. \tag{6}$$

Before evaluating (6) quantitatively we would like to note the following two items:

First, at high pressures it is the so-called Lorentz–Lorenz function LL which is independent of temperature, pressure, and density ρ (Lorentz, 1909). LL is given by

$$\mathrm{LL} = \frac{n^2 - 1}{n^2 + 2} \frac{1}{\rho} . \tag{7}$$

For $n \approx 1$ the Lorentz–Lorenz function can be approximated to

$$\mathrm{LL} = \frac{2}{3} \frac{n - 1}{\rho} \tag{8}$$

and, hence, yields (3). The application of (7) instead of (3) to our HAD experiment and inclusion of the compressibility of real gases has been studied in detail by Schulte (1983).

Second, so far we have neglected the small amount of trace constituents in the Jovian atmosphere. In light of the desired accuracy of our helium measurement, this neglect needs further consideration. In fact, we have either to remove the trace gases from the Jovian gas sample before the refractive index is measured or we have to apply numerical corrections to the result obtained from (6). These corrections would depend on information from sources other than the HAD about the ambient abundances of NH_3, H_2O, and CH_4. We have elected to remove these trace gases before measuring the refractive index (Schütze, 1986), because in the alternative approach it appears difficult to establish how much of the ambient NH_3 and H_2O in fact enters the interferometer gas cells.

3.2. Instrument Design

The Galileo Helium Abundance Detector uses a two-arm, double-pathlength interferometer or Jamin–Mascart interferometer (Mascart, 1874). This type of interferometer allows for a particularly compact and simple design (Figure 2). The light source (1) is a light emitting diode (LED) operating at a wavelength of 900 nm. An interference filter (2) with a 15 nm passband aids in producing near-monochromatic light. A Jamin plate (4) produces two parallel and coherent light beams (5 and 6). Four cells, each of length $l = 100$ mm, house the Jovian gas (7) and the reference gas (8). Additional optical elements are the collimator (3), the inversion prism (9), and the objective (10). The inversion prim is very slightly tilted about an axis parallel to the incoming light beams. This feature, in combination with the objective produces a well-defined interference pattern of consecutive equidistant bright and dark fringes at a linear array of nine photodetectors (12). This pattern does not change if both cells are filled with gas mixtures having the same refractive index. However, any differences between the refractive indices of the Jovian and the reference gas causes a continuous shifting of the pattern with increasing pressure (that is, as the entry probe penetrates deeper into the Jovian atmosphere). The detector allows measurement of the position and motion of the interference fringes in multiples of $\frac{1}{8}$ of the fringe separation.

The instrument carries a simple optical test device (11) which allows a measurement of the contrast of the interference fringes and a verification of the operation of the fringe counter during Earth-based tests and the interplanetary cruise of the Galileo spacecraft. It consists of a plane-parallel glass plate mounted between the objective (10) and the

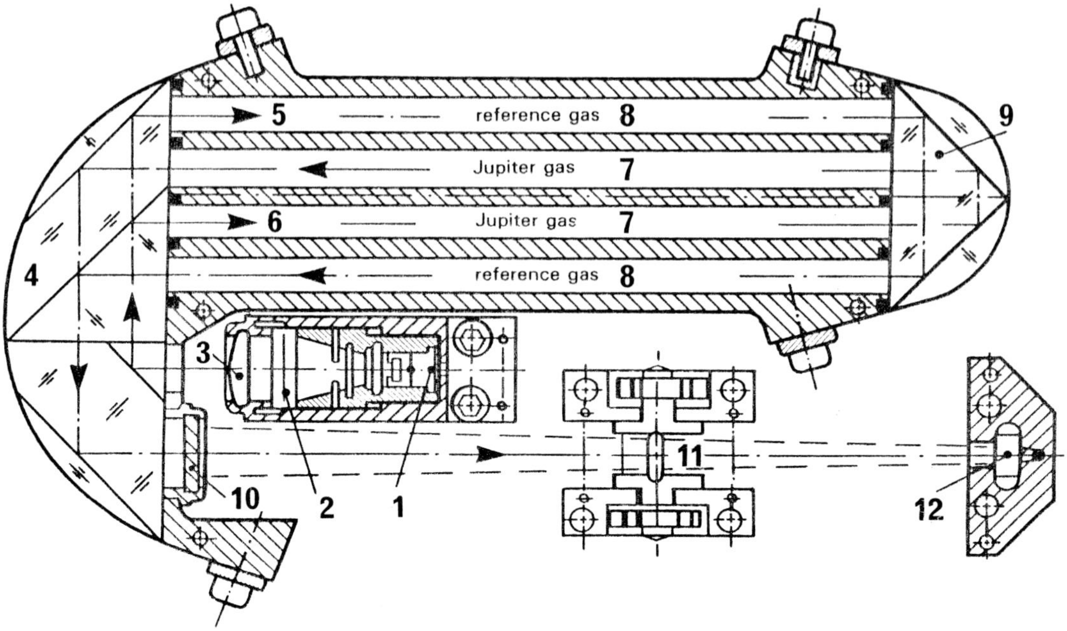

Fig. 2. Optical layout of HAD interferometer (to scale). The length of one gas cell is $l = 10$ cm. 1 = light emitting diode; 2 = interference filter; 3 = collimator; 4 = (single) Jamin plate; 5, 6 = parallel light beams reflected from the Jamin plate; 7 = Jovian gas cells; 8 = reference gas cells; 9 = inversion prism; 10 = objective; 11 = test device; 12 = photodetector array.

detector array (12). By telemetry command this plate can be slowly tilted up to about 30° about an axis parallel to the interference fringes at the detector. This causes a lateral shift of the interference pattern across the detector array.

Figure 3 shows a schematic of the gas flow system of the HAD instrument. After entering the Instrument at *I*, the Jovian gas is passed through a two-stage chemical absorber (A1 and A2) to be scrubbed first of traces of NH_3 and H_2O and then of CH_4. In addition, immediately before entering the gas cells (CS, CR) the Jovian and the reference gases are each passed through heat exchangers (H) made of stainless steel wool to fully accommodate the gas temperatures to that of the surrounding metal structure.

The reference gas consists of a mixture of argon and neon having the same refractive index as a mixture of 11.1% He and 88.9% H_2. The reference gas is carried within the instrument in a storage volume (*R*) of about 20 cm^3 at a pressure of 25 bar. During the descent into the Jovian atmosphere the reference gas is released into its interferometer cells (CR) by means of a membrane valve (V). It keeps the differential pressure between the Jovian gas cell (CS) and the reference gas cell (CR) near 75 mbar. The latter value is nearly independent of the total pressure because opening of the valve is largely determined by the pressure *difference* across the membrane (and the elastic constant of the membrane). This pressure difference is measured by a pressure sensor (PD) within

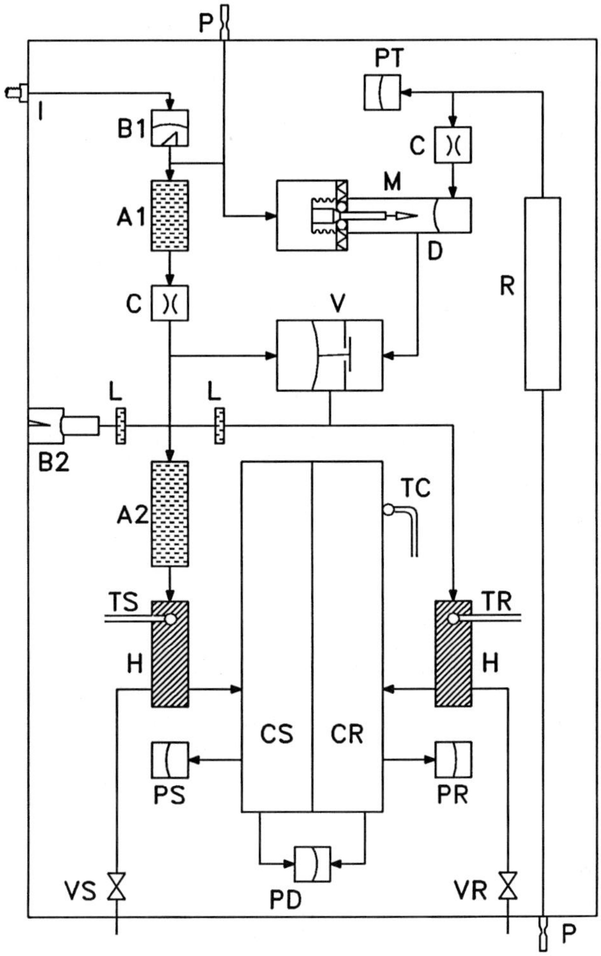

Fig. 3. Schematic of gas flow system, with Jovian gas on the left side, the reference gas on the right side. I = inlet for sample gas; B1, B2, D = burst diaphragms; A1, A2 = absorbers No. 1 and No. 2; C = capillaries; M = release mechanism for reference gas; L = microleaks; H = heat exchangers; CS = interferometer cell(s) for sample gas; CR = interferometer cell(s) for reference gas; R = storage volume for reference gas; V = membrane regulated valve; TS, TC, TR = temperature sensors; PT, PS, PR, PD = pressure sensors; P = pinch-off tubes; VS, VR = valves for laboratory use.

a few millibar to fully account for the influence of this pressure differential on the observed fringe motion.

During launch and cruise of the Galileo spacecraft towards Jupiter the entrance orifice of the instrument for Jovian gas is closed by a thin metal diaphragm (B1). This diaphragm is designed to burst upon reaching an outside pressure of 2.5 bar. Subsequently, the ambient pressure actuates a needle device (M) which pinches a hole in a second diaphragm (D) which previously had closed off the reference gas in its storage volume (R). Both gases are then passed into the interferometer through capillaries (C)

which limit the initial rate of pressure increase inside the interferometer cells to 50 mbar s^{-1}. Measurements are to continue until the reference gas is expanded to the local ambient pressure which should occur near 12 bar ambient pressure. Recent calculations of the descent profile of the entry probe predict that it will take the entry probe

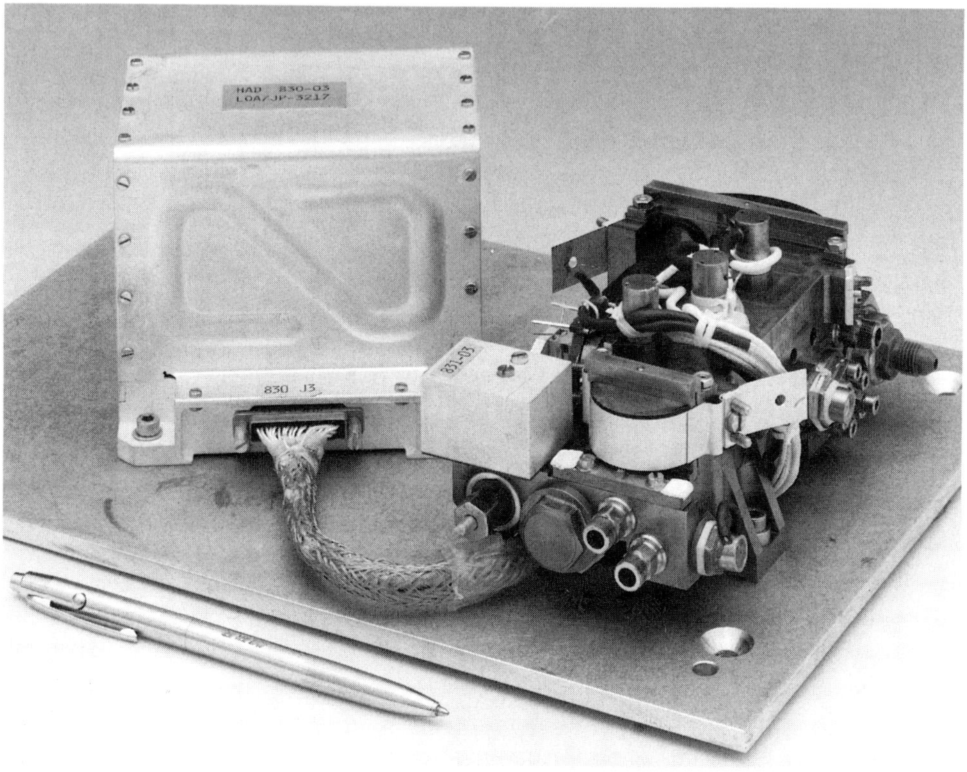

Fig. 4. The HAD flight instrument (without thermal cover). The right unit constitutes the interferometer and gas flow system; the box on the left contains the electronics. The small box on the top left side of the interferometer structure contains the photodetector array and preamplifiers; 3 small cylinders on top of the gas cells are pressure transducers.

about 28 min to descend from 2.5 to 12 bar ambient pressure. The fringe counter measures the motion of the interference pattern starting from vacuum conditions through the 'in-rush' period near 2.5 bar ambient pressure and up to 12 bar.

The structure carrying the optical elements of the interferometer, the gas flow elements, the storage volume for the reference gas, 3 pressure sensors, and 4 temperature sensors is machined from beryllium. This material was chosen for its high mechanical rigidity, low specific mass, and good thermal conductivity. A photograph of the fully assembled flight instrument is shown in Figure 4. A concise list of its important parameters is given in Table III.

To save energy, the LED is powered only for 0.5 ms at 64 Hz. Also, the fringe position is measured 64 times per second which allows the fringe counter to follow the fringe motions for pressure surges of up to 750 mbar s^{-1}. The average power consumption of the HAD instrument is 0.9 W.

One telemetry data frame of the HAD instrument consists of 256 bits and is transmitted every 64 s. It contains the content of the fringe counter, the readings from 3 precision pressure sensors, 4 precision temperature sensors, a number of housekeeping channels and the analog signal of one of the photodetectors. The latter should enable us to obtain a reasonable result from the HAD experiment even if the logic of the fringe counter fails.

TABLE III

Parameters of the helium interferometer

Length l of individual gas cell	100 mm
Pathlength L of light beams in gas cells	200 mm
Wavelength λ	900 nm
Range of Jovian pressure p_j	up to 12 bar
Reference gas	27.64% Ar, rest Ne
Interferometer structure	beryllium
Mass of instrument	1.4 kg
Internal measuring speed	64 fringe positions per s
Telemetry data rate	1 sample per 64 s (= 4 bit per s)
Power consumption	0.9 W
Sensitivity $\Delta m = \frac{1}{8}$ corresponds to	$\Delta q_{He} = 0.0006$
Accuracy	$\delta q_{He} = \pm 0.0015$

3.3. Sensitivity and Accuracy

Equation (6) can be used to quantitatively evaluate the sensitivity and accuracy of the HAD instrument for determining the helium mole fraction q_{He}. These evaluations are based on the input parameters as given in Table IV.

TABLE IV

HAD parameters and uncertainties

Parameter	Value	Uncertainty	Unit
Center wavelength	900.5	± 3	nm
Length of gas cells	2 × 100.0	± 0.05	mm
$(n_{H_2,0} - 1) \times 10^6$	137.026	± 0.04	
$(n_{He,0} - 1) \times 10^6$	34.7196	± 0.04	
$(n_{ref,0} - 1) \times 10^6$	125.682	± 0.05	
Δm	≤6	± 0.0625	
$T_{j,e}$	~280	± 0.5	K
ΔT_e	0	± 0.1	K
$p_{j,e}$	~10000	± 100	mbar
Δp_e	~75	± 3	mbar

The sensitivity S of the HAD instrument is given by the change Δm of the order of the interference pattern for a given change Δq_{He} of the He mole fraction: $S = \Delta m / \Delta q_{\mathrm{He}}$. S is obtained directly from the second term of (6)

$$S = (n_{\mathrm{H}_2,0} - n_{\mathrm{He},0}) \frac{p_{j,e} T_0 L}{p_0 T_{j,e} \lambda} . \tag{9}$$

Inserting the parameters as given in Table IV yields $S = 222$. Hence, if the genuine helium abundance deviates by 1% ($\Delta q_{\mathrm{He}} = 0.01$) from the value simulated by the reference gas ($q_{\mathrm{He}} = 0.111$), the fringe counter would register a change Δm of 2.22 fringes between 0 and 10 bar ambient pressure. For any value of q_{He} within the error bar of the q_{He} measurement of Conrath *et al.* (1984) we would obtain $\Delta m \leq 7$. As will be shown in the section on HAD calibrations, the coherence length of the LED/filter combination makes the HAD instrument capable of producing more than 50 interference fringes.

The accuracy expected from the HAD experiment for determining the helium mole fraction is evaluated using (6). The error bar on the measured q_{He} depends to some measure on the uncertainty of each of the 10 variables entering (6). However, different variables affect the final result of q_{He} in quite different ways. Here it should suffice to point out only the most critical ones.

We will denote with ${}^1q_{\mathrm{He}}$ the first term on the right-hand side of Equation (6)

$${}^1q_{\mathrm{He}} = \frac{n_{\mathrm{H}_2,0} - n_{\mathrm{ref},0}}{n_{\mathrm{H}_2,0} - n_{\mathrm{He},0}} . \tag{10}$$

Substituting $n_{\mathrm{ref},0}$ from our standard reference gas we obtain

$${}^1q_{\mathrm{He}} = 0.110883 .$$

An error $\delta {}^1q_{\mathrm{He}}$ calculated by using the uncertainties listed in Table IV has no direct relevance for the HAD experiment, because the value of ${}^1q_{\mathrm{He}}$ was verified by laboratory calibrations of the HAD. Each calibration consisted of a complete simulation of the descent of the HAD flight instrument into a hydrogen + helium atmosphere to pressure levels of at least 10 bar. All of these calibrations employed the same standard reference gas mixture in the HAD and hydrogen + helium mixtures surrounding the HAD instrument. Three hydrogen + helium mixtures were commercially procured and employed in the calibrations, each of them having a helium mole fraction of about 0.11, known with an accuracy of $\delta q_{\mathrm{He}} \leq 6 \times 10^{-5}$. In many dozens of such calibration runs, the HAD instrument measured with great precision the correct q_{He} of the commercial $\mathrm{H}_2 + \mathrm{He}$ mixtures with an accuracy δq_{He} better than 1×10^{-3}, where ${}^1q_{\mathrm{He}} = 0.110883$ was used to calculate the final result. This proves that when we use a value of 0.110883 for this ratio (in combination with our standard reference gas) its uncertainty has no significant effect on our q_{He} determination.

We will denote with ${}^2q_{\mathrm{He}}$ the second term on the right-hand side of Equation (6). Again using the parameters of Table IV, one obtains

$${}^2q_{\mathrm{He}} = 4.511 \times 10^{-3} \Delta m . \tag{11}$$

The error $\delta^2 q_{\rm He}$ vanishes entirely in the case that Δm goes to zero. In the idealized case of $\Delta p_e = \Delta T_e = 0$, this happens when $n_j = n_{\rm ref}$. Under these circumstances the uncertainties of $T_{j,0}$, $p_{j,0}$, λ, L, and the two refractive indices do not enter at all, which is the basic advantage and incentive for performing a 'differential' measurement.

Of all the parameters entering ${}^2q_{\rm He}$ it is the error $\delta p_{j,e}$ which has the largest effect on $\delta^2 q_{\rm He}$. In Table IV, $\delta p_{j,e}/p_{j,e}$ is (conservatively) given as ± 0.01. Hence, considering (11), $\delta p_{j,e}$ alone causes an error $\delta^2 q_{\rm He} = \pm 4.5 \times 10^{-5}\,\Delta m$. Taking the other parameters in ${}^2q_{\rm He}$ into account as well, this error rises slightly to

$$\delta^2 q_{\rm He} = \pm 5.1 \times 10^{-5}\,\Delta m\,. \tag{12}$$

As outlined above, Δm will be always smaller than 7 for $0.08 \le q_{\rm He} \le 0.14$. Hence, in this range of $q_{\rm He}$ we find

$$|\delta^2 q_{\rm He}| \le 3.5 \times 10^{-4}\,. \tag{13}$$

We will denote with ${}^3q_{\rm He}$ the third term on the right-hand side of Equation (6). Again using the parameters of Table IV, one obtains

$$\delta^3 q_{\rm He} = \pm 5.7 \times 10^{-4}\,. \tag{14}$$

In this case, both $\delta\Delta p_e$ and $\delta\Delta T_e$ contribute about equally to the error sum. Unfortunately, it is not possible to actually measure the gas temperatures in the gas cells with the required accuracy (though the wall temperatures can be). We have to rely on the effectiveness of the heat exchangers and continued intensive contact of the gas with the walls of the gas cells to maintain the required thermal equilibrium between the two gases.

As discussed in Section 4.2, we also have to include in the final estimate of the overall error $\delta q_{\rm He}$ due to the effects of the chemical absorbers which introduce an uncertainty

$$\delta^a q_{\rm He} = \pm 5 \times 10^{-4}\,.$$

Adding quadraticaly the terms $\delta^2 q_{\rm He}$, $\delta^3 q_{\rm He}$, and $\delta^a q_{\rm He}$ we obtain the calculated overall error of the measured He mole fraction

$$\delta q_{\rm He} = \pm 8 \times 10^{-4}\,.$$

This error still meets with our originally announced goal of an error smaller than ± 0.001 (von Zahn and Hoffmann, 1976), but it comes uncomfortably close to the latter. This is because during the instrument development phase the effects of the required chemical absorbers turned out to be more significant than estimated. To be conservative we now prefer to state that our overall error on the helium mole fraction will eventually turn out to be smaller than ± 0.0015.

4. Tests and Calibrations

4.1. Tests

The instrument subsystems and critical components underwent an intensive series of developmental tests. The major objectives addressed in these tests were:

- radiation hardening of selected components, in particular the IR emitting diode and the photodiode array detector (Mett, 1980);
- overpressure testing of electronics components;
- temperature testing of critical electronics circuits;
- testing of the entrance burst diaphragm, opening device for the reference gas, and of the membrane valve controlling the pressure differential between the Jovian and reference gas cells;
- characterizing and quantifying the effectiveness of the absorbers employed to remove H_2O, NH_3, and CH_4 from the measured Jovian gas sample;
- verification of a sufficient coherence length of the IR light source.

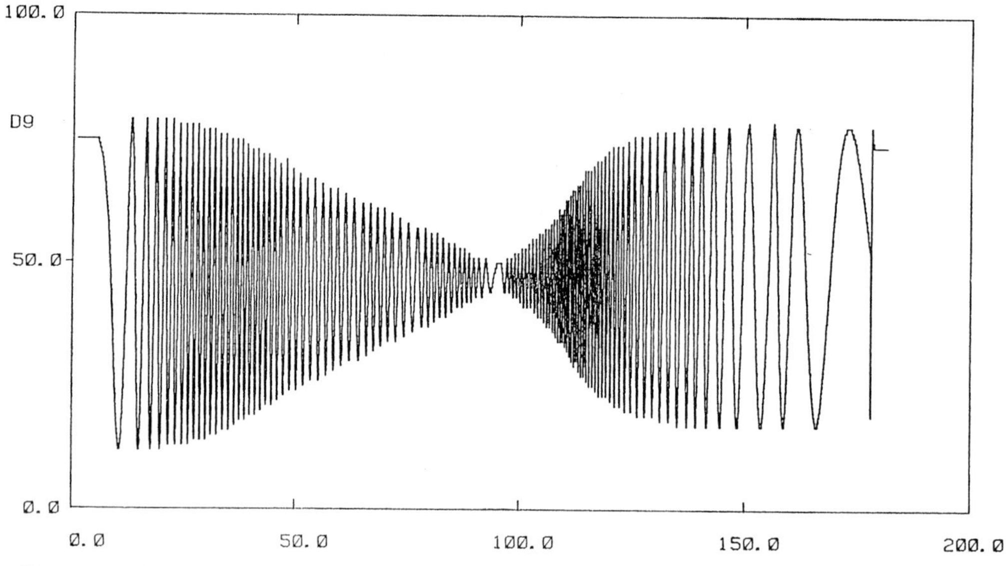

Fig. 5. Interference fringes caused by first filling the Jovian gas cell with N_2 and then pumping it down, while the reference gas cell stays evacuated. Along the ordinate is shown the analog signal of the diode in the detector array, along the abscissa time (= No. of major frames of the telemetry). More than 50 fringes can be counted.

From the large number of individual test results we show in Figure 5 an example of a test for sufficient coherence length of the LED. This test is started by evacuating all gas cells of the interferometer. Then nitrogen is slowly introduced in the 'Jovian' gas cell(s) which causes a shift of the interference fringes at the detector. This shift is registered by the photodiodes. The analog signal from one of the 9 diodes is shown in

Figure 5 (along the ordinate). Along the abscissa, time is increasing (represented by the number of telemetry major frames). After counting 55 fringes the pressure rise is stopped and the nitrogen pumped out of the gas cell (with the final level of the diode signal matching nicely its initial level). As was noted above the maximum number of interference fringes to be counted during the descent into the Jovian atmosphere is 7. It follows that the bandwidth of the IF filter, and hence the coherence length of the LED assembly, is chosen very conservatively and does not pose any limitations on the instrument performance.

The entire HAD instruments underwent the following tests according to specifications laid down in the document NASA JP-512.03 (1979):

- temperature tests (in 1 bar of dry nitrogen);
- thermal-vacuum tests;
- high pressure tests;
- vibration tests (sine and random accelerations);
- acoustics tests;
- steady-state deceleration tests;
- shock tests;
- electromagnetic interference tests.

The HAD engineering instrument was tested to design qualification levels, the HAD flight instrument to the expected flight levels. During these tests the performance of the HAD optics gave, in general, only pleasant surprises. Before the design of the instrument could be successfully qualified, however, we were required to spend a considerable effort isolating the instrument from thermal and mechanical stresses induced by the mounting platform of the Galileo Probe on the interferometer. At the same time, it is obviously necessary to provide a rigid mounting for the HAD instrument inside the Probe in order for it to survive without harm the extreme mechanical loads experienced during its entry into the Jovian atmosphere.

We might add that after all this elaborate testing and only a few weeks after 'final' delivery of the flight instrument to NASA for integration into the Galileo entry probe, the sensor *PT*, which measured the pressure of the reference gas in its storage vessel, started to leak. This event raised serious concerns about the reliability of this particular type of pressure sensor. Since there was no time to qualify and integrate a new type of pressure sensor, the leaky sensor was substituted by a dummy. This, in fact, ended our ability to monitor the pressure p_r in the storage vessel for the reference gas. As this quantity does not enter in the evaluation of the Jovian helium abundance, no loss in accuracy of our results are expected from this mishaps.

4.2. Calibrations

The HAD instrument carries 4 precision thermistors *TS*, *TR*, *TC*, and *TF* for temperature measurements in the range from -25 °C to $+40$ °C. With the help of the following electronics the read-out from the sensors is made nearly linear between -10 °C and $+15$ °C, the range in which the telemetry resolution is 0.2 °C (except for TF which measures the IR filter temperature with a resolution of 0.8 °C). The absolute calibration of all four sensors is performed jointly to an accuracy of ± 0.5 °C. During the first checkout of the HAD instrument on its cruise towards Jupiter each of the TS, TR, and TC sensors read a temperature within the range of $+7.20$ °C $\pm$ 0.12 °C. The spread of values is fully accounted for by the telemetry resolution.

The sensors *PS* and *PR* measure the pressures in the gas cells in the range from 0 to 20 bar, while the sensor *PD* has 400 mbar full range. The absolute sensitivity of each sensor was calibrated against a rotating piston gage to within 0.1% of its reading, as was the temperature dependence of these sensitivities. It turned out that the temperature dependencies of the *PS* and *PR* sensors are negligible for our experiment. The temperature dependence of the *PD* sensor, however, needs to be taken into account in deriving the helium mole fraction. At an indicated pressure of 100 mbar it amounts to change of about 3 mbar in the temperature range from -30 °C to $+35$ °C. Accounting for this temperature dependence, the reproducibility of the pressure readings are within 1 mbar throughout the expected operating range of the instrument.

The spectral intensity distribution of the LED/filter combination was measured and the temperature shift of its centroid wavelength determined. The latter is

$$\lambda(T) = (900.5 + 0.041T)\ \mathrm{nm}\,, \tag{15}$$

with T in °C. The temperature of the IR filter is measured by the sensor *TF*, but its value is obviously not critical.

We tested a large number of absorber materials but unexpectedly could not find one which absorbed methane efficiently, but did not absorb hydrogen. The effect of hydrogen absorption is not large, but measurable with the accuracy of our instrument. We measured this effect for many types of absorbers in a large number of descent simulation tests (see below). These were performed with the HAD instrument and covered the temperature range from -15 °C to $+25$ °C. We finally selected 1.3 g of silica gel as absorber for water and ammonia (absorber No. 1) and 1.15 g of activated carbon as absorber for methane (absorber No. 2). Additional tests using this absorber combination quantified the required correction $\Delta^a q_{\mathrm{He}}$ of the measured helium mole fraction for gas mixtures having q_{He} in the range between 0.008 and 0.014. For a true He mole fraction $q_{\mathrm{He}} = 0.11$ the correction is

$$\Delta^a q_{\mathrm{He}} = 4.0 \times 10^{-3} - 7.9 \times 10^{-5} T_a \pm 5 \times 10^{-4}\,, \tag{16}$$

with T_a being the absorber temperature, again measured in °C. The quoted uncertainty of this correction term is to be taken independent of temperature and helium mole fraction in the tested ranges of those parameters.

Verification of the value of ${}^{1}q_{\mathrm{He}}$, the ratio of refractive index differences, has been discussed in Section 3.3. To this end a great number of laboratory simulations of the descent of the instrument into the Jovian atmosphere have been performed. These consisted of mounting the HAD instrument in a high-pressure chamber in which the chamber pressure, the temperature of the mounting platform, and the temperature of the chamber gas could be programmed to follow the values expected during the actual Jovian descent. Many of these tests were performed with all of the burst diaphragms in place inside the HAD instrument, but in the majority of the tests no burst diaphragms were installed. For the development and calibration of HAD instruments, more than 500 such descent simulations were performed and evaluated.

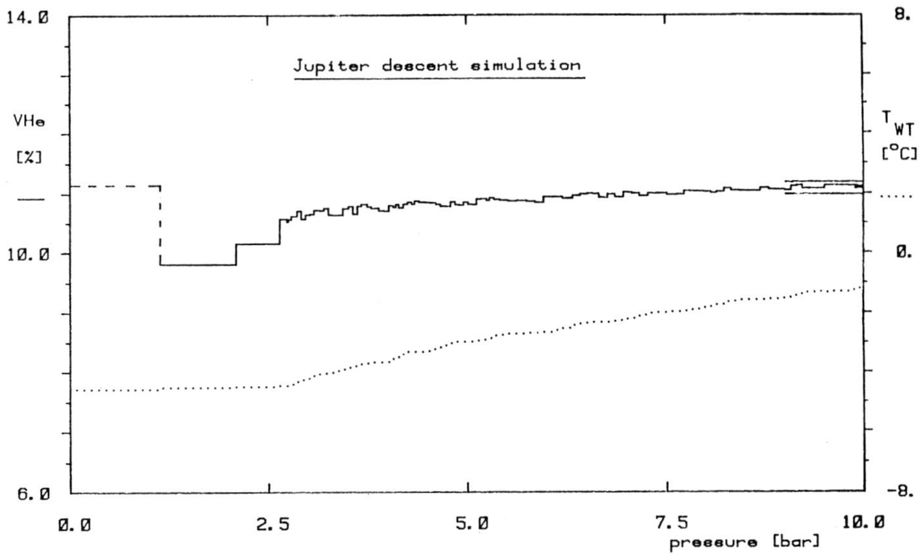

Fig. 6. Laboratory simulation of descent into the Jovian atmosphere (test No. 131): Flight instrument equipped with burst membranes; Jovian gas simulated by binary gas mixture of 11.08% He + the rest H_2. The solid line gives helium mole fraction (V_{He}, in %) calculated from measured fringe motion. Double lines near 10 bar indicate range of helium mole fraction 0.1108 ± 0.001. The dotted line and right ordinate scale gives temperature T_{WT} of the simulated Jovian gas.

Figure 6 shows a typical result of a laboratory simulation of descent into the Jovian atmosphere (test No. 131), performed with the HAD flight instrument equipped with burst membranes. The Jovian gas is simulated by a binary gas mixture of 11.08% He + rest H_2. The solid line gives the helium mole fraction (V_{He}, in %) calculated from the measured fringe motion. The abscissa gives the pressure p_s measured in the Jovian gas cell. After bursting the entrance diaphragm it rises comparatively rapidly. This is the reason why the data points, which are sampled at fixed time intervals, stretch out for $p_s \leq 2.5$ bar. The double lines near 10 bar indicate the exact helium mole fraction ($= 0.1108$) of the sampled gas with a 'permitted' measurement error of ± 0.001. It is evident that the measured helium mole fraction (solid line) gives the correct result and does so well within the predicted uncertainty. The dotted line and right ordinate scale gives the temperature T_{WT} of simulated Jovian gas.

Figure 7 shows a second example of such a descent simulation (test No. 341). Here the HAD engineering instrument No. 2 is tested (without inlet burst membranes). The Jovian gas is simulated by a mixture of 10.92% He, 0.198% CH_4, and the rest H_2. Again, the solid line gives the helium mole fraction (V_{He}, in %) calculated from the measured fringe motion and applying the correction for hydrogen absorption in the chemical absorbers. The double lines near 10 bar indicate the exact helium mole fraction (= 0.1092) of the sampled gas mixture with the 'permitted' measurement error of

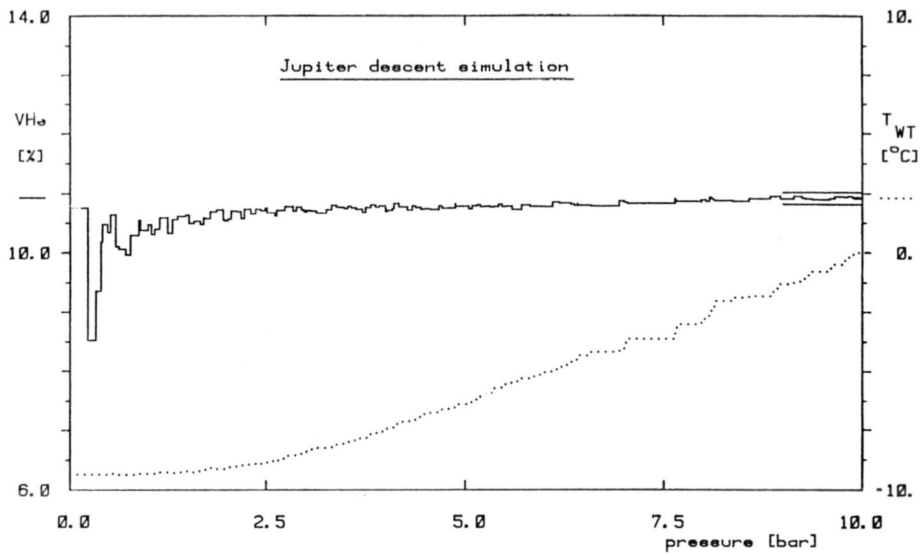

Fig. 7. Laboratory simulation of descent into the Jovian atmosphere (test No. 341): Engineering instrument No. 2 (without inlet burst membranes); Jovian gas simulated by mixture of 10.92% He, 0.198% CH_4, and rest H_2. The solid line gives helium mole fraction (V_{He}, in %) calculated from measured fringe motion. Double lines near 10 bar indicate range of helium mole fraction 0.1092 ± 0.001. The dotted line and right ordinate scale gives temperature T_{WT} of simulated Jovian gas.

$\pm$ 0.001. It is evident that the measured helium mole fraction (solid line) again gives the correct result and does so well within the predicted uncertainty. As in Figure 6, the dotted line and right ordinate scale gives the temperature T_{WT} of simulated Jovian gas.

Acknowledgements

We are most grateful to many of our colleagues and students for their diligent and untiring efforts on behalf of the HAD experiment. Over the years H. J. Hoffmann contributed heavily to the instrument development effort, W. Mett was responsible for radiation hardening of all the instrument subsystems, W. Schulte developed the apparatus for laboratory simulations of the instrument descent into the Jovian atmosphere, and H. Schütze performed the calibration and environmental tests of the various units of the HAD instrument. Integration of the instrument into the spacecraft and systems tests were supported by H. Schütze and G. Lehmacher. K. Pelka assisted during various phases of the project through software development. The interferometer part of the HAD instrument was developed by Carl Zeiss (Oberkochen), the other portion of the

HAD by Messerschmitt–Bölkow – Blohm (Ottobrunn). This experiment was supported by the Bundesministerium für Forschung und Technologie, Bonn, Germany through grant WRK 01 QJ 107 0.

References

Conrath, B. J., Gautier, D., Hanel, R. A., and Hornstein, J. S.: 1984, 'The Helium Abundance of Saturn from Voyager Measurements', *Astrophys. J.* **282**, 807.

Conrath, B., Gautier, D., Hanel, R., Lindal, G., and Marten, A.: 1987, 'The Helium Abundance of Uranus from Voyager Measurements', *J. Geophys. Res.* **92**, 15003.

Gough, D. O.: 1983, in P. A. Shaver, D. Kunth, and K. Kjär (eds.), 'The Protosolar Helium Abundance', *Proceedings of the ESO Workshop on Primordial Helium*, ESO, ISBN 3-923524-16-1, pp. 117–136.

Hubbard, W. B.: 1989, in S. K. Atreya, J. B. Pollack, and M. S. Matthews (eds.), 'Structure and Composition of Giant Planet Interiors', *Origin and Evolution of Planetary and Satellite Atmospheres*, University of Arizona Press, Tucson, pp. 539–563.

Jenkins, F. A. and White, H. E.: 1976, *Fundamentals of Optics*, 4th ed., McGraw-Hill, New York.

Lewis, J. S.: 1974, 'The Chemistry of the Solar System', *Sci. Amer.* **230**(3), 50.

Lorentz, H. A.: 1909, *The Theory of Electrons*, Teubner, Leipzig.

Mascart, M.: 1874, 'Sur les modifications qu'éprouve la lumière par suite du mouvement de la source lumineuse et du mouvement de l'observateur', *Ann. Sci. l'Ecole Normale Sup.* **3**, 363.

Mett, W.: 1980, 'Entwicklung strahlenresistenter opto-elektronischer Bauteile für den Einsatz am Jupiter', Ph.D. Thesis, University of Bonn, Bonn.

NASA Document JP-512.03: 1979, *Science Instruments and Related Requirements; Test Program*, Revision 4, 1 November, 1979.

Orton, G. S. and Ingersoll, A. P.: 1976, in T. Gehrels (ed.), 'Pioneer 10 and 11 and Ground-Based Infrared Data on Jupiter: The Thermal Structure and He–H_2 Ratio', *Jupiter*, University of Arizona Press, Tucson, pp. 207–215.

Pagel, B. E. J.: 1989, in J. Beckman and B. E. J. Pagel (eds.), 'Evolution of ^{4}He in Galaxies', *Evolutionary Phenomena in Galaxies*, Cambridge University Press, Cambridge.

Paul, W., Anton, F., Paul, L., Paul, S., and Mampe, W.: 1989, 'Measurement of the Neutron Lifetime in a Magnetic Storage Ring', *Z. Phys.* **C45**, 230.

Podolak, M. and Cameron, A. G. W.: 1975, 'Further Investigations of Jupiter Models', *Icarus* **25**, 627.

Salpeter, E. E.: 1973, 'On Convection and Gravitational Layering in Jupiter and in Stars of Low Mass', *Astrophys. J.* **181**, 183.

Schütze, H.: 1986, 'Ein Refraktometer zur interferometrischen Bestimmung der Heliumhäufigkeit in der Jupiteratmosphäre', Ph.D. Thesis, University of Bonn, Bonn.

Schulte, W.: 1983, 'Ein Jamin-Interferometer zur Bestimmung der Häufigkeit von Helium in der Atmosphäre des Jupiters', Ph.D. Thesis, University of Bonn, Bonn.

Shaver, P. A., Kunth, D., and Kjär, K. (eds.): 1983, *Proceedings of the ESO Workshop on Primordial Helium*, ESO, ISBN 3-923524-16-1.

Smoluchowski, R.: 1967, 'Internal Structure and Energy Emission of Jupiter', *Nature* **215**, 691.

Stix, M.: 1989, *The Sun*, Springer-Verlag, Berlin.

von Zahn, U. and Hoffmann, H.-J.: 1976, *Jupiter Helium Abundance Interferometer Experiment*, Proposal to NASA for participation in the Juper Orbiter Probe 1981/1982 Mission.

Yang, J., Turner, M., Steigman, G., Schramm, D., and Olive, K.: 1984, 'Primordial Nucleosynthesis: A Critical Comparison of Theory and Observation', *Astrophys. J.* **281**, 493.

THE PLASMA INSTRUMENTATION FOR THE GALILEO MISSION

L. A. FRANK, K. L. ACKERSON, J. A. LEE, M. R. ENGLISH, and G. L. PICKETT

Department of Physics and Astronomy, The University of Iowa, Iowa City, IA 52242, U.S.A.

Abstract. The plasma instrumentation (PLS) for the Galileo Mission comprises a nested set of four spherical-plate electrostatic analyzers and three miniature, magnetic mass spectrometers. The three-dimensional velocity distributions of positive ions and electrons, separately, are determined for the energy-per-unit charge (E/Q) range of 0.9 V to 52 kV. A large fraction of the 4π-steradian solid angle for charged particle velocity vectors is sampled by means of the fan-shaped field-of-view of 160°, multiple sensors, and the rotation of the spacecraft spinning section. The fields-of-view of the three mass spectrometers are respectively directed perpendicular and nearly parallel and anti-parallel to the spin axis of the spacecraft. These mass spectrometers are used to identify the composition of the positive ion plasmas, e.g., H^+, O^+, Na^+, and S^+, in the Jovian magnetosphere. The energy range of these three mass spectrometers is dependent upon the species. The maximum temporal resolutions of the instrument for determining the energy (E/Q) spectra of charged particles and mass (M/Q) composition of positive ion plasmas are 0.5 s. Three-dimensional velocity distributions of electrons and positive ions require a minimum sampling time of 20 s, which is slightly longer than the spacecraft rotation period. The two instrument microprocessors provide the capability of inflight implementation of operational modes by ground-command that are tailored for specific plasma regimes, e.g., magnetosheath, plasma sheet, cold and hot tori, and satellite wakes, and that can be improved upon as acquired knowledge increases during the tour of the Jovian magnetosphere. Because the instrument is specifically designed for measurements in the environs of Jupiter with the advantages of previous surveys with the Voyager spacecraft, first determinations of many plasma phenomena can be expected. These observational objectives include field-aligned currents, three-dimensional ion bulk flows, pickup ions from the Galilean satellites, the spatial distribution of plasmas throughout most of the magnetosphere and including the magnetotail, and ion and electron flows to and from the Jovian ionosphere.

1. Introduction

Although the first direct detection of the presence of plasmas in the vicinity of Io's orbit was reported by Frank *et al.* (1976) with measurements from the plasma analyzer on Pioneer 10, the first definitive measurements of Jovian magnetospheric plasmas were acquired during the Voyager flybys. The Voyager plasma observations were used to define the required capabilities for the Galileo plasma instrumentation. Briefly we summarize here the plasma domains of the Jovian magnetosphere. This information is largely taken from the review by Belcher (1983). For more recent work the reader is referred to further analysis of the torus ions (Bagenal, 1985; Bagenal *et al.*, 1985), the torus electrons (Sittler and Strobel, 1987), and the middle magnetosphere (Sands and McNutt, 1988). Measurements of medium-energy charged particles, $E \gtrsim 30$ keV, are summarized by Krimigis and Roelof (1983).

The heart of the Jovian magnetosphere is the great torus of plasmas that encompasses the orbit of Io. This torus is fed by the ionization of neutral gases from Io's atmosphere and may respond to the sporadic injection of gases from this moon's volcanic activity.

Space Science Reviews **60:** 283–307, 1992.

The composition of the ion plasmas in this torus is rich in heavy ions, e.g., S^+, O^+, S^{2+}, O^{2+}, and Na^+. The plasma torus is divided into two regimes, a cold torus inside Io's orbit and a hot torus at greater Jovicentric distances. The maximum ion densities, $\sim 3000\ \mathrm{cm}^{-3}$, are located near Io's orbit. The ion temperature decreases severely from ~ 40 eV at 6 R_J (Jovian radii) to ~ 1 eV at 5 R_J. This temperature decrease is due to radiative cooling. Ion temperatures in the hot torus at radial distances ~ 6 to 8 R_J are in the range of 40 to 100 eV. The electron temperatures can be described in terms of a two-temperature Maxwellian distribution, i.e., a cold and hot distribution. At the inner edge of the torus the electron temperatures decrease to ~ 0.5 eV with decreasing radial distances whereas the cold electron temperatures beyond ~ 6 R_J are typically ~ 10 to 100 eV. Characteristic temperatures of the hot electron velocity distributions are ~ 1 keV and the number densities are less than those for the cold electrons. The torus plasmas corotate with the planet. The corresponding corotational energy of an S^+ ion is 960 eV at equatorial radial distance 6 R_J. The deflection of plasma bulk flow near the Io flux tube is consistent with that expected for incompressible flow around a cylinder and is evidence for an Alfvén wave associated with the plasma flow past Io. The estimated current in the Io flux tube is $\sim 3 \times 10^6$ A, presumably carried in large part by electrons.

At distances beyond the plasma torus, $\gtrsim 10$ R_J, a plasma sheet extends to the dayside magnetopause. At 15 R_J the typical thickness of the plasma sheet is ~ 2 R_J. These plasmas are observed to corotate more-or-less rigidly with Jupiter's rotational motion to radial distances of about 20 R_J. At distances of 20 to 40 R_J this azimuthal bulk speed of the plasma is less than that expected from rigid corotation by factors of 2 or more. Beyond 40 R_J the plasmas are again observed to rigidly corotate at frequent times as inferred from measurements with the medium-energy charged particle detector. The corotational energy of an S^+ ion at 40 R_J is 43 keV. Whereas the density of the hot electrons is only $\sim 1\%$ of the total density at 8 R_J, the hot electron density is similar to that for the cold electrons at 40 R_J. Ion temperatures are also higher in the plasma sheet relative to those in the torus, ~ 20 to 40 keV at radial distances 30 to 100 R_J. Plasma densities in the plasma sheet are ~ 1 to $10\ \mathrm{cm}^{-3}$ at 10 to 20 R_J and vary from $\sim 10^{-3}$ to $1\ \mathrm{cm}^{-3}$ at larger radial distances. Above and below the plasma sheet the densities can be as low as 10^{-5} to $10^{-4}\ \mathrm{cm}^{-3}$.

Beyond radial distances of 130 R_J in the dawn side of the Jovian magnetosphere the ion bulk flows become generally anti-sunward with a strong component along directions that are radially outward from the planet. This region was detected with the medium-energy charged particle detector and is called the magnetospheric wind.

2. Advantages of the Galileo Plasma Measurements

The Galileo Mission advantages for plasma investigations in the Jovian magnetosphere are (1) the spinning section of the spacecraft, (2) an instrument microprocessor to restructure the instrument operation by ground command, and (3) a series of orbits that allow close flybys of the Galilean satellites, a survey of the Jovian magnetotail, and a substantial local-time survey of the magnetosphere. The spinning section of the spacecraft provides the important capability for a suitably designed instrument to view the entire 4π-steradian solid angle for particle velocity vectors at the spacecraft position. The instrument microprocessor can be used to tailor the operation of the plasma instrument for the most effective measurements in each of the diverse plasma regimes of the magnetosphere and its environs, e.g., magnetosheath, plasma sheet, satellite wake or flux tube, or magnetospheric wind. Targeted encounters with the satellites and a tour of the magnetosphere and magnetotail offer exceptional opportunities for studies of most of the important plasma regions and their temporal responses to variations of Iogenic and solar wind plasmas, and the interactions of magnetospheric plasmas with the satellites.

The Galileo plasma instrumentation (PLS) is substantially more capable for measurements of the Jovian plasmas than those of the Pioneer and Voyager spacecraft because it is specifically designed for this purpose. The basic advantages are in the performance areas of (1) extended energy range, (2) coverage of the angular distributions of plasmas, (3) angular resolution, (4) temporal resolution, and (5) ion composition.

The energy-per-unit charge ranges of the Pioneer and Voyager plasma instruments are 100 to 4800 V and 10 to 5920 V, respectively. The corresponding range of the Galileo plasma analyzer is 0.9 to 52 000 V. This extended energy range spans the important energy gap between 5920 and 30 000 V in the combined performances for the Voyager plasma instrument and medium-energy particle detectors. The 4π-steradian solid angle for particle velocity vectors at the spacecraft position is sampled adequately to provide determinations of the three-dimensional velocity distributions for positive ions and electrons. Thus such important plasma parameters as field-aligned currents, cross-field currents, plasma bulk flow velocities, heat fluxes, and free energy are to be determined for the first time with the Galileo instrument. The angular resolution is sufficient to provide definitive measurements of the above plasma parameters. Temporal resolutions for obtaining electron and positive ion spectra are about 200 s for the Pioneer analyzer (ions only) and about 100 s for the Voyager Faraday cups. The corresponding temporal resolution for the Galileo plasma analyzer is about 0.5 s; complete three-dimensional velocity distributions for positive ions and electrons can be telemetered once each 20 s. These improved temporal resolutions are particularly important during the brief encounters with the satellites and the traversals of plasma boundaries such as those of the plasma sheet and current sheet in the middle and outer magnetospheres.

Three miniature mass spectrometers which are positioned at the exit apertures of the electrostatic analyzers in the Galileo instrument provide determinations of the positive ion composition. The Voyager determinations of ion composition from E/Q spectra are model-dependent and are possible when the Mach number of the corotational flow is greater than 5 or 6. This method is acceptable generally near the Io orbit but as the Jovian radial distance increases, ion thermal speeds rapidly increase and prevent decisive identification of ion species. The Galileo mass spectrometers provide a direct determination of ion composition, specifically the mass-per-unit charge.

In addition to the above performance features, the Galileo plasma analyzer can be operated flexibly via electronic reconfiguration by ground command. The operational configuration of energy-per-unit charge (E/Q) passbands, mass-per-unit charge (M/Q) channels, sensors, and angular sectors can be tailored for a specific plasma region. The temporal resolution for a given measurement can also be selected. The Galileo plasma analyzer is equipped with sufficient onboard hardware and software to implement automated beam capture modes for ion velocity distributions and for determination of ion composition.

3. Several Anticipated Scientific Results

The capabilities of the Galileo plasma instrumentation are demonstrated here by application to several plasma regimes in the Jovian magnetosphere.

As the Galileo spacecraft crosses the plasma sheet in the middle and outer magnetospheres the magnitudes of field-aligned and cross-field currents are determined. Their values and location are correlated with the position of the current sheet as found with the magnetometer. The motions of the plasma sheet are directly determined from the three-dimensional bulk flow vector and the azimuthal component is separated from the radial outflow or inflow. Angular distributions and ion compositions are examined in order to discern the contributions of electrons and ions from the ionosphere, the solar wind via the magnetosheath, and Io in the inner magnetosphere. Thus the formation and dynamics of the plasma sheet can be understood. The mechanism for the unusual heating of plasma with increasing radial distance is expected to be identified.

The encounters with the Galilean satellites offer exciting opportunities for observing plasma phenomena. Examination of the ion velocity distributions in the wakes of these satellites is used to determine the mechanism for ion loss from these bodies. The effectiveness of ion pickup by the magnetospheric plasma flow is derived from the signatures in the velocity distributions of these ions. The mass spectrometers are used to identify the major ions produced in the vicinity of the satellite. For Io these ions include O^+, S^+, and SO_2^+, and for icy satellites perhaps H^+, C^+, and H_2O^+ can be found. Such measurements give the rate of mass loss from each satellite. Perturbations of the plasma flow can be identified in terms of the conductivity of the satellite. During the closest satellite encounters it is possible that a magnetopause or ionopause is detected, thus providing further information concerning the magnetic and atmospheric properties of that body. If the flyby of the satellite is polar, detection of strong field-aligned currents to and away from the Jovian ionosphere might be expected. Field-aligned acceleration of ions and electrons by electrostatic double layers or anomalous resistivity is possible. The relative contributions of the various Galilean satellites for providing the ions in the plasma torus and sheet are assessed during the encounters.

The substantial periods of time that the Galileo spacecraft is located in the plasma sheet offer the unique opportunity to view the responses of the Jovian magnetosphere to the volcanic activity on Io. If specific Io volcanic eruptions can be identified with temporal fluctuations in densities, composition, and motions of the plasma sheet, remarkable advances in our knowledge of the transport of mass and momentum in the Jovian magnetosphere are envisioned.

Simultaneous observations of three-dimensional plasma velocity distributions and of plasma waves with the Galileo spacecraft allow the first studies of wave-particle interactions in the wide-ranging types of plasmas in the Jovian magnetosphere. A discussion of measurements of plasma waves during the Voyager encounters has been given by Gurnett and Scarf (1983). For example, the velocity distributions of ions can be examined to determine whether or not resonant acceleration by ion-cyclotron waves is an important mechanism for ion heating in the torus and plasma sheet. Further the amplitudes of broadband electrostatic noise can be compared with plasma velocity distributions to determine the importance of the anomalous resistivity in plasma heating. Free-energy sources, e.g., ring distributions in the electron velocity distributions, for the generation of electron cyclotron or upper hybrid waves may be identified and related to the wave amplitudes observed with the plasma wave instrument. In general the direct measurement of the plasma density and other parameters gives the growth, propagation and resonance conditions for plasma waves in wave-particle interactions. Thus the mechanisms for providing Jupiter with intense radio sources and particle precipitation into the auroral ionosphere can be further understood.

The existence of the magnetospheric wind at radial distances $> 130\ R_J$ in the dawn sector of the magnetosphere offers exciting goals for the orbit into the distant magnetotail. The origins of this wind are unknown. It is possible that the magnetospheric wind develops near the Alfvén point, where the corotational speed is equal to the Alfvén speed. The actual position must be determined from considerations of the tangential stress balance (cf. Vasyliunas, 1983). Thus magnetic bubbles could be slung radially outwards into the magnetotail. The low pressures in the magnetotail would produce super-Alfvénic radial outflow. On the other hand, the outflow wind might be thermally powered by the hot plasma in the plasma sheet inside the Alfvén point. A third possibility is that the magnetospheric wind is the signature of reconnection of magnetotail lines in a convection pattern controlled by dayside magnetic merging rates. The response of the magnetotail to fluctuating internal plasmas, e.g., Iogenic plasmas, or to a varying solar wind are unknown. In the magnetotail characterized by spectacular, explosive activity or a mere quiescent outflow of plasmas? The exploratory orbit into the magnetotail will indeed answer many questions concerning the origins and dynamics of this immense and little understood plasma region.

4. Overview of the Plasma Instrument

A pictorial view of the plasma instrument (PLS) is shown in Figure 1. The instrument is divided into two analyzers, A and B. Each electrostatic analyzer comprises three 70° spherical-segment plates. The outer and inner plates are grounded and the center plate is supplied with a programmed series of voltages to effect analyses of the energy spectra of electrons (E) and positive ions (P). The inner and outer channels between the plates are the positive ion and electron analyzers, respectively. A charged particle successfully passes through the channel on the basis of its energy-per-unit charge (E/Q). Continuous-channel electron multipliers, or Spiraltrons, are employed as sensors and are positioned

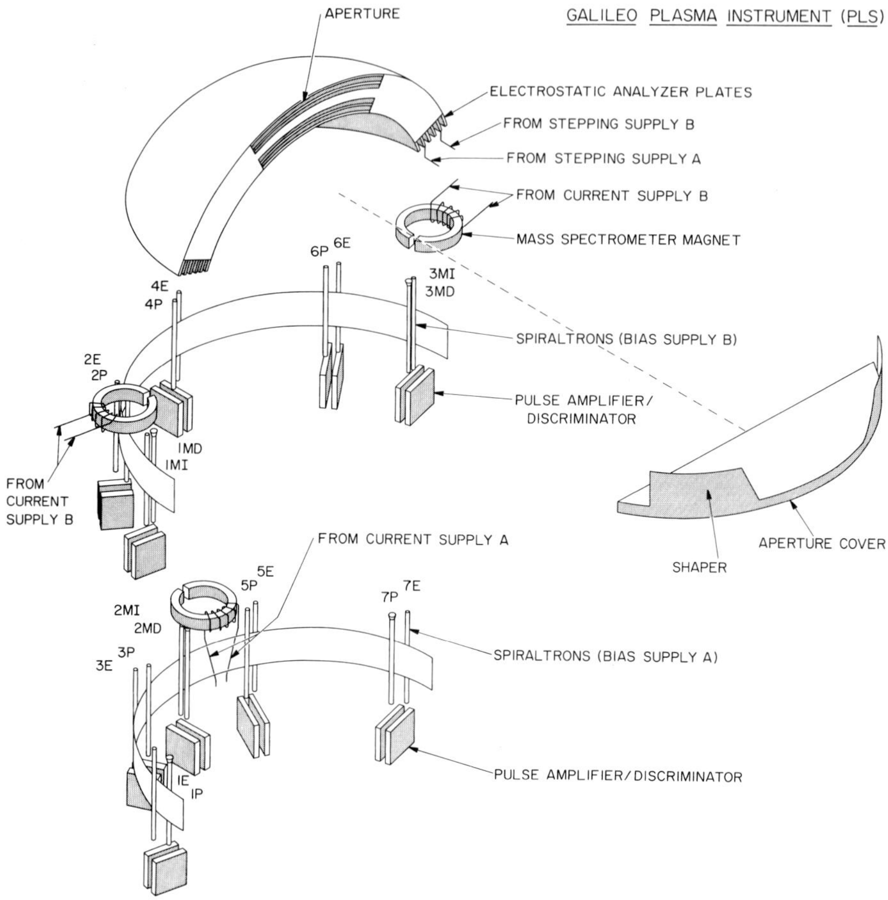

Fig. 1. An exploded view of the electrostatic analyzers, miniature mass spectrometers, and sensors of the plasma instrumentation (PLS) for the Galileo Mission.

at the exit apertures of the electrostatic analyzers. Charged particles arrive at positions at the exit aperture according to their direction of arrival at the entrance aperture. The analyzers are mounted on the instrument (magnetometer) boom of the spacecraft such that charged particles moving perpendicular to the spacecraft spin axis arrive at sensors $4E$ and $4P$, and particles generally moving parallel and anti-parallel to the spin axis are detected with sensor pairs $7E$, $7P$ and $1E$, $1P$, respectively. Thus the fan-shaped fields-of-view are divided into segments by the use of multiple sensors. The instantaneous fields-of-view for the seven ion sensors as projected onto the unit sphere for velocity vectors are shown in Figure 2. Rotation of the spacecraft spinning section allows coverage of almost the entire unit sphere and angular distributions are obtained by electronically sectoring the sensor responses as a function of spacecraft rotation angle. The angular sampling of electron velocity distributions is similar. The instrument

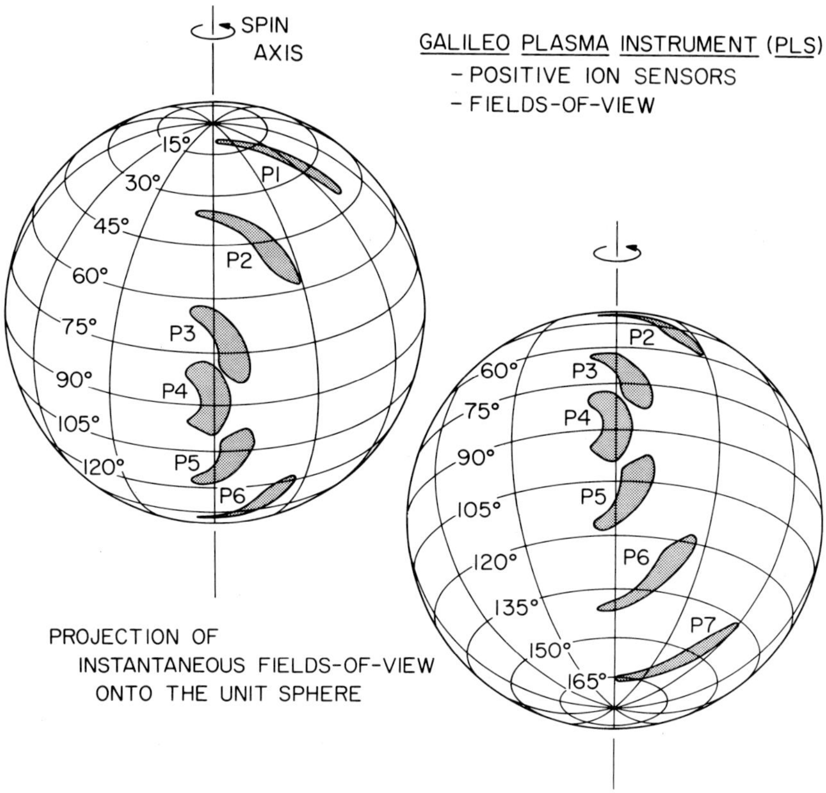

Fig. 2. The instantaneous fields-of-view for the seven positive ion sensors as projected onto the unit sphere. Rotation of the spacecraft spinning section provides coverage of 80% of the unit sphere for ion velocity vectors. The coverage of this solid angle with the seven electron sensors is similar.

is placed at a sufficient distance out along the boom to avoid obstruction of the fields-of-view by the large dish antenna of the spacecraft.

Three miniature mass spectrometers are included in the instrument for determining the composition, i.e., mass-per-unit charge (M/Q), of the positive ion plasmas. As shown in Figure 1 two of these mass spectrometers are positioned at the exit aperture of electrostatic analyzer *B*, the third spectrometer is in analyzer *A*. Each of these mass spectrometers is equipped with two Spiraltrons as sensors and an electromagnet. One of these sensors is placed behind the electromagnet such that it accepts ions not deflected by the gap magnetic field. These 'integral flux' sensors are shown as 1MI, 2MI, and 3MI. The second sensor in each mass spectrometer is displaced from the undeflected path and accepts ions with M/Q values that are a function of the gap magnetic field. These 'differential flux' sensors are 1MD, 2MD, and 3MD. A programmed series of currents is fed to the electromagnet. If the polar angle is taken as $0°$ in the direction of the spacecraft spin axis, then the fields-of-view are $11°$–$38°$, $87°$–$93°$, and $142°$–$169°$ for spectrometers 1, 2, and 3, respectively. The instantaneous fields-of-view

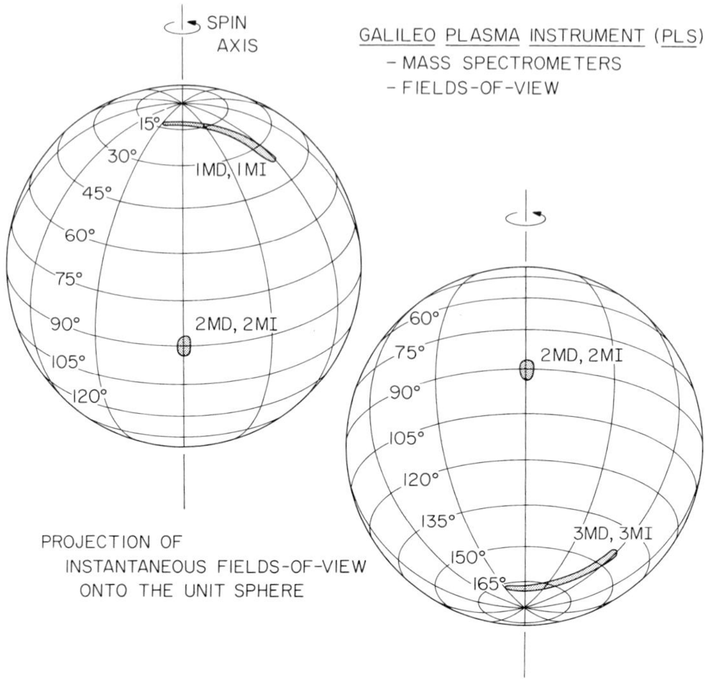

Fig. 3. The instantaneous fields-of-view for the three mass spectrometers as projected onto the unit sphere.

of the three mass spectrometers as mapped onto the unit sphere are shown in Figure 3.

The aperture cover shown in Figure 1 serves two purposes. Prior to and during launch the cover in its closed position prevents contamination of the sensors from dust and condensable vapors. After the launch sequence, the cover is opened and is employed to tailor the fields-of-view of the sensors viewing at small angles to the spin axis of the spacecraft. The corresponding obstructions are identified as shapers in Figure 1.

A photograph of the Galileo plasma instrument is shown in Figure 4, along with a 6-inch ruler. The aperture cover has been opened to show the two arc-shaped openings that are the entrance apertures for the two analyzers, *A* and *B*. A thermal blanket is yet to be installed around the instrument.

5. Design of the Instrumentation

5.1. Electrostatic analyzer

The spherical-segment analyzer plates are precision-machined from solid blocks of magnesium. The radii of the inner and outer surfaces, respectively, of the four electrostatic analyzers are 9.68 and 9.95, 10.08 and 10.36, 11.77 and 12.10, and 12.23 and 12.57 cm. These plates are concentric. Thus the analyzer constant $C \simeq 18.2$, where

Fig. 4. Photograph of the plasma instrument. The two arc-shaped openings are the apertures for the two sets of electrostatic analyzers. A 6-inch ruler is shown for comparison.

$E/Q = CV$ and V is the plate voltage. The angle from the center of the entrance aperture to the exit aperture of the analyzer is 70° as referenced to the common center of curvature. Each of the concave surfaces within the analyzers has been machined with 140 saw-tooth serrations. The interior surfaces of the analyzer plates have been also electrodeposited with platinum black over gold electroplate. The two latter measures are taken in order to suppress the scattering of ultraviolet radiation and charged particles within the analyzers into the sensors. The entrance aperture is 60° wide. Because the entrance aperture is wide in order to provide good angular coverage in spacecraft latitude (see Figure 2), the fields-of-view shapers are used on the protective cover to limit excessive spreading in the azimuthal direction for the polar sensors. The center plate of each analyzer pair is supplied with voltages ranging from 0.05 to 2880 V in order to provide energy (E/Q) spectra of positive ion and electron intensities. The E/Q range is 0.9 eV q^{-1} to 53 keV q^{-1}. There are 64 plate voltages that cover this energy range in logarithmically equal increments. The averaged full-width at half maximum responses (FWHM) of the ion and electron passbands are equal, $\Delta E/E = 0.11$. The range and sequence of plate voltages can be selected by ground command.

A total of seven sensors are used for the two electron analyzers, and seven for the two positive ion analyzers. These continuous channel multipliers are Spiraltrons, model SEM 4211 with 1-mm diameter apertures and model SEM 4213 with 3-mm diameter apertures, manufactured by Galileo Electro-Optics Corporation. Entrance apertures of these sensors are positioned at a distance 16 mm from the exit aperture of their respective electrostatic analyzers. The Spiraltrons with larger apertures are used for the two ion sensors that view closest to the spin axis of the spacecraft, i.e., the polar sensors, in order to offset the reduced projected area of the entrance aperture. The sensors are screened for stability by operation for $\sim 2 \times 10^9$ accumulated counts at a gain $\gtrsim 10^8$. Grounded mesh screens are mounted in front of the entrance apertures of the sensors to shield the sensor post-acceleration electric fields for the prevention of the collection of secondary charged particles produced in the interior of the instrument. The post-acceleration voltage for the ion sensors is approximately the bias voltage, and about + 150 V for the electron sensors. The nominal gain of the Spiraltrons is 5×10^7 to 3×10^8 in the saturated pulse counting mode. The output charge is collected by small plates and the collection efficiency is improved by a potential difference of about 120 V for the electron sensors and 200 V for the ion sensors. This charge is received by hybrid amplifiers and discriminators manufactured by AMPTEK Inc., model A101. The threshold for these amplifiers was conservatively set at 4×10^6 electrons. The high voltage for sensor bias is programmable by ground command in 32 increments spanning the range 2200 to 3800 V in order to maximize the operating lifetime of the sensors against degradation by using the minimum charge per pulse. The pulse pair resolution of the amplifier/discriminator is nominally 250 ns (4 mHz), and about 1.4 μs (700 kHz) after modification for use in the instrument.

5.2. Miniature mass spectrometers

Three miniature mass spectrometers are included in the plasma instrument, one spectrometer in analyzer A and two spectrometers in analyzer B as shown in Figure 1. A more detailed diagram of one of these small mass spectrometers is shown in Figure 5. After passage through the electrostatic analyzer the positive ions enter two collimating slits. The dimensions of the first slit are 11.1×0.15 mm and for the second slit, 8.5×0.15 mm. These two slits are separated by 9.5 mm. The paths of the positive ions are then deflected according to their M/Q by the magnetic field in the gap of a small electromagnet. The gap dimension is 3.0 mm and the length and width of the pole pieces are 9.9 and 4.0 mm, respectively. The magnet core is fabricated from a material similar to HY MU 80 and wound with about 5000 turns of $33\frac{1}{2}$-gauge silver wire. Overlapped sheets of Permalloy 80 with thickness 0.010 inch are used to encase the plasma instrument to reduce the maximum stray field to 16 nanotesla (nT) at a distance of 1 m. The mass of the electromagnet is 150 g. The electromagnet is supplied with a programmed series of 64 currents ranging from 0.6 to 105 mA. The sequence of current values can be controlled by ground command. The corresponding range of gap magnetic fields is 0.0014 to 0.225 T. The ions are detected with two Spiraltrons, one Spiraltron (integral) with a 1-mm aperture for undeflected ions, and one Spiraltron (differential) with a 3-mm aperture that is offset from the path for undeflected ions. The magnet is non-focussing and the 3-mm aperture Spiraltron is used to achieve approximately equal geometric factors for the differential and integral channels. The sensor apertures are positioned at a distance 20.1 mm from the exit face of the electromagnet. A slit with width 0.76 mm

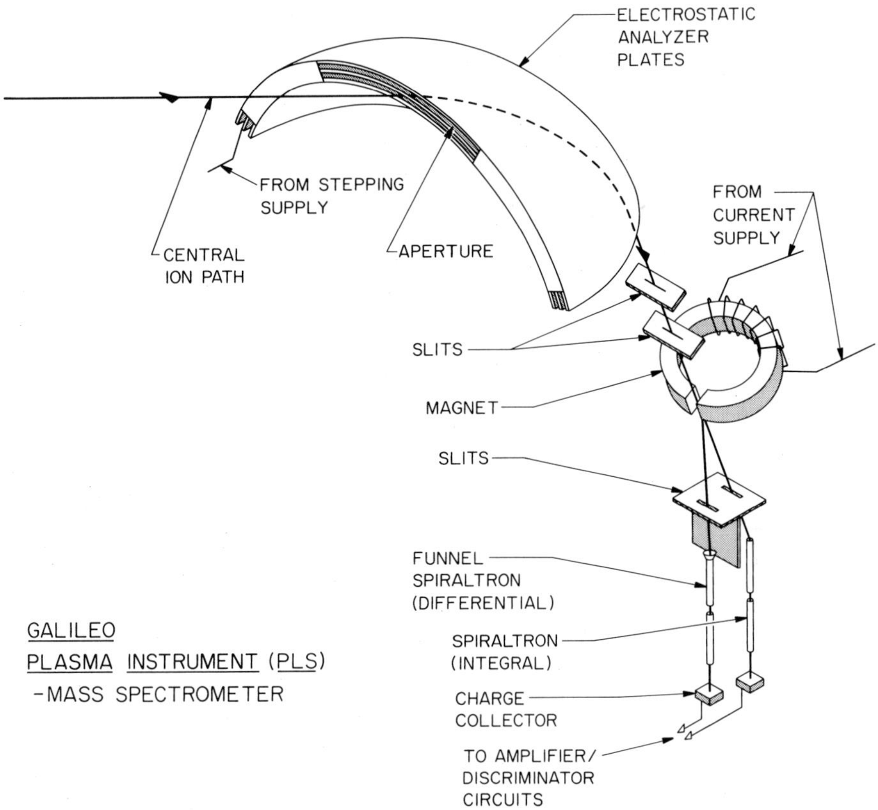

Fig. 5. Configuration of one of the three miniature, magnetic mass spectrometers in the plasma instrument.

is placed in front of each of the two Spiraltrons. The centers of these slits are separated by 3.30 mm. The Spiraltrons are operated in a similar manner as previously described for the sensors for the electrostatic analyzers.

An example of the mass spectrometer performance taken from laboratory calibrations of the flight instrument is shown in Figure 6. The value E/Q is 947 V for the two ions in the beam, H_2^+ and OH^+. The sweeping of the ions from the integral sensor as a function of M/Q and mass channel (magnet current) is clearly evident. At higher mass channels (larger current) these ions are deflected sufficiently to be detected with the differential sensor. A summary of the measured performance of the miniature mass spectrometer as functions of M/Q, E/Q, and electromagnet current is given in Figure 7. The M/Q value for the integral sensor is taken at a fraction 0.5 of the undeflected responses. For a given current step of the mass spectrometer, the averaged FWHM for the three mass spectrometers in terms of ion energy is $\Delta E/E = 0.06$. In general the differential channel is used for the detection of trace fluxes of light ions and the integral channel for abundant heavy ions in the Jovian magnetosphere. The mass resolutions

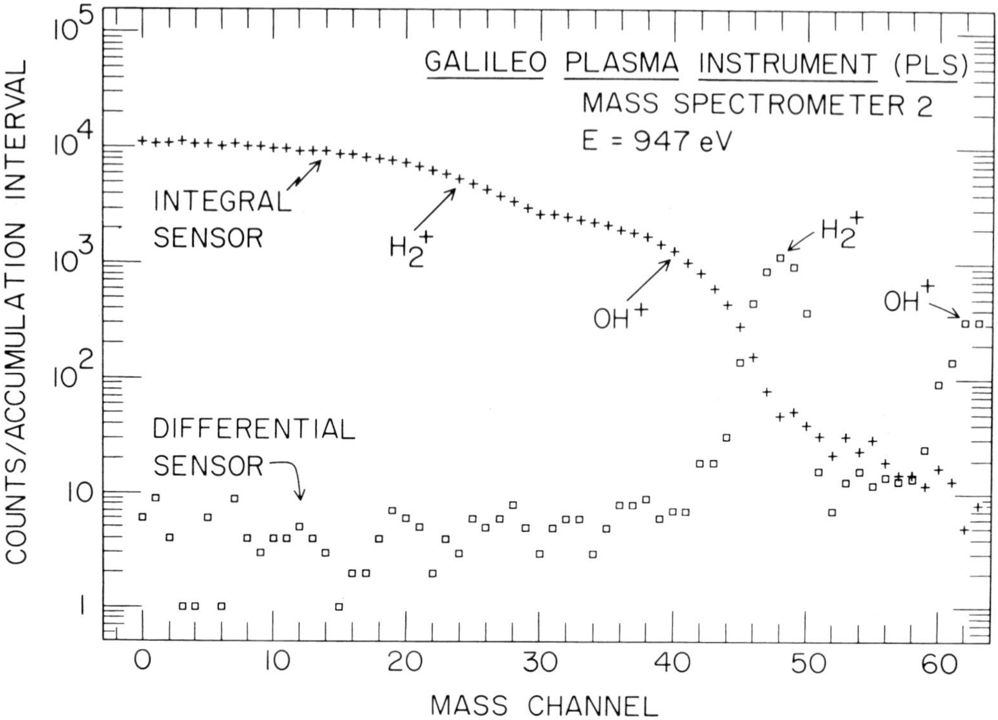

Fig. 6. Calibration of a miniature mass spectrometer with an ion beam (H_2^+, OH^+). The unresolved water group (O^+, OH^+, H_2O^+) is designated as OH^+. The beam energy is 947 eV and the beam current is constant. The responses of the integral and differential channels are shown as a function of electromagnet current (mass channel). The current increases by a logarithmically equal increment for each mass step.

of the mass spectrometers are $M/\Delta M = 4.1$ at full-width at 50% responses (FWHM) for the differential sensors (*MD*) and $M/\Delta M \simeq 2$ for the integral sensors (*MI*). This resolution has been chosen to allow identification of the species H^+, H_2^+ (He^{++}), He^+, O^{++}, O^+, Na^+, S^+, and K^+ with the *MD* sensors and H^+, H_2^+ (He^{++}), O^{++}, O^+, S^+, and SO_2^+ with the *MI* sensors. The E/Q ranges vary with the M/Q of the ion species, e.g., for the *MD* sensors, 0.9 V to 20 keV for H^+, 0.9 V to 3 kV for O^+, and 0.9 V to 800 V for S^+. For the *MI* sensors, these ranges are 10 V to 52 kV for H^+, 0.9 V to 52 kV for O^+, and 0.9 V to 14 kV for S^+. The mass spectrometers cannot distinguish between two ions with the same M/Q, e.g., O^+ and S^{++}. The mass spectrometers are designed in part with the criterion that corotating SO_2^+ ($M/Q = 64$ amu, $E/Q \simeq 2$ kV) can be identified at Io's orbit.

5.3. Geometric Factors

A summary of the latitudinal coverage, energy resolutions, and geometric factors of each of the twenty sensors in the plasma instrument is given in Table I. The averaged geometric factors for the electron and positive ion sensors of the electrostatic analyzers

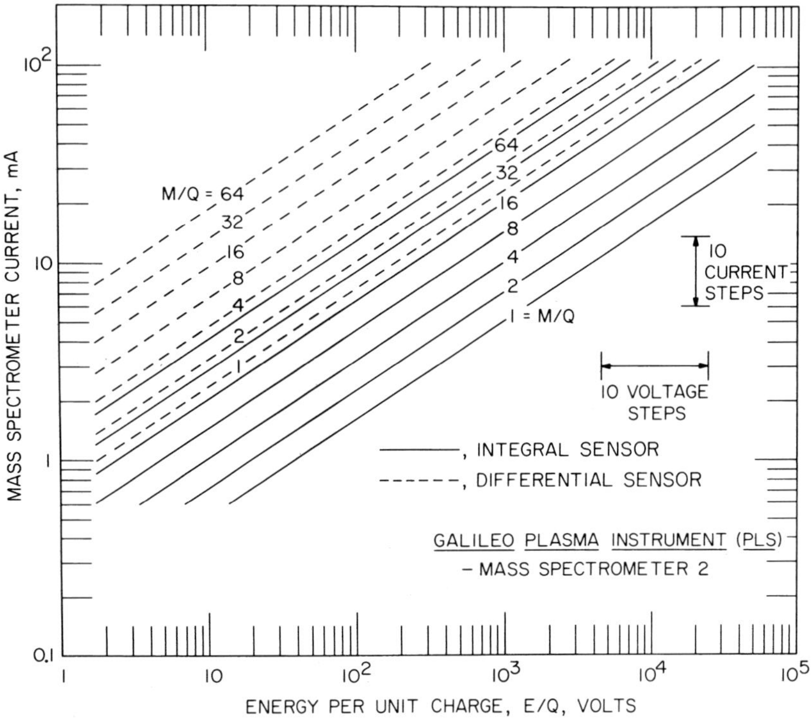

Fig. 7. The relationship of energy-per-unit charge (E/Q), mass spectrometer current, and mass-per-unit charge (M/Q) for the detection of positive ions with the miniature mass spectrometers. The E/Q value is 960 V for a corotating S^+ ion at equatorial radial distance 6 R_J in the Jovian magnetosphere.

and the positive ion sensors of the mass spectrometers are 3.4×10^{-5}, 6.4×10^{-5}, and 3.2×10^{-6} cm^2 sr eV eV^{-1}, respectively.

These values are computed by comprehensive ray tracing of trajectories through the electrostatic and magnetic analyzers and with the nominal entrance area of the sensor. In practice both the efficiency and this area vary with individual sensors and final values of the geometric factors are derived from laboratory measurements and inflight responses in an isotropic plasma such as that in the plasma sheet during Earth1 encounter. These geometric factors are tailored to provide effective measurements of both the dense plasmas in the torus and the sparse plasmas of the outer Jovian magnetosphere.

The sensitivities for detecting these plasmas are summarized in Figure 8. The maximum responses of a single sensor to several representative plasmas are shown as functions of the plasma temperature, bulk flow speed V, and species. The bulk speed of 100 km s^{-1} has been chosen as scale-wise representative for the corotational speeds in the torus. The densities of all the plasmas are each assumed to be 1 cm^{-3}. For example, if the density of S^+ ions is 1000 cm^{-3}, V is 100 km s^{-1}, and the temperature kT is 100 eV, the maximum responses of the ion sensors of the electrostatic analyzer

TABLE I

Galileo PLS performance parameters

Sensor	Polar angle coverage, θ	Energy resolution, $\Delta E/E$ at FWHM	Geometric factor[b], cm^2 sr eV eV^{-1}
Electrons			
Energy range: $0.9\ \mathrm{V} \le E/Q \le 52\ \mathrm{kV}$			
1*E*	14°–41°	0.14	1.9×10^{-5}
2*E*	38°–62°	0.12	3.7×10^{-5}
3*E*	58°–80°	0.10	4.1×10^{-5}
4*E*	81°–102°	0.08	5.0×10^{-5}
5*E*	100°–122°	0.10	4.1×10^{-5}
6*E*	121°–146°	0.12	3.6×10^{-5}
7*E*	142°–171°	0.14	1.3×10^{-5}
Positive ions			
Energy range: $0.9\ \mathrm{V} \le E/Q \le 52\ \mathrm{kV}$			
1*P*[a]	9°–41°	0.15	9.8×10^{-5}
2*P*	35°–59°	0.12	3.5×10^{-5}
3*P*	62°–84°	0.09	4.1×10^{-5}
4*P*	78°–99°	0.07	5.0×10^{-5}
5*P*	97°–119°	0.09	4.0×10^{-5}
6*P*	118°–141°	0.11	3.6×10^{-5}
7*P*[a]	136°–166°	0.15	1.5×10^{-5}
Ion composition			
Energy range: species dependent			
Differential (*D*) sensor:	0.9 V to 20 kV (H^+) 0.9 V to 800 V (S^+)		
Resolves:	H^+, H_2^+, He^+, O^+, Na^+, S^+, K^+ with $M/\Delta M = 4.1$		
Integral (*I*) sensor:	10 V to 52 kV (H^+) 0.9 V to 14 kV (S^+)		
Resolves:	H^+, H_2^+, He^+, O^+, S^+, SO_2^+ with $M/\Delta M \simeq 2.0$		
1*MD*[a], 1*MI*	11°–38°	0.03	2.4×10^{-6}
2*MD*[a], 2*MI*	87°–93°	0.03	4.7×10^{-6}
3*MD*[a], 3*MI*	142°–169°	0.03	2.4×10^{-6}

[a] 3-mm entrance diameter, others are 1 mm.

[b] Preliminary values based upon ray tracing (see text).

(*P*) and of the ion sensors of the mass spectrometers (*M*) are 4×10^6 and 2×10^5 counts s^{-1}, respectively, when viewing in the bulk flow direction. The geometric factor of the ion sensor (*P*) is sized such that these responses are somewhat above the saturation values for the sensor/amplifier, $\sim 10^6$ counts s^{-1}. The ion sensors in the mass spectrometers are employed to extend the dynamic range of these ion measurements to the larger ion densities by means of their lesser geometric factors. On the other hand, the large geometric factor of the ion sensors for the electrostatic analyzers provides the capability of the determining densities of hot ($\sim$ tens of keV), isotropic ions as low as 10^{-3} to 10^{-2} cm^{-3} in the outer regions of the magnetosphere. Thus the combined geometric factors of the electrostatic analyzers and mass spectrometers accommodate a large range of ion densities. If the electron densities in the center of the plasma torus are 3000 cm^{-3}, then the maximum responses for the electron sensors are $\sim 2 \times 10^5$ and 6×10^5 counts s^{-1} for electron temperatures $kT = 1$ and 10 eV, respectively. For an electron temperature of 10 keV in the outer magnetosphere, densities as low as 10^{-4} to 10^{-3} cm^{-3} can be well determined.

Considerable attention in the design of the instrument was directed toward minimizing the sensor responses to the intense fluxes of energetic electrons in the inner Jovian magnetosphere. The Spiraltrons are shielded in all directions by a minimum of 2.5 g cm^{-2} equivalent of aluminum. This corresponds to an electron range of ~ 5 MeV. In addition the Spiraltrons used for ion sensors are operated at a sufficiently low voltage that two or more initial secondary electrons at their entrance apertures are necessary to yield an electron pulse above the discriminator level of the amplifiers. This mode of operation reduces the sensor efficiency for the detection of ions by 50% ($\pm 10\%$), with a corresponding decrease in the geometric factors cited in Table I. Such operation of the sensors at bias voltages ~ 2400 V allows discrimination against detection of penetrating electrons. The omnidirectional geometric factors for detection of penetrating, $\gtrsim 5$ MeV electrons are $\sim 10^{-4}$ cm^2 for the ion sensors with 1-mm apertures, and $\sim 10^{-3}$ cm^2 for the 3-mm ion sensors (see Table I). The corresponding geometric factors for the Spiraltrons used in the electron analyzers are $\sim 10^{-3}$ cm^2. At the orbit of Io the electron intensities with $E \gtrsim 5$ MeV are $\sim 2 \times 10^7$ cm^{-2} s^{-1} (Van Allen, 1976). Thus the background counting rates are $\sim 2 \times 10^3$, 2×10^4, and 2×10^4 counts s^{-1} for the 1-mm ion sensors, the 3-mm ion sensors, and the 1-mm electron sensors, respectively. For comparison, the sensor responses in the direction of flow (S^+, 1000 cm^{-3}, 50 eV, 100 km s^{-1} from Figure 8) are $\sim 5 \times 10^6$ counts s^{-1} for the ion channels of the electrostatic analyzer and $\sim 3 \times 10^5$ counts s^{-1} for the sensors in the mass spectrometer. The analyzer responses to electrons (e^-, 1000 cm^{-3}, 50 eV from Figure 8) are expected to be $\sim 6 \times 10^5$ counts s^{-1}. The corresponding S/N ratios are 2500, 150 (I), and 15 (D), and 30 for the ion sensors, mass spectrometer sensors, and electron sensors, respectively.

At larger radial distances, $\gtrsim 20$ R_J, the intensities of electrons with $E \gtrsim 5$ MeV are typically $\lesssim 10^3$–10^4 cm^{-2} s^{-1} within and near the plasma sheet (Baker and Van Allen, 1976). The corresponding maximum background rates are then $\lesssim 1$ and 10 counts s^{-1} for the 1-mm positive ion and electron sensors, respectively. For these maximum rates, the densities for which $S/N = 1$ for an isotropic, H^+ plasma are 3×10^{-3} cm^{-3} at $kT = 10$ keV and 5×10^{-3} cm^{-3} for electrons at 1 keV (see Figure 8). The corresponding densities for the mass spectrometer sensors are ~ 0.1 cm^{-3} (I) and 1 cm^{-3} (D). These above examples for H^+ give the most pessimistic values because we have assumed worst-case background rates and because the ion plasmas are partially corotating. The S/N ratios will be typically larger by factors of ~ 10 to 100.

The spacecraft potential is expected to be important at the lower energy range of the analyzer. A quantitative assessment of anticipated spacecraft potentials is given by the Voyager plasma measurements. In the outer magnetosphere, typical Voyager spacecraft

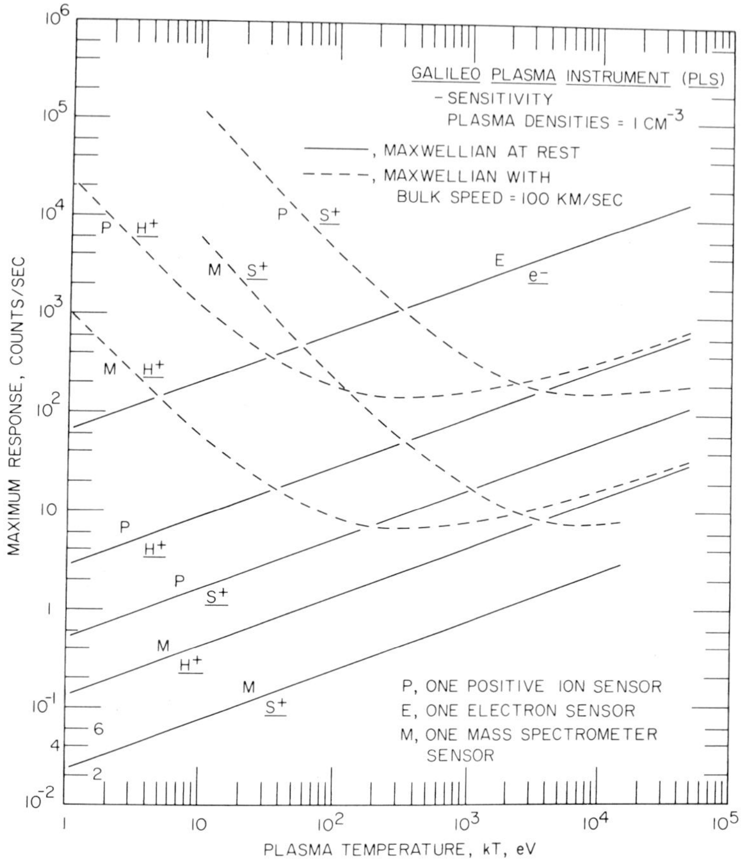

Fig. 8. Maximum responses of several individual sensors to a Maxwellian plasma distribution as a function of plasma temperature. Responses to H^+ and S^+ plasmas at rest and with bulk flow speed 100 km s^{-1} are shown. The densities of electron and ion plasmas are each taken as 1 cm^{-3}.

potentials were positive in the range of several volts to 10 V (Scudder *et al.*, 1981). Because the plasmas are generally hot, temperatures ~ keV, in the outer magnetosphere the plasma measurements should not be greatly impaired. On the other hand, in the highest density regions of the Io torus, Voyager spacecraft potentials were negative with magnitudes up to 25 V (Sittler and Strobel, 1987). In this region electron temperatures are tens of eV or less and the observations of thermal electron plasmas may be precluded if the Galileo spacecraft potential is similar. The energy range of the Galileo plasma instrument is sufficient to determine this spacecraft potential. Determination of the magnitude of the potential will have to await the *in-situ* observations. The potentials along the boom on which the plasma instrument is mounted and those of the spacecraft body will also affect the trajectories of low-energy particles as viewed by the plasma analyzer. This effect will have to be modeled in detail in order to determine the deflections of the observed angular distributions as a function of the particle energy.

5.4. Instrument Electronics

The plasma instrument is divided into two analyzer systems *A* and *B* as shown in Figure 9 (see also Figure 1). This configuration of the instrument has been chosen in

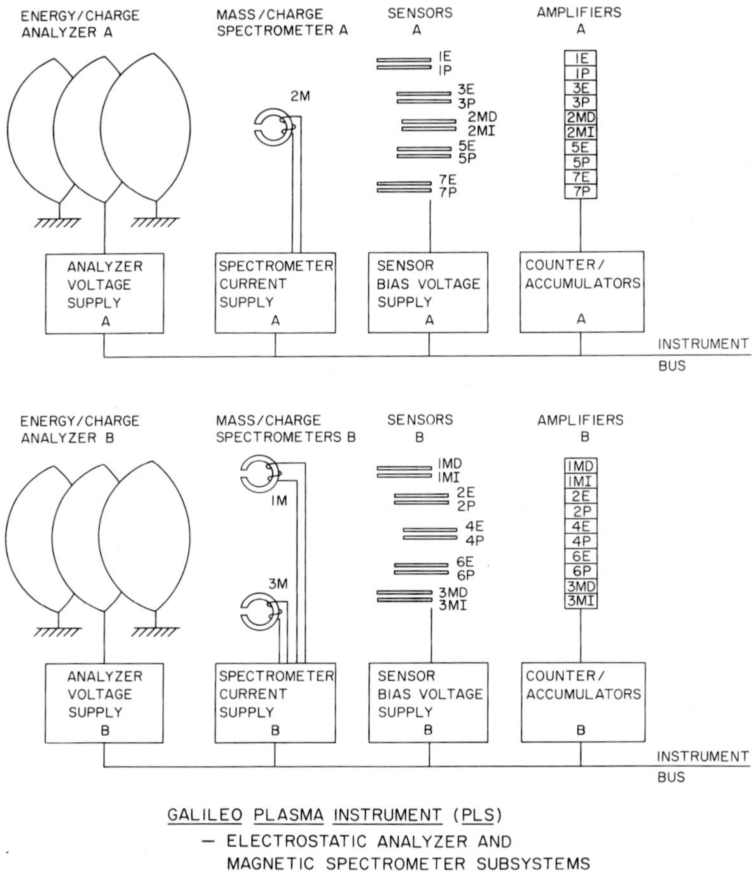

Fig. 9. Block diagram for the two analyzers, *A* and *B*, in the plasma instrument.

order to reduce the number of possible single-point failures that could result in the total loss of the scientific objectives. Each analyzer is equipped with a set of electrostatic analyzer plates, at least one miniature mass spectrometer, and a partial set of the sensors for the measurements of the three-dimensional velocity distributions of positive ions and electrons. A dedicated plate voltage supply, magnet current supply, and sensor bias voltage supply are provided for each analyzer. Each sensor is serviced by a 16-bit accumulator. The electronics for both analyzers are controlled from the instrument bus.

The reduction of single-point failures of the instrument proved to be considerably more difficult for the data handling and control subsystem (DHCS). The configuration that was chosen for the microprocessors and associated electronics is shown in Figure 10. There are two separate buses, *A* and *B*, that can singly operate the two analyzers shown in Figure 9. Similarly there are two RCA 1802 microprocessors, 1 and 2, that are each equipped with 4 kbytes of read-only memory (ROM) and 4 kbytes of

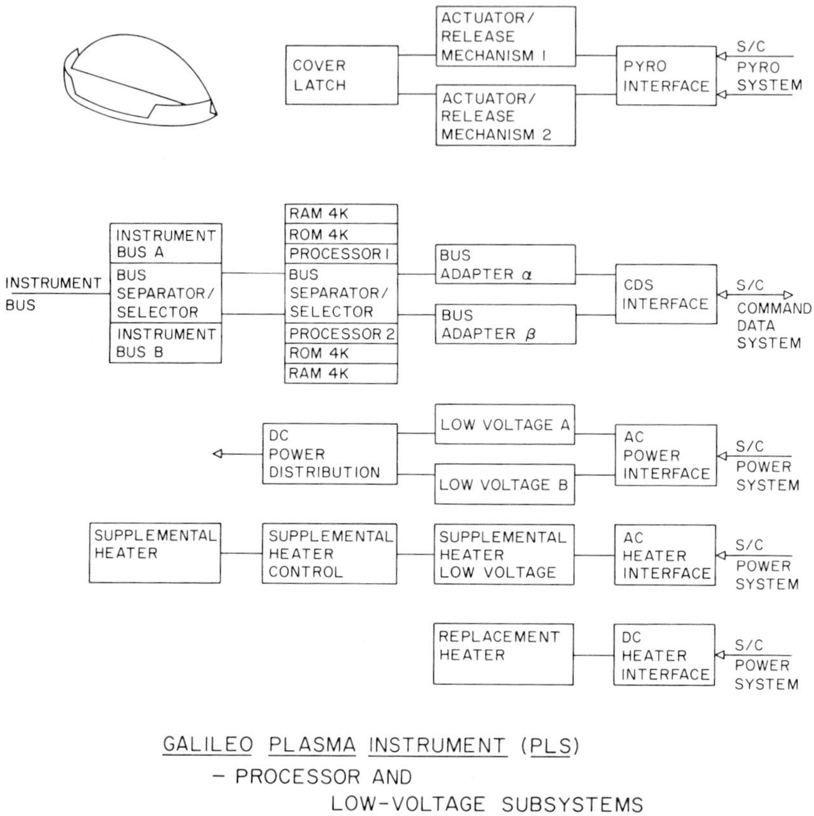

Fig. 10. Block diagram for the electronics in the plasma instrument. Two microprocessors are used to improve the reliability.

read/write memory (RAM). Two bus adapters, α and β, couple the microprocessors with the command data system (CDS) of the spacecraft. The instrument is operated with one bus adapter, one microprocessor, and one bus. The bus separator/selector allows the use of any combination of these electronic elements, e.g., bus adapter α, processor 2, and instrument bus *A*. This configuration for the DHCS is set via a hardware bus command (HBC) that transfers the necessary information in the address portion of the packet header from the spacecraft CDS. The HBC is executed regardless of which processor and bus adapter are currently selected. If the currently selected bus adapter fails, the HBC can be used to select the other bus adapter.

Each of the two microprocessors is provided with identical I/O electronics that include an analog-to-digital (ADC) converter (model AD571, Analog Devices, Inc.), three digital-to-analog (DAC) converters, and a digital status input port. A 16-input multiplexor is used with the ADC to monitor voltages within the instrument. The DACs provide the control voltages for the programmable high voltage (plate and bias) and current (electromagnet) supplies.

Two low-voltage power supplies, A and B, are included within the plasma instrument. By means of a power distribution system, failure of a single low voltage supply does not result in the loss of the DHCS or instrument bus. Analyzer A or B becomes inoperable with the failure of the one of the low-voltage power supplies, A or B. A power switching circuit that is controlled by ground command is used to select the analyzer to be operated with the functioning low-voltage power supply. The replacement and supplemental heaters shown in Figure 10 are used for thermal control during the mission. The latch for releasing the protective cover over the instrument aperture is a one-shot redundant device with two electrically fired, black powder Unidynamics bellows actuators.

5.5. Mass, size, and power

The overall dimensions of the plasma instrument are 8.00 × 15.00 inch (mounting surface) and 13.68 inch (height). The total mass is 13.2 kg, of which 0.33 kg is used for magnetic shielding and 3.57 kg is invested in radiation shielding of the sensors and electronics with tantalum. The average power, without heaters, is dependent upon the electronic configuration of the instrument and is in the range of 6.5 to 10.7 W.

6. Inflight Operation of the Instrument

The operating modes of the plasma instrument are designed to accommodate the diverse plasmas in the Jovian magnetosphere. We provide here a brief introduction to those capabilities. The instrument cycle time is 243 s and is subdivided into 12 equal intervals, or instrument spin modes. Each spin mode is a separate instrument operations and data collection cycle. The duration of a spin mode is 20.3 s and thus slightly longer than the range of anticipated rotation periods for the spacecraft spinning section, 18.3 to 19.8 s. By ground command the plasma instrument can be configured to sample a combination of a given set of sensors, a range of energy passbands, a range of mass channels, and a set of angular sectors as the fields-of-view rotate. The operations of analyzers A and B can be programmed independently. Limitations of these analyzers are imposed by the minimum dwell time for the energy passbands and mass channels of 8.3 ms, a service time of 1 ms for the processing of the contents of a count accumulator, and the telemetry rate allocated to the instrument of 612 bits s^{-1} (72 sensor samples s^{-1} plus overhead). Each sample of sensor responses is quasi-logarithmically compressed into an 8-bit word. Internal buffers can allow rapid bursts of $\lesssim 1500$ measurements to be trickled into the telemetry stream.

Consider the measurement cycle time of the plasma instrument if onboard software were not available to improve the operational efficiency. If all energy passbands, mass channels, and sensors were sampled in each of 16 angular sectors, then the time for this complete plasma measurement (1.3×10^6 samples) would be 5.1 hours. Such instrument operation is ineffective and wasteful of the capabilities for obtaining plasma parameters, e.g., individual 64-point energy or mass spectra in 0.5 s. Thus the spin modes are each designed to obtain a specific type of plasma measurement during one spacecraft rotation, e.g., a three-dimensional velocity distribution, high angular and energy resolutions of an ion beam, and the mass composition of an ion beam. A spin mode is constructed of nested control loops. These loops control (1) the number of angular sectors sampled during a spacecraft rotation, (2) the number of energy passbands or mass channels in a sector, (3) the duration of an energy passband or mass channel, (4) the readout of the selected sensors, (5) the sequence of energy passbands, and (6) the sequence of mass channels. Four sequencing tables are used to determine the operation of the instrument during a spin mode: (1) sensor, (2) mass channel, (3) energy passband, and (4) angular sector. The angular sectors are referenced to a fixed position on the celestial sphere by means of information from the spacecraft attitude control system. Instrument software is available for five basic types of spin modes. Default values for the sequence tables are also included in read-only memory in the instrument processor in lieu of values from ground commands. We briefly illustrate below the capabilities of the various spin modes.

Spin mode 1. Survey of positive ion and electron velocity distributions. All electron, ion, and integral ion sensors (spectrometers) are sampled. The number of angular sectors, the energy range, and the number of energy passbands are selected by ground command. The product of the numbers of passbands and angular sectors is 64. For example, the responses of all of the above sensors for 64 passbands sampled in a single angular sector of 45° can be telemtered each spacecraft spin period. Alternately 16 passbands (every fourth passband) in each of our 90°-sectors can be telemetered during a single rotation period in order to obtain the principal features of the three-dimensional velocity distributions of positive ions and electrons once each 20.3 s.

Spin mode 2. Determination of the velocity distribution of a positive ion beam. Electron and ion sensors corresponding to those nearest the direction of the ion beam are selected. These sensors and the spacecraft rotation angle for the beam are determined with measurements from a preceding spin mode 1. The rapid energy scans in the direction of the beam are limited to the energy range of the beam as determined during spin mode 1. For example, during one spacecraft rotation, energy passbands 8 through 23 can be sampled with three sensors for positive ions and two or three sensors for electrons for each of five contiguous 22.5°-sectors in the direction of the beam. Two electron sensors are used for analyzers *A* and *B*, with the exception of three for analyzer *B* if the beam is nearly perpendicular to the spacecraft spin axis. Again angular size of the sectors and the number of energy passbands can be selected by ground command.

Spin mode 3. Survey of ion composition. Mass spectrometers 1 and 2 are sampled for a selected range of gap magnetic fields. During one spacecraft rotation a single energy passband of the electrostatic analyzer is used and the gap magnetic fields are incremented over a selected series of values. Thus for a given energy passband and a single spacecraft rotation it is possible to sample the entire M/Q range in 64 current steps in each of four 90°-sectors.

Spin mode 4. Survey of ion composition. This spin mode is identical to spin mode 3 with the exception that mass spectrometer 3 replaces 1.

Spin mode 5. Determination of the composition of an ion beam. The mass spectrometer with direction of field-of-view nearest to that of the ion beam is chosen on the basis of previous measurements with spin mode 1. The energy passband and angular sectors for the ion beam are similarly identified. For example, during one spacecraft rotation in the plasma sheet or torus of the Jovian magnetosphere full coverage of the M/Q range in 64 channels can be sampled in each of five contiguous 22.5° sectors.

The instrument cycles for analyzers *A* and *B* are each selected as a sequence of 12 spin modes. The order of the spin modes and their operating parameters such as energy and mass ranges, angular resolution, etc., are controlled by the sequence tables. As an example, a sequence of spin modes during an instrument cycle for analyzer *A* can be 1, 1, 2, 1, 1, 5, 1, 1, 4, 1, 1, 3. Thus the various operating modes of the plasma instrument can be implemented and cycled automatically with minimal demand for command uplinks to the Galileo spacecraft. Major command sequences are used to restructure the spin modes and their sequencing for special events such as the close encounters with the Galilean satellites and the exploratory survey into the distant magnetotail.

Acknowledgements

At The University of Iowa L. A. Frank is principal investigator and K. L. Ackerson is a co-investigator for the plasma investigation. The other co-investigators are F. V. Coroniti of the University of California at Los Angeles and V. M. Vasyliunas of the Max-Planck-Institüt für Aeronomie, Lindau, Germany. E. C. Stone of the California Institute of Technology is a co-investigator whose responsibility is the Heavy Ion Composition (HIC) investigation which employs a separate instrument on the Galileo Orbiter. The authors wish to express their appreciation to the following personnel of the Jet Propulsion Laboratory for their assistance in the implementation of the plasma instrumentation: J. R. Casani, W. G. Fawcett, H. W. Eyerly, C. M. Yeates, M. S. Spehalski, W. J. O'Neil, R. F. Ebbett, and T. V. Johnson. This research was supported in part at The University of Iowa by the Jet Propulsion Laboratory under contract 958778.

References

Bagenal, F.: 1985, 'Plasma Conditions Inside Io's Orbit: Voyager Measurements', *J. Geophys. Res.* **90**, 311.

Bagenal, F., McNutt, R. L., Jr., Belcher, J. W., Bridge, H. S., and Sullivan, J. D.: 1985, 'Revised Ion Temperatures for Voyager Plasma Measurements in the Io Plasma Torus', *J. Geophys. Res.* **90**, 1755.

Baker, D. N. and Van Allen, J. A.: 1976, 'Energetic Electrons in the Jovian Magnetosphere', *J. Geophys. Res.* **81**, 617.

Belcher, J. W.: 1983, in A. J. Dessler (ed.), 'The Low-Energy Plasma in the Jovian Magnetosphere', *Physics of the Jovian Magnetosphere*, Cambridge University Press, Cambridge, p. 68.

Frank, L. A., Ackerson, K. L., Wolfe, J. H., and Mihalov, J. D.: 1976, 'Observations of Plasmas in the Jovian Magnetosphere', *J. Geophys. Res.* **81**, 457.

Gurnett, D. A. and Scarf, F. L.: 1983, in A. J. Dessler (ed.), 'Plasma Waves in the Jovian Magnetosphere', *Physics of the Jovian Magnetosphere*, Cambridge University Press, Cambridge, p. 285.

Krimigis, S. M. and Roelof, E. C.: 1983, in A. J. Dessler (ed.), 'Low-Energy Particle Population', *Physics of the Jovian Magnetosphere*, Cambridge University Press, Cambridge., p. 106.

Sands, M. R. and McNutt, R. L., Jr.: 1988, 'Plasma Bulk Flow in Jupiter's Dayside Middle Magnetosphere', *J. Geophys. Res.* **93**, 8502.

Scudder, J. D., Sittler, E. C., Jr., and Bridge, H. S.: 1981, 'A Survey of the Plasma Electron Environment of Jupiter: A View from Voyager', *J. Geophys. Res.* **86**, 8517.

Sittler, E. C. and Strobel, D. F.: 1987, 'Io Plasma Torus Electrons: Voyager 1', *J. Geophys. Res.* **92**, 5741.

Van Allen, J. A.: 1976, in T. Gehrels (ed.), 'High-Energy Particles in the Jovian Magnetosphere', *Jupiter*, University of Arizona Press, Tucson, p. 928.

Vasyliunas, V. M.: 1983, in A. J. Dessler (ed.), 'Plasma Distribution and Flow', *Physics of the Jovian Magnetosphere*, Cambridge University Press, Cambridge, p. 395.

THE GALILEO HEAVY ELEMENT MONITOR

T. L. GARRARD[1], N. GEHRELS[2], and E. C. STONE[1]

Abstract. The Heavy Ion Counter on the Galileo spacecraft will monitor energetic heavy nuclei of the elements from C to Ni, with energies from ~ 6 to ~ 200 MeV nucl^{-1}. The instrument will provide measurements of trapped heavy ions in the Jovian magnetosphere, including those high-energy heavy ions with the potential for affecting the operation of the spacecraft electronic circuitry. We describe the instrument, which is a modified version of the Voyager CRS instrument.

1. Introduction

The Heavy Ion Counter (HIC) is included on the Galileo spacecraft primarily for the purpose of monitoring the fluxes of energetic heavy ions in the inner Jovian magnetosphere and high-energy solar particles in the outer magnetosphere in order to characterize the ionizing radiation to which electronic circuitry is most sensitive. The measurements performed will also be of scientific interest, since the instrument's large geometry factors and extended energy range will provide spectral information for ions from ${}_6$C to ${}_{28}$Ni with energies of ~ 6 to $\gtrsim 200$ MeV nucl^{-1}. In this article we will concentrate on Jovian magnetospheric science. We review here previous scientific results concerning trapped high-energy heavy ions, describe anticipated new findings with Galileo, and provide a brief instrument description.

2. Previous Results

During the Voyager encounters with Jupiter, it was discovered (Krimigis *et al.*, 1979a, b; Vogt *et al.*, 1979a, b) that a major component of the trapped radiation in the inner Jovian magnetosphere is energetic heavy ions. The dominant heavy species inside $\sim 10\ R_J$ in the MeV nucl^{-1} energy range were found to be oxygen and sulfur, with sodium also present. This is illustrated in Figure 1 which compares the elemental composition between 4.9 and 5.8 R_J with the more normal solar-like composition found in the same energy range in the middle and outer magnetosphere. The abundances in the inner magnetosphere indicate that the source of the ions is the surface or atmosphere of Io.

The phase space density of the energetic oxygen and sulfur ions has a positive radial gradient (i.e., increasing outward) in the inner magnetosphere (Gehrels *et al.*, 1981), implying that the diffusive flow at these energies is inward. On the other hand, the density of the oxygen- and sulfur-rich plasma in the inner magnetosphere is highest at the Io plasma torus and decreases outward (Siscoe *et al.*, 1981). Barbosa *et al.* (1984) have proposed that charge exchange in the plasma torus produces fast neutrals which

[1] California Institute of Technology, Pasadena, CA 91125, U.S.A.
[2] NASA Goddard Space Flight Center, Greenbelt, MD 20771, U.S.A.

Space Science Reviews **60:** 305–315, 1992.

escape to the outer magnetosphere where ~0.2% are re-ionized by solar ultraviolet or electron impact and recaptured Some of the newly created ions are subsequently energized by stochastic acceleration caused by magnetohydrodynamic waves, producing ions with large magnetic moments that adiabatically diffuse inward, some attaining energies in excess of 30 MeV nucl^{-1} in the inner magnetosphere.

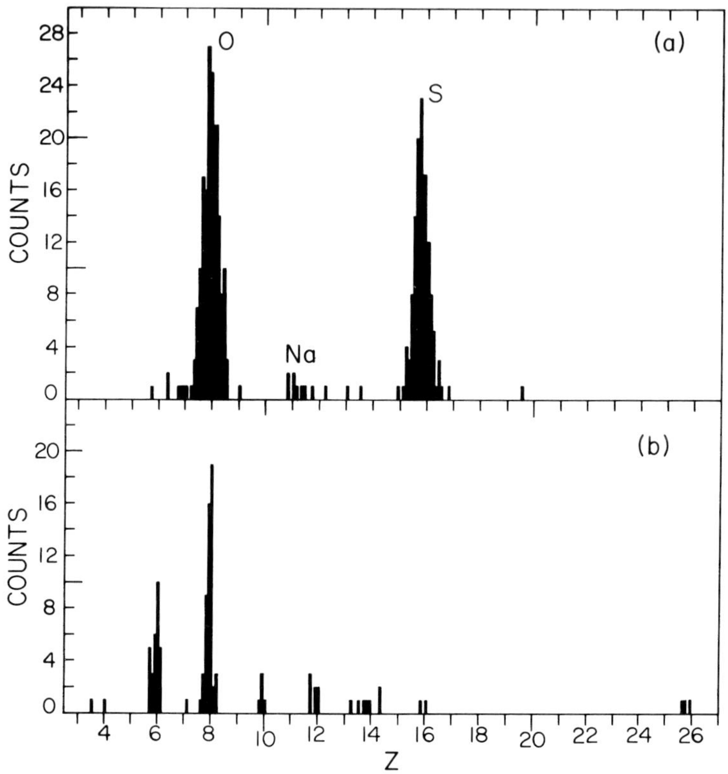

Fig. 1. Measured element (Z) distribution for heavy ions in the Jovian environment with energies from 7 MeV nucl^{-1} to typically ~18 MeV nucl^{-1}. (a) 4.9 to 5.8 R_J, ~1 hour elapsed time. (b) Outside ~11 R_J, ~9 days elapsed time. From Vogt *et al.* (1979a). Because the measurement efficiency is independent of Z, the observed distributions directly reflect the relative abundances of the energetic nuclei in each panel.

As the energetic ions flow inward, they are lost from the magnetosphere, presumably by pitch-angle scattering of the mirroring particles into the loss cone (see, e.g., Thorne, 1982). The strength of this loss mechanism determines the rate at which the heavy ions precipitate into the Jovian atmosphere, exciting ultraviolet and X-ray auroral emissions. At the time of the Voyager 1 encounter in March 1979, it appeared losses due to pitch-angle scattering were occurring at nearly the maximum rate, resulting in the precipitation of ~10^{24} ions s^{-1} above ~70 MeV nucl^{-1} G^{-1}, with an extrapolation down to 10 MeV nucl^{-1} G^{-1} suggesting the possibility of a loss rate of ~10^{26} ions s^{-1} and an auroral power of ~10^{13} W (Gehrels and Stone, 1983). The evidence that major losses are occurring in the magnetic moment (energy) range appropriate to HIC is shown in Figure 2, where the total inward flow rate of oxygen ions with magnetic moments greater than several thresholds is given as a function of radial distance (L) from Jupiter. The number of inflowing ions first decreases inside 15 R_J and falls off sharply inside 10 R_J.

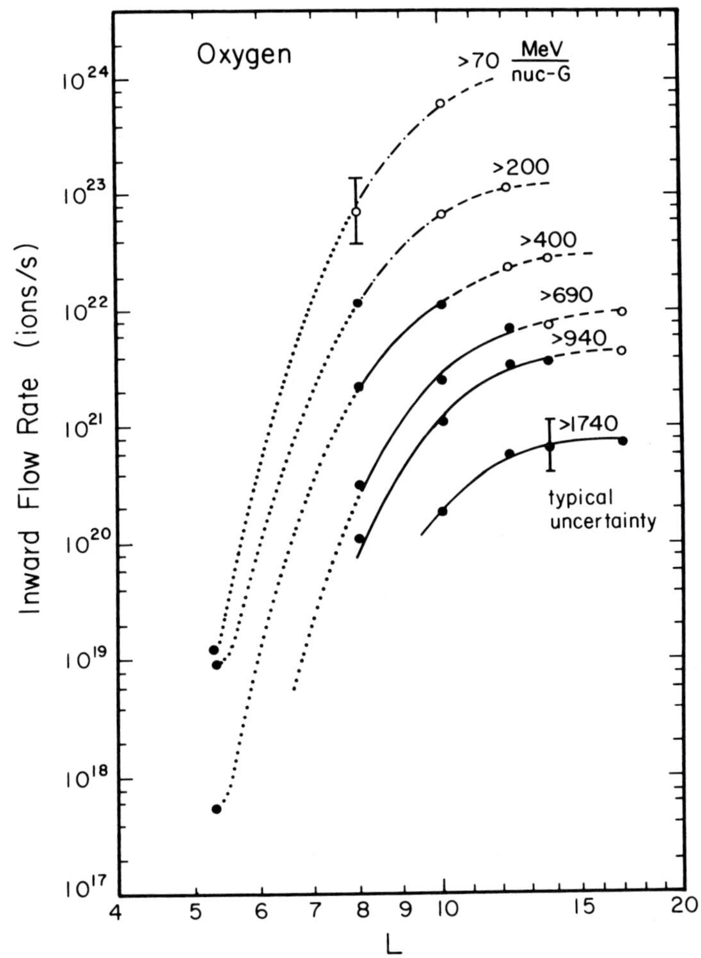

Fig. 2. The inward flow rate of oxygen ions with magnetic moments greater than the indicated values as a function of distance, L, in R_J. The filled circles are from measured spectra and the open circles from extrapolation of the spectra. Typical uncertainties in the measured and extrapolated points are dominated by uncertainties in the diffusion coefficients. There is, in addition, a factor of 3 absolute uncertainty that applies to all points. From Gehrels and Stone (1983).

The curves in Figure 2 also illustrate that only a small fraction of the trapped particles survive to produce the trapped radiation at 5 R_J. As a result, changes in the rate of pitch-angle scattering can result in significant changes in the flux of particles reaching

$5\ R_J$. It is therefore important to better characterize not only the flux at $5\ R_J$, but also the nature and any long-term variability in the loss mechanism operative outside of $5\ R_J$.

3. Anticipated New Results

The HIC instrument should provide new information on the spectra of heavy ions at higher energies than previously possible. As illustrated in Figure 3, fluxes at these

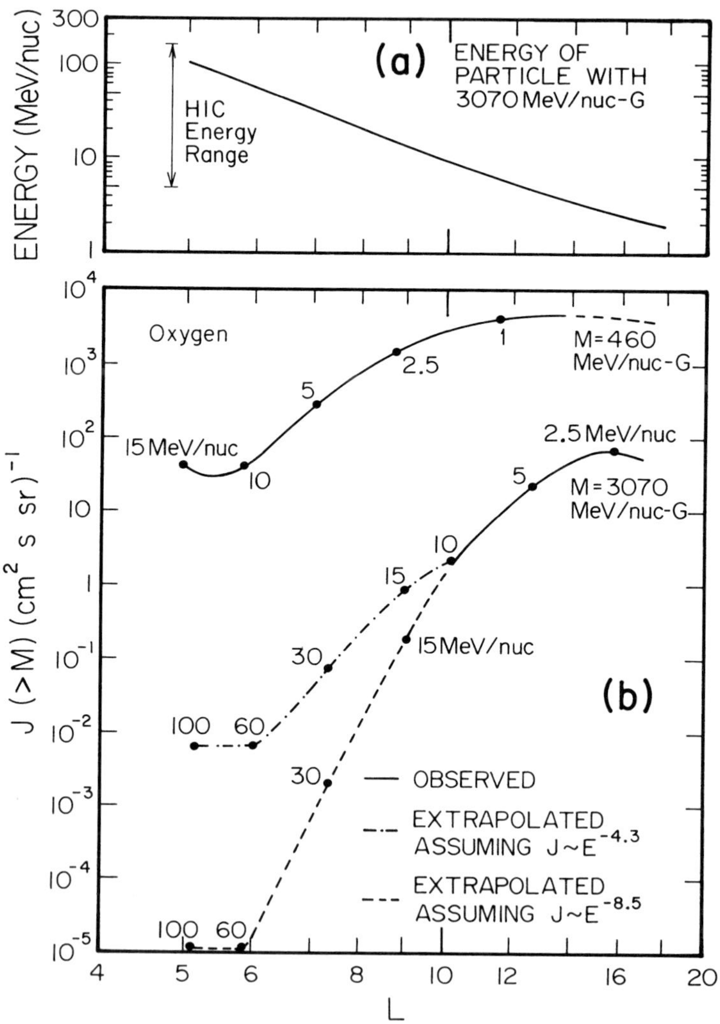

Fig. 3. (a) The energy of a particle with magnetic moment 3070 MeV nucl^{-1} G^{-1} as a function of radial position compared with the HIC energy range. (b) The integral flux of oxygen ions with magnetic moments greater than 460 MeV nucl^{-1} G^{-1} and 3070 MeV nucl^{-1} G^{-1} as functions of radial position. The solid lines represent data from the Voyager CRS instrument, and the dashed lines are possible extrapolations. The energy of the particles is indicated by the numbers at specific radial locations along the intensity curves.

energies are based on extrapolations of spectra measured at lower energies and as a result are very uncertain.

Two different estimates of the expected fluxes of oxygen ions are shown in Figure 3(b) for magnetic moments of ≥ 3070 MeV nucl^{-1} G^{-1} and ≥ 460 MeV nucl^{-1} G^{-1}, corresponding to particles with $E \geq 100$ MeV nucl^{-1} and ≥ 15 MeV nucl^{-1} at 5 R_J. The ≥ 15 MeV nucl^{-1} flux was directly measured by the Cosmic-Ray Science instrument (CRS) on Voyager 1 and is reasonably certain. The ≥ 100 MeV nucl^{-1} flux, however, has been determined by extrapolation, and is therefore quite uncertain as indicated by the two different profiles. The dot-dashed line is the result of an assumption that $J(>E) \propto E^{-4.3}$ which is based on a least-squares fit of a power law to Voyager 1 data in the energy interval from 7 to 20 MeV nucl^{-1}. Because of limited lifetime and geometrical factor, Voyager 1 observed no particles with energies ≥ 30 MeV nucl^{-1}, so a much softer spectrum is also possible at higher energies as indicated by the dashed line corresponding to $J(>E) \propto E^{-8.5}$. These two extrapolations differ by a factor of ~ 500 in the predicted of flux of oxygen with $E \geq 100$ MeV nucl^{-1} at 5 R_J.

Measurements with the HIC instrument will significantly reduce the uncertainty in the high-energy fluxes. As shown in Figure 3(a), the energy of a 3070 MeV nucl^{-1} G^{-1} oxygen ion is within the HIC energy range into 5 R_J. The geometry factor for $E > 20$ MeV nucl^{-1} oxygen ions (inside 8.5 R_J for 3070 MeV nucl^{-1} G^{-1} ions) is ~ 4 cm^2 sr compared with 0.88 cm^2 sr for the Voyager CRS instrument. More importantly, the livetime for measuring oxygen and sulfur ions with $E \geq 50$ MeV nucl^{-1} will be essentially 100% because a polling priority system (see instrument description section) will discriminate against the large fluxes of protons and electrons that dominate the CRS analysis rate. As a result, a flux of only 10^{-5} cm^{-2} s^{-1} sr^{-1} should produce 1 analyzed event in 10 hours, the time that Galileo is inside 7 R_J.

Figure 3 indicates that the oxygen ions which have ≥ 100 MeV nucl^{-1} at 5 R_J have ≥ 10 MeV nucl^{-1} at 10 R_J. Thus, as Galileo makes repeated orbital passes in the vicinity of Europa, it will be possible to monitor the fluxes of particles with $M \geq 3000$ MeV nucl^{-1} G^{-1} throughout the mission. Such long term information is especially important, since there are a number of reasons why the fluxes might vary. For example, the flux of high-energy ions could be measurably affected by changes in the density of the Io plasma torus which is the source of the escaping neutral atoms and by changes in the solar ultraviolet which re-ionizes the neutrals in the outer magnetosphere. The HIC instrument will measure any time dependence of the energetic heavy ion fluxes and permit correlative studies of any associated changes in auroral emissions or diffusion processes.

Figure 3 also illustrates that the fluxes at 5 R_J depend strongly on the loss processes occurring inside of ~ 15 R_J. It is postulated that the losses are due to pitch-angle scattering of the mirroring particles into the loss cone, and that the rate of scattering is close to the strong pitch-angle diffusion limit in the inner magnetosphere (Thorne, 1982; Gehrels and Stone, 1983). At this limit there is sufficient pitch-angle scattering to refill the loss cone as rapidly as particle precipitation empties it, resulting in a nearly isotropic pitch-angle distribution. Since the HIC detectors will view nearly perpendicular to the spacecraft spin axis, essentially complete coverage of the pitch-angle distribution will be possible. Investigation of any time dependence of the loss process will also be possible as Galileo repeatedly passes through the radial range between 15 and 10 R_J where the radial profiles in Figure 2 indicate that significant losses occur.

4. Intrument Description

The Galileo Heavy Ion Counter (HIC) consists of two solid-state detector telescopes called Low Energy Telescopes or LETs. Use of these two telescopes over three energy intervals provides the geometry factor and energy range necessary to determine the fluxes of the heavy, penetrating radiation to which solid state memories are most sensitive. Heavy collimation and high discrimination thresholds on all detectors provide the necessary immunity to accidental coincidences from the large proton background. Three-parameter analysis provides additional rejection of background.

Two separate telescopes are included in order to cover a wide energy range while minimizing pulse pile-up through the optimum selection of detector and window thicknesses. The LET E is optimized for the detection of nuclei with energies as high as 200 MeV nucl^{-1}, requiring thicker detectors. Thick windows protect this system from low-energy proton pile-up, but also exclude lower energy oxygen and sulfur nuclei. The second telescope, LET B, has a substantially thinner window so that it can detect lower energy nuclei (down to ~6 MeV nucl^{-1}), especially in the outer magnetosphere. The properties of the two telescopes are indicated in Table I and discussed below.

TABLE I
Nominal LET parameters

Detector	Radius (cm)	Thickness (μm)	Threshold (MeV)
LB1	0.95	32.1	0.3
LB2	0.95	29.6	0.4
LB3	1.13	421	3.7
LB4	1.13	440	2.0
LE1	0.95	30.4	9.3 (1.4)[a]
LE2	0.95	33.4	2.0 (0.3)[a]
LE3	1.13	463	25 (5.0)[a]
LE4, 5	1.66	≃2000	117 (23)[a]

[a] High gain.

The LET B telescope in shown in Figure 4. In this telescope ions which penetrate LB1 and LB2 and stop in the thicker LB3 detector are analyzed. Detector LB4 is used in anti-coincidence. (If desired, the command system can be used to allow analysis for ions which penetrate LB1 and stop in LB2, or for ions which stop in LB1. Also detector LB4 can be turned off.) Lighter nuclei (especially hydrogen and helium) are rejected by 'slant' discrimination with a weighted sum of the signals from the front three detectors,

$$SLB = LB1 + 0.42\ LB2 + 0.2\ LB3$$

required to be above about 9.6 MeV.

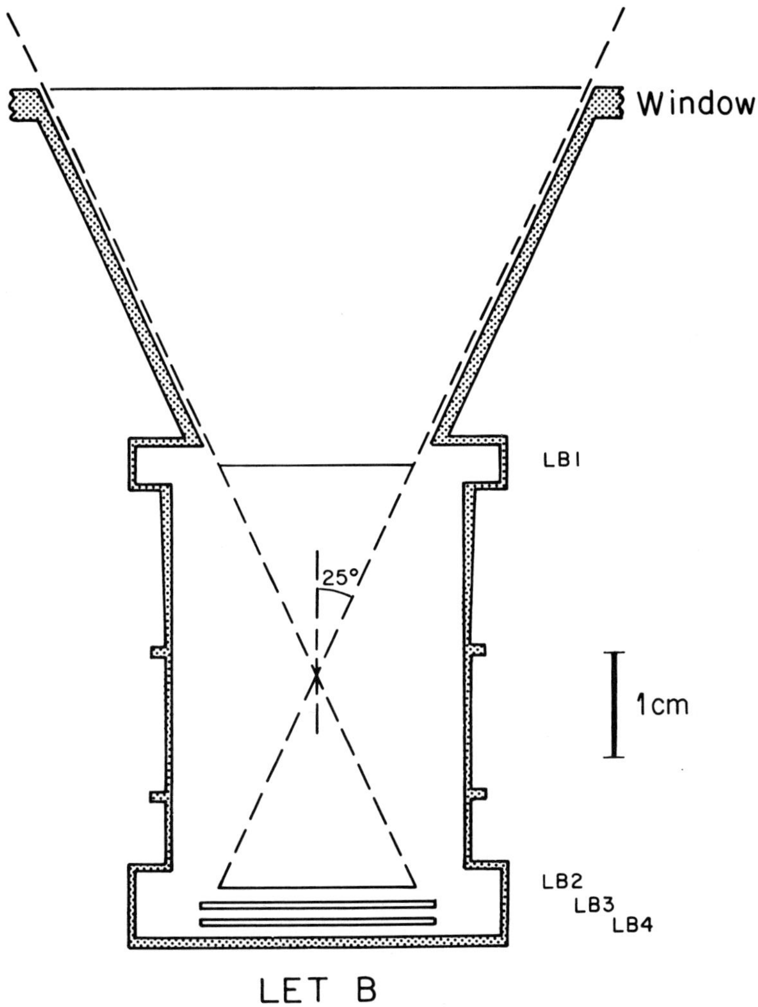

Fig. 4. Schematic cross section of the LET B telescope. The dotted housing is Al. Only the active regions of the detectors are shown.

The thin window (25 μm Kapton) shown in the figure serves for thermal control and for protection from sunlight. All detectors are surface-barrier type. An important feature is the use of keyhole detectors for LB1 and LB2, which define the event geometry. The active area of these detectors excludes the nonuniform edge of the silicon wafer through the use of keyhole-shaped masks during the deposition of the Au and Al contacts.

This telescope is an improved version of the Voyager CRS LETs (Stone *et al.*, 1977) which have demonstrated charge (Z) resolution of 0.1 charge units at oxygen under solar flare conditions (Cook *et al.*, 1980) and 0.2 charge units at $\sim 5\ R_J$, Voyager 1's deepest penetration of the Jovian magnetosphere (Gehrels, 1982). Additional collimation and the thicker window decrease the HIC LET's response to background protons in the Jovian magnetosphere. The flux of protons in detector L1, for example, should be reduced by at least a factor of 10 from that observed on Voyager 1.

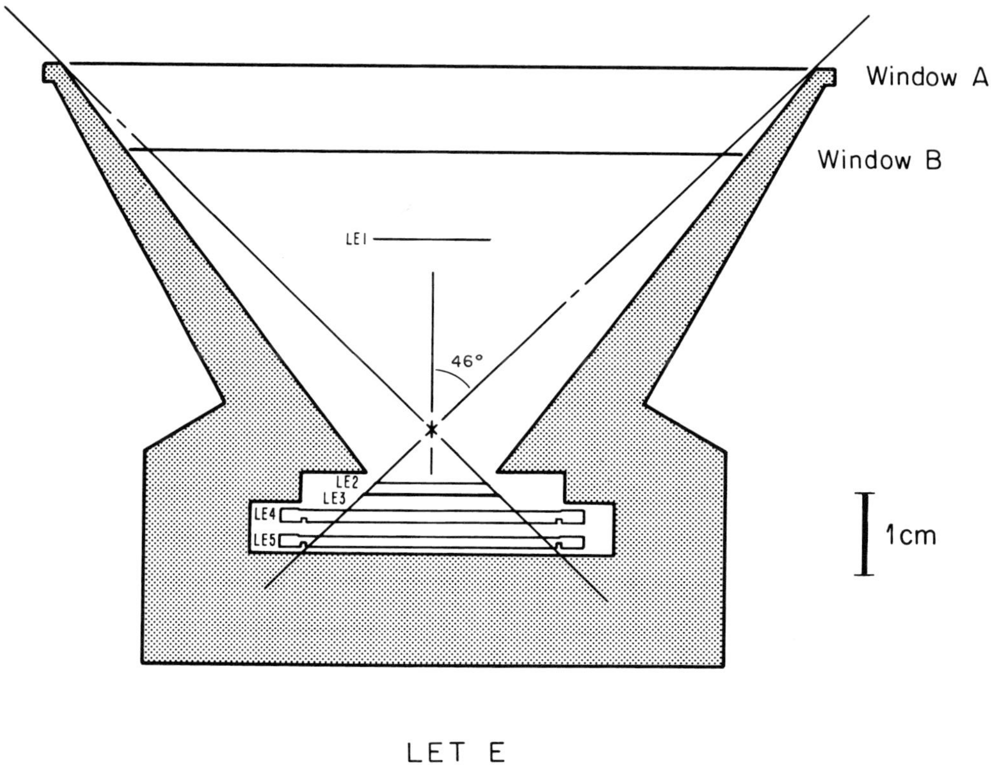

Fig. 5. Schematic cross section of the LET E telescope. The dotted housing is Al. Only the active regions of the surface-barrier detectors (LE1, 2, 3) are shown. The LiD detectors (LE4, 5) are active inside the groove on the silicon wafer. LE1 is supported by a very light-weight spider arrangement.

The higher energy LET E telescope is illustrated in Figure 5. In order to measure the flux of heavy nuclei at higher energies, LE4 and LE5 are each 2000-μm lithium-drifted detectors. The front detectors, LE1, LE2, and LE3, are identical to their LET B counterparts, providing spectral continuity and overlap. The collimator housing is relatively thick to provide background immunity and has a large opening angle to provide a large geometrical factor. The windows are 76 μm Kapton and 254 μm Al. As in LET B, low charge (low Z) events are recognized and rejected by a slant discriminator,

$$SB = LE1 + 0.5\,LE2 + 0.1\,LE3 + 0.04\,(LE4 + LE5)\,,$$

where SB must exceed 9.6 MeV to allow analysis.

For intermediate energies, a narrow-angle geometry is defined by LE1 and LE2 for particles stopping in either LE2 or LE3. For the highest energies, where the fluxes are exceedingly small, measurements are made with the wide-angle geometry defined by LE2, LE3, LE4, and the collimator. LE1 is not required but its discriminator is recorded as a tag bit. LE5 distinguishes stopping and penetrating events. The maximum energy observed is determined by the LE2 discriminator threshold. This maximum is ~ 185 MeV nucl^{-1} for oxygen as noted in Table II. To allow detection of penetrating galactic cosmic-ray nuclei at higher energies in the outer magnetosphere, the LET E pre-amplifier gains can be increased by a factor of 5 to 7 by command.

With the exception of spacecraft interface circuitry, all of the electronics were originally part of the Proof Test Model of the Voyager Cosmic-Ray Science Instrument (CRS), and additional detail may be found in Stone *et al.* (1977) and Stilwell *et al.* (1979). With the adjustment of amplifier gains and discriminator thresholds and the incorporation of thicker detectors, collimators, and windows, it has been possible to develop an instrument which is optimized for the measurement of high-energy heavy ions trapped in the Jovian magnetosphere. Minor modifications to the logic allow the instrument to recognize events of the various types mentioned above, which are summarized in Table II.

Event data are stored in buffers which are read according to a polling scheme which prevents domination of the telemetry by any one type of event. The buffer polling logic cycles through the five buffers listed in Table II, reading out one each minor frame ($\frac{2}{3}$ s) and stepping to the next non-zero buffer on the subsequent minor frame. If a particular type of event occurs less than 0.3 times per second (i.e., is rare) then all of that type will be transmitted regardless of activity in other event types. If a particular type of event occurs more often than 0.3 times per second, it will be readout at least 0.3 times per second and more of ten if the other event buffers are empty.

Telemetry of counting rates and pulse height analyzed events is rapid compared to the nominal 3 rpm spin rate of the spacecraft; thus pitch-angle distributions of the trapped radiation can be measured. Both telescopes have their axes oriented near the spin plane for this purpose. The time resolution of the HIC is in the range from $\frac{2}{3}$ to 2 s, implying an angular resolution in the range from 12° to 36°, which is to be compared to the telescope opening angles of 25° in narrow geometry mode and 46° in wide geometry mode.

Many of the functions of the coincidence logic and the buffering/readout scheme can be modified by command to optimize the instrument for changing environments or partial failures. As noted above, commands can also be used to change gain on the LET E pre-amplifiers.

TABLE II

Analysis modes

Name	Requirement	Geometry factor (cm^2 sr)	Z range	Oxygen energy range (MeV $nucl^{-1}$)	Sulfur energy range (MeV $nucl^{-1}$)	Signals telemetered
LETB	LB1. LB2. LB3. $\overline{\text{LB4}}$	0.44	C to Ni	6 to 18	9 to 22	LB1, LB2, LB3
DUBL	LE1. LE2. $\overline{\text{LE3}}$	0.44	C to Fe	16 to 17	24 to 25	LE1, LE2
TRPL	LE1. LE2. LE3. $\overline{\text{LE4}}$	0.44	C to Ni	17 to 27	25 to 38	LE1, LE2, LE3
WDSTP	LE2. LE3. LE4. $\overline{\text{LE5}}$	4.0	C to Fe	26 to 46	37 to 70	LE2, LE3, LE4
WDPEN	LE2. LE3. LE4. LE5	4.0	C to Fe	49 to $\sim$185	$\geq$70	LE2, LE3, LE4 + LE5
WDPEN (High gain)	LE2. LE3. LE4. LE5. HG	4.0	Li to O	49 to $\sim$500	–	LE2, LE3, LE4 + LE5

Acknowledgements

A special acknowledgement is due A. W. Schardt, whose untimely death prevented his co-authorship of this paper. This project was made possible by the development of the CRS instrument under the leadership of R. E. Vogt. W. E. Althouse (Caltech) and D. E. Stilwell (GSFC) have provided invaluable engineering and programmatic support and advice. We also are pleased to acknowledge the contributions of A. C. Cummings at Caltech; M. Beasley, W. D. Davis, J. H. Trainor, and H. Trexel at GSFC; and D. R. Johnson at JPL. This work was supported by a number of NASA contracts and grants.

References

Barbosa, D. D., Eviatar, A., and Siscoe, G. L.: 1984, *J. Geophys. Res.* **89**, 3789.

Cook, W. R., Stone, E. C., and Vogt, R. E.: 1980, *Astrophys. J.* **238**, L97.

Gehrels, N.: 1982, 'Energetic Oxygen and Sulfur Ions in the Jovian Magnetosphere', CIT Ph.D. Thesis.

Gehrels, N. and Stone, E. C.: 1983, *J. Geophys. Res.* **88**, 5537,

Gehrels, N., Stone, E. C., and Trainor, J. H.: 1981, *J. Geophys. Res.* **86**, 8906.

Krimigis, S. M., Armstrong, T. P., Axford, W. I., Bostrom, C. O., Fan, C. Y., Gloeckler, G., Lanzerotti, L. J., Keath, E. P., Zwickl, R. D., Carbary, J. F., and Hamilton, D. C.: 1979a, *Science* **204**, 998.

Krimigis, S. M., Armstrong, T. P., Axford, W. I., Bostrom, C. O., Fan, C. Y., Gloeckler, G., Lanzerotti, L. J., Keath, E. P., Zwickl, R. D., Carbary, J. F., and Hamilton, D. C.: 1979b, *Science* **206**, 977.

Siscoe, G. L., Eviatar, A., Thorne, R. M., Richardson, J. D., Bagenal, F., and Sullivan, J. D.: 1981, *J. Geophys. Res.* **86**, 8480.

Stilwell, D. E., Davis, W. D., Joyce, R. M., McDonald, F. B., Trainor, J. H., Althouse, W. E., Cummings, A. C., Garrard, T. L., Stone, E. C., and Vogt, R. E.: 1979, *IEEE Trans. Nuc. Sci.* **NS-26**, 513.

Stone, E. C., Vogt, R. E., McDonald, F. B., Teegarden, B. J., Trainor, J. H., Jokipii, J. R., and Webber, W. R.: 1977, *Space Sci. Rev.* **21**, 355.

Thorne, R. M.: 1982, *J. Geophys. Res.* **87**, 8105.

Vogt, R. E., Cook, W. R., Cummings, A. C., Garrard, T. L., Gehrels, N., Stone, E. C., Trainor, J. H., Schardt, A. W., Conlon, T., Lal, N., and McDonald, F. B.: 1979a, *Science* **204**, 1003.

Vogt, R. E., Cummings, A. C., Garrard, T. L., Gehrels, N., Stone, E. C., Trainor, J. H., Schardt, A. W., Conlon, T. F., and McDonald, F. B.: 1979b, *Science* **206**, 984.

THE GALILEO DUST DETECTOR

EBERHARD GRÜN[1], HUGO FECHTIG[1], MARTHA S. HANNER[2], JOCHEN KISSEL[1], BERTIL-ANDERS LINDBLAD[3], DIETMAR LINKERT[1], DIETER MAAS[1,4], GREGOR E. MORFILL[5], and HERBERT A. ZOOK[6]

Abstract. The Galileo Dust Detector is intended to provide direct observations of dust grains with masses between 10^{-19} and 10^{-9} kg in interplanetary space and in the Jovian system, to investigate their physical and dynamical properties as functions of the distances to the Sun, to Jupiter and to its satellites, to study its interaction with the Galilean satellites and the Jovian magnetosphere. Surface phenomena of the satellites (like albedo variations), which might be effects of meteoroid impacts will be compared with the dust environment. Electric charges of particulate matter in the magnetosphere and its consequences will be studied; e.g., the effects of the magnetic field on the trajectories of dust particles and fragmentation of particles due to electrostatic disruption. The investigation is performed with an instrument that measures the mass, speed, flight direction and electric charge of individual dust particles. It is a multicoincidence detector with a mass sensitivity 10^6 times higher than that of previous *in-situ* experiments which measured dust in the outer solar system. The instrument weighs 4.2 kg, consumes 2.4 W, and has a normal data transmission rate of 24 bits s^{-1} in nominal spacecraft tracking mode. On December 29, 1989 the instrument was switched-on. After the instrument had been configured to flight conditions cruise science data collection started immediately. In the period to May 18, 1990 at least 168 dust impacts have been recorded. For 81 of these dust grains masses and impact speeds have been determined. First flux values are given.

1. Introduction

The Galileo Mission provides opportunity for a thorough investigation of the interplanetary dust complex and the Jovian dust environment. The scientific objectives of the Galileo dust investigation were stated in our proposal dated November 1976: "The overall objective of the proposed investigation is the exploration of the physical and dynamical properties of small dust particles (with masses between 10^{-19} and 10^{-9} kg) in the Jovian environment. The parameters to be studied include mass, speed, flight direction, and charge of individual particles. The impact rate, size distribution, angular distribution, and charge will be determined with respect to the Jovian distance from the satellites and the magnetospheric coordinates. The primary specific objectives are:

(a) to investigate the interaction of the Galilean satellites with their dust environment in order to study the relationship between dust influx on satellites and their surface properties, and to perform direct measurements of ejecta particles from the satellites;

(b) to study the interaction between dust particles and magnetospheric plasma, high-energy electrons and protons, and magnetic field, to determine the relationship

[1] Max-Planck-Institut für Kernphysik, Heidelberg, Germany.
[2] Jet Propulsion Laboratory, Pasadena, California, U.S.A.
[3] Lund Observatory, Lund, Sweden.
[4] Kernforschungszentrum Karlsruhe, Karlsruhe, Germany.
[5] Max-Planck-Institut für Extraterrestrische Physik, Garching, Germany.
[6] NASA Johnson Space Center, Houston, Texas, U.S.A.

Space Science Reviews **60:** 317–340, 1992.

between dust concentrations and attenuation of the radiation belts, and to investigate the effects of the Jovian magnetic field on the trajectories of charged dust particles;

(c) to investigate the influence of the Jovian gravitational field on the interplanetary dust population and to search for rings around Jupiter."

Since the approval of the Galileo Mission in 1977, two Voyager spacecraft have flown by Jupiter and have increased our knowledge about the Jovian system considerably (Burns *et al.*, 1980; Grün *et al.*, 1980; Morfill *et al.*, 1980a; Showalter *et al.*, 1987). The existence of a dust ring of μm-sized particles around Jupiter, which had previously only been suspected (Acuña and Ness, 1976; Fillius, 1976) was confirmed by photographs taken by both spacecraft (Smith *et al.*, 1979a, b). Since *in-situ* dust detectors can detect local dust concentrations much better than cameras, new information on the Jovian environment can be expected.

The reduced capability of the launching system required the Galileo spacecraft to cruise through the inner Solar System for several years and to have multiply flybys of Venus and Earth before it reaches Jupiter. This prolonged interplanetary flight trajectory provides the opportunity of investigating the interplanetary dust complex in a unique way from about 0.7 to 5 AU. During this six-year long cruise measurements by the Galileo dust detector could also serve as in-ecliptic base line for the out-of-ecliptic measurements taken by the twin dust detector on the Ulysses Mission (Grün *et al.*, 1983).

The Galileo and Ulysses dust detectors are descendants of the dust detector flown on the HEOS-2 satellite (Dietzel *et al.*, 1973). This instrument carried out measurements in the near-Earth space and observed interesting effects of Earth's magnetosphere and the Moon on the interplanetary dust population (Hoffmann *et al.*, 1975, 1976; Fechtig *et al.*, 1979). All these dust detectors are based on the impact ionization phenomenon (Friichtenicht and Slattery, 1963; Fechtig *et al.*, 1978), which provides extremely high sensitivty for recording small dust particles. Table I compares the mass sensitivities of different interplanetary dust detectors.

A team of experts has been assembled to develop, to build and to test the dust detector

TABLE I

Mass sensitivities and measurement ranges of different interplanetary dust detectors. The mass thresholds refer to 20 km s^{-1} impact speed. The Pioneer 10 and 11 detectors are threshold detectors.

Spacecraft	Mass threshold	Dynamic range	Sensitive area	References
Pioneer 8/9	2×10^{-16} kg	10^2	0.010 m^2	Grün *et al.*, 1973
Pioneer 10	2×10^{-12} kg	–	0.26 m^2	Humes *et al.*, 1974
Pioneer 11	10^{-11} kg	–	0.26 (0.57) m^2	Humes, 1980
HEOS 2	2×10^{-19} kg	10^4	0.010 m^2	Hoffmann *et al.*, 1975
Helios 1/2	9×10^{-18} kg	10^4	0.012 m^2	Dietzel *et al.*, 1973
Ulysses	2×10^{-18} kg	10^6	0.10 m^2	Grün *et al.*, 1983
Galileo	2×10^{-18} kg	10^6	0.10 m^2	This work

instrument and its data-processing system and to carry out scientific investigations based on the data. Table II lists the present principle team members and their responsa-

TABLE II

Galileo Dust Detector Team

Principal investigator	Eberhard Grün
Co-investigators	Hugo Fechtig Martha S. Hanner Jochen Kissel Bertil-Anders Lindblad Gregor E. Morfill Herbert A. Zook
Project engineer	Dietmar Linkert
Electronics engineers	Günter Baust Rainer Hofacker Günter Matt Werner Schneider Alfred Zahlten
Data analysts	Gudrun Linkert Dieter Maas Nadeem Siddique
Clerical support	Ortrud Kress

bilities. A more complete account of people whose contributed significantly to the development of the Galileo dust investigation over the past years is given in the Acknowledgements.

2. Scientific Background

2.1. Interplanetary Dust

Galileo's interplanetary cruise to Jupiter comprises different stages: between Earth and Venus (two times), between Earth and the asteroidal belt (three times), within the asteroidal belt, and beyond the asteroidal belt until the arrival at the planet Jupiter.

The space between the Earth and Venus was scanned by a dust detector on the Helios space probe (Grün *et al.*, 1980, 1985a). The Helios sensor's sensitivity was similar to that of the Galileo detector. There was a large variety of new results. In agreement with the zodiacal light experiment, which was also flown on Helios, the number of interplanetary dust grains were found to vary with $r^{-1/3}$ (r = distance from the Sun; Leinert *et al.*, 1981). The Helios dust detector identified two different types of interplanetary particles orbiting the Sun: particles of densities $\rho = 3$ to $8\ \mathrm{g\ cm^{-3}}$ on low excentric trajectories and particles of low densities ($\rho < 1\ \mathrm{g\ cm^{-3}}$) orbiting on high excentric ellipses around the Sun (Grün *et al.*, 1980). For a general survey see McDonnell (1978) and Leinert and Grün (1990).

For dust particles outside the Earth's orbit we have the results of two dust experiments on Pioneers 10 and 11. There the mass sensitivity was considerably lower than in the Galileo experiment. The Galileo dust detector can record particles as small as 0.1 μm diameters, i.e., 2 orders of magnitudes smaller in size than those measured on Pioneer 10/11. Between 1 and 3.3 AU these experiments detected a decrease of the dust abundance proportional to $r^{-1.5}$ (Humes *et al.*, 1974; Hanner *et al.*, 1976).

Within the asteroidal belt, the Pioneer 10/11 dust experiments (Figure 1) did not find

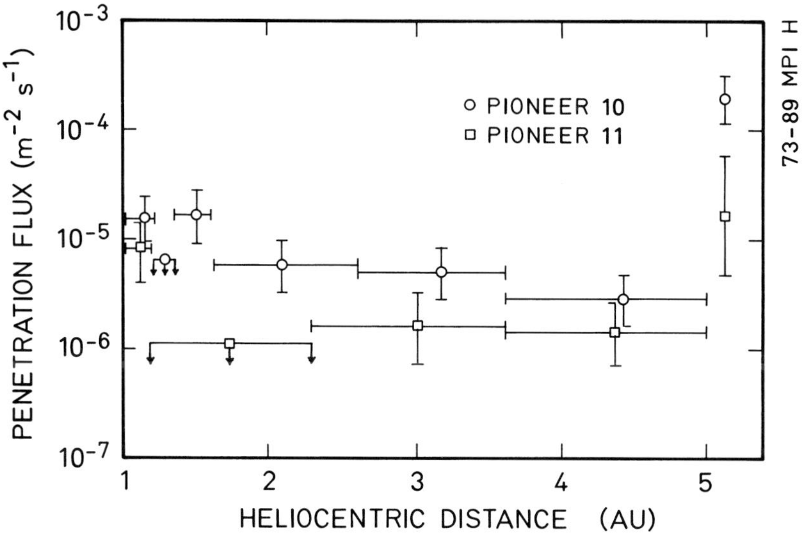

Fig. 1. Dust particle fluxes between Earth and Jupiter measured by the Pioneer 10 and 11 penetration experiments (Channel 0; Humes, 1980). The walls of the penetration cells were 25 μm and 50 μm thick stainless steel for Pioneer 10 and 11, respectively. The fluxes shown at Jupiter correspond to the average flux recorded within 50 planetary radii; the near-Jupiter peak fluxes is about an order of magnitude higher.

any increase of the dust abundances as had previously been expected by some investigators. The conclusion of the experimentalists (Humes *et al.*, 1974) was that no dust is produced in the asteroid belt. A significant increase, however, of the particles in the sensitivity range of these experiments (> 20 and 40 μm diameter) cannot necessarily be expected since the dust particles are spiralling towards the Sun and thus leaving the asteroid belt. Observation of asteroidal dust bands by the infrared satellite IRAS (Hauser *et al.*, 1984) confirm that the asteroid belt is a significant source of dust in the solar system. Measurements of the dust population between 0.1 and 1 AU show that most of the mass of the dust grains in the inner solar system are of cometary origin (Grün *et al.*, 1985b). At 1 AU, however, the study of lunar microcraters clearly shows that $\frac{2}{3}$ of the projectiles which have produced the lunar microcraters by high velocity impacts are of densities between 3 and 8 g cm^{-3} and only $\frac{1}{3}$ of the projectiles are of low densities (< 1 g cm^{-3}). The higher density particles are most likely of asteroidal origin (Nagel *et al.*, 1975, 1976a, b; Fechtig *et al.*, 1975; Brownlee *et al.*, 1973; Fechtig, 1989).

The two dust experiments on Pioneers 10/11 observed particles larger than approximately 20 μm in diameter in the outer solar system (Humes *et al.*, 1974; Hanner *et al.*, 1976). The most surprising result was that the photometer did not record any scattered sunlight beyond the asteroidal belt (Hanner *et al.*, 1976) while the penetration experiment recorded dust grains at up to about 20 AU distance from the Sun at almost constant impact rate on highly eccentric orbits (Humes, 1980). Within the asteroid belt and inward to 1 AU presumably $\frac{2}{3}$ of the dust grains are of asteroidal origin and hence recorded by both experiments in agreement with the results from lunar microcrater studies. The situation outside the asteroid belt, however, may be explained in such a way that the registered dust grains which are on high inclination orbits (Humes, 1980) are most likely of cometary origin. From the Halley investigations it is known that cometary grains have low albedos (Keller *et al.*, 1986). With the Pioneer photometer it was not possible to distinguish the scattered sunlight of these particles from the integrated starlight background. Therefore, an explanation for the dust population outside the asteroid belt is that these dust grains are comparatively young cometary dust grains of extremely low albedos on 'cometary' orbits. For further information see Fechtig (1989).

Dust grains in interplanetary space are expected to have rather varied, and often quite complicated, dynamical histories. They will be emitted under gas pressure from comets or ejected as collisional debris from asteroids. Radiation pressure will reduce the solar gravitational force, and the orbits of the particles will spiral in towards the Sun under Poynting–Robertson and solar wind drag; these latter forces differ from particle to particle and depend on the cross-section-to-mass ratio of each individual grain. Mutual collisions between meteoroids will destroy some particles and create others from the collision fragments. Finally, the gravitational forces of the planets can have strong effects on the orbital evolutionary paths of dust grains on low inclination orbits. Numerical modeling by Jackson and Zook (1989) shows that the inner planets, and especially the Earth, are expected to trap dust grains of larger than 10 μm radius into orbits that are in orbital period resonance with the planet. They predict, in particular, that there is a heliocentric ring of dust near 1 AU that is due to dust grains trapped into exterior resonances with the Earth. The Galileo dust detector should be suited to confirm the existence, and explore the extent of this hypothetical ring if it is a significant feature of the interplanetary dust complex.

2.2. Jovian Dust Environment

Within the Jovian system dust plays a major role in several ways. The strong gravitational field of Jupiter locally increases the interplanetary dust flux considerably. Io's volcanic activity is most likely the source for tiny particles in the Jovian magnetosphere. The many Jovian satellites are potential dust sources for ejecta particles, which are produced by meteoroid impacts. The smaller the satellites, the more effective they are as sources, because ejecta particles can more easily escape from those smaller satellites. Finally the strong Jovian magnetosphere is a potential production source for fine dust by electrostatic fragmentation of larger bodies.

The first and so far only recordings of dust in the Jovian system were obtained by the meteoroid penetration detectors on board the Pioneer 10 and 11 spacecraft. Pioneer 10 registered 11 impacts of dust particles with masses $m \geq 2 \times 10^{-12}$ kg and Pioneer 11 registered 2 dust particles with masses $m \geq 2 \times 10^{-11}$ kg (Humes, 1976). The corresponding peak dust flux is 2 to 3 orders of magnitudes higher than the flux in the neighboring interplanetary space.

In the neighborhood of Jupiter one has to expect the above-mentioned sources for dust. The following sections discuss these sources closer.

Gravitational focusing of interplanetary meteoroids is expected due to velocity increase within the gravitational field of Jupiter. According to Öpik (1951) the spatial flux $F(R)$ at a distance R from a planet is given by

$$F(R) = N_0 u \left(1 + \frac{v_e^2(R)}{u^2}\right), \tag{1}$$

with N_0 is the spatial density of dust particles in interplanetary space; $v_e(R)$, escape speed from planet at distance R; u, unaccelerated speed of meteoroids relative to the planet.

Compared to the situation near the Earth where the HEOS-2 dust experiment (Hoffmann *et al.*, 1975) measured a gravitational increase of roughly a factor 2 one could expect an increase by a factor of 10 to 20 at Jupiter.

The Voyager spacecrafts discovered a tenuous ring around Jupiter between 1.72 and 1.81 R_J (R_J is the Jovian radius) (Owen *et al.*, 1979; Jewitt and Danielson, 1981). Showalter *et al.* (1985, 1987) confirm this result and report a toroidal halo at the ring's inner edge extending inwards roughly half way to the cloud tops with a vertical thickness of about 10^4 km. For further details of the ring and its halo see Burns *et al.* (1984) and Showalter *et al.* (1987). Since the small ring particles interact with the magnetospheric plasma (Grün *et al.*, 1984), they have a limited lifetime of approximately 10^3 years. They are also transported out of the ring by plasma drag on a similar time scale. Because of these destruction and dispersion mechanisms, dust particles in the visible ring have to be replenished on time scales of 10^3 years. As sources for micron-sized dust particles Morfill *et al.* (1980b) and Burns *et al.* (1980) proposed the numerous large parent bodies residing within the bright ring. Small ring particles are produced by impact erosion of these large bodies. The projectiles stem from the volcanos on Io (Morfill *et al.*, 1980b) or are interplanetary micrometeoroids (Burns *et al.*, 1980). A 'gossamer' ring has been discovered in the Voyager images (Showalter *et al.*, 1985) outward of the bright ring up to a distance of $\sim 3\ R_J$. It is quite possible that this tenuous ring extends all the way through the Jovian system.

The dust experiment on HEOS-2 (Hoffmann *et al.*, 1975) investigated the dust population around the Earth between 5000 and 240 000 km altitudes. Apart from the 'interplanetary' dust population, the experiment recorded a near-Earth increase of the dust flux. Parts of this increase – the so-called 'groups' – have been associated with lunar ejecta produced by meteoroid impacts on the surface of the Moon (Fechtig *et al.*, 1979). Although considerable doubts have been raised as to whether or not the Moon can lose material, Zook *et al.* (1984) have experimentally shown that oblique impact on the Moon produces large quantities of ejecta at velocities well above the lunar escape velocities.

Jupiter with its numerous satellites – many of them smaller than the Moon – may produce considerable numbers of dust particles which contribute to the several orders of magnitude increase in dust population recorded by the Pioneer 10/11 dust experiments.

Dust particles in interplanetary space or within a planetary magnetosphere carry electrical charges. There are two competing mechanisms which charge-up particulates: photoemission of electrons by absorption of solar ultraviolet radiation and capture of impinging electrons from the ambient plasma. In interplanetary space the photoeffect generally dominates and yields a surface potential $U \approx +3$ to $+6$ V (Rhee, 1967; Wyatt, 1969). This charge has some effects on the dynamics of the interplanetary dust cloud (Morfill *et al.*, 1986). In regions of planetary magnetosphere where the plasma density is sufficiently high dust particles are charged-up negatively to a potential which corresponds to the electron energy E_e (cf. Grün *et al.*, 1984):

$$U \approx -\frac{E_e}{e}, \tag{2}$$

where e is the electronic charge (1.6×10^{-19} C). The charge Q on a spherical particle is then given by

$$Q = 4\pi\varepsilon_0 U s, \tag{3}$$

with permittivity $\varepsilon_0 = 8.859 \times 10^{-12}$ C/Vm and grain radius s. Figure 2 shows dust charges and the Galileo dust detector's range of measurement.

The HEOS-2 dust experiment recorded so-called 'dust swarms' which have been interpreted as fragmentation products due to electrostatic disruption of solids within the Earth's auroral zones (Fechtig *et al.*, 1979). Figure 2 shows the limiting tensile strengths of the electrostatic disruption process. If such processes occur already within the Earth's magnetosphere one has to expect a much stronger effect within the Jovian magnetosphere. Depending on the location one might expect temporal and local increases up to 5 orders of magnitudes.

3. Instrument Description

3.1. Experimental approach

The Galileo Dust Detector System (DDS) detects individual particles impacting on the sensor and measures their mass, impact speed, electric charge, and determines the impact direction. The instrument consists of an impact ionization detector and the appropriate electronics. The impact ionization detector is similar to the HEOS-2 cosmic dust experiment, which was in Earth orbit from 1972 to 1974. The modifications to the previous experiment are (1) an increase in the sensitive area from ~ 0.01 m^2 to 0.10 m^2, (2) the addition of a measurement channel for the electrical charge of dust particles, and (3) the installation of an electron multiplier in order to obtain an additional, independent signal of dust particle impacts.

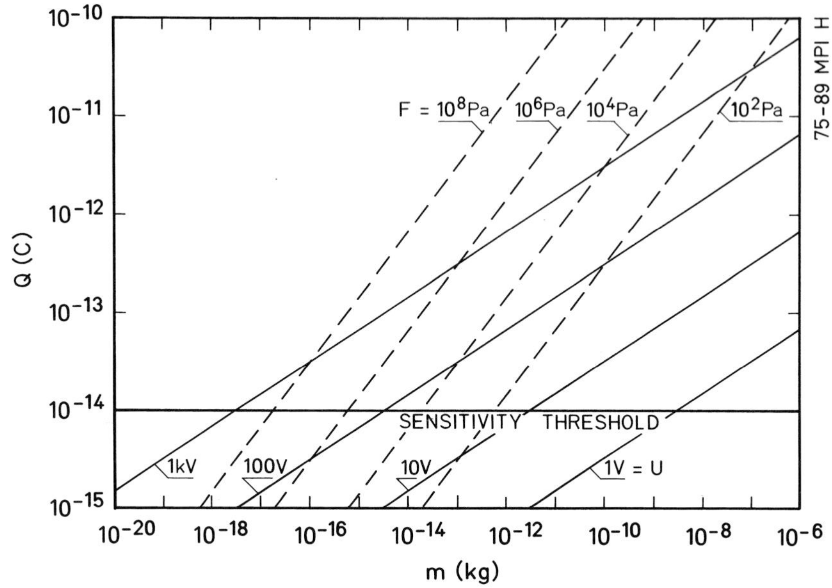

Fig. 2. Charges Q on dust particles as a function of their surface potential U (solid lines). In interplanetary space the surface potential is typically several volts positive, whereas in planetary magnetospheres it may reach a few kV negative. A particle density of $\rho = 1$ g cm^{-3} is assumed. If the electrostatic stress $F = \varepsilon_0 U/s$ (broken lines) exceeds the tensile strength of the material, the dust particle will be disrupted. Typical tensile strengths are 10^3 Pa for fluffy aggregates, 10^5–10^7 Pa for ice, and 10^6–10^8 Pa for silicates, respectively.

The basic sensor is shown in Figure 3. Positively or negatively charged particles entering the sensor are first detected via the charge which they induce to the charge grid while flying between the entrance and shield grids. The grids adjacent to the charge pick-up grid are kept at the same potential in order to minimize the susceptibility of the charge measurement to mechanical noise. All dust particles – charged or uncharged – are detected by the ionization they produce during the impact on the hemispherical impact sensor. After separation by an electric field, the ions and electrons of the plasma are accumulated by charge sensitive amplifiers (CSA), thus delivering two coincident pulses of opposite polarity. The rise times of the pulses, which are independent of the particle mass, decrease with increasing particle speed. From both the pulse heights and rise times, the mass and impact speed of the dust particle are derived by using empirical correlations between these four quantities. The thresholds and the dynamic ranges of the different measurements are given in Table III.

A third independent signal originates from part of the positive impact charge which is detected and amplified (approx. × 100) by an electron multiplier (channeltron). This signal serves as a control for the identification of dust impacts. The multiplier has a low dark current at the required amplification. Channeltron gain loss will be compensated by changes in the channeltron voltage, which will be effected by telecommand. An impact event just exceeding the detection threshold of the instrument (corresponding to

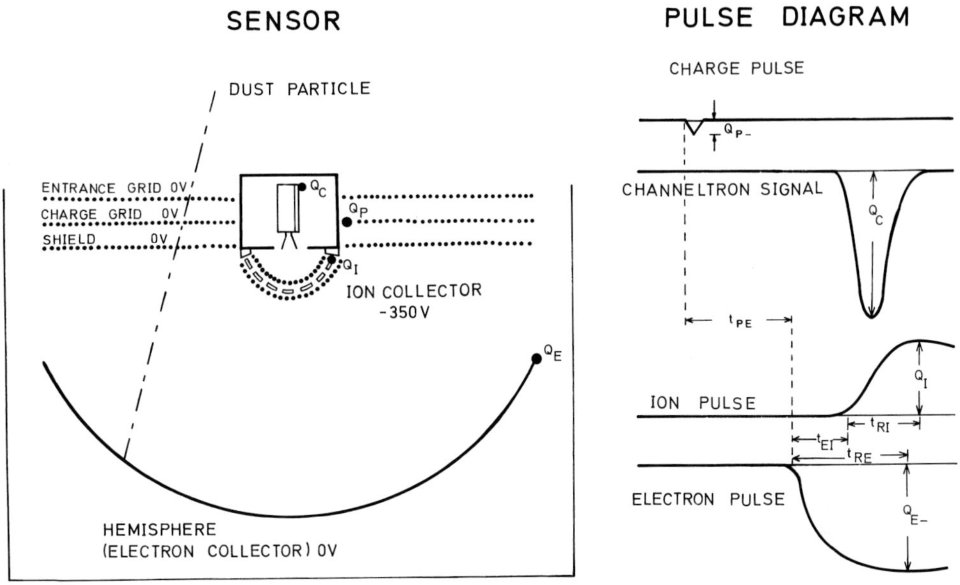

Fig. 3. Sensor configuration (schematic) and measured signals upon impact of a negatively charged dust particle.

10^{-14} C) will result in an output signal from the channeltron of $\sim 10^{-12}$ C. On one hand, noise events (microphonics or electromagnetic interferences) that exceed the CSA threshold will produce smaller output signal at the channeltron. On the other hand, energetic particles which may stimulate the channeltron are less effective at the channels which measure the charge by means of CSAs. Therefore one has an umambiguous method of distinguishing noise events from impact events by measuring the channeltron output signal and deriving coincidence signals between both CSAs and the channeltron. This information is used by the on-board processor to give higher priority to the transmission of data from impact events than to those from noise events (see below).

From Table III it can be seen, that the amplitude of the positive and negative charge pulses, the channeltron signal and the corresponding rise times are redundant measures of mass and impact speed of the dust particle. This redundancy gives a further check on the identification of an impact event and increases the accuracy of the measurement considerably. In order to further increase the reliability of impact identification coincidences between the different signals are derived and evaluated in an event classification scheme. The instrument and its operation are designed to reliably suppress noise and allow detection of as few as one impact event per month, the charge of which is close to the detection threshold. This is about a factor of 10 below the lowest expected dust particle flux.

TABLE III

Signals measured by the Galileo Dust Detector upon impact of a dust particle onto the sensor and related particle parameters

Signal designation	Measured quantity	Range	Accuracy (number of logarithmic steps)	Related particle parameters
Q_E	Negative charge generated upon impact (electrons)	10^{-14}–10^{-8} C	48	Mass, speed
Q_I	Positive charge generated upon impact (ions)	10^{-14}–10^{-8} C	48	Mass, speed
Q_C	Positive charge generated upon impact (partially)	10^{-13}–10^{-9} C (channeltron output)	32	Impact identfication (mass, speed)
Q_P	Induced charge positive	10^{-14}–10^{-12} C	16	Electric charge
	Induced charge negative	10^{-14}–10^{-10} C	32	
t_E	Rise time of negative charge	10–100 μs	16	Speed
t_I	Rise time of positive charge	10–100 μs	16	Speed
t_{EI}	Time difference negative positive charge signals	1–50 μs	16	Impact identification
t_{PE}	Time difference induced negative charge signals	1–400 μs	32	Speed

3.2. Angular sensitivity and sensor pointing

The geometric detection probability is defined by the sensitivity of the detector for particles impinging from different directions in an isotropic flux of particles. Directions are determined by an angle ϑ to the sensor axis.

The angular sensitivity was calculated by a computer simulation program that 'fired' a particle beam onto the sensor from different directions. The cross-section A of the beam, which hit the target, was calculated. This is predominantly a cosine function; for larger incidence angles ϑ it is modified by the shielding of the detector's side walls. The maximum area (0.10 m^2), of course, is found for $\vartheta = 0$ (Figure 4). Because of the axial symmetry of the detector, the solid angle interval is $d\Omega = 2\pi \sin\vartheta \, d\vartheta$. The relative sensitivity $I(\vartheta)$ is therefore given by

$$I(\vartheta)\, d\Omega = \frac{A(\vartheta)}{A(\vartheta = 0)}\, 2\pi \sin\vartheta \, d\vartheta \,. \tag{4}$$

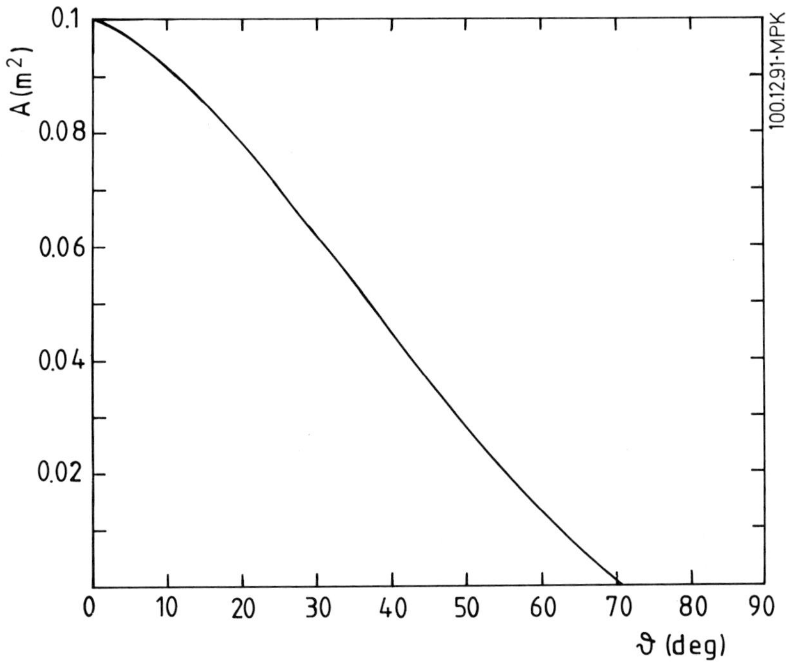

Fig. 4. Angular sensitivity. The sensitive area A is shown as a function of angle ϑ with respect to the sensor axis. In an isotropic flux, 50% of all particles will impact the sensor at $\vartheta < 32°$.

Integration of this function leads to the effective solid angle interval covered by the detector, which is 1.45 sr (cf. Göller and Grün, 1989).

The distribution of the impact angles onto the target was determined in a similar way. Impact angles were measured between particle direction and the normal of the target at the impact point. As the target was curved they were generally different from the incidence angles ϑ (measured to the sensor axis). The average impact angle was 36°. As the calibration differed slightly for different impact angles their distribution was used for finding the proper average calibration curve.

In addition the distribution of flight paths inside the detector was determined. Depending on incidence angle and point of impact at the target, the particles had different flight paths. This part of the calculation was done to determine the average pathlength and its deviations, to find a conversion from time of flight (t_{PE}) to particle speed. The average flight path of a particle in the detector was 20 cm ± 5 cm. The uncertainty given refers to 68% of all trajectories.

The sensor is mounted to the science boom on the spinning section of the spacecraft. For reasons of the experiment's simplicity and reliability, as well as for keeping the weight low, the pointing direction of DDS was fixed at a single angle with respect to the spacecraft spin axis. The chosen angle of 55° with respect to the positive spin axis (z-axis of the spacecraft) appeared to provide optimal conditions for detecting, and then analyzing a variety of known and potential cosmic dust families and their trajectories, especially in the Jovian system.

3.3. Technical Description

The whole instrument comprises a sensor with channeltron and pre-amplifiers, signal conditioning, and spacecraft interface electronics. The latter two are separated from the sensor and pre-amplifiers, and included in the electronics box (Figure 5).

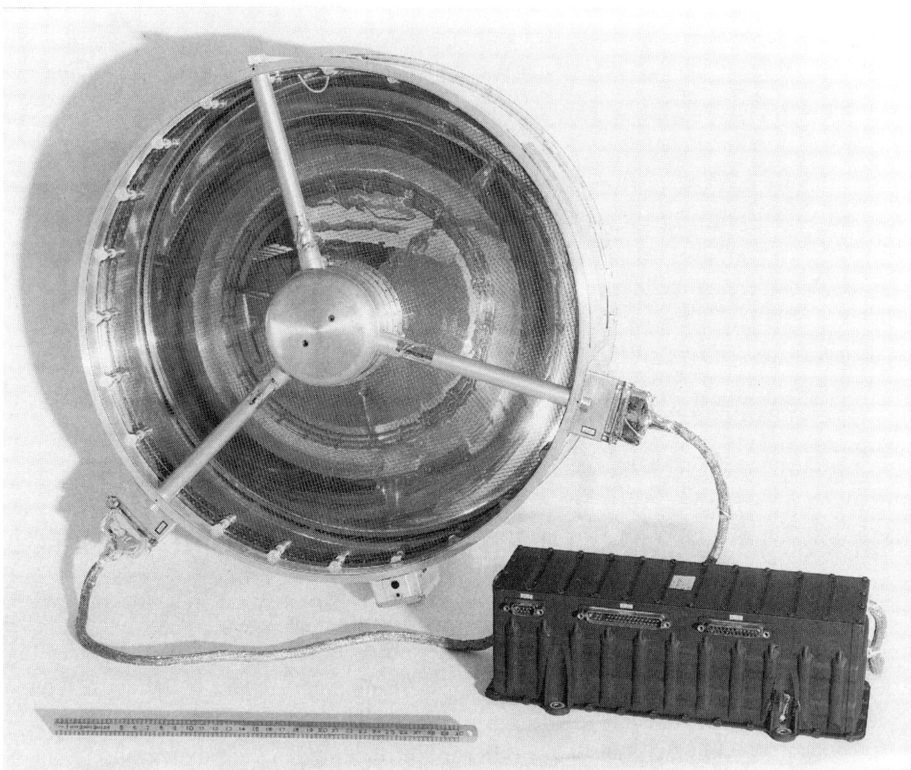

Fig. 5. Sensor and electronics box of the Galileo Dust Detector.

The sensor consists of a grid system for the measurement of the particle charge, an electrically grounded target (hemisphere) and a negatively biased ion collector (Figure 3). A charged dust particle entering the sensor will induce a charge to the charge grid, which is connected to a charge sensitive amplifier. The output voltage of this amplifier rises until the particle passes this grid, and falls off to zero when it reaches the shield grid. The peak value (Q_P) is stored for a maximum of 600 μs and is only processed if an impact is detected by the impact ionization detector within this time. A dust particle hitting the hemispherical target produces electrons and ions, which are separated by the electric field between hemisphere and ion collector into negative charges (electrons and negative ions) and positive ions. The negative charges are collected at the hemisphere and measured by a charge sensitive amplifier (Q_E). Positive ions are collected and measured at the negatively biased ion collector with a charge sensitive amplifier (Q_I). Some of the ions penetrate the ion collector, which is partly transparent (total transmission approximately 40%), are further accelerated, and hit the entrance cone of an electron multiplier (channeltron). Secondary electrons are produced, amplified, and measured by a charge sensitive amplifier (Q_C). Other quantities measured, are the rise times (t_I, t_E) of both the positive and negative charge pulses (Q_I and Q_E). The measurement of the time delay (t_{EI}) between electron pulse and ion pulse serves as a means for distinguishing impact events from noise. Impact events have time delays of 2–50 μs, while mechanical noise has a time delay of milliseconds. These signal amplitudes and times of a single recorded event are digitized and stored in an Experiment Data Frame (EDF).

A measurement cycle is initiated if either the negative charge Q_E on the hemispherical target, or the positive charge on the ion-collector Q_I, or Q_C exceeds a threshold. Since the hemisphere has a large area which is directly exposed to interplanetary plasma and high-energy radiation, we expect that this may cause some interferences for the Q_E measurement. To avoid these interferences during high activity times, it is possible to switch by command to a mode in which a measurement cycle is initiated if only the charge on the ion collector Q_I (small area and not directly exposed) or channeltron signal Q_C exceeds the threshold. If more than one event occur within the transmission time of one EDF, then these events are counted by several amplitude-dependent counters. The dead-time caused by the measurement cycle is 5 ms.

Figure 6 shows a block diagram of the instrument. The signals from the sensor are conditioned and analysed. The microprocessor coordinates the experiment measurement cycle, collects the buffered measurement data and processes the data according to a program stored in the memory.

Each event recorded by the instrument is classified according to its ion signal amplitude into 6 amplitude intervals of about one decade width. In addition each event is categorized by one out of four event classes. Class 0 are all events (noise events and some unusual impact events – e.g., hits of the sensor's internal structure) which are not categorized in a higher class. For class 1 through class 3 events the measured parameters and their relation to each other are increasingly restricted, so that class 2 represents 'potential' and class 3 represents 'probable' dust impact events. Data from events of all classes are buffered in the *A*-range of the instrument memory. This range contains complete data sets from 6 events of different ion signal amplitudes. In addition up to eight class 3 events are stored in the *E*-range of the instrument memory. This classification scheme is initially optimized to laboratory noise conditions but it is also adaptable to the in-flight noise situation by changing the classification parameters (set points) or by adjusting the onboard classification program by commands. Detailed information on noise is mandatory in order to evaluate the reliability of impact detection as well as to determine the effect on the dead-time of the instrument.

Each event is counted in one out of 24 counters according to its ion amplitude and

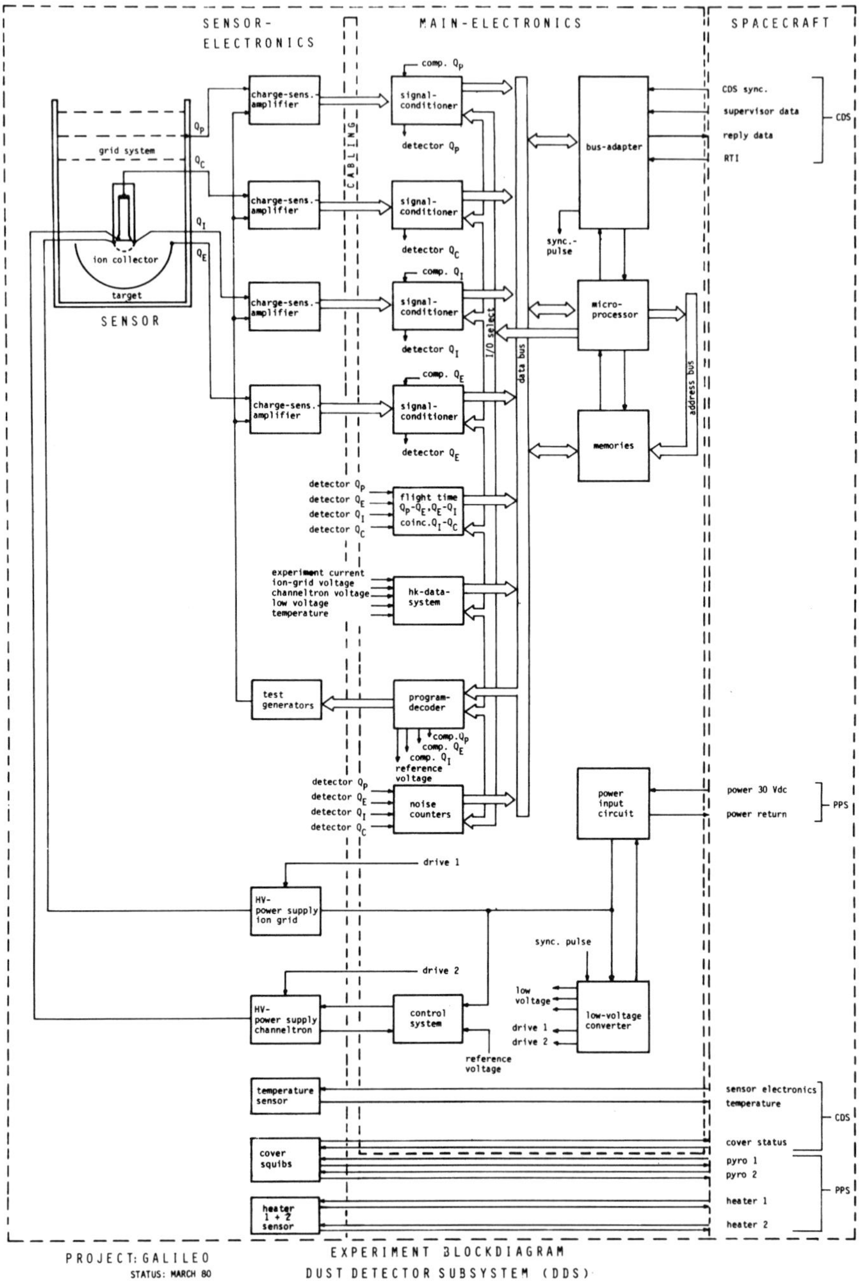

Fig. 6. Functional block diagram of the Galileo Dust Detector.

event class. The scheme of the 24 accumulators (ACs) is shown in Table IV. Fourteen EDFs (each containing the full information on a single event and the corresponding accumulator contents) are stored in the memory and are shifted sequentially through the

TABLE IV
Scheme of DDS accumulators

Class number		C0	C1	C2	C3
Amplitude range	A1	AC01	AC11	AC21	AC31
	A2	AC02	AC12	AC22	AC32
	A3	AC03	AC13	AC23	AC33
	A4	AC04	AC14	AC24	AC34
	A5	AC05	AC15	AC25	AC35
	A6	AC06	AC16	AC26	AC36

bus adapter to the spacecraft. The exchange of data with the spacecraft and instrument modes are controlled by the microprocessor. An internal test generator stimulates and calibrates the electronics periodically or by command and generates 'test pulse-data'.

3.4. Instrument Data

In this paragraph DDS data, its data transmission modes and its initial on-ground processing steps are described. All information on events (dust impacts or noise) recorded by the instrument is contained in data packages of 16 bytes (1 byte = 8 bits) of science data which are supplemented by 10 bytes of engineering housekeeping data (Table V). DDS data are generally transmitted within the Low Rate Science (LRS) mode.

One EDF consists of 26 bytes and is read out from the instrument to the spacecraft within 13 Low Rate Science frames (LRS) and with 2 bytes of data contained in each LRS. Apart from science data there are also other types of DDS data which are transmitted (cf. Table V), depending on the mode selected:

– Mode 1 is the science collection mode, the first 16 bytes of data contain science data and the next 10 bytes housekeeping information,

– Mode 2 is the memory read-out mode, in this mode the first 2 bytes contain the 16 bits starting address of the memory readout. The next 13 bytes of data contain readout data from any part of the instrument memory (ROM or RAM, depending on the starting address). The 10 bytes housekeeping data and status information are the same as describe in mode 1, and

– Mode 3 is the instrument set point mode; two packets of DDS data will contain the actual set point data information.

In addition to mode 2, DDS has another capability to read out the memory data by using the direct memory transfer to the spacecraft. In this case it is possible to read out the memory data in packets of 32 bytes.

During normal Low Rate Science data transmission one EDF is sent every 8.7 s, which corresponds to an average DDS data transmission rate of 24 bits s^{-1}. During interplanetary cruise and during orbital cruise around Jupiter one or two passes (of about 8 hr duration) of LRS data per week will be transmitted. This rate corresponds to an average DDS data rate of 1–2 bits s^{-1}. During the first phase of the mission, i.e.,

TABLE V

Data of the Galileo Dust Detector. One Experiment Data Frame (EDF) contains 208 bits: 128 bits of science data recorded for one event and 80 bits of housekeeping data. In special modes the memory content or instrument set points are read-out.

Name	Bits	Description
		Science data
CA	5	Amplitude of channeltron signal Q_C
CCP	2	Channeltron (Q_C) threshold setting
CL0	8	Class 0 counter, 6 amplitude ranges
CL1	8	Class 1 counter, 6 amplitude ranges
CL2	8	Class 2 counter, 6 amplitude ranges
CL3	8	Class 3 counter, 6 amplitude ranges
CLN	2	Class number of transmitted event
CN	4	Channeltron noise counter, which counts all threshold exceedings. The counter is reset every 81 ms
EA	6	Amplitude of negative charge Q_E measured at the target
ECP	2	Negative charge (Q_E) threshold setting
EIC	1	Coincidence between negative and positive charge signals: if both pulses exceed the thresholds in a time interval of 200 ns
EIT	4	Time t_{EI} measured between negative and ion charge signals
EN	4	Negative charge noise counter, which counts all threshold exceedings. The counter is reset every 41 ms
ERS	3	*E*-range frame number
ET	4	Rise time t_E of negative charge signal
EVD	3	Event definition status, indicates the actual setting for the initiation of data acquisition by one or more measuring channels (negative charge-, ion- or channeltron signal)
FN	3	Data frame number, indicator for EDF content: *A*-range or *E*-range science data, set point data 1 or 2, commanded or automatic testpulse data or memory content
IA	6	Amplitude of ion charge Q_I measured at the ion grid
ICC	1	Coincidence between ion and channeltron signals: if both pulses exceed the thresholds in a time interval of 20 μs
ICP	2	Ion charge (Q_I) threshold setting
IN	4	Ion noise counter, which counts all threshold exceedings. The counter is reset every 81 ms
IT	4	Rise time t_I of ion charge signal
PA	6	Amplitude of induced charge Q_P, most significant bit indicates polarity
PCP	2	Induced charge (Q_P) threshold setting
PET	5	Time difference t_{PE} between induced and negative charge signals
PN	4	Induced charge noise counter, which counts all threshold exceedings. The counter is reset every 10 ms
SEC	8	Spacecraft sector at time of event
SES	1	Sector data status, sector data valid or not valid
THS	1	Threshold status, indicator bit for threshold settings by command or automatically by each individual noise counter
TIME	8	Spacecraft event time; short time interval: 121.33 s, 1.33 s step size; long time interval: 138 hr 32.35 min step size
TS	1	Transmit status of EDF: first time or repeated transmission

Table V (continued)

Name	Bits	Description
		Housekeeping data
ADD–HK	2	Address counter, determines whether the transmitted byte contains the sub-commutated word PARN, CMD–ACC, or CMD–REJ
CMD–ACC	8	Command accepted, counts all commands which are identified as DDS commands
CMD–REJ	8	Command rejected, counts all unrecognized commands
CMP–S	8	Computer status, contains information about computers and memories and a 1-bit indicator for the spacecraft event time interval
CUR	8	Experiment current, range 15 to 100 mA
HVC	8	Channeltron high voltage, range 0 to 2500 V
HVI	8	Ion grid high voltage, range 0 to − 512 V
PARN	8	Parity error number, counts all supervisory data parity errors
SYNC	6	Synchronization word
TEMP	8	Main electronics box temperature, range − 30° to + 80 °C
VA +	8	Low voltage + 7.5 V, range 0 to 10 V
VA −	8	Low voltage − 7.5 V, range 0 to − 10 V
VD +	8	Low voltage + 10 V, range 0 to 15 V
		Memory readout and Set Point data
MHAD	8	Memory high address
MLAD	8	Memory low address
MEMC	8	Memory content
SP1 to SP18	8	Instrument set point data, 18 words

to Venus and back to Earth, when the high gain antenna cannot be used, DDS data are fed into the engineering data stream at an average rate of 0.01 bits s^{-1}. For this period *E*-range data (class 3 events) and accumulator data are transmitted with high priority.

Data received from DDS on ground will be initially evaluated according to the following scheme:

(1) Probable impact events, potential impact events, and noise data are extracted from the raw data and put into separate files. Count rates of probable events (class 3 counters) are accumulated over fixed time intervals and transferred to the Common Data File, which is shared among the Fields and Particles investigators.

(2) Statistical analysis of the noise data will result in a Dead-Time History File, with the help of which impact rates can be accurately determined.

(3) Identification of all dust impact data and calibration of this data will result in a Physical Parameter File, which then can be used for further analysis by the DDS investigator team.

3.5. Calibration

Extensive calibration tests have been performed with the Galileo Dust Detector (Göller and Grün, 1985, 1989) at the Heidelberg dust accelerator. Calibration tests were performed with iron, carbon, and silicate particles. The particles were in the speed range from 1 km s^{-1} to 70 km s^{-1} and in the mass range from 10^{-18} kg to 10^{-13} kg. In addition to the projectile material variation, we carried out calibration tests for iron particles, and varied the impact angles.

There are three possibilities to determine a particle's speed: the rise times t_I, t_E, and the ratio Q_C/Q_I. Figure 7 shows the dependency of the rise time of the ion charge signal

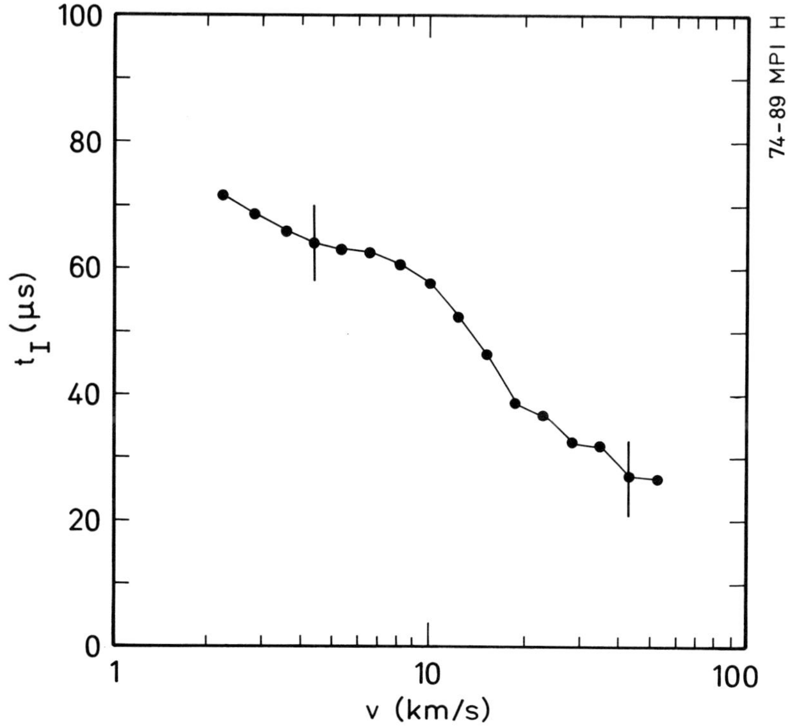

Fig. 7. Rise time t_I of the ion charge signal as a function of the impact speed v for iron particles (from Göller, 1988).

as a function of the impact speed. If we use all three measurements the speed can be determined with an accuracy of a factor of 1.6. Using only one measurement the accuracy is about a factor 2.

When the particle speed is known, the mass can be determined from the charge yields Q_I/m and Q_E/m. If the speed is known within a factor of 1.6 and both yields are used for mass determinations the value can be measured with an uncertainty of a factor of 6. The main part of this error is caused by the limited accuracy of the speed measurement.

From the charge yields the sensitivity of the sensor can be calculated. The negative charge pulses have higher yields than the positive pulses. This fact is mainly caused by some positive ions leaving the detector after impact. Thus they are not collected by the ion collector.

The speed-dependent measurable mass range is shown in Figure 8. Under flight conditions, especially if the noise level is high, the electronic detection threshold can be set to a higher value. In this case the sensitivity will be reduced. Since the charge sensitive amplifiers cover a range of six orders of magnitude, the upper limit for particle measurement is also indicated (saturation range). For larger particles the detector operates as a threshold detector. The calibration covered a speed interval ranging from about 2 km s^{-1} to 70 km s^{-1} (see Figure 8).

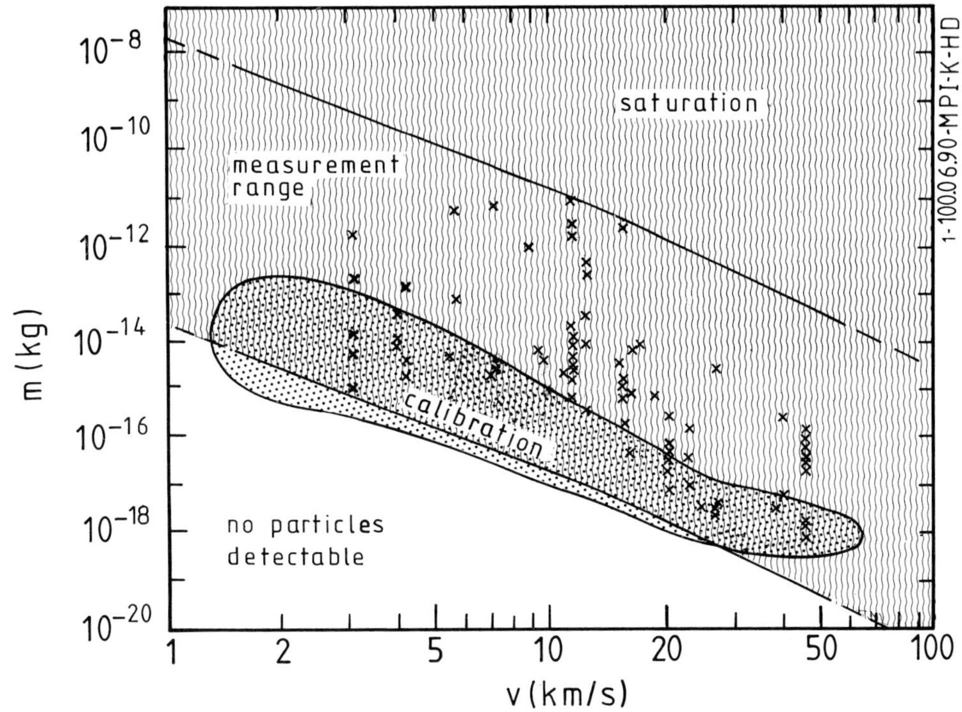

Fig. 8. Dust particle masses and speeds which can be measured by the Galileo Dust Detector. In the region marked 'saturation' the instrument operates as a threshold detector. The mass and speed range is indicated, in which the instrument was calibrated with silicate, carbon, and iron particles. Shown are also mass and speed values for the 81 discrete impacts detected in all but the lowest amplitude/class ranges until May 18, 1990. The accuracies are factors 2 and 10 for speed and mass values, respectively.

3.6. Measurement ranges and payload parameters

The impact rate will be measured from 3×10^{-7} s^{-1} to 10^2 s^{-1} in order to cover both the impact rates expected in interplanetary space and during times of high activity (e.g., in the Jovian magnetosphere).

The measurable particle mass ranges from 10^{-19} kg $< m < 10^{-13}$ kg at 40 km s^{-1} to 10^{-16} kg $< m < 10^{-10}$ kg at 5 km s^{-1} impact speed. The detectable impact speed is $v \geq 1$ km s^{-1}. The particles' electric charges are measured from 10^{-14} C to 10^{-10} C

for negative charges and 10^{-14} C to 10^{-12} C for positive charges. The maximum angle of incidence for dust particles with respect to the sensor axis will be 70°.

The volumetric envelope of the sensor is cylindrical: 442 mm in diameter and 301 mm long (without cover). The electronics box has a volume of 283 × 100 × 100 mm^3. The total weight of the instrument is 4.2 kg and its power consumption is 2.4 W plus up to 3.0 W for heaters. The average data transmission rate is 24 bit s^{-1} in spacecraft tracking mode. The operating temperature limits are −20 °C to +40 °C.

4. Initial Performance

The following report provides a summary of the performance of the Galileo DDS instrument as observed during the 4-day science check-out period from December 27 to 30, 1989 (during this period about 8 hours per day of LRS data was available) and includes data from 26 instruments memory read-outs until May 18, 1990.

On December 27 the DDS cover was successfully deployed as indicated by the microswitch status change and the sensor temperature decrease which was as expected. On December 28 DDS was switched on for the first time. All functions were successfully checked-out. When the instrument was switched to the measuring mode, low amplitude noise on three (EA, CA, and PA) out of four charge measuring channel was detected. Such noise has never been observed during ground tests. An internal or external noise source could not be identified during the 4-day check-out and subsequent periods. By help of real-time commanding DDS was brought into a state where it was immune against the types of noise encountered initially.

Around Venus encounter high noise rate was detected also at the ion channel (IA). This noise sems to have caused increased dead-time starting at least February 8, 1990. By commanding (on February 16, 1990) the detection threshold was increased to step one (IA ≥ 3) and the noise rate was reduced to very low values thereafter.

Noise events have been observed only in amplitude/class ranges AC01, AC11, and AC02 (cf. Table IV), although some impacts of these types have been identified as well in the discrete data. Only impact events were recorded by all other accumulators. Therefore, the contents of AC01, AC11, and AC02 are used for determining the maximum noise rate, whereas all other accumulators are added-up in order to determine the impact rate.

A summary of the discretely recorded dust impacts is displayed in Figure 9. The positions of Galileo, Venus, and Earth for May 18, 1990 are indicated. Recorded dust impacts are overlaid on Galileo's orbit. The impact direct direction distribution shows the predominance of impacts from the apex direction, an effect which was observed also by previous dust detectors (Berg and Grün, 1973; Grün *et al.*, 1985a; Hoffmann *et al.*, 1975). Gaps apparent in the distribution of impacts, are due to two about 4-weeks long gaps in the data transmission during which an overflow in the internal memory assigned for discrete impact data occured.

Preliminary particle parameters have been calculated for the recorded discrete impacts. The physical parameters are based on IA-, EA-, IT-, and ET-data only and

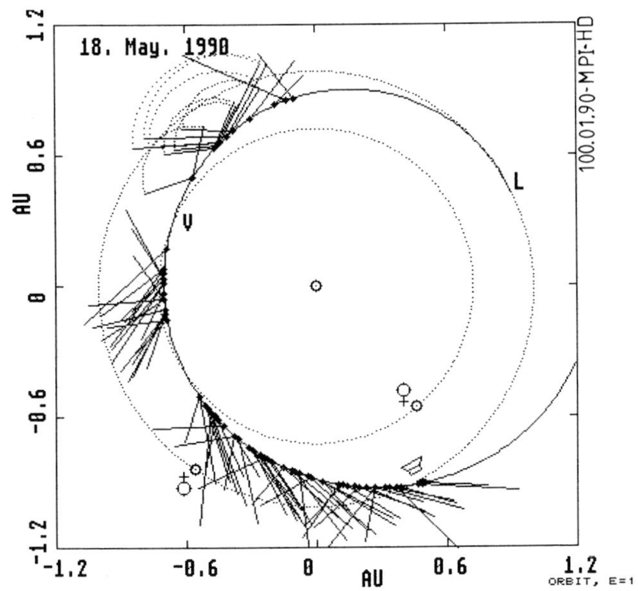

Fig. 9. Orbit of the Galileo spacecraft and impacts recorded until May 18, 1990. The positions for this date of Galileo, Venus and Earth are indicated. Dust impacts (dots) are overlaid on Galileo's orbit. Ecliptic projections of the sensor pointings are indicated by attached bars. No pointing information was obtained for impacts Nos. 7 to 13, here the full range of sensor pointings is shown. The length of the bars represents the magnitude of the impact charge (IA).

on calibration with iron particles. CA-information has not yet been used because we do not have accurately determined the channeltron amplification yet. An effect of the low channeltron voltage is the increased amplitude range in which class number 0 events occur. Therefore, small events (IA ≤ 15) are not complete.

During the first 141 days of DDS operation, the dust experiment has detected and accumulated at least 168 impacts in all but the lowest amplitude/class ranges. Eighty-one of these dust particles have been recorded by measuring the individual masses and speeds relative to the sensor. Preliminary impact speeds are determined sofar only in the range 3 to 50 km s^{-1}. Figure 8 shows the masses and speeds of the recorded dust particles in relation to the measurement range. The accuracy of the speed is about a factor two and that of the masses is about an order of magnitude. In the lower $\frac{1}{3}$ of the measurement range (about 2 orders of magnitude in mass) the true particle distribution is significantly under-represented because of instrumental limitations. No electric charge carried by dust particles has been identified sofar.

From the accumulated impact events it was possible to calculate total fluxes. For comparison, three flux values are given: (a) flux Φ between 0.875 AU (DDS switch-on) and perihelion (0.698 AU), (b) between perihelion and 0.875 AU, and (c) between 0.875 AU and 1.035 AU. These fluxes are (refering to 1.45 sr effective solid angle):

(a) $\Phi = 8.5 \times 10^{-5}$ m^{-2} s^{-1} between 0.875 AU and perihelion.

(b) $\Phi = 2.1 \times 10^{-4}$ m^{-2} s^{-1} between perihelion and 0.875 AU.

(c) $\Phi = 1.3 \times 10^{-4}$ m^{-2} s^{-1} between 0.875 and 1.035 AU.

It is remarkable that the inbound and outbound fluxes between 0.875 AU and perihelion are almost a factor 2.5 different from each other. This has to do with the orientation of the viewing cone of the dust sensor relative to the orbit of the spacecraft. In the outbound section the fluxes are comparatively higher because the spacecraft motion is outward towards the hemisphere the sensor is facing (cf. Figure 9). The decrease of the flux around 1 AU in the third section in comparison with the flux close to the Sun is due to the decrease of the dust population with increasing distance from the Sun (in qualitative agreement with relevant from the Helios dust experiment, Grün *et al.*, 1985a).

As shown in Figure 8 the recorded dust grains are generally between 10^{-18} kg and 10^{-11} kg. However, the mass threshold depends on the relative speed of the dust particles and the dust sensor. 10^{-18} kg is correct at a speed of approximately 30 km s^{-1}, but for 3 km s^{-1} the smallest mass observed is only 10^{-15} kg. The mean mass threshold of the reported flux is about 10^{-17} kg (about 0.3 μm diameter) and corresponds to a speed of roughly 20 km s^{-1}. These flux values are comparable to the in-ecliptic flux observed at 1 AU (1.1×10^{-4} m^{-2} s^{-1} for $m \geq 10^{-17}$ kg and 1.4 sr; Grün *et al.*, 1985b). Conclusions concerning the orbit distribution of the interplanetary dust particles have to wait for further analysis.

Acknowledgements

The development, implementation, and testing of this instrument and its data processing system was only made possible by the efforts of a great number of individuals from various departments of the Max-Planck-Institut für Kernphysik and from outside institutions and companies. The authors are especially indebted to G. Baust, H. Geldner, J. R. Göller, R. Hofacker, O. Kress, G. Linkert, G. Matt, C. Nehls, D. Roemer, H. Sallmutter, G. Schäfer, W. Schneider, N. Siddique, O. Titze, and A. Zahlten for their personel involvement during that campaign. Important contributions came from P. Gammelin, F. Eckl, and D. Semder of ARGE PEES, W. Fischer and G. Pahl of Walter Fischer Gerätebau, and A. Kaiser of Ingenieurbüro Kaiser. Careful attention and effective coordination and support has been provided by W. Kempe of DLR and by M. Agabra, C. J. Coppock, W. Fawcett, G. McSmith, C. Polanskey, and the Galileo Instrument Team of JPL. We are grateful for the support received from the Galileo project scientist T. V. Johnson. We thank E. Schwarzenbacher for the careful preparation of this manuscript. This work has been supported by the Bundesminister für Forschung and Technologie, under the grands 01 QJ 087-WRF 7701 and 01 QJ 900 400.

References

Acuña, M. H. and Ness, N. F.: 1967, 'The Complex Main Magnetic Field of Jupiter', *J. Geophys. Res.* **81**, 2917.

Berg, O. E. and Grün, E.: 1973, in 'Evidence of Hyperbolic Cosmic Dust Particles', *Space Research XIII*, Akademie Verlag, Berlin, pp. 1047–1055.

Brownlee, D. E., Hörz, F., Vedder, J. F., Gault, D. E., and Hartung, J. B.: 1973, 'Some Physical Parameters of Micrometeoroids', *Proc. Lunar Sci. Conf.* **4**, 3197.

Burns, J. A., Showalter, M. R., Cuzzi, J. N., and Pollack, J. B.: 1980, 'Physical Processes in Jupiter's Ring: Clues to Its Origin by Jove!', *Icarus* **44**, 339.

Burns, J. A., Showalter, M. R., and Morfill, G. E.: 1984, in R. Greenberg and A. Brahic (eds.), 'The Ethereal Rings of Jupiter and Saturn', *Planetary Rings*, University of Arizona Press, Tucson, pp. 200–272.

Dietzel, H., Eichhorn, G., Fechtig, H., Grün, E., Hoffmann, H.-J., and Kissel, J.: 1973, 'The HEOS 2 and Helios Micrometeoroid Experiments', *J. Phys. (E) Sci. Instr.* **6**, 209.

Fechtig, H.: 1989, 'Dust in the Solar System', *Z. Naturforsch.* **44a**, 877.

Fechtig, H., Gentner, W., Hartung, J. B., Nagel, K., Neukum, G., Schneider, E., and Storzer, D.: 1975, 'Microcraters on Lunar Samples', *Proc. Soviet–American Conference on Cosmochemistry*, Moscow, pp. 453–472 (translation: NASA SP-370, pp. 585–603).

Fechtig, H., Grün, E., and Kissel, J.: 1978, in J. A. M. McDonnell (ed.), 'Laboratory Simulation', *Cosmic Dust*, Wiley and Sons, Chichester, pp. 607–669.

Fechtig, H., Grün, E., and Morfill, G.: 1979, 'Micrometeoroids within Ten Earth Radii', *Planetary Space Sci.* **27**, 511.

Fillius, W.: 1976, in T. Gehrels (ed.), 'The Trapped Radiation Belts of Jupiter', *Jupiter*, University of Arizona Press, Tucson, pp. 896–927.

Friichtenicht, J. F. and Slattery, J. C.: 1963, *Ionization Associated with Hypervelocity Impact*, NASA Technical Note D-2091.

Göller, J. R.: 1988, 'Kalibrationsmesungen an Mikrometeoriten für die Missionen Ulysses, Galileo und Giotto', Ph.D. dissertation, University of Heidelberg, Heidelberg.

Göller, J. R. and Grün, E.: 1985, in R. H. Giese and P. Lamy (eds.), 'Calibration of the Galileo/ISPM Dust Detectors with Iron Particles', *Properties and Interactions of Interplanetary Dust*, D. Reidel Publ. Co., Dordrecht, Holland, pp. 113–115.

Göller, J. R. and Grün, E.: 1989, 'Calibration of the Galileo/Ulysses Dust Detectors with Different Projectile Materials and at Varying Impact Angles', *Planetary Space Sci.* **37**, 1197.

Grün, E., Berg, O. E., and Dohnanyi, J. S.: 1973, in M. J. Rycroft and S. K. Runcorn (eds.), 'Reliability of Cosmic Dust Data from Pioneers 8 and 9', *Space Research XIII*, Akademie Verlag, Berlin, pp. 1057–1062.

Grün, E., Morfill, G., Schwenn, G., and Johnson, T. V.: 1980, 'A Model of the Origin of the Jovian Ring', *Icarus* **44**, 326.

Grün, E., Fechtig, H., Giese, R. H., Kissel, J., Linkert, D., McDonnell, J. A. M., Morfill, G. E., Schwehm, G., and Zook, H. A.: 1983, in K. P. Wenzel, R. G. Marsden, and B. Battrick (eds.), 'The ISPM Dust Experiment', *The International Solar Polar Mission – Its Scientific Investigations*, ESA SP-1050, pp. 227–241.

Grün, E., Morfill, G. E., and Mendis, D. A.: 1984, in R. Greenberg and A. Brahic (eds.), 'Dust-Magnetosphere Interactions', *Planetary Rings*, University of Arizona Press, Tucson, pp. 275–332.

Grün, E., Fechtig, H., and Kissel, J.: 1985a, in R. H. Giese and P. Lamy (eds.), 'Orbits of Interplanetary Dust Particles Inside 1 AU as Observed by Helios', *Properties and Interactions of Interplanetary Dust*, D. Reidel Publ. Co., Dordrecht, Holland, pp. 105–111.

Grün, E., Zook, H. A., Fechtig, H., and Giese, R. H.: 1985b, 'Collisional Balance of the Meteoritic Complex', *Icarus* **62**, 244.

Hanner, M. S., Sparrow, J. G., Weinberg, J. L., and Beeson, D. E.: 1976, in 'Pioneer 10 Observations of Zodiacal Light Brightness Near the Ecliptic: Changes with Heliocentric Distances', *Interplanetary Dust and Zodiacal Light, Lecture Notes in Physics* **48**, 29.

Hauser, M. G., Gillett, F. C., Low, F. J., Gautier, T. N., Beichmann, C. A., Neugebauer, G., Aumann, H. H., Band, B., Boggess, N., Emerson, J. P., Houck, J. R., Soifer, B. T., and Walker, R. G.: 1984, 'IRAS Observations of the Diffuse Infrared Background', *Astrophys. J.* **278**, L15.

Hoffmann, H.-J., Fechtig, H., Grün, E., and Kissel, J.: 1975, 'Temporal Fluctuation and Anistropy of the Micrometeoroid Flux in the Earth–Moon System', *Planetary Space Sci.* **23**, 985.

Hoffmann, H.-J., Fechtig, H., Grün, E., and Kissel, J.: 1976, in B. Donn *et al.* (eds.), 'Particles from Comet Kohoutek Detected by the Micrometeoroid Experiment of HEOS 2', *The Study of Comets*, NASA SP-393, pp. 949–961.

Humes, D. H.: 1976, in T. Gehrels (ed.), 'The Jovian Meteoroid Environment', *Jupiter*, University of Arizona Press, Tucson, pp. 1052–1067.

Humes, D. H.: 1980, 'Results of Pioneer 10 and 11 Meteoroid Experiments: Interplanetary and Near-Saturn', *J. Geophys. Res.* **85**, 5841.

Humes, D. H., Alvarez, J. M., O'Neal, R. L., and Kinard, W. H.: 1974, 'The Interplanetary and Near-Jupiter Meteoroid Environment', *J. Geophys. Res.* **79**, 3677.

Jackson, A. A. and Zook, H. A.: 1989, 'A Solar System Dust Ring with the Earth as Its Shepherd', *Nature* **337**, 629.

Jewitt, D. C. and Danielson, G. E.: 1981, 'The Jovian Ring', *J. Geophys. Res.* **86**, 8691.

Keller, H. U., Arpigny, C., Barbieri, C., Bonnet, R. M., Cazes, S., Coradini, M., Cosmovici, C. B., Delamare, W. A., Huebner, W. F., Hughes, D. W., Jamar, C., Malaise, D., Reitsema, H. J., Schmidt, H. U., Schmidt, W. K. H., Seige, P., Whipple, F. L., and Wilhelm, K.: 1986, 'First Halley Multicolour Camera Imaging Result from Giotto', *Nature* **321**, 320.

Leinert, C., Richter, I., Pitz, E., and Planck, B.: 1981, 'The Zodiacal Light from 1.0 to 0.3 AU as Observed by the Helios Space Probes', *Astron. Astrophys.* **103**, 177.

Leinert, C. and Grün, E.: 1990, in R. Schwenn and E. Marsch (eds.), 'Interplanetary Dust', *Physics of the Inner Heliosphere*, in the series *Physics and Chemistry in Space*, Springer-Verlag, Heidelberg, pp. 207–275.

McDonnell, J. A. M. (ed.): 1978, in 'Microparticle Studies by Space Instrumentation', *Cosmic Dust*, Wiley and Sons, Chichester, pp. 337–426.

Morfill, G. E., Grün, E., and Johnson, T. V.: 1980a, 'Dust in Jupiter's Magnetosphere: Physical Processes', *Planetary Space Sci.* **28**, 1087.

Morfill, G. E., Grün, E., and Johnson, T. V.: 1980b, 'Dust in Jupiter's Magnetosphere: Origin of the Ring', *Planetary Space Sci.* **28**, 1101.

Morfill, G. E., Grün, E., and Leinert, C.: 1986, in R. G. Marsden (ed.), 'The Interaction of Solid Particles with the Interplanetary Mediu', *The Sun and the Heliosphere in Three Dimensions*, D. Reidel Publ. Co., Dordrecht, Holland, pp. 455–474.

Nagel, K., Neukum, G., Eichhorn, G., Fechtig, H., Müller, O., and Schneider, E.: 1975, 'Dependencies of Microcrater Formation on Impact Parameters', *Proc. Lunar Sci. Conf.* **6**, 3417.

Nagel, K., Neukum, G., Dohnanyi, J. S., Fechtig, H., and Gentner, W.: 1976a, 'Density and Chemistry of Interplanetary Dust Particles Derived from Measurements of Lunar Microraters', *Proc. Lunar Sci. Conf.* **7**, 1021.

Nagel, K., Neukum, G., Fechtig, H., and Gentner, W.: 1976b, 'Density and Composition of Interplanetary Dust Particles', *Earth Planetary Sci. Letters* **30**, 234.

Öpik, E.: 1951, 'Collision Probabilities with the Planets and the Distribution of Interplanetary Matter', *Proc. Roy. Irish Acad.* **54A**, 165.

Owen, T., Danielson, G. E., Cook, A. F., Hansen, C., Hall, V. L., and Duxbury, T. C.: 1979, 'Jupiter's Rings', *Nature* **281**, 442.

Rhee, J. W.: 1967, in J. C. Weinberg (ed.), 'Electrostatic Potential of a Cosmic Dust Particle', *The Zodiacal Light and the Interplanetary Medium*, NASA SP-150, pp. 291–297.

Showalter, M. R., Cuzzi, J. N., and Pollack, J. B.: 1985, 'Discovery of Jupiter's Gossamer Ring', *Nature* **316**, 526.

Showalter, M. R., Burns, J. A., Cuzzi, J. N., and Pollack, J. B.: 'Jupiter's Ring System: New Results on Structure and Particle Properties', *Icarus* **69**, 458.

Smith, B. A., Soderblom, L. A., Beebe, R. F., Boyce, J., Briggs, G. A., Carr, M. H., Collins, S. A., Cook, A. F., II, Danielson, G. E., Davies, M. E., Hunt, G. E., Ingersoll, A. P., Johnson, T. V., Masursky, H., McCauley, J. F., Morrison, D., Owen, T., Sagan, C., Shoemaker, E. M., Strom, R. G., Suomi, V. E., and Veverka, J.: 1979a, 'The Galilean Satellites and Jupiter: Voyager 2 Imaging Science Results', *Science* **206**, 927.

Smith, B. A., Soderblom, L. A., Johnson, T. V., Ingersoll, A. P., Shoemaker, E. M., Hunt, G. E., Masursky, H., Carr, M. H., Davies, M. E., Cook, A. F., II, Boyce, J., Danielson, G. E., Owen, T., Sagan, C., Beebe, R. F., Veverka, J., Strom, R. G., McCauley, J. F., Morrison, D., Briggs, G. A., and Suomi, V. E.: 1979b, 'The Jupiter System through the Eyes of Voyager 1', *Science* **204**, 951.

Wyatt, S. P.: 1969, 'The Electrostatic Charge of Interplanetar Grains', *Planetary Space Sci.* **17**, 155.

Zook, H. A. and Berg, O. E.: 1975, 'A Source for Hyperbolic Cosmic Dust Particles', *Planetary Space Sci.* **183**, 183.

Zook, H. A., Lange, G., Grün, E., and Fechtig, H.: 1984, 'Lunar Primary and Secondary Microcraters and the Micrometeoroid Flux', *Lunar and Planetary Science*, Vol. XV, Houston, pp. 965–966.

THE GALILEO PLASMA WAVE INVESTIGATION

D. A. GURNETT[1], W. S. KURTH[1], R. R. SHAW[†], A. ROUX[2], R. GENDRIN[2], C. F. KENNEL[3], F. L. SCARF[†], and S. D. SHAWHAN[†]

Abstract. The purpose of the Galileo plasma wave investigation is to study plasma waves and radio emissions in the magnetosphere of Jupiter. The plasma wave instrument uses an electric dipole antenna to detect electric fields, and two search coil magnetic antennas to detect magnetic fields. The frequency range covered is 5 Hz to 5.6 MHz for electric fields and 5 Hz to 160 kHz for magnetic fields. Low time-resolution survey spectrums are provided by three on-board spectrum analyzers. In the normal mode of operation the frequency resolution is about 10%, and the time resolution for a complete set of electric and magnetic field measurements is 37.33 s. High time-resolution spectrums are provided by a wideband receiver. The wideband receiver provides waveform measurements over bandwidths of 1, 10, and 80 kHz. These measurements can be either transmitted to the ground in real time, or stored on the spacecraft tape recorder. On the ground the waveforms are Fourier transformed and displayed as frequency-time spectrogams. Compared to previous measurements at Jupiter this instrument has several new capabilities. These new capabilities include (1) both electric and magnetic field measurements to distinguish electrostatic and electromagnetic waves, (2) direction finding measurements to determine source locations, and (3) increased bandwidth for the wideband measurements.

1. Introduction

This paper describes the Galileo plasma wave investigation. The basic objective of this investigation is the study of plasma waves and radio emissions in the magnetosphere of Jupiter. The Voyager 1 and 2 flybys of Jupiter have now clearly shown that many complex types of plasma wave and radio emission phenomena occur in the Jovian magnetosphere (Scarf *et al.*, 1979; Warwick *et al.*, 1979a, b; Gurnett *et al.*, 1979). These include electromagnetic whistler mode emissions called chorus and hiss, continuum radiation trapped in the magnetospheric cavity, electrostatic waves associated with harmonics of the electron cyclotron frequency, electrostatic and electromagnetic ion-cyclotron waves, and a wide variety of escaping radio emissions. Some of these waves, such as the whistler mode emissions and ion cyclotron waves, are believed to play an important role in the dynamics of the magnetosphere by controlling the pitch-angle scattering and loss of magnetically trapped radiation belt particles. Certain types of plasma waves, such as lightning generated whistlers and upper hybrid resonance emissions, provide important diagnostic tools, from which fundamental plasma parameters such as the electron density can be computed. For an overview of plasma waves in the Jovian magnetosphere see, for example, the review by Gurnett and Scarf (1983).

Since the Galileo spacecraft will be the first orbiter of Jupiter, this spacecraft will provide much better spatial coverage of the Jovian magnetosphere than was possible

[1] Department of Physics and Astronomy, The University of Iowa, Iowa City, IA 52242, U.S.A.
[2] Centre National d'Etudes des Telecommunications/Centre de Recherches en Physique de l'Environment Terrestre et Planetaire, 92131 Issy-les-Moulineaux, Cedex, France.
[3] Department of Physics, UCLA, Los Angeles, CA 90024, U.S.A.
[†] Deceased

Space Science Reviews **60:** 341–355, 1992.

with the previous Pioneer and Voyager flybys of Jupiter. Specifically, the orbit of Galileo will provide a survey of the magnetotail near local midnight at distances of up to 150 R_J, a region that has never previously been explored; repeated passes through the plasma sheet, the magnetopause, and the tail lobes; and numerous close flybys of the Galilean satellites. Of particular importance will be a very close pass by the satellite Io. The Voyager flybys showed that volcanic gases escaping from Io are the main source of plasma in the Jovian magnetosphere and that the primary energization occurs in the dense torus of plasma that surrounds Jupiter near Io's orbit.

In addition to exploring regions never previously investigated, Galileo, by virtue of its long lifetime in orbit around Jupiter, also provides a unique new capability for carrying out studies of temporal variations on time scales that cannot be investigated with a single flyby. For example, it is known that the kilometric and decametric radio emissions associated with Io and its plasma torus have temporal variations on time scales of weeks and longer. With Galileo these temporal variations can be monitored over periods of several years and compared with other remote sensing instruments. These measurements should be able to tell us, for example, whether the variations are associated with volcanic eruptions on Io or some other cause. Considerable interest also exists in searching for evidence of magnetospheric substorm phenomena, possibly comparable to auroral substorms in the Earth's magnetosphere (Akasofu, 1976). With the Galileo plasma wave instrument, it should be possible to provide remote sensing of substorms in a manner comparable to the remote sensing of terrestrial auroral kilometric radiation, which is known to be closely associated with terrestrial substorms. Remote detection of lightning, from whistler measurements, can also provide information on convective storms in Jupiter's atmosphere.

To carry out comprehensive studies of plasma waves and radio emissions at Jupiter, the Galileo plasma wave instrument incorporates several new features that provide improvements over the previous Voyager 1 and 2 measurements. These improvements include (1) both electric and magnetic field measurements to distinguish electrostatic waves from electromagnetic waves, (2) direction finding measurements to determine source locations, and (3) wideband measurements with increased bandwidth to resolve fine structure in the plasma wave and radio emission spectrum. The main instrument package and the electric antenna system were designed and constructed at the University of Iowa, and the search coil magnetic antenna was provided by the Centre de Recherches en Physique de l'Environnement Terrestre et Planetaire (CRPE). A summary of the principal instrument characteristics is given in Table I and a detailed description of the instrument is given in Sections 2 and 3. Section 4 describes the instrument calibration and Section 5 describes the in-flight performance.

2. Electric and Magnetic Field Sensors

The plasma wave sensors on Galileo consist of one 6.6 m tip-to-tip electric dipole antenna and two search coil magnetic antennas. The electric dipole antenna is mounted at the end of the magnetometer boom approximately 10.6 m from the spacecraft, as

TABLE I

Plasma wave instrument characteristics

Parameter	Characteristic
Frequency range, electric	5.62 Hz to 5.65 MHz
Frequency range, magnetic	5.62 Hz to 160 kHz
Frequency resolution	(Low freq.) 5.62 Hz to 31.1 Hz, $\Delta f/f \simeq 67\%$ (Med. freq.) 40 Hz to 160 kHz, $\Delta f/f \simeq 8\%$ (High freq.) 100 kHz to 5.65 MHz, $\Delta f/f \simeq 10\%$
Time resolution	(Low freq.) 5.62 Hz to 31.1 Hz, Δt = 2.67 s (Med. freq.) 40 Hz to 160 kHz, Δt = 18.67 s (High freq.) 100 kHz to 5.65 MHz, Δt = 18.67 s
Sensitivity, electric	$E/\sqrt{\Delta f} \simeq 15\ \text{nV m}^{-1}\ \text{Hz}^{-1/2}$ at ~ 10 kHz
Sensitivity, magnetic	$B/\sqrt{\Delta f} \simeq 50\ \mu\gamma\ \text{Hz}^{-1/2}$ at ~ 100 Hz decreasing to $\simeq 3\ \mu\gamma\ \text{Hz}^{-1/2}$ at 20 kHz
Dynamic range	5.62 Hz to 31.1 Hz, 110 db 40 Hz to 5.65 MHz, 100 db
Wideband waveform modes	Mode 1, 50 Hz to 10 kHz Mode 2, 50 Hz to 80 kHz Mode 3, 5 Hz to 1 kHz
Waveform resolution	Mode 1, 4-bits, 25 200 samples s^{-1} Mode 2, 4-bits, 201 600 samples s^{-1} Mode 3, 4-bits, 3150 samples s^{-1}
Mass	Main electronics box 3.94 kg Search coil 1.52 Electric antenna 1.68 Total 7.14 kg
Power	6.80 W, heater power 3.0 W

shown in Figure 1, and the search coil magnetic antennas are mounted on the high gain antenna feed. The electric antenna consists of two graphite epoxy elements with a root diameter of 2.0 cm, tapering to a diameter of 0.3 cm at the tip. To minimize electric field asymmetries induced by the spacecraft structure the dipole elements are mounted perpendicular to the magnetometer boom. The antenna axis is also oriented perpendicular to the spacecraft spin axis in order to permit direction finding. Each element is hinged 1.8 m from the tip so that the antenna can be folded for launch. A housing at the base of the dipole elements contains two pre-amplifiers. These pre-amplifiers provide low impedance signals to the main electronics package, one for each element. Each element is grounded to the spacecraft structure through a 250 MΩ resistance to limit differential charging effects. A photograph of the electric antenna assembly is shown in Figure 2.

The search coil magnetic antenna consists of two high permeability rods, 25.5 and 27.5 cm long, one optimized for low frequencies, 5 Hz to 3.5 kHz, and the other optimized for high frequencies, 1 to 50 kHz. The winding on the low frequency search

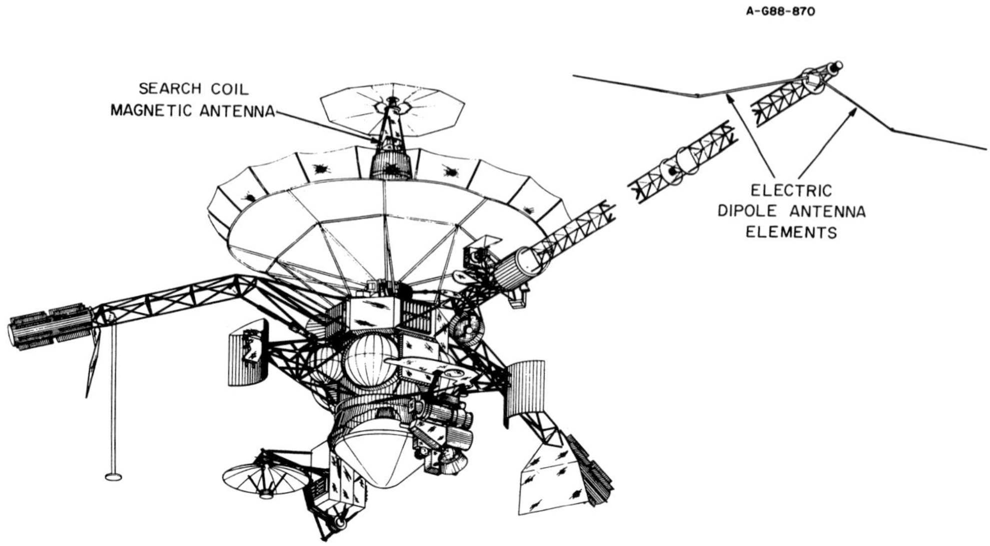

Fig. 1. A sketch of the Galileo spacecraft showing the location and orientation of the electric dipole and search coil antennas used by the plasma wave instrument.

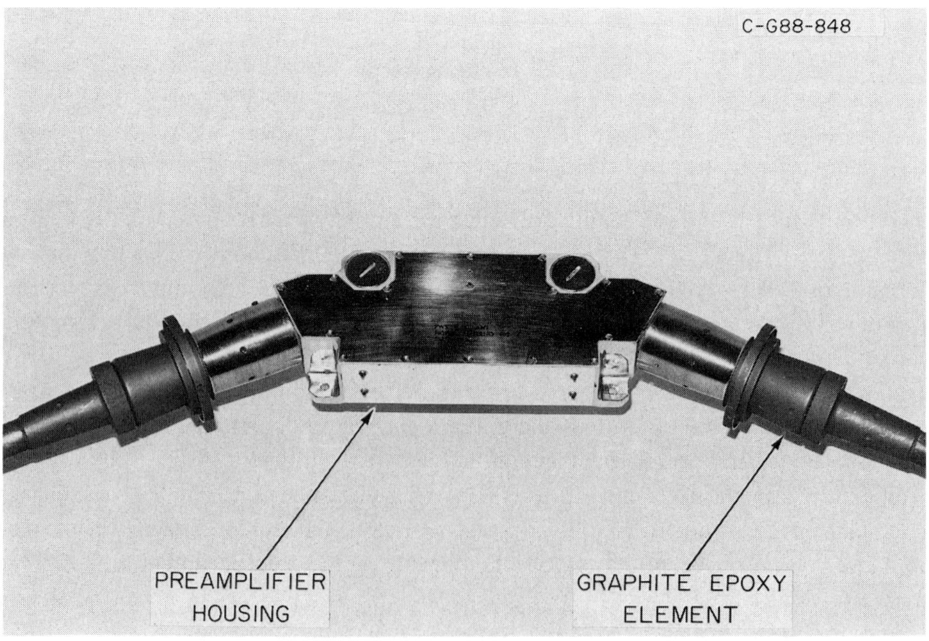

Fig. 2. A photograph of the electric dipole antenna and preamplifier assembly. This assembly is mounted on the end of the magnetometer boom as shown in Figure 1.

coil consists of 50000 turns of 0.07 mm diameter copper wire and the winding on the high-frequency search coil consists of 2000 turns of 0.14 mm diameter copper wire. The two search coils are mounted orthogonally to minimize the electrical coupling between the sensors. Both search coils are mounted perpendicular to the spacecraft spin axis. The high-frequency sensor is oriented perpendicular to the electric dipole antenna and the low-frequency sensor is oriented parallel to the electric dipole antenna. Two pre-amplifiers are mounted in a housing near the search coils in order to provide low impedance signals to the main electronics package. Photographs of the search coil and pre-amplifier assemblies are shown in Figure 3.

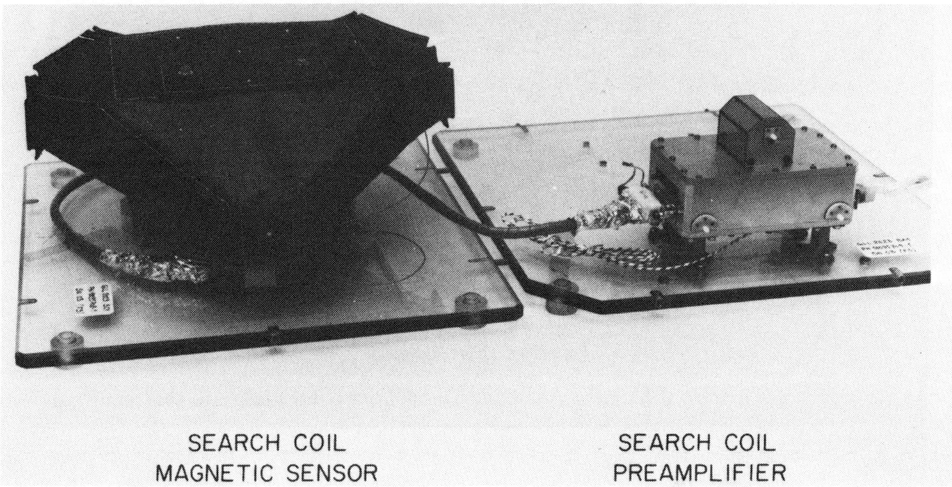

Fig. 3. A photograph of the search coil magnetic sensor assembly. These assemblies are mounted on the high gain antenna as shown in Figure 1.

3. Main Electronics Package

All of the signal processing for the plasma wave experiment is performed in a single main electronics package. The main electronics package is mounted in the spacecraft body near the base of the magnetometer boom. A photograph of the main instrument housing is shown in Figure 4, and a simplified block diagram of the instrument is shown in Figure 5. The signal processing is performed by two main systems, a low rate system to provide survey spectrums, and a high rate system to provide high resolution (wide-band) spectrums. The low rate system consists of three spectrum analyzers: a high-frequency spectrum analyzer, a medium-frequency spectrum analyzer, and a low-frequency spectrum analyzer. The high-frequency spectrum analyzer provides 42 frequencies from 100.8 kHz to 5.645 MHz with a fractional frequency spacing of $\Delta f/f \simeq 10.0\%$ and a bandwidth of 1.34 kHz. The dynamic range of the high-frequency analyzer is 100 db, and a complete spectral sweep requires 28 minor telemetry frames, which corresponds to 18.67 s. The medium-frequency spectrum analyzer provides 112 frequencies from 40 Hz to 160 kHz with a fractional frequency spacing of $\Delta f/f \simeq 8.0\%$. The 112 frequencies of this analyzer are divided into four bands, each with 28

Fig. 4. A photograph of the main electronics housing. This housing is mounted in the spacecraft body near the base of the magnetometer boom.

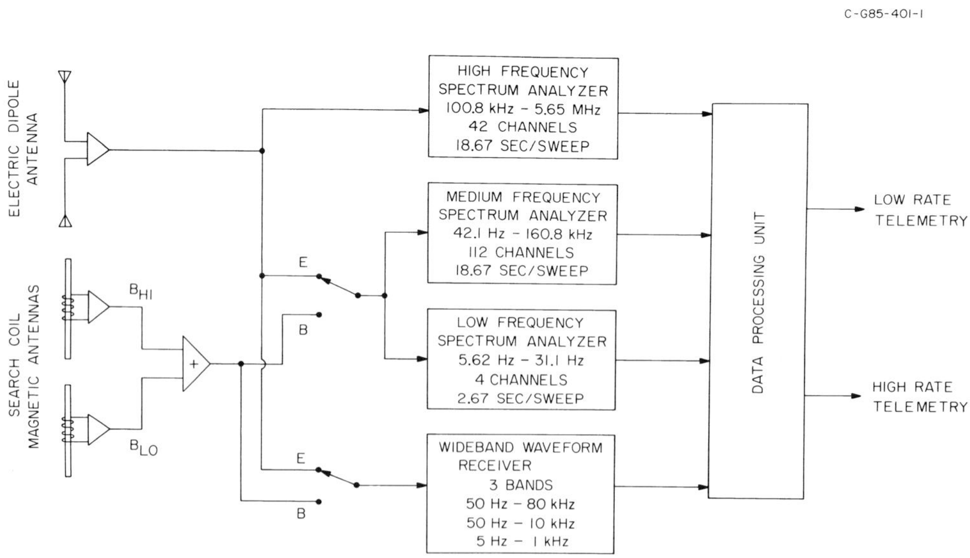

Fig. 5. A block diagram of the plasma wave instrument electronics. Signals from the electric and magnetic antennas are processed by three spectrum analyzers (at low, medium, and high frequencies) and a wideband waveform receiver. All analog to digital conversions and instrument control functions are handled by a data processing unit which interfaces with the spacecraft Command and Data System (CDS).

frequencies. The filter bandwidths associated with each of these bands (in order of increasing frequency) are 4.26, 6.67, 120, and 1510 Hz. The medium-frequency analyzer gives one spectral sweep every 18.67 s with a dynamic range of 100 db. The low-frequency analyzer provides 4 logarithmically-spaced frequency channels from 5.62 to 31.1 Hz with filter bandwidths of 0.83, 1.86, 2.75, and 4.79 Hz. These 4 channels are sampled once every 2.67 s and have a dynamic range of 110 db. The exact sampling scheme for the high-, medium-, and low-frequency spectrum analyzers is somewhat complex and is summarized in Figure 6. As can be seen the basic cycle time required

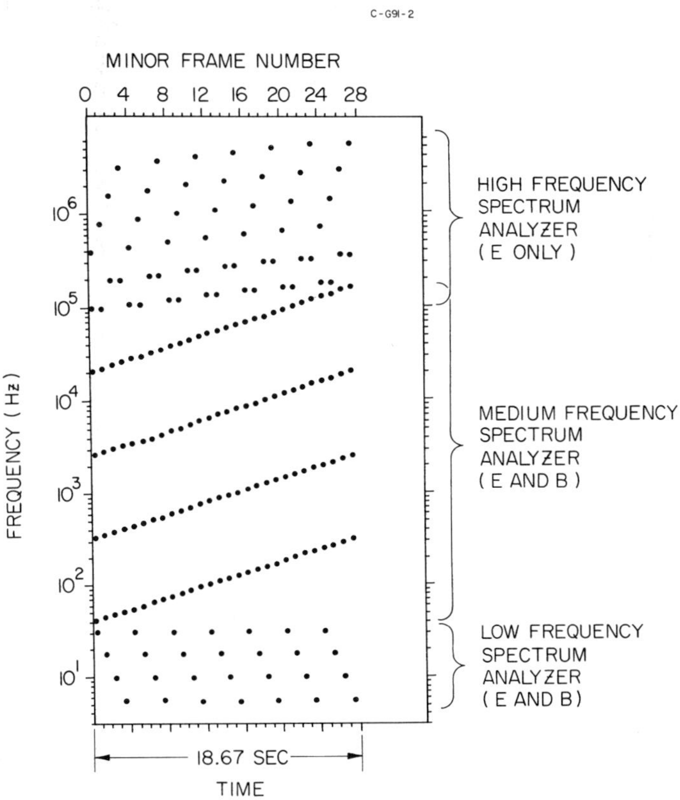

Fig. 6. A diagram showing the scheme for sampling the outputs from the various receiver channels. The basic sampling cycle requires 28 minor frames, which corresponds to 18.67 s.

to obtain a complete spectrum from one antenna is 18.67 s. The signals from the spectrum analyzer filters are logarithmically compressed and detected using a circuit that provides a piecewise-linear approximation to a logarithmic response. The outputs from the logarithmic compressors are then digitized with 8-bit resolution and transmitted to the ground via the low rate telemetry system. The effective data rate for the low rate portion of the instrument is 240 bits s^{-1}.

The high rate wideband receiver provides waveform measurements in three frequency bands, 5 Hz to 1 kHz, 50 Hz to 10 kHz, and 50 Hz to 80 kHz. The frequency band used

TABLE II

Representative Hi-rate waveform modes

Spacecraft mode	Instrument Hi-rate mode	Frequency coverage	Frequency resolution	Size of data block	Time interval between data blocks	Duty cycle
MPW (*R*/*T* or record)	Mode 3 (12.6 kb s^{-1})	5 Hz–1 kHz	25 Hz (64 frequencies)	512 bits	$66\frac{2}{3}$ ms	61%
MPP (*R*/*T* or record)	Mode 3 (12.6 kb s^{-1})	5 Hz–1 kHz	any desired resolution	840 bits	$66\frac{2}{3}$ ms	100%
HCJ, HPJ (*R*/*T*)	Mode 3 (12.6 kb s^{-1})	5 Hz–1 kHz	any desired resolution	840 bits	$66\frac{2}{3}$ ms	100%
XPW, XRW (*R*/*T*)	Mode 1 (100.8 kb s^{-1})	50 Hz–10 kHz	49 Hz (256 frequencies)	3104 bits	$66\frac{2}{3}$ ms	46%
HPW, HRW (*R*/*T* or record)	Mode 1 (100.8 kb s^{-1})	50 Hz–10 kHz	25 Hz (512 frequencies)	6304 bits	$66\frac{2}{3}$ ms	94%
PW4 (Record)	Mode 2 (806.4 kb s^{-1})	50 Hz–80 kHz	394 Hz (256 frequencies)	3104 bits	$8\frac{1}{3}$ ms	46%
PW8 (Record)	Mode 2 (806.4 kb s^{-1})	50 Hz–80 kHz	197 Hz (512 frequencies)	6400 bits	$8\frac{1}{3}$ ms	95%
WF Survey	Mode 4 (in LRS data)	50 Hz–10 kHz and 5 Hz–1 kHz	98 Hz 12 Hz (128 frequencies)	1120 bits	$18\frac{2}{3}$ s	–

[a] Based on using FFT techniques on 1 block of data only.

is controlled by signals from the spacecraft Command and Data Subsystem (CDS). Waveforms from the wideband receiver are digitized by a 4-bit analog to digital converter. Since 4-bit digitization only provides a dynamic range of about 24 db, an analog automatic gain control (AGC) circuit is used to control the amplitude of the waveforms into the analog to digital converter. The AGC time constant is 0.1 s in the two high frequency bands and 1.0 s in the low-frequency band. The sample rate is either 3150, 25 200, or 201 600 samples per second, depending on the frequency band selected. The waveform data can be either transmitted in real time or recorded on the spacecraft digital tape recorder.

The plasma wave instrument has several modes of operation and methods of data transmission. These modes are also controlled by the spacecraft CDS. The medium and low-frequency spectrum analyzers and the wideband waveform receiver can be connected to either the electric dipole antenna or the search coil magnetic antennas. In the normal mode of operation, the medium- and low-frequency spectrum analyzers are cycled between the electric and magnetic antennas so that alternate electric and magnetic spectrums are obtained. Since the search coils do not provide signals in the frequency range covered by the high-frequency analyzer, this analyzer is always connected to the electric antenna. In the cycling mode of operation, the time required for

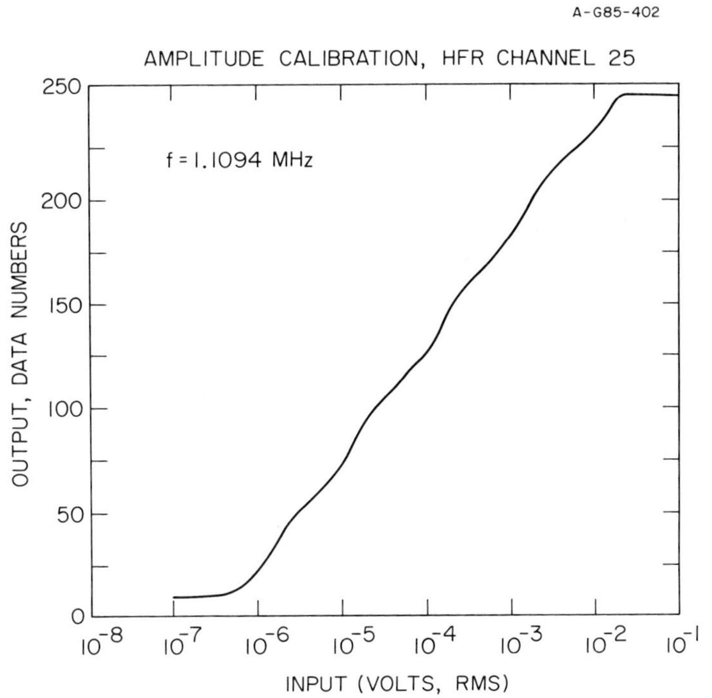

Fig. 7. A representative transfer function for a single channel of the high-frequency spectrum analyzer. The deviation from a straight line is caused by the piecewise-linear stages in the logarithmic compressor.

a complete set of electric and magnetic field spectrums is 37.33 s. The medium- and low-frequency analyzers can also be locked on either the electric or magnetic antennas to provide improved time resolution at the expense of complementary electric and magnetic field coverage. In the ground data processing the spectrum analyzer data are displayed in the form of color frequency-time spectrograms. The frequency scale of the Galileo spectrograms extends from 5.6 Hz to 5.65 MHz, and variable time scales are available, ranging from 30 min to more than 24 hours, depending on the application. Normally, 24-hour spectrograms are used for the low rate survey data. These survey spectrograms are used to select specific intervals for more detailed analysis, such as comparison with charged particle or magnetic field data, or direction-finding analyses.

The greatest flexibility in the operation of the plasma wave instrument is available in the wideband waveform receiver. This receiver provides very high-resolution measurements of electric and magnetic field waveforms during times of special interest, such as the pass through the Io torus and satellite encounters. The waveform data provide the highest possible frequency and time resolution, subject only to the constraints of Fourier analysis, $\Delta f \Delta t \simeq 1$. Although the waveform receiver has only three frequency bands, with bit rates of 12.6, 100.8, and 806.4 kbits/sec, several spacecraft modes are available for recording and transmitting the data to the ground. These modes of wideband data transmission and the corresponding time resolution for the most conveniently usable "blocks" of wideband data are summarized in Table II. In the highest time resolution mode, a continuous sample of the electric or magnetic field waveform can be obtained over a bandwidth of 50 Hz to 80 kHz for periods of up to 18 min (the time required to fill the spacecraft tape recorder).

On the ground the wavefrom data are Fourier transformed in discrete packets, usually consisting of 1024 samples, and displayed in the form of a frequency-time spectrogram. These frequency-time spectrograms provide the highest time resolution data available from the Galileo plasma wave instrument. In certain modes of operation, such as MPW, XPW, and PW4, the duration of the wideband recording can be extended at the expense of reduced duty cycle, frequency coverage, or analysis bandwidth. To provide waveform measurements when the high rate telemetry link is not available, a waveform survey output is included in the regular low rate telemetry data. This waveform survey output provides one block of 280 waveform samples every 18.67 s in any of the three wideband frequency ranges, 5 Hz to 1 kHz, 50 Hz to 10 kHz, or 50 Hz to 80 kHz.

4. Instrument Calibration

An extensive series of calibrations and performance checks were carried out on the plasma wave instrument both before and after integration on the spacecraft. Since the

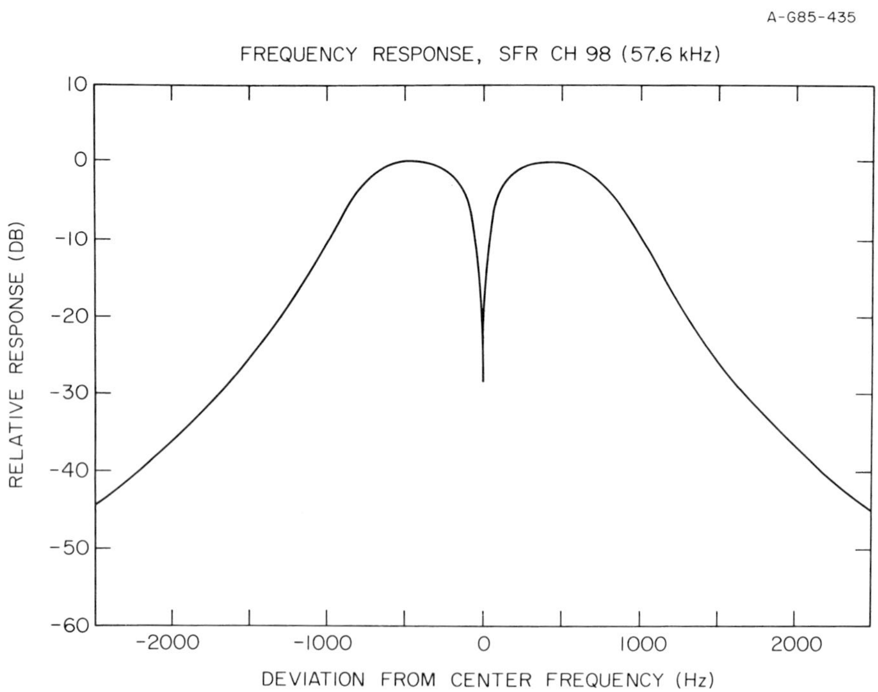

Fig. 8. A representative frequency response of a single channel of the medium frequency spectrum analyzer. The notch at the center of the band is caused by the frequency mixing scheme, which has the intermediate frequency (IF) at zero frequency.

piecewise-linear logarithmic compressors used in the spectrum analyzers do not give a true logarithmic response, the transfer function of the logarithmic compressors must be calibrated. Since a large number of channels are used, it was not practical to measure the transfer function for each frequency channel separately. Since the receivers are organized into distinct frequency bands, each of which uses the same filter and logarithmic compressor, the calibration can be broken into two steps. First, within a given band the transfer function from antenna input to the telemtry output is measured by progressively increasing the signal amplitude at a fixed frequency. Second, within this same band the frequency response is measured, from channel to channel, at a fixed input amplitude. By combining these amplitude and frequency response measurements a complete calibration can be obtained for all signal intensities and frequencies.

A representative transfer function is shown in Figure 7. This curve is for channel 25 (1.1094 MHz) of the high-frequency receiver. This transfer function gives the amplitude response of the logarithmic compressor, including all gain factors from the antenna (electric, in this case) to the logarithmic compressor. The slight deviation from a true logarithmic response is caused by the piecewise-linear response of the logarithmic compressor. Using calibration curves such as this, a look-up table can be constructed which converts the telemetry data number to input signal strength. When combined with the channel-to-channel gain, these calibrations are sufficient to determine the signal strength in all channels of the high-frequency receiver.

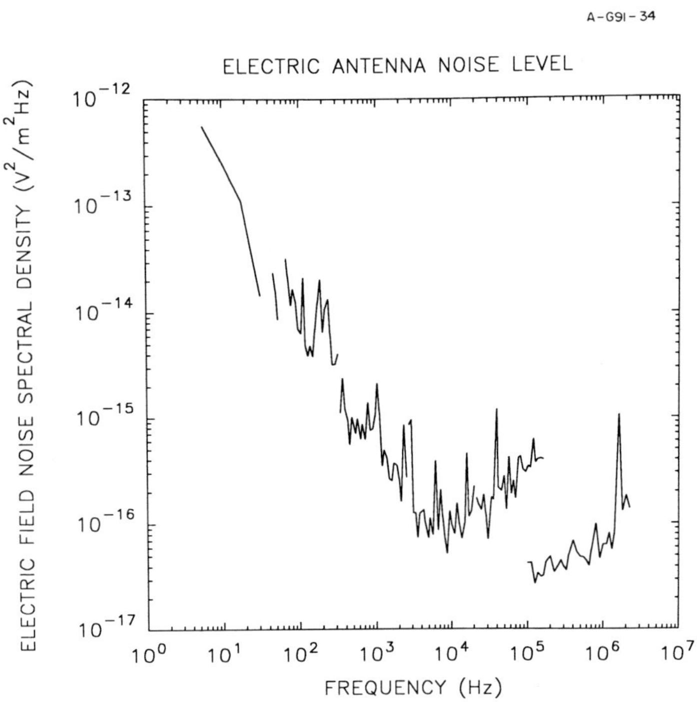

Fig. 9. The system level (unity signal-to-noise ratio) for the electric antenna system.

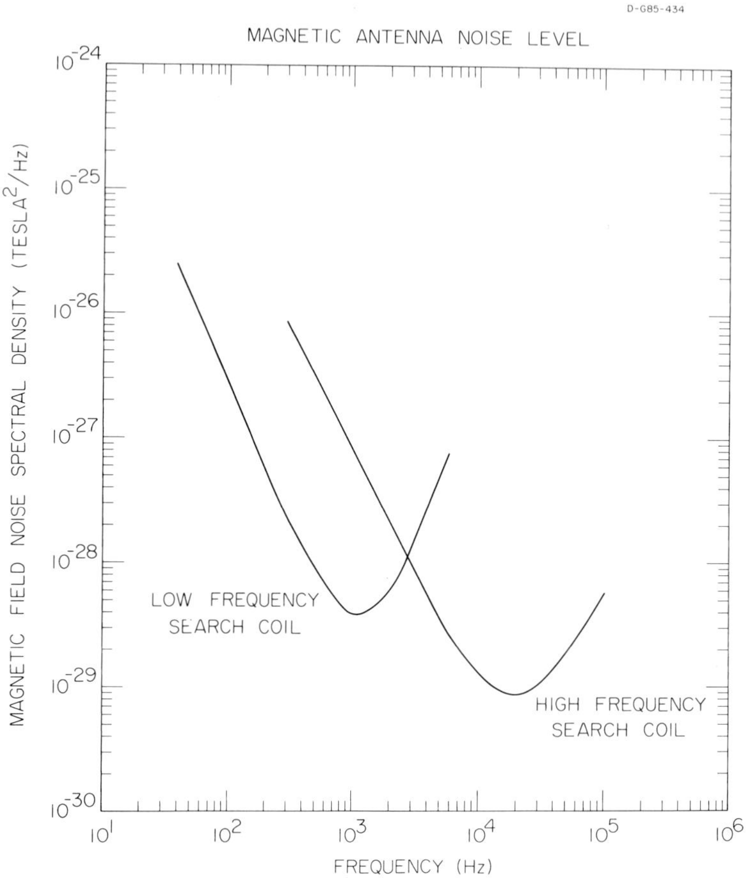

Fig. 10. The system level (unity signal-to-noise ratio) for the magnetic antenna system.

In addition to measuring the transfer function and channel-to-channel gain, a detailed frequency response measurement was also performed for each filter. These measurements confirm that the filters have the proper shape and no spurious responses. A typical filter frequency response is shown in Figure 8. This frequency response curve is for channel 98 (57.6 kHz) of the medium-frequency spectrum analyzer. The sharp notch at the center of the filter band is characteristic of the type of receiver used (zero frequency IF). Whenever possible the notch is centered on harmonics of the spacecraft 2.4 kHz power supply in order to reduce spacecraft generated interference. The filter frequency response measurements provide a good first-order estimate of the noise bandwidth of the filters. The effective noise bandwidths are also independently determined by stimulating the instrument with a white noise signal of known spectral density.

As can be seen from the transfer function curve in Figure 7, a level exists below which no signal can be detected. For our purposes the noise level of the instrument is defined as the signal level which gives a unity signal to noise ratio ($S/N = 1$). Plots showing the noise levels for the electric and magnetic antennas are shown in Figures 9 and 10. These noise levels have been converted to units of electric and magnetic spectral densities. For the electric field antenna, the electric field strength is computed by assuming that the antenna has an effective length of $l_{eff} = 3.5$ m. This length is the distance between the geometric centers of the two dipole elements. The discontinuities between various parts of the curve are due to differences in the internal noise levels of the high-, medium-, and low-frequency spectrum analyzers. For the search coil magnetic antennas, the magnetic field sensitivity and frequency response were calibrated in the IPG magnetic field observatory at Chambon La Foret, France. The transfer function measurements were performed using a Helmholtz coil driven by a known AC current source. The absolute accuracy of the magnetic field calibration is estimated to be about 3%. The magnetic noise levels were measured by placing the search coils in a μ-metal chamber, which shields the sensors from external noise sources.

5. Inflight Performance

The Galileo spacecraft was successfully launched on October 18, 1989. The plasma wave instrument was turned on shortly after launch and operated satisfactorily. During a period known as the '4-day checkout', which occurred from December 27 to 31, 1989, a detailed checkout of the instrument was performed, including tests to determine if any interference was present from other spacecraft systems or instruments. The instrument operation proved to be nominal in all respects. The electric field interference levels also proved to be very low. Only a few very minor interference lines could be detected, mostly in the high-frequency receiver channels. Many type III solar radio burst were detected

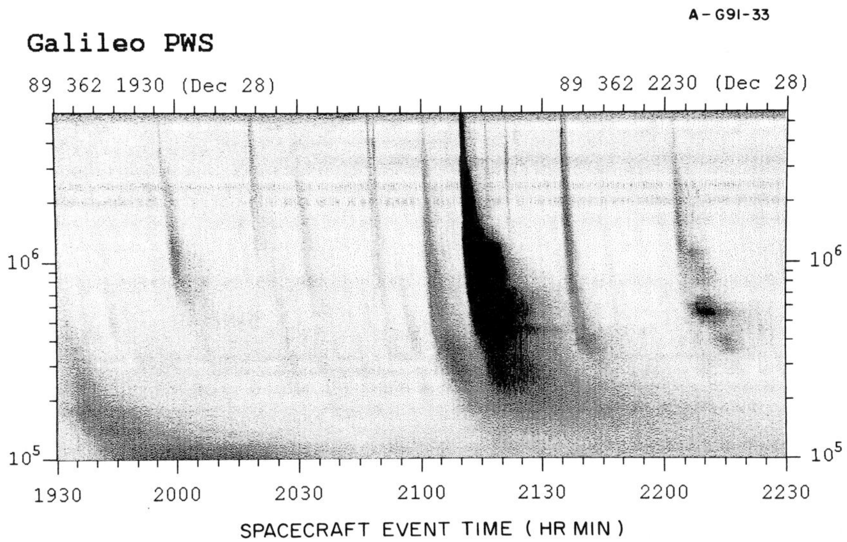

Fig. 11. A spectrogram showing numerous type III solar radio bursts detected during the 4-day checkout. Note the almost complete absence of spacecraft generated interference.

during the 4-day checkout, which demonstrates the excellent sensitivity of the instrument. A frequency-time spectrogram showing some of these radio bursts is shown in Figure 11. A persistent elevation of the electric field noise levels above the pre-launch noise levels exists in the low-frequency channels, below about 100 Hz. This noise is essentially structureless, and appears to be caused by the plasma.

The magnetic field data showed a series of strong interference lines at the fundamental and low order harmonics of the 2.4 kHz spacecraft power supply. These magnetic interference lines extend as much as 20 db above the search coil noise level, and were comparable to the magnetic interference levels observed in the pre-launch testing. The strong magnetic interference is a consequence of the close proximity of the search coils to the spacecraft body, which has substantial power system currents at 2.4 kHz and its harmonics. It would have been much better to have mounted the search coil on a boom far from the spacecraft body. This option was considered early in the spacecraft design, but was rejected because of certain critical spacecraft weight and balance considerations which were of importance at that stage of the project. Although the power supply interference is undesirable, adequate measurements can still be made at frequencies between the harmonics. Therefore, the objectives of the magnetic field sensors are not seriously compromised. Strong interference was also detected from the ultraviolet spectrometer (UVS) at frequencies up to a few hundred Hz. This interference apparently originates from an unshielded stepper motor that was not identified as an interference source during the pre-launch testing. From a second compatibility test conducted during the Earth 1 flyby we demonstrated that the scientific impact of this interference can be minimized by duty cycle operation of the UVS instrument.

6. Conclusion

The Galileo plasma wave experiment is a capable instrument which should provide valuable new information on wave-particle interactions in the Jovian magnetosphere. The electric field noise levels are comparable to the Voyager plasma wave instrument and the magnetic field noise levels are comparable to those used on previous terrestrial plasma wave investigations. The frequency resolution is substantially better than the previous Voyager measurements, and the direction-finding and magnetic field measurements will provide an entirely new capability for measuring plasma waves and radio emissions in the Jovian magnetosphere.

Acknowledgements

The authors would like to express their thanks to Steve Remington, Don Kirchner, Bob Barrie, Elwood Kruse, and Dave Tomash at the University of Iowa, for their role in designing, assembling, and testing the instrument; to Alain Meyer of CRPE for his role in designing the search coil magnetic antennas; to Terry Averkamp, Larry Granroth and Scott Allendrof of the University of Iowa and Claude de Villedary of CRPE for their role in preparing the data processing routines; and to Gaylon McSmith and Bill Fawcett at the Jet Propulsion Laboratory for their help in coordinating our effort with JPL. Unfortunately three of our co-investigators, Fred Scarf from TRW, Bob Shaw from the University of Iowa, and Stan Shawhan from NASA Headquarters have died during the course of this investigation. This instrument was funded by NASA through the Jet Propulsion Laboratory under contracts 955234 and 958779.

References

Akasofu, S.-I.: 1976, *Physics of Magnetospheric Substorms*, D. Reidel Publ. Co., Dordrecht, Holland.

Gurnett, D. A. and Scarf, F. L.: 1983, in A. J. Dessler (ed.), 'Plasma Waves in the Jovian Magnetosphere', *Physics of the Jovian Magnetosphere*, Cambridge University Press, Cambridge, pp. 285–316.

Gurnett, D. A., Kurth, W. S., and Scarf, F. L.: 1979, 'Plasma Wave Observations Near Jupiter: Initial Results from Voyager 2', *Science* **206**, 987.

Scarf, F. L. and Gurnett, D. A.: 1977, 'A Plasma Wave Investigation for the Voyager Mission', *Space Sci. Rev.* **21**, 287.

Scarf, F. L., Gurnett, D. A., and Kurth, W. S.: 1979, 'Jupiter Plasma Wave Observations: An Initial Voyager 1 Overview', *Science* **204**, 991.

Warwick, J. W., Pearce, J. B., Riddle, A. C., Alexander, J. K., Desch, M. D., Kaiser, M. L., Thieman, J. R., Carr, T. D., Gulkis, S., Boischot, A., Harvey, C. C., and Pedersen, B. M.: 1979a, 'Voyager 1 Planetary Radio Astronomy Observations Near Jupiter', *Science* **204**, 995.

Warwick, J. W., Pearce, J. B., Riddle, A. C., Alexander, J. K., Desch, M. D., Kaiser, M. L., Thieman, J. R., Carr, T. D., Gulkis, S., Leblanc, Y., Pedersen, B. M., and Staelin, D. H.: 1979b, 'Planetary Radio Observations from Voyager 2 Near Jupiter', *Science* **206**, 991.

THE GALILEO MAGNETIC FIELD INVESTIGATION

M. G. KIVELSON, K. K. KHURANA, J. D. MEANS, C. T. RUSSELL, and R. C. SNARE

UCLA Institute of Geophysics and Planetary Physics, Los Angeles, CA 90024-1567, U.S.A.

Abstract. The Galileo Orbiter carries a complement of fields and particles instruments designed to provide data needed to shed light on the structure and dynamical variations of the Jovian magnetosphere. Many questions remain regarding the temporal and spatial properties of the magnetospheric magnetic field, how the magnetic field maintains corotation of the embedded plasma and the circumstances under which corotation breaks down, the nature of magnetic perturbations that transport plasma across magnetic shells in different parts of the system, and the electromagnetic properties of the Jovian moons and how they interact with the magnetospheric plasma. Critical to answering these closely related questions are measurements of the dc and low-frequency magnetic field. The Galileo Orbiter carries a fluxgate magnetometer designed to provide the sensitive measurements required for this purpose. In this paper, the magnetometer is described. The instrument has two boom-mounted, three-axis sensor assemblies. Flipper mechanisms are included in each sensor assembly for the purpose of offset calibration. The microprocessor controlled data handling system produces calibrated despun data that can be used directly without further processing. A memory system stores data for those periods when the spacecraft telemetry is not active. This memory system can also be used for storing high time-resolution snapshots of data.

1. Introduction

The Jovian magnetosphere envelopes the planet in a vast cocoon that extends more than seven million kilometers from the planet in all directions and deflects the solar wind in its outward flow from the Sun. The gargantuan scale of the system is readily understood as arising from a combination of factors: the low dynamic pressure of the solar wind at the orbit of Jupiter, down by a factor of more than 25 from its level at 1 AU; the large planetary radius, 11 times that of the Earth; the strong dipole magnetic moment of the planet, more than four orders of magnitude larger than that of the Earth; and the existence of a relatively dense, centrifugally accelerated plasma that acts to inflate the magnetosphere from within.

The properties of this extraordinary magnetosphere were first revealed in the 1970s when it was explored in flyby passes by four spacecraft: Pioneer 10 and 11, and Voyager 1 and 2. For full reviews of the observations and interpretations of the processes of importance in the magnetosphere, the reader should consult *The Physics of the Jovian Magnetosphere*, edited by Dessler (1983) and *Jupiter*, edited by Gehrels (1976) as well as review articles by Birmingham (1983) and Connerney (1987). The first missions showed that the magnitude of the magnetic dipole moment is 4.2 $R_J^3 \times 10^{-4}$ T, or 4.2×10^5 nT at the level of the cloud tops near the equator (Acuna *et al.*, 1983). This is an order of magnitude larger than the magnetic field at the terrestrial surface. As at Earth, the Jovian plasma contains ions of solar wind origin (see, for example, Krimigis and Roelof, 1983) and ions of ionospheric origin (see, for example, Belcher, 1983). Unique to Jupiter is the large population of heavy ions whose source is the

Space Science Reviews **60:** 357–383, 1992.

Galilean moon, Io; these ions form a torus of relatively high plasma density near Io's orbit but they also diffuse out to form an important component of the plasma present throughout the magnetosphere. The heavy ions, spun up to corotation velocity by magnetic ($\mathbf{j} \times \mathbf{B}$) forces related to the closure of field-aligned currents flowing between the ionosphere and the magnetosphere (Hill, 1979), carry cross-field currents that distort the magnetic field into a disk-like geometry at large distances (see the discussion in Vasyliunas, 1983).

Although many features of the magnetosphere were identified by the first four spacecraft to make measurements within it, there are still many puzzles. For example, it is not clear how the plasma, whose ultimate source is Io, is transported through the magnetosphere. Over large spatial regions, the magnetic field imposes approximate corotation, but it is thought that corotation breaks down at large distances in the evening sector, and possibly more generally when new plasma is added at an unusually large rate. The system appears to display large-scale temporal changes, but it is not clear whether the changes are internally or externally driven and the time scales for many of the changes are not understood. The Galileo mission will provide new data with which it will be possible to map the Jovian magnetosphere, monitor magnetospheric dynamics, and investigate magnetosphere-ionosphere coupling. The presence of Jupiter's large moons perturbs the plasma, and Galileo spacecraft investigations of the plasma interaction with the satellites will yield information about properties of both the plasma and the moons themselves. For example, during the flyby of Io it will be possible to determine whether or not Io has an internally generated magnetic field (Kivelson *et al.*, 1979). By remaining within the magnetosphere for years instead of weeks (as for a flyby mission), the Galileo Orbiter will be able to study the structure and the time variability of the system on both large and small scales. Finally, the spacecraft will make measurements in the magnetotail near the midnight meridian where it may be possible to find evidence of a planetary wind flowing away from Jupiter.

The lengthy (six year) interval of interplanetary cruise will also provide opportunities for carrying out scientific investigations. During the flyby of Venus in 1990, particles and fields instruments on the spacecraft acquired data of value to the study of the cloud-covered planet. The magnetometer, in particular, provided data on the structure of the bow shock and on upstream waves. Solar wind data in the vicinity of Venus will be studied in conjunction with data from the Pioneer Venus Orbiter to provide insight into the time constants for transport of plasma and fields through the ionosphere of Venus. Studies of pick-up ions and of the electromagnetic signals from lightning on Venus will also be carried out. The first flyby of Earth whose magnetic properties are well-known, provided an excellent opportunity for calibration of the instrument. In addition, the data collected promise to provide valuable insight into substorm processes and tail geometry. They will be used for coordinated studies with observations from ground-based instruments and from spacecraft operating in Earth orbit. During the interplanetary cruise period, the solar wind will be monitored. Galileo measurements will provide a valuable baseline for the study of the large-scale structure and evolution of the solar wind. For this purpose, comparison with the data of the Ulysses spacecraft which will be in the solar wind at different distances and latitudes will be of special interest.

Magnetometer measurements are central to accomplishment of some of the primary objectives of the mission. For example, properties of the satellites and their atmospheres determine the strength of the field-aligned currents that flow in the magnetospheric plasma surrounding them. During close flybys of the Galilean moons, the magnetometer will need to identify relatively small changes (of order of nT) of field magnitude and orientation in large background fields (roughly 2000 nT near Io) in order to characterize the currents and thereby contribute to understanding the structure of the moons. Not only must the magnetometer identify slowly varying features of the magnetic geometry, it must also be capable of characterizing fluctuating fields of order 0.1 nT at rates below and just within the low-frequency bands of the search coil magnetometer. Small amplitude magnetohydrodrynamic waves are important in establishing plasma equilibrium in local regions of the magnetosphere while larger amplitude waves serve to couple energy and momentum among its widely separated parts. Complete spectral coverage will augment direct plasma measurements in providing the information needed to understand plasma sources, transport and losses. To obtain useful data, it is essential that the spacecraft noise level and the noise level of the magnetometer itself be low enough not to mask the ambient signals. As well, linear response is required so that measurements of fluctuating fields can be made accurately independent of the strength of the average field (as high as 6000 nT near closest approach to Jupiter).

In addition to providing data of direct importance for scientific investigations, the magnetometer performs a critical service function, providing the reference direction essential to the meaningful analysis of the plasma and energetic charged particle measurements. As the interpretation of all fields and particle measurements relies on knowledge of the magnetic field direction and/or magnitude, it is essential that physically meaningful data from the magnetometer be available without delay following the transmission of data from the spacecraft.

The design and development of the magnetometer instrument and its data-processing system that provides scientifically useful data required the cooperation of many people. The scientific investigations will rely on the contributions of many others. Table I lists the principal team members and their responsibilities.

The magnetometer has been designed to provide highly accurate and stable measurements of field vectors over an exceptionally broad dynamic range from the low values characteristic of the distant solar wind and the distant magnetotail (~ 0.1 to 1 nT) to values larger than the largest anticipated ($\sim 6 \times 10^3$ nT) near closest approach to Jupiter. So that small perturbations can be measured with great accuracy, the stray field of all spacecraft systems were constrained to produce less than 0.1 nT magnetic perturbations at the outboard sensor. Knowledge of sensor orientation after in-flight calibration is better than $0.1°$.

The design of the magnetometer on the Galileo Orbiter incorporates several innovations that will enhance the scientific value of the measured data. In particular, the Galileo magnetometer can provide calibrated data in 'despun coordinates' using on-board data-processing. Such data can be used by investigators immediately upon receipt without further delay for data analysis and proved extremely useful during the first flyby of Earth. The magnetometer also provides for storage of time-averaged data. Thus, even during intervals when the spacecraft is not transmitting data to Earth, a complete record of magnetic field data will exist.

TABLE I

Magnetometer Team
Institute of Geophyscis and Planetary Physics, UCLA

Principal investigator	Margaret G. Kivelson
Co-investigators	Paul J. Coleman Charles F. Kennel Robert L. McPherron Christopher T. Russell David J. Southwood Raymond J. Walker
Project engineers	Robert C. Snare Joseph D. Means
Project computer specialists	Neal Cline Steven Joy Todd King
Project science support	Krishan K. Khurana

The design of the Galileo magnetometer evidently presented many challenges. The requirement of high sensitivity is difficult to achieve on a complex spacecraft. Reliability, essential because of the prolonged duration of the mission with a six-year cruise preceding the beginning of the two-year mission, posed additional problems. Special requirements on the data coverage and format required development of a new approach to data-handling capabilities as noted above.

2. Critical Aspects of the Instrument Design

The design challenges described in the introduction were met in a variety of ways. We chose to fly a boom-mounted dual triaxial fluxgate magnetometer with ring core sensors in a flipper assembly.

We chose a fluxgate magnetometer because of its low power and its robustness, as well as because our laboratory has had experience on many missions with successful fluxgate operation. The ring core sensor was chosen because of its proven stability and low noise. The flipper assembly was included so that the sensor along the spin axis can periodically be rotated into the spin plane where its zero level can be checked. This feature also offers some redundancy. If the sensor along the spin axis were to cease operation, one of the two sensors could be rotated 90° to replace it. Then the three vector components could still be measured, albeit only once per spin period.

The need for extreme sensitivity was recognized not only in the instrument design but also in the design of the spacecraft which has a long boom to isolate the primary set of sensors from the magnetic fields generated by the complex spacecraft and its subsystems. A second set of sensors at a location roughly $\frac{2}{3}$ as far from the spacecraft spin axis as the primary set was incorporated to measure the large fields of the inner Jovian magnetosphere and to assist in the identification of spacecraft fields (Ness *et al.*, 1971). Spacecraft fields will decrease in magnitude with distance from the spacecraft and will often be related to specific spacecraft activities. Such magnetic fields can be characterized to first order by using data from both the inboard and outboard sensors if it is assumed that the field sources are localized dipoles; the data can then be corrected by removing the contributions of locally-generated fields.

The sensitivity of the magnetometer is needed not merely to measure small magnetic fields but also to measure small variations of the field in a large background magnetic field. This requirement cannot be met with an instrument that has low fractional resolution in any one range but many operating ranges; instead the instrument must have both a large dynamic range and high resolution. This need was met by using a high precision A/D converter with 12-bit digitization and 15-bit accuracy. Averaging data with this accuracy prior to telemetry increases the effective resolution of the measurements. Our careful plans for achieving high sensitivity could be frustrated by the presence of significant sources of oscillating spacecraft fields in the bandwidth of the magnetometer. A careful program of magnetic testing, shielding, compensation, and proper cable design for the entire spacecraft was undertaken to insure spacecraft magnetic cleanliness and thus to preserve magnetometer sensitivity.

The requirement for extreme reliability was met in several ways. The use of separate inner and outer sensor assemblies adds some redundancy and so does the flipper, as mentioned above. In addition to guarding against failure, one also has to be aware of changes in instrument operation. As the spacecraft spins, the gains and relative phases of the sensors in the spin plane can be intercompared. The flipper can be used to bring the spin axis-aligned sensor into the spin plane so its gain and phase can also be determined. An internal calibrate signal referenced to the instrument's A/D converter is also included. This signal reveals any changes in the operation of the analog portion of the magnetometer. The pointing of the sensors may change during the mission because the sensors are mounted on the end of a very long boom whose orientation may be temperature sensitive and may be affected by spacecraft maneuvers. Corrections for such changes can be made by using the clearly measurable magnetic signal produced by a fixed coil (called the MAG Cal Coil) mounted on the side of the spacecraft at the end of the boom, as we discuss in a section on instrument calibration.

In many respects the design problems of the digital portion of the magnetometer are similar to those found in building a high fidelity sound system. Signal fidelity must be preserved. To preserve the signal, two aliasing problems must be addressed before the signals are transmitted. First, signals outside the telemetry bandwidth, as given by the Nyquist frequency (half the sample rate), must not enter the analog to digital converter. This is accomplished by analog filtering before digitization. Secondly, on a spinning spacecraft, signals in the ambient medium polarized clockwise and counterclockwise about the spin axis are detected at different frequencies. Depending on signal conditioning in the telemetry process, the amplitude and phase relationship of these two circular components of a single ambient signal can be altered. The signal processing procedures adopted minimize the distortion of the low-frequency signal, a point to which we return following a description of the on-board data processing.

The service function of the magnetometer requires that pertinent data be readily available to other experimenters. Our solution is to calculate physically meaningful data on-board the spacecraft and put those data into the telemetry stream. We can, by command, despin our data on-board to produce high time-resolution despun data required by other spacecraft instruments. We note that in order to obtain accurate despun data, the instrument operation must be monitored carefully and calibration constants in the instrument memory updated as necessary. As mentioned, we also routinely obtain despun vector data averaged over approximately 30-s intervals and transmit this information in the subcommutated data. When these data arrive at Earth, they can be immediately displayed and used by all experimenters without further computer processing. This feature supports rapid analysis of the data both for mission planning purposes and for public information purposes.

Finally, the scientific requirement for continuous data in the face of possible intermittent spacecraft tracking led us to store 200 vectors of averaged data for later transmission. The available storage registers also permit us to use the full bandwidth of the magnetometer to store limited intervals (7 s per hour) of 30 samples per second data that we call *snapshots* for later transmission. The snapshot memory can store 210 vectors. In short, the design requirements for the Galileo mission were met with the development of a sophisticated magnetometer with redundancy, cross-checks, internal computational power, and data storage.

3. The Instrument

The magnetometer has a separate electronics box containing digital and analog electronics and power supplies, located in electronics Orbiter bay number three, and two flipper assemblies and their ring core sensors mounted on a boom or 'astromast' deployed from a canister mounted on the science boom. Below we describe in more detail these various sub-assemblies, the instrument modes, calibration methods, and the on-board data processing.

3.1. Sensors

The sensors (Figure 1) in each triad are oriented so that one is closely aligned along the spin axis of the spacecraft and two sensors lie with orthogonal orientation in the spacecraft spin plane. The individual sensors contain drive and sense coils surrounding a magnetically permeable core. Currents in the drive coils drive the permeable core into saturation twice during each drive cycle, symmetrically positive and negative. An external field breaks the symmetry of the periodic core saturation, and sense coils detect the resultant harmonics of the drive frequency. The amplitude of the even harmonics are proportional to the component of the ambient magnetic field along the axis of the sense coils. The instrument senses the second harmonic signal, i.e., the harmonic of largest amplitude. A feedback current that nulls the external field is applied to a coil around the sensor so that the core operates in a zero field condition. Thus, the amplitude of the feedback current is proportional to the component of the external field along the sensor axis. The data system senses the analog voltage proportional to the feedback current and converts it to a digital signal which is transmitted to Earth.

Fig. 1. Sensor assembly with cover removed showing mechanism. The bi-metallic springs with heaters that actuate the flipper mechanism are at the top of the assembly.

The basic sensors are ring core fluxgate sensors similar to those described by Gordon and Brown (1972). Each sensor assembly contains three orthogonally mounted fluxgate sensors and a flipping mechanism. The flipper mechanism rotates two of the sensors so that the non-spinning sensor parallel to the spacecraft spin axis may be exchanged with one of those mounted in the spin plane of the spacecraft. As mentioned above, such an exchange of sensor positions provides data for determination of the offset of the sensor parallel to the spacecraft spin axis.

The flipping mechanism is similar in design to that developed at NASA Ames Research center and used on Explorer 35, Pioneer 9, ISEE 1 and 2, and UK-AMPTE. The mechanism uses electrically heated bi-metallic springs and a series of levers to rotate the sensor shaft by 90° and back again. A photograph of one of the sensor assemblies is shown in Figure 1.

The outboard sensor, mounted at the end of a deployable boom, is located 11.03 m from the spacecraft spin axis. The inboard sensor is mounted on the same boom at a distance of 6.87 m from the spacecraft spin axis. The locations can be seen in Figure 2.

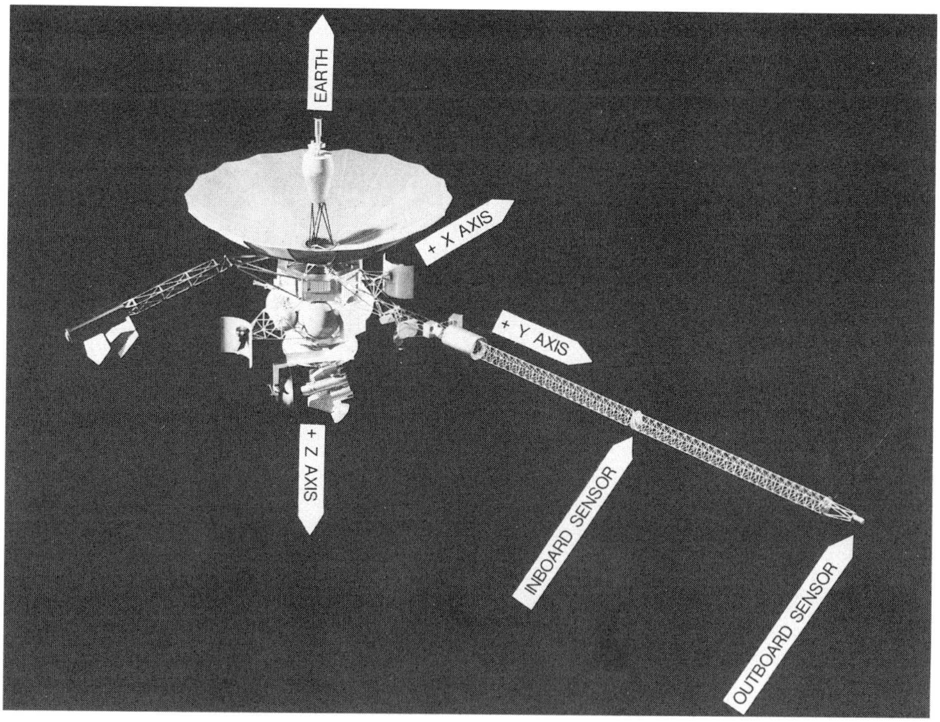

Fig. 2. A model of the Galileo spacecraft with labels defining axes.

The outboard sensors have dynamic ranges of ± 32 nT and ± 512 nT. The inboard sensors have dynamic ranges of ± 512 nT and $\pm 16\,384$ nT. The limitation to 2 ranges for each magnetometer assembly allowed us to optimize the design to achieve low noise levels and small offsets. The lower range of the outboard sensor was designed for use in the solar wind and in the distant Jovian magnetosphere and magnetotail. The higher range of the inboard sensor was designed for use in the inner magnetosphere of Jupiter, i.e., within $\sim 12\ R_J$ and to the closest approach at $\sim 4\ R_J$. The common ± 512 nT range, with resolution of 0.03 nT, is appropriate for the middle and outer magnetosphere or the solar wind within 1 AU. This is the range that will be used for dual magnetometer estimation of the spacecraft field. The magnetometer to be operated, i.e., inboard or outboard and the desired dynamic range are selected by commands from the Earth.

3.2. Electronics

The electronics unit housed within the spacecraft contains the drive and sense electronics of the sensors, analogue-to-digital converters, a microprocessor, and power conditioning and control circuitry. The electronic system is constructed of semiconductor circuits that will withstand the radiation environment of the Jovian magnetosphere. The analog circuits are constructed on double-sided printed circuit boards with plated-through holes. The data system is constructed on planar stitchwire circuit cards. The electronic circuit cards are mounted on a standard sub-chassis shelf which is mounted in the Galileo electronic bay. The physical characteristics of the system are given in Table II.

TABLE II
Instrument characteristics

Sensors	
Mass	0.72 kg
Dimensions	8.1 cm (diameter), 16.0 cm (length)
Power	4.35 W for 4 min to activate flip
Electronics	
Mass	3.98 kg
Volume	$14.6 \times 17.8 \times 41.8\ (\text{cm})^3$
Power	3.9 W
Dynamic Range	
Inboard magnetometer	± 512 nT and $\pm 16{,}384$ nT
Outboard magnetometer	± 32 nT and ± 512 nT

The electronics unit contains an analog circuit board for each of the six fluxgate sensors. These circuits provide sensor excitation currents, amplification, filtering, and second harmonic detection. The magnetometer circuits are similar to those used for the Pioneer Venus Magnetometer, reported by Snare and Means (1977). Changes were made to adapt the circuits to radiation-tolerant semi-conductors and to reduce power consumption. The output of each circuit is an analog voltage that is linear with the measured magnetic field at the sensor. The data system contained in the electronics unit is microprocessor controlled. Figure 3 shows a photograph of the electronics system.

Figure 4 shows the block diagram of the system. In the block diagram those blocks residing above the internal bus, i.e., central processor (CPU), read only memory (ROM), and the random access memories (RAM), form a standard configuration micro-

Fig. 3. Electronic chassis of the magnetometer. The six analog magnetometer boards are on top of the chassis. The digital data handling circuits are on the lower side.

processor system using the 1802 family of integrated circuits. The hardware multiplier is a separate logic circuit that performs 8-bit multiplications at a high rate thus enhancing the capability of the microprocessor to process data rapidly. The flipper control, analog power control, and internal calibrate circuits are switches that are set by the microprocessor in response to ground commands. The power regulator and memory 'keep alive' circuit are power conditioning circuits that receive power from the spacecraft's

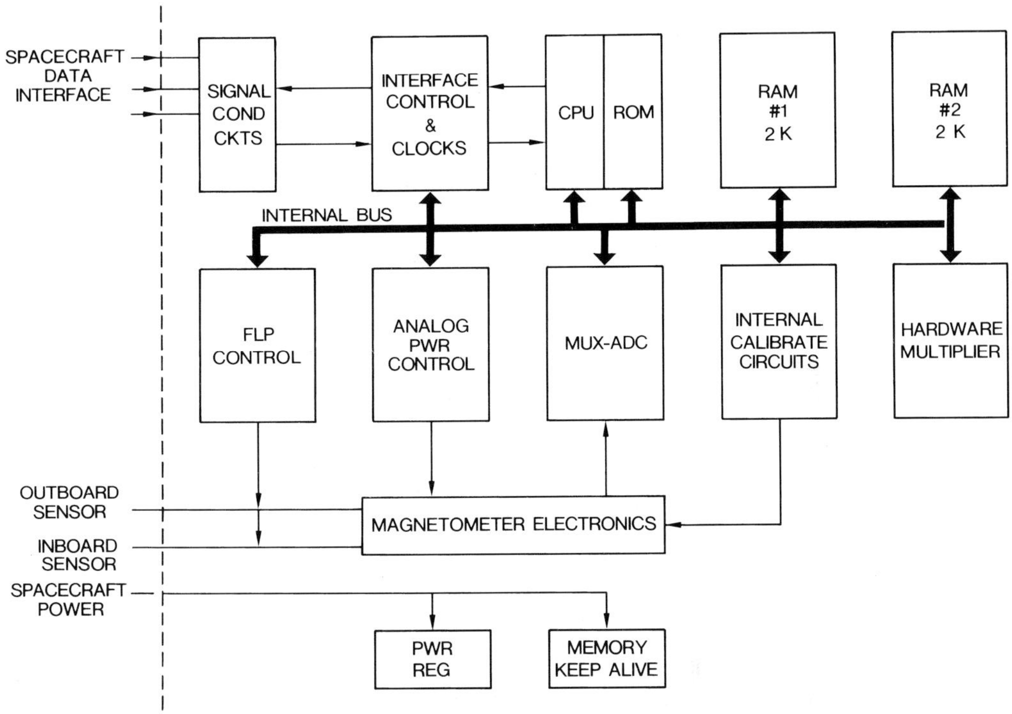

Fig. 4. Functional block diagram of the experiment.

alternating current power bus. The multiplexer, analog to digital converter (MUX-ADC) collects analog data under microprocessor control from the magnetometer circuits and from internal monitors of voltage and temperatures.

4. Magnetometer Commands and Modes of Instrument Operation

Commands can be sent to the spacecraft to control instrument operation. Table III lists the commands and the frequency with which routine commands will be sent. Interactive commands are those that affect other subsystems on the spacecraft whereas non-interactive commands are internal to the MAG subsystem. Critical commands are those which select either the inboard or outboard sensor to make the measurements and determine whether the high-gain (low-range) or low-gain (high-range) measurements are desired. These selections are made by specific command, not selected automatically, in order to avoid a situation where changes in instrument mode are made in the middle of an important measurement. The selection of optimal averaging or snapshot mode determines which sort of data will be stored in the instrument buffer (see the discussion of these modes in the section on data processing). The memory load provides an updated table for the coordinate transformation that converts measurements into absolute field vectors in a specified orthogonal coordinate system. The remaining commands are related to special procedures that give the parameters needed to convert measurements into absolute field values in a known coordinate system. They are discussed in the next section.

TABLE III

Instrument commands

Instrument mode	Time or frequency
Interactive commands	
Instrument power on/off	On prior to boom deployment
Flipper power on/off	Start/end flipper checks
Flip right, Flip left	For flipper checks
Flip (a toggle command)	(affects power consumption when activated)
Non-interactive commands	
Inboard power on/off	Few/orbit
Outboard power on/off	
Inboard/Outboard sensor select	Few/orbit
Hi range/low range select	Few/orbit
Despin select (on/off)	On change of sensor, flip position, gain, averaging interval
Optimal averaging (on/off)	When no real time or taped low rate data
Snapshot mode (on/off)	off/on command every ~40 min
Memory load as needed to update	The coordinate transform matrix/averaging interval
Internal calibration coil signal (on/off)	Weekly during cruise, twice per orbit at Jupiter
Interactive (calibration) commands using programmed sequences	
Offset checks/Flipping sensors	Weekly
Constant slew once/orbit	
MAG Cal Coil (on/off)	Following major engine burns and twice per orbit at Jupiter

The instrument processes data in two ways. In normal operation, the data processor demodulates the data to remove the effect of spacecraft spin and resolves it into an orthogonal system before transmitting the averages to the spacecraft data system. This process is described in greater detail below. Alternatively, direct measurements from each sensor can be sampled and averaged in even increments of time for transmission to the spacecraft data system. The latter method would require 'despinning' of the data on the ground and would be used only as a backup procedure or for instrument calibration as discussed below.

5. Determination of Instrument Offsets, Sensor Orientation and Gain

The internal calibrate signals drive prescribed currents through internal coils around the sensors and are used to determine instrument gain. In addition, the measurement must be corrected for slight misalignment of the sensors from the nominal orthogonal orientations. For this purpose, the data are used without despinning. A small amplitude modulation at the spacecraft spin period of the field in the sensor nominally aligned with the spin axis indicates misalignment. The measurements can readily be corrected to remove the oscillatory signal and obtain the true spin-axis component of the ambient field. Sensors precisely in the spin plane will record fields modulated at the spacecraft spin frequency. If offsets have been corrected and stray spacecraft fields are negligible, such signals average to zero over integral numbers of spins for steady background fields or if averaged over long times. Any non-vanishing average field in the sensors nominally in the spin plane can be attributed to the rotation of spin plane sensors out of the spin plane. By confirming that this spin average is proportional to the component of **B** along the spin-axis, it is possible to confirm that corrections for offsets and stray fields from the spinning part of the spacecraft have been made properly. The data can then be corrected to yield the spin plane components of the ambient field. If the sensors in the spin plane are perpendicular to one another, the signals are in quadrature. Misalignment within the spin plane results in phase shifts. They can be readily identified and the data further corrected. The procedure outlined here provides components of the field along three orthogonal directions, but the absolute orientation of the axes in the spin plane remains uncertain. The remaining uncertainty is removed by measurement of a reference signal that identifies a direction within the spin plane. The signal is produced by a coil rigidly mounted on the spacecraft.

The MAG Cal Coil (MCC) is mounted at the center of the lower ($+z$) edge of the boom with its axis aligned with the center line of the boom axis. It is a 20″ × 18″ roughly rectangular coil of 117 turns. During MAG Cal Coil calibrations, the coil is driven at one of several selected frequencies with amplitude (outboard) of 4.5 nT; (at 0.5 and 0.75 Hz the signal is a square wave, and at 1.5 Hz it is switched on for $\frac{2}{3}$ of each cycle). Because the frequency is known, the signal can be identified even if the field is not entirely quiet. The flexible choice of frequency allows for the possibility of noise sources on the spacecraft that may make it undesirable to operate in certain frequency bands. All else being equal, higher frequencies are desirable because natural background noise falls off with increasing frequencies. The MCC signal defines a reference direction in space, thus providing the required information to allow the sensor orientations to be well constrained.

The above procedure must be carried out for each sensor triad and for both flip positions of the sensor triad.

Fields produced by systems mounted on the despun part of the spacecraft cannot readily be differentiated from ambient fields. A special procedure has been designed to permit measurements of the magnetic signal of the despun portion of the spacecraft. Starting in dual spin mode, the relative spin rate of the despun portion of the spacecraft is gradually reduced until the spacecraft is in all-spin mode and the procedure is reversed until the spacecraft is once again in the dual spin mode. The procedure takes approximately four hours. The procedure is initiated by the *constant slew* command and will be carried out once per orbit at Jupiter.

Finally, offsets in the instrument itself can be identified by measurements made in the known field of the Earth's magnetosphere. This will be possible during the initial post-launch period and during the two subsequent flybys of Earth. Standard methods of identifying offsets through statistical analysis of tangential discontinuities in the solar wind will also be used.

6. On-Board Data Processing

A flow diagram of the data processing is shown in Figure 5. The system uses a 12-bit analog-to-digital converter that has been trimmed such that each bit is accurate to at least $\frac{1}{4}$ of the value of the least significant bit. The data, sampled at 30 samples s^{-1}, are filtered to provide 16-bit words, tests have demonstrated that these words are accurate to the 15th bit. This filtering is performed by the microprocessor system using a recursive

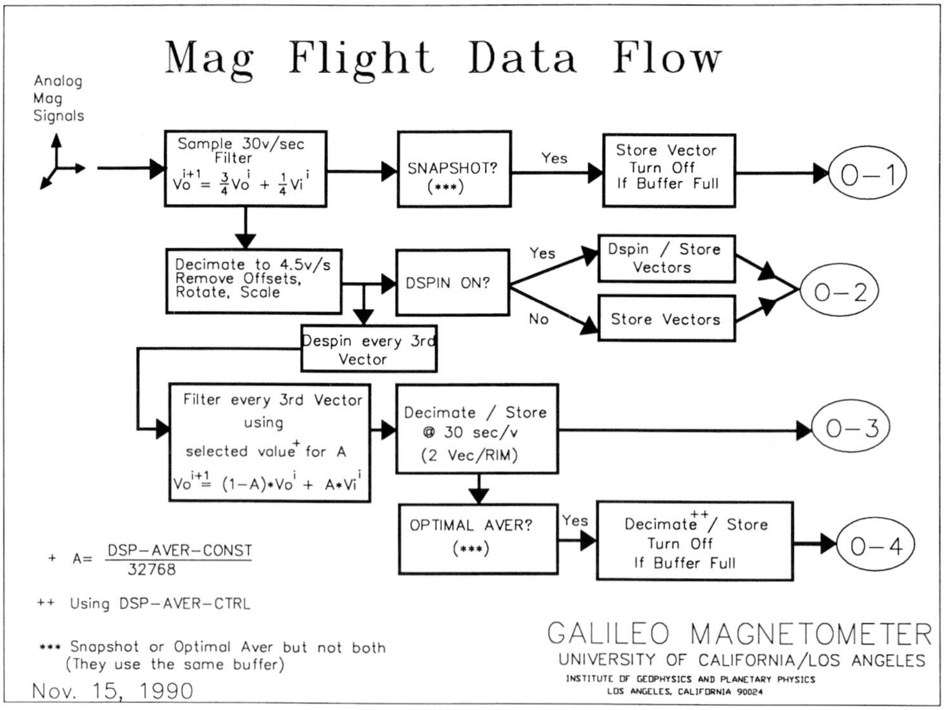

Fig. 5. Data processing flow diagram. O-1 to O-4 are types of data readout available. When tracking is available, O-2 and O-3 readouts are both transmitted and O-1 is optional. Readouts O-4 are collected during intervals when tracking is not available and stored for later transmission. The abbreviation v/s is vectors/second. DSP-AVER-CONST and DSP-AVER-CTRL are parameters defined in the instrument command log; their use is discussed in the text.

filter, a form selected for economy of implementation. The recursive filter algorithm is of the form,

$$V_0^j = (1 - A)\,V_0^{j-1} + A V_i^j\,, \tag{1}$$

where V_0^j is the (output) value calculated from measurements prior to measurement j and V_i^j is the (input) value read in at measurement j. The value of A is set at $\frac{1}{4}$ for the filter at this stage of the data processing. This filter gives a transfer function (not used in on-board processing) of the form,

$$T(f) = \frac{\pi\Omega_c}{(1 + \pi\Omega_c) - \exp(-i\pi f/f_n)}\,. \tag{2}$$

The relationship of A in Equation (1) to Ω_c of Equation (2) is given by the expression,

$$A = \frac{\pi\Omega_c}{1 + \pi\Omega_c}\,, \tag{3}$$

where $\Omega_c = f_c/f_n$, and f_n = Nyquist frequency = $(2DT)^{-1}$, f_c = the filter corner frequency and DT is the data sampling interval, $\frac{1}{30}$ s. This gives $f_c \approx 1.6$ Hz, i.e., the filter cuts off frequencies above the telemetry Nyquist frequency, 2.25 Hz. At frequencies well below f_c (including the range from *dc* through low multiples of the spin frequency) T is close to unit amplitude and has a small imaginary part.

The special problems associated with detection of signals in the ambient medium that are circularly polarized about the spin axis was alluded to in the introduction. A filter that is not flat for frequencies in a range between *dc* and low multiples of the spin frequency can distort ambient signals, especially near the spin frequency. The signal processing of Equation (1) does not distort signals in this range which lies well below f_c. Next the data are corrected by subtracting the sensor offsets from the data and multiplying the data by a coordinate transformation matrix. This matrix contains sensor gain and pointing information. The correction routine yields accurate data at 3 vectors each $\frac{2}{3}$ s (4.5 v/s) in true spacecraft coordinates. The information for the corrections, i.e., offsets and matrix values, is computed on the ground and sent via the command system as a direct memory load to the microprocessor memory.

In Figure 6, one hour of data is used to illustrate the results obtained by the different paths to the read-outs (O-1 to O-4) represented on the right side of Figure 5. The data are decimated to 3 vectors each $\frac{2}{3}$ second for the low rate science data output which follows the path to O-2. These data may be commanded to be either in spacecraft coordinates or to be despun data in inertial coordinates. The spacecraft coordinates are defined with the spin axis $+z$ pointing in the anti-Earth direction; the magnetometer boom is the $+y$ axis, and the $+x$-axis completes the right-hand triad as shown in Figure 2. Figure 6(a) illustrates the non-despun data for the one hour sample interval. In this form, the data in the two spin plane directions (x and y) are modulated at the spacecraft spin period. These data were acquired at an early phase of operations and the spin-axis-aligned component, z, and the field magnitude show very small amplitude ripples at the spin period. Even this small amplitude ripple is later removed in ground data processing. Improved parameters have been developed and are now in use for on-board calibration.

To obtain magnetic field data in inertial (I) coordinates, the data are despun on board by multiplying the spacecraft (sc) frame B_x and B_y by sine and cosine functions generated from spacecraft roll information.

$$B_{xI} = B_{xsc} \cos \theta - B_{ysc} \sin \theta , \tag{4}$$

$$B_{yI} = B_{xsc} \sin \theta + B_{ysc} \cos \theta , \tag{5}$$

$$B_{zI} = B_{zsc} . \tag{6}$$

The angle θ, the spacecraft roll angle, is measured from the projection onto the spacecraft spin plane of a southward pointing vector normal to the ecliptic plane.

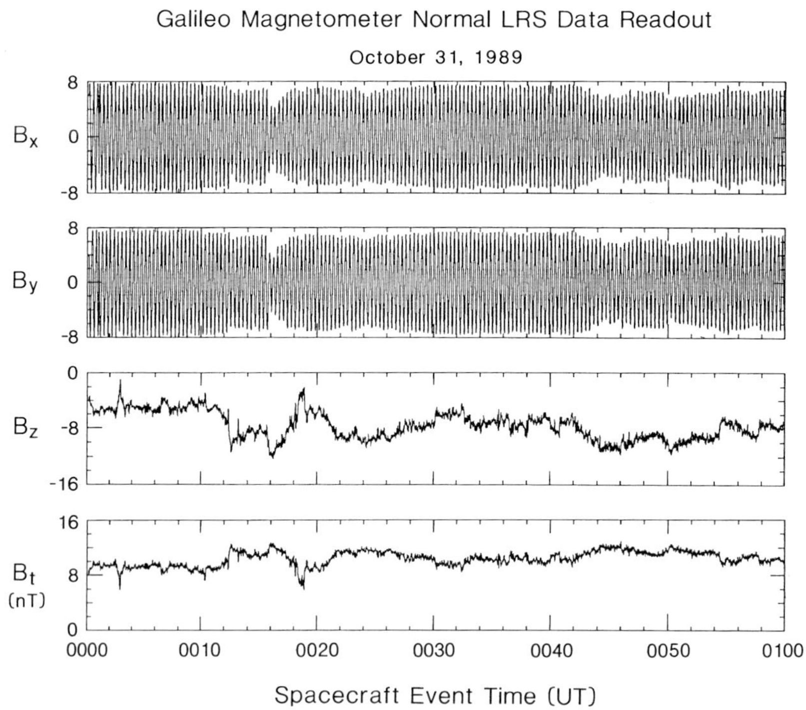

Fig. 6a.

Fig. 6. One hour of data taken shortly after Galileo launch (00:00 to 01:00 UT on 31 October, 1989) illustrating different data formats available. (a) The low rate science data (4.5 vectors per second) in the spacecraft coordinate system. (b) The low rate science data of (6a) rotated into an inertial coordinate system. The despinning could have been done on board. (c) The despun averaged data processed on the spacecraft and transmitted as part of the low rate science data at a rate of 1 vector/($30\frac{1}{3}$ s). (d) The same data as in (a) presented as if the instrument had been in the optimal average mode with a sampling rate of 1 vector/(8 min). (e) A 7-s 'snapshot' within the hour. Here the samples are at $\frac{1}{30}$ s intervals.

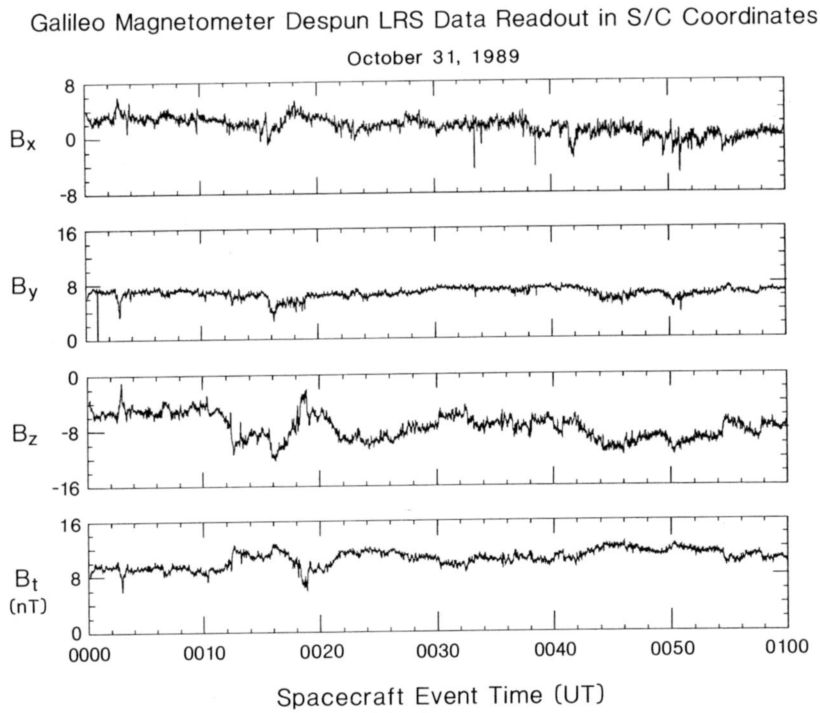
Galileo Magnetometer Despun LRS Data Readout in S/C Coordinates
October 31, 1989
Bx
By
Bz
Bt
(nT)
8
0
-8
16
8
0
0
-8
-16
16
8
0
0000
0010
0020
0030
0040
0050
0100
Spacecraft Event Time (UT)

Fig. 6b.

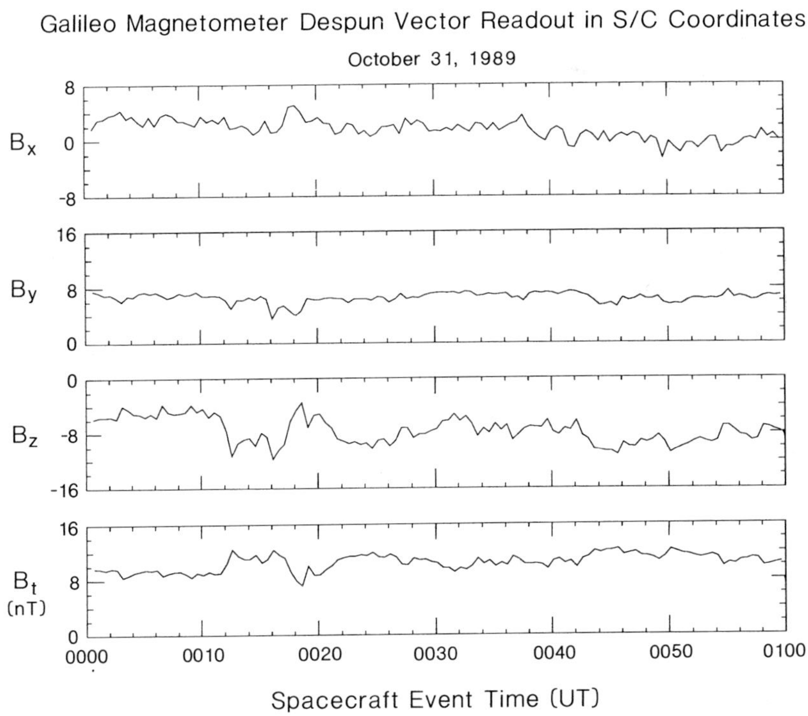
Galileo Magnetometer Despun Vector Readout in S/C Coordinates
October 31, 1989
Bx
By
Bz
Bt
(nT)
8
0
-8
16
8
0
0
-8
-16
16
8
0
0000
0010
0020
0030
0040
0050
0100
Spacecraft Event Time (UT)

Fig. 6c.

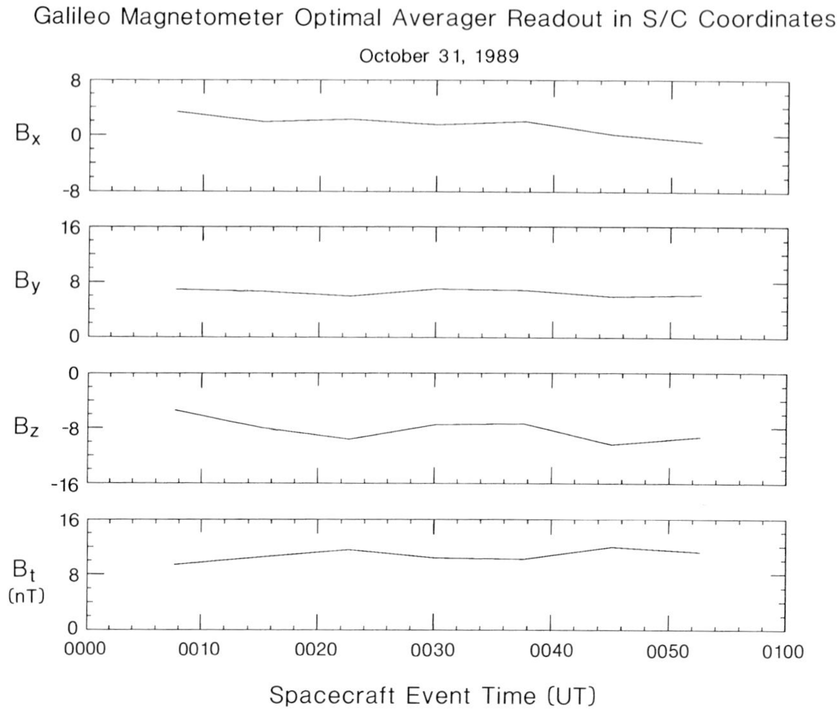
Galileo Magnetometer Optimal Averager Readout in S/C Coordinates
October 31, 1989
Bx
By
Bz
Bt
(nT)
8
0
-8
16
8
0
0
-8
-16
16
8
0
0000
0010
0020
0030
0040
0050
0100
Spacecraft Event Time (UT)

Fig. 6d.

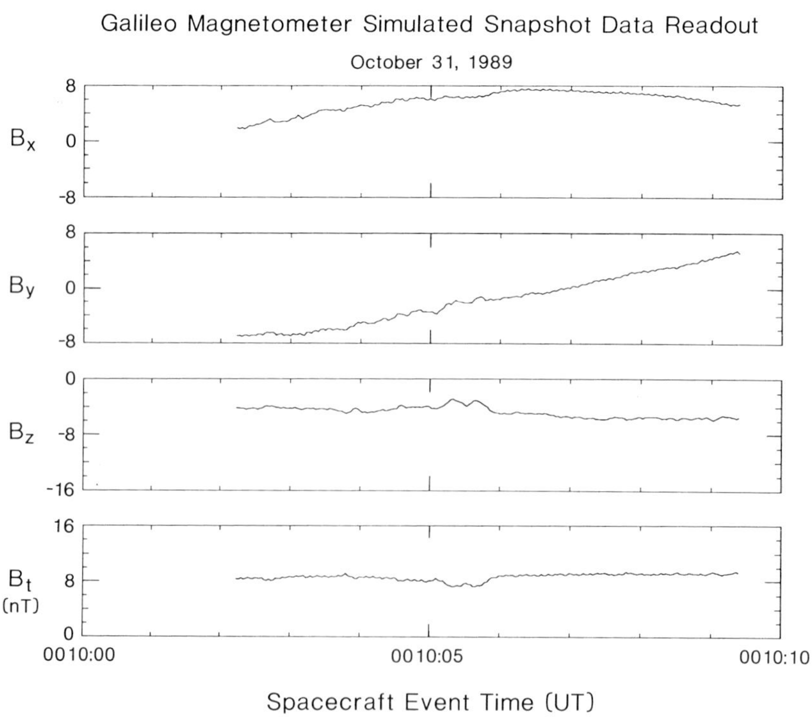
Galileo Magnetometer Simulated Snapshot Data Readout
October 31, 1989
Bx
By
Bz
Bt
(nT)
8
0
-8
8
0
-8
0
-8
-16
16
8
0
0010:00
0010:05
0010:10
Spacecraft Event Time (UT)

Fig. 6e.

Therefore, in inertial coordinates positive B_{xI} points south, normal to the ecliptic plane, positive B_{yI} lies in the ecliptic plane in the sense of Jupiter's orbital motion and positive B_{zI} is in the anti-earth direction. Figure 6(b) illustrates the results for the same input data if the instrument were operating in the despin mode. Note that the z-component, virtually aligned with the spin axis, and the total field have not changed but the large modulation at the spin period has disappeared.

Following the path in Figure 5 to O-3, despun data are again filtered and 2 vectors each minute (actually each $60\frac{2}{3}$ s) are provided in the subcommutated data. The filter is of the form given in Equation (1), with $A = 512/32768$. Figure 6(c) illustrates this lower resolution record for the same time interval. These subcommutated despun data are provided continuously, even when the main instrument output data are provided in spacecraft coordinates and give physically useful information directly from the experiment data record without any further processing.

For periods of up to a week during the cruise mode on the journey to Jupiter, the spacecraft data system may not be collecting data from the science instruments. During such times, the magnetometer will be left on. Then, following the path of Figure 5 to O-4, despun data will be averaged over times longer than the basic 30 s intervals and stored in memory. The averaging constant, A in Equation (1), is selected so that the memory will contain time and attitude information and 200 evenly spaced data points for the planned period between telemetry readouts. Commands set the parameter DSP-AVER-CONST of Figure 5, where A = (DSP-AVER-CONST)/32768 and the parameter ranges from 1024 for 1-min time resolution through 1 for 16-hr time resolution. This mode of operation is referred to as the *optimal averaging* mode. Figure 6(d) illustrates the output of the optimal average mode at 8-min time resolution for the same time interval.

The instrument memory can alternatively be used for recording short bursts of high speed data at 30 samples per second. This mode, called the *snapshot*, will be used for trying to capture transient events such as shocks or to increase the Nyquist frequency of the magnetometer measurements in order to provide a continuous spectrum of plasma waves from DC to frequencies that overlap the lowest frequency band of the Plasma Wave System (PWS). In Figure 5, this mode follows the path to O-1. The snapshot data come from the output of the digital averager at 30 samples per second. By inverting the operation of Equation (1), the original unfiltered input data are retrieved. The highest time resolution samples are not equispaced in time, the time-interval separating digitized values of the vector components is approximately 0.2 ms and the time lapse between successive vector samples is fixed at 33.3 ms. The buffer memory which contains the snapshot data (210 48-bit vectors and a time record) is continuously read out at 16 words per minute. The memory readout appears in the subcommutated MAG data. A seven second burst within the sample time interval is illustrated in Figure 6(e). As the spacecraft spin period is 20 s, the x- and y-components are modulated by the spacecraft spin, evident as a portion of a sinusoid.

7. Data Format

The low rate science data format contains three vector sets of x, y, z samples and a sub-commutator word. Each word is 16 bits in length and the readout provides 160 bits each $\frac{2}{3}$ of a second for a total readout of 240 bps. The time interval between the first and second, and second and third vectors is 233 ms. The time interval between the third vector and the first vector of the next minor frame is 200 ms. This slight asymmetry of sampling could not be avoided because of incongruity between telemetry and instrument data rates. Ground processing can be used to remove the undesirable features of non-uniform sampling when necessary for refined analysis. The approach is to break the non-equispaced time series of n elements into three equispaced time series of lower time resolution (666 ms). The elements of the three series are composed of elements $3i$-2, $3i$-1, or $3i$, respectively, where $i = 1, 2, \ldots, n/3$. Each series is Fourier transformed. The concept of frequency folding is used to relate these Fourier transforms to the transform of an equispaced time series of 222 milliseconds resolution. The procedure requires the inversion of a set of 6×6 equations. Inverse Fourier transformation then produces the desired equispaced time series. Details are provided in Appendix 1.

The subcommutated data consist of 91 words, corresponding to a data rate of $(60\frac{2}{3}\ \mathrm{s})^{-1}$. The sub-commutator contains 2 readouts of despun averaged data, 16 words of stored buffer data; the remainder is housekeeping information to determine the status of the instrument. Included in the housekeeping data are various power supply voltages, a readout of all commands, matrix values and all memory-stored constants that are transferred to the instrument from the command system, and memory checksums.

8. Performance

8.1. Instrument noise levels

The magnetometer sensors and sensor electronics have been selected to achieve the lowest noise possible. However, as is true with many magnetometers, the limiting factors in instrument noise are the digitization noise of the analog-to-digital converter and electronic noise. The technique of averaging 12-bit data to obtain 16-bit words was used

TABLE IV

Instrument noise performance

Range (nT)	Measured Resolution (nT)		Digitization noise ($nT^2\,Hz^{-1}$)	Measured noise ($nT^2\,Hz^{-1}$)	
	12 bit	16 bit		at 1 Hz	at 0.01 Hz
± 16384	8	$\frac{1}{2}$	1.9×10^{-2}	1×10^{-1}	$< 10^{-1}$
± 512	$\frac{1}{4}$	$\frac{1}{64}$	5.8×10^{-4}	1×10^{-3}	$< 4 \times 10^{-2}$
± 32	$\frac{1}{64}$	$1/1024$	3.6×10^{-5}	5×10^{-5}	$< 2 \times 10^{-2}$

on magnetometers for ISEE 1 and 2 and successfully reduced digitization noise as described further by Russell (1978).

Table IV shows the resolution for each dynamic range of the magnetometer. The third column displays the calculated digitization noise for the 16-bit average. The noise measured in the laboratory at 1 Hz and at 0.01 Hz is shown in the last columns. The instrument noise may be lower than the table entries as ambient noise sources cannot be completely eliminated in the laboratory tests. The noise level at 1 Hz is close to the calculated digitization noise in the low dynamic ranges of the instrument. For the larger dynamic range of $\pm 16\,348$ nT, the electronic noise of the system raises the measured noise above the digitization noise at 1 Hz by approximately one half decade. In this range, 1 nT corresponds to 300 μV which is of the order of the electronic noise of the circuits. Appropriately selected data from flight operation will be used to provide improved estimates of instrument noise levels.

Khurana *et al.* (1991) have shown that the power in the quietest spectra obtained from data in the lobes of the Jovian magnetosphere (where the ± 32 nT range is relevant) is of order 1 $nT^2\,Hz^{-1}$ near 0.01 Hz. The lobe power falls off approximately like the inverse frequency, consistent with levels near 10^{-2} $nT^2\,Hz^{-1}$ near 1 Hz, which will be readily detectable above instrument noise levels.

8.2. In-flight Calibration

The magnetometer assemblies were calibrated repeatedly in the laboratory prior to launch, providing excellent initial calibration parameters which were loaded into the instrument memory prior to launch. Data from flight operation revealed that the calibration parameters in the flight configuration could be further improved. Appendix 2 describes the approach developed for in-flight calibration and gives examples of its application. Calibration must be done separately for the different operating modes of the instrument of which there are 8 (two sensor assemblies each with two flip orientations and two ranges). The calibration uses low rate science data ($\frac{2}{9}$ s sample^{-1}) of which only limited intervals were available during the first year of operation; only two of the operating modes were fully calibrated before the first flyby of the Earth; for these modes, the calibrated data have achieved the desired levels of accuracy. The MAG Cal Coil wave form has been studied but the analysis of absolute sensor orientation has not been completed at this time.

Acknowledgements

At UCLA we are grateful to F. R. George and D. Dearborn who did the preliminary circuit design and final calibrations, and R. J. Walker who provided expert advice on many instrument-related issues. Excellent low noise sensors were supplied by R. Lundsten of Naval Surface Weapons Center, White Oak. The final design, packaging and fabrication was skillfully directed by D. Frankel and R. Mack of Westinghouse. We benefitted much from the advice of M. Larson of the Office of Naval Research and Ulrika Williamson, Douglas Clay, Maria Etchetgaray, Ralph Parish, Howard Eyerly, and Carol Polanskey of JPL.

Appendix 1. Redigitization of Magnetometer Data

Because of a mismatch between the instrument and spacecraft clock cycles, the magnetometer data received from the spacecraft in the low-rate science mode are not equispaced. The spacecraft returns 3 vectors every 0.666 s. The time spacing (Δt) between vectors 1 and 2 and vectors 2 and 3 is equal to 0.233 s, but the spacing between vectors 3 and 4 is equal to 0.200 s. Many applications (e.g. fast Fourier transforms, auto- and cross-correlations, convolution, etc.) require equispaced data. If one incorrectly assumes that the data are equispaced then the input signal is distorted. The distortion of the input signal increases with increasing input frequency, the most severe distortion occurring near the Nyquist frequency (2.25 Hz).

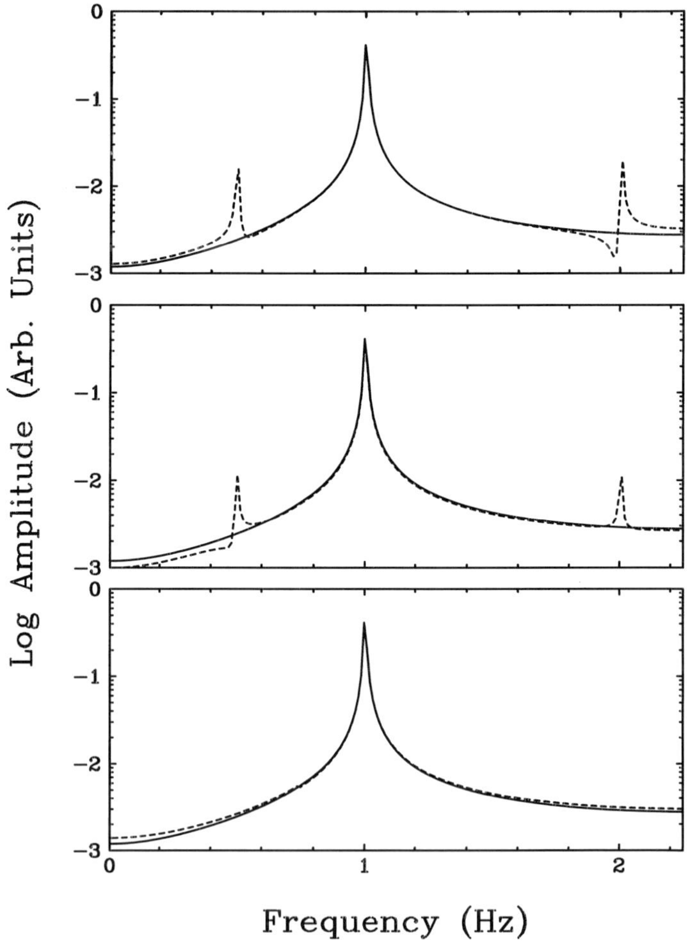

Fig. A1-1. *Top panel:* Amplitude spectra of the unequispaced series $F(t)$ (broken line) and the equispaced time series digitized at 0.222 s resolution (solid line). *Middle panel:* Amplitude spectra of the linearly interpolated redigitized time series (broken line) and the equispaced time series (solid line). *Bottom panel:* Amplitude spectra of the time series redigitized by the present scheme (broken line) and the equispaced time series (solid line).

To study this effect, we sampled a sinusoidal signal of frequency $f = 1$ Hz with $\Delta t = 0.222$ s and at the irregular spacings of our magnetometer. The fast Fourier transforms of the two time series (treated as if the data were equispaced) are shown in the top panel of Figure A1.1. It can be seen that in the transform of the irregularly spaced series (dashed curve), there are ghosts of the input frequency at 0.5 Hz and 2 Hz in addition to the peak at 1 Hz. The transform of the equispaced data (solid curve) show only a peak at 1 Hz. If one linearly interpolates the irregularly spaced data at $\Delta t = 0.222$ s, ghosts still appear in the Fourier transform (Figure A1.1 middle panel), though their amplitude is diminished.

We have devised a redigitization scheme that takes advantage of certain properties (frequency folding, phase shifting) of the Fourier transforms to minimize the signal distortion. The technique consists of breaking the original time series into three equispaced lower resolution (0.666 s) time series and recombining them in the frequency domain so that the inverse Fourier transform would yield an equispaced time series at $\Delta t = 0.222$ s. In our process we split the original time series $F(t)$ into three time series $G_i(t)$, $i = 1, 2\ 3$ such that

$G_1(t)$ contains points 1, 4, 7, 10 ... ...,

$G_2(t)$ contains points 2, 5, 8, 11 ... ...,

and

$G_3(t)$ contains points 3, 6, 9, 12 ... .

Notice that series G_2 and G_3 are equispaced with $\Delta t = 0.666$ s but are shifted from G_1 by 0.233 s and 0.466 s, respectively. The objective is to combine the series G_1, G_2, and G_3 to obtain a redigitized equispaced ($\Delta t = 0.222$ s) time series that we denote by $G(t)$. If the Nyquist frequency of $G(t)$ is $f_N = 2.25$ Hz then the Nyquist frequency of the lower time resolution series is f_{N*} ($= 0.75$ Hz). If $G(t)$ had power in frequencies above f_{N*}, then that information would be folded into frequencies below f_{N*} in the spectra of $G_i(t)$. In general the aliased power at a frequency $f \leq f_{N*}$ can be written as (Kanasewich, 1975)

$$P_a(f) = \sum_{n=-\infty}^{\infty} P(f - 2nf_{N*}) =$$

$$= P(f) + \sum_{n=1}^{\infty} P(f - 2nf_{N*}) + \sum_{n=1}^{\infty} P(f + 2nf_{N*}) . \tag{1}$$

As we are interested only in frequencies up to $3f_{N*}$, we will include terms only up to $|n| = 1$ in the above equation. Then $G_1(f)$ (the Fourier transform of $G_1(t)$) is given by

$$G_1(f) = G(f) + G(f - 2f_{N*}) + G(f + 2f_{N*}) . \tag{2}$$

For Fourier transforms $G_2(f)$ and $G_3(f)$, the frequency folding occurs in the same way but their elements also suffer a phase shift in accordance with the fact that for a Fourier transform pair, a shift of τ in the time domain translates into a phase change of $2\pi f \tau$

in the frequency domain. Therefore,

$$G_2(f) = G(f)\, e^{i2\pi f\tau} + G(f - 2f_{N*})\, e^{i2\pi(f - 2f_{N*})\tau} + G(f + 2f_{N*})\, e^{i2\pi(f + 2f_{N*})\tau}, \tag{3}$$

$$G_3(f) = G(f)\, e^{i4\pi f\tau} + G(f - 2f_{N*})\, e^{i4\pi(f - 2f_{N*})\tau} + G(f + 2f_{N*})\, e^{i4\pi(f + 2f_{N*})\tau}. \tag{4}$$

Complex equations (2) to (4) can be rewritten in the matrix form as

$$\begin{pmatrix} \mathcal{R}(G_1(f)) \\ \mathcal{I}(G_1(f)) \\ \mathcal{R}(G_2(f)) \\ \mathcal{I}(G_2(f)) \\ \mathcal{R}(G_3(f)) \\ \mathcal{I}(G_3(f)) \end{pmatrix} = \tag{5}$$

$$= \begin{pmatrix} 1 & 1 & 1 & 0 & 0 & 0 \\ 0 & 0 & 0 & 1 & 1 & 1 \\ \cos\alpha & \cos\beta & \cos\gamma & -\sin\alpha & -\sin\beta & -\sin\gamma \\ \sin\alpha & \sin\beta & \sin\gamma & \cos\alpha & \cos\beta & \cos\gamma \\ \cos 2\alpha & \cos 2\beta & \cos 2\gamma & -\sin 2\alpha & -\sin 2\beta & -\sin 2\gamma \\ \sin 2\alpha & \sin 2\beta & \sin 2\gamma & \cos 2\alpha & \cos 2\beta & \cos 2\gamma \end{pmatrix} \begin{pmatrix} \mathcal{R}(G(f)) \\ \mathcal{R}(G(f - 2f_{N*})) \\ \mathcal{R}(G(f + 2f_{N*})) \\ \mathcal{I}(G(f)) \\ \mathcal{I}(G(f - 2f_{N*})) \\ \mathcal{I}(G(f + 2f_{N*})) \end{pmatrix},$$

where $\alpha = 2\pi f\tau$, $\beta = 2\pi(f - 2f_{N*})\tau$, $\gamma = 2\pi(f + 2f_{N*})\tau$, and $\mathcal{R}$ and $\mathcal{I}$ denote the real and imaginary parts of the functions. There are six unknowns and six equations. By inverting the set of Equations (5) one can reconstruct the spectrum of $G(f)$ which will have frequencies up to $f_N = 3f_{N*}$. The bottom panel of Figure A1.1 shows (dashed curve) the Fourier transform constructed in this way. Notice that the ghosts of the primary input frequency have vanished in the reconstructed spectrum and that the aliased power changes the amplitude below the Nyquist frequency by a negligible amount. An inverse Fourier transform of the reconstructed spectrum yields the desired equispaced data.

Appendix 2. In-flight Calibration of the Magnetometer Assemblies by Non-linear Optimization

Galileo is scheduled to reach Jupiter in December 1995. The long idle time between the fabrication of the magnetometer and the launch of the spacecraft provided ideal opportunities to calibrate the magnetometer assemblies in the laboratory repeatedly with very high accuracy. Three different calibrations were performed over a period of four years and the gains and the relative angles of the magnetometer assemblies were obtained with an accuracy of 1 part in 10^4.

With the launch of the spacecraft in October 1989, operational data became available and the matrices available from the ground calibrations were tested and improved further. A scheme based on non-linear optimization through a gradient search method (Bevington, 1969) was adopted to obtain the elements of the coupling matrix and the offsets. To illustrate the scheme, let us assume that the data from the spacecraft are obtained in a rotating coordinate system (1, 2, 3) so that the measured components of the magnetic field are (B_1, B_2, B_3). We identify directions 1, 2, and 3 as the 'look' directions of the sensors which are measuring field components almost along the two orthogonal spin plane directions of the spacecraft coordinate system (x and y) and along the spin axis (z), respectively. The objective is to calculate the magnetic field in a rotating coordinate system that is truly orthogonal with $\hat{z}$ strictly along the spin axis. The calculated field $\mathbf{B} = (B_z, B_y, B_z)$ must also be corrected for offsets and gains. This requires us to know the elements of a 3×3 matrix (called a coupling matrix) and the offsets (O_1, O_2, O_3) in the three sensors so that we can calculate

$$\begin{pmatrix} c_{11} & c_{12} & c_{13} \\ c_{21} & c_{22} & c_{23} \\ c_{31} & c_{32} & c_{33} \end{pmatrix} \begin{pmatrix} B_1 - O_1 \\ B_2 - O_2 \\ B_3 - O_3 \end{pmatrix} = \begin{pmatrix} B_x \\ B_y \\ B_z \end{pmatrix} . \tag{1}$$

The data are then despun into an inertial coordinate system (X, Y, Z) to obtain (B_X, B_Y, B_Z).

Data in this frame should normally not contain power at f_{spin}, the spin frequency, or its harmonics at levels above background. The elements of the coupling matrix and the offsets are varied in order to minimize the amplitude of the spin ripple and its harmonics in one or more components of the vector field $\mathbf{B}$ in the inertial coordinate system (X, Y, Z) using enough data that power from fortuitous ambient signals at f_{spin} is very small. One starts with some assumed values (either 0 for the minor elements and the offsets and 1 for the diagonal elements, or the parameters obtained from laboratory calibrations) and calculates the harmonic power. Then using a gradient search routine, one searches nearby regions of parameter space for the parameter set that minimizes the power at the spin frequency and twice the spin frequency in the data in the inertial coordinates. It can be shown that only 8 of the 12 unknowns in the above equation can be determined from schemes that minimize power at spin-related frequencies. For example, small changes in the element c_{33} which multiples the term $(B_3 - O_3)$ make no first-order change in the amplitude of the spin ripple in B_z because the term $(B_3 - O_3)$ itself contains very little spin ripple. Similarly, it can be shown that by using this technique, one can improve the estimate of c_{12} or c_{21} or a linear combination of them, but not both. The same is true of the pair c_{11} and c_{22}. Offset O_3 also cannot be determined from this method because an offset in the spin-direction sensor does not produce any spin ripple in the inertial coordinate system. This means that 4 parameters must be determined by other methods. Our initial approach is to set c_{12} and O_3 to zero and c_{11} and c_{33} to values obtained from laboratory calibrations. We optimize the other 8 parameters. The Galileo magnetometer can flip sensor 3 from the spin axis into the spin plane. Data from the flipped configuration enables us the improve estimates of O_3. It can be shown that the effect of fixing c_{11} but letting c_{22} vary is equivalent to varying the gain of sensor 2 in terms of the known gain of sensor 1. Calibration coils internal to the magnetometer will be used to test the constancy of the sensor gains over time.

Once the elements of the coupling matrix have been obtained, we know the relative angles of the three sensors, but the information on the absolute orientation of the sensor assembly around the spin axis cannot be obtained by reducing spin-related power. To

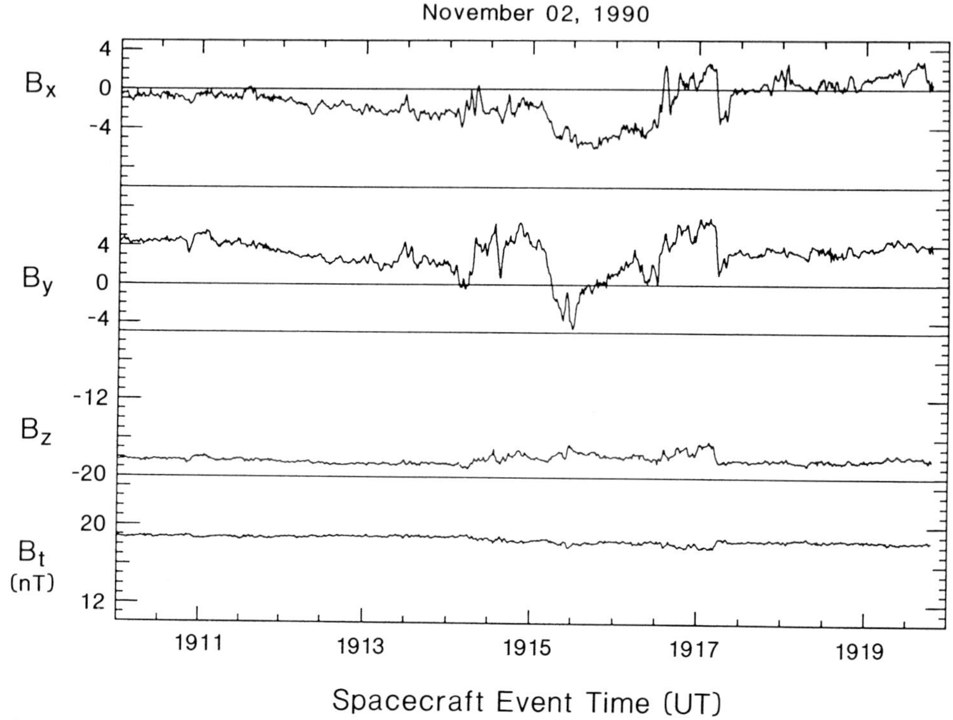

Fig. A2-1. Despun data obtained using the matrix and the offsets from the non-linear optimization technique.

obtain the orientation of the assembly relative to the spacecraft axes, an external calibration coil mounted on the spacecraft is used. This coil generates a precisely known square wave that provides a spacecraft-fixed reference direction in the spin plane. This procedure yields further improvements in the elements c_{12} and c_{21}.

We have successfully tested this scheme for several modes of Galileo magnetometer operations. The relative angles of the sensors obtained from this scheme are within 0.1° of the angles obtained from the laboratory calibrations. The optimization program itself is very simple and less than 10 min of computer time is required on a Microvax work station to obtain all 8 parameters from 5 hours of input data. To illustrate the quality of the output despun data, we show the application of the method to a data set obtained from Galileo about two weeks after the launch when the spacecraft was in the solar wind. Figure A2.1 shows the despun data with the best matrix and offsets obtained from the optimization scheme described above. The spin ripple in all three components is below the ambient background noise in the data. This accuracy was achieved when only a few days of magnetic data with spacecraft attitude information were available to determine the matrix elements and the offsets. Further improvements in the parameters can be anticipated when quieter intervals become available for examination.

References

Acuna, M. H., Behannon, K. W., and Connerney, J. E. P.: 1983, in A. J. Dessler (ed.), *Physics of the Jovian Magnetosphere*, Cambridge University Press, Cambridge, MA.

Belcher, J. W.: 1983, in A. J. Dessler (ed.), *Physics of the Jovian Magnetosphere*, Cambridge University Press, Cambridge, MA.

Bevington, P. R.: 1969, *Data Reduction and Error Analysis for the Physical Sciences*, McGraw-Hill Book Co., New York.

Birmingham, T. J.: 1983, *Rev. Geophys.* **21**, 375.

Connerney, J. E. P.: 1987, *Rev. Geophys.* **25**, 615.

Dessler, A. J. (ed.): 1983, *Physics of the Jovian Magnetosphere*, Cambridge University Press, Cambridge, MA.

Gehrels, T., (ed.): 1976, *Jupiter*, University Arizona Press, Tucson.

Gordon, D. I. and Brown, R. E.: 1972, *IEEE Transactions on Magnetics* **MAG-8**, 76.

Hill, T. W.: 1979, *J. Geophys. Res.* **84**, 6554.

Kanasewich, E. R.: 1975, *Time Series Analysis in Geophysics*, University Alberta Press, Edmonton, Canada.

Khurana, K. K., Chen, S.-H., Hammond, C. M., and Kivelson, M. G.: 1991, *Adv. Space Res.*, (in press).

Kivelson, M. G., Slavin, J. A., and Southwood, D. J.: 1979, *Science* **205**, 491.

Krimigis, S. M. and Roelof, E. C.: 1983, in A. J. Dessler (ed.), *Physics of the Jovian Magnetosphere*, Cambridge University Press, Cambridge, MA.

Ness, N. F., Behannon, K. W., Lepping, R. P., and Schatten, K. H.: 1971, *J. Geophys. Res.* **76**, 3564.

Russell, C. T.: 1978, *IEEE Transactions on Geoscience Electronics* **GE-16**(3).

Snare, R. C. and Means, J. D.: 1977, *IEEE Transactions on Magnetics* **MAG-13**(5).

Vasyliunas, V. M.: 1983, in A. J. Dessler (ed.), *Physics of the Jovian Magnetosphere*, Cambridge University Press, Cambridge, MA.

THE GALILEO ENERGETIC PARTICLES DETECTOR

D. J. WILLIAMS[1], R. W. McENTIRE[1], S. JASKULEK[1], and B. WILKEN[2]

1 *The Johns Hopkins University, Applied Physics Laboratory, Laurel, MD 20723, U.S.A.*

[2] *Max Planck Institute für Aeronomie, Katlenburg/Lindau, Germany*

Abstract. Amongst its complement of particles and fields instruments, the Galileo spacecraft carries an Energetic Particles Detector (EPD) designed to measure the characteristics of particle populations important in determining the size, shape, and dynamics of the Jovian magnetosphere. To do this the EPD provides 4π angular coverage and spectral measurements for $Z \geq 1$ ions from 20 keV to 55 MeV, for electrons from 15 keV to > 11 MeV, and for the elemental species helium through iron from approximately 10 keV nucl^{-1} to 15 MeV nucl^{-1}. Two bi-directional telescopes, mounted on a stepping platform, employ magnetic deflection, energy loss versus energy, and time-of-flight techniques to provide 64 rate channels and pulse height analysis of priority selected events. The EPD data system provides a large number of possible operational modes from which a small number will be selected to optimize data collection during the many encounter and cruise phases of the mission. The EPD employs a number of safeing algorithms that are to be used in the event that its self-checking procedures indicate a problem. The EPD has demonstrated its operational flexibility throughout the long evolution of the Galileo program by readily accommodating a variety of secondary mission objectives occasioned by the changing mission profile, such as the Venus flyby and the Earth 1 and 2 encounters. To date the EPD performance in flight has been nominal. In this paper we describe the instrument and its operation.

Preface

We wish to acknowledge the major contribution of three EPD team members who did not live to see the results of their efforts. Wolfgang Stüdemann, a member of the EPD science team, fulfilled major responsibilities in detector head design, verification testing, and instrument calibration. Wolfgang provided a professional competence and a personal warmth that was felt throughout the EPD team. He met an untimely and tragic death on December 7, 1987. James Cessna, a member of our engineering team, was responsible for the design and construction of the original time-of-flight circuitry for the EPD composition telescope. Jim performed this task with a quiet excellence that commanded the respect of all EPD team members. Jim's untimely death occurred on March 11, 1989. Eric Bubla, a member of our technical team, participated in the mechanical design of the EPD detector heads and was responsible for their construction and assembly. Eric performed these tasks as a master craftsman and his work was respected and appreciated by the EPD team. Sadly, Eric passed away on April 1, 1989. These three talented people were positive influences not only on the EPD team but also on all who were fortunate enough to know them. We will miss them. We dedicate this paper to our late colleagues, Wolfgang Stüdemann, James Cessna, and Eric Bubla.

1. Introduction

Jupiter possesses the largest planetary magnetosphere in the solar system. It is the largest in spatial dimension, has the highest trapped particle energies and intensities, has

Space Science Reviews **60:** 385–412, 1992.

the greatest compositional variety in its major particle populations, displays the largest co-rotational effects and has the largest number of moons within the magnetosphere that provide both strong sources for and losses of the observed particle populations. These characteristics, uncovered by the Pioneer and Voyager flybys (see for example, the articles in the following special issues Pioneer 10 – *Science* **183**, 301–324, 1974; Pioneer 11 – *Science* **188**, 445–477, 1975; Voyager 1 – *Science* **204**, 945–1008, 1979; *Nature* **280**, 725–806, 1979; Voyager 2 – *Science* **206**, 925–996, 1979; Voyager 1 and 2 – *J. Geophys. Res.* **86**, 8123–8841, 1981), demand an instrument design capable of accommodating the great range in parametric values established by these extremes.

Within the Jovian magnetosphere, the energetic (≥ 20 keV) particle populations play an important dual role. First, they represent a major factor in determining the size, shape, and dynamics of the system. For example, observations of energetic particle intensities and corresponding energy densities show that these populations are important in (1) standing off the solar wind and thereby determining magnetopause position; (2) determining the general magnetic field configuration in the evening magnetosphere; and (3) establishing the bulk of the ring current responsible for the magnetodisk configuration of the middle-Jovian magnetosphere (see for example, Krimigis and Roelof, 1983; Mauk and Krimigis, 1987).

Secondly the energetic particles play an important diagnostic role in the determination of energization, transport, and loss processes active in the Jovian magnetosphere. In this role they also provide a remote sensing capability for identifying magnetospheric structures through finite gyroradius effects and for diagnosing remote processes through field-aligned flow, $\overline{\mathbf{E}} \times \overline{\mathbf{B}}$ drift, and magnetic drift effects.

The Galileo EPD will provide major extensions to the Jovian energetic particle data base obtained from the Pioneer and Voyager flybys. For example:

(1) Galileo will be placed into a highly elliptical orbit around Jupiter. The nominal two-year mission lifetime will allow both a direct measure of time variations in the Jovian magnetosphere and a significantly larger spatial sample of the system than has been possible with the previous flybys.

(2) The nominal mission includes several close (≤ 1000 km) flybys of the Galilean satellites thereby providing the best opportunity to date to observe details of the satellite/magnetospheric interactions.

(3) The EPD provides the first 4π steradian angular coverage for Jovian energetic particles, thereby assuring that the necessary energetic particle measurements will be obtained independent of satellite orientation and magnetic field direction.

(4) The low-energy thresholds of the EPD effectively close the energy gap between plasma and energetic particle measurements that has existed in previous observations and assures that processes thought to operate in that gap will be tested by direct observation. For example, it has been suggested (Gehrels and Stone, 1983) that the particles powering Jovian aurora are ions of energies ≤ 100 keV nucl^{-1}, a composition energy range to be measured by Galileo instrumentation at Jupiter.

The EPD science team, shown in Table I, will be the primary users of EPD data and will work with the other Galileo instrument teams to maximize this unique opportunity

TABLE I

EPD Science Team

Principal investigator	D. J. Williams	JHU/APL
Co-investigators	T. P. Armstrong	University of Kansas
	T. A. Fritz	Los Alamos National Laboratory
	S. M. Krimigis	JHU/APL
	L. J. Lanzerotti	Bell Laboratories
	R. W. McEntire	JHU/APL
	J. G. Roederer	University of Alaska
	E. C. Roelof	JHU/APL
	†W. Stüdemann	Max Planck Institute für Aeronomie (MPAe)
	B. Wilken	MPAe
Theoretical consultants	W. I. Axford	MPAe
	A. Hasegawa	Bell Laboratories
	L. R. Lyons	Aerospace Corporation
	R. M. Thorne	University of California, LA

† Deceased (December 7, 1987).

to extend our knowledge of the Jovian magnetosphere in particular and planetary magnetospheres in general. The following sections of this paper describe the EPD and its operation.

2. EPD Overview

The EPD instrument is the result of a joint effort between The Johns Hopkins University Applied Physics Laboratory (JHU/APL), The Max-Planck-Institute für Aeronomie (MPAe), and The National Oceanic and Atmospheric Administration Space Environment Laboratory (NOAA/SEL). Proposed in 1976 with initial funds received in late 1977, the EPD was launched onboard the Galileo spacecraft on October 12, 1989. The MPAe was responsible for the detector heads and three analog circuit boards associated with those heads. The NOAA/SEL was responsible for the original time-of-flight (TOF) circuitry. The TOF circuitry employed in the upgraded TOF detector actually flown (and described in the composition measurement system, CMS, section) was the joint responsibility of MPAe and JHU/APL. The JHU/APL was responsible for all remaining electronics, the scanning motor, the data system, instrument power, structure, test, instrument integration, and spacecraft integration. Calibrations were performed by JHU/APL and MPAe.

The general characteristics of the EPD are listed in Table II. A picture of EPD is shown in Figure 1 and a schematic representation of the instrument is shown in Figure 2. The two bi-directional solid-state detector telescopes are the Low Energy Magnetospheric Measurement System (LEMMS) and the Composition Measurement System (CMS). These detector heads are mounted on a platform and rotated as shown in Figure 2 by a stepper motor contained in the main electronics box. Figure 3 shows the stepper motor positions as seen looking down on the top of the EPD along the motor

TABLE II

Galileo Energetic Particle Detector (EPD) characteristics

Mass: 10.5 kg Size: 19.5 cm × 27 cm × 36.1 cm	Power: 6W electronics; 4W heaters;	Bit rate: 912 bps
Two bi-directional telescopes mounted on stepper platform		
4π steradian coverage with 52 to 420 samples every 7 S/C spins ($\sim$140 s)		
Geometric factors: 6×10^{-3}–5×10^{-1} cm^2 ster, dependent on detector head		
Time resolution: 0.33–2.67 s, dependent on rate channel		
Magnetic deflection, $\Delta E \times E$, and time-of-flight systems		
Energy coverage: (MeV nucl^{-1})		
0.02–55	$Z \geq 1$	
0.025–15.5	Helium	
0.012–10.7	Oxygen	
0.01–13	Sulfur	
0.01–15	Iron	
0.015 → 11	Electrons	
64 rate channels plus pulse height analysis		

Fig. 1. Photograph of the Energetic Particles Detector (EPD) with the dust covers open and with no thermal blanket on the detector heads and time-of-flight electronics box.

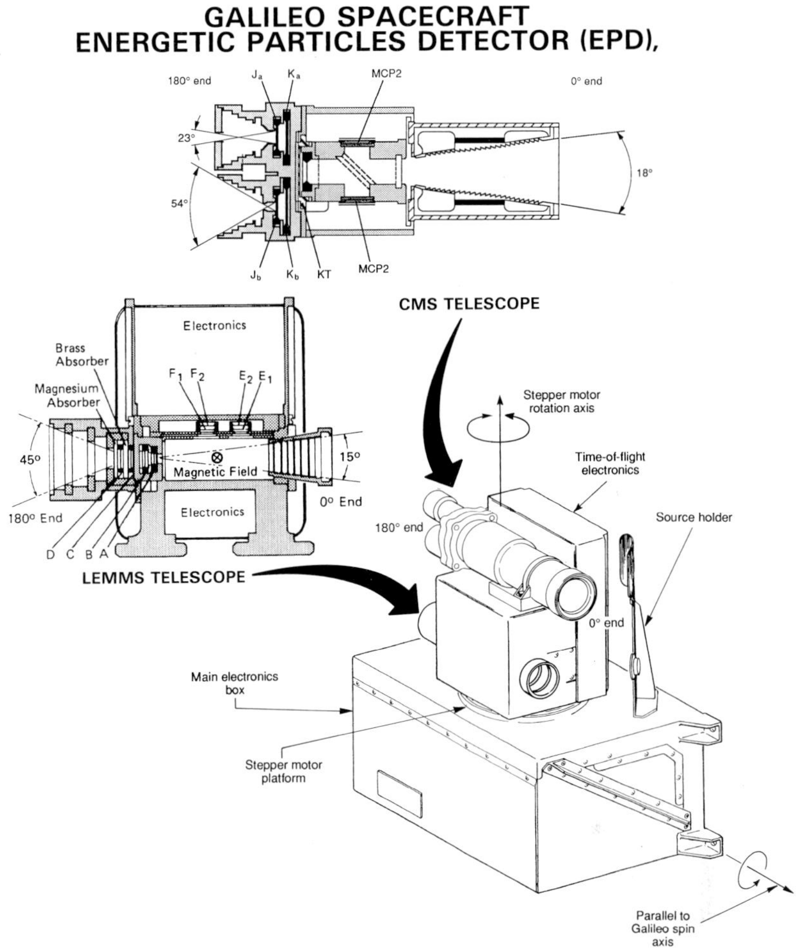

Fig. 2. Schematic showing the EPD telescope heads, the overall EPD configuration, the EPD stepper motor rotation axis, and the Galileo spin axis direction.

rotation axis. The combination of the satellite spin and the stepper motor rotation (nominally stepping to the next position after each spacecraft spin) provides 4π steradian coverage of the unit sphere. The $0°$ ends of the two telescopes (see Figure 3) have a clear field of view over the unit sphere and also can be positioned behind a foreground shield/source holder for background measurements and in-flight calibrations. The $180°$ ends experience obscuration effects in motor positions 4, 5 and 6 caused by the magnetometer boom and foreground shield.

The $0°$ end of the LEMMS unit uses magnetic deflection to separate electrons from ions and provides, from detectors A and B, total-ion energy spectra above ~ 20 keV and from detectors E_1, E_2 and F_1, F_2, electron spectra above ~ 15 keV. The $180°$ end of LEMMS uses absorbers in combination with detectors C and D to provide measurements of ions $\gtrsim 16$ MeV and electrons $\gtrsim 2$ MeV.

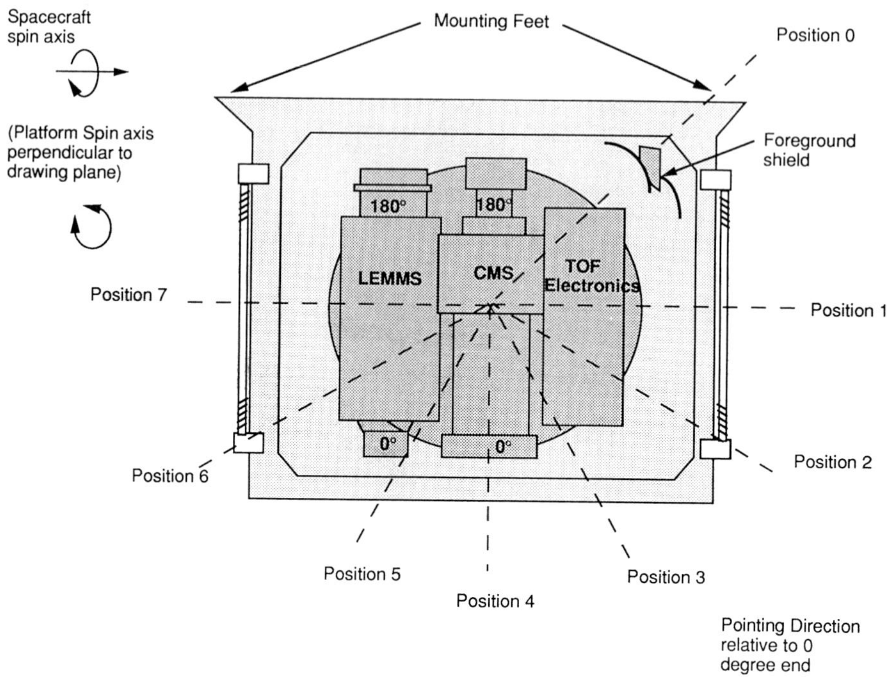

Fig. 3. View looking down on EPD along stepper motor rotation axis showing EPD viewing positions. Galileo spin axis direction is parallel to plane of figure and from right to left. Up to 60 samples per spin for the nominal spin period of 20 s are obtained, giving good angular resolution over the full 4π steradian of the unit sphere.

The 0° end of the CMS telescope employs a time-of-flight (TOF) versus total energy technique to measure elemental energy spectra above ~ 10 keV nucl^{-1} for helium through iron. A sweeping magnet in the entrance collimator prevents electrons with energies $\lesssim 265$ keV from entering the system. TOF start and stop pulses are generated as the incoming ions pass, respectively, through a thin entrance foil and impinge on the detector KT. Electrons released from the foil and the detector are accelerated and deflected through a series of grids and are detected by the microchannel plates, MCP1 and MCP2. The time difference between the start pulse, MCP1, and the stop pulse, MCP2, is then obtained, along with the ion total energy from KT. Knowing the ion total energy and its travel time through the system (which gives its velocity), the ion mass is determined.

The 180° end of the CMS telescope measures the ion energy loss, ΔE, as the ions pass though detectors J_a and J_b and the ion residual energy $E = E_{\text{total}} - \Delta E$, as they impact detectors K_a and K_b. The resulting ΔE and E measurement provides a measure of ion composition for energies $\gtrsim 200$ keV nucl^{-1}.

The planned normal mode of EPD operation is to have both telescopes powered and to step the stepper platform once each satellite spin. This will yield a 4π scan of the unit sphere approximately every 140 s. Many other scanning modes are available and will be used for special circumstances. For example, during satellite encounters the EPD will be configured to scan particular directions such as the expected direction of the magnetic flux tube, the direction of the Galilean satellite wakes as they travel through the Jovian magnetosphere, and the direction of $\overline{\mathbf{E}} \times \overline{\mathbf{B}}$ drift paths.

A more detailed description of the EPD hardware follows. The primary elements of this system are the LEMMS/CMS detector heads with their analog electronics, the motorized scanning system, the digital and support electronics, and the data system. Figure 4 shows a functional block diagram of the EPD to be referred to when reading the following sections.

3. The Low-Energy Magnetospheric Measurements System (LEMMS)

The LEMMS telescopes are designed to measure low to medium energy ion and electron fluxes with wide dynamic range and high angular ($<20°$) and temporal ($\frac{1}{3}$ to $\frac{4}{3}$ s) resolution. As stated previously, a full 4π coverage of the unit sphere is obtained by the use of a stepping platform in conjunction with satellite spin.

The LEMMS detector head (shown in Figure 5) is a double-ended telescope containing eight heavily shielded silicon solid state surface barrier totally depleted detectors providing measurements of electrons from 15 keV to >11 MeV, and ions from 22 keV to ~ 55 MeV, in 32 rate channels that are summarized in Figure 6 and Table III. The LEMMS design uses a baffled collimator and shaped-field magnetic deflection to provide clean separation between ions and electrons in detectors A, E, and F (0° end in Figure 5), and extends measurements to much higher energies via detectors C and D in the oppositely-directed double-absorber-detector telescope stack on the 180° end. The functions of the different LEMMS detector and mechanical subsystems are discussed in the following sections.

3.1. Low-energy ion ($Z \geq 1$) telescope: detectors A, B

This is the primary set of detectors for measuring low energy ions ($\gtrsim 22$ keV) in the Jovian magnetosphere. Particles entering through the 15° full angle collimator of LEMMS pass through a series of eleven baffle plates that define seven hexagonal entrance aperture channels converging toward detector A (Figure 5). This geometry provides a relatively wide aperture with the narrowly-defined total acceptance angle necessary for the focused deflection of electrons in the shaped magnetic field beyond the aperture plates. Electrons $\lesssim 1$ MeV are deflected away from detector A, and focused onto detectors E and F for analysis. Ions pass through the field region with relatively little deflection and impact on detector A, a 25 mm^2, 102 μ thick surface barrier detector mounted with the aluminium contact out to minimize radiation damage. Detector B is 35 mm^2 in area and 984 μ thick. The total geometry factor for detector A and the A/B pair is 0.006 cm^2 sr. Ions ($Z \geq 1$) impacting A are measured in 8 differential rate channels covering the energy range from 22 keV to 3.2 MeV. Additional high energy $Z = 1$, $Z \geq 2$, and electron channels are defined using coincidence logic between A and B. In addition to these fast discriminator and accumulator-based rate channels the

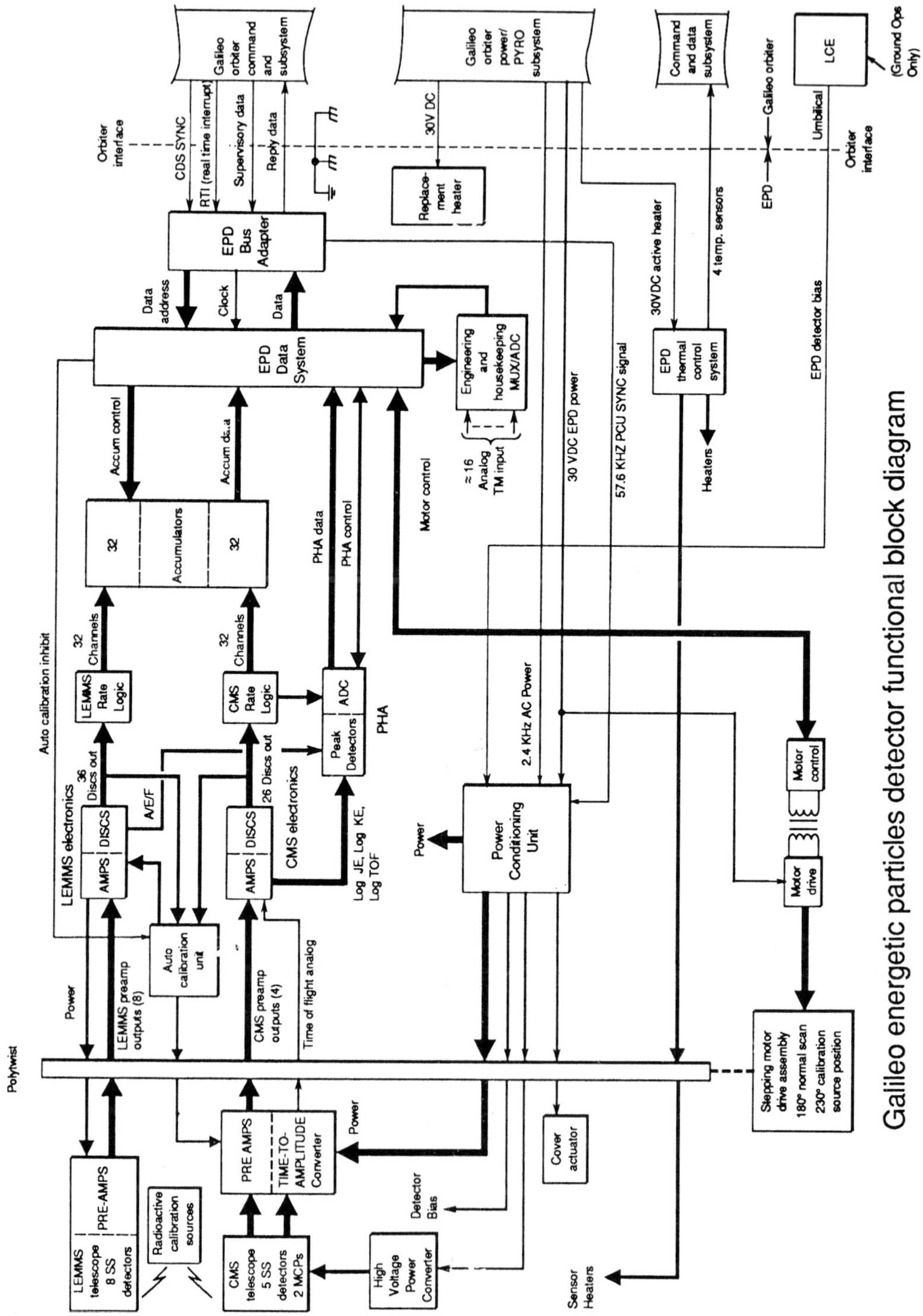

Fig. 4. EPD functional block diagram.

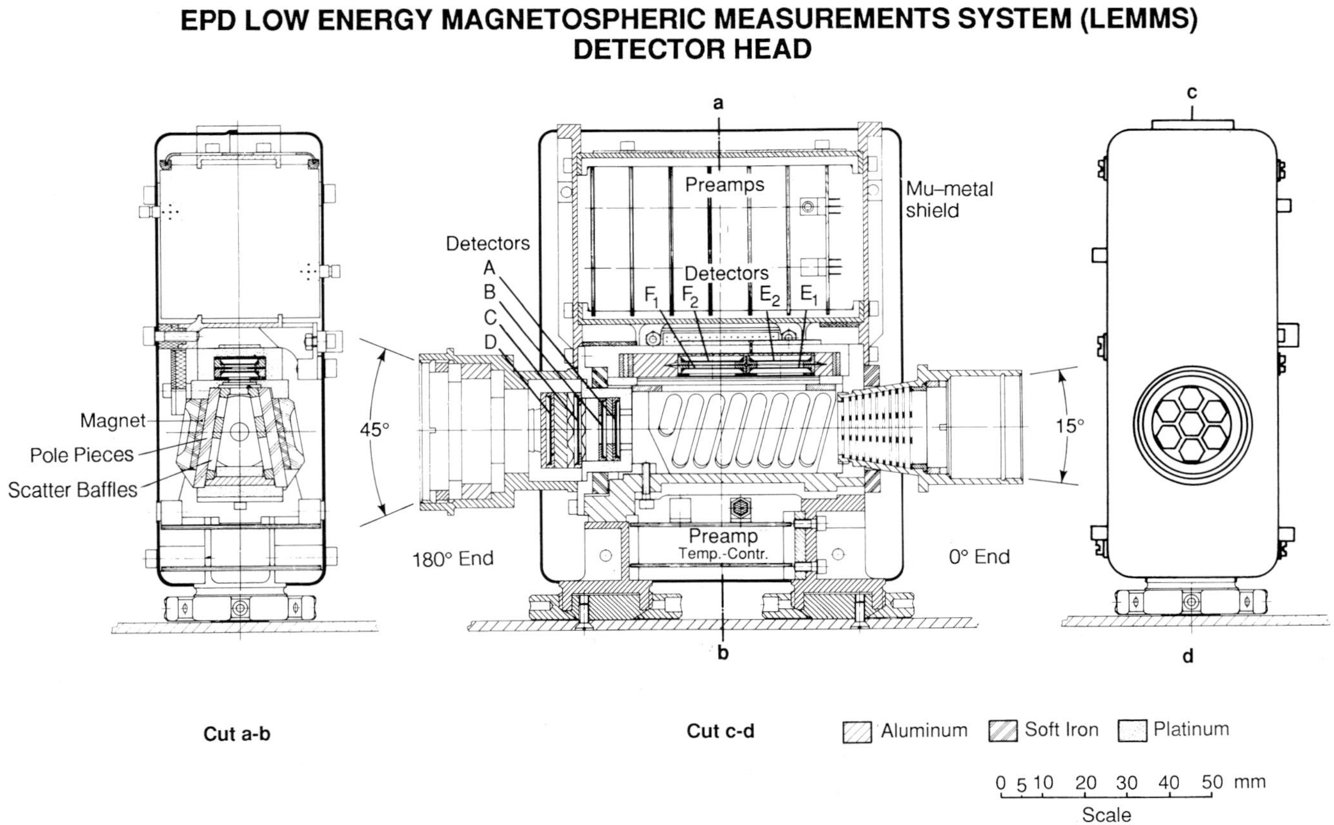

Fig. 5. Detail of the EPD LEMMS detector head.

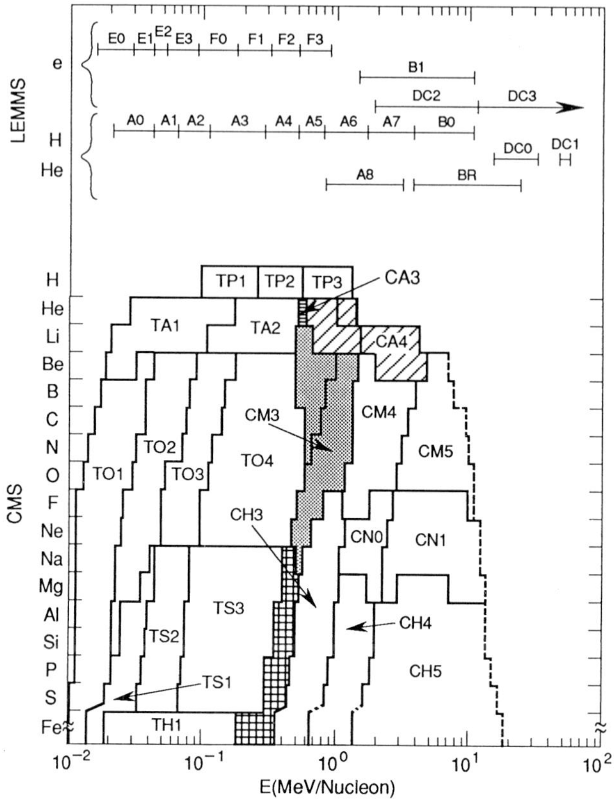

Fig. 6. Graphical summary of EPD energy and species rate channel coverage. Additional high energy resolution and detailed species identification are obtained from pulse height analysis data gathered simultaneously with the rate channel data shown.

output of detector A is digitized to provide a 46-channel energy spectrum (22–290 keV) approximately once every $4\frac{2}{3}$ s. Background contamination at Jupiter is minimized by coincidence/anti-coincidence conditions between detectors, and by surrounding platinum-iridium shielding $\geq 12\ \text{g cm}^{-2}$ thick. The lowest energy discriminator on the output of A can be raised to 27 keV by command in the event of a system noise increase.

3.2. Low-energy electrons: detectors E_1, E_2; F_1, F_2

Figure 5 shows the collimator and magnetic deflection region of the low energy end of the LEMMS telescope. Co–Sm permanent magnets generate an inhomogeneous magnetic field between inclined pole pieces with a maximum center line strength of about 650 G. Entering electrons are focused onto detectors E_1 (≥ 15 to 200 keV) and F_1 (~ 100 to > 1000 keV), which provide the low-energy electron measurements for the EPD experiment. Both detectors are rectangular, 45 mm^2 surface barrier silicon detectors. E_1 is 303 μ thick, F_1 is 1097 μ thick, and both are backed by 300 μ thick detectors (E_2, F_2) of equal area operated in anticoincidence to actively reduce any background

TABLE III

LEMMS channel descriptions

Channel name	Species	Energy range (MeV)	Readout interval[b] (in $\frac{1}{3}$s)
A0	$Z \geq 1$	0.022– 0.042	1
A1	$Z \geq 1$	0.042– 0.065	1
A2	$Z \geq 1$	0.065– 0.120	2
A3	$Z \geq 1$	0.120– 0.280	2
A4	$Z \geq 1$	0.280– 0.515	2
A5	$Z \geq 1$	0.515– 0.825	2
A6	$Z \geq 1$	0.825– 1.68	2
A7	$Z \geq 1$	1.68 – 3.20	2
A8	$Z \geq 2$	3.50 – 12.4	4
B0	$Z = 1$	3.20 – 10.1	4
B1	electrons	~1.5 – 10.5	4
B2	$Z = 2$	16.0 –100.	4
DC0	$Z \geq 1$	14.5 – 33.5	4
DC1	$Z \geq 1$	51. – 59.	4
DC2	electrons	$\gtrsim$ 2.	4
DC3	electrons	$\gtrsim$ 11.	4
E0	electrons	0.015– 0.029	1
E1	electrons	0.029– 0.042	1
E2	electrons	0.042– 0.055	2
E3	electrons	0.055– 0.093	2
F0[a]	electrons	0.093– 0.188	2
F1	electrons	0.174– 0.304	2
F2	electrons	0.304– 0.527	2
F3	electrons	0.527– 0.884	2
AS	singles	All counts in detector	8
BS	singles	All counts in detector	4
CS	singles	All counts in detector	4
DS	singles	All counts in detector	4
EB1	background	Sidewise penetrators	4
EB2	background	E1E2 coincidences	4
FB1	background	Sidewise penetrators	4
FB2	background	F1F2 coincidences	4

[a] Channel F0 is the sum of ~100–200 keV events seen in detectors E1 and F1.

[b] At nominal 3 RPM: $\frac{1}{3}$ s = 60 readouts/spin, 6° 'sectors'
$\frac{2}{3}$ s = 30 readouts/spin, 12° 'sectors'
$\frac{4}{3}$ s = 15 readouts/spin, 24° 'sectors'

LEMMS PHA is accumulated for $\frac{2}{3}$ s every 4.66 s. There are 46 linear channels and 1 high-energy channel.

Sequence is: 11 spectra from Det. A (20–290 keV @ 6.4 keV/ch)
1 spectra from Det. E1 (15–190 keV @ 4.5 keV/ch)
1 spectra from Det. F1 (80–890 keV @ 20.1 keV/ch)

This sequence is $60\frac{2}{3}$ s long, and repeats.

penetrating the surrounding shielding of ≥ 6 g cm^{-2} of platinum-iridium alloy. No electron ≤ 1 MeV can directly reach detector A, and no non-deflected particle can directly reach detectors E_1, F_1. The inside of the deflection volume (inside of the yoke, pole pieces, and magnets) is lined with aluminium baffles to minimize scattering.

The efficiencies of the electron detectors have been measured as a function of energy and input direction relative to the collimator central axis. Sample results from detector E are shown in Figures 7(a) and 7(b) at electron energies of 40 and 80 keV. At these energies detector F efficiencies effectively are zero. The effect of the honeycomb aperture (Figure 5) clearly is seen.

Also the geometric factor of the collimator and magnetic optics system has been analyzed by extensive computer modeling (Wu and Armstrong, 1988). The energy-dependent aperture-integrated geometry factors for detectors E_1 and F_1 are shown in Figure 7(c) and 7(d). The detector efficiencies and geometric factor calculations, as represented by Figure 7, provide a complete determination of the response of the LEMMS electron channels to electrons entering the aperture. The outputs of detectors E_1 and F_1 are registered in eight differential accumulator-based fast rate channels and are separately pulse-height analyzed to provide 46 channel energy spectra for each detector (Table III).

3.3 High-energy electrons and ions: detectors C, D

The C, D detector telescope is used to extend the energy range of the LEMMS electron and ion measurements to higher energies. Both detectors are ~ 100 mm^2, ~ 500 μ thick devices separated by 0.48 cm. The detectors are shielded from the side by > 7.24 g cm^{-2} of platinum and the rear-facing collimator provides a directional field of view of $\sim 45°$, resulting in a geometric factor of ~ 0.5 cm^2 sr. In the look direction detector D is shielded by a magnesium disk 2 mm thick (0.36 g cm^{-2}), and between D and C there is an additional absorber of 3.2 mm of brass (2.8 g cm^{-2}). The detectors each have two discriminator thresholds and operate both in coincidence/anticoincidence and in singles modes that, combined with the magnesium and brass absorbers, define directional thresholds of > 2 and > 11 MeV for electrons in D and C, respectively, and ion (proton) thresholds of 14.5 and 51 MeV (Table III).

As shown in Figure 5, all of the LEMMS preamplifiers are packaged in close proximity to their detectors to minimize noise, and all of LEMMS except the mounting feet and collimators is enclosed in a μ-metal shell to reduce magnetic field leakage. All LEMMS detector energy signals except for A and E_1 have bipolar pulse shaping with a 1 μs full width. Detectors A and E_1 have unipolar Gaussian shaping (0.9 μs full width) and baseline restoration to achieve the lowest possible thresholds for measuring low energy ions and electrons.

3.4. LEMMS analog electronics

Figure 8 shows the block diagram for the LEMMS analog electronics. The unbuffered, preamp tail-pulse output signals are sent down the polytwist wiring harness to three LEMMS analog boards, where the signals are amplified and shaped. All signal process-

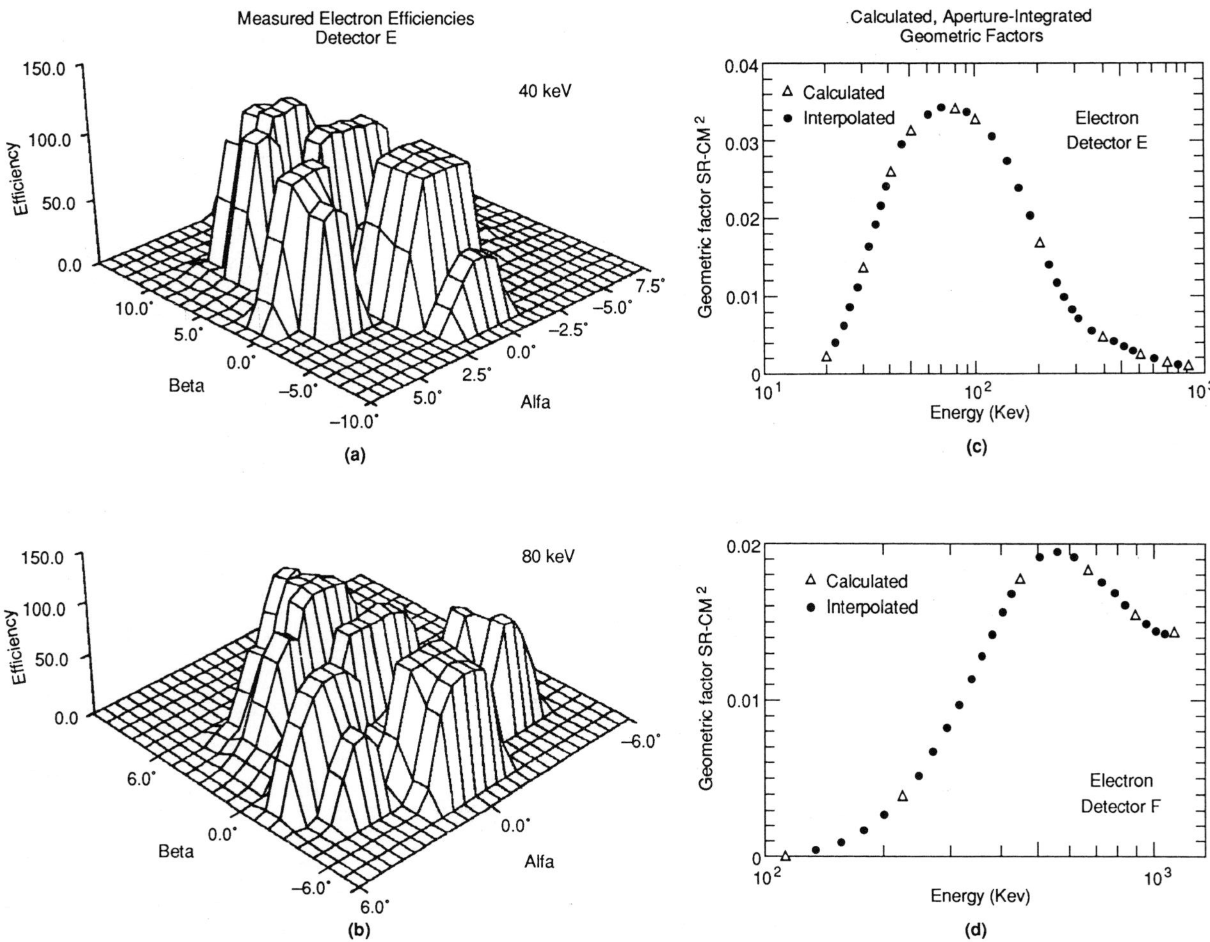

Fig. 7. Response of the EPD LEMMS low energy electron detectors. The geometric factor, as calculated from a detailed simulation of the LEMMS head, is shown as function of energy.

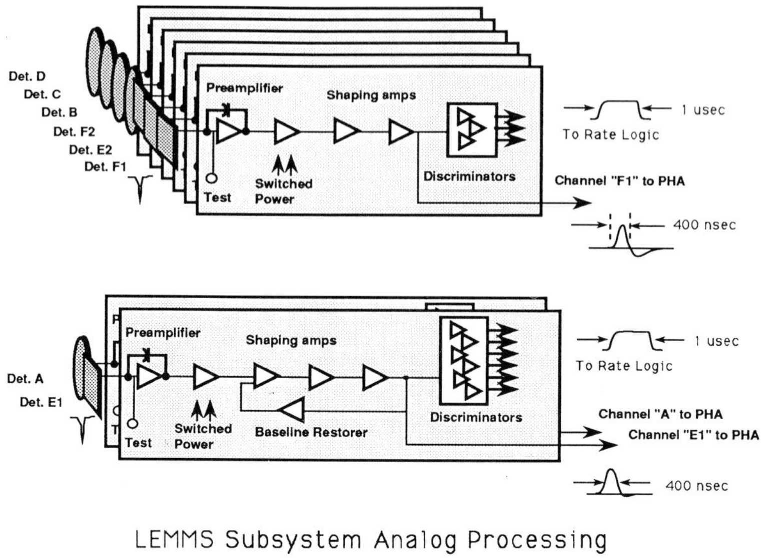

Fig. 8. Block diagram of LEMMS analog electronics.

ing in the LEMMS channels is done in the linear domain. Comparator-like discriminator circuits assigned to each of the eight channels fire when pre-defined threshold levels are exceeded.

The resulting discriminator outputs are sent to the rate logic electronics, where combinatorial logic is used to define 32 rate channels (event bins). Logic definitions for these channels include requirements for coincidence of some discriminators and anti-coincidence for others; this is used to define desired events. Those events which fall into the rate channels (by having the correct combination of discriminator levels) are counted in custom APL-designed high rate 24-bit accumulator hybrids. The rate data is periodically read into the telemetry stream by the data system. Each channel preamplifier and amplifier electronics can be powered ON/OFF by command.

LEMMS data also are processed through a pulse height analyzer (PHA) that produces 13 46-channel energy spectra per major frame ($13 \times 4\frac{2}{3}$ s). The detector outputs analyzed are selected via command from detectors A, E_1, and F_1 with the normal selection being 11 spectra from detector A and one each from detectors E_1 and F_1.

4. The Composition Measurement System (CMS)

The CMS is designed to measure the composition of ions in the Jovian environment from energies of $\gtrsim 10$ keV nucl^{-1} to > 10 MeV nucl^{-1}. This system contains two types of energetic particle telescopes, shown schematically in Figure 9. A small time-of-flight (TOF) telescope looks in the 0° direction while a pair of $\Delta E \times E$ solid state detector

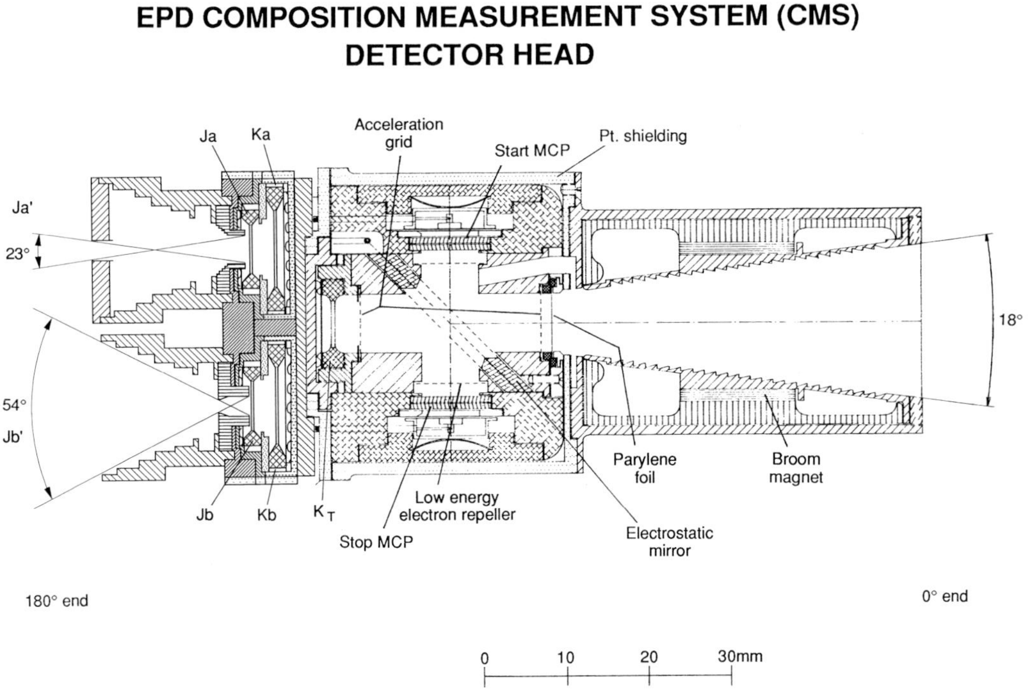

Fig. 9. Detail of EPD CMS detector head.

telescopes (*JK* detectors) covering higher energies are oriented in the 180° direction. The location of CMS on EPD is shown in Figure 2, and energy coverage for the combined system is summarized in Figure 6 and Table IV. The CMS subsystems are discussed below.

4.1. CMS TIME-OF-FLIGHT (TOF) TELESCOPES

The TOF portion of CMS shown in Figure 9 is a replacement of the original EPD TOF telescope built for launch in early 1986. JPL, NASA, and the Bundesministerium für Forschung und Technologie allowed us the opportunity to take advantage of new technology in order to lower our energy threshold for composition measurements by at least a factor of ten. This technology was the basis of a new generation of TOF telescopes developed for and/or successfully flown on AMPTE/CCE, VIKING, GIOTTO, and Ulysses (Stüdemann and Wilken, 1982; Gloekcler *et al.*, 1983, 1985; Wilken and Stüdemann, 1984; McEntire *et al.*, 1985; Stüdemann *et al.*, 1987; Wilken *et al.*, 1987). The new TOF telescope, its high voltage power supply, the new TOF circuitry, and the major modifications to the electronics subsystem were designed, built, tested, and integrated into the EPD in 1987 and 1988. All instrument and spacecraft tests and calibrations were successful and were indicative of an instrument response that exceeded expectations.

The CMS TOF telescope, as shown in Figure 9, consists of a collimator containing

TABLE IV

CMS channel descriptions

Channel name	Species	Energy range (MeV nucl^{-1})	Channel name	Species	Energy range (MeV nucl^{-1})
TOF × E			Delta E × E		
TP1		0.08–0.22			
TP2	Protons	0.22–0.54			
TP2		0.54–1.25			
TA1	Alphas	0.027–0.155	CA1		0.19– 0.49
TA2		0.155–1.00	CA3	Alphas	0.49– 0.68
			CA4		0.68– 1.4
TO1		0.012–0.026			
TO2	Medium	0.026–0.051	CM1	Medium	0.16– 0.55
TO3	Nuclei (O)	0.051–0.112	CM3	Nuclei (O)	0.55– 1.1
TO4		0.112–0.562	CM4		1.1 - 2.9
			CM5		2.9 –10.7
TS1	Intermediate	0.016–0.030	CN0	Intermediate	1.0 - 2.2
TS2	Nuclei (S)	0.030–0.062	CN1	Nuclei (Na)	2.2 –11.7
TS3		0.062–0.31			
TH1 [a]	Heavy		CH1	Heavy	0.22– 0.33
	Nuclei (Fe)	0.02–0.20	CH3	Nuclei (Fe)	0.33– 0.67
			CH4		0.67– 1.3
			CH5 [a]		1.3 - 15.0
TACS [a]	Singles		JaS [a]	Singles	
STARTS [a]	Rates		JbS [a]	Rates	
KtS [a]			KS [a]		

[a] These channels are read out every $\frac{8}{3}$ s. All other CMS channels are read out every $\frac{4}{3}$ s.

a sweeping magnet followed by a thin-foil solid state detector telescope with an effective geometric factor of 0.007 cm^2 sr. The thin (36 μg cm^{-2} total thickness) parylene front foil is doubly aluminized to reject both light and scattered Lα background. It is mounted on a 90% transparency stainless steel grid that is followed by a 3.77 cm time-of-flight path to the rear solid state detector K_T (50 mm^2, 12.6 μ thick). Secondary electrons are emitted from each surface when an incident ion traverses the front foil and impacts K_T. The electrons from the inner surface of the foil and the aluminium front surface of K_T are electrostatically accelerated and reflected through 90° onto microchannel plates (MCP's) that provide (respectively) 'start' and 'stop' fast timing pulses to measure the ion time-of-flight between the front foil and K_T. The ion energy is measured in K_T and the incident ion mass is determined by the measured velocity and energy (with corrections applied for the K_T pulse height defect and dead layer). The microchannel plates are 1.4 cm diameter, 12 μ pore size, 80/1 ratio, 200 MΩ plates operated in chevron configuration at a gain of $\sim 5 \times 10^6$. Because these devices are sensitive to contamination, the EPD telescopes were purged with N_2 up until launch. The MCP's and electrostatic optics are powered by a single HV power supply, adjustable over the operational range in 14 steps (up to a maximum of 4.6 KV) to maintain constant MCP gain over the mission life.

Predicted track locations of several elements in the TOF versus energy space measured by the telescope are shown as dashed lines in Figure 10. The resolution limit of the sensor is set primarily by two factors; at higher Z and lower energies it is set by pulse-height-defect spread in the energy measured in K_T, while at high energies the limiting factor is the ~ 0.6 ns (FWHM) timing resolution of the electron optics and electronics. For example the telescope can resolve H_1, H_2, He^3, He^4, C, O, and the major Jovian species and species groups with a mass resolution that is a function of both mass and energy. Vertical energy discriminators and slanted mass discriminators (defined on the logarithmic sum of the energy and TOF pulse height) are defined (solid lines) to separate TOF-E space into 13 composition channels covering the species or species groups listed in Table IV. These accumulator-based composition rate channels (and associated singles channels) provide good flux and angular information on abundant species with (in some modes) 100% duty cycle but with moderate energy and species resolution. A much more precise spectral and elemental resolution is obtained from 256-channel pulse-height analysis (PHA) of the energy and TOF signals for individual events. An average of ~ 4 of these fully-analyzed events are transmitted per second (selected by a 4-level adaptive priority system used to obtain balanced coverage over species and energy and shared between the CMS TOF and $\Delta E \times E$ telescopes). These PHA data, normalized by the rate-channel count rates, allow detailed fluxes and spectra to be obtained for each element measured. Figure 10 contains PHA data from sample calibration runs at the Goddard Space Flight Center and University of Rutgers accelerators where the telescope was calibrated with beams of H, He, C, O, Ne, Na, S, Ar, Fe, and Kr from energies below 100 keV to over 10 MeV. The TOF head is resistant to electron pile-up and over most of the orbit will be insensitive to penetrating background due to shielding and coincidence requirements. A valid TOF head event requires a triple coincidence (a start MCP event followed within 60 ns by a stop MCP event, and a coincident K_T energy pulse). The sweeping magnet prevent electrons $\lesssim 185$ keV from reaching the front foil, and electrons $\lesssim 265$ keV from directly reaching the rear detector. This detector, K_T, has a lower threshold of 66 keV (which can be increased to 122 keV by command) and due to its 12.6 μ thickness is insensitive to energetic electrons in its field of view. All active components are shielded by a minimum of 3.8 g cm^{-2} of platinum.

4.2. CMS $\Delta E \times E$ TELESCOPES

The CMS $\Delta E \times E$ telescopes are designed to extend the CMS ion composition ($Z \geq 2$) measurements to significantly higher energies than are attained by the TOF telescope. These rear (180° direction) facing telescopes are shown in cross section in Figure 9. Each consists of a thin front solid state detector ($J_{a,b}$) followed by a thick rear detector ($K_{a,b}$). J_a is 44 mm^2, 5.5 μ thick, J_b is 23 mm^2, 5.6 μ thick, and $K_{a,b}$ are each ~ 200 mm^2, 185 μ thick. Apertures for each telescope are defined by an aluminium collimator and a field stop immediately in front of each J detector. This yields geometry factors of

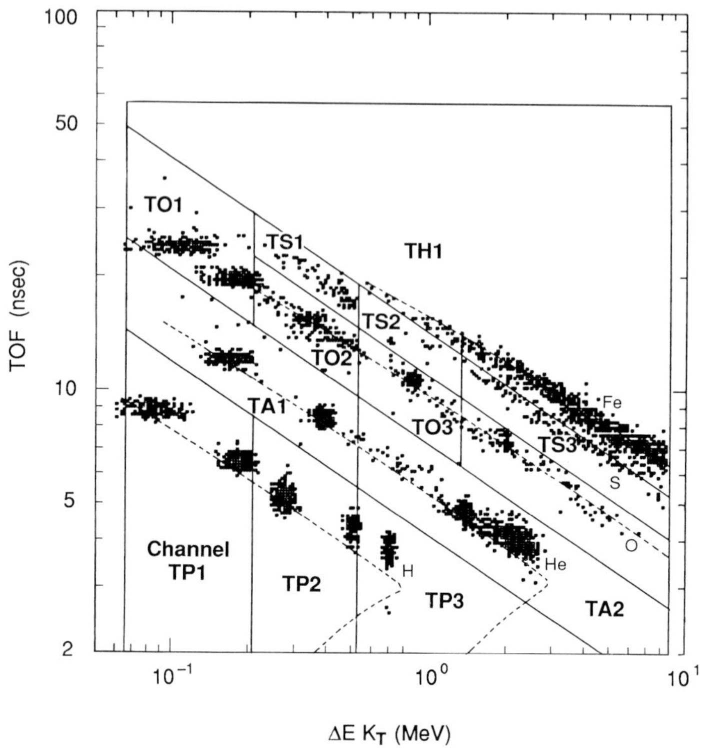

Fig. 10. TOF versus energy observed in detector K_T. Discriminator levels defining the indicated TOF rate channels (solid lines), predicted TOF vs E curves (dashed lines), and data from accelerator calibrations (points) are shown.

0.008 cm^2 sr for J_a and 0.025 cm^2 sr for J_b. The two telescopes provide redundancy and the geometry factors are chosen to minimize radiation damage accumulated over the Galileo mission.

In each telescope the J and K detectors are primarily operated in coincidence, with dE/dx measured in J and total residual energy measured in K. The $\Delta E \times E$ energy space measured is shown in Figure 11, with dashed lines showing predicted $\Delta E_J \times E_K$ tracks for a number of elements over the telescope range of energies. Energy discriminators on J (horizontal lines) and K (vertical lines), and two discriminators defined on the logarithmic sum of the J and K energy measurements (slanted lines) define 10 discrete composition channels covering the species shown in Figure 6 and summarized in Table IV. The three single parameter ($J\bar{K}$) channels CA1, CM1, and CH1 (shown below the lowest K discriminator in Figure 11 but not shown in Figure 6) extend measurements ($Z \gtrsim 2$, 3, 12, respectively) to lower energies in order to increase overlap with the TOF telescope. The dashed line on the right in Figure 6 indicates the energy of penetration of the K detectors. The actual sensitivity of the telescope extends to higher energies. As with the TOF telescope, the accumulator-based $\Delta E \times E$ rate channels can provide coverage with essentially 100% duty cycle and very wide dynamic range. Detailed

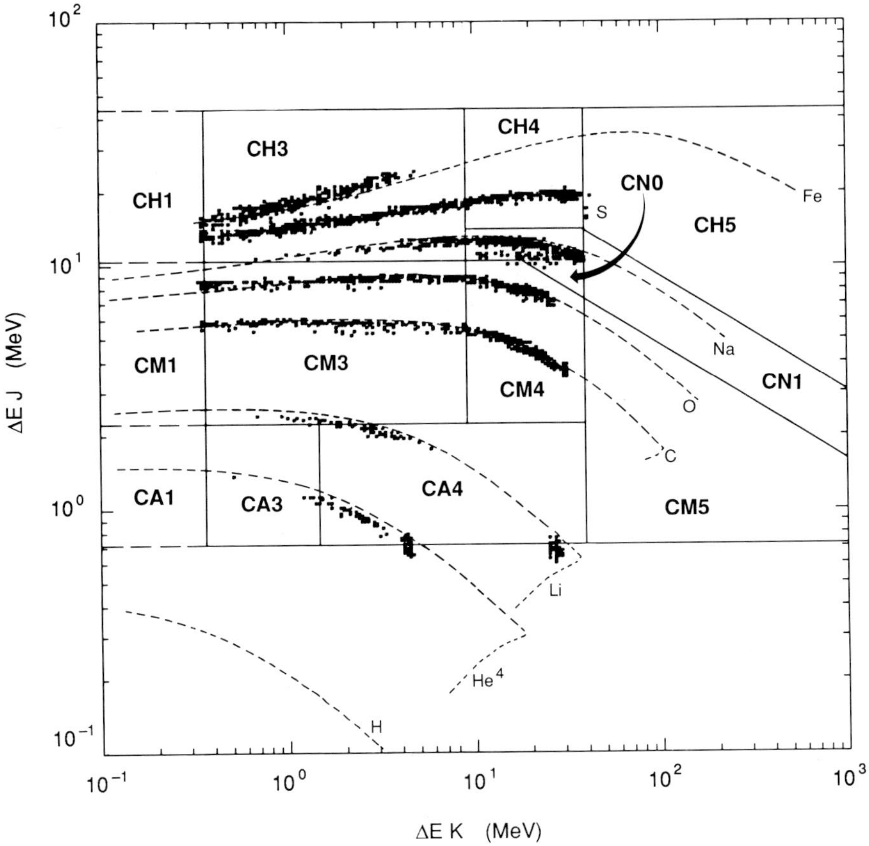

Fig. 11. Energy loss in detector J, (ΔE), versus energy measured in detector K, (E). Discriminator levels defining the indicated rate channels, predicted $\Delta E \times E$ curves, and data from accelerator calibrations are shown.

spectral and species resolution is derived from PHA events transmitted at ~4 events s^{-1}. Figure 11 contains sample PHA data taken during calibration runs at the Rutgers Tandem Van de Graff accelerator where the $\Delta E \times E$ telescopes were calibrated with beams of He, Li, C, O, F, Na, Si, S, and Fe over a wide range of energies. As can be seen from the figure, elemental resolution is possible (depending on relative abundance) through sodium and all major Jovian species and species groups will be distinguished.

Each of the JK detector pairs is surrounded by $\geq 2.5 \text{ g cm}^{-2}$ of platinum and aluminium shielding. Coincidence and anti-coincidence requirements on valid events, and discriminator thresholds set well above expected energy depositions for foreground electrons and most penetrating particles, allow the $\Delta E \times E$ telescopes to make clean measurements in all but the most intense portions of the Jovian radiation environment.

4.3. CMS ANALOG ELECTRONICS

Figure 12 shows a block diagram of the CMS analog electronics. The velocity measurement is made via the Time-to-Amplitude Conversion (TAC) electronics, which measure ion velocity by converting microchannel plate-produced start and stop charge pulses into a bipolar linear signal whose amplitude is proportional to the time-of-flight of the ions over a known distance. The analog signal shaping times are similar to those used on the other CMS channels.

Unlike the LEMMS processing, the energy and time signals are logarithmically compressed after initial amplification. This enables direct calculation of mass from time and energy data using linear analog circuitry and extends the dynamic range of the channels to three decades. These advantages come at the expense of rate capability; the log amp recovery time can be as great as 40 μs (for the worst case situation of very large energy signals in the K detector), and represents the limiting rate factor in the CMS electronics chain. The TOF telescope requires significant rate corrections at input count rates of $\gtrsim 150$ kcps. The $\Delta E \times E$ electronics can experience pile-down effects at $\gtrsim 50$ kcps, a value that is energy dependent.

As in LEMMS, discriminator circuits are assigned to each of the analog channels, allowing energy and species categorization to be performed. The 26 discriminator outputs are sent to the rate logic board, where combinatorial logic is used to define 32 rate channels (event bins). The data are accumulated and read into the data system as in LEMMS.

In addition to the rate channel measurements, the instrument is able to perform higher resolution analysis of events using a pulse height analyzer (PHA). This circuit performs an 8-bit A/D conversion of the relevant energy and time parameters. In LEMMS, this is used to produce a 47-bin single channel energy spectrum for the three low energy ion and electron channels. In CMS a 256 × 256 dual parameter measurement is made of $\Delta E \times E$ and total energy × time-of-flight.

Due to processing and telemetry limitations, only a small fraction of all events are analyzed by the PHA. Approximately 7 LEMMS energy spectra and 230 CMS dual parameter measurements are produced every minute.

5. Rate Capability and System Limitations

The LEMMS electronics can be modeled as 8 independent, non-paralyzable channels with 1.4 μs dead-times. Events rates up to 600 kHz can be measured in each channel without requiring significant rate corrections. Figure 13 shows the measured true versus apparent count rate response of the LEMMS.

The CMS electronics is more rate restricted, however. The log amplifier recovery time is a function of both the signal level and overall event rate. Performance in each channel, while independent up to and beyond the log amplifiers, is affected by the common event timing logic. As mentioned above, energy piledown occurs in the CMS $\Delta E \times E$ energy channels at rates of $\gtrsim 50$ kcps (depending upon event energy distribution). The time-of-flight system is less sensitive to these problems and will operate to rates well above 150 kcps.

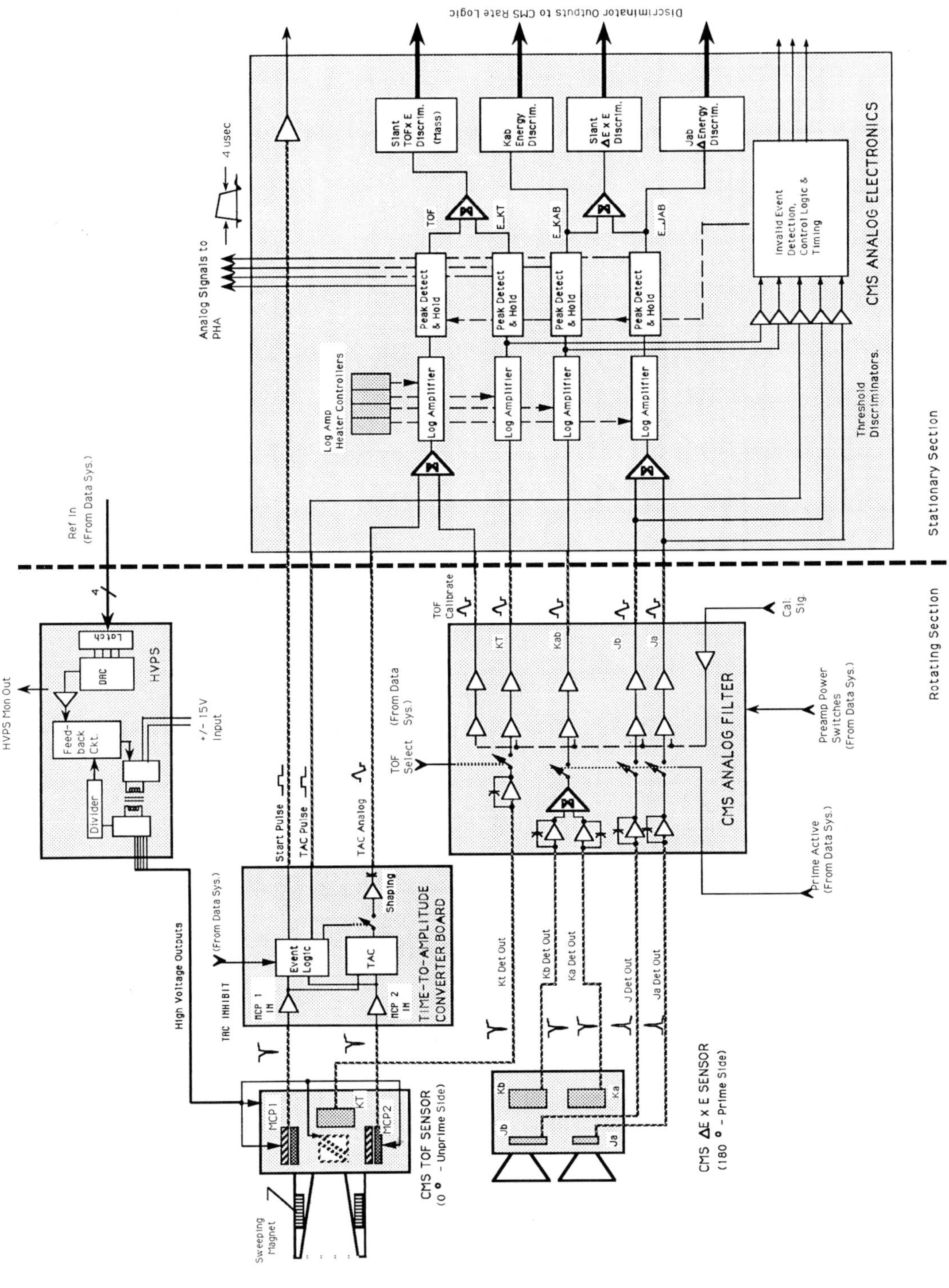

Fig. 12. Block diagram of CMS analog electronics.

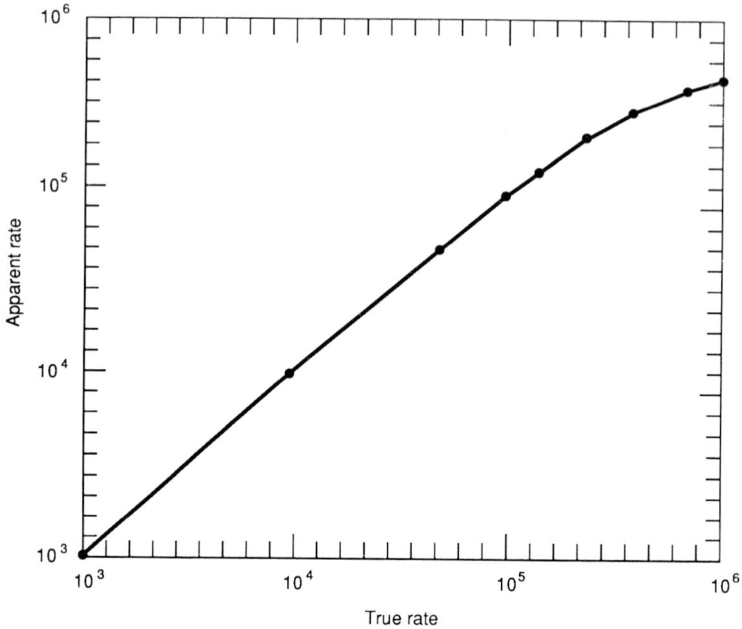

Fig. 13. Apparent versus true count rate measured for LEMMS rate channels.

6. Scanning System

Full 4π solid angle coverage is provided to the EPD sensors by using a motor subsystem to step the detector assemblies through 228° of rotation (see Figure 3). This stepping, in conjunction with the approximately 3 revolution-per-min rotation of the spacecraft science boom, provides three dimensional angular resolution of the measured particle distributions, as indicated in Figure 2.

The scanning system is a microprocessor-based positioning system designed to be automatic, self-correcting, and self-protecting. Optional modes provide extensive flexibility, allow adaptation to most single-point failures, and provide for in-flight program modification. A detailed description of the scanning system is given by Fort (1985).

The system is comprised of several components, including control and driver electronics, opto-isolated position sensors, rotating wiring harness, and the motor itself. The heart of the system is a 4-phase 90° stepper motor acting through a 50 : 1 reduction gear to yield 1.8° rotational steps at the 78 000 g cm^{-2} load. The motion is translated to the sensor packages via a beryllium copper shaft and platform. All major sensor structures are mounted to the platform.

A special wiring assembly, called a polytwist, carries 116 power, control, and analog signals between the rotating platform and the main electronics assembly through a 2.26 cm shaft. The polytwist was designed and an identical unit successfully tested for a 500 000 cycle lifetime.

The motor is powered from the energy stored in a 0.0127 μF capacitor bank. The capacitors are trickle charged when the system is at rest to reduce the peak power demands on the spacecraft. The motor, drive electronics, and capacitor bank are transformer-isolated from the control circuitry, and are powered separately off the spacecraft power bus.

A set of five LED/phototransistor pairs provide positioning information to the RCA 1802 microprocessor-based controller hardware. These sensors identify each of the eight regions, or defined positions within the scan range, as well as provide centerline and control feedback information. Figure 3 shows the motor position definitions.

During normal operation, the system steps through 228.6° and back, stopping at 8 positions to allow for the accumulation of science data and the recharging of the capacitor bank. The angle between motor position centerlines is 30.6°, except for position 0 (calibration position), which is 45° from sector 1. In normal mode, the motor position is changed every 20 s, although this may be varied by command to better match a non-standard spacecraft spin rate.

The motor controller receives its commands through the normal instrument command interface via the data system electronics. Motor system status is provided in the instrument telemetry via the data system as indicated in Figure 4.

A great deal of flexibility has been incorporated into the motor control hardware and software. Amongst other capabilities, the system may step at various rates, scan between commandable position limits, operate open loop (in case the position information is not available), change recharge rates, or remain at specified locations. The motor may also be single-stepped between defined positions, using the full energy stored in the capacitors for the 1.8° movement.

Autonomous error recovery routines are available to allow graceful degradation of the system should mechanical or electrical problems be encountered. The control software may be modified in flight to handle unforeseen system limitations.

9. Data System and Support Electronics

The EPD instrument contains a highly flexible command, control, and telemetry system. The 'brains' of the system is a radiation hardened, RCA 1802 microprocessor-based computer with 6 kbyte mask programmed ROM and 2.25 kbyte RAM, and 22 8-bit input/output ports. The system is responsible for all instrument control, command decoding, and telemetry processing and formatting. While this system (as well as other parts of the instrument hardware) may not be impressive by today's VLSI standards, at the time it was designed and built, it was state-of-the-art.

The command and telemetry interface with the spacecraft is made over 4 dedicated serial data lines, three going to the instrument and one going to the spacecraft. These data lines connect to high-speed direct memory address control electronics within the instrument. Through this electronics, the spacecraft is able to directly read and write into the contents of the EPD data system memory.

Commands intended for the EPD are thus written to a specific area in RAM, and formatted telemetry packets are likewise read from alternating buffers in RAM. This whole process is transparent to the instrument microprocessor electronics. Special instrument registers count the number of invalid bus transactions. Each instrument on the spacecraft is connected to the spacecraft bus in a similar manner, thus forming a very flexible distributed processing network.

In addition, there is a low voltage power supply which converts the spacecraft 30 V DC power into 10 isolated output voltages needed in the instrument. The converter, based upon a buck/boost regulator, operates at 57.6 kHz, and contains overcurrent crowbar protection (350 ma), load switching, and dual detector bias output levels. It operates at approximately 78% efficiency under normal load.

The converter also switches power to the instrument's cover release mechanism. The mechanism, based upon a bi-phase wax actuator and nine-watt heater element, successfully released two clamshell covers which protected the CMS and LEMMS detectors from chemical contamination and physical damage. The covers can not be reclosed during the mission.

The instrument may be calibrated in flight in two ways. Alpha particle and electron radioactive calibration sources are mounted on the foreground shield positioned in motor sector 0. These sources can be seen by both the CMS and LEMMS. An electronic pulser circuit also can be used to generate signals for the LEMMS and CMS analog channels. These pulses are injected after the preamplifier electronics, and are intended to verify channel operation and discriminator settings. A sophisticated feedback circuit enables the data system to automatically cycle the pulsers to each discriminator, measuring the 12% and 88% trigger levels in each to measure channel noise.

Instrument status is monitored through the use of an 8-bit analog to digital converter and multiplexer. Five temperatures, nine voltages, and the instrument input current are measured and reported in the telemetry stream once a minute.

10. Thermal Control

Instrument temperatures are regulated via a system of ten autonomous heater circuits; six are proportionally-controlled linear heaters and four are discretely-controlled switching heaters. The combined maximum capability of the heaters is 10.1 W, although the power use at Jupiter is expected to be 3.9 W. The temperature setpoint for most of the heaters is -18 to -20 °C. To keep the temperature-sensitive CMS log amplifier circuit gain constant under varying thermal conditions, four of the heaters (one for each amplifier) have setpoints of $+20$ °C.

In addition to the heaters, the instrument is almost completely enclosed in conductively coated thermal blankets, and is mounted to the science boom via thermal isolators. A spacecraft-provided retractable sunshade will protect the instrument from the Sun during the first three years of the mission.

11. Telemetry and Command Processing

The system is able to receive and execute 6 command bytes per second via a 32-byte command buffer. There are 145 different instrument commands, 54 of which are passed through to the motor via a dedicated command interface. The motor control electronics, discussed above, are operated in a slave mode to the data system, receiving commands and sending motor status information upon request. Four of the instrument commands are actually macro commands, putting the electronics into pre-defined operating modes. The number of received valid and invalid EPD commands is reported in the instrument telemetry.

Telemetry is sent to the ground via the spacecraft at a fixed 912 bit/sec rate. A new telemetry packet is produced every $\frac{2}{3}$ s by the instrument (see Figure 14). As this packet is being read by the spacecraft, the next packet is formated in a different location in RAM. The EPD telemetry packet includes information on instrument status via a subcommutated housekeeping channel; the data is reported for each of the 91 channels once every $60\frac{2}{3}$ s.

Various EPD telemetry modes may be selected by command, including multiplexing of LEMMS and CMS PHA measurements and simultaneous, alternating, or dedicated measurements from the two CMS detector heads. Many of the subsystems, including the motor, PHA, calibrator, and rate logic may be operated in a number of configurations, allowing reallocation of telemetry to specific channels of interest.

One byte of motor status is generated every packet, including operating mode, position, error conditions, etc. This information, together with spacecraft generated navigation and spin parameters, will be used on the ground to sort the received event data into 7 motor position-based sets of 64 spin-based sector bins.

12. Fault Protection

The instrument must be able to adapt to unforeseen problems during the mission. Two ways that this is done are via software changes to the data system and an internal alarm monitor. The EPD operating software can be modified in-flight through the use of RAM software patches. This can be used to 'patch' around sections of memory that fail, or add new control algorithms whose need was unforeseen when the ROM-based operating system was created.

The instrument's internal alarm monitor checks 4 temperatures, one voltage and one current against defined alarm thresholds approximately once every subcommutation cycle ($60\frac{2}{3}$ s). If the measured value exceeds the limits, an alarm flag is set in the instrument telemetry, and the instrument configures itself in a Power-On-Reset state. The spacecraft, upon receiving the alarm, will turn off the instrument power. The alarm function can be disabled by command for any or all individual thresholds.

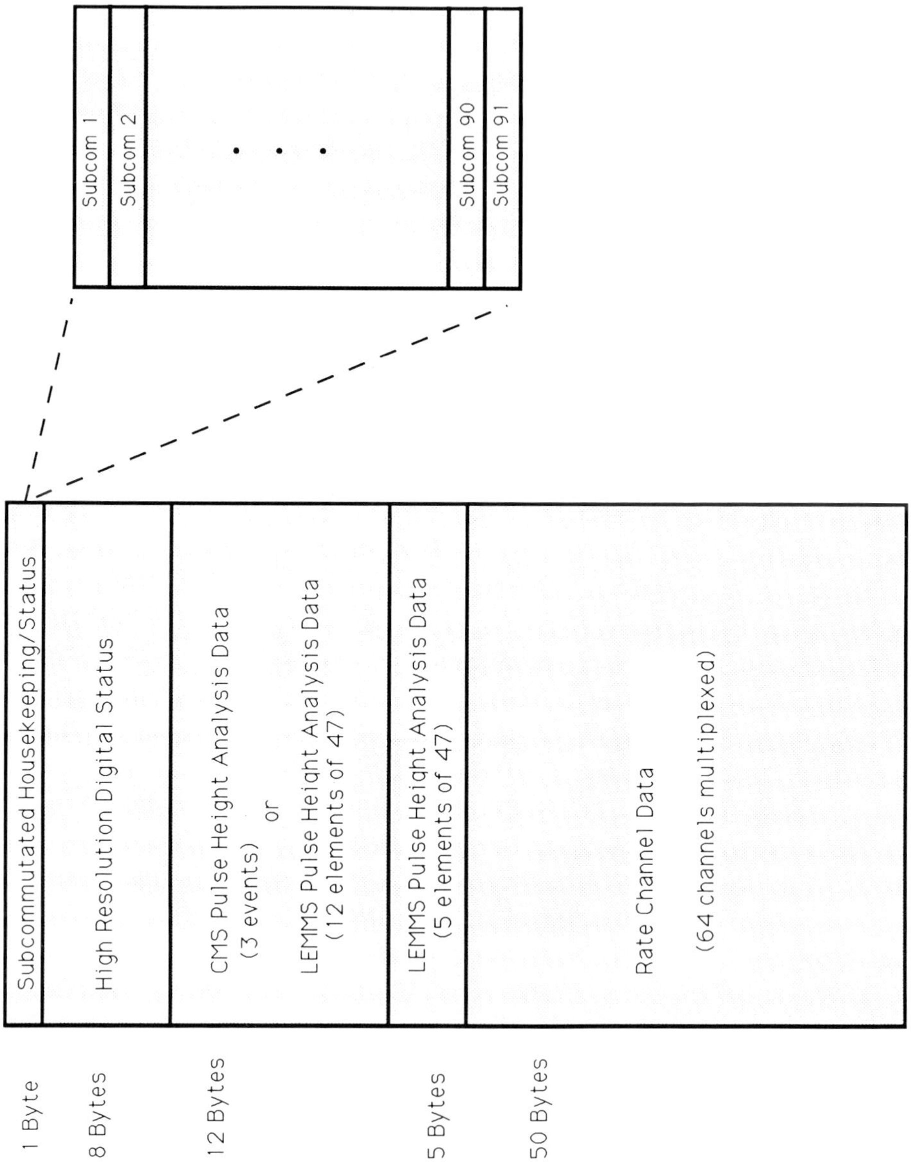

Fig. 14. **EPD telemetry packet allocation.**

13. Summary

The EPD was launched successfully on board the Galileo spacecraft on October 12. 1989. A full instrument check performed in December 1989 showed completely nominal operation. All data received and operations performed to date indicate continuing nominal operation. A more complete set of data will become available during the first Earth encounter period whose closest approach occurs on December 8, 1990.

Acknowledgements

The successful construction, test, and launch of the EPD has required the continuing efforts of many dedicated professionals since proposal submission in 1976 and receipt of initial funding in late 1977. Space does not permit the recognition of everyone connected with the EPD over this 14 year time interval. However, with apologies to inadvertent omissions, we wish to thank and acknowledge those many people who had major and/or continuing responsibilities in the EPD effort. The following major contributors, exclusive of the authors of the present paper and those named therein, are listed with the organizational affiliation existing at the time of their association with EPD: J. Crawford, J. Dassoulas, D. Fort, S. Gary, J. Heiss, B. J. Hook, J. Kohl, H. Malcom, R. Moore, T. Mueller, M. Puritz, S. Purwin, N. Rothman, P. Schwartz, R. Thompson, B. Tossman, J. Townsend, Jr., C. Wingate and H. Wong, all from The Johns Hopkins Applied Physics Laboratory; W. Boeker, W. Klemme, H. Sommer, W. Weiss, and H. Wirbs, from the Max Planck Institute für Aeronomie; R. Dayhoff, and C. Holmes, from the NOAA Space Environment Laboratory; W. Fawcett, R. Gibbs, G. McSmith, R. Parrish, J. Taylor, and J. Willett, from the Jet Propulsion Laboratory; S. Brown from Goddard Space Flight Center; and J. Burke and R. Martin, contract employees. Thanks to all of you who have helped make the EPD such a successful effort to date. This work was supported by a National Aeronautics and Space Administration contract to The Johns Hopkins Applied Physics Laboratory under the Department of Navy Task IAYX910X; contract N00039–89–C–001.

References

Fort, D. E.: 1985, JHU/APL Space Systems Report CP087.

Gehrels, N. and Stone, E. C.: 1983, *J. Geophys. Res.*, **88**, 5537.

Gloeckler, G., Geiss, J., Balsiger, H., Fisk, L. A., Gliem, G., Ipavich, F. M., Ogilvie, K. W., Stüdemann, W., and Wilken, B.: 1983, ESA **SP-1050**, 75.

Gloeckler, G., Ipavich, F. M., Stüdemann, W., Wilken, B., Hamilton, D. C., Kremser, G., Hovestadt, D., Gliem, F., Lundgren, R. A., Rieck, W., Tums, E. O., Cain, J. C., Masung, L. S., Weiss, W., and Winterhof, P.: 1985, *IEEE Trans. Geosci. Remote Sensing* **GE-23**, 234.

Krimigis, S. M. and Roelof, E. C.: 1983, in A. J. Dessler (ed.), *Physics of the Jovian Magnetosphere*, Cambridge University Press, Cambridge, p. 106.

Mauk, B. H. and Krimigis, S. M.: 1987, *J. Geophys. Res.* **92**, 9931.

McEntire, R. W., Keath, E. P., Fort, D., Lui, A. T. Y., and Krimigis, S. M.: 1985, *IEEE Trans. Geosci. Remote Sensing* **GE-23**, 230.

Stüdemann, W. and Wilken, B.: 1982, *Rev. Sci. Instr.* **53**.

Stüdemann, W. and Wilken, B., Kremser, G., Korth, A., Fennell, J. F., Blake, B., Koga, R., Hall, D., Bryant, D., Soraas, F., Bronstad, K., Fritz, T. A., Lundin, R., and Gloeckler, G.: 1987, *Geophys. Res. Letters* **14**, 455.

Wilken, B. and Stüdemann, W.: 1984, *Nuc. Inst. Methods* **22**, 587.

Wilken, B., Weiss, W., Stüdemann, W., and Hasebe, H.: 1987, *J. Phys. E: Sci. Inst.* **20**.

Wu, Y. and Armstrong, T. P.: 1988, *Nucl. Instr. Methods Phys. Res.* **A265**, 561.

THE GALILEO SOLID-STATE IMAGING EXPERIMENT

MICHAEL J. S. BELTON

National Optical Astronomy Observatorires, Tucson, AZ, U.S.A.

KENNETH P. KLAASEN, MAURICE C. CLARY, JAMES L. ANDERSON

Jet Propulsion Laboratory, Pasadena, CA, U.S.A.

CLIFFORD D. ANGER

ITRES Research Ltd, Calgary, Canada

and

Institute for Space and Terrestrial Sciences, York University, Ontario, Canada

MICHAEL H. CARR

United States Geological Survey, Menlo Park, CA, U.S.A.

CLARK R. CHAPMAN

Planetary Science Institute, SAIC, Tucson, AZ, U.S.A.

MERTON E. DAVIES

RAND Corp, Santa Monica, CA, U.S.A.

RONALD GREELEY, DONALD ANDERSON, LAWRENCE K. BOLEF,
TIMOTHY E. TOWNSEND

Arizona State University, Tempe, AZ, U.S.A.

RICHARD GREENBERG

University of Arizona, Tucson, AZ, U.S.A.

JAMES W. HEAD III

Brown University, Providence, RI, U.S.A.

GERHARD NEUKUM

DLR, Oberpfaffenhofen, Germany

CARL B. PILCHER

NASA Headquarters, Washington, DC, U.S.A.

JOSEPH VEVERKA, PETER J. GIERASCH

Cornell University, Ithaca, NY, U.S.A.

FRASER P. FANALE

Institute for Geophysics, Honolulu, HI, U.S.A.

ANDREW P. INGERSOLL

California Institute of Technology, Pasadena, CA, U.S.A.

HAROLD MASURSKY

United States Geological Survey, Flagstaff, AZ, U.S.A.

DAVID MORRISON and JAMES B. POLLACK

NASA Ames Research Institute, Moffet Field, CA, U.S.A.

Space Science Reviews **60:** 413–455, 1992.

Abstract. The Solid State Imaging (SSI) experiment on the Galileo Orbiter spacecraft utilizes a high-resolution (1500 mm focal length) television camera with an 800 × 800 pixel virtual-phase, charge-coupled detector. It is designed to return images of Jupiter and its satellites that are characterized by a combination of sensitivity levels, spatial resolution, geometric fiedelity, and spectral range unmatched by imaging data obtained previously. The spectral range extends from approximately 375 to 1100 nm and only in the near ultra-violet region (~350 nm) is the spectral coverage reduced from previous missions. The camera is approximately 100 times more sensitive than those used in the Voyager mission, and, because of the nature of the satellite encounters, will produce images with approximately 100 times the ground resolution (i.e., ~50 m lp^{-1}) on the Galilean satellites. We describe aspects of the detector including its sensitivity to energetic particle radiation and how the requirements for a large full-well capacity and long-term stability in operating voltages led to the choice of the virtual phase chip. The $F/8.5$ camera system can reach point sources of $V(\text{mag}) \sim 11$ with $S/N \sim 10$ and extended sources with surface brightness as low as 20 kR in its highest gain state and longest exposure mode. We describe the performance of the system as determined by ground calibration and the improvements that have been made to the telescope (same basic catadioptric design that was used in Mariner 10 and the Voyager high-resolution cameras) to reduce the scattered light reaching the detector. The images are linearly digitized 8-bits deep and, after flat-fielding, are cosmetically clean. Information 'preserving' and 'non-preserving' on-board data compression capabilities are outlined. A special "summation" mode, designed for use deep in the Jovian radiation belts, near Io, is also described. The detector is 'preflashed' before each exposure to ensure the photometric linearity. The dynamic range is spread over 3 gain states and an exposure range from 4.17 ms to 51.2 s. A low-level of radial, third-order, geometric distortion has been measured in the raw images that is entirely due to the optical design. The distortion is of the pincushion type and amounts to about 1.2 pixels in the corners of the images. It is expected to be very stable.

We discuss the measurement objectives of the SSI experiment in the Jupiter system and emphasize their relationships to those of other experiments in the Galileo project. We outline objectives for Jupiter atmospheric science, noting the relationship of SSI data to that to be returned by experiments on the atmospheric entry Probe. We also outline SSI objectives for satellite surfaces, ring structure, and 'darkside' (e.g., aurorae, lightning, etc.) experiments. Proposed cruise measurement objectives that relate to encounters at Venus, Moon, Earth, Gaspra, and, possibly, Ida are also briefly outlined. The article concludes with a description of a 'fully distributed' data analysis system (HIIPS) that SSI team members intend to use at their home institutions. We also list the nature of systematic data products that will become available to the scientific community. Finally, we append a short 'historical' note outlining the responsibilities and roles of institutions and individuals that have been involved in the 14 year development of the SSI experiment so far.

1. Introduction

Remote sensing experiments on the Voyager mission laid the basis for modern studies of Jupiter and its satellites some 11 years ago (Stone and Lane, 1979a, b). In 1995 a new package of experiments on the Galileo orbiter and its atmospheric Probe will reach the planet and provide a second opportunity for a substantial advance in our understanding. Along the way Galileo will have collected new data on Venus, the Moon, Earth, and one or two asteroids. Among the experiments is a high-resolution, slow-scan television camera called the Solid State Imager (SSI), which is designed to give the same angular resolution as the 'narrow-angle' camera (Snyder, 1980) on Voyager. It features a solid-state, charged-coupled device, detector of a radically different design to the SeS_2 vidicons that were used previously. As a result, the sensitivity of the camera has been increased by a factor of roughly 100; spectral coverage is extended by almost a factor of two into the near-infrared (~1 μm), although the sensitivity in curtailed in the ultra-violet; the photometric response is linear and free of residual image; and the geometric fidelity of the images is ultra-stable and free of any dependency on scene brightness gradients. Because Galileo's orbit will reduce the Galilean satellite encounter distances from the order of 100 000 km to 1000 km (some will be larger, some will be smaller), the ground resolution achievable by the camera will be increased by up to a factor of ~100 over Voyager. Repeated flybys of the satellites and a nominal 2-year orbital mission will allow the SSI to greatly extend the existing imaging data base in time, space, and spectral coverage.

During interplanetary cruise, the SSI experiment will return new data on Venus' clouds, extend the existing high-resolution multi-spectral imaging database on the Moon, and provide the first close-up images of at least one asteroid. At Jupiter it should take our knowledge of atmospheric phenomena into previously unexplored combinations of time and spectral coverage. Galileo's orbital tour will allow the long-term evolution of atmospheric structures to be followed at high spatial resolution, while the infra-red spectral sensitivity is expected to allow us to obtain information on the vertical structure of such features at the same time. Moreover, combined with observations from the atmospheric Probe, the SSI will extend some of the locally determined properties at the region of descent to the atmosphere as a whole. SSI observations of Jupiter's rings will not only improve knowledge of their structure and the source regions that produce the short-lived ring particles but, with extended time coverage, they may also yield new insights into their dynamics.

The SSI camera will be operated as a facility instrument in the fullest sense. It will be used to provide 'optical' navigation for essential refinements to the spacecraft trajectory and to calibrate precisely the pointing of the scan-platform. The SSI camera team also expects to provide essential pointing data to other remote sensing experiments on the Orbiter, and cooperative sequences and observations with other experiments are being emphasized.

2. The SSI Camera

A detailed description of many aspects of the camera system design and its operating modes has previously been published by Klaasen *et al.* (1984), and technical aspects of the detector and its response to energetic particle radiation can be found in Janesick *et al.* (1981). Also a preliminary report on the early inflight performance of the system is contained in Belton *et al.* (1992). In this section our intention is to provide a quantitative overview of the system for future reference, while including additional information not published elsewhere. The overall properties of the camera system are displayed in Table I.

2.1. Optical System

The optical system of the Galileo camera (Figure 1) is a modified flight spare of the 1500-mm focal length, all-spherical, catadioptric, 'narrow-angle' telescope (Snyder, 1980) flown on Voyager. Its basic design is that of the two telescopes flown on Mariner 10 (cf. Anderson *et al.*, 1978). Modifications include improved baffling to further suppress off-axis scattered light, changes to the design of the thermal coatings

TABLE I

Overview of the properties of the Galileo SSI camera

General	
Mass	28 kg
Peak power	23 W
Dimensions	$90 \times 25 \times 30$ cm^3
Optical	
Focal length	1500 mm
Field-of-view	8.1 mrad
Angular resolution	10.16 μrad pixel^{-1}
Camera speed	$F/8.5$
	T/10.8 at 576 nm
Filter positions	8 (see Table II)
Minimum exposure time	4.167 ms
Maximum exposure time	51.2 s
Detector	
Type	Silicon Virtual Phase CCD
Dimensions	12.19×12.19 mm
Pixel density	65.6 pixels mm^{-1}
Pixel format	800×800
Pixel full-well capacity	108,000 e^- (normal modes)
	432,000 e^- (summation mode)
Noise floor	$\pm 30\ e^-$
Signal digitization	8 bits, 256 DN
1 DN (low gain state)	318 e^-
Gain state factors	1:4.8:9.8
Nonlinearity	0.2 DN (normal modes)
	0.45 DN (summation mode)
Charge transfer efficiency	0.99997
Dark counts	$< 10\ e^-$ s^{-1} pixel^{-1}
Frame times for operating modes	
Summation mode (2×2 pixel)	2.333 s
Normal modes[a]	60.667 s
	30.333 s
Full-resolution record mode	8.667 s
Tape recorder characteristics for SSI data	
Transfer rate (bits s^{-1})	8.064×10^5
Maximum storage (images):	
Summation mode	478
$60\frac{2}{3}$	128
$30\frac{1}{3}$ compressed	257
$8\frac{2}{3}$	128
$8\frac{2}{3}$ compressed	257

[a] Normal modes can either be recorded or channelled directly for real-time transmission. The summation mode must be recorded. In the normal modes the data can, if necessary, be compressed by a factor of ≈ 2.5 in either an information preserving fashion (lines may be truncated), or in a non-information preserving (lines are complete but pixel values may lose accuracy).

Fig. 1. The SSI camera. The entrance aperture is to the left and is shown with the protective front aperture cover in position. The SSI experiments at Venus and Earth–Moon 1 encounters were performed with the cover in place. The white cylindrical appendage to the right is the radiative cooler for the detector. Camera electronics are in the rectangular box below the main body of the telescope.

to allow for the greater spectral range of the detector, the addition of a front aperture cover to protect exterior optical surfaces from contamination in the launch environment, improved shielding of the detector from high-energy particle radiation, and the addition of a 'pre-flash' system used to prepare the detector for an exposure and ensure photometric linearity in the images. A 'wide-angle' camera was not included in the Galileo payload due to weight and budget restrictions. Thus there is no redundancy in the instruments able to make high-resolution images.

The SSI camera contains five transmissive elements plus the two mirrors (Figure 2). A front aperture cover, featuring a transparent annulus, was installed to prevent contamination of the optics during the launch phase when the camera was in the space shuttle bay. The cover will be removed before the encounter with the asteroid 951 Gaspra if flight experience indicates negligible contamination rates of exterior optical surfaces during interplanetary cruise. All optical surfaces have anti-reflective coatings, and the camera operates at $F/8.5$ with a transmission factor of 62% at 576 nm. The front

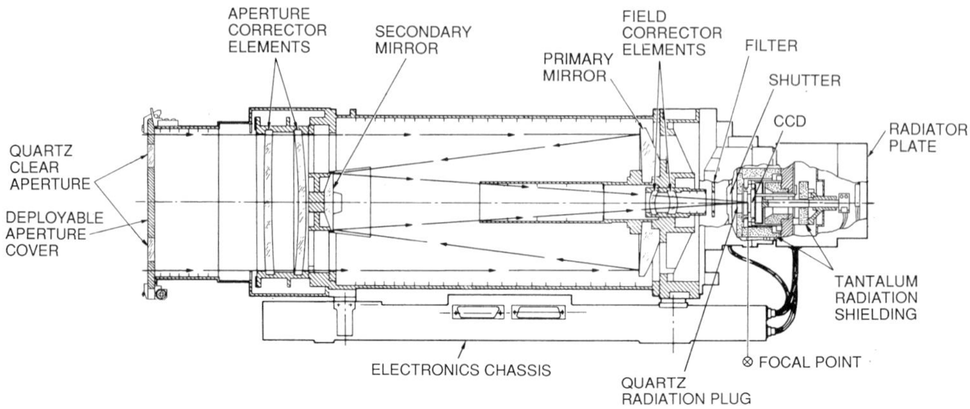

Fig. 2. Cut-away schematic of the SSI camera showing the location of the principal optical components, the particle radiation shield, and the front aperture cover.

optical element is coated to minimize radiative heat loss from the telescope, and heaters are provided to maintain thermal stability across the front aperture and within the body of the telescope. The long wavelength sensitivity limit (≈ 1100 nm) is set by the detector. The short wavelength limit (≈ 375 nm) is set by the properties of the anti-reflection coatings. Only in its short wavelength sensitivity does the previous Voyager camera have superior performance to the Galileo design. There is no analog to the Voyager 'ultra-violet' passband in the Galileo camera.

An eight-position filter wheel was also inherited from the Voyager project (Snyder, 1980). The filters were fabricated by Barr Associates and Microcoatings, Inc., and are of the interference type with out-of-bandpass absorption blocking components. The design of these filters was complicated by the combination of the wide spectral sensitivity of the detector, the desire for narrow spectral passbands, and a requirement for an identical optical thickness in all filters (the camera has a fixed focus). As a result, the filter passbands, while well defined, are far from smooth. The properties of the filters are listed in Table II and illustrated in Figure 3. The choice of filter passbands was based on the following considerations. A clear quartz filter was included to provide maximum sensitivity. The three broad-band filters – violet (404 n), green (559 nm), and red (671 nm) provide for visible color reconstruction and compatibility with the passbands utilized on Voyager. The remaining filters were chosen to investigate phenomena in the near-infrared spectral region. There are two narrow-band filters centered at 727 nm and 889 nm that are designed to sample Jovian methane absorption bands, and there is a 'continuum' filter centered at 756 nm. Finally, a filter at 986 nm is included to both provide spectral overlap with the Near-Infrared Mapping Spectrometer (NIMS) experiment and to sample any moderately strong electronic absorption features near 1000 nm, such as are often seen in the reflection spectra of common Fe-bearing minerals.

The camera shutter is also a Voyager flight spare (Snyder, 1980). The minimum exposure time is about 4.167 ms; and 28 discrete exposures are available up to a maximum of 0.8 s in the 'normal' mode, and beginning at 1.067 s up to 51.2 s in the 'extended' mode.

TABLE II

Filter data and SSI absolute sensitivity

Filter	Effective[c] wavelength (nm)	Passband[d] (nm)	Sensitivity[a] (W cm^{-2} ster^{-1} nm^{-1})	Sensitivity[b] (DN ms^{-1})
Clear	611	440	1.9×10^7	35.0
Violet	404	45	1.2×10^6	2.0
Green	559	65	2.5×10^6	5.5
Red	671	60	2.8×10^6	4.9
'Methane' (7270)	734	10	2.9×10^5	0.44
'Continuum'	756	19	5.4×10^5	0.78
'Methane' (8890)	887	16	1.5×10^5	0.17
'Infrared'	986	50	1.7×10^5	0.16

[a] Radiance needed to produce a signal of 1 DN ms^{-1} in the low gain state, assuming constant spectral radiance across the filter bandpass.
[b] Sensitivity at 5.2 AU for a scene of unit reflectivity illuminated by sunlight.
[c] Assumes a solar spectrum.
[d] Full-width at half maximum.

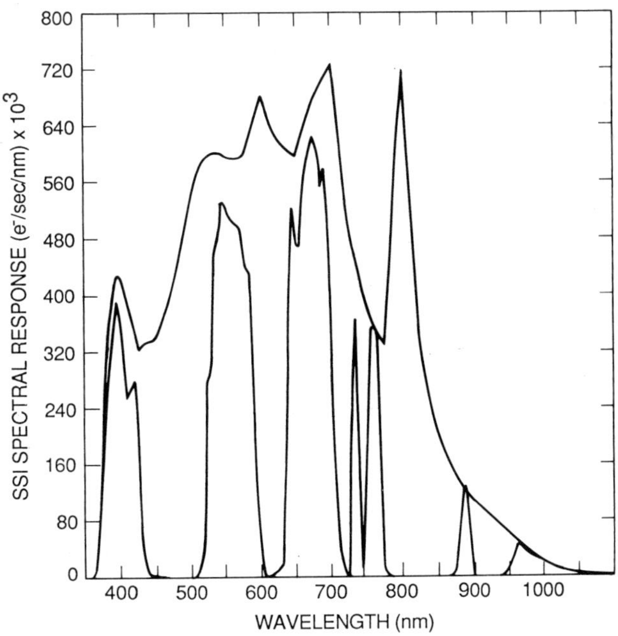

Fig. 3. SSI spectral response to a constant source radiance of π^{-1} watts m^{-2} ster^{-1} nm^{-1} through each of its eight filters.

2.2. The detector

The sensor (cf. Table I) is a virtual-phase ($V\phi$), buried channel, thick, frontside illuminated, 800 line by 800 column, charge-coupled, silicon device (CCD), fabricated by Texas Instruments Corporation in an experimental development program (Janesick *et al.*, 1981) that was directed by the staff of Division 38 of the Jet Propulsion Laboratory (JPL). The initial Galileo design called for a 3ϕ detector of the type that was subsequently incorporated in the Wide Field/Planetary Camera on the Hubble Space Telescope. However, at the time of the Galileo flight hardware development (1978–1982), difficulties in fabricating 3ϕ chips with full-well capacities exceeding $\approx 30\,000\ e^-$, and the discovery that the 3ϕ technology was susceptible to irreversible damage at the levels of particle radiation expected in orbit about Jupiter (see below), caused the SSI to switch to the development of a chip based on $V\phi$ technology. Many thousands of experimental $800 \times 800 V\phi$ CCDs were built for Galileo; however, the yield was such that only two chips were finally judged acceptable for flight. One of these, No. 29, is now on its way to Jupiter, and its characteristics are summarized in Table I.

The $V\phi$ CCD employs a polysilicon gate structure with 15.2 μm center-to-center spacing between pixels. During image read out, all 800 lines are simultaneously shifted one line in the column (parallel) direction causing the first image line to be shifted through a transfer gate into the serial output register. The first image line is then read out through an on-chip amplifier. The line read out rate (and therefore associated read out noise) is identical for all of the SSI operating modes. This procedure is repeated until all 800 lines have been readout. Important characteristics of the $V\phi$ CCD are:

(1) Image geometry which is independent of scene brightness gradients. Unlike earlier planetary camera systems, the SSI will have no reseaux (fiducial marks) impressed on the images.

(2) Linear photometric response to incident light. This property greatly simplifies the systematic calibration of the raw images.

(3) High charge transfer efficiency of ~ 0.99997 (Klaasen *et al.*, 1984), which ensures photometric linearity at low light levels.

(4) A high quantum efficiency and low read out noise which makes the SSI approximately 100 times more sensitive than the equivalent Voyager camera.

(5) Wide spectral range. The $V\phi$ CCD has appreciable sensitivity to light out to 1100 nm. This compares favorably with the long wavelength cut-off of earlier planetary cameras near ≈ 650 nm.

During ground calibration tests it was found that read out of the chip did not completely remove the effects of a previous exposure to light at wavelengths > 650 nm. A faint 'residual image' was impressed on the photometric response of the chip to a succeeding exposure. In order to prevent this effect from corrupting the photometric accuracy of the data, the camera operation was modified so that prior to each exposure the chip is 'pre-flashed', i.e., bathed in near-IR (~ 930 nm) light from several photodiodes placed around the chip, and then read out several times at high speed. This preparation cycle is found to remove all traces of previous exposure thistory. Details and an explanation of this phenomenon can be found in Janesick *et al.* (1984).

Much of the hardware incorporated into the SSI camera is equipment left over from the Voyager project. While this approach, which was dictated to the Galileo project, has possibly saved some expenditures – it has also restricted the scientific potential of the SSI experiment. For example, eight filter passbands are insufficient to adequately cover the spectral range of the SSI detector with the result that there are gaps in the spectral coverage; the thermal design of the telescope is in conflict with the requirements for a cold detector (SeS_2 vidicons were operated near 'room' temperature) and this is a contributory reason why the SSI lacks the ultra-violet sensitivity of the Voyager cameras; in addition the 8-bit linear encoding of the signal is not well matched to the dynamic range available with a CCD and will lead to the presence of digital 'contouring' in certain types of low contrast images. Nevertheless, in spite of these difficultires, the basic sensitivity, linearity, and geometric fidelity of the SSI camera is so great that it is expected to create images that are vastly superior to those obtained in previous planetary missions.

2.3. Particle Radiation Hazards

The effects of the irradiation of the camera by high-energy particles in the Jovian magnetosphere and by the flux of neutrons from the Galileo radioisotope thermoelectric generators (RTG's) are expected to be severe (Janesick *et al.*, 1981; Klaasen *et al.*, 1984) and, during the development of the SSI, several tests were performed to quantify them. In the first, which used a Co^{60} source, the effects of γ-ray and high-energy electron impacts on candidate 3ϕ devices were investigated at fluence levels expected to be reached during the orbital tour. The CCD, as shielded in the SSI, is predicted to receive a dose of ~ 900 rad (from fluences of $\sim 4 \times 10^{10}\, e^-\ \text{cm}^{-2}$ and $\sim 2 \times 10^7$ protons cm^{-2} with energies above 100 keV) by the end of five orbits about Jupiter. Passage of a high-energy charged particle through the SiO_2 epitaxial layers, which function as insulators between the various 3ϕ gate structures, produces local ionization and possible breakdown in their insulating properties. The observed result was an unacceptable, dose dependent, change in the applied 'flatband' operating voltages which control the full-well capacity and the charge transfer efficiency characteristics of the device. In one case, major physical damage occurred and resulted in a large permanent blemish that covered thousands of pixels. It was on the basis of this experiment that the decision was made to switch from the 3ϕ to $V\phi$ chip. Since the latter has only a single gate structure, it is expected that it will be far less susceptible to these effects. $V\phi$ CCD's were tested to fluences of 10^8 protons cm^{-2} (at 10 MeV) and $7.5 \times 10^{10}\, e^-\ \text{cm}^{-2}$ (at 2.5 MeV) without suffering any unacceptable permanent damage. In addition, a modification to the camera design was made to increase the radiation shielding around the detector. The CCD is encapsulated in a tantalum cavity with 1 cm thick walls. A 0.5 cm thick quartz 'plug' was also placed in the optical path directly in front of the sensor.

Although well protected, the $V\phi$ CCD is still an excellent particle detector, and calculations showed that within 10 R_J (Jupiter radii) the chip might easily approach full-well saturation in the time (60.667 s) it takes to read it out in the normal, real-time mode. Tests were therefore made to quantify the effects of the expected particle fluxes on the quality of the images and, in particular, to understand how the camera might operate at 5 R_J, near Io (cf. Figures 4(a) and 4(b)). A Sr^{90} source of energetic electrons was used, and it was found necessary to increase the maximum read out rate of the detector by a factor of four in order to assure satisfactory operation at Io. The frame time in this special mode is 2.333 s, and in order to match the read-out rate of the camera to the on-board tape recorder, it was necessary to reduce the image format by the same factor. The SSI team chose an option in which adjacent pixels in the image are summed (one 'summed' pixel equals four mutually adjacent pixels; the resulting image is then in a 400 × 400 pixel format) during the read-out of the chip. This is referred to as the 'summation mode' (Table I). The images at Io are expected to contain a background signal, formed by radiation 'hits' that reaches a maximum of about 8 data numbers (DN) in the summation mode. At 15 R_J, which is the orbital radius of Ganymede, individual radiation 'spikes' will be present in the images at a rate of about 10^4 frame^{-1} in the 8.667 s mode. The amplitude of the average spike is expected to be about 4 DN in the low gain state.

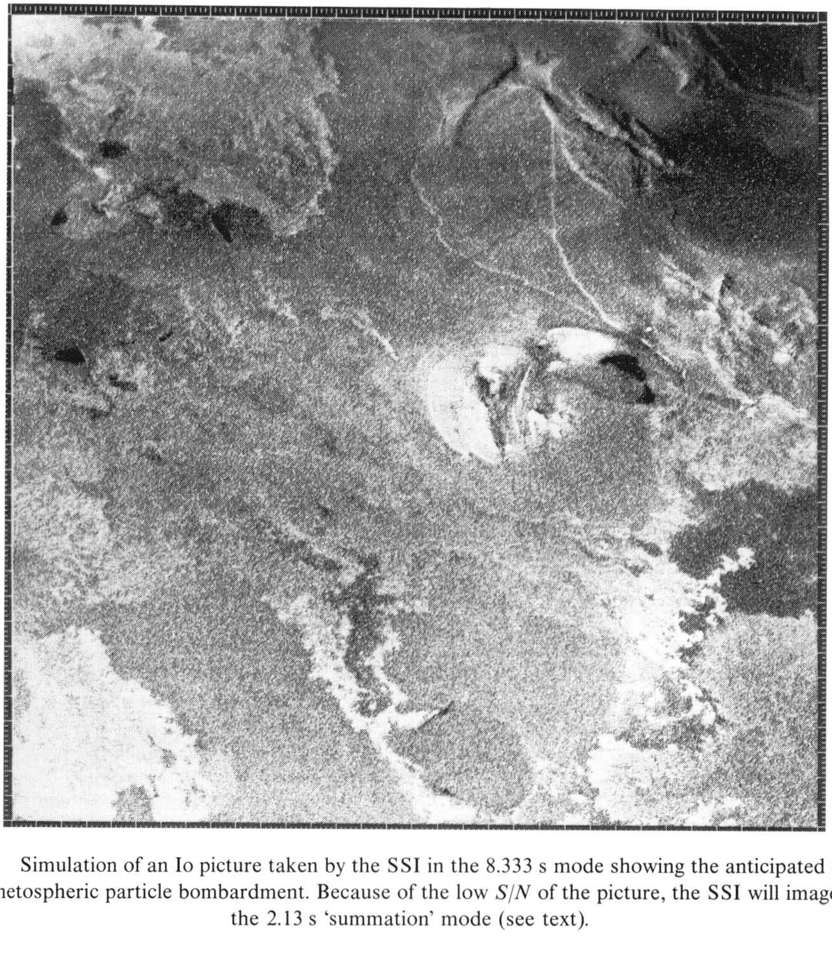

Fig. 4a. Simulation of an Io picture taken by the SSI in the 8.333 s mode showing the anticipated effects of magnetospheric particle bombardment. Because of the low S/N of the picture, the SSI will image Io in the 2.13 s 'summation' mode (see text).

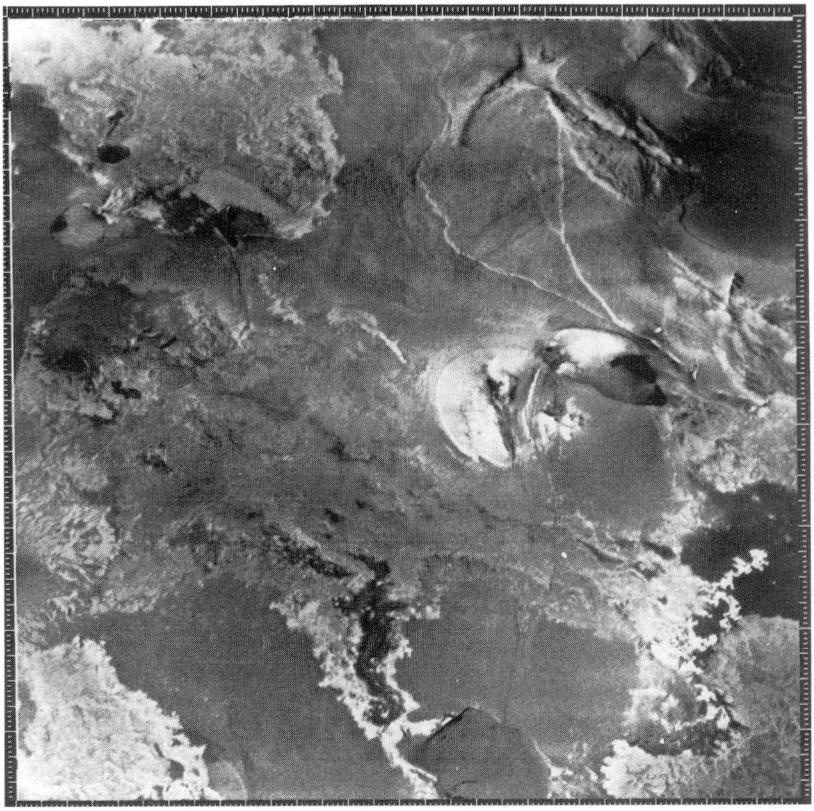

Fig. 4b. Same as Figure 4(a) but without the effects of energetic particle irradiance.

A further, insidious source of particle radiation is the spacecraft RTGs, which produce a steady flux of neutrons. These particles cause irreparable dislocations in the lattice structure of the detector material which can then act as sites for the copious thermal production of charge. These sites are sometimes referred to as 'hot pixels' or 'dark spikes' in the images. To quantify this effect, the Northrop Corporation's Triga reactor was used as a neutron source to simulate the dose expected from Galileo's RTGs. As expected, lattice displacement damage was observed to occur and permanent 'dark spikes' were produced in the output images. To ameliorate the effects of this unavoidable phenomenon, the operating temperature of the detector was reduced from -70 °C to -110 °C by enlarging the thermal radiator. Lattice dislocations are expected to occur at a rate of about 10^5 yr^{-1} and are expected to slowly reduce the charge transfer efficiency of the flight chip. However, its operating lifetime is expected to far exceed the duration of the nominal Galileo mission.

The fact that the $V\phi$ CCD is sensitive to particle radiation is the price that must be paid for increased quantum efficiency and spectral range. Already during an inflight checkout of the camera, and in a later 'mini-calibration sequence', we have seen the deleterious effects of cosmic rays and energetic particles of solar origin in the images (Belton *et al.*, 1992). However, because of the tests and modifications described above, the hardware is expected to return excellent images throughout the nominal mission.

2.4. Ground calibration and performance

The process of calibrating the images involves three basic steps. First the zero-exposure bias in the signal is subtracted. This is followed by the removal of small pixel-to-pixel variations in sensitivity (fixed-pattern noise) by division of the image by a second image (a 'flat-field') of a uniformly illuminated, extended source. Finally, the signal in the images is placed on an absolute scale through the application of factors that are determined in ground calibrations. It is intended to update the absolute calibration during the flight mission through a program of inflight calibration sequences.

The performance characteristics of the camera have been thoroughly evaluated. Figure 5 shows a typical flat-field image. The dark circular 'donuts' are shadows of small, ~10 μm in diameter, dust specks, most of which are located on the surfaces of

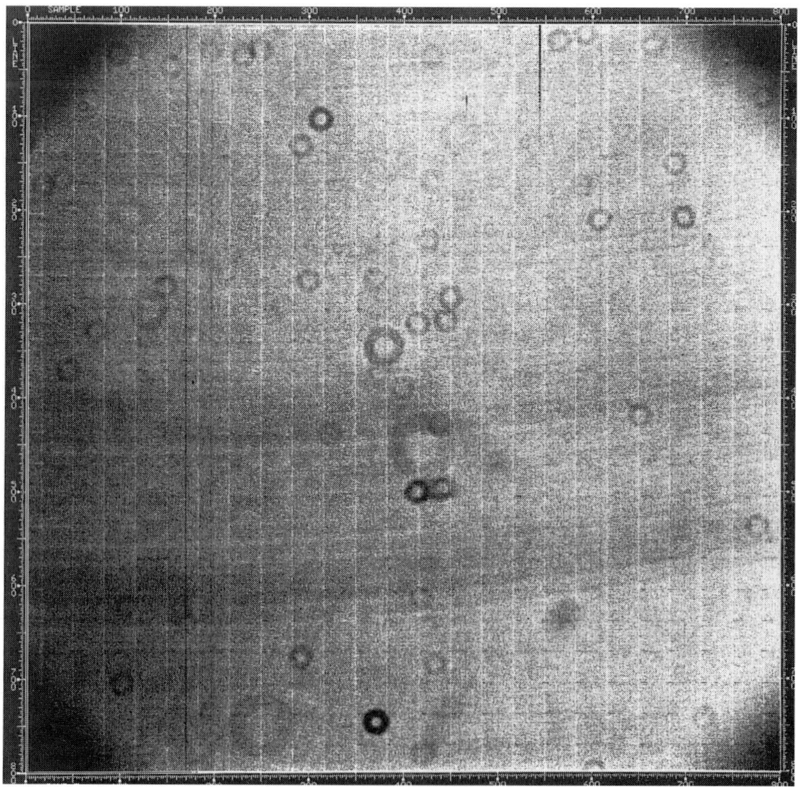

Fig. 5. Contrast enhanced image of a flat field target which shows the non-uniformities in the camera response. The circular features are shadows of dust particles on the window of the detector package. These features are completely removed from the images after flat-field calibration (see text). The picture has been contrast enhanced by a factor of 25 in order to bring out the non-uniformities.

the window of the detector package. The darkening at the corners of the image is due to a slight vignetting of the field-of-view by the filter aperture. The light vertical columns, which occur every 33 pixels, are due to an increased pixel area in these columns. This, in turn, is an artifact of the step-and-repeat photolithographic technique that is used in fabricating the $V\phi$ CCD.

The mechanical shutter was shown to be very repeatable. Actual exposure times were found to be slightly less than the commanded time by about 1.2 ms, averaged across the frame. This shutter time offset varies from about 1.5 ms at the top of the frame (line 1) to about 1.05 ms at the bottom (line 800). When the shutter offset is properly taken into account, the SSI response to a fixed source radiance over a 50×50 pixel test area in the center of the image was found to be linear with exposure time to better than 1% over its entire dynamic range.

The absolute sensitivity of the system through each filter was determined to an accuracy of about 10% and is shown in Table II. Also listed is the SSI sensitivity to a scene normally illuminated by sunlight at a distance of 5.2 AU (Jupiter's distance from the Sun) at zero phase, assuming unit reflectivity. The signal from each pixel is encoded linearly into 256 DN (data numbers). The portion of the CCD dynamic range that is actually encoded is determined by the settings of two commandable gain switches in the SSI signal processing electronics. Table II is computed for the 'low' gain state in which one DN equals 380 e^-. The gain can be increased (i.e., the number of electrons per pixel decreased) by factors of 4.8 or 9.8 on command. Full-well is at about 108 000 e^- pixel^{-1}. Figure 3 shows the overall SSI system spectral response function for each filter position.

The camera sensitivity is such that, for unresolved objects, it can detect a solar-type star with $V = 11$ mag at a $S/N \sim 10$, through the clear filter in the high gain state. On extended targets, it can reach surface brightnesses down to 20 kilorayleighs at 60 nm using the longest available exposure.

The radiometric response has been calibrated for every pixel in the CCD array and with the exception of a few blemishes, the spatial nonuniformities in response across the array can be corrected using a linear calibration function. Figure 6 shows the calibrated version of the flat-field image in Figure 5. Note that all traces of dust shadows, corner vignetting, and the step-and-repeat pattern are completely removed. Most of the remaining blemishes are located in columns 1, 2, and 170 and rows 799 and 800. Also, nine columns were found to contain pixels that have lower than average full-well capacities. These, under certain conditions, can lead to 'bad' columns in the output image. Because they have a limited charge transfer capability (in the range from 30 000 to 85 000 e^-), images that have high average signal levels may show bad, or partially bad, columns as a result of these low-full-well pixels.

Raw 'dark' (i.e., unexposed) images have a random noise level of about ± 0.95 DN in the highest gain state (equivalent to about $\pm 40\ e^-$) and a coherent noise level of about ± 0.35 DN. The random component is primarily due to spurious charge generation produced in the $V\phi$ CCD during clocking. The level of this component stays at the same level in all of the readout modes (for a particular gain state) since the line

Fig. 6. Contrast enhanced image of a flat-field target after calibration. In this case (compare Figure 5) the picture has been stretched by a factor of 65.

clocking rate is independent of the mode. Thermal generation of charge at SSI operating temperatures is negligible. The primary source of coherent noise is the spacecraft power supply (at 2400 Hz). This shows up as a very faint pattern of vertical stripes in the images spaced 42 pixels apart. However, the amplitude of the coherent noise is small enough that it will not normally be visible in the images. The random noise component noted above includes the effects of CCD spatial sensitivity variations and blemishes. The noise level in a high-gain dark frame was found to be reduced to ± 0.82 DN by the calibration process, and the coherent noise is reduced to ± 0.17 DN. This is equivalent to the theoretical level expected for a CCD with a read noise of $\pm 31\ e^-$, combined with a quantization error of ± 0.37 DN, a residual uncalibrated component equal to 0.78% of the mean signal level. In the lower gain states, the uncalibrated component is found to be smaller (around 0.3%).

Quantization noise was found to be larger than the theoretical level of ± 0.29 DN expected for a perfect analog-to-digital converter (ADC). The flight ADC exhibits some unevenness in the widths of its DN bins, particularly for DN levels that are multiples of 8. By adjusting the SSI output DN levels by fractions of a DN in the calibration process, the quantization error can be reduced to about ± 0.31 DN.

The SSI shutter was tested for light leakage and was found not to be perfectly light tight; approximately 10^{-7} of the incident light is transmitted when the blades are closed. This is equivalent to a signal level of about 300 e^{-} min^{-1} if it is assumed that the camera is pointing directly at Jupiter at 0° phase angle in the clear filter.

The achievable spatial resolution is illustrated in Figure 7. Unaliased contrast is

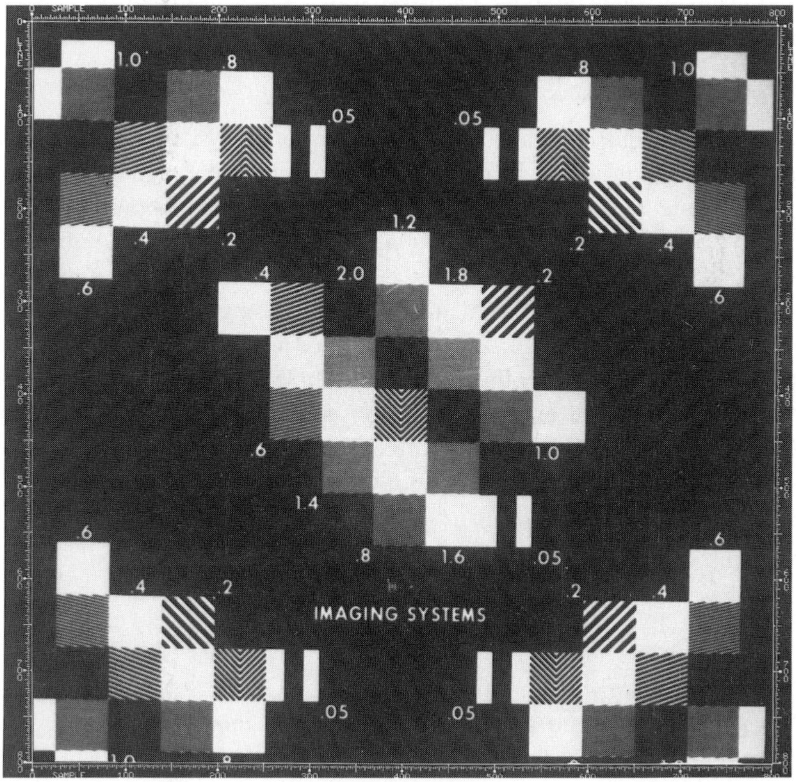

Fig. 7. Slant-bar test target showing the spatial resolution of the camera. The numbers next to the boxes indicate the fraction of the SSI Nyquist frequency (33 cycles mm^{-1}) represented by the slant bar pattern in the box.

visible down to 80% of the Nyquist frequency (0.049 cycles μrad^{-1}). SSI inflight performance is expected to be at least as good as this as the test setup which was used to acquite Figure 7 results in some degradation. Images of wide horizontal and vertical bar targets were also acquired and analyzed to determine the modulation transfer function (MTF) of the system. At the Nyquist frequency, the MTF ranges systematically from about 0.2 in the violet filter, to about 0.08 in the infrared filter. This is consistent with the expected diffraction-limited performance of the telescope. The point-spread response was also characterized by imaging targets containing pinholes. The observed standard deviation of the resulting charge distributions about the location of the center of the distribution ranged between 0.75 and 1.1 pixels. As a rough 'rule-of-thumb', the camera has a useful resolution of about 1 km (0.5 km pixel^{-1}) at a range of 50000 km.

Geometric distortion in SSI images was investigated by measuring the locations of intersections of a set of regular horizontal and vertical lines in an image of a regular target projected through a distortion-free collimator. A low level of pincushion distortion was observed, as expected from analysis of the optical design. This distortion is azimuthally symmetric and increases as the cube of the distance from the center of the image. It reaches a maximum value of about 1.2 pixels at the corner of the array. The coefficient of the cubic term is about 6.58×10^{-9} pixel^{-3}.

2.5. Inflight calibration

Because of possible detector drift during the very long flight time to Jupiter and the potential for changes in the performance of the camera while immersed in the Jovian magnetosphere, the SSI experiment includes a major inflight calibration effort, already underway. We are using photometric standard stars, star clusters, and observations of a 'calibration plaque' (permanently mounted on the spinning section of the spacecraft) to acquire absolute calibrations and flat-fields, and thereby follow the evolution of the SSI calibration parameters. It is planned to perform a major inflight calibration sequence just prior to each encounter and, if possible, twice every orbit in the Jovian magnetosphere. Early data on the inflight performance of the system *with the front aperture cover in place* are reported by Belton *et al.* (1992). The camera performance was found to be nominal, although the aperture cover introduced anomalies.

3. Scientific Objectives and Measurement Strategy

The scientific objectives for the orbital mission were determined very early in the mission planning and were are derived directly from high-level mission goals (Yeates *et al.*, 1985). After the mission was reconfigured to the VEEGA trajectory in 1987, additional science objectives were included that cover the cruise encounters. In this section we give an overview of the SSI objectives for the orbital tour and a brief outline of the scientific focus of the SSI experiment at each of the planetary bodies that we expect to encounter.

There are many possibilities for cooperative investigations with the other orbiter experiments, and reasonable efforts have been made to identify them. Several of Galileo's interdisciplinary scientists (IDSs) have become intimately involved in the design of the SSI experiment with this purpose in mind. Particularly strong cooperative investigations are expected between the SSI and NIMS, the Photo-Polarimeter and Radiometer (PPR), and UltraViolet Spectrometer (UVS) experiments.

3.1. Satellite observations

(1) For each Galilean satellite, determine the morphology of at least 50% of the surface at a resolution of 1 km or better (in particular in those areas for which Voyager resolution is poorer than 5 km) and selected areas of the surface at 100 m or better.

(2) Map the spatial variation in color and albedo of each Galilean satellite over at least 50% of its surface at a resolution of 2 km or better and monitor any temporal variations.

(3) Determine the photometric function at several locations on each Galilean satellite.

(4) Determine the location of the spin axis, the rotation rate, and the geometric figures of the Galilean satellites to an absolute accuracy of $\pm$ 3 km.

(5) Determine what photometric phenomena result from the interaction of the Galilean satellites with Jupiter's magnetosphere.

(6) Determine the surface morphology, color, and albedo, and measure the global properties of the non-Galilean satellites as opportunities arise.

3.2. Jupiter atmospheric observations

(7) Characterize the state of the atmosphere in the vicinity of the Probe descent region as closely as possible to the time of the descent of the Probe through the atmosphere.

(8) Separate atmospheric mass motions from wave motions on global scales. Identify any wave propagation modes that are present and the level of the atmosphere in which they exist. Characterize fields of atmospheric flow, cloud structures, and waves.

(9) Investigate the vertical structure of a few specific types of atmospheric features by center-to-limb, and multi-spectral observations. Characterize any relationships between vertical cloud structure, color, and morphology.

(10) Measure the mean vertical wind profile through relative cloud motions at the boundaries of major belts and zones and in the region of descent of the Galileo atmospheric Probe.

(11) Measure the scattering properties of individual cloud features at several wavelengths and at as wide a range of phase angles as possible.

3.3. Observations of magnetospheric interactions

(12) Map and characterize auroral phenomena in the Jovian atmosphere and on the satellites. Correlate them with *in situ* particle and field observations made from the Orbiter.

3.4. Jovian ring observations

(13) Map the ring structure and search for any temporal variations.

(14) Search for new small satellites in the vicinity of the Jovian ring and observe all small inner satellites at the highest possible resolution to improve our knowledge of their sizes, shapes, colors, albedos, surface textures, and spin states.

(15) Measure the color, albedo, and column densities and infer the size distribution of the ring particles.

3.5. Cooperative observations and targets of opportunity

(16) Pursue cooperative observations with other Galileo investigations.

(17) Make observations of objects which can be expected to present special target opportunities during the course of the mission.

(18) Make observations of the photometric, geometric, and geological character of at least one asteroid.

(19) Make 'best effort' investigations of Venus, Earth, and the Moon during encounters that occur during interplanetary transit to Jupiter.

4. Measurement Strategy at Major Encounters

4.1. Overview of the Galileo Trajectory

The mission trajectory has the following components: (1) Interplanetary cruise, which includes major encounters at Venus, the Earth–Moon system (twice), and, hopefully, two asteroids (951 Gaspra and 243 Ida); (2) Approach to Jupiter and Jupiter Orbital Insertion (JOI), which includes an 'observatory phase' study of Jupiter, encounters with Europa and Io, a detailed characterization of the Probe descent region, and observations of the planet's darkside; (3) The orbital tour (Yeates *et al.*, 1985) at distances $> 10\ R_J$ from Jupiter, which includes 10 orbits each characterized by an encounter with the planet itself, a close (100–1000 km) encounter with a specific Galilean satellite, and in some cases an 'untargeted' (20000–100000 km) encounter with one of the other satellites. Major opportunities exist for the SSI in all these mission phases. As the orbital tour progresses, Io and Europa will be monitored continuously. Also, in suitable orbital configurations, the rings, darkside phenomena on the planet, and non-Galilean satellites will be targeted. The Galileo spacecraft will reach Jupiter in late 1995 and JOI is expected to occur on December 7, 1995. The subsequent orbital tour will nominally conclude on October 7, 1997.

4.2. Venus

The Galileo spacecraft encountered Venus on UT February 10, 1990 and, at the time of writing, has returned three images of the planet. One of these, taken through the violet filter (with the front aperture cover in place), is shown in Figure 8. A preliminary description of these images has been given by Belton *et al.* (1992). The full tapeload of 81 Venus frames will be played back to the Earth when the telecommunications link improves in late November, 1990.

The spacecraft trajectory was favorable for an investigation of small-scale motions in the Venus clouds. The spacecraft flew behind Venus to receive acceleration from the planet, and then receded from Venus over the afternoon quadrant of the atmosphere. Previous imaging experiments (e.g., Belton *et al.*, 1976; Rossow *et al.*, 1980) have shown that small-scale motions are most intense and most clearly visible in the afternoon quadrant.

The observational strategy was to track a region of clouds as it passed from morning to afternoon across the disk and follow the evolution of specific markings as the Sun's elevation changed. Internal waves should be detectable by phase propagation relative to the general drift of features. Waves are of special interest because they are capable of transporting momentum for large distances, producing accelerations where they are

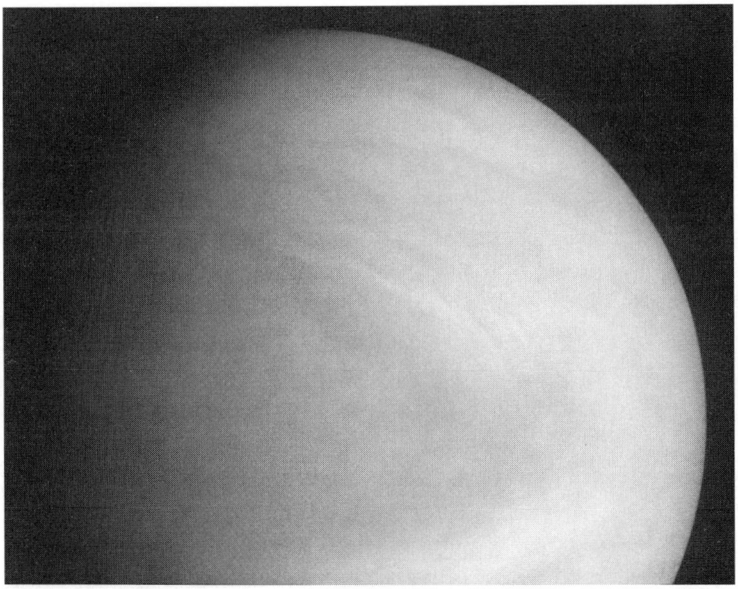

Fig. 8. Violet (404 nm) image of Venus taken by the SSI on UT February 13, 1990. The picture has been calibrated and stretched but not filtered. It was taken with the front aperture cover in place. Cosmetic removal of dust shadow residuals has been performed (cf. Belton *et al.*, 1990).

absorbed. Convective activity should be detectable by the appearance of time-dependent cellular structures. In addition top the study of small-scale motions by feature tracking, images were sequenced at larger ranges to determine the general circulation over the entire planet. This should be particularly useful for comparison with the infrared spectral-imaging data from the NIMS experiment, which will determine the rate of motion of the newly discovered thermal emission features deep in the Venus cloud layers (Allen, 1987).

Other SSI objectives included a search for optical evidence of lightning flashes, a survey of limb haze structure from the equator to the pole, and an attempt to image cloud structure in the near-infrared near 1 μm. One of the three Venus test images indicated that cloud markings, with a contrast of 3%, do exist at 1 μm. The morphology of these markings is different from those seen in violet and ultraviolet images.

4.3. The Moon

Lunar observations will have the highest priority in the SSI experiment during the Earth Gravity Assist (EGA) encounters. Combined with NIMS observations, the data should permit ground-truth calibrations by observing regions on the nearside from which samples have been returned by Apollo and Luna missions (Adams and McCord, 1971, 1972). We then plan multispectral imaging of the north polar region, limbs, and farside that should allow extension of our knowledge of nearside mineralogical sample and unit compositions to these regions for the first time. The SSI spectral coverage, particularly

Fig. 9. The Orientale basin on the Moon, which is the highest priority target in the SSI experiment at EGA 1. It is about 900 km in diameter and is thought to have been formed slightly less than 4 billion years ago from the impact of a large projectile. The small white patch is the outline of a map of the State of Rhode Island and gives some idea of the scale of the basin.

in filters at 404, 756, and 986 nm, will allow *ratio* images (e.g., 404/756, 986/756) to be formed that track the spatial distributions of minerals with differing Fe, Ti, or Ca contents.

During EGA-1 the SSI will obtain excellent coverage of the Orientale Basin (Figure 9), the youngest major basin on the Moon (Head, 1974; Wilhelms, 1987). This impact feature has excavated and ejected crustal and perhaps mantle material from considerable depths. Multispectral images will provide information on the distribution

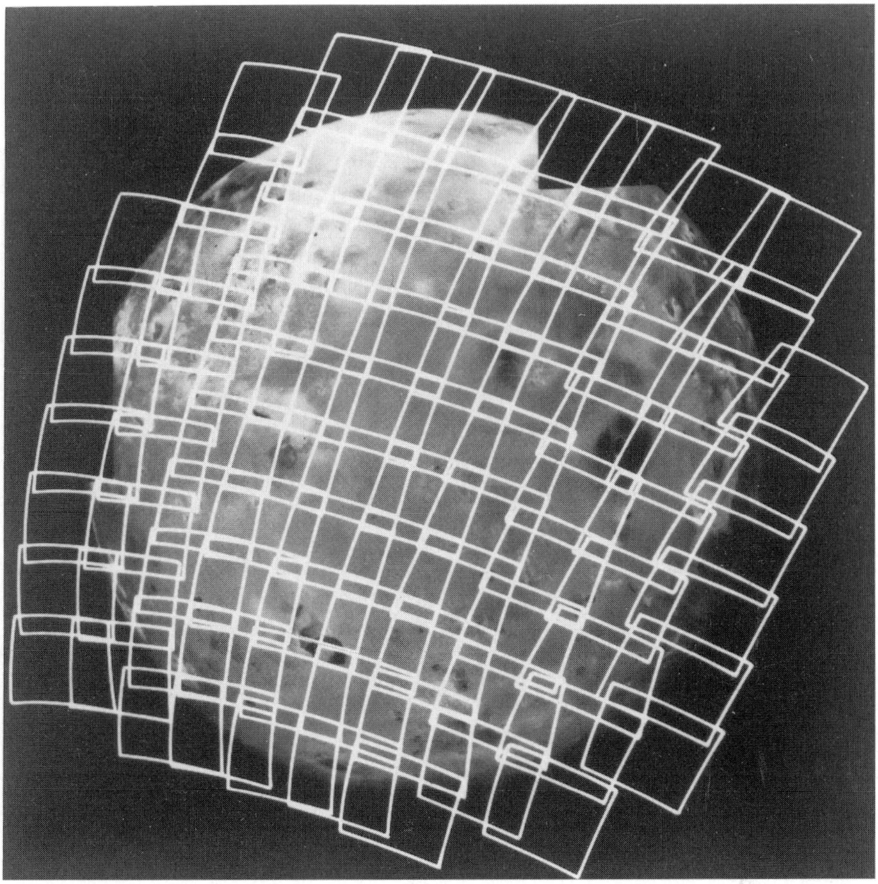

Fig. 10. The presently planned last, full disk, highest resolution SSI mosaic of Io superimposed on a Voyager color mosaic.

of different compositional units associated with Orientale and other basins, possibly providing clues to crustal stratigraphy. Together with NIMS data, it may be possible to correlate compositional units with known rock types. Analysis of the images is expected to yield new information on problems of impact basin formation (e.g., do large basins penetrate through the crust and into the mantle?), on the composition of the lunar farside and polar crust (are the nearside and farside crusts of different compositions?), and on the stratigraphy of the lunar crust (is it layered or has it been largely homogenized by impact basin reworking (Pieters *et al.*, 1985; Pieters, 1986)?).

Other questions that we will address include a determination of the global distribution of lunar maria, the melt products of early differentiation (Pieters, 1978; Pieters *et al.*, 1980). It is well known that there is a nearside-farside asymmetry in distribution of the maria, but there has been insufficient compositional data to determine the distribution mare types on the farside. SSI will provide new information on the farside and in north polar regions.

In addition, we intend to make observations of previously unexplored areas of the south polar region and to provide geodetic control points between the front and back sides to improve the mapping control network.

4.4. Earth

Galileo's two flybys of the Earth are expected to generate considerable public interest. However, our planet is already under constant surveillance by an array of specialized spacecraft, which are specially instrumented for Earth studies. New measurements by the SSI are therefore not expected to add a great deal that is of high scientific interest. For this reason SSI measurement priorities at the two EGAs are being directed towards the Moon. When opportunities permit, SSI will acquire sequences of multi-spectral images of planet Earth from a variety of distances and, in essence, build-up short movie clips of cloud motions and planetary rotation. These activities will focus on developing the team's expertise in creating similar, but far more extensive, Jupiter sequences. In addition, we intend to obtain images of well understood geological features on the Earth (chiefly in Australia and in the western hemisphere) at a range of resolutions comparable to those that will be acquired on the Galilean satellites. In this way we expect to be able to understand the value and limitations of Galileo images by making comparisons of Galileo data with observations made from the air or directly from the ground.

4.5. Asteroids

As soon as it was generally appreciated that flybys of mainbelt asteroids were possible as part of missions to the outer planets at a very modest cost of resources, NASA established a policy that such flybys would be a part of all future missions to the outer solar system. This policy is to be implemented for the first time by the Galileo mission, and the spacecraft will encounter 951 Gaspra on October 29, 1991, and, probably, 243 Ida on August 28, 1993. The known physical characteristics of these asteroids are summarized in Table III. The encounter distance remains to be decided, but is expected to be in the range of 500–5000 km. Two major factors, spacecraft safety and expected science return, need to be considered to determine the optimum distance for an asteroid

TABLE III

Characteristics of asteroids to encountered by Galileo

	951 Gaspra	243 Ida
Diameter (km)	16	32
Taxonomic type	S[a]	S
Rotational period (hr)	7.04	4.1
Light curve amplitude (mag)	0.7	>0.4
Possible family membership	Flora	Koronis
Flyby Speed (km s^{-1})[b]	8	12.5
Approach phase angle	32°	19°

[a] Gaspra is an 'unusual' S-type (C. Chapman).

[b] The encounter distance is not yet decided (see text).

flyby. After much effort by several study groups, the Galileo project has adopted the position that no 'circum-asteroid' dust hazard will be encountered beyond one hundred radii of the target.

For an asteroid as small as 951 Gaspra, the Radio Science Experiment requires a flyby distance of 200 km or less to achieve a mass determination to an accuracy of about 10%. Scan platform instruments, especially NIMS and SSI, currently prefer larger flyby distances (800 and 2000 km, respectively) in order to optimize the competing requirements for high-resolution and substantial areal coverage at a wide range of phase-angles. As both of the flybys involve approach at low phase angles, which is optimum for NIMS spectral mapping, but not for high-resolution imaging, the SSI attaches great significance to targeting frames successfully near the time of closest approach when surface features will be illuminated at low Sun angles, i.e., high incidence angles were shadows accentuate the topographic relief. Given the possibility of significant downtrack errors in the position of the asteroid, the SSI desires to encounter the asteroids at larger distances than desired by NIMS, in order to ensure that at least one frame at moderate phase angle has a high probability of containing the asteroid. At the time of writing, a consensus on the optimum targeting strategy has yet to be achieved.

The SSI asteroid strategy has two major aims. The first is to obtain as much global, multicolor coverage of the asteroid as possible during the approach phase. As the camera achieves a resolution of about 1 km lp^{-1} at 10^5 km, it should be possible to resolve Gaspra at about one day before closest approach for a flyby speed near 10 km s^{-1} thus defining the useful duration of the approach phase. The second goal is to obtain as much coverage of the illuminated part of the asteroid near the time of closest approach when the range is minimum and the phase angle is near 90°. This produces the low-Sun illumination conditions needed to bring out surface morphology. Most of the visible, illuminated face of the asteroid can be imaged at 100–200 m $pixel^{-1}$ and some data may be obtained in the terminator regions at resolutions approaching 20 m $pixel^{-1}$.

The SSI asteroid science objectives are outlined in Table IV. It is our intention that

TABLE IV

SSI asteroid measurement and scientific objectives

Global properties	Surface morphology
Accurate size/shape Volume estimate Period and pole position Cratering statistics and 'age' of surface Search for satellites	Crater morphology as function of diameter Ejecta patterns Evidence of past internal activity Search for spallation features
Compositional properties	**Regolith properties**
Surface composition Compositional heterogeneity	Ejecta dispersal Stratigraphy Photometric properties

the objectives relating to surface composition will be carried out in conjunction with the NIMS investigation. While the SSI spectral range (Table I) is more restricted than that of NIMS (Yeates *et al.*, 1985), the two instruments clearly complement one another as the SSI extends the 1–5 μm spectral coverage of NIMS down to 0.42 μm. More important, the far greater spatial resolution of SSI should make it sensitive for detecting the small-scale albedo/color differences that can be used to define compositional units and relate them to small-scale topography.

Both Gaspra and Ida are classified as S-asteroids (Chapman *et al.*, 1975). But the precise nature of S-asteroids remains a hotly contested issue. On one hand, several pieces of evidence suggest that S-objects are elated to ordinary chondrites (Wetherill and Chapman, 1988); on the other, the observed spectra of S-asteroids and ordinary chondrites do not match precisely (Feierberg *et al.*, 1982; Gaffey, 1984). The differences are considered significant by some who believe S-types may be differentiated cores of broken up parent bodies (Bell *et al.*, 1989). But others argue that poorly understood regolith processes could possibly modify the spectral reflectance properties of ordinary chondrite materials on S-asteroid surfaces (Britt and Pieters, 1988). The issue is a fundamental one: if S-asteroids are not related to ordinary chondrites, then we have the problem of finding a source for these most abundant meteorites. Worse still, we must explain why those S-asteroids, which are placed best to deliver fragments to Earth (i.e., those in specific locations in the asteroid belt, e.g., near the 3:1 resonance zone with Jupiter, and those in Earth-approaching orbits), fail to contribute significantly to the meteorite population.

An important clue to the nature of S-asteroids would result if it proves possible to determine masses to $\sim 10–20\%$ during the flybys. Such a measurement, combined with an accurate volume determination from the SSI images, will give mean densities that may eliminate, or at least constrain, some existing models. For instance, in one model S-asteroids are viewed as collisionally stripped-down cores of metallic asteroids (Chapman, 1976). Mean densities of such objects should be significantly greater than 3.5 g cm^{-2}.

The SSI images will certainly reveal what the surfaces of small asteroids look like. Will they be as densely cratered as current ideas predict? Will they look something like the surface of Phobos – a body intermediate in size to Gaspra and Ida (Veverka and Thomas, 1979), and will we perhaps find evidence of surface fractures associated with large craters (Thomas and Veverka, 1979)? Or will we find evidence for downslope movement of regolith similar to that discovered on Deimos (Thomas and Veverka, 1980) and on the walls of young craters on Phobos (Avanesov *et al.*, 1991)? Will the surface be spectrally bland or heterogeneous, as in the case of Phobos where the crater Stickney appears to have excavated material with color properties distinct from that of surrounding regolith (Murchie *et al.*, 1990a)?

4.6. Jupiter

In interpreting the mission objectives for Jupiter's atmosphere (Yeates *et al.*, 1985), it is important to remember that the atmosphere covers an enormous area, extends to great depths, varies on all horizontal scales, and changes with time. The Galileo atmospheric Probe will satisfy the objectives at one latitude, one longitude, and one instant in time. The SSI, in combination with other remote sensing experiments on the orbiter, will provide the spatial and temporal coverage (motion picture sequences with 10 times better resolution than Voyager, for example) and long time base (10 orbits over 2 years plus a possible extended mission) that are needed to complete the picture.

Imaging is the most straightforward way to measure the spatial and temporal structure of the clouds. The SSI experiment is capable of resolving features smaller than half a scale height (~ 10 km) and as large as the radius of the planet. The appropriate temporal resolution depends on how fast the smallest resolvable features are changing or moving relative to each other. Analysis of Voyager images shows that relative velocities, v, are of order $L^{-1/2}$, where v is in m s^{-1} and L is the size of the features in kilometers. For example, $v = 10$ m s^{-1} for $L = 100$ km and $v = 32$ m s^{-1} for $L = 1000$ km. The largest sampling interval capable of resolving these motions is $\sim L/2v$, or $500L^{-1/2}$ s. Thus for spatial scales L of order 10 and 100 km, the appropriate sampling intervals are 0.4 hr and 1.4 hr, respectively.

At a resolution of 10 km, we expect to see vertical structure (clouds, haze, molecular scattering) on the planetary limb. On the planet's disk, SSI will observe processes such as convection, gravity waves, and shear instability whose horizontal scale is comparable to the atmospheric scale height. For example. Flasar and Gierasch (1986) have shown that an interesting class of mesoscale waves occurs near the cores of atmospheric jets. They were unable to establish phase speeds with Voyager images. This will be possible with Galileo data, and analysis of these waves should yield information about vertical structure of the mean wind fields. At a resolution of 100 km, the latitudinal bands and large ovals will be imaged, as well as intermediate-scale features (100 to 1000 km in size, 1 to 10 days in lifetime) that coexist with the long-lived structures but do not destroy them. For example, Dowling and Ingersoll (1989) have shown that mapping of the vorticity field at cloud top level can give information about the latitudinal variations in motions that exist much deeper in the atmosphere.

The wavelengths of the SSI filters, ranging from violet to near-infrared, should enable the effects due to molecular and aerosol scattering, different chromophores in the clouds, and absorption by gases such as methane to be separated out. By observing features at different positions on the disk (morning, noon, evening) and under different viewing geometries (forward and backward scattering, low and high emission angles), it is expected that the effects of vertical structure, aerosol abundance, particle size, and composition can be separated. It may also be possible to measure vertical shear in the zonal wind fields. Coordination of SSI sequences with those of the other orbiter instruments, and those on the Probe, is essential to this process.

SSI's observational strategy is to be highly selective as regards the atmospheric features to be studied. We will attempt to gather as much information as possible in order to ensure a substantial advance in our understanding of a few distinct classes of features (e.g., the great red spot, equatorial plumes, white ovals, barges, etc.). The planet is just too large and bewildering in its complexity to be studied with full comprehension by a single narrow-angle camera on a single spacecraft with a limited picture budget (budgeting for the experiment has been based on a limitation of 60 000 frames for the nominal mission excluding cruise science).

In what follows, we highlight a sample of the measurement sequences that the SSI team plans to use for atmospheric studies. While we highlight the role of the SSI, cooperative experiments with the other remote sensing instruments will be the central focus of the measurement strategy at Jupiter.

Probe entry site (PES) observations. Basically a combination of the sequences outlined below in which the evolution of the PES and its environment are observed, first at low and then at high resolution, starting two months before Probe entry. The Probe data will provide *in situ* truth to help in the interpretation of these observations. The Orbiter data, in turn, will provide information on the larger scale processes that are occurring at the PES and help extrapolate them to global conditions.

Full feature track (FFT). This sequence follows a selected region from limb to limb during one half rotation (5 hr) in order to obtain vertical structure information on clouds, winds and temperature. Our intent is to schedule FFT's at various longitudes and phase angles throughout the mission, all in the range of distances of 10–40 R_J. The inversion of the data to obtain information on vertical structure is the primary objective.

Feature velocity map (FVM). A variant of the FFT, the objective is to measure the velocity (SSI), temperature (PPR), composition and clouds (NIMS and UVS) over an area roughly 20° in latitude, 60° in longitude. The sequence takes about 4 h and will be conducted at phase angles less than 35° and a range inside of 20 R_J. A larger area is covered near the center of the disk than for the FFT, which samples more frequently in time.

High-resolution north–south scan (HRNSS). This is a cooperative sequence with the PPR which will be used to characterize spatial structures at 1000 to 2000 km scales for temperature and cloud microphysical properties and for which the SSI provides contextual imaging. Full longitudinal coverage will require a duration of 10 h. To maximize the spatial resolution (especially for scan platform instruments other than SSI) the measurement will, when possible, be conducted at minimum range ($\sim 15\ R_J$). HRNSS can be done at phase angles greater than 90°, but SSI will either have to look 'through' the spinning RTG booms or rely on observations from an earlier rotation of the planet.

High-resolution cylindrical map (HRCM). This is a cooperative sequence with NIMS, in which we intend to acquire high-resolution spectral data on the aerosol opacity and minor species throughout the near equatorial (~ 20°–30° N–S) region. Primarily a search for hot spots, the emphasis will be on obtaining data on a number of latitudinal regions with later repeats to look for temporal variations. The measurement requires 10 hr of observing time and must be conducted within 15 or 20 R_J to obtain the required spatial resolution.

Global velocity map (GVM). This is a moderate resolution (30 km pixel^{-1}) sequence in which we intend to measure the velocity field over the entire planet from a range of 50 R_J, at phase angles less than 110°, over two full rotations (20 hr). The measurement is built on imaging mosaics that are frequently repeated (every 40–50 min) over the duration of the sequence. Displacements (motions) will be measured using images obtained 10 hr apart.

High-resolution velocity map (HRVM). A combined SSI and PPR sequence designed to measure the meridional transport of heat and momentum across a latitude circle. The SSI data will be used to measure velocities by feature displacement over 2 hr time increments, and temperature via overlapping PPR field-of-views. The duration of the measurement will be about 12 hr, the range must be less than 20 R_J, and the phase angle less than 40°.

4.7. Io

Io is the most volcanically active body in the solar system, and the SSI experiment will focus primarily, although not exclusively, on this aspect of the satellite. Voyager observed several volcanic plumes extending up to 300 km above the surface, numerous volcanic hot spots on the surface, and a wide variety of landforms suggestive of both sulfur and silicate volcanism (Smith *et al.*, 1979a). Dissipation of tidal energy (Peale *et al.*, 1979) is believed to be the cause of the volcanism, but precisely how and where within the satellite the tidal energy is released is not known. Nor is it known what fraction of the tidal energy released within the satellite is conducted to the surface and what fraction is dissipated by transport of magma. Another major issue is the chemical nature of the volcanism. Are the eruptions primarily of sulfur, as is suggested by the spectral properties of the surface, or are they primarily of silicates as is suggested by the apparent strength of the surface materials (Carr *et al.*, 1979)?

Voyager was able to place some lower limits on resurfacing rates from the absence of impact craters and from modelling of the plumes (Johnson *et al.*, 1979), but the estimates are quite uncertain. While almost everything on Io's surface is probably the result of some form of volcanism, some of the landforms observed in the Voyager images are strongly suggestive of crustal deformation and erosion. Yet how does erosion occur on what is effectively an airless body? Possibilities include thermal erosion and sublimation, but other, as yet undiscovered, processes may also be involved. The SSI experiment will include a variety of observations designed to address all of these issues.

The single Galileo encounter at Io, at the beginning of the orbital tour, is typical of the satellite flybys that will occur later in the orbital tour. During approach to JOI there will be a period at relatively small phase angles, which will be devoted to multi-spectral mapping of the surface with moderate resolution (<2–4 km pixel^{-1}). Just prior to the encounter a full-disk mosaic of images will be acquired at the highest resolution possible (~ 1–2 km pixel^{-1}) in the clear filter (Figure 10). The actual encounter distance for Io will be ~ 1000 km, i.e., 20 times closer than the closest Voyager 1 encounter, and will allow images to be taken of specific localities with ground resolutions ~ 20 m pixel^{-1} (Figure 11). By comparison, the highest resolution achieved by Voyager on Io was ~ 200 m pixel^{-1}. Because Io is deep in the Jovian magnetosphere (5 R_J), the SSI must operate in a hostile particle radiation environment described above. As a consequence, the entire sequence will be performed in the lower resolution 'summation' mode (cf. Table I) so that the quality of the images is maximized.

Fig. 11. An example of the anticipated coverage by the SSI of a specific region on Io at the time of encounter when the ground resolution will approach ~ 20 m pixel^{-1}.

High-resolution imaging will help assess the nature of volcanism on Io. For example, estimates of long-term global volcanic rates will be possible from the presence or absence of impact craters in the highest resolution images. The nature of the volcanic activity will be assessed primarily from the highest resolution images, those with 20 m pixel^{-1} to 1 km pixel^{-1} resolution. At such high resolutions a variety of volcanic features, such as lava channels, levees, flow fronts, vents, and fumarolic deposits should be visible, and key measurements of flow thicknesses, caldera depths, slopes of walls, and so forth can be made. In this way we expect to place much more precise limits on the properties of the erupted materials and their probable compositions. In addition, high-resolution images will give an indication of what types of volcanic activity are occurring, particularly the relative importance of flow eruptions, pyroclastic activity, and various forms of remobilization of near surface volatiles.

Although the high-resolution imaging of Io will take place during only a single encounter, we will be able to monitor Io with spatial resolution comparable to that of Voyager 2 (5–10 km pixel^{-1}) during most of the two year orbital phase of Galileo. A program of continual monitoring will allow us to explore time variations, in addition to what we can learn from comparisons over the eighteen-week interval between the Voyagers and the 16-year interval between Voyager and Galileo.

The longer interval is likely to reveal significant surface changes due to the active and changing volcanism. Substantial portions of the surface may have been resurfaced with volcanic plume deposits. Earth-based observations over the past decade show that volcanic hotspots have bloomed and faded so that individual volcanic source sites, including vents and hot lava lakes, may be quite different in the mid-1990's from what was imaged by Voyager (McEwen *et al.*, 1989; Howell and Sinton, 1989; Greenberg, 1989). Plume activity will likely be changed considerably as well. Less dramatic, but still possible, may be changes in lava flow features. Another category of change to be explored over this longer time-scale is the rotational orientation of Io with respect to the direction of Jupiter. The lunar analogy would predict that Io rotates synchronously, but Greenberg and Weidenschilling (1984) have shown that flaws in that analogy admit the possibility of some deviation.

Monitoring during the orbital phase of the mission will allow us to explore the more rapid variations in volcanic activity. A strategy for this monitoring has been developed, taking into account our understanding of the various temporal and spatial scales of expected changes. Based on Voyager data, volcanic plumes of the Pele class (order of 300 km in height) have life spans less than a few weeks, while the Prometheus-class plumes (75 km) may endure for several years (Strom *et al.*, 1981; McEwen and Soderblom, 1983). Monitoring for the existence of plumes will be appropriate over the time-scale of the orbital mission and will be possible from distances as great as 100 R_J. Morphology of the plumes may well change on much shorter time-scales. The Loki plume showed significant variation over the course of hours, although Loki does not fit into either of the above plume classes. Monitoring such short term changes will only be possible at ranges $< 30\ R_J$ for Pele-class plumes and only at close approach for the smaller Prometheus class. The monitoring strategy will include sequences that may provide a movie of plume variation.

Monitoring will also be used to investigate the phenomenon of post-eclipse brightening, which has been observed with mysterious inconsistences from the Earth. Repeated observations of Io's emergence from eclipse with moderate resolution may help reveal the mechanism for this phenomenon.

4.8. The Io torus

Io is unusual not only for its active volcanism, but also for the material it contributes to the Jovian environment. The Io torus is the ring of neutrals and plasma that surround Jupiter at approximately Io's orbital distance. The neutral clouds of sodium and potassium atoms and the torus of sulfur and oxygen ions that make up the optically observable torus all originate on Io (Krimigis *et al.*, 1979; Brown *et al.*, 1983)).

The neutrals are concentrated around Io itself. Within a few Io radii of the satellite's surface, the maximum sodium D-line brightness probably exceeds 100 kilorayleighs and should be easily accessible to the SSI camera giving signals of several tens of DN in images taken in highest gain state with extended exposure. Although the sodium cloud is bright relative to other components of the torus, observing this emission will be made difficult by the proximity of the bright Io surface and the absence of a narrow-band filter to optically isolate the sodium emission. Nonetheless, we expect that the high photometric accuracy of the CCD detector and our efforts to minimize scattering in the SSI optics should allow us to subtract the scattered light from Io, enabling measurements of the near-Io sodium. Such measurements might reveal source regions or streamlines in the outflow of sodium and contribute greatly to our understanding of the mechanisms by which neutrals are removed from Io.

The plasma component of the torus is distributed in a large circum-Jovian ring, with the brightest emission often occurring at great distances from Io. However, the brightest of the plasma emissions in the wavelength range of the SSI are less than 1 kilorayleigh. These low brightness levels are not expected to be detectable owing to the absence of narrow-band filters and the 54-s limit on SSI exposures.

4.9 Europa

Europa is probably the least understood of the Galilean satellites due to comparatively poor resolution in the Voyager images, but it certainly is one of the most intriguing. It is believed to have a relatively thick, icy crust, and a silicate interior. The most prominent feature on its surface is a global system of intersecting dark and light linear markings. Several recent studies have investigated the origin of these marking; possible generic mechanisms include tidal deformation (Helfenstein and Parmentier, 1983), nonsynchronous rotation of a decoupled ice shell (Helfenstein and Parmentier, 1985), and polar wander of a decoupled ice shell (Ojakangas and Stevenson, 1989). Impact craters are extremely rare, so the surface is young and Europa is likely to have been actively resurfaced throughout its recent history (Lucchitta and Soderblom, 1982) and is, perhaps, continuing today. Although Voyager observed no obvious volcanic plumes, some equivocal evidence of brightening above its limb has been interpreted as indicating very low-level emissions of material from the surface (Helfenstein and Cook, 1984), and results of photometric studies indicating a very high surface porosity have been interpreted as being more consistent with 'pyroclastic' resurfacing than with an impact-gardened regolith (Domingue *et al.*, 1989).

The best Voyager images have a resolution of about 4 km pixel^{-1}. The best SSI images should approach $\sim$10 m pixel^{-1}, i.e., two orders of magnitude improvement in ground resolution. With this improved resolution, discrimination between different theories for the formation of the network of linear markings, should be possible. For example, at SSI resolutions, we should be able to characterize the morphology of the different types of markings in detail and determine the extent to which each originated by tectonic or volcanic activity. High-resolution images may also enable us to detect small craters and, through counts, establish the relative rates of resurfacing in different areas. With such a dramatic improvement in resolution, we may expect a whole new panoply of Europian features and phenomena to study. We anticipate that Europa may turn out to be the source of some of the most exciting and novel discoveries to be made by Galileo orbiter. The surface and limb will also be monitored throughout the mission in order to detect any changes that might occur and look for any indication of active 'ice-volcano' plumes.

4.10. Ganymede and Callisto

Ganymede and Callisto, by virtue of their similar sizes and densities (indicative of a mixture of rock and ice) present an extremely important case study in comparative planetary evolution. The surface of Callisto consists almost entirely of heavily cratered terrain exhibiting little obvious evidence of volcanic and tectonic modification (Smith *et al.*, 1979a, b); the most notable exception is the vicinity of the Valhalla multi-ring structure, where at least minor volcanic resurfacing has occurred (Remsberg, 1981). About half of Ganymede's surface constists of dark terrain which is superficially similar in appearance, but is pervasively cut by systems of linear and curvilinear furrows (Smith *et al.*, 1979a, b). Early interpretations of the geology of dark terrain were that it is an impact-fractured primordial surface, perhaps capping a deep, differentiated water-ice mantle (e.g., Passey and Shoemaker, 1982; Shoemaker *et al.*, 1982; McKinnon and Parmentier, 1986). More recent investigations of the stratigraphic relations of furrows with deposits of dark material and the crater size-frequency distribution in dark terrain suggest a volcanic surface deformed at least in part by endogenic processes (e.g., Croft and Goudreau, 1987; Murchie *et al.*, 1989, 1990b).

The remaining half of Ganymede's surface consists of lighter-colored material pervasively deformed by sets of linear grooves (Smith *et al.*, 1979a, b). Studies of crater morphology and size-frequency distribution suggest that the light material forms a layer undreds of maters to a few kilometers thick burying underlying dark terrain (Schenk and McKinnon, 1985; Murchie *et al.*, 1989). The grooves themselves are widely thought to be extensional in origin, formed as tension fractures or graben modified by mass wasting, volcanism, or other processes (cf. McKinnon and Parmentier, 1986, for a review). Large areas of grooved light terrain appear to have been formed by downdropping of blocks of lithosphere followed by resurfacing and fracturing (Parmentier *et al.*, 1982). This process occurred as a sequence of initial fracturing of the blocks followed by formation of different morphologic classes of grooved terrain, for which a generalized sequence of events have been worked out (Murchie *et al.*, 1986). On a global scale these structures form a coherent pattern, with both global 'fabric' and regionally dominant groove orientations in evidence (Bianchi *et al.*, 1986). Although the dominant form of deformation in grooved terrain clearly was extensional, in some locations there is evidence for shear and compressional deformation (Murchie and Head, 1988). More detailed examinations of these locations and assessment of the roles of different types of deformation in the formation of grooved terrain could be some of the more important geologic observations by SSI.

The difference in the geology of Ganymede and Callisto despite their similar bulk properties, i.e., the 'Ganymede–Callisto dichotomy', is thought to be a consequence of divergent thermal and dynamical evolutions (McKinnon and Parmentier, 1986). Evidence for the cause of the divergence occurs in the spatial and temporal patterns of volcanic and tectonic surface modification, but also in the form of other indicators of the satellites' thermal evolutions. One such indicator, applicable to Ganymede, is the evolution of the widths of tectonic troughs, which is a function of the evolution of the thickness and thermal gradient in the lithosphere (Golombek and Banerdt, 1986). Another indicator is the history of viscous relaxation of craters and palimpsests. Passey and Shoemaker (1982) have attempted to reconstruct the thermal history of Ganymede using such observations, but this technique is complicated by uncertainty about the roles of impact melting and subsequent resurfacing in also degrading crater morphologies (Croft, 1983; Murchie *et al.*, 1989).

SSI imaging of these satellites will emphasize multi-phase-angle coverage of selected areas, multi-spectral images and the acquisition of the very highest spatial resolution image data of selected regions (down to ~ 10 m pixel^{-1}). Details of structural features, such as morphologic modification of furrow, groove, and crater rims, and walls and floors (possibly by resurfacing or mass wasting), should be easily visible. New features may be detected, e.g., further subdivision of undivided material units, detection of extrusion vents, flow features, and possibly, ancient volcanic constructs. The spectrum of crater forms will be enlarged by details involved with ejecta material emplacement (pedestal craters, bright and dark rays), central peak and pit modification, rim morphology, terraces, and secondary crater chains. These investigations will be aided by photometric material classification at all SSI wavelengths. Thus, the aim of the SSI experiment at Ganymede and Callisto is to greatly refine the preliminary Voyager reconnaissance of geological features as well as to investigate some types of features more intensively.

With the spatial resolution capability noted above, it should be possible to easily discriminate between small primary craters and secondary craters, as well as differentiate impact craters from craters of other origin. It should be possible to improve our knowledge of the crater size-frequency distribution over all crater sizes and for all the different geologic units on the surfaces of Ganymede and Callisto. From this, improved comparisons will be possible with the cratering record in the inner solar system. These investigations should yield further insight into the nature of the impactor populations which are responsible for the cratering record both in the Jovian system as well as the inner solar system, and may clarify the relative abundance of heliocentric impactors versus planetocentric impactors as a function of time (Chapman and McKinnon, 1986; Murchie *et al.*, 1989). Thus we may learn about early accretional and post-accretional processes in the satellite system of Jupiter and, especially, about the bombardment history of the satellites in terms of possible catastrophic impacts that may have led to destruction and re-accretion of the satellites.

Rheologic properties derived from morphology of surface features will also be an important tool to better determine the thermal histories of both satellites. Detailed analysis of crater morphology and the age sequence of widths of tectonic features may provide important tools for establishing a time scale for any changes in the rheologic properties of Ganymede's lithosphere. Also, detailed mapping of furrows and grooves may help in determining regional or local stress domains associated with conditions of the interior.

4.11. Jupiter's rings

The main ring of Jupiter extends from a very sharp edge at 1.8086 R_J inward to a fuzzy boundary at about 1.71 R_J and lies in the equatorial plane of the planet. The slant optical thickness of the brightest component of the rings is about 10^{-5} to 10^{-6}, which gives some idea of how ethereal these rings are. Embedded in the main ring are (at least) two small satellites, both near the outer edge (Metis at 1.792 R_J and Adrastea at 1.807 R_J). The ring is probably sustained by ejecta from these satellites due to meteoroid impact, even as radiation drag drives ring particles in towards the planet.

Just interior to the main ring, from about 1.71 R_J to a very fuzzy inner boundary at 1.3 to 1.4 R_J, is the toroidally shaped 'halo' component of the ring system, whose thickness is comparable to its radial extent. The halo is thought to be produced by the interaction of electric charges on small ring particles with the Jovian magnetic field: As ring particles reach the inner edge of the main ring, Lorentz forces drive them into increasingly inclined orbits, generating the thickness of the swarm inside 1.71 R_J (Schaffer and Burns, 1987). The total amount of material in the halo is thought to be comparable to that in the main ring, although the halo appears to have a much lower surface brightness in edge-on view because the material has a much greater vertical extent. Voyager images also revealed a 'gossamer ring', which stretches from the main ring out to $\sim 3\ R_J$, where it became too faint to detect. The size of the optically important particles in the rings lies in the range from 0.1 to 10 μ.

We plan to obtain multi-spectral images to map the details of any structures in the main ring and better characterize its particle size distribution. We will search for time-dependencies and for presently undiscovered satellites that may be substantially contributing to the source of the particles. The intriguing structure of the gossamer ring, which has an intensity maximum near the distance where Keplerian orbits would be synchronous with Jupiter's rotation, will be mapped more comprehensively than was possible from Voyager.

Other issues that we plan to address include the radial and vertical structure and extent of the halo. Does it, for example, really agree with the Lorentz resonance hypothesis? A logistical complication is that observations at high phase angle will require the SSI to be looking back towards the Sun through the spinning RTG and magnetometer booms and scattered sunlight may mask the signature of the rings. The solution will probably be to make such high-phase-angle observations only during occasional 'science turns' or while the spacecraft is in the shadow of Jupiter, which may, in turn, substantially limit the coverage of the rings that is possible.

An equally important part of the SSI ring experiment will be to obtain image data that will further elucidate the nature of the larger objects that presumably serve as the basic sources of ring material. These range from the satellites that have already been discovered, down through a possibly continuous range of intermediate and smaller sizes. These source objects will be best seen at low-phase-angles in backscattered light from as small a range as possible. Multi-spectral images in backscattered light may help characterize the surface composition of the larger objects.

4.12. Auroral, airglow and dark-side phenomena

As a result of the high sensitivity of the $V\phi$ CCD, it will be possible to observe a variety of low-light-level phenomena on the dark faces of Jupiter and its satellites. The Voyager imaging experiment (and other investigations) has already shown that intense optical auroras are present on Jupiter. However, their global patterns, distribution, and dynamics remain poorly understood. Two-dimensional maps of Jovian aurorae made with SSI images will provide unique information on the occurrence of precipitating plasma; therefore accurate mapping of their rate of occurrence and locality is a critical measurement for a deeper understanding of Jovian magnetospheric dynamics. The SSI has a sensitivity to extended sources that is some 20 times that of the Voyager *wide angle* camera (clear filter, $S/N \sim 10$). It is, therefore, expected that the SSI will be able to make unique measurements of these phenomena.

Some key questions which we expect to address are: What is the time scale for changes in the Jupiter aurora? What is the relative importance of particles in the torus compared with those in the magnetotail plasma sheet for producing auroral phenomena? What differences, if any, in auroral intensities and structure occur in the vicinity of the magnetic anomaly ('active sector')? What differences exist between visible aurorae in the northern and southern hemispheres? Can we characterize the auroral emissions in visible wavelengths at the foot of the Io fluxtube?

Although we believe the SSI is capable of making significant contributions to magnetospheric science on its own, the prospects are multiplied when its capability is combined with those of other Galileo instruments. For example, cooperative observations with the UVS and with *in situ* magnetometer, plasma, and energetic particle experiments should make it possible to trace magnetospheric field lines back to their source in the ionosphere.

Other 'darkside', low-surface-brightness phenomena that we plan to investigate include: atmospheric lightning discharges that were originally discovered (Cook *et al.*, 1979) during the Voyager mission (is it possible to trace out the global distribution of convective activity deep in tropospheric clouds with the spatial distribution of visible lightning flashes?); glows associated with volcanic activity on Io; and meteor fireballs.

5. Data Analysis System

5.1. Overview of the SSI data set

We estimate that the SSI will generate ~ 70 000 useful images during the entire mission. Of these we expect 81 of Venus, 5000 of the Earth–Moon system, 2000 of asteroids, 40 000 of Jupiter's atmosphere, and 20 000 of Jupiter's satellites and its rings. It is our intention to rapidly (see below) and systematically produce calibrated versions of these images and to perform a preliminary analysis of them as the mission proceeds. In order to constrain the substantial costs which are involved in producing photographic hardcopy, we will rely on modern computer techniques to interact with the vast bulk of the data. Photographic hardcopy will be produced for only a fraction of the images.

The data are expected to be made available to the general scientific community within one year of their receipt from the spacecraft and will be distributed through NASA's Planetary Data System and through the NASA Regional Planetary Image Facilities.

5.2. HIIPS: Home Institution Image Processing System

The 1980's saw a rapid spread of computer-based image processing capabilities into the hands of individual research scientists. What was once the domain of a relatively few large research centers has become commonplace throughout the scientific community.

The scientific data analysis for the SSI experiment will largely be accomplished during the flight mission using a 'fully distributed', networked computing system called HIIPS, which features nodes at each team members' home institutions. When HIIPS was first conceived in the early 1980s, only a few institutions were capable of conducting large-scale image processing. It was argued that it would cost much less for each team member to operate a dedicated image processing system rather than to be supported by a large cadre of computer analysts at JPL, the production of large quantities of 'hard-copy' photographic prints, and the support of extensive travel and subsistence in residence at mission control, as had been the case in previous missions. These arguments became stronger when it was recognized that each team member would require research associates who would also need support in residence at mission control. Thus, it was found not only less expensive, but more efficient to develop a system of dedicated image processing capabilities at each team member's home institution. As capabilities of image-processing computers has improved during the mission delays of the 1980's, these arguments seem all the more cogent.

A typical HIIPS station consists of a microcomputer, tape drive, mass storage, display processor, and various peripherals such as digitizers, hard copy devices, etc. (Figure 12). Most are configured for multiple users. All HIIPS are networked and linked to the Jet Propulsion Laboratory (Figures 13(a–c)). The linkage is currently over the SPAN network and works at a rate of 9.6 kbaud. During mission operations it is planned to increase this to 56 kbaud. The image processing software consists of Jet Propulsion Laboratory's VICAR routines (LaVoie, 1989) running under the Transportable Applications Executive (TAE). Both VMS and UNIX versions are employed. Other packages, e.g., IRAF (Tody, 1986), are also expected to be used for special applications. Special software to address specific scientific problems is also being developed by individual team members, by staff at MIPL, and as a joint SSI Team/MIPL activity.

We anticipate that well before the time of the primary mission a medium other than magnetic tape will be available for the distribution of SSI data. The rapid development of CD-ROM's and erasable optical disks holds great potential for fast and relatively inexpensive, distribution of the large Galileo data set.

5.3. Systematic processing

SSI systematic products differ from those made for specific scientific investigations. Their primary purpose is to meet the operational and archival requirements of the

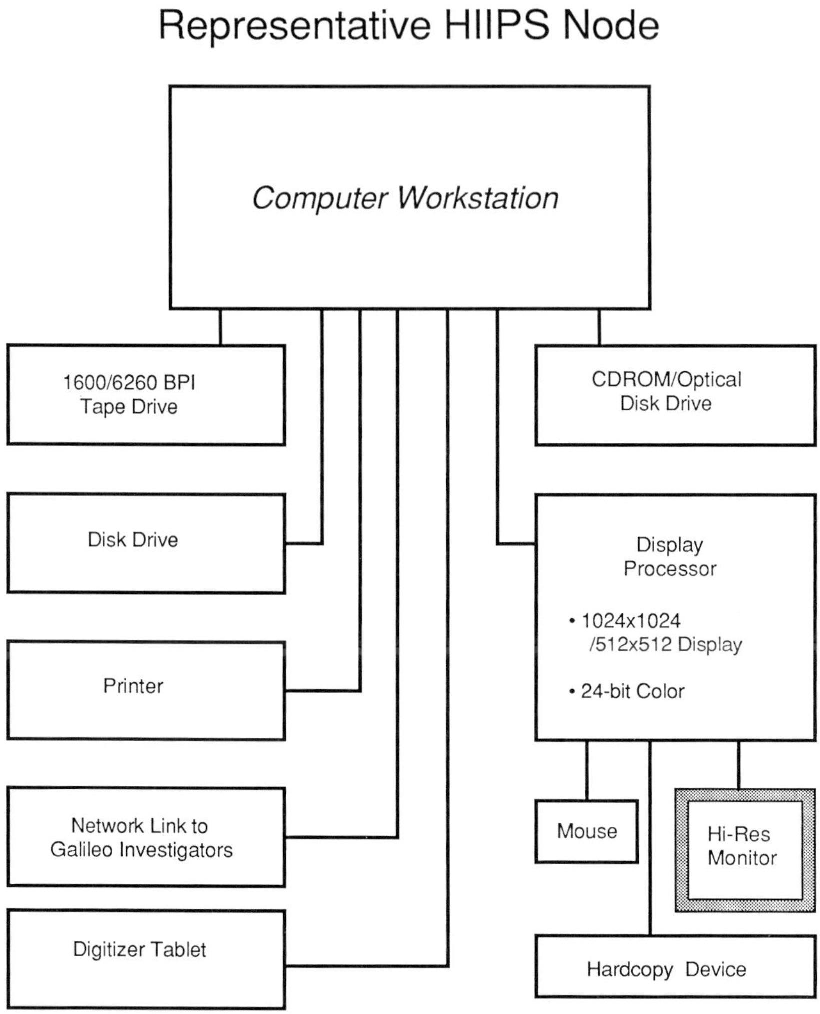

Fig. 12. Representative configuration at a typical HIIPS node. Current workstations are DEC Micro VAX II/III, DEC VAX 750 (VMS Operating System), or Sun 3/4 (UNIX Operating System). The standard image analysis software is VICAR, available from MIPL (VMS version) and Arizona State University (UNIX version). Individual HIIPS may use other packages, such as IRAF. All HIIPS sites are involved in the design and production of special purpose software for specific research requirements.

Galileo project. In doing this they also represent the materials on which much future research will be based. These products will be produced in great volume and according to a strict schedule in order to support ongoing mission operations, e.g., a combination of approximately 5000 monochrome, color, and mosaicked images are expected to be produced on short time scales ($\sim$ 30 days) every orbit in order to recognize changes that may have occurred from one orbit to the next and plan modifications in encounter and orbital sequences.

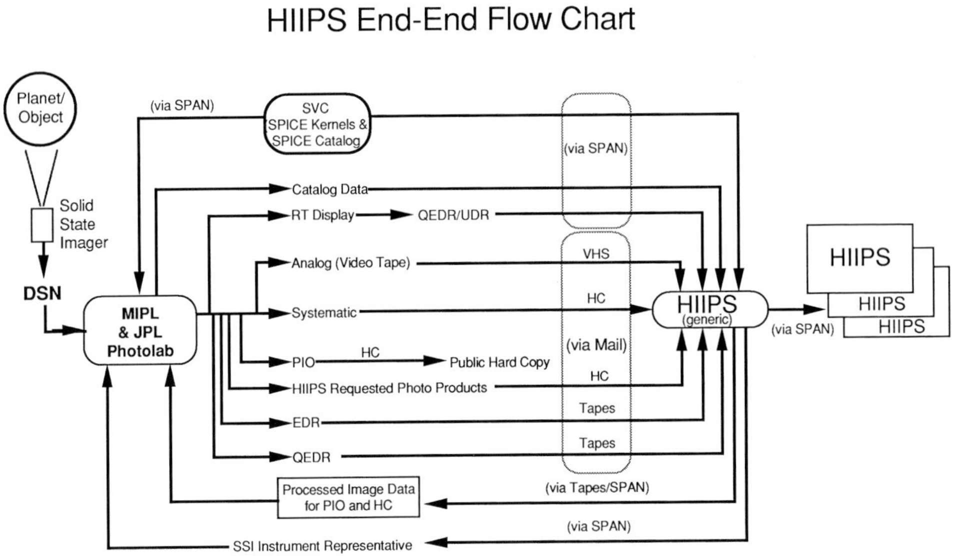

Fig. 13a. The SSI HIIPS network showing the relationship of HIIPS to the computers and data flow functions at mission control (JPL). The acronyms are defined in Figure 13(b).

Various photoproducts required for inflight sequence generation and updating, will also be systematically produced. These include duplicate negatives, duplicate positives, paper positive contact prints and paper positive enlargements of a small fraction of the data. All SSI images of solid surfaces will be enhanced and film-recorded in black and white. However, only about 10% of atmospheric images are expected to be enhanced and film recorded in this way. Atmospheric investigations will rely largely on the volatile image displays at the HIIPS workstations.

These products also include a 'reduced' data set, or EDR (Experiment Data Record), that will be made available in the scientific community for subsequent research analysis within a year of the receipt of the data. This data set will be consistently processed, will be radiometrically calibrated, and will include all requisite calibration files. Since these operations are all linear the original, or 'raw', form of the data will always be recoverable.

An introductory explanation to some of the basic systematic enhancement operations that will be performed on the SSI images is given by Cutts (1974). Systematic enhancement operations that we intend to apply to the SSI data include: blemish removal, replacement of invalid pixels by an average computed from their near neighbors; map projection and registration to allow accurate mosaicking and color, or time-sequence, registration; photometric function removal; contrast enhancement; high-pass spatial filtering; modulation transfer function compensation; and color enhancement.

For the EDR, the SSI images will be radiometrically calibrated and distributed as 16-bit files in a special VICAR, project defined, format. During early cruise the distribution media will be magnetic tape. However, we expect that by the time of orbital operations at Jupiter, CD-ROM's will be used.

Glossary for HIIPS End-End Flow Chart

Abbreviation	Full Name	Definition
Analog	Analog (Video Tape) Data	VHS format video tape with still frames of EDR's and "movies" from Encounter Time Lapse sequences.
DSN	Deep Space Network	A world-wide system of sensitive radio receivers and antennas built to capture faint signals transmitted from NASA spacecraft.
EDR	Experimental Data Record	Scientifically useful digital data provided by MIPL on magnetic tape to all HIIPS sites on a best effort basis.
HC	Hard Copy	Photographic negatives and prints, movies on video or film, and paper documents.
HIIPS	Home Institution Image Processing System	An image processing hardware and software system located in a scientist's own local laboratory.
MIPL	Multi-Mission Image Processing Laboratory	An organization at JPL (the NASA Jet Propulsion Lab, Pasadena) which processes spacecraft data received by the DSN into a scientifically useful form.
Photolab	Photographic Laboratory	A group at JPL that accepts exposed negatives and produces photographic prints and negatives.
PIO	Public Information Office	The office at JPL that distributes information to the press and public. Commonly PIO refers to hard copy officially released to the press by the JPL Public Information Office.
QEDR	Quick-look Experimental Data Record	Scientifically useful digital data which is partially processed and incompletely characterized, provided to the HIIPS sites by request of individual scientists before EDR's are available.
RT Display	Real-Time Display	A MIPL system which displays spacecraft data as soon as practical after it is received and processed.
SPAN	Space Physics Analysis Network	The general-purpose, world-wide NASA science network. SPAN is used by all NASA science, not just physicists. SPAN uses the DECnet protocol.
SVC	Galileo Science VAX Cluster	System available for some Galileo science data analysis, where the SPICE Kernels are stored. The SPICE Kernels provide navigation and pointing information for the Galileo instruments.
Systematic	Systematic Processing	Pre-planned sequence of processing operations by which MIPL produces hard copy (PIO) and digital products (EDR's and QEDR's)
UDR	Unprocessed Data Record	Digital data on disk at MIPL, available through SPAN, but not yet processed to remove noise, fill in missing lines, etc.

Fig. 13b. Definition of the acronyms used in Figure 13(a).

Calibration files and other information that will be included in the EDR are: radometric calibrations for each camera operational mode; statistics on signal chain noise; dark current files for each camera readout rate and for a range of radiation environments; the modulation transfer function as determined both on the ground and inflight; and, finally, blemish files that also include information on any changes that occur.

Another systematic product will be a set of VHS video tapes to provide an informal visual catalog of the images. These are intended to allow rapid visual searches through the data using inexpensive VCR/TV equipment immediately after receipt of the data.

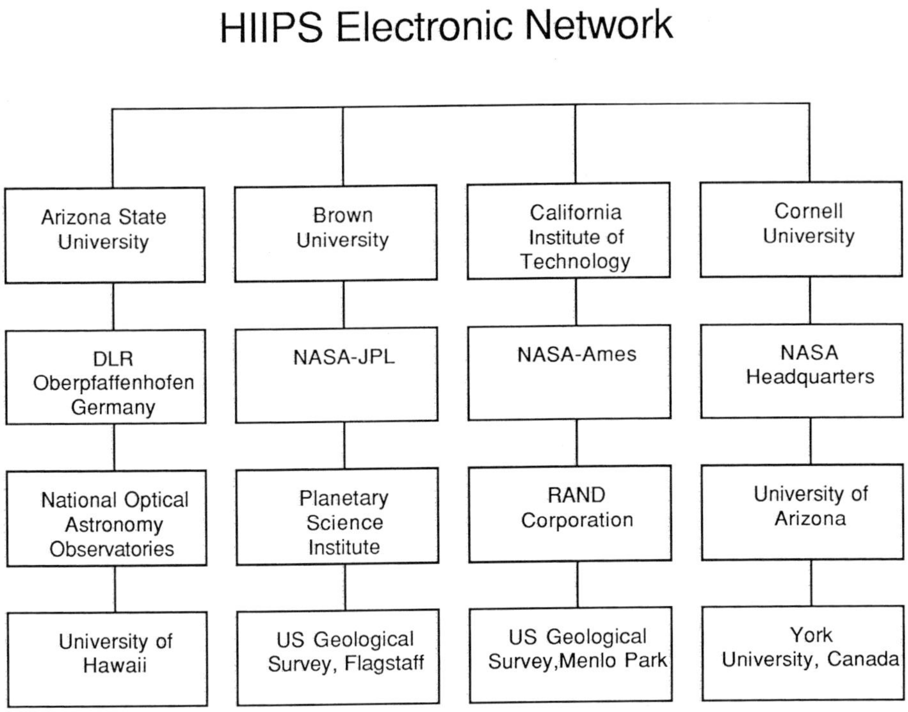

Fig. 13c. The SSI HIIPS network identifying the names and locations of the nodes. The sixteen HIIPS sites are electronically linked via the NSI (NASA Science Internet), which uses both the DECnet (SPAN network) and TCP/IP (NSN network) protocols. Individual HIPPS also connect to other scientific networks such as NSFnet and ARPANET. Network link speeds range from 9.6 kbaud to 1.5 Mbaud.

Information specifying the camera pointing geometry and other instrument and spacecraft conditions at the time of the creation of each image will be generated using the NAIF/SPICE system (Acton, 1990). This essential ancillary data will be provided in both predict and 'final' form on mission computers and on the HIIPS system. A set of records, companion to the EDRs, will also be produced as a systematic product on CD–ROMs.

Historical Notes and Acknowledgements

The SSI team was formed by NASA in 1977. The original members were M. J. S. Belton (Team Leader), C. D. Anger, M. H. Carr, C. R. Chapman, M. E. Davies, R. Greeley, R. Greenberg, J. W. Head III, G. Neukum,C. B. Pilcher, G. Schubert, J. Veverka, and J. Wellman. The Jet Propulsion Laboratory (JPL) science coordinator is K. P. Klaasen. The original JPL hardware manager was F. Landauer. The hardware was designed and fabricated in JPL Division 38. The SSI team acknowledges the work of D. Norris, J. Janesick, M. Blouke, F. Vescelus, L. Snyder, R. Locke, D. Johnson (who was responsible for the image compressor). The ground calibration was done under the direction of K. P. Klaasen in conjunction with the staff of the MIPL, and the SSI team wishes to acknowledge the work done by D. Alexander, E. Korsmo, M. Pestana, L. Wainio, L. Wynn, and G. Yagi.

In the intervening years J. Wellman and G. Schubert have left the team to follow other pursuits – however G. Schubert remains as an interdisciplinary scientist in the Galileo project. In 1979 the hardware management passed to F. Vescelus and then to M. C. Clary in 1980. Shortly after launch M. C. Clary retired and was replaced by the present hardware manager, W. Cunningham. In 1988 K. P. Klaasen was formally added as a member of the team. The SSI team has striven for close associations with other Galileo experimenters and, as a result, special arrangements have been made with several Galileo interdisciplinary scientists (IDS's) who have formally associated themselves with the SSI experiment and have accepted major responsibilies in the experiment. These *Associated IDS's* are in effect regular members of the team and form an intellectual and experimental bridge between the SSI and other experiments. They are F. P. Fanale, P. J. Gierasch, A. P. Ingersoll, D. Morrison, and J. B. Pollack. The late Dr H. Masursky was also an Associated IDS and contributed greatly to the development of the experiment. His responsibilities to the SSI experiment have now been undertaken by A. McEwen.

The team was structured into several subgroups during the pre-launch project development. These were *Scientific Data Processing* (R. Greeley); *Systematic Data Processing* (K. P. Klaasen); *Hardware Oversight* (M. E. Davies); *Satellite Science* (M. H. Carr); *Asteroid Science* (J. Veverka); and *Atmospheric Science* (C. Chapman and M. J. S. Belton). The Scientific Data Processing Group has been largely responsible for the development of the HIIPS concept, and its implementation was managed by D. Anderson. The ongoing effort to maximize the power and reliability of the HIIPS and to develop scientific applications software in the team was first managed by the late T. E. Townsend, who has now been replaced by L. Bolef.

The interface between the team, the project, and the resources at JPL is managed by K. P. Klaasen and the members of his 'photoscience' group. The team acknowledges the work done on its behalf by B. Paczkowski, H. Breneman, S. Edberg, G. Johnson, C. Kahn, D. Speer, and W. James.

A second interface of great significance to the SSI experiment is that between the team and the Multimission Image Processing Laboratory (MIPL). This responsibility is undertaken by J. L. Anderson and L. Wainio. The team acknowledges the work done on its behalf by R. Bambery, G. Bothwell, R. Brill, M. Girard, P. Jepsen, K. Jones, L. Kamp, S. LaVoie, S. Pohorsky, N. Sirri, S. Tews, and P. Zamani. This group is also responsible for stripping the SSI data from the real-time telemetry steam, and for making it rapidly available at JPL and to the HIIPS at the time of an encounter.

For the flight mission some of the responsibilities in the team have been changed. The Hardware group will oversee the continuing inflight calibration of the SSI. It is planned to delegate the prime responsibility for planning each encounter to two or three team members per encounter plus one lead member fo the photoscience team. The encounters during the orbital mission are so close in time that we are planning for a great deal of parallel activity. The planning of the Venus encounter was undertaken by P. J. Gierasch, M. J. S. Belton and H. Breneman; the team activities for the Moon 1 encounter are being lead by J. W. Head III, M. H. Carr, and B. Paczkowski; Earth 1 by C. Chapman, C. D. Anger and B. Paczkowski. J. Veverka and C. Chapman will be the team members primarily responsible for the Gaspra encounter. Other future encounters remain to be assigned.

The Imaging Team acknowledges support from the NASA/JPL Galileo Project for the performance of this work. In particular the team would like to specially acknowledge the support given it by the Galileo Science Manager, Dr Clayne Yeates. His recognition of the future value of the HIIPS concept and his continued support in its implementation has been invaluable. We also acknowledge the help of the first Galileo project manager, Dr J. Casani, who at least listened to our pitch for a wide-angle camera and an additional tape-recorder.

Individuals who have made important contribution to the experiment include E. Alvarez, T. Colvin, C. Cunningham, D. Godfrey, P. Helfenstein, S. Howell, C. Pieters, P. Thomas, T. Thompson, C. Anger acknowledges support from the Canadian National Science and Engineering Research Council and from the Ontario Institute for Space and Terrestrial Science. G. Neukum acknowledges support from the German government. The Planetary Science Institute is a division of Science Applications International Corporation. The National Optical Astronomy Observatories are operated by AURA Inc. under a cooperative agreement with the National Science Foundation.

References

Acton, C.: 1990, *Proc. AIAA/NASA 2nd Int. Symp. on Space Information Systems.*

Adams, J. B. and McCord, T. B.: 1971, *Lunar Planetary Sci. Conf.* **II**, 2183.

Adams, J. B. and McCord, T. B.: 1972, *Lunar Planetary Sci. Conf.* **III**, 3021.

Allen, D. A.: 1987, *Icarus* **69**, 221.

Anderson, J. L., Belton, M. J. S., Danielson, G. E., Evans, N., and Soha, J. M.: 1978, *Astrophys. J. Suppl.* **36**, 275.

Avanesov, G. B., Zhukov, B., Ziman, Ya., Kostenko, V., Murav'ev, V., Fedotov, V., Bonev, B., Mishev, D., Petkov, D., Krumov, A., Simeonov, S., Boycheva, V., Uzunov, Yu., Kempe, F., Rebel, B., Weide, G-G., Halmann, D., Neumann, W., Petuchova, I., Possel, W., Head, J., Murchie, S., Lumme, K., Muinonen, K., Peltoniemi, J., Duxbury, T., Murray, B., Herdenoff, K., Fanale, F., Irvine, W., and Smith, B.: 1991, *Planetary Space Sci.* (in press).

Bell, J. F., Davis, D. R., Hartmann, W. K., and Gaffey, M. J.: 1989, in R. P. Binzerl, T. Gehrels, and M. S. Mathews (eds.), *Asteroids II*, University of Arizona Press, Tucson, p. 921.

Belton, M. J. S., Smith, G. R., Schubert, G., DelGenio, A. D.: 1976, *J. Atmospheric Sci.* **33** 1394.

Belton, M. J. S., Gierasch, P., Klaasen, K. P., Anger, C. D., Carr, M. H., Chapman, C. R., Davies, M. E., Greeley, R., Greenberg, R., Head, J. W., Neukum, G., Pilcher, C. B., Veverka, J., Fanale, F., Ingersoll, A. P., Pollack, J. B., Morrison, D., Clary, M. C., Cunningham, W., and Breneman, H.: 1992, *Adv. Space Res.* (in press).

Bianchi, R. R., Casacchia, R., Lanciano, P., Pozio, S., and Strom, R.: 1986, *Icarus* **67**, 237.

Britt, D. T. and Pieters, C. M.: 1988, *Astron. Vestnick* **22**, 229.

Brown, R. A., Pilcher, C. B., and Strobel, D. F.: 1983, in A. J. Dessler (ed.), *Physics of the Jovian Magnetosphere*, Cambridge University Press, Cambridge, p. 197.

Carr, M. H., Masursky, H., Strom, R. G., and Terrile, R. J.: 1979, *Nature* **280**, 729.

Chapman, C. R.: 1976, *Geochim. Cosmochim. Acta* **40**, 701.

Chapman, C. R. and McKinnon, W.: 1986, in J. Burns and M. Matthews (eds.), *Satellites*, University of Arizona Press, Tucson, p. 492.

Chapman, C. R., Morrison, D., and Zellner, B.: 1975, *Icarus* **25**, 104.

Cook, A. F., Duxbury, T. C., and Hunt, G. E.: 1979, *Nature* **280**, 794.

Croft, S.: 1983, *J. Geophys. Res.* **88**, *Suppl.* **B71**.

Croft, S. and Goudreau, B.: 1987, *Lunar Planetary Sci. Conf.* **XVIII**, 209.

Cutts, J. A.: 1974, *Mariner Mars 1971 Television Picture Catalog, Experiment Design and Picture Data*, Technical Memorandum 33-585 **1**, Jet Propulsion Laboratory, Pasadena.

Domingue, D., Hapke, B., Lockwood, W., Thompson, D.: 1989, *Bull. Am. Astron. Soc.* **21**, 984.

Dowling, T. E. and Ingersoll, A. P.: 1989, *J. Atmospheric Sci.* **46**, 3256.

Flasar, F. M. and Gierasch, P. J.: 1986, *J. Atmospheric Sci.* **43**, 2683.

Feierberg, M. A., Larson, H. P., and Chapman, C. R.: 1982, *Astrophys. J.* **257**, 361.

Gaffey, M. J.: 1984, *Icarus* **60**, 83.

Golombek, M. and Banerdt, W.: 1986, *Icarus* **68**, 252.

Greenberg, R. and Weidenschilling, S. J.: 1984, *Icarus* **58**, 186.

Greenberg, R.: 1989, in M. J. S. Belton, R. West, and J. Rahe (eds.), *Time Variable Phenomena in the Jovian System*, NASA SP-494, Ch. 7, p. 100.

Head, J. W.: 1974, *The Moon* **91**, 327.

Helfenstein, P. and Parmentier, E. M.: 1983, *Icarus* **53**, 415.

Helfenstein, P. and Cook II, A. F.: 1984, *Lunar Planetary Sci. Conf.* **XV**, 354.

Helfenstein, P. and Parmentier, E. M.: 1985, *Icarus* **61**, 175.

Howell, R. R. and Sinton, W. M.: 1989, in M. J. S. Belton, R. West, and J. Rahe (eds.), *Time Variable Phenomena in the Jovian System*, NASA SP-494, Ch. 3, p. 47.

Janesick, J. R., Hynecek, J., and Blouke, M. M.: 1981, *Proc. SPIE* **290**, 165.

Janesick, J. R., Elliot, T., Collins, S., Marsh, H., Blouke, M., and Freeman, J.: 1984, *Proc. SPIE* **501**, 2.

Johnson, T. V., Cook II, A. F., Sagan, C., and Soderblom, L. A.: 1979, *Nature* **280**, 746.

Klaasen, K. P., Clary, M. C., and Janesick, J. R.: 1984, *Opt. Eng.* **23**, No. 3, 334.

Krimigis, S. M., Armstrong, T. P., Axford, W. I., Bostrom, C. O., Fan, C. Y., Gloeckler, G., Lanzerotti, L. J., Keath, E. P., Zwickl, R. D., Carbary, J. F., and Hamilton, D. C.: 1979, *Science* **204**, 998.

LaVoie, S.: 1989, *Vicar User's Guide*, Jet Propulsion Laboratory, Pasadena.

Lucchitta, B. K. and Soderblom, L. A.: 1982, in D. Morrison (ed.), *Satellites of Jupiter*, University of Arizona Press, Tucson, p. 521.

McEwen, A. and Soderblom, L.: 1983, *Icarus* **55**, 191.

McEwen, A. S., Lunine, J., and Carr, M. H.: 1989, in M. J. S. Belton, R. West, and J. Rahe (eds.), *Time Variable Phenomena in the Jovian System*, NASA SP-494, Ch. 2, p. 11.

McKinnon, W. and Parmentier, E. M.: 1986, in J. Burns and M. Mathews (eds.), *Satellites*, University of Arizona Press, Tucson, p. 718.

Murchie, S. and Head, J.: 1988, *J. Geophys. Res.* **93**, 8795.

Murchie, S., Head, J., Helfenstein, P., and Plescia, J.: 1986, *J. Geophys. Res. Suppl.* **91**, E222.

Murchie, S., Head, J., and Plescia, J.: 1989, *Icarus* **81**, 271.

Murchie, S. D., Britt, D., Head, J., Pratt, S., Fisher, P., Zhukov, B., Kuzmin, A., Ksanformality, L., Nikitin, G., Zharkov, A., Fanale, F., Blaney, D., and Robinson, M.: 1990a, *Proc. Lunar Planetary Sci.* **XXI**, 825.

Murchie, S., Head, J., and Plescia, J.: 1990b, *J. Geophys. Res.* **95**, 10743.

Ojakangas, G. and Stevenson, D.: 1989, *Icarus* **81**, 242.

Parmentier, E. M., Squyers, S., Head, J., and Allison, M.: 1982, *Nature* **295**, 290.

Passey, Q. and Shoemaker, E.: 1982, in D. Morrison (ed.), *The Satellites of Jupiter*, University of Arizona Press, Tucson, p. 379.

Peale, S. J., Cassen, P., and Reynolds, R. T.: 1979, *Science* **203**, 892.

Pieters, C. M.: 1978, *Lunar Planetary Sci. Conf.* **IX**, 2825.

Pieters, C. M.: 1986, *Rev. Geophys.* **24**, 557.

Pieters, C. M., Head, J., Adams, J. B., McCord, T. B., Zisk, S. H., and Whitford-Stark, J.: 1980, *J. Geophys. Res.* **85**, 3913.

Pieters, C. M., Adams, J. B., Mouginis-Mark, P., Zisk, S. H., Head, J. W., McCord, T. B., and Smith, M.: 1985, *J. Geophys. Res.* **90**, 12393.

Remsberg, A.: 1981, *Lunar Planetary Sci. Conf.* **XII**, 874.

Rossow, W. B., DelGenio, A. D., Limaye, S. S., and Travis, L. D.: 1980, *J. Geophys. Res.* **85**, 8107.

Schaffer, L. E. and Burns, J. A.: 1897, *J. Geophys. Res.* **92**, 2264.

Schenk, P. and McKinnon, W.: 1985, *J. Geophys. Res.* **90**, C775.

Shoemaker, E., Lucchitta, B., Plescia, J., Squyres, S., Wilhelms, D.: 1982, in D. Morrison (ed.), *The Satellites of Jupiter*, University of Arizona Press, Tucson, p. 435.

Smith, B. A., Soderlom, L. A., Johnson, T. V., Ingersoll, A. P., Collins, S. A., Gunt, G. E., Masursky, H., Carr, M. H., Davies, M. E., Cook II, A. F., Boyce, J., Danielson, G. E., Owen, T., Sagan, C., Beebe, R. F., Veverka, J., Strom, R. G., McCauley, J. F., Morrison, D., Briggs, G. A., and Suomi, V. E.: 1979a, *Science* **204**, 951.

Smith, B. A., Soderblom, L. A., Beebe, R. F., Boyce, J., Briggs, G. A., Carr, M. H., Collins, S. A., Cook II, A. F., Danielson, G. E., Davies, M. E., Hunt, G. E., Ingersoll, A. P., Johnson, T. V., Musursky, H., McCauley, J. F., Morrison, D., Owen, T., Sagan, C., Shoemaker, E. M., Strom, R. G., Suomi, V. E., and Veverka, J.: 1979b, *Science* **206**, 927.

Snyder, L.: 1980, *Opt. Eng.* **19**, 566.

Stone, E. C. and Lane, A. L.: 1979a, *Science* **204**, 945.

Stone, E. C. and Lane, A. L.: 1979b, *Science* **206**, 925.

Strom, R. G., Schneider, N. M., Terrile, R., Cook, A. F., and Hansen, C.: 1981, *J. Geophys. Res.* **86**, 8593.

Thomas, P. and Veverka, J.: 1979, *Icarus* **40**, 395.

Thomas, P. and Veverka, J.: 1980, *Icarus* **42**, 234.

Tody, D.: 1986, *Proc. SPIE* **627**, 733.

Veverka, J. and Thomas, P.: 1979, in T. Gehrels (ed.), *Asteroids*, University of Arizona Press, Tucson, p. 628.

Wetherill, G. W. and Chapman, C. R.: 1988, in J. F. Kerridge and M. Mathews (eds.), *Meteorites and the Early Solar System*, University of Arizona Press, Tucson, p. 35.

Wilhelms, D.: 1987, *USGS Prof. Paper*, No. 1348.

Yeates, C. M., Johnson, T. V., Colin, L., Fanale, F. P., Frank, L., and Hunten, D. M.: 1985, in *GALILEO: Exploration of Jupiter's System*, NASA SP-479, U.S. Government Printing Office, Washington, D.C.

NEAR-INFRARED MAPPING SPECTROMETER EXPERIMENT ON GALILEO

R. W. CARLSON*, P. R. WEISSMAN*, W. D. SMYTHE*, J. C. MAHONEY*,

and

THE NIMS SCIENCE AND ENGINEERING TEAMS†

Abstract. The Galileo Near-Infrared Mapping Spectrometer (NIMS) is a combination of imaging and spectroscopic methods. Simultaneous use of these two methods yields a powerful combination, far greater than when used individually. For geological studies of surfaces, it can be used to map morphological features, while simultaneously determining their composition and mineralogy, providing data to investigate the evolution of surface geology. For atmospheres, many of the most interesting phenomena are transitory, with unpredictable locations. With concurrent mapping and spectroscopy, such features can be found and spectroscopically analyzed. In addition, the spatial/compositional aspects of known features can be fully investigated. The NIMS experiment will investigate Jupiter and the Galilean satellites during the two year orbital operation period, commencing December 1995. Prior to that, Galileo will have flown past Venus, the Earth/Moon system (twice), and two asteroids; obtaining scientific measurements for all of these objects.

The NIMS instrument covers the spectral range 0.7 to 5.2 μ, which includes the reflected-sunlight and thermal-radiation regimes for many solar system objects. This spectral region contains diagnostic spectral signatures, arising from molecular vibrational transitions (and some electronic transitions) of both solid and gaseous species. Imaging is performed by a combination of one-dimensional instrument spatial scanning, coupled with orthogonal spacecraft scan-platform motion, yielding two-dimensional images for each of the NIMS wavelengths.

The instrument consists of a telescope, with one dimension of spatial scanning, and a diffraction grating spectrometer. Both are passively cooled to low temperatures in order to reduce background photon shot noise. The detectors consist of an array of indium antimonide and silicon photovoltaic diodes, contained within a focal-plane-assembly, and cooled to cryogenic temperatures using a radiative cooler. Spectral and spatial scanning is accomplished by electro-mechanical devices, with motions executed using commandable instrument modes.

Particular attention was given to the thermal and contamination aspects of the Galileo spacecraft, both of which could profoundly affect NIMS performance. Various protective measures have been implemented, including shades to protect against thruster firings as well as thermal radiation from the spacecraft.

* Jet Propulsion Laboratory, California Institute of Technology, Pasadena, CA 91109, U.S.A.

† The Near Infrared Mapping Spectrometer (NIMS) Engineering and Science Teams consist of I. Aptaker (Instrument Manager), G. Bailey (Detectors), K. Baines (Science Coordinator), R. Burns (Digital Electronics), R. Carlson (Principal Investigator), E. Carpenter (Structures), K. Curry (Radiative Cooler), G. Danielson (Co-Investigator), T. Encrenaz (Co-Investigator), H. Enmark (Instrument Engineer), F. Fanale (Co-Investigator), M. Gram (Mechanisms), M. Hernandez (NIMS Orbiter Engineering Team), R. Hickok (Support Equipment Software), G. Jenkins (Support Equipment), T. Johnson (Co-Investigator), S. Jones (Optical-Mechanical Assembly), H. Kieffer (Co-Investigator), C. LaBaw (Spacecraft Calibration Targets), R. Lockhart (Instrument Manager), S. Macenka (Optics), J. Mahoney (Instrument Engineer), J. Marino (Instrument Engineer), H. Masursky (Co-Investigator), D. Matson (Co-Investigator), T. McCord (Co-Investigator), K. Mehaffey (Analog Electronics), A. Ocampo (Science Coordinator), G. Root (Instrument System Analysis), R. Salazar (Radiative Cooler and Thermal Design), D. Sevilla (Cover Mechanisms), W. Sleigh (Instrument Engineer), W. Smythe (Co-Investigator and Science Coordinator), L. Soderblom (Co-Investigator), L. Steimle (Optics), R. Steinkraus (Digital Electronics), F. Taylor (Co-Investigator), P. Weissman (Co-Investigator and Science Coordinator), and D. Wilson (Manufacturing Engineer).

Space Science Reviews **60:** 457–502, 1992.

This paper is dedicated to the memories of Gary Bailey and Hal Masursky. Gary would have been very pleased with the excellent performance of his detectors and Hal would have enjoyed the Galileo flyby of Venus, one of his favorite planets. Their crucial contributions to NIMS and Galileo will continue to be apparent throughout the mission, and will be appreciated in whatever future success we may enjoy.

Table of Contents

1. Introduction

The Near-Infrared Mapping Spectrometer (NIMS) experiment on the Galileo Orbiter Spacecraft represents a combination of imaging and spectroscopic methods. The advantage of spectral imaging, as opposed to pure imaging systems or single-field-of-view spectrometers, is the ability to simultaneously find, identify, and map compositional units on planetary surfaces. With such information one can investigate the geochemical evolution of satellite surfaces as well as the dynamical and compositional properties of atmospheres.

The NIMS instrument possesses modest spatial and spectral resolution, and operates in the near-infrared range of 0.7–5.2 μ. This spectral range is particularly diagnostic of minerals known or suspected to occur on planetary and satellite surfaces, and also includes many observable features of atmospheric species. The spectral resolution was designed for investigating the relatively broad bands seen in surface reflectance, yet is adequate for identifying several major and minor atmospheric constituents.

Galileo will be inserted into Jupiter orbit in December of 1995, commencing a nearly two year investigation of the Jovian system, performing eleven orbits around the planet during this period. Prior to Jupiter arrival, Galileo has flown by Venus (February 1990) and will fly by the Earth and Moon twice (December of 1990, 1992) and two asteroids – Gaspra (October 1991) and Ida (August 1993). NIMS will investigate all of these bodies. The scientific objectives for these measurements are briefly presented in Section 2.

The purpose of this paper is to describe and document the NIMS instrument design and development, which posed many unique design and spacecraft integration challenges, owing largely to the low temperatures required by the detectors and optics. The instrument itself is described in Section 3, followed by calibration aspects (Section 4), instrument operating modes (Section 5), and spacecraft considerations (Section 6). Mission design aspects which relate to NIMS are presented in Section 7. General aspects of the NIMS experiment have been briefly documented by Carlson (1981), and a description of the Galileo mission and its complement of experiments is contained in a volume by Yeates *et al.* (1985).

Although this paper is devoted to a discussion of the instrument – i.e., the hardware – the corresponding analysis tools – the software – must also be mentioned. NIMS generates a great amount of such data in a short amount of time, and this large volume of data can only be digested using highly-efficient, visual, interactive computer methods. These computer tools are therefore an integral part of the NIMS experiment, and a separate discussion is warranted. These aspects are not discussed here. In the interim, refer to Torson (1989) for an overview of NIMS data visualization capabilities.

2. Scientific Objectives

2.1. Venus

During the Venus flyby (February, 1990), NIMS measurements concentrated on spectral features which arise from surface and deep-atmosphere thermal emission. These features occur in spectral regions where CO_2 is relatively transparent, allowing one to perform chemical probing far below the cloud region and to also measure the spatial variations of the intervening cloud extinction. This infrared radiation is observable on the nightside of Venus and was only recently discovered (Allen and Crawford, 1984). Two types of measurements were obtained from NIMS: (1) multiple spatial images at selected infrared wavelengths for dynamical studies, and (2) detailed spectra at a variety of latitudes and longitudes for chemical abundance information, specifically H_2O and CO in the deep atmosphere. The results from this flyby are described by Carlson *et al.* (1991).

2.2. Earth

During the first flyby of the Earth–Moon system, NIMS performed both atmospheric and surface spectral-imaging of the Earth and similar geological investigations of the Moon. For the Earth measurements, NIMS investigated mesospheric airglow emission features and obtained geological maps of Australia and Antarctica. Exploratory global mapping of the Earth was also performed; the first time ever in this spectral region. Comprehensive lunar measurements were also obtained by NIMS at multiple phase angles, but with relatively poor spatial resolution.

One of the primary goals of NIMS during both the 1990 and 1992 flybys is investigation of mesospheric water, observable through limb scans of infrared fluorescence in the 2.7-μ band. It has been recently proposed (Thomas *et al.*, 1989) that noctilucent clouds, and upper-atmosphere water in general, has increased over the past century due to a larger amount of biological and anthropogenic emission of methane, a photochemical source of water in the mesosphere. The abundance of water in the upper atmosphere is poorly determined. Microwave measurements are in general agreement with photochemical models, but rocket ion spectroscopy and infrared measurements indicate a greater water abundance, perhaps implying an additional source of mesospheric water (cf. Garcia, 1989). NIMS may provide an independent measurement of the vertical water profile at several latitudes, extending to the summertime south polar region.

Other limb airglow emission features that were investigated are the O_2 infrared bands, the well-known infrared hydroxyl bands, the strong $CO_2\,\nu 3$ band, ozone emissions at 4.8 μ, and the tail of the NO (1–0) band, which is a prime cooler of the thermosphere. The NIMS limb measurements occur within roughly $\pm\frac{1}{2}$ hour of closest approach, and sample both the night- and day-side airglow.

As Galileo receded from the Earth, NIMS investigated specific geographic regions as determined by the timing of the flyby, illumination geometry, and the gain characteristics of the instrument. These gain characteristics were established for our ultimate goal, Jupiter at 5.2 AU from the Sun, and tend to be too sensitive for general Earth observations. The Earth at 1 AU is simply too bright. Nevertheless, by choosing favorable geometry, we were able to perform spectral mapping of both Australia and Antarctica during the 1990 pass and plan to investigate other regions in 1992. Using ground-truth measurements, we can corroborate and extend our remote sensing measurements to both continental and planetary scales.

2.3. Moon

On the first pass through the Earth–Moon system, the closest approach of Galileo was ~ 350 000 km, limiting NIMS resolution to ~ 170 km, in contrast to 10 km typical of spot spectrometer measurements obtained from ground-based observations. This initial pass provided viewing of roughly a quarter of the lunar surface (selenographic longitudes 90° to 180°) not previously observed spectroscopically. The phase angle range of observations (30° to 150°) extended the range of viewing geometry available from Earth, particularly for sub-Earth meridians, where Earth observations are limited to 90° phase angles. The wavelength coverage was also extended by NIMS; terrestrial observations being generally limited to below 2.5 μ by the atmosphere. However, because the NIMS dynamic range is designed for measurements at Jupiter, the lunar observations were saturated at many wavelengths, particularly for small incidence angles; unsaturated spectra to 2.5 μ generally require incidence angles greater than 80°. At wavelengths beyond about 3 μ, thermal emission becomes important. Because of this dynamic-range limitation, full NIMS spectra are available for only narrow angular regions. Nonetheless, with multiple observations taken at many geometries, some important lunar questions can be addressed, including: (1) an initial search for hydrated minerals – there is a remote possibility that hydrated minerals may be present near the polar regions where low temperatures may occur. The high sensitivity of NIMS allows a search for the 3 μ hydration feature during the second Earth–Moon encounter which passes over the Moon's north polar region, with a spatial resolution of about 60 km. (2) Spectral characterization of additional lunar areas – with the two passes, about 20% of the Moon not visible from the Earth can be mapped with surface resolution from 200 to 500 km. (3) Obtaining the first spectra in the 2.5 to 5.2 μ region. The NIMS wavelength range will allow an accurate determination of local surface temperatures and a correction for thermal emission for that portion of the spectrum that contains reflected radiation.

2.4. Asteroids

Galileo will fly by the main belt asteroids Gaspra and Ida in October 1991 and August 1993, respectively. Ground based studies have identified both asteroids as S-type. The NIMS goal at each encounter will be to acquire spectral-reflectance and thermal-emission images of the asteroid at maximum possible spatial resolution. Because of the high velocity of the flybys, it will be possible to obtain resolved images of only one hemisphere of each object. Full disc, unresolved spectra of the other hemisphere will be obtained during approach. NIMS will identify and map mineral species on the surface of each asteroid, and will seek to determine if the surfaces are chemically heterogeneous.

2.5. Jupiter atmosphere

Although NIMS was originally conceived for satellite surface spectral reflectance measurements, the experiment is well-suited for a variety of investigations of the Jupiter atmosphere. This is due to: (1) The spectral range available to NIMS, which includes signatures from several minor species such as germane, phosphine, and water, which are produced in the deep atmosphere of Jupiter and may serve as tracers of motions in this unexplored altitude region of the planet. (2) The NIMS spectral range also includes absorption features of the more abundant molecules: CH_4, NH_3, and H_2. Due to variations in absorption strengths for these molecules, NIMS can probe a large altitude range, ranging from the high-altitude polar aerosols, far above the ammonia cloud deck, extending down into the water cloud region at the several-bar-level. (3) The spatial resolution of the experiment, about 300 km, is sufficient to resolve the many dynamical features of the atmosphere and to investigate their temporal changes. Investigation of changes over time scales of hours to years are possible with the repeated observations available from Galileo. Finally (4) a full range of phase angle coverage is available with Galileo, enabling one to optically investigate the microphysical properties of the diverse cloud layers. Prior discussions of the NIMS measurements at Jupiter can be found in Taylor and Calcutt (1984) and Hunten *et al.* (1986).

2.6. Jovian Satellites

For the three large icy Galilean satellites, Callisto, Ganymede, and Europa, the primary NIMS science objectives are to map the various surface compositional units and to identify their elemental and mineralogical composition. An important aspect is to study these three objects as a collection. Voyager multispectral data suggests that there are common compositional units across the three objects. For example, the crust of Callisto, saturated with scars of ancient impact structures, has the same albedo and color as the remnants of the oldest terrain on Ganymede. Likewise, the much younger grooved terrains on Ganymede appear to be similar to the linear markings on Europa.

A primary question concerning these units is the composition of the dark components that are mixed with the dominant water-ice crusts. Are these materials silicates or organic-rich materials derived from primitive objects such as comet nuclei? If silicates, NIMS may detect bands due to olivines, pyroxenes, or a range of iron-bearing minerals. C–H features might be present if the satellite surfaces contain dark organic components such as those found on some asteroids. Magnetospheric sources of implanted material (e.g., sulfur) may be an important process, providing chemically reactive species which can modify the surfaces (e.g., generating S–O from the S implanted in H_2O).

Another set of questions pertain to the formation and history of these compositional units. Are there systematic correlations of dark component abundance or nature with geologic age and setting? Were the dark materials added early during the accretion or late during subsequent impacts? Methods to address these questions involve studying the global distribution of these units, examining leading- versus trailing-hemispheres as well as the geologic setting of such units. For example, dark-rayed craters on Ganymede may expose units of concentrated dark materials that exist as layers and lenses in the subsurface. Ejecta may show compositional correlations with size, age, latitude, terrain type, or longitude that will provide insights into their nature and origin. Another important question concerns other volatile species. These might be involved in active processes of eruption and transport, in particular on Europa, or be found as cold-trapped species, for instance in Ganymede's icy polar caps.

The innermost Galilean satellite, Io, is of great interest to the NIMS investigation, exhibiting not only a wide range of volcanic processes, some of which are continuously active, but also an equally bizarre mixture of surface chemistry, composition and mineralogy. Active volcanic processes include: (1) two classes of eruptive plumes thought to be driven by superheated volatiles in the form of sulfur and sulfur dioxide vapor, (2) eruption of icy clouds of material along scarps and fractures that have been proposed to be due to the escape of liquid sulfur dioxide to the surface, where it explosively forms gases and ice, and (3) dark hot spots, ranging in temperature up to

400 K, which have been proposed to be crusted pools of liquid sulfur. All of the processes likely involve a wide range of chemical components other than S and SO_2. Other volatiles may be cold-trapped in the polar regions. H_2S could exist in regions on the surface; it has been identified in ground-based spectra. Not only is there a plethora of potential spectral reflection features from the exotic compositional components, but gaseous absorption features may be detected over hot volcanic regions and the thermal output of the volcanic activity itself will be monitored and mapped. There is even the possibility that molten silicates will occasionally breach through to the surface.

3. Instrument Description

3.1. General description and operation

Performing simultaneous spectral and spatial infrared mapping at Jupiter necessitates use of high light-gathering power optics and sensitive detectors, both operating at low temperatures to minimize detector and background noise. The resulting NIMS instrument, shown in a photograph in Figure 1 and schematically in Figure 2, consists of the

Fig. 1. Photograph of the NIMS instrument. The telescope is shown on the right, with one of the athermalizing Invar rods, separating the primary and secondary mirrors, visible outside the telescope housing. The spectrometer portion is at the rear, while the radiative cooler is facing at the left. The black portion at the center of the cooler is the radiating surface. This is surrounded by a shield which obscures radiation from the scan platform and much of the spacecraft.

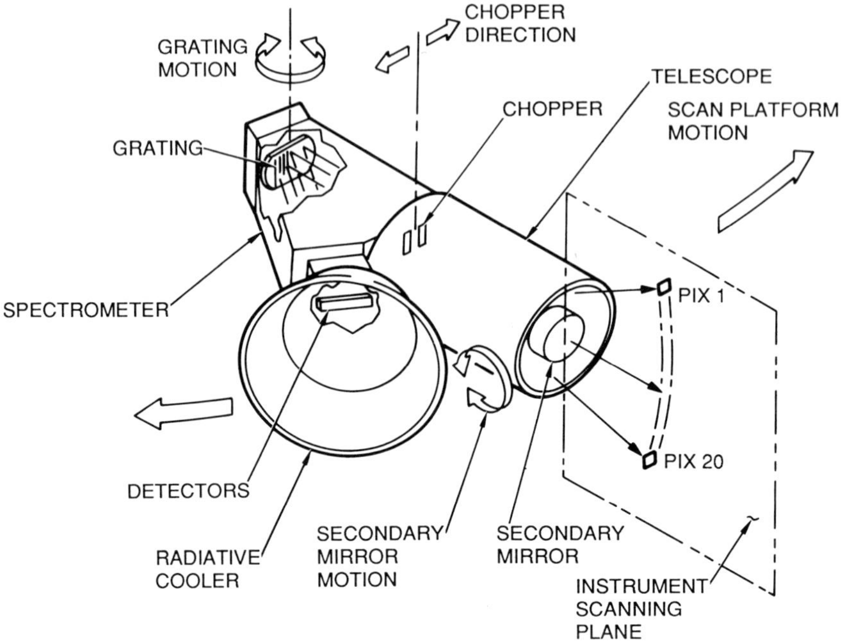

Fig. 2. Schematic diagram of the instrument and scanning motions. The instrument is shown as it would appear sitting on the spacecraft scan platform. The cone angle motion of the scan platform is illustrated by the large arrows. The internal 20-position spatial scan, produced by motion of the telescope secondary mirror, is in the orthogonal cross-cone direction. The position of the initial pixel in a major frame (RIM) is shown as pixel 1. The plane of spectral dispersion is parallel to the plane of cone angle motion.

following major elements: (1) a telescope with one dimension of spatial scanning, (2) an optical chopper to provide a dark reference, (3) a diffraction grating spectrometer which disperses the radiation onto the focal plane assembly, (4) a focal plane assembly consisting of 17 individual detectors, optical filters, and preamplifier circuitry, (5) a passive radiative cooler which cools the focal plane to cryogenic temperatures, and (6) signal processing and control electronics, not shown.

The typical operating mode of the instrument can be described with the aid of Figure 2. One dimension of spatial scanning is provided within the telescope employing a 'wobbling' secondary mirror, giving 20 contiguous pixels, each with 0.5×0.5 mrad resolution, for a total angular field of 0.5×10 mrad. The other dimension of spatial scanning is provided by slowly slewing the spacecraft scan platform in the cone direction. During one half of the up/down mirror scan, the grating is set at a fixed angle, with a corresponding set of 17 wavelengths striking the 17 individual detectors of the focal plane assembly. At the extremes of the mirror scan, the grating is stepped to a new setting, and a new set of wavelengths are measured during the second half of the mirror scan. The chopper frequency is synchronized to the 63 Hz spacecraft timing, and the mirror and grating motions are synchronized to the chopper, with motions taking place during the dark portion of the chopper cycle. The size and number of grating steps can be adjusted for specific encounter conditions and scientific objectives, and the scan platform motion is matched to the resulting spectral scan time. Details of the instrument timing are given in Section 3.6 and the instrument modes are described in Section 5. Additional information on many aspects of the instrument are contained in a series of papers by Aptaker (1982a, 1982b, 1983, 1987). A tabulation of instrument parameters is given in Table I.

3.2. Optical design

A variety of considerations led to the ultimate design of the NIMS optics, which primarily involved signal and instrument noise aspects. In order to minimize the latter, one must use small-area detectors, as the noise varies as the square root of detector area. At the same time, maximizing the signal on a detector requires large acceptance angles, i.e., use of a low *f*-number camera system illuminating the detectors. Similarly, adequate signal is obtained through use of a large diameter telescope, consistent with the system 'etendue' and angular resolution requirements. The resulting optical design of the instrument is illustrated in Figure 3 and consists of all-reflective telescope and spectrometer sections. Much of the optical design and fabrication were performed by the Perkin–Elmer Corporation. Development and testing of the NIMS optics have been previously described by Macenka (1983).

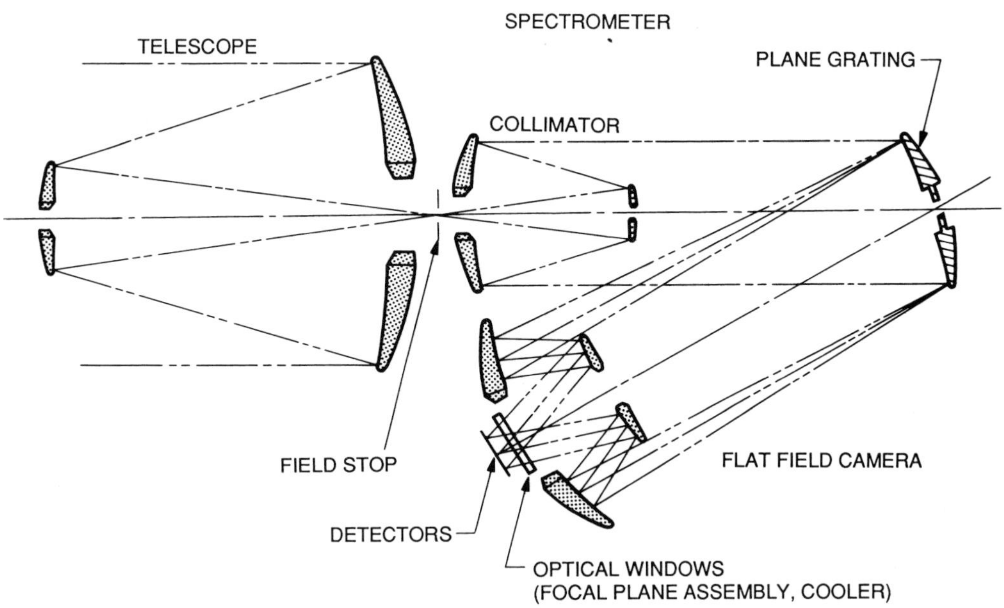

Fig. 3. Schematic of the NIMS optical design. A telescope of 228 mm aperture forms an image at the entrance slit of the spectrometer. The telescope secondary mirror pivots around an axis which is in the plane of the paper, giving a 20-position spatial scan. The spectrometer employs a dual-blazed diffraction grating. The dispersed radiation is detected by 17 detectors in the camera focal plane, each detector located for a particular spectral region. Rotation of the grating changes the wavelength striking each detector. A complete spectrum, at full spectral resolution, is accomplished with 24 grating positions.

TABLE I

Instrument parameters for the near-infrared mapping spectrometer

Angular resolution	0.5 mrad × 0.5 mrad (individual pixel size)
Angular field	10 mrad × 0.5 mrad (20 pixels, cross-cone direction)
Spectral range	0.7–5.2 μ
Spectral resolution	0.0250 μ ($\lambda > 1$ μ), 0.0125 μ ($\lambda < 1$ μ)
No. of spectral samples	Variable; 408, 204, 102, ..., 1
Spectral scan time	Variable: $8\frac{2}{3}$ s (408 spectral samples) $4\frac{1}{3}$ s (204 spectral samples) $\frac{1}{3}$ s (17 spectral samples)
Telescope	22.8 cm diameter *f*/3.5 Ritchey–Chrétien 800 mm equivalent focal length Spatial scanning via moving secondary mirror Operating temperature ~150 K
Telescope etendue	1.1×10^{-4} cm^2 steradian
Spectrometer	39 lines mm^{-1} plane grating spectrometer 400 mm focal length, *f*/3.5, Dall–Kirkham collimator 200 mm focal length, *f*/1.8, flat-field camera, Bipartite diffraction grating, 30% blazed for 1.9 μ, 70% blazed for 3.8 μ Operating temperature ~150 K
Detectors	Seventeen individual photovoltaic diodes, 15 indium antimonide, 2 silicon Active area = 0.2 mm × 0.2 mm Quantum efficiencies > 70%
Radiative cooler	Passive radiative cooler Achieves 64 K, with detectors energized
Noise equivalent radiance	7×10^{-9} W cm^{-2} sterad^{-1} per pixel and per spectral resolution element (0.025 μ), at 3 μ and 70 K FPA temperature
Noise equivalent albedo	0.0002 at 5 AU
Mechanisms	Torque motor drives for spatial and spectral scans Tuning fork chopper Deployable covers for telescope, radiative cooler
Electronics	Seventeen channel signal chain Microprocessor controlled (RCA 1802)
Gain states	4 ground-commandable gain states, detectors 1–14 Automatic gain switching, detectors 15–17
Protective devices	Covers for pre- and post-launch protection Continuous instrument purging through launch Heaters for continuing contamination control
Mass	18 kg
Power	12 W (average), 13 W (peak)
Dimensions	83 × 37 × 39 cm (optics) 20 × 25 × 13 cm (electronics)
Data rate	11.52 kbps
Data encoding	10 bits (0–1023)
Mounting	Scan platform Co-aligned with SSI, UVS, PPR
On-board calibration	Photometric Calibration Target (PCT, a solar reflectance target), and a Radiometric Calibration Target (RCT-NIMS, a blackbody radiator)

The telescope is a 228-mm diameter, $f/3.5$ Ritchey–Chrétien design, with an equivalent focal length of 800 mm. The secondary mirror steps in 20 equal increments, sweeping the image in the plane of the field stop in 0.5 mrad increments and thereby providing one dimension of spatial scanning. The field stop is a 400 μ wide slit, which defines a 0.5 mrad field-of-view, normal to the mirror scan direction and parallel to the plane of dispersion of the spectrometer. The angular resolution in the other direction is defined by the projection of the detectors at the field stop, and is approximately 0.5 mrad. Spatial response measurements are given in Section 4.3. The slit was made longer than required in order to ensure against misalignments caused by thermal or vibration induced shifts. For a distant point source, the telescope forms an image with 90% of the energy contained within a circle of angular diameter 0.05 mrad, which is quite satisfactory when compared to the aforementioned 0.5 mrad resolution. An InGaAs light emitting diode is mounted on the telescope spider and is used for inflight wavelength verification of the spectrometer.

The spectrometer employs a plane diffraction grating, illuminated by a Dahl–Kirkham collimator and followed by a wide-angle flat-field camera which focuses the entrance slit (the telescope field stop) onto the detectors. Stepwise rotation of the grating allows the complete NIMS spectrum to be generated.

The collimator has an effective focal length of 400 mm and a focal ratio of $f/3.5$. The smaller mirror in the collimator has a slightly toroidal surface to correct some system aberrations, primarily astigmatism from grating anamorphism.

The grating is a 39 lines mm^{-1} dual-blazed grating, with 30% of the area blazed for 1.9 μ and the remainder for 3.8 μ. The first order of the grating is used for wavelengths greater than 1 μ (the InSb detectors), and the second order for shorter wavelengths (Si detectors). Between the blaze configuration and use of multiple orders, reasonable efficiencies can be obtained over the relatively large wavelength range of NIMS. Measurements of the blaze efficiency of the flight grating are presented by Macenka (1983). Ruling of the master NIMS grating, replication, and measurement of efficiencies were performed by the Perkin–Elmer Corporation.

The detectors are widely spaced in the focal plane, requiring a wide-angle, flat-field camera. The two mirrors comprising the camera are both rather extreme aspheric surfaces, with an effective focal length of 200 mm and a focal ratio of $f/1.75$.

The linear dispersion in the focal plane is 8 mm μ^{-1} in first order. For the active area of the detectors (0.2 mm × 0.2 mm) the spectral width of each detector is 0.025 μ. This matches the width of the entrance slit, yielding a triangular spectral bandpass, slightly broadened due to finite spot sizes (see Section 4.2). The grating can be stepped through minimum increments of one-half of a spectral resolution element. The spectrometer exhibits some residual astigmatism and was aligned for the best spectral focus, consequently the vertical spatial resolution profile is somewhat broadened (see Section 4.3).

All of the mirrors and the grating were fabricated from fused-silica. Because of the large number of reflections, efficient infrared-reflective surfaces are required; NIMS uses pure gold with no protective covercoats, obtaining reflectivities of R = 96 to 98% in the spectral interval 0.7 to 5.2 μ. Obscuration, mainly by the camera secondary mirror, reduces the incident energy with a transmission factor of $T_{obs} = 60\%$, giving an 'etendue' for the optics, exclusive of reflection losses and grating efficiencies, of $A\Omega T_{obs} = 6.1 \times 10^{-5}$ cm^2 steradian. Stray light is reduced with baffles and an interior finish of matte black (Bostik–Finch Catalac Black).

The optics and their housing must operate at low temperatures in order to minimize photon shot noise from background thermal emission. Furthermore, initial alignment was performed at room temperature, thus thermal aspects are an important part of the design, and are discussed in Section 3.7.

3.3. Detectors and the Focal Plane Assembly

There are seventeen individual detectors (15 InSb and 2 Si) contained within the focal plane assembly (FPA), along with their associated spectral filters and electronic preamplifier components. Shielding from high energy particles, necessary in the Jovian magnetosphere, is provided by a hermetically sealed Tantalum case, which includes a sapphire window for optical input. A platinum resistance thermometer provides a measure of the FPA temperature. Figure 4 shows photographs of the FPA while Figure 5 illustrates the various components and their packaging. The FPA was manufactured by Cincinatti

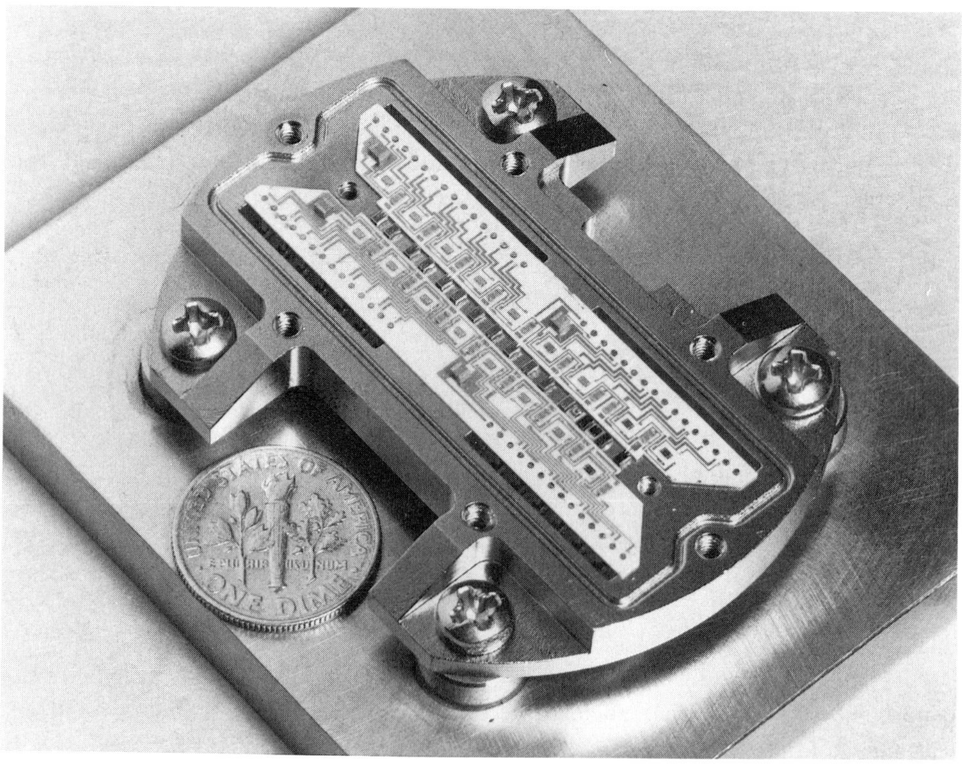

Fig. 4a.

Fig. 4b.

Fig. 4. Photographs of the focal plane assembly. The lower portion of the tantalum case and the wire-bonded hybrid circuitry containing the detectors, dual field effect transistors, feedback resistors, and decoupling capacitors. Not shown are the spectral blocking filters, field-of-view limiters, and the upper portion of the tantalum case containing the sapphire optical input window. The completed unit is evacuated and hermetically sealed using an indium gasket.

Electronics Corporation, and various aspects are described by Bailey (1979) and Smith *et al.* (1982).

Dispersed radiation from the spectrometer enters the FPA through two sapphire windows, the first contained within the radiative cooler (see Section 3.4), while the second is integral with the FPA and is anti-reflection coated for the near infrared. The radiation then passes through optical filters, whose purpose is to reject higher-order radiation and to limit the amount of thermal radiation incident on the detectors. These filters are at the cryogenic temperatures of the FPA and emit negligible amounts of thermal radiation that can be sensed by the detectors. A cold field-of-view limiting aperture is placed just behind the filters, further limiting thermal radiation on the detectors.

Each of the seventeen photodiode detectors contained in the FPA has a photo-active area of 0.2×0.2 mm, and each is anti-reflection coated for its own individual spectral

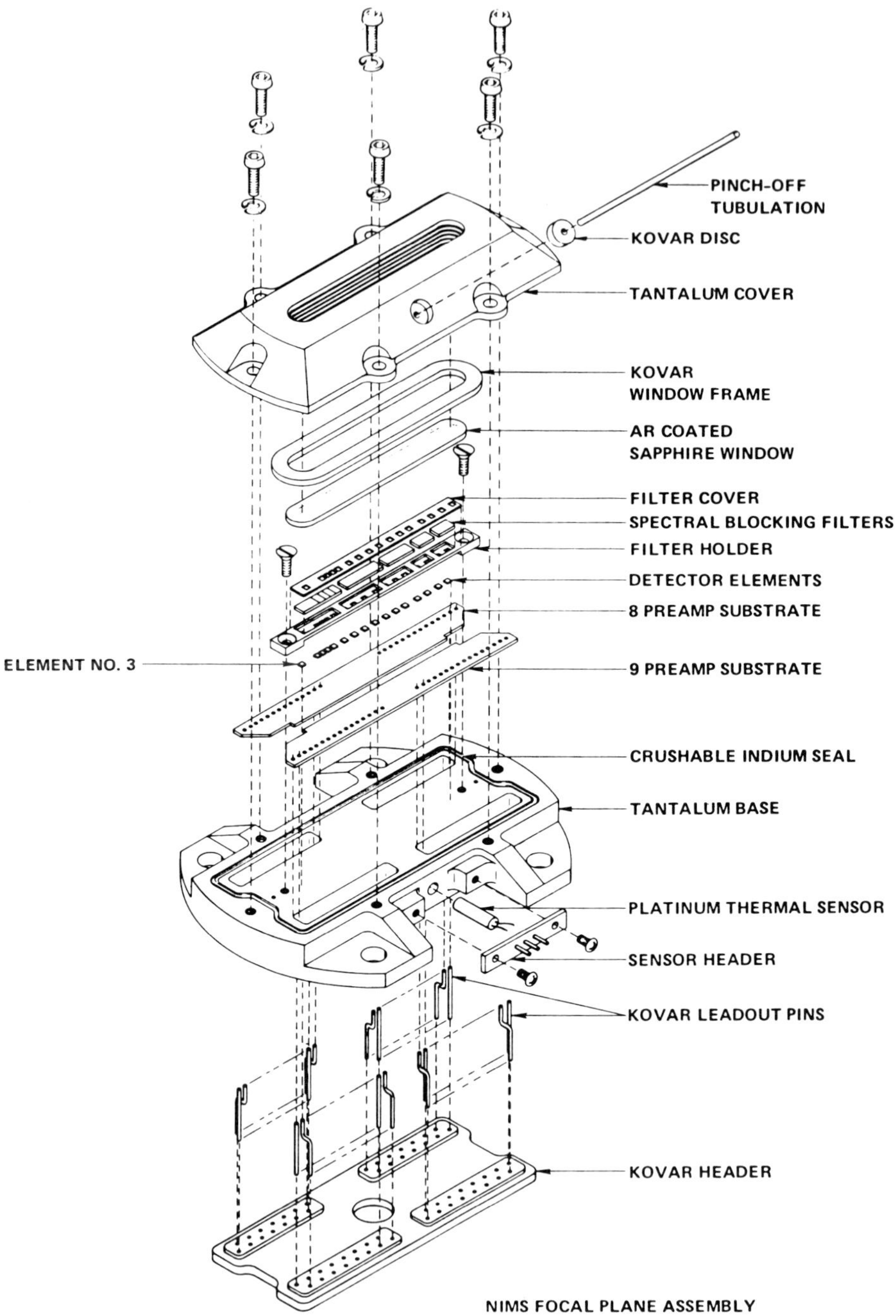

Fig. 5. Focal plane assembly components. The Focal Plane Assembly (FPA) contains seventeen individual photovoltaic diodes (15 InSb, 2 Si) and their associated pre-amplifier components, optical filters, field limiters, an optical window, a sealed Ta housing, a Pt temperature sensor, and feed-through connectors which couple to a low thermal conductivity ribbon cable (not shown).

region. Detector quantum efficiencies of 70% or greater were measured for the coated photodiodes. By operating the detectors at a constant bias voltage, near zero, the detectors act as high impedance current sources, with linear response to incident photon flux.

Noise performance of the instrument is determined by the detectors and their pre-amplifiers, each of which can be characterized by a noise-current spectral density. When the detectors are operated at near-zero bias, only Johnson noise contributes, with a noise-current spectral density of

$$i = \sqrt{4kT\Delta f/R}\,,$$

where R is the junction resistance (Hall *et al.*, 1975). T is the temperature, k is Boltzmann's constant, and Δf is the frequency range. R exhibits a rapidly varying temperature dependence (Kruse *et al.*, 1962), varying as

$$R \propto T^{-1/2} \exp(eV/2kT)\,,$$

V being the detector band gap (0.22 V for InSb). From this, the InSb detector noise is expected to show roughly a factor of three change for a temperature change of 10 K, and this dependence was indeed observed for the NIMS instrument. Measurements of individual NIMS InSb photodetectors at 77 K show a noise current density of 1.8×10^{-15} Amps $\mathrm{Hz}^{-1/2}$ with 0.9×10^{-15} Amps $\mathrm{Hz}^{-1/2}$ for the pre-amplifiers. This implies that NIMS will be pre-amplifier-noise limited for FPA temperatures less than about 70 K. NIMS noise levels measured in flight, as well as in the laboratory, are entirely consistent with the above values.

A hybrid dual junction field effect transistor (JFET) pre-amplifier front-end and $10^{10}\ \Omega$ feedback resistors are mounted in close proximity to each detector. Locating these components within the FPA minimizes circuit noise contributions by providing minimum capacity at the detector and by providing a low impedance interface to the external electronics assembly. The pre-amplifier circuitry contains a differential source-follower circuit which was experimentally found to give the best electromagnetic interference (EMI) performance. A commandable automatic bias circuit is incorporated to maintain the detectors at the afore-mentioned constant near-zero bias. This feature allows higher temperature operation of the InSb detectors by enforcing nonsaturation of the preamplifiers due to increased dark current and was included for the unlikely event of high FPA temperatures. Experience in flight has shown that there is no need for this precaution, although possible inflight contamination of the radiative cooler could change this situation. Additional details of the preamplifier electronics are given in Section 3.6.

Radiation shielding is provided by a 3 mm thick tantalum enclosure and by shielding the back of the camera secondary mirror, providing complete angular shielding. The shields were designed to limit the integrated exposure to < 10 krad during the nominal Galileo mission in order to prevent radiation damage, while simultaneously reducing noise caused by penetrating magnetospheric particles, mainly electrons. Tests were performed on the developmental model using energetic gamma rays; from this one can predict the resulting performance at Jupiter: Jovian radiation noise will be most severe for the observations taken during the Io close encounter just prior to Jupiter orbit insertion, where a predicted signal-to-noise ratio of 10:1 is predicted. Corresponding results for flyby observations of Europa and Ganymede are 30:1 and 100:1 respectively. Measurements from Calisto's distance are unaffected by ambient magnetospheric radiation.

High-energy magnetospheric protons can also cause displacement damage, with loss of sensitivity for InSb detectors, but such loss can be recovered by annealing the detectors at roughly 300 K. For this reason, the radiative cooler contains a commandable heater which can elevate the detectors to annealing temperatures.

3.4. Radiative cooler

The indium antimonide detectors require cryogenic temperatures for operation and this is achieved with a single-stage passive radiative cooler, illustrated in Figure 6. The cold stage, containing the FPA, has a 627 cm^2 aluminum honeycomb plate which radiates energy to space and cools the detectors to 64 K. The ultimate temperature is determined by input power to the cold stage and arises from many sources. The FPA dissipates 9 mW of electrical power, and thermal conduction from the cable and support mechanism provides another thermal path. Incident radiation is another source, and can arise from the instrument and the cooler housing and shield, from the spacecraft and other instruments, and from the planet itself. Minimization of these sources is discussed in the following paragraphs and in Section 6.1. The cooler was provided by the Santa Barbara Research Center.

Mechanical support of the cold stage is accomplished through a suspension system of fiberglass bands, similar to the mounting of a bicycle wheel hub. This arrangement not only provides an extremely low thermal conductance path, but also very stable mechanical positioning stability (0.01 mm). Electrical access to the FPA is provided by a 60-conductor ribbon cable with narrow stainless steel plated conductors in a thin polyimide sandwich, with a Ni coating for shielding.

Radiative loading from the instrument is minimized using a surrounding outer shroud with very low-emittance gold surfaces. The outer surfaces of the cold stage are similarly treated giving very low radiative coupling. Radiation from the spacecraft is reduced using a shield around the radiator plate to optically block emissions from the scan platform and portions of the spacecraft, the latter dependent upon the scan platform cone angle. Radiative loading to the honeycomb plate by the inner surface of the shield itself is reduced by use of a low emissivity (<0.03), highly specular (99% at 7 μ) gold coating. Additionally, the shield is thermally insulated from the cooler body, and radiatively cools to ~ 120 K. The surface of the shield which faces the scan platform is thermally isolated with multiple layers of thermal blanketing, while the outward-facing radiating surface is painted for high emissivity.

In order to maintain the cooler performance, it is crucial that the low emissivity of the inner shield surface be maintained. Any contamination would seriously increase the radiative loading; consequently the shield contains a 26 W strip heater to elevate the temperature and thereby avoid condensation of spacecraft outgassing and thruster products. This heater is used nearly continuously, except during observation periods, and elevates the shield to about 300 K. Additional contamination protection, for the pre-launch and the early post-launch period, was provided by inclusion of a deployable radiator cover and continuous dry nitrogen purging (see Sections 3.8 and 6.2). A heater is also included for the cold stage, for use in annealing the detectors in the unlikely event of radiation damage (see Section 3.3).

The radiator is mounted at 62.5° to the telescope optical axis, in the plane of rotation of the scan platform. Its full field-of-view is circular, with an angular radius of 71.5°. The total mass is 1.9 kg. The thermal mass of the cold stage is about 250 J deg^{-1} (at 90 K, temperature dependent), giving a cool-down time from 300 to 65 K of 30 hours. The initial cooling rate is very rapid, but slows at lower temperatures, requiring 24 hours to cool from 125 to 65 K. Additional in-flight performance values are given in Section 6.1.

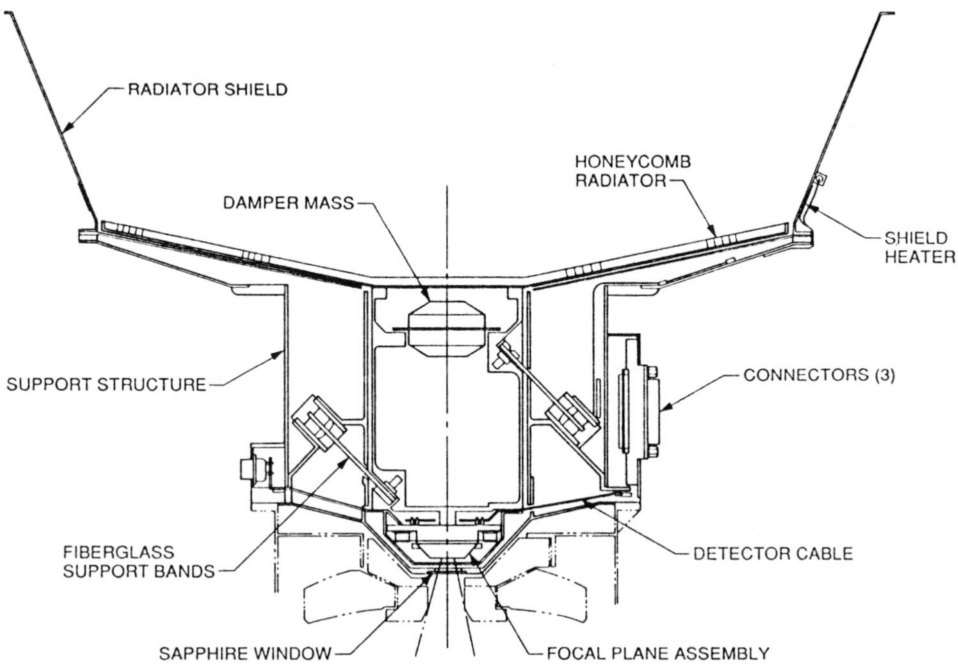

Fig. 6. Radiative cooler assembly. The Focal Plane Assembly (FPA) is thermally coupled to a honeycomb radiator. Operating temperatures of 64 K are achieved by radiation and by minimizing thermal loads to the FPA-radiator stage through use of low thermal conductivity fiberglass support bands and low emittance surfaces. A shield surrounds the radiator plate to block thermal emission from the scan platform and other portions of the spacecraft. The shield can be heated to 300 K in order to avoid contamination of the low-emissivity inner surface of the shield. The outer surface, on the side facing away from the scan platform, is painted with high-emissivity paint to allow the shield to radiatively cool, while the surface facing the platform is thermally blanketed to reduce scan platform radiative loads to the shield. The outer diameter of the shield periphery is ~40 cm.

3.5. Mechanisms

There are three mechanisms in the NIMS instrument: an optical chopper and two nearly identical, mirror and grating drives. In addition, the telescope and cooler had protective covers, since deployed in flight.

The chopper serves to modulate the detected radiation, allowing the dark current level of the detector to be subtracted on a pixel-by-pixel basis. It is a tuning fork with a resonant frequency slightly above the spacecraft 63 Hz timing signal, and contains two moving blades, each mounted on opposing tines, which chop the light. The chopper is located immediately in front of the telescope field stop (the spectrometer entrance slit) and modulates the radiation with an approximate 50% duty cycle. The back surfaces of the blades are coated with a black finish, while the front is coated with reflective gold. When the instrument is off, the blades are in the open position, and behind a pre-slit. This protects the low-thermal-mass blades from damage by inadvertent Sun pointing.

The chopper is driven at the spacecraft 63 Hz timing rate. This rate is slightly below the natural resonant frequency; in the event of a drive circuit failure the fork can be operated at its natural frequency, with overall synchronization with spacecraft timing being maintained by NIMS instrument software. Motion sensing and drive power are provided by magnetic pickup coils and a phase comparator feedback loop. The amplitude is kept at a constant level through an amplitude comparison circuit. The tuning fork and electronic design were provided by the American Time Company.

Mirror and grating motion is accomplished with direct-current magnetic motors, similar to a loudspeaker drive, but adapted for angular rather than linear motion. The on low-loss flexures and contains a samarium-cobalt magnet. An enclosing drive solenoid is attached to the stator. Motion is sensed through a pair of linear voltage differential transformers (LVDTs), one for position/velocity feedback control, the second as an independent, telemetered measure of the grating and mirror positions. Each LVDT consists of a permeable core, mounted on the rotor, and three stator-mounted coils – two for the drive current and one for position sensing. Motion of the core within the coils varies their coupling, and the induced EMF is linearly proportional to displacement.

In operation, the desired position (mirror or grating) is generated in the microprocessor and converted to an analog voltage. Error signals generate the motion drive, which is further controlled by rate information. The mechanism thus steps from one position to the next, settling to a stable position within a short interval (less than the dark half of a chopper cycle for the mirror, somewhat more for the grating).

Two ejectable covers are used to protect the NIMS optics and radiator (see Section 3.8). Each cover consists of a lightweight aluminum frame with a multi-layer thermal blanket closure. They are mounted onto the telescope and cooler apertures using uncaptured hinges, and locked into place with removable latch pins. Conical dowels ensure against launch-induced translation. Both the 25 cm optics cover and the 41 cm cooler cover are ejected at the same time by a common release device. The cover eject command is performed by an electric signal that fires redundant pyrotechnic squibs.

These 'bellows actuators' remain hermetically sealed after firing to prevent any contamination to the instrument, and through a lever-and-piston mechanism cause a pair of steel cables to be pulled. Each cable simultaneously unlatches a cover by pulling the constraining latch pin. Cover ejection is accomplished using torsion springs mounted at the hinge, causing a cover to rotate open when unlocked, and to slide free under its own momentum once it has opened approximately 100°.

3.6. Electronics design

The electronics assembly is mounted on the spacecraft scan platform near the optical assembly and contains the following circuits: analog, digital, scan mechanism and chopper drivers, and power supplies. A block diagram of the electronics showing the interaction of the various sub-modules and their external connections to the Command and Data Subsystem (CDS) and the Power/Pyro Subsystem (PPS) is presented in Figure 7.

The analog subassembly consists of 17 signal processing circuits (one for each detector), a multiplexer, analog-to-digital converter and miscellaneous circuitry such as engineering telemetry and a calibration lamp driver.

To maintain detector bias voltage and dark current stability with temperature and time, each InSb detector amplifier (channels 3–17) incorporates a bias correction servo loop which samples the amplifier output signal during the dark-signal portion of the chopper cycle and holds the amplifier output, and thus the detector bias voltage, at a predetermined level.

There are two types of signal processing circuits employed in the NIMS (see Figure 9). For most of the NIMS wavelength interval, the signals are determined by surface or atmospheric albedo and their range can be accurately predicted. Thus one can use ground commands to accomplish the infrequently required gain changes. On the other hand, there are transient, localized hot spots in the Jovian atmosphere which arise from unpredictable cloud clearings. These features allow one to probe the deep atmosphere, showing variable and often intense thermal emission in the 5 μ region, and require a much larger dynamic range than those required for the lower wavelength channels.

Commandable gain state switching is used in channels 1 through 14 (albedo channels). These commandable gain states are achieved by switching resistors at the input to the gated integrators. The instrument may be commanded to one of four possible gain states which then selects the appropriate resistors for each of the 14 albedo channels. The gain of each channel has been set for that detector's predicted response at Jupiter. In particular, the nominal gain state (gain state 2) corresponds at full scale (1023 data numbers, DN) to an albedo of ~ 1.2 at Jupiter's solar distance. Gain states 3 and 4 are each more sensitive by factors of two and four, respectively. Gain state 1 is similar to gain state 2, except channels 10–14 are each reduced in order to obtain measurements from the spacecraft Radiometric Calibration Target (RCT–NIMS, see Section 4.5).

Channels 15 through 17 (the thermal channels) use a dual-gain amplifier which automatically switches gain with input level and achieves a dynamic range of 10 000. The

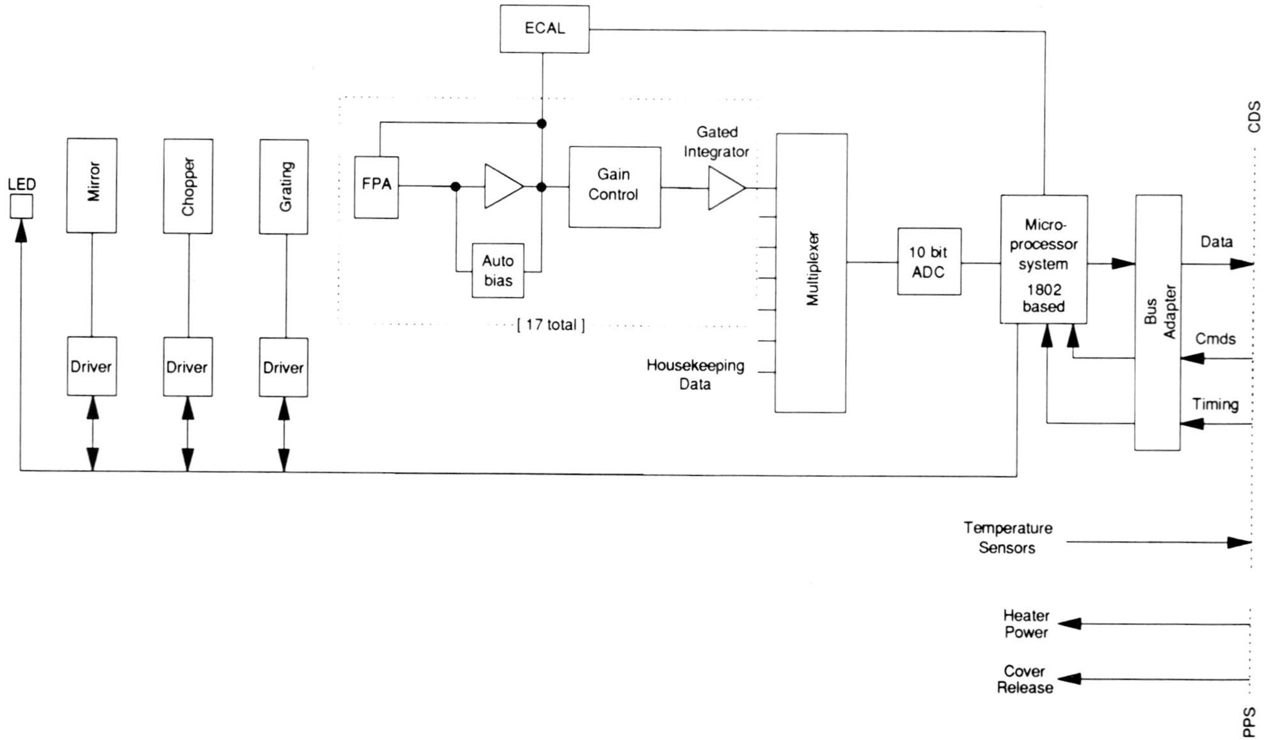

Fig. 7. Electronics block diagram. The instrument is controlled by an 1802 microprocessor which communicates with the spacecraft Command and Data Subsystem (CDS) through a bus adaptor. The microprocessor controls the mirror, grating, and chopper mechanisms, the timing of the signal chains, and calibration and housekeeping measurements. There are individual signal chains for each detector, and their outputs are sampled with an analog multiplexer and digitized to 10 bit precision. There are six temperatures monitored within the instrument (telescope, chopper, spectrometer, radiator shield, focal plane, and electronics). These measurements are obtained directly by the CDS. Heater and cover release functions are performed by commandable functions in the Power and Pyrotechnic Subsystem (PPS).

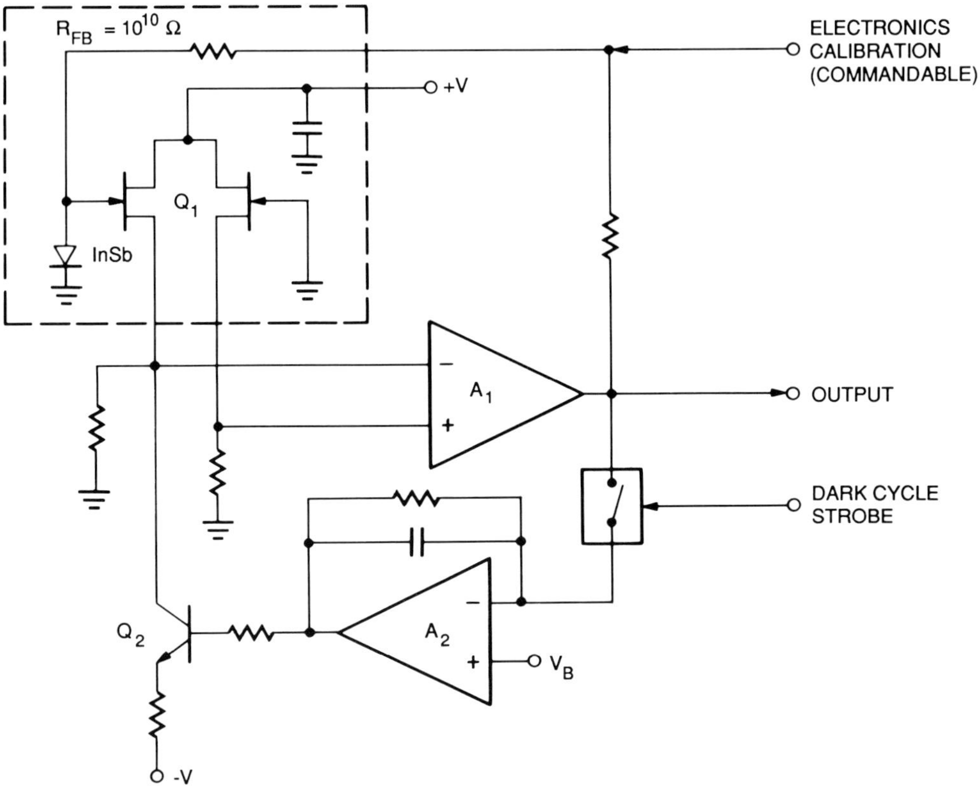

Fig. 8. Pre-amplifier and autobias circuitry. The components (detectors, field effect input transistors, feedback resistors, and bypass capacitors) contained within the Focal Plane Assembly (FPA) are shown in the outlined rectangle in the upper left of the Figure. A differential source follower drives operational amplifier A_1, which forms a transresistance amplifier using feedback resistor R_{FB}. The output of this pre-amplifier feeds the remaining signal processing circuitry shown in Figure 9. Automatic biasing of the detectors is accomplished by sampling the output of A_1 during the dark portion of the chopper cycle; this signal is sampled and averaged using amplifier A_2. The output of A_2 controls the drain current of the input transistor, which controls its gate-source voltage V_{GS}, and holds the bias of the InSb photodiode detector at zero bias voltage. With this zero bias condition, dark current changes due to changes in detector temperature remain small, yielding a large temperature operating range before amplifier saturation occurs. This bias condition also ensures detector linearity and minimizes detector noise.

two possible gains (low signal level = high gain, high signal level = low gain) for each of the thermal channels has also been optimized for Jovian system measurements, with brightness temperatures of 230 to 325 K measurable in the low gain configuration.

A commandable electronic calibration signal can be introduced to verify the gain (in gain state 2) for each of the seventeen signal chains.

The digital subassembly contains the central controller and mode sequencing functions necessary to operate the mechanisms in response to external commands, to acquire and format science and engineering data, and to output engineering telemetry and science data via the spacecraft data bus. Science data is transferred at a rate of

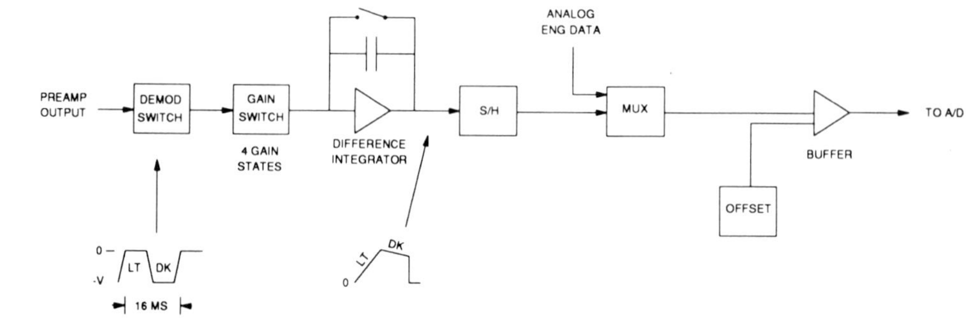

Block Diagram - signal chain for albedo channels 1 - 14

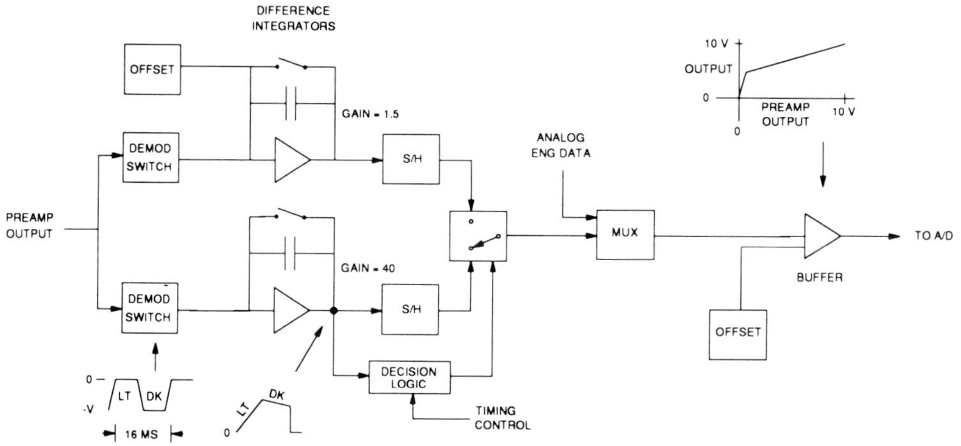

Block Diagram - signal chain for thermal channels 15 - 17

Fig. 9. Analog signal chain electronics. The upper diagram refers to detectors 1–14, termed the albedo channels, each of which has 4 commandable linear gain states. Pre-amplifier signals are synchronously demodulated and the current, which is determined by gain switched resistors, is integrated over one light/dark chopper cycle by the difference integrator. The integrated light-minus-dark signal is sampled at the end of each chopper cycle and stored by a sample-and-hold (S/H) circuit. It is then sequentially sampled by an analog multiplexer (MUX) and sent via an analog buffer to a 10-bit analog-to-digital converter (ADC). An offset is added in the buffer to ensure positive output values, >0 DN. The signal path for each of the thermal channels, detectors 15–17, is shown below. The circuitry are similar, but there are no commandable gain states. Instread, there are two independent paths, each with differing gains. The value sent to the MUX depends upon signal levels, yielding a large dynamic range.

11.52 kilobits s^{-1} and engineering telemetry is transferred at a rate of 36 bits s^{-1}. Commands to the mechanisms are initiated at the closing of the chopper blades. Since the chopper is phase-locked to the instrument timing chain, which in turn is phase-locked to the spacecraft real-time interrupt (RTI), all mechanism and other instrument operations are accurately synchronized to the spacecraft clock.

The digital subassembly is based on the RCA 1802 microprocessor. Its operating software is interrupt-driven and permits a flexible selection of instrument operating sequences. Mode defining parameter tables are used to control the operation of each sequence. Ground commands permit modification of the parameter tables which allows the instrument operation to be uniquely tailored to a specific science opportunity (see Section 5.3).

The chopper driver circuitry operates on commands from the digital subassembly and phase-synchronizes the chopper blades to the instrument and spacecraft clocks. The analog signal chain demodulator signal can be derived directly from the chopper (Chopper Reference mode) or from the 63 Hz drive signal (63 Hz mode). There are slight phase differences between the two; the former is the preferred mode, and is slightly more sensitive. The relative timing of the mechanisms is shown in Figure 10. The

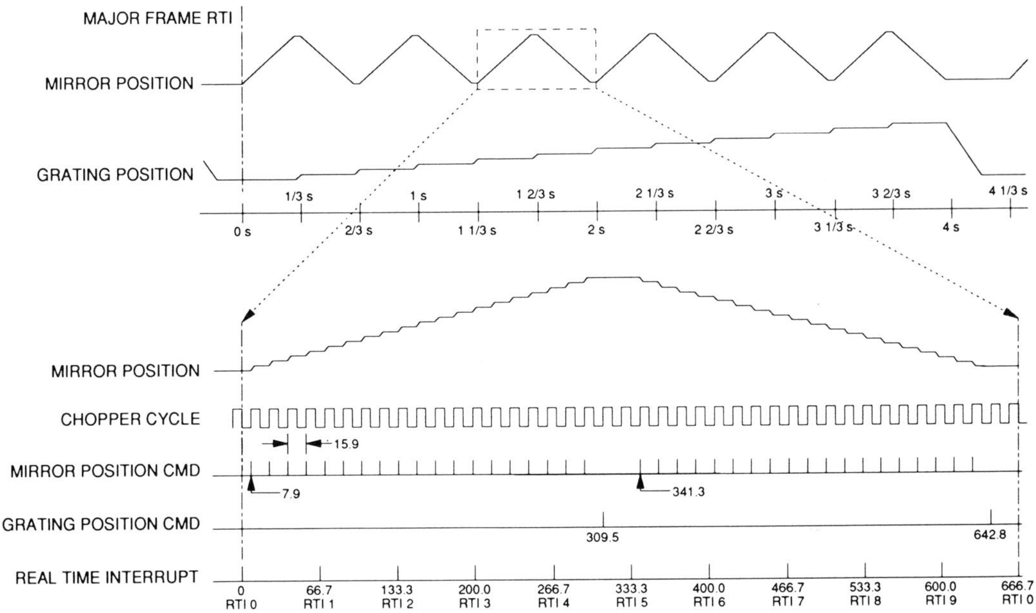

Fig. 10. Instrument timing. This timing diagram is representative of most NIMS modes, and particularly describes the FULL MAP mode, which has 12 grating positions and a grating-flyback period, all occurring during a period of $4\frac{1}{3}$ s (or $6\frac{1}{2}$ minor frames). During the grating flyback, the mirror rests, so that the next spectral cycle has the same mirror motion for the same wavelengths. The mirror scan consists of 20 positions, each corresponding to a single chopper cycle of $\frac{1}{63}$ s duration. During the dark cycle of a chopper cycle, the mirror steps from one position to the next. At the end of a mirror scan, the mirror is stationary for one cycle, during which the grating steps to a new position.

circuitry also regulates the physical amplitude of the chopper vane movement. The chopper's resonant frequency is slightly greater than the 63 Hz drive frequency. This allows a third chopper mode, wherein the chopper oscillates at its own natural frequency, with synchronism maintained by software control, occasionally slipping chopper cycles to maintain lock.

Both the scan mirror and the grating are actuated by torque motors. Each torque motor is driven by a closed-loop servo circuit which uses the output of a linear voltage differential transducer (LVDT) as the position sensor. The LVDT output is continuously compared against the position requested by the microprocessor. An error signal is generated which is then applied to the torque motor for position correction. To minimize instantaneous power demand, the software sequences commands to the mirror and grating drivers such that the two will not be in motion simultaneously.

The power supply consists of current-limited power converters and regulators necessary to supply all voltages needed by the various subassemblies. It also generates a power-on reset signal which controls the initialization of the instrument. At power-on, the instrument is placed in a quiescent mode, with neither the mirror nor grating scanning; the signal chain is set to the nominal gain state 2 and the instrument is synchronized to the 63 Hz clock instead of the chopper.

3.7. Thermal design

The low temperatures required for the FPA and optics posed many unique problems in both instrument design and thermal loads from external sources; the latter is discussed in Section 6.1. There are two general categories of design problems, the first is in achieving the required low temperatures, the second is ensuring that alignment shifts and mechanical stresses are reduced to tolerable levels.

The optical assembly is passively cooled by radiation, attaining ~ 150 K during both laboratory measurements as well as inflight. This is accomplished by having an unblanketed instrument and using narrow stainless steel mounting struts with low thermal conduction. The three struts form a kinematic mount, consisting of a monopod, bipod, and tripod. A prestress was applied during mounting to allow for thermal contraction of the instrument relative to the warm scan platform. Thermal conduction by the electrical cabling was reduced using long, minimum diameter wiring. Radiative coupling to the scan platform and nearby instruments is minimized by their thermal blanketing. The optical assembly was painted with a low solar absorptance, high emissivity white paint (zinc orthotitanate). This material is a moderate electrical conductor, providing protection against electrostatic discharge. Optical alignment was carried out at room temperature and it is important that it be maintained at the much lower operating temperature. Most critical are the primary-secondary mirror spacings, which are athermalized using invar rods. Thermal gradient distortion is avoided by directly mounting the optics, except for the grating, to a central optical 'bench' of high-thermal-conductivity material (aluminum). The mirror mount design is crucial to the optical quality, since differential thermal contraction can distort, or even break, the optical elements. NIMS uses an inherently athermal mounting scheme. The back surface of each mirror, in the central mounting region, is ground flat, while the front has a conical shape, with the apex of the cone in the plane defined by the back (cf. Figure 3). Flat and conical aluminum retainers were machined with the same geometry. Since there is no change in shape with temperature, clearances and clamping forces remain constant. Only a relative sliding motion results, facilitated by 0.001 inch thick mylar sandwiched between the surfaces.

3.8. Instrument Contamination Protection

There are a multitude of contamination sources which can degrade, or destroy, NIMS instrument performance. Ground-based sources include water-vapor absorption and deposition of particulate matter on the optics and radiator. These sources are especially pronounced during the launch phase, where vibration frees many particles, and outgassing of organics and other molecules also occurs. During the interplanetary injection IUS burn, the entire spacecraft is subject to impingement by motor products. During cruise, the spacecraft is a source of outgassing water vapor and organics, which can condense on the cold NIMS instrument. Thruster byproducts from trajectory and attitude correction maneuvers are continuing sources of contamination. For all of these reasons, NIMS has adopted a dual approach for contamination protection: first, to incorporate within the instrument as many protective measures as possible (discussed below), and second, to minimize external contamination sources insofar as possible (see Section 6.2).

The first NIMS protective measure was to use only low-outgassing materials and to further subject them to high temperature bakeout, prior to assembly. Upon final assembly, the instrument was kept in a dry environment, either in a dessicated container, or it was purged with dry nitrogen. Purge protection was nearly continuously maintained, even during Shuttle operations and launch, finally terminating two minutes prior to release from the Shuttle bay.

There were two separate purge paths within the instrument, dictated by relative contamination sensitivities. The low-emissivity surfaces within the cooler are most sensitive to contamination, consequently the cooler has its own purge path, separate from the optics purge.

Protective covers were installed over the telescope and cooler apertures, and remained in place, except during calibration and thermal-vacuum testing. They were deployed 77 days after launch, hopefully after most spacecraft outgassing had occurred.

During launch and cruise, NIMS is the coldest object on the spacecraft, and is subject to condensation of water and other volatiles onto sensitive surfaces. Heaters were included in order to minimize any immediate condensation, and to subsequently drive off condensates that might be deposited during unheated periods. The optics assembly has two 40 W heaters, operated simultaneously, which produce temperatures of 240–250 K. This suffices to remove water, but thruster products may be less easily removed, particularly if they have remained on the surfaces and have reacted to form less-volatile species. A heater was also incorporated in the radiator shield (see Section 3.4). It serves the same purpose, in this case protecting the most contamination-sensitive surface on NIMS – the inner surface of the radiator shield.

4. Instrument Calibration

4.1. Introduction

For the NIMS instrument, there are three broad calibration categories: spectral, radiometric, and spatial. The majority of the calibration measurements were performed in the laboratory, however there are several important calibration verification activities that will be performed in flight, and discussions of the relevant spacecraft hardware is included below.

Unless otherwise noted, the measurements reported here were obtained in the NIMS thermal-vacuum facility, which is a large stainless steel vacuum chamber, evacuated with a liquid-nitrogen-baffled diffusion pump. The NIMS instrument is mounted on an internal table which simulates thermal properties of the spacecraft scan platform. A liquid-nitrogen-cooled shroud surrounds the instrument, allowing the optical assembly to radiatively cool to its flight temperature – roughly 150 K. An additional space-background-simulator, cooled with liquid nitrogen or liquid neon, was used to cool the radiator. A large area blackbody source was installed in the vacuum chamber and can be rotated into the field-of-view of the telescope. External optical access is provided by two interchangeable window assemblies, one being a single, large diameter quartz window, the second consisting of a mosaic of small diameter calcium fluoride windows.

Calibration measurements were performed for a variety of operating conditions and instrument modes. The focal plane temperature, optics temperature, electronics temperatures, and input power voltages were all varied over expected operating ranges. In addition, all appropriate instrument gain states and modes were investigated.

4.2. Spectral calibration

The goal of the spectral calibration is to establish the wavelengths sensed by a detector for each of the 32 possible grating positions over the range of conditions expected in flight. This involves calibrating the NIMS spectrometer itself, prior to launch, and also characterizing an internal spectral light source, to be used in flight to detect any spectral shifts. In addition, when investigating highly detailed atmospheric spectra, it is necessary to know spectral bandpass profiles in order to convolve theoretical spectra to the NIMS resolution.

The grating equation, when expressed for the NIMS optical geometry, reads

$$m(\lambda/d) = \sin(\gamma + \phi) - \sin(\gamma - \phi + \chi)\,,$$

where m is the order of diffraction, λ the wavelength, d the grating constant, γ is half of the angle separating the collimator and camera optical axes, ϕ is the grating rotation angle, and χ is the angular displacement of each detector from the optical axis,

$$\chi_i = \arctan(x_i/f)\,.$$

Here, x_i is the linear displacement of detector i and f is the effective focal length, and includes refraction effects by the cooler and FPA windows. The grating rotation angle

$$\phi = \phi_0 + \varepsilon(p - \delta p)\,,$$

with p being the grating position ranging from 0 to 31, ε is the grating rotation increment, and ϕ_0 is the grating offset for grating position zero. δp is included to account for any launch induced shifts; it is defined as zero for the laboratory calibration. Determination of these constants, and any variation in them due to thermal and electrical effects, constitutes one portion of the spectral calibration. Several of these quantities, in particular γ and x_i were well determined during fabrication and assembly, leaving only ε, f, and ϕ_0 to be determined.

In order to accomplish these measurements, an auxiliary monochromator was used, which itself was calibrated using the Hg I λ5461 Å green line in various orders. This monochromator, used in conjunction with an incandescent source, a diffuser, and a collimator, was used to illuminate the NIMS telescope, forming a diffuse, spectrally narrow image at the entrance slit of the NIMS spectrometer. For a given setting of the external monochromator, a spectrum was obtained by the NIMS instrument. The wavelength setting of the auxiliary monochromator was then changed by an interval small compared to the NIMS spectral bandpass and the process repeated. Thus, for each detector and grating position, one can determine the spectral position and bandpass profile; an example is shown in Figure 11. Least squares fitting of such data for

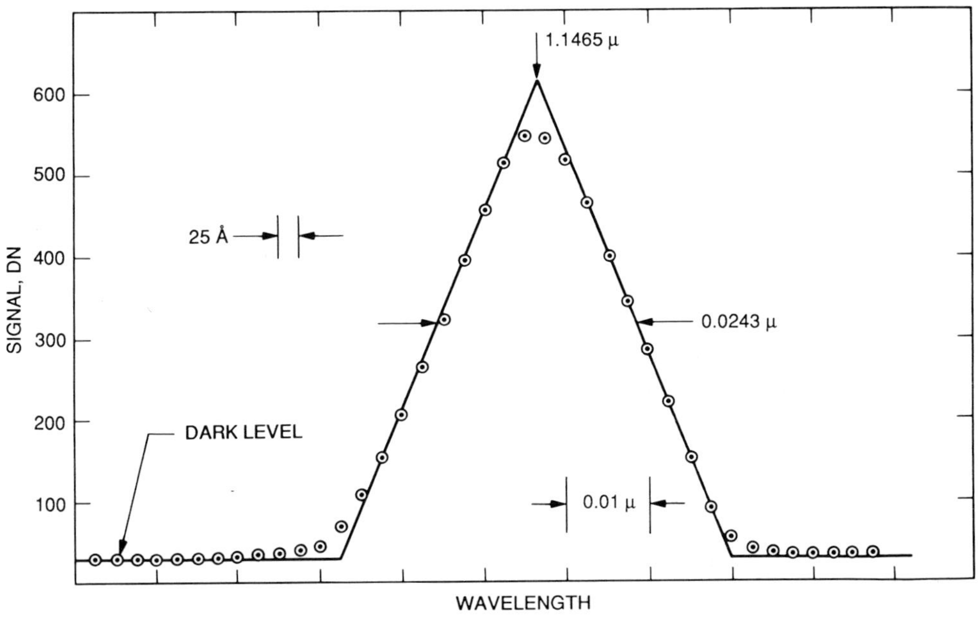

Fig. 11. Spectral bandpass response. The spectral response of the instrument for a given detector and grating position, in this case Det. 3 and Grating Position 13. An input monochromator, with a spectral bandpass of 25 Å, was varied in wavelength to determine this instantaneous spectral response. The ideal response of the NIMS instrument is expected to be triangular, with a full-width-at-half-maximum (FWHM) bandwidth of 0.0250 μ. The experimental results are seen to be close to expectation. Roughly one half of the broadening is due to the finite bandwidth of the input monochromator, with the remainder arising from the finite spot sizes (abberations) of the NIMS spectrometer optics.

a large number of grating positions and detectors allows an accurate determination of the calibration parameters, with an observed variance in wavelength of 0.001 μ, or less than one tenth of a grating step. Wavelength checks were also performed using indene and polystyrene absorption features.

Temperature variations of the wavelength positions were determined using Hg lines and by varying the optics assembly temperature using the instrument heaters. Comparing the relative displacement of the positions of atomic lines shows that the grating step value, ε, varies by 0.03% per degree centigrade, an insignificant amount considering the observed constancy of inflight temperature values.

Spectral shifts could arise from launch induced vibration or thermal effects. In order to quantify any such deviations, a spectral lamp is contained within the instrument, mounted on the telescope spider, which can be exercised by ground command. The lamp is an InGaAs light emitting diode, which emits a relatively narrow band of radiation, about 0.025 μ wide, centered at 0.8500 μ (at 130 K). The center wavelength is slightly temperature dependent, and there is a platinum resistance thermometer located nearby on the spider which can be used to accurately determine the temperature and wavelength. Using this internal source, any spectral shifts can be measured to an accuracy of better than 0.05 of a grating step.

4.3. Radiometric calibration

The purpose of the preflight radiometric measurements is to determine the parameters which relate the signal S received from the NIMS instrument (in data numbers, DN, ranging from 0 to 1023) to the radiance I of the target. The response of the instrument was found to be linear, with $S = S_0 + \sigma I$, where the sensitivity σ depends upon several parameters, including detector, wavelength, detector temperature, instrument gain, chopper mode, and polarization of the source. The dark value offset, S_0, depends upon detector, gain state, and other variables. These dark values were simultaneously determined from laboratory measurements but will also be determined in-flight, before and after an encounter sequence.

In order to cover the NIMS wavelength range of 0.7 to 5.2 μ, we used two types of light sources, the first being an incandescent tungsten-filament spectral irradiance standard which allows calibration to 2.5 μ. The second source is the aforementioned extended blackbody source mounted within the NIMS vacuum chamber, and provides useful spectral radiance for wavelengths longer than 2 μ. Some details of the two different sources are given below.

The shorter wavelength measurements used a 1 kW filament lamp with a quartz envelope containing halogen gas. The spectral radiance of this source was calibrated by EG&G Inc. and this calibration is directly traceable to the National Institute of Standards and Technology (NIST). The lamp was powered by regulated direct current at the prescribed amperage, measured using NIST-traceable instruments. In order to produce an extended source of known radiance, a large-area Halon target was constructed according to the prescription of Weidner and Hsia (1981), illuminated by the standard lamp. We use Weidner and Hsia's (1981) measured directional/hemispheric spectral reflectance values and their bidirectional reflectance data to find the reflectance for our particular geometry: normal incidence and $\sim 20°$ emission angle. This target was viewed by the NIMS instrument through the quartz window, whose transmission was independently measured. The lamp was placed at various distances from the target, allowing a test of the linearity of response and yielding a precise determination of the instrument sensitivity. The precision of the sensitivity determination was found to be a fraction of a percent. The accuracy of the derived values is estimated to be about 10%, partly due to uncertainties in the original lamp calibration (< 5%) and partly due to uncertainties in the bidirectional reflectance of the Halon target. This target is currently being compared to a Labsphere Spectralon standard, itself calibrated with traceability to NIST.

Measurement of the instrument sensitivity for the longer wavelength region was performed using an extended blackbody source, with a diameter larger than the NIMS telescope aperture. This source is a *V*-groove radiator, electrically heated and regulated to maintain a constant, preset temperature. The physical temperature of the radiating surface, which exhibits an emissivity of 0.99, is measured by two copper-constantan thermocouples, each relative to the ice point established by a distilled water ice bath. The thermocouples and potentiometer used were all calibrated to NIST-traceable standards. The heated target is loosely thermally-coupled to a liquid nitrogen-cooled heat sink, allowing a controllable temperature range of 200 to 350 K to be achieved. As with the previous lamp, measurements were obtained for a variety of source settings, in this case different temperatures, and the data fit in the sense of least squares to find sensitivity values. The precision in this determination is generally a fraction of 1%. Thermal gradients of ~ 0.5 K occur over the surface, limiting the accuracy of the derived sensitivity to 5%. In the overlap region, 2.3 to 2.5 μ, the difference in sensitivities found using the two different sources is 12.5%.

The instrument uses a diffraction grating, which are known to show efficiency differences for different polarizations. The polarization sensitivity was checked throughout the entire operating range using dichroic and wire grid polarizers. The maximum difference in sensitivity for two orthogonal polarizations was found to be only 5%, and this occurs only in the vicinity of the grating blaze wavelengths, as predicted by scalar diffraction theory of gratings (Strong, 1958).

4.4. Spatial Calibration

In this section we discuss aspects of the spatial calibration, which includes the pointing geometry, the angular resolution and angular sensitivity profile, and scattered light rejection. The NIMS spatial scan pattern is twenty pixels aligned along the cross-cone direction of the scan platform and formed by stepping the telescope secondary mirror through twenty positions. Measurement of the angular location of each pixel was accomplished using an illuminated slit, mounted in the focal plane of an external collimator. By translating this horizontally oriented slit in the vertical (cross-cone) direction one can find the point of maximum response, and thus the angular location for each of the twenty mirror positions. It was found that the angular locations are well represented as a linear function of the mirror position number, with an angular step size of 0.528 mrad.

Using a very narrow slit, either horizontally or vertically oriented, and translating it in the orthogonal direction, one can obtain a measure of the angular sensitivity profile in the spacecraft scan platform cone and cross-cone directions, respectively. An example is shown in Figure 12. It can be seen that the response in cone angle approximates a

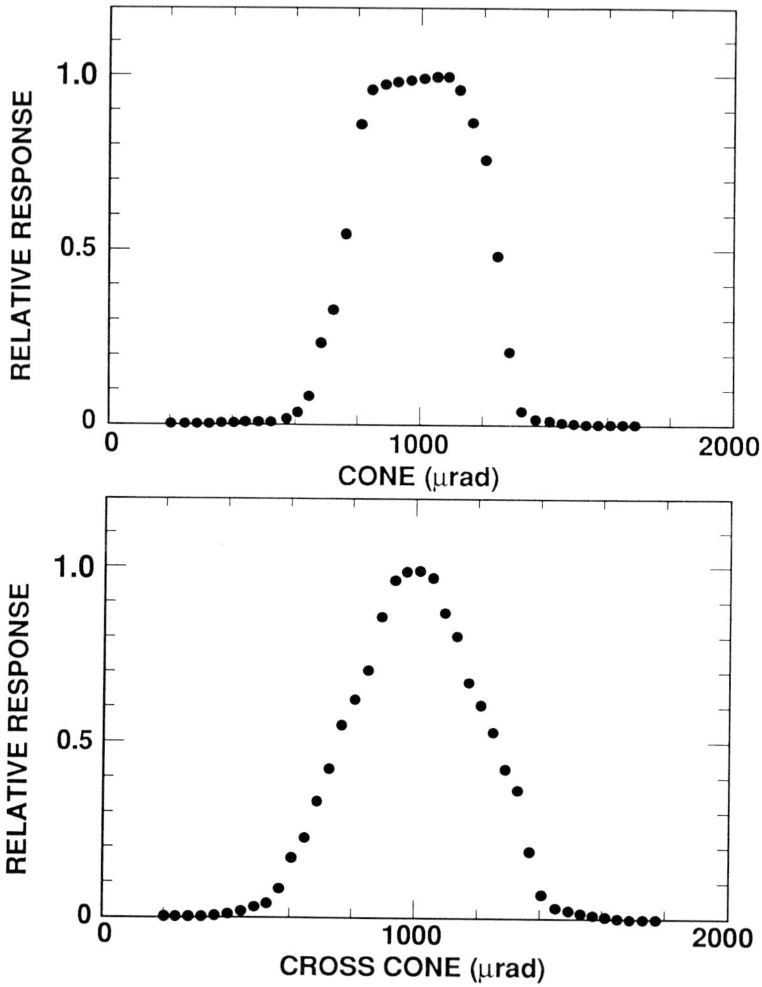

Fig. 12. Spatial response. The cone and cross-cone response of the NIMS instrument; see Figure 2 for definitions. In the cone-angle direction (shown above), the response is rectangular, whereas the cross-cone response (below) is broadened by the natural astigmatism of the spectrometer.

rectangle, as would be expected since this response is determined by the entrance slit of the NIMS spectrometer which is in the telescope focal plane. The response in the cross-cone direction is more triangular, and is the natural result of spectrometer astigmatism, which gives a slightly defocused vertical image of the detectors at the telescope focal plane.

The co-alignment of the Galileo scan platform experiments (NIMS, SSI, UVS, PPR) was measured prior to launch. The co-alignment of the NIMS and SSI experiment was found to be within 0.3 mrad, well within specifications. Simultaneous tests using stellar sources will be used to verify post-launch alignment.

For a variety of NIMS measurements, the instrumentally scattered light is of concern. Examples of observations which could be affected by scattered light include limb scans, dark feature measurements, terminator scans, and Jupiter dark side imaging. Although the best time to measure the scattered light properties of the instrument is during flight, at which time any launch and contamination influences will be included, prelaunch measurements of stray light effects were also performed. These tests, reported here, are only upper limits to the actual performance, since additional contributions from the collimator and both surfaces of the quartz window are included. Two different tests were performed to evaluate the scattered light, the first being a 'knife edge' test and the second a 'black hole' test.

In the 'knife edge' test, the mirror motion scans across a sharp boundary in the focal plane of the external collimator, the upper half-plane being a brightly illuminated surface while the lower unilluminated portion is made as dark as possible using a black surface oriented to form a cavity. For this geometry, the near-field (1 to 5 mrad from the boundary) scattered light contribution is detector dependent, varying from 0.1 to 0.4%. The disadvantage of this test is not knowing just how dark the lower boundary really is, and is partially solved using the following test.

In the 'black hole' tests, the instrument and collimator are focused at the center of a hole contained within a large, diffusely reflecting plate. An absorbing conical blackbody cavity is placed behind the plate, and the entire assembly is illuminated by a source placed in front and displaced from the axis of the conical absorber. Radiation reflected by the plate, which does not enter the collimator, is absorbed by optically black baffles. By using plates with the same hole size but with varying albedos, radiation emanating from the center of the hole is constant, while the instrumentally produced stray light varies with the surface albedo. Using plates with various albedos and hole sizes, the near-field scattered light response was found to be 30% less than that for the 'knife edge' tests. Again, all of these measurements include extraneous contributions from the three intervening optical elements. Definitive tests will be performed inflight, however preliminary results based upon the Venus and Earth limb scans indicate excellent scattered light rejection.

4.5. Spacecraft Radiometric Calibration Target

Calibration targets are provided on the Galileo spacecraft for inflight calibration verification of the remote sensing instruments, and to monitor the relative response throughout the mission by performing periodic calibration observations. All of the remote sensing instruments can use the Photometric Calibration Target, discussed in the next Section. In addition, there is a Radiometric Calibration Target (RCT–NIMS) which is intended for verification of NIMS performance in the long wavelength region, and is described in the following paragraphs.

The NIMS Radiometric Calibration Target is a near-field, extended, blackbody source, mounted on the scan platform sunshade and in front of the NIMS telescope when the scan platform is in the 0° cone angle position. When used as a calibration source, the target is heated with 25 W of electrical power, elevating the target surface to ~315 K. This provides a known radiance which can be used for radiometric calibration for wavelengths longer than 2.5 μ. The emitting surface, slightly larger than the telescope aperture, consists of a mosaic of hexagonal honeycomb cavities, each 0.25 inch in width and 0.50 inch in depth. An infrared black paint (Chemglaze Z004) was applied, which, with the cavity geometry, gives a normal emissivity of greater than 0.98 (Sparrow and Heinisch, 1970). Thermally insulating rods are used to mount the target, and a strip heater is employed to provide uniform heating and temperatures. Temperature differences across the surface were measured using thermocouples, with a root-mean-square deviation of 0.7 K being found. Upon heating, the target reaches equilibrium temperatures, within 0.5 K in a time period of 6 hours.

The physical temperature of the target is measured using the spacecraft Command and Data System (CDS) engineering telemetry. Two temperature sensors are employed, consisting of platinum and nickel resistance thermometers, and both were calibrated with NIST traceability. At the target operating temperature, both sensors exhibit nearly the same resistance, roughly 550 Ω. In order to calibrate the CDS circuitry itself, a temperature insensitive resistance of 562 Ω is also measured. All three measurements use the same current source and measurement circuitry. Consequently, temperatures can be measured to an accuracy limited only by the digitization interval, $\sim \pm 0.3$ K. If we assume a combined temperature error from all sources of ± 1 K, then radiometric accuracies of 7 and 4% will be achieved at 2.5 and 5 μ, respectively. Comparison of flight RCT data with pre-launch measurements indicates stable instrument performance.

4.6. Spacecraft Photometric Calibration Target

The Photometric Calibration Target (PCT) and an associated optical element, the Photometric Calibration Mirror (PCM) together form a source of diffusely reflected solar radiation which can be used by the remote sensing experiments for intra- and inter-instrument comparisons. These two elements are mounted on the Science Boom; the mirror reflects solar radiation onto the diffusing target surface which is placed outboard from the mirror and in a position that can be viewed from the scan platform. In order that the target be illuminated over the nominal range of solar cone angles, the mirror is convex, with a radius of curvature of 46 cm and located 53 cm from the target. The reflecting surface is vacuum deposited aluminum, with a protective overcoat formed by its natural oxide. The target surface is similar to the Voyager diffuser plate, consisting of sand blasted aluminum. The combination produces a spectrally gray diffuse surface, with an effective albedo of roughly 0.05. They were calibrated over the spectral range of 0.3 to 5.2 μ and for a variety of incidence angles.

5. Operating Modes and Data Acquisition

5.1. Instrument modes and operation

For most Galileo observations, the time available is limited, and one must tailor each observation for specific scientific goals. For the NIMS experiment, this translates into optimization of the spatial and spectral sampling aspects. For example, atmospheric measurements usually require the best available spectral resolution, whereas surface reflectance spectra are generally broader, allowing coarser spectral sampling. In addition, the spatial coverage and resolution demands are quite different for Jupiter and satellite measurements, the latter requiring much more rapid spatial sampling during the short amount of time available. Not only are there internal spatial/spectral tradeoffs to be considered but, in addition, it has been a longstanding goal among the Galileo remote sensing experiments to perform coordinated and compatible observations through simultaneous use of the scan platform. With these considerations in mind, we have developed a flexible set of instrument modes, described in this section.

The relevant NIMS instrument parameters that can be adjusted for differing observations are mainly spectral, determining the number of wavelengths to be sampled and their relative placement. In one extreme, the entire spectrum is obtained at full resolution, at the other extreme, the grating is fixed and only one wavelength band is sampled for each of the seventeen detectors. Intermediate combinations are possible, each with differing times required to complete a spectrum. Throughout this time, mapping is being accomplished by scan platform motion, with spatial and spectral sampling occurring simultaneously. During the time required to form a complete grating motion cycle, the scan platform will have moved some fraction of a NIMS spatial resolution element (0.5 mrad). In the following, contiguous spatial sampling corresponds to a motion of one spectral sample per spatial resolution element. Nyquist sampling is twice as frequent.

NIMS modes are implemented in the instrument software using parameter tables (PTABs). There are two such tables, each describing a specific spectral measurement sequence. Use of two PTAB tables allows for hybrid combinations, giving flexibility to instrument sequencing. The assignment of an individual parameter table (PTAB) is shown below (Figure 13) and described as follows: N is the number of grating positions per cycle, ranging from 1 to 24, D is the grating angle step size, unity corresponding to a single step of one-half of a spectral resolution element, and S is the grating start position. The number of times to repeat a given spectral sequence is given by the parameter R. Additional parameters include M, mirror scanning (on or off), and A, autobias (on or off).

There are a total of sixteen modes available to NIMS, twelve of which are pre-defined in the instrument read-only-memory (ROM). An additional four modes, yet undefined, can be placed in the instrument's random-access-memory (RAM) via uplink commands. All mode parameters can be changed by ground commands. The standard ROM modes are summarized in Table II.

The NIMS instrument modes were designed to be synchronous with spacecraft timing and its various time units. The largest unit is a RIM (or MAJOR FRAME) which

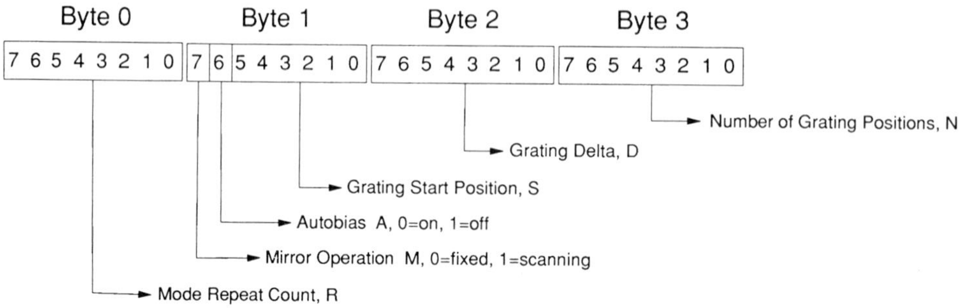

Fig. 13. Parameter table assignments. Spectral modes are derived from the number of grating positions N, the increment between successive grating positions D, and the starting position of the grating, S. Spatial scanning is controlled by the mirror operation bit, M.

TABLE II
NIMS standard ROM modes.

Mode number	Name	N	D	M	Execution time	Nyquist slew rate
0	Safe	0	0	Off	$\frac{1}{6}$	15000
1	Full Map	12	2	Scanning	$4\frac{1}{3}$	60
2	Full Spectrometer	12	2	Off	$4\frac{1}{3}$	60
3	Long Map	24	1	Scanning	$8\frac{2}{3}$	30
4	Long Spectrometer	24	1	Off	$8\frac{2}{3}$	30
5	Short Map	6	4	Scanning	$2\frac{1}{3}$	110
6	Short Spectrometer	6	4	Off	$2\frac{1}{3}$	110
7	Fixed Map	1	0	Scanning	$\frac{1}{3}$	750
8	Band Edge Map	2	–	Scanning	$1\frac{1}{3}$	190
9	Band Edge Spectrometer	2	–	Off	$1\frac{1}{3}$	190
10	Stop & Slide Map	24, 12	1, 2	Scanning	Variable	Variable
11	Stop & Slide Spectrometer	24, 12	1, 2	Off	Variable	Variable

The execution time, given in seconds, is the time required to complete a grating cycle. The Nyquist slew rate is the scan platform cone angle rate necessary to move through one-half of a NIMS pixel ($\frac{1}{2}$ of 0.5 mrad) in the execution time. It is given in microradians s^{-1}.

is $60\frac{2}{3}$ s, subdivided into 91 MINOR FRAMES, having whole fractions of $\frac{1}{91}$ ($\frac{2}{3}$ s), $\frac{1}{26}$ ($2\frac{1}{3}$ s), $\frac{1}{14}$ ($4\frac{1}{3}$ s), $\frac{1}{7}$ ($8\frac{2}{3}$ s), and $\frac{1}{2}$ ($30\frac{1}{3}$ s). There are NIMS modes synchronous with all of these fractions except $\frac{1}{2}$, which is that used for a compressed imaging mode for the SSI instrument.

The instrument steps through a grating cycle (N positions of the grating) then repeats R times (Mode Repeat Count). When the repeat count is consumed, then the other PTAB controls the instrument mode. This process continues as long as the instrument is powered on.

The instrument software has specific branching instructions which depend mainly on the value of N, the number of grating positions. These instructions result in somewhat different timing for the scan mirror, as well as the number of steps required to complete a cycle. $N = 12$ results in 13 spatial scan cycles ($\frac{1}{3}$ s each) and a total grating cycle time of $\frac{13}{3} = 4\frac{1}{3}$ s. The scan mirror rests at the starting point during the 13th interval, during which time the grating returns to its starting position. If $N = 24$, then this number of grating positions are positioned, followed by two scan mirror rest times, during which the grating is reset to the origin, giving a total cycle time of $\frac{26}{3} = 8\frac{2}{3}$ s.

The NIMS instrument modes in ROM are named according the following conventions. The MAP mode indicates the spatial mirror is scanning, SPECTROMETER indicates the spatial mirror is off. LONG, FULL, SHORT (which is sometimes called PARTIAL) and FIXED refer to the number of grating positions in a single grating cycle (24, 12, 6, and 1, respectively). The FIXED mode has been embedded in spacecraft documentation as FIXED GRATING; however the systematic name consistent with the definition in ROM would be FIXED MAP. The BANDEDGE mode has one grating position per PTAB, with a different grating position in each PTAB, alternating between PTABS each $\frac{2}{3}$ s. The STOP&SLIDE mode is a hybrid mode, combining the FULL mode in one PTAB with a FIXED mode in the other.

In the MAP modes the instrument has the spatial mirror turned on. This yields an effective field of view of 20 pixels, arranged in a linear stripe of 10 mrad $\times$ 0.5 mrad in the cross-cone direction. These 20 pixels are measured sequentially, and all within $\frac{1}{3}$ s. The initial scan motion, at the beginning of a RIM, is downward (toward the scan platform), then up for the next grating step.

The SPECTROMETER mode, for which there is no internal spatial scanning, is used for certain observations, for example atmospheric limb scans, when only one-dimensional image scanning using the scan platform motion is required, or when redundant sampling of a given location can be used to advantage. In general, the scan platform rates are the same as those for the MAP modes, since the same amount of time is required to complete a grating cycle. The only exception is between the FIXED MAP and SAFE modes.

Among the various grating modes, the LONG mode provides the best spectral resolution, with two spectral samples per spectral resolution element (spectral Nyquist sampling). This mode requires $8\frac{2}{3}$ s to complete, with seven spectra contained in a RIM. For this mode, Nyquist spatial sampling requires ~ 30 microrad s^{-1} slew rates, yielding two complete spectra while crossing through a NIMS field of view (0.5 mrad). It may be difficult to match this mode with simultaneous imaging (SSI) observations when time is the limiting factor, since the slew rate for a nominal SSI frame is about 120 microrad s^{-1}. One strategy is to use alternate imaging filters in a sliding mosaic.

The FULL mode gives lower spectral resolution, with only one spectral measurement per spectral resolution element, and is useful for diffuse solid surface reflection features. This mode permits a coordinated observation with other scan platform instruments using a slew rate of about 110 microrad s^{-1}, which is twice the preferred spatial Nyquist sampling rate for NIMS.

The SHORT mode under-samples the NIMS spectral capability by a factor of 4, using only 6 grating positions. This mode is a compromise which was included to work with the SSI camera's compressed mode and to permit obtaining some spectral information when the time to complete a mosaic is extremely limited. Some optimization is possible by matching the starting grating position for a cycle (and thus the spectral 'comb') to spectral bands expected on the observed surface.

The FIXED mode reduces the number of grating positions to 1. This mode is spectrally minimal, but ensures excellent spectral registration for the wavelengths that are measured. It proved to be very useful in Venus darkside observations.

The BANDEDGE mode alternates between two grating positions, providing two stripes (20 samples at one grating position over the 17 detectors which are approximately evenly spaced over the spectral range of the instrument) at different grating positions. This mode, for instance, alternates sampling between the continuum and the maximum of a spectral feature. It is useful for spatial mapping a particular spectral feature at a high spatial scan rate. This mode samples one stripe every $\frac{2}{3}$ s – $\frac{1}{3}$ s for the stripe and $\frac{1}{3}$ s preparing for the next grating position (defined by the next PTAB).

The STOP&SLIDE mode is a combination mode – normally a combination of the FULL mode and the LONG mode. Its purpose is to provide a mode compatible with multi-color sequences for the SSI framing camera. The sequence works as follows: The scan platform is stopped and NIMS enters its highest spectral resolution mode (LONG) while the SSI instrument is acquiring frames in several filters. The scan platform then slews slowly over to the next picture position (one overlapped SSI FOV) with NIMS in a mode compatible with the slew rate (nominally FULL mode), in a time period which corresponds to the readout time for the framing camera. The Mode Repeat Count parameter is set to accomplish this compatible sequence. The result for SSI is a multi-color image. The result for NIMS is a spatially complete spectral map of the target at modest spectral resolution, and spatially sparse high spectral resolution samples with a best spectral resolution. This optimizes the use of scarce scan platform time, characteristic of close encounter geometries, when multi-color framing mosaics are being performed.

In addition to the ROM modes discussed above, four of the 16 instrument modes may be loaded from RAM. The PTABs for these modes are defined by ground commands to the instrument and are stored in RAM. The RAM mode may then be loaded into the active area with a single command. The RAM mode remains valid as long as the instrument is powered on and the instrument is not reset.

After the two PTABS which define a mode are loaded into the active area, the active area may be modified to change the characteristics of the mode. The appropriate command modifies one of the four PTAB values in each loaded PTAB (independently) with each invocation of the command. This capability is particularly important for the BANDEDGE and STOP&SLIDE modes. The BANDEDGE grating start positions and the STOP&SLIDE duration in the high spectral resolution mode often need to be modified from those defined in the instrument Read-Only Memory (ROM).

5.2. Instrument commands

In addition to the various modes that the NIMS instrument is capable of executing (described above), there are a variety of instrument states which can be used for measurement or calibration. Of prime concern is the instrument gain state, for which there are four (see Section 3.6). In addition, two different calibration sequences can be executed: an OPCAL command, for which an internal electroluminescent diode provides a wavelength reference, and an ECAL command, which injects a known signal into the electronic amplification chain.

The commands are labeled by instrument number (NIMS = 37) and a mnemonic, which indicates the type of command sent. One NIMS command will be executed in a given spacecraft RIM; in the event of multiple commands, the last command loaded will be executed. The command must be loaded into the instrument command buffer by minor frame 89 of the previous RIM. The following is a brief description of available commands.

37IOP – Instrument Operation: This command loads a NIMS mode (set of PTABS) from ROM or RAM to the active area. It also permits specification of the grating start position (loaded into both active PTABS).

37IST – Instrument Status: This command modifies the gain state (4 available), the chopper state (63 Hz, chopper off, chopper reference, and free run), and can invoke the electronics calibration or the optical calibration. The electronics calibration should be carried out in gain state 2. The calibration lamp (a light emitting diode) measurement should be done in gain state 4 and LONG SPECTROMETER mode.

37MPT – Modify Parameter Table: This command modifies one of the 4 parameters in the PTAB for both of the PTABS in the active area. The value for each PTAB is specified independently in the command (but the same parameter is modified in each PTAB). It is possible to turn off the thermal channel autobias with this command – a capability which is intended for use only where the focal plane is approximately at room temperature.

37GO – Grating Offset: This command sets the grating offset. Acceptable values are 0–7; the default is 4 which is entered whenever the instrument is turned on. Note that this grating offset is different than the PTAB grating start position parameter.

37SS – Special Sequence: This command programs the RAM modes. It must be invoked twice, once for each PTAB in a mode.

37IRT – Instrument Reset: This command permits resetting the instrument without cycling instrument power.

37MN – Memory Normal: This sets the instrument CPU to use the normal ROM address space.

37MRL – Memory Reallocate: This sets the instrument CPU to branch to the RAM address space.

37PL – Program Load: This permits loading programs into the spare RAM.

5.3. Typical encounter operations

During each of the eleven Jupiter orbits, there will be satellite flyby opportunities which occur on time scales of hours, demanding efficient usage of the various instrument capabilities in order to maximize the scientific return. Obervations of the Jupiter atmosphere occur over longer time scales, measured in days, but the number of important features to be measured, their angular sizes, and temporal measurement frequency also demands efficient instrument mode usage. The following discussion illustrates, in a simplified fashion, the NIMS use of modes during a typical orbit at Jupiter. General sequencing priorities are (1) keep NIMS in LONG mode as much as possible to optimize spectral resolution, (2) maximize spatial coverage and resolution, and (3) include as many scan platform instruments as possible in an observation – to maximize synergistic science return.

On a typical Jovian orbit, Jupiter becomes visible over the spacecraft sunshade at about 25–40 Jovian radii (R_J). At this distance it is possible to mosaic the entire planet – at full spectral resolution – in a reasonably short time. At periapsis, occurring at about 10 R_J, the angular size of Jupiter is quite large, and it is not possible to mosaic the entire planetary disc at full spectral resolution. Similarly, for a close Galilean satellite encounter, it is possible to fully map the satellite at highest spectral resolution at about 4 hours out, but at closest approach the angular size and surface-relative smear rates become very high, forcing a choice between spatial and spectral mode coverage.

On the inbound portion of a Jupiter orbit, the NIMS instrument would be in LONG MAP mode to mosaic the Jupiter day and night side, occasionally executing the STOP&SLIDE mode for compatibility with SSI multi-color images of the planet. Many of the mosaics will be oversampled vertically, which is required for overlapping fields-of-view for the thermal instrument (PPR). As perijove approaches, the LONG MAP mode is maintained for atmospheric measurements, but the areal coverage is reduced to include only a few specific features. These features will be consistently measured throughout the orbital pass, yielding their temporal and phase-angle variations. It may be useful to develop some RAM modes for these atmospheric measurements, particularly if one is interested in only a portion of the complete spectrum available to a detector. It is possible to maintain high resolution, i.e., using a grating delta parameter of $D = 1$, but using a lesser number of grating positions N and an appropriate grating start position S to choose the spectral region of interest. In doing so, the observations will encompass less time, allowing more features to be examined.

For the satellite encounters, a full-disk mosaic will be acquired each time the spatial resolution changes by a factor of ~ 2. Thus, a LONG MAP mosaic would be followed by a FULL MAP mosaic, which would be followed by a SHORT MAP mosaic, as spectral resolution is traded for spatial resolution. The STOP&SLIDE mode would be invoked as the framing camera acquires it's multi-color mosaics. At closest approach, the FIXED mode would be utilized to compensate for the high angular rate of the satellite surface motion.

6. Spacecraft Interactions

6.1. Thermal control

Integration of the NIMS instrument into the Galileo spacecraft design involved a number of new and unique problems for a remote sensing instrument. The greatest problem was minimizing spacecraft thermal loads on the instrument and its radiator, so that it would be able to cool passively to the very low operating temperatures desired. Also, because the Galileo configuration changed several times to accommodate Shuttle and upper-stage delays and different launch opportunities, it was necessary to pay constant attention to changes which might have negative impacts on the instrument's performance.

One problem involved the location of NIMS on the despun scan platform, relative to the other remote sensing instruments. The desired location for NIMS was at the right edge of the platform so that the side-mounted radiator had a clear view towards space. In addition, the end position prevented NIMS from being sandwiched between two warm, thermally blanketed instruments, and allowed the instrument body to radiate to space both above and to the side of it. However, this desire conflicted with the Photopolarimeter/Radiometer (PPR) which also wanted the outboard platform position so that it would be able to view around spacecraft structure down to relatively small angles from the Sun. The solution was to mount the PPR on a downward extension of the scan platform so that it was to the right of, and below NIMS, satisfying the requirements of both instruments.

During spacecraft assembly it was discovered that the PPR telescope barrel extended into the field-of-view of the NIMS radiator. Located at the end of the Photopolarimeter/Radiometer (PPR) barrel was a thermally actuated hinge mechanism which would have provided a significant thermal source for the NIMS focal plane. The PPR cover was rotated so that its actuating mechanism was out of the FOV of the NIMS radiator. The small fraction of the blanketed telescope barrel still viewed by the radiator was estimated to cause an increase in focal plane temperature by approximately 0.5 K.

The greatest thermal problem for NIMS came from the two radio-isotope thermoelectric generators (RTGs) used to produce the spacecraft's electricity. Mounted on booms on the spun side of the spacecraft, each RTG radiates more than 4000 W of heat. The NIMS instrument body and radiator had clear views of the RTGs. Thermal modeling showed that when the radiator viewed the RTG's spinning through its field of view, it would be heated to at least 120 K, well beyond the operating range of InSb detectors. The solution was to implement RTG shades mounted on the booms, blocking direct views to NIMS. The shades had to be carefully designed because the RTG's depended on a clear view to space for cooling; if the shades reflected back too much thermal energy the electrical power of the spacecraft would be degraded.

Solar loads on NIMS at Jupiter were not expected to be a significant thermal source. Flight rules required that the NIMS radiator not be illuminated by the Sun prior to observation sequences. The scan platform sunshade provided protection when the platform was stowed at 0° cone angle. However, after construction it was discovered that a small slice of the NIMS radiator was not shadowed by either the radiator shield or the sunshade, when the platform was at 0° cone. The solution of this problem would nominally have been not to leave the platform parked at low cone angles. However, the spacecraft bus sunshade added for the 1989 VEEGA mission provided shading of the radiator.

Early flight experience with NIMS showed that the attention given to minimizing spacecraft thermal loads was successful. However, Venus operations revealed a new problem. Venus darkside observations by the PPR and NIMS required the instruments to 'shoot through the booms' of the spinning section, and allowed the NIMS radiator to view a significant amount of warm spacecraft structure. Because of the small heliocentric distance, some spacecraft structures were much warmer than expected. The first PPR observation, 16 hours before Venus closest approach, left the scan platform at a cone angle of 27°, giving the radiator a view of the back of the bus sunshade. Apparently, the back of the shade was warmer than predicted and the NIMS focal plane was heated from 64 K to about 95 K, beyond its desired operating temperature. Real-time commands were sent to the spacecraft, which moved the platform to a safe position, enabling the radiator to view deep space. The focal plane cooled to 86 K by the time of the first NIMS observation. Because the NIMS observations of Venus were at larger cone angles than the initial PPR observation, the radiator no longer viewed the bus sunshade, and the focal plane did not heat beyond 88 K during the critical nightside observations of Venus.

Flight rules were subsequently changed to require that the scan platform not be left at low cone angles where the radiator could view spacecraft structure, unless science observations were actually underway. Tests were also planned for calibrating the thermal load from the various spacecraft structures, using NIMS itself to measure the temperature of each spacecraft element.

6.2. Contamination control

A major concern of the NIMS experimenters was possible contamination of the instrument from various spacecraft sources. Because it was unblanketed and because it operated at such low temperatures, it was feared that NIMS would serve as a cold trap for volatiles outgassed from the spacecraft and from thruster plume byproducts. In addition, the performance of the NIMS radiator shield was closely tied to maintaining the emissivity and reflectivity of the shield surface – contamination could seriously threaten the ability to cool the NIMS focal plane.

An examination of past experience with Viking and Voyager showed several unexpected problems of this type. For example, outgassing of the Viking lander capsule was so severe that it resulted in nongravitational accelerations on the spacecraft orbit for the first six months after launch. On Voyager, thruster plume impingement on spacecraft structures resulted in a 20% inefficiency in spacecraft maneuvers.

Most alarming to the NIMS experimenters was data that came from the Infrared Thermal Mapper (IRTM) and Mars Atmospheric Water Detector (MAWD) on Viking. These instruments had a common calibration target which consisted of a small sandblasted aluminum plate, illuminated by the Sun. It was not possible for either instrument to view the target prior to deployment of the Viking landers. When they were finally used, it was found that each target had decreased in albedo by several percent, and continued to decrease over the life of the mission. Spectral data from the two instruments allowed the investigators to determine that the contaminating material was reddish in color.

Numerous sources for the contamination were considered, such as dust in Mars orbit, but were rejected. The studies concluded that the most likely contaminant was byproducts from the Viking main propulsion engine. This engine used a bipropellant combination of monomethyl-hydrazine and nitrogen tetroxide, and one of the byproducts was monomethyl-hydrazine nitrate, a dark reddish material with extremely low volatility, particularly after exposure to sunlight. Surprisingly, the IRTM/MAWD cal target did not have a direct view of the Viking main engine, and was located about 160° from the engine thrust centerline. However, plume expansion in vacuum was known to carry contaminants to locations behind the main engine, and even able to expand around spacecraft structures.

Galileo used the same type bipropellant system as Viking, not only for the main engine but also for the attitude control thrusters; Viking used a cold gas attitude control system. Extensive studies were undertaken by the Galileo spacecraft team to quantify the expected contamination from thrusters and the main engine. These studies showed that significant contamination could occur on the scan platform instruments and on the calibration targets. As a result, a number of protective measures were instituted.

The first protection was shields around the thruster clusters to prevent a direct line-of-sight to the scan platform, and a similar shield between the platform and the 400 newton main engine. Secondly, minimal contamination positions were determined for the platform for either thruster or main engine firings. The platform was commanded to the respective safe positions prior to any thruster use or maneuver. Third, the thrusters on the spinning section were limited to firing during passage through relatively narrow arcs centered approximately 90° from the despun scan platform position. Fourth, the NIMS and calibration target heaters were kept on whenever possible to prevent condensation of volatiles on the surfaces, and to sublimate any materials that had condensed. This latter measure also provided protection from spacecraft outgassing early in the mission. The primary outgassing product is water from thermal blankets.

Deployable covers were incorporated into the NIMS design for both the telescope and the radiator, fired by a single, redundant pyro device. The covers were retained until 77 days post-launch, providing protection during the initial period of spacecraft outgassing, the first two trajectory correction maneuvers, and during many other early pyro events on the spacecraft. In addition a dry nitrogen purge was provided for the instrument during ground testing and spacecraft assembly. This purge was maintained during the Shuttle launch and in Earth orbit, until a few minutes before deployment of the Galileo/IUS stack from the Shuttle bay.

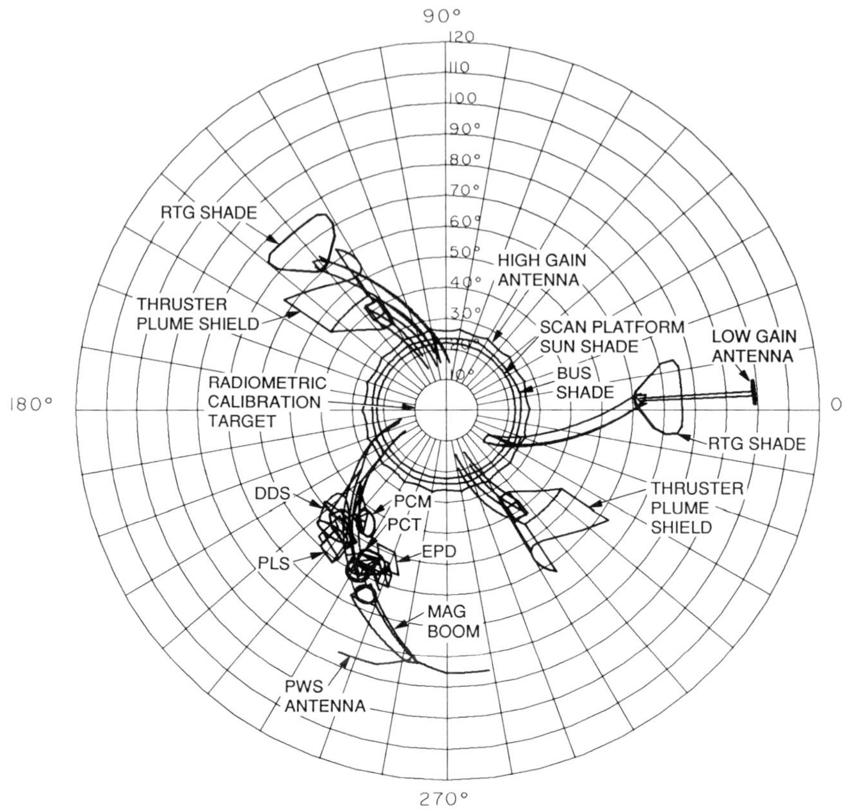

(a)

Fig. 14.

6.3. Instrument pointing

NIMS is a whisk-broom imager with one spatial dimension being provided by its scanning secondary mirror, and the other dimension provided by either spacecraft motion or movement of the scan platform. The ability to move the scan platform very smoothly at a predetermined rate relative to a target was thus very important for the success of the experiment. This was a somewhat new concept to the attitude control engineers on Galileo whose previous experience was with imaging systems that typically worked in a 'stop-and-shoot' mode.

Numerous discussions were held with spacecraft engineers to perfect the pointing requirements for NIMS and the other remote sensing instruments. In addition, discussions were held with the other remote sensing science teams to determine compatible instrument modes where all instruments could take data simultaneously. Both the UVS and the PPR were also interested in taking data in a continuous slew mode, whereas the Imaging system still preferred a 'stop-and-shoot' system. Scan platform capabilities and instrument modes were developed to accommodate both situations.

The nominal FULL MAP operating mode for NIMS called for a continuous Nyquist

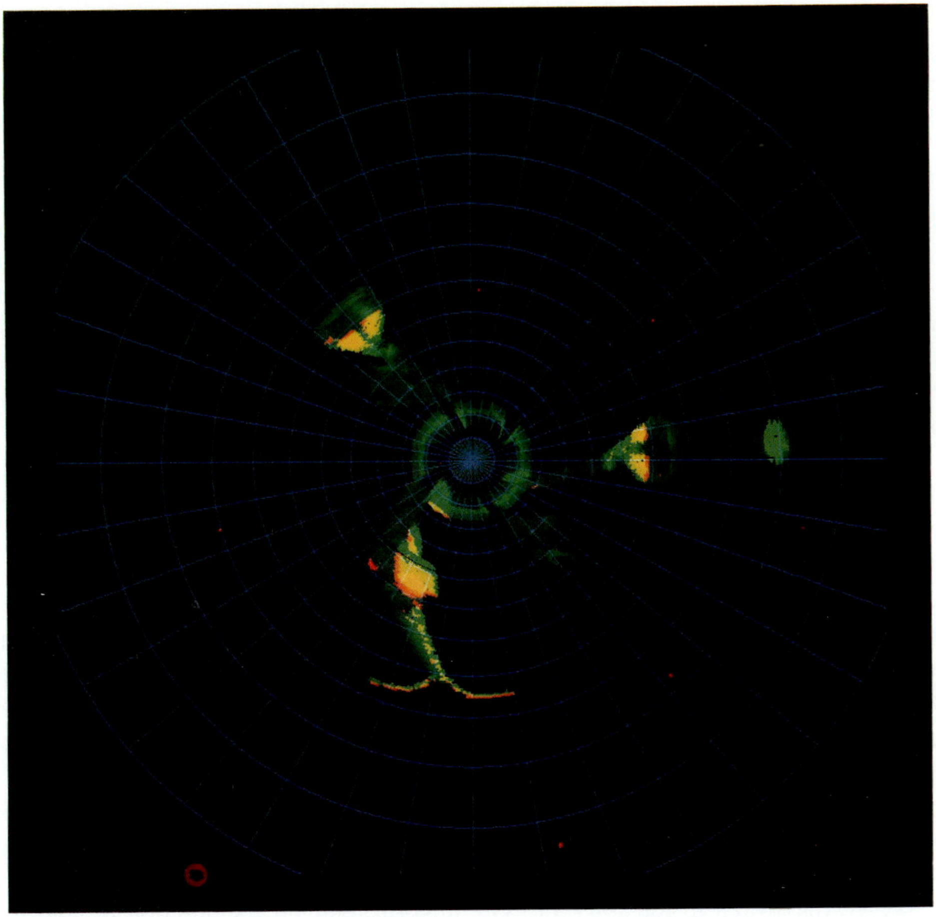

(b)

Fig. 14. Spacecraft obscuration. The obscuration and Sun glint properties of the spacecraft, as seen by the NIMS instrument, are shown above. (A) is a computer-generated plot showing the angular field which is completely clear of any obscuration within the NIMS field, taking into account the size of the telescope and the angular divergence of the field-of-view. (B) shows an experimentally measured obscuration plot, obtained by NIMS during flight. It is a combination of Sun glint, shown as red, and thermal emission, shown as green. Regions of relative warmth (~ 240 K) which also include solar glint contributions appear as yellow. Note the bright Photometric Calibration Target (PCT), which is intentionally illuminated by the Sun.

slew rate of 60 μrad s^{-1}. Other desired rates were 30 μrad s^{-1} for LONG MAP, 110 μrad s^{-1} for SHORT MAP, and 750 μrad s^{-1} for FIXED MAP. These are all relatively slow rates compared to the maximum scan platform capability of 17500 μrad s^{-1} (i.e., a degree per second). Thus, the scan platform performance was optimized for slew rates less than 3000 μrad s^{-1}. Maximum allowed deviation from a desired slew path was set at 125 μrad, or one-quarter of a NIMS pixel.

Since some satellite flybys could result in fairly large target smear rates, target motion compensation was added to the pointing system. This allowed the spacecraft attitude control system to take out relative target motion without the need for extensive ground design of sequences.

6.4. Spacecraft obscuration

The dual spinner design of Galileo led to a situation where much of the sunward hemisphere of the sky is often obscured by spacecraft structure. The obscuration of NIMS by the spacecraft is shown in Figure 14. It is not possible to synchronize instrument operation with the spinning booms, so observations that view through the spinning structures must be repeated to fill in gaps caused by boom obscuration. Although NIMS can view down to less than 30° cone angle, limited by the bus shade, obscuration by other structures is considerable at that angle.

Considerable effort was devoted to keeping the anti-sunward hemisphere completely free of obscuring structure. The only spacecraft part that extended into this hemisphere was the ends of the Plasma Wave Spectrometer (PWS) antenna, located at the end of the magnetometer boom. However, the switch to the 1989 VEEGA mission required a second low gain antenna on the aft end of the spacecraft. This was accomplished by hanging a deployable boom off one of the RTG booms, and extending down to about 114° cone angle. This additional obscuration was accepted because LGA-2 would not be needed after the second Earth gravity assist flyby, and the boom would be folded out of the way and not used again. However, Venus and Earth observations near 90° cone angle were degraded by this additional obscuration.

Careful attention was paid to minimizing sources of glint from spacecraft structures, to minimize scattered light during observations. Since most of the spacecraft was expected to be covered in black thermal blankets, this was not a very difficult problem. However, the switch to the 1989 VEEGA mission required the addition of gold foil blankets to many Sun-facing surfaces to minimize heating at the expected post-Venus perihelion distance of 0.69 AU. Integration of these changes with science concerns was not as optimal as it had been earlier in the mission, though science suggested changes were accommodated in many cases. During the Venus flyby, glint was observed from the edges of some of the spacecraft booms, however it was weak and from very narrow regions, and did not degrade any of the data.

7. NIMS Mission Design Aspects

Design of the NIMS experiment required the opportunity to view targets at optimum geometry for obtaining spectral images. For solid surfaces, this meant viewing at low phase angles, preferably less than 30° phase, where shadowing of target surfaces would be minimal. This conflicted sharply with Imaging Team desires to view surfaces at phase angles near 75° where shadowing would help to highlight surface topography. Another conflict involved the NIMS team desire to emphasize global mapping of satellite surfaces at resolutions of 25 km nimsel^{-1} or better, whereas the Imaging Team was interested in obtaining the highest possible resolution of surface features, at better than 1 km line-pair^{-1}.

The approach and departure hyperbolae to satellite flybys afforded good opportunities for global mapping by both investigations. However, close flybys, desired by both Imaging and by the fields and particles investigations, made it impossible to view the entire sub-spacecraft hemisphere. In addition, there was not sufficient time to image the sub-spacecraft hemisphere at closest approach because of the high flyby velocity and high smear rates.

The solution to many of these problems involved 'non-targeted' or 'accidental' flybys of satellites. Because of the resonant motion of the inner three Galilean satellites, the spacecraft would often fly close, within about 10^5 km, to one satellite when being targeted to a gravity assist encounter with another. If these passes occurred on the sunlit side of the satellite they provided good opportunities for global mapping at resolutions of better than 50 km nimsel^{-1}, or 2 km line-pair^{-1} for the SSI. In some ways these non-targeted flybys were even better than global mapping on the approach hyperbolae to targeted encounters, because the range to the target changed more slowly, and there was a greater possibility for viewing at relatively low phase angles.

Again, interaction with the engineers designing the satellite tour at Jupiter was an important factor in optimizing the mission for NIMS. Analysis programs were written to estimate mapping coverage for each prospective satellite tour, and these results were compared and suggestions for improvements made. This was a difficult exercise because of the limits of the trajectory changes possible using gravity assists, and because of the large number of science requirements from the different investigation teams.

Jupiter observation geometries were largely dictated by the satellite tour selected, though they were also considered in the tour design. Because the planet covered such a large solid angle near spacecraft periapsis, special sequences needed to be worked out to allow all the remote sensing instruments to cover a maximum area of the planet simultaneously.

Flyby trajectories at Venus and the Earth were optimized for gravity assist and had to be used by the science instruments without modification. Even with these restrictions, it was possible to construct valuable science sequences. These sequences proved to be very useful in conducting complete tests of the Galileo sequencing, commanding, downlink, and data reduction systems. Many problems were discovered and solved which might have not been detected until much later in the nominal mission. The value of exercising the spacecraft and instruments on real targets, prior to arrival at Jupiter, cannot be overstated.

Acknowledgements

Development of the NIMS instrument, along with incorporation of necessary spacecraft accommodations, involved the work of many people, whose unique contributions ranged from project management to craftmanship in soldering. All of these skills are necessary to NIMS' success, and it is unfortunate that we cannot cite each individual for their own contributions. We would, however, like to specifically recognize and thank members of the initial Galileo Project management team who provided encouragement and advice during the development stages of NIMS: John Casani, Ron Draper, Norm Haynes, and Bill Fawcett. This work was supported by NASA Contract NAS 7–100 to the California Institute of Technology, Jet Propulsion Laboratory.

References

Aptaker, I.: 1982a, 'Near-Infrared Mapping Spectrometer for Investigation of Jupiter and Its Satellites', *SPIE Vol. 331, Instrumentation in Astronomy IV* **182**.

Aptaker, I.: 1982b, 'A Near-Infrared Mapping Spectrometer for Investigation of Jupiter and Its Satellites', *SPIE Vol. 366, Modern Utilization of IR Technology VIII* **96**.

Aptaker, I.: 1983, 'A Near-Infrared Mapping Spectrometer for Investigation of Jupiter and Its Satellites', *SPIE Vol. 395, Advanced IR Sensor Technology* **132**.

Aptaker, I.: 1987, 'A Near-Infrared Mapping Spectrometer for Investigation of Jupiter and Its Satellites', *SPIE Vol. 834, Imaging Spectroscopy II* **196**.

Bailey, G.: 1979, 'Design and Test of the Near Infrared Mapping Spectrometer (NIMS) Focal Plane for the Galileo Jupiter Orbiter Mission', *SPIE Vol. 197, Modern Utilization of Infrared Technology* **210**.

Carlson, R. W.: 1981, 'Spectral Mapping of Jupiter and the Galilean Satellites in the Near Infrared', *SPIE Vol. 268, Imaging Spectroscopy*.

Carlson, R. W. *et al.*: 1991, 'Galileo Infrared Imaging Spectroscopy Measurements at Venus', *Science* **253**, 1541.

Garcia, R. R.: 1989, 'Dynamics, Radiation, and Photochemistry in the Mesosphere: Implications for the Formation of Noctilucent Clouds', *J. Geophys. Res.* **94**, 14605.

Hall, D. N. B., Aikens, R. S., Joyce, R., and McCurnin, T. W.: 1975, 'Johnson Noise Limited Operation of Photovoltaic InSb Detectors', *Appl. Optics* **14**, 450.

Hunten, D. M., Colin, L., and Hansen, J. E.: 1986, 'Atmospheric Science on the Galileo Mission' *Space Sci. Rev.* **44**, 191.

Kruse, P. W., McGlauchlin, L. D., and McQuistan, R. B.: 1962, 'Elements of Infrared Technology: Generation, Transmission, and Detection', Wiley and Sons, New York.

Macenka, S. A.: 1983, 'Near-Infrared Mapping Spectrometer Optical Subsystem Development and Testing', *SPIE Vol. 430, Infrared Technology IX* **313**.

Smith, D. S., Wimmers, J. T., Hermann, J. A., and Bailey, G. C.: 1982, 'Hybrid Packaging Approach to Improved Low-Noise Operation of Photovoltaic InSb Detectors', *SPIE* **331**, 29.

Sparrow, E. W. and Heinisch, R. P.: 1970, 'The Normal Emittance of Circular Cylindrical Cavities', *Appl. Optics* **9**, 2569.

Strong, J.: 1958, 'Concepts of Classical Optics', W. H. Freeman and Co., San Francisco.

Taylor, F. W. and Calcutt, S. B.: 1984, 'Near-Infrared Spectroscopy of the Atmosphere of Jupiter', *J. Quant. Spectrosc. Radiat. Transfer* **32**, 463.

Thomas, G. E., Olivero, J. J., Jensen, E. J., Schroeder, W., and Toon, O. B.: 1989, 'Relation between Increasing Methane and the Presence of Ice Clouds at the Mesopause', *Nature* **338**, 490.

Torson, J. M.: 1989, *Interactive Image Cube Visualization and Analysis*, Proceedings Chapel Hill Workshop on Volume Visualization, Chapel Hill, North Carolina, May 1989.

Weidner, V. R. and Hsia, J. J.: 1981, 'Reflection Properties of Pressed Polytetrafluoroethylene Powder', *J. Opt. Soc. Am.* **71**, 856.

Yeates, C. M., Johnson, T. V., Colin, L., Fanale, F. P., Frank, L., and Hunten, D. M.: 1985, *Galileo: Exploration of the Jupiter System*, NASA SP-479, National Aeronautics and Space Administration, Washington, D.C.

GALILEO ULTRAVIOLET SPECTROMETER EXPERIMENT

C. W. HORD, W. E. McCLINTOCK, A. I. F. STEWART, C. A. BARTH, L. W. ESPOSITO, G. E. THOMAS,

Laboratory for Atmospheric and Space Physics, University of Colorado at Boulder, Boulder, CO 80309-0392, U.S.A.

B. R. SANDEL, D. M. HUNTEN, A. L. BROADFOOT, D. E. SHEMANSKY,

Lunar and Planetary Laboratory, 901 Gould-Simpson Building, University of Arizona, Tucson, AZ 85721, U.S.A.

J. M. AJELLO, A. L. LANE, and R. A. WEST

Jet Propulsion Laboratory, California Institute of Technology, Pasadena, CA 91109, U.S.A.

Abstract. The Galileo ultraviolet spectrometer experiment uses data obtained by the Ultraviolet Spectrometer (UVS) mounted on the pointed orbiter scan platform and from the Extreme Ultraviolet Spectrometer (EUVS) mounted on the spinning part of the orbiter with the field of view perpendicular to the spin axis. The UVS is a Ebert-Fastie design that covers the range 113-432 nm with a wavelength resolution of 0.7 nm below 190 and 1.3 nm at longer wavelengths. The UVS spatial resolution is 0.4 deg × 0.1 deg for illuminated disc observations and 1 deg × 0.1 deg for limb geometries. The EUVS is a Voyager design objective grating spectrometer, modified to cover the wavelength range from 54 to 128 nm with wavelength resolution 3.5 nm for extended sources and 1.5 nm for point sources and spatial resolution of 0.87 deg × 0.17 deg. The EUVS instrument will follow up on the many Voyager UVS discoveries, particularly the sulfur and oxygen ion emissions in the Io torus and molecular and atomic hydrogen auroral and airglow emissions from Jupiter. The UVS will obtain spectra of emission, absorption, and scattering features in the unexplored, by spacecraft, 170-432 nm wavelength region. The UVS and EUVS instruments will provide a powerful instrument complement to investigate volatile escape and surface composition of the Galilean satellites, the Io plasma torus, micro- and macro-properties of the Jupiter clouds, and the composition structure and evolution of the Jupiter upper atmosphere.

1. Introduction

The Galileo Ultraviolet Spectrometer investigation will use data obtained by two instruments. The Ultraviolet Spectrometer (UVS) covers the wavelength range from 113 to 432 nm and was the original instrument selected for the Galileo Orbiter. The Extreme Ultraviolet Spectrometer (EUVS) was added to the Orbiter payload after the Challenger accident in 1986. A joint effort by the Universities of Colorado and Arizona culminated in the EUVS instrument, a modified version of the Voyager Ultraviolet Spectrometer, covering the wavelength region from 54 to 128 nm.

Primary objectives of the investigation are to understand physical processes occurring in: (1) the upper atmosphere of Jupiter; (2) the Io plasma torus; (3) the volatile gases escaping from Galilean satellites. The surface scattering properties of Jupiter's moons will be investigated. The spectral features of H and H_2 in the Jupiter aurora and airglow investigated by the Voyager Ultraviolet Spectrometer form a solid point of departure for the investigation of the composition and thermal structure of the upper atmosphere. Sulfur and oxygen ions discovered by Voyager provide the tracers to understand the time evolution of volcanic gases from Io in their transit through the plasma torus and ultimate

Space Science Reviews **60:** 503-530, 1992.

deposition in the Jupiter auroral zone. Spectra from the International Ultraviolet Explorer and rocket-borne spectrometers, along with data from the UV channel of the Voyager Photopolarimeter have revealed photochemical products of CH_4 and auroral zone darkening phenomena that will be addressed by emission and absorption spectroscopy in the 115–340 nm wavelength region of the UVS. The new, longer wavelength region available on Galileo will provide the opportunity to study the micro- and macroscopic properties of the cloud particles in the atmosphere of Jupiter. Satellite surface scattering properties will be explored in concert with the Near Infrared Mapping Spectrometer (NIMS) investigation to understand the evolutionary history of these surfaces.

2. Scientific Objectives

2.1. Interplanetary medium

The interplanetary medium is supplied with atoms of hydrogen and helium by the interstellar wind. Both species are detectable by the UVS through their resonance scattering of solar photons at 121.6 and 58.4 nm (Thomas, 1978). Gravitational focussing by the Sun produces a cone of enhanced helium abundance downwind of from the Sun; the density and shape of this cone reflects the velocity distribution of atoms in the ISW. In contrast, a cavity in the hydrogen medium is created by charge exchange with solar wind protons, and the atoms of hydrogen experience a repulsion due to radiation pressure that is comparable to solar gravity. The size and shape of the cavity depends on variables such as the solar wind, Lα and EUV fluxes (all of which vary with solar longitude and latitude), as well as the characteristics of the ISW itself. Processes occurring at the heliopause may also affect the cavity. By carrying out a systematic program of H and He measurements over the course of the mission, UVS will improve our knowledge of the ISW and of the processes that affect its passage through the solar system. Of interest early in the mission will be the search for evidence of an inner-solar-system source of hydrogen, perhaps from subvisual comets. Observations by UVS and EUV should be a more sensitive search than those of the Voyager UVS, from which Hall and Shemansky (1988) have derived an upper limit to the possible Lα signal.

2.2. Venus

Very little is known about the EUV spectrum of Venus, while the UV spectrum has been extensively studied. EUV measurements will tell us about the ratio of O^+ to O and H. The geometry of the Galileo flyby will permit pole-to-pole and dawn-to-dusk measurements of the abundance of SO_2 in the cloud-top region, and of the abundances of O and CO in the thermosphere. The latter quantities are determined by the balance between global-scale solar-driven thermospheric wind systems and vertical transport induced by the dissipation of tides and gravity waves. The wind systems carry atoms of oxygen and nitrogen to the night side, where they undergo chemiluminescent recombination to form O_2 and NO (Lawrence *et al.*, 1977; Stewart and Barth, 1979). The broad wavelength coverage of UVS will allow the simultaneous measurement of the resulting NO and O_2 emissions, whose distribution over the night side reflects the high- and low-altitude components of these winds. The altitudes of the emitting layers are also of dynamical interest. In the far UV, the altitude of emissions from O will constrain the energy of auroral electrons that presumably derive either from the daytime topside ionosphere or from the solar wind itself.

2.3. Earth and Moon

Although the UV and even the EUV emissions from the Earth have been extensively studied, the two Galileo flybys offer opportunities to address unanswered questions. As on Venus the post-encounter passage near the subsolar point at long range allows the near-simultaneous measurement of pole-to-pole and dawn-to-dusk variations in the UV airglow and in reflected sunlight. From the former we can investigate the global O/N_2 ratio through their respective UV emissions, while from the latter we can investigate the distribution of ozone. It is also of interest to establish the Earth's UV albedo in the Schumann–Runge band region near and below 200 nm; the absorption of solar photons in this spectral range produces photodissociation in the mesosphere and is an important energy source for the photochemistry of that region.

A search for a tenuous lunar atmosphere using the resonance emissions of H, O, and OH will address the question of the rate of bombardment of the Moon by small bodies, and of the fate of solar wind protons that strike the surface. Previous searches of this kind have had negative results (Fastie *et al.*, 1973), but the possibility of episodic events remains. The flybys also allow the Earth–Moon system to be mapped, and these data will also contain an image of the hydrogen geocorona from a unique sunward vantage point.

2.4. Asteroids

The UVS will measure the albedo of the asteroids Gaspra and Ida during flyby. Spatial resolution on the bodies' surfaces will not be possible, but their scattering properties as a function of phase angle will be measured, and the presence of absorption features at wavelengths longer than 200 nm will be determined. At these and at shorter wavelengths (down to Lα) the asteroid's albedo may be directly compared to that of the Moon measured during the two Earth encounters.

2.5. Jovian atmosphere

2.5.1. *Clouds and Hazes*

The Galileo orbiting mission offers the opportunity to observe Jupiter's clouds and hazes repeatedly over a wide range of phase angle and wavelength; the imaging and mapping capabilities of the optical instruments will allow these studies to be made on individual cloud features as well as on planetary-scale features. Because its ability to examine small scattering angles is restricted by solar protection considerations, the contribution of UVS will be to determine the imaginary parts of the aerosols' refractive indices by obtaining the single-scattering albedo from photometric measurements. It will sample the lower end of the aerosol size distribution due to its sensitivity down to 200 nm. The distribution of aerosols with altitude will be measured in the stratosphere by measuring limb radiance profiles and in the troposphere by making nadir-to-limb scans.

Regarding time variations, optical observations from Earth, Earth orbit, and flyby missions indicate great changes on many spatial and time scales (Beebe *et al.*, 1989). No clear seasonal patterns have been identified, but changes in the color, structure, and albedo of belts and zones and of smaller features have been seen on timescales of years and sometimes months. Galileo's tour designs favor observations of the morning hemisphere, and offer the opportunity for detailed spectroscopic and photometric studies at timescales ranging from days to the duration of the mission (about two years, less than a complete Jovian season).

2.5.2. *Stratospheric Composition and Chemistry*

Jupiter has a reducing atmosphere whose photochemistry is dominated by the photolysis of methane (Strobel, 1975). Its derivatives, C_2H_6, C_2H_4, and C_2H_2 are removed by vertical transport. C_2H_2 (acetylene) initiates a reaction chain leading to aerosol formation, and these aerosols are also removed by transport. Methane and ethane (C_2H_6) are detectable in UV reflectance by broad absorption continua below 160 nm, and C_2H_4 and C_2H_2 have characteristic strong absorption bands in the same region. UVS will use reflectance spectroscopy during disc and limb scans to compile an inventory of these hydrocarbons as a function of location and altitude; the models of photochemistry, vertical transport, and aerosol formation can be tested against these data, and horizontal variations will indicate the zonal and meridional variations in vertical transport processes. UVS limb scans will yield stratospheric temperatures through the scale height of the signal from Rayleigh scattered sunlight. These scale heights are a factor in the retrieval of vertical transport parameters, and the temperatures may affect the photochemistry.

Stratospheric temperatures show considerable variations (Beebe *et al.*, 1989). At the time of the Voyager flybys, which occurred near northern autumnal equinox, stratospheric temperature contrasts of order 10 K were noted between the northern and southern hemispheres. Galileo will observe Jupiter near northern vernal equinox, and a reversal of the contrast, which has been attributed to the greater summer insolation in the northern summer due to proximity to perihelion (Caldwell *et al.*, 1979), will be looked for. The Voyagers also noticed a 2 K equatorial warming over the 4 months between the flybys (Geirasch *et al.*, 1986) such variations will be readily studied by Galileo, whose mission will consist of orbits with periods between 1 and 4 months.

2.6. THERMOSPHERE

The thermosphere of Jupiter is characterized by unexpectedly high temperatures (of order 1100 K in the upper thermosphere) and by unexpectedly bright UV emissions from molecular hydrogen. Lα emission from H shows an equatorial bulge that sometimes extends across the morning terminator. None of these phenomena have been totally explained. Diffuse fluxes of soft electrons have been proposed (the 'electroglow' concept) but the existence, the energy source, and the need for such electrons all remain controversial (Clarke *et al.*, 1989a). The ionospheric electron density profiles measured by radio occultation techniques from the Pioneer and Voyager spacecraft are also poorly understood (Eshleman *et al.*, 1979). Emission, heating, and ionization are slightly linked processes, and a careful study of spectral, horizontal, vertical, and diurnal and other time variations is an important objective for the Galileo UVS and EUV experiments.

2.7. Aurora

The Jovian aurora has been observed at X-ray, EUV, UV, and IR wavelengths (Clarke *et al.*, 1989a). There is considerable uncertainty about the nature of the primaries, and indeed there may be several different types of aurora, produced by the precipitation of energetic electrons, protons, and heavy ions (oxygen and sulfur from the Io torus). In the UV the dominant emissions are H Lα and the Werner and Lyman bands of H_2. The spectrum of the H_2 bands clearly show the effects of absorption by acetylene, indicating that the primaries penetrate at least as deep as the homopause (Gladstone and Skinner, 1989). The Lα emission shows none of the Doppler-shifted component to be expected from precipitating protons or hydrogen atoms (Clarke *et al.*, 1989b). The X-rays may be bremsstrahlung or emissions from torus ions (Metzger *et al.*, 1983). The IR emissions are attributed to CH_4, C_2H_2, and other trace gases; they indicate enhanced temperatures in auroral 'hot spots'. These hot spots are located on the traditional 'auroral zone' lying at the feet of magnetotail field lines, whereas the UV and EUV emissions are observed at the feet of lines intersecting the Io torus. In both cases the emissions are preferentially seen near 180° System III longitude. In the southern hemisphere, there is a broader clumping of UV emissions around 0° longitude.

Galileo's mostly equatorial orbits mean that the aurora will be observed near the northern or southern limbs, allowing excellent longitudinal resolution at the cost of lesser latitude resolution. The spectral effects of atmospheric absorption will be enhanced. Jupiter's rapid rotation will facilitate the determination of longitudinal dependences of the emissions on each orbit. The possibility of correlations between the aurora and conditions in the Io torus will be explored.

2.8. Satellites

While close-range observations of Io and Europa will be prevented by the radiation environment, the outer two Galilean satellites will be visited many times in close (from a few hundred km up) encounters. Io's surface is dominated by the effects of its vulcanism, and it appears to possess an atmosphere of sulfur dioxide. Europa has a young, icy, frequently resurfaced exterior, marked by light and dark lineaments. Callisto's surface is dark and heavily cratered with very little evidence of volcanic activity. Ganymede has a surface part of which resembles Callisto's and part of which is light-colored and extensively marked by extensional events. The entire surface shows systems of grooves.

The Galileo UVS will measure and map the UV albedos of these moons. The measurements will be compared with those of the Moon and of the asteroids Gaspra and Ida. The rich variety of surface terrain and materials will greatly expand our knowledge of the UV scattering properties of satellite surfaces. The UVS will also look for evidence of tenuous and possibly sporadic atmospheres that might be produced by sublimation, sputtering by co-rotating plasma, or even eruptive events.

2.9. Io Torus

The torus of Io is produced by the ejection of material from Io, its subsequent ionization mostly by charged particle bombardment, and the pickup of the resultant ions by the rotating magnetic field of Jupiter (Strobel, 1989). Prior to ionization, the neutral material leaving Io forms a 'neutral cloud' under the influences of Jovian gravity and, for some species, radiation pressure. Conditions in the torus thus reflect the surface and atmospheric composition of Io and the nature and efficiency of the escape and ionization processes, as well as the complex interactions of the ionized material with the magnetic and gravitational fields of Jupiter and with the rest of the magnetosphere. Information on these materials and processes can be gained by measuring the abundance and distribution of the neutral and ionized species. The EUV instrument will perform this task for the ionized states of sulfur and oxygen, whose origin is ultimately the venting of sulfur dioxide from Ionian volcanoes. The UV instrument will measure neutral sulfur and oxygen. The relative abundances of the neutral and ionized states of these gases reflect local plasma density and especially the electron temperature. The EUV will observe the torus on inbound orbital legs, for a period equal to several Jovian rotations and comparable to a revolution of Io; thus the data will reveal many dynamical aspects of the torus in addition to its composition. Spatial resolution will be best at the sunward ansa which will be viewed from about 20 R_J on the early orbits. The absolute brightness of the emissions, especially those from ionized sulfur and oxygen in the EUV, are of great importance because the emissions are a major cooling process for the heated torus plasma, whose energy balance is not fully understood (Strobel, 1989). The range of variability in torus conditions with time is also inadequately explored; the combination of an approximately two-year timebase and high spatial resolution offered by Galileo will be invaluable.

2.10. Magnetosphere

There are many processes in the exosphere of Jupiter, on the constantly irradiated satellites, in the Io torus, and in the magnetosphere in general, that might provide sources of neutral atoms in the magnetosphere. We might expect to find hydrogen and even OH in addition to oxygen and sulfur. The UV instrument will search for such material at times when the radiation noise in the instrument is at a minimum.

2.11. Joint investigations

Collaborative studies are planned with the fields and particles investigators. Our goal is to improve our understanding of the transportation of sulfur and oxygen ions from the Io plasma torus to their ultimate precipitation in the Jupiter auroral region. The *in-situ* particle measurements will describe the ion and electron populations in the Io torus. Ultraviolet emissions from torus ions, and possibly neutrals, in the 55–430 μm range provide temperature, radiative energy loss, and compositional information that will help constrain physical conditions of the torus. *In-situ* measurements of the torus can be made on only the two passages of the spacecraft through the torus region. Ultraviolet data can obtain torus data on every orbit of Jupiter extending and measuring variability over the full length of the mission. Ultraviolet measurements of auroral emissions from atomic and molecular hydrogen will depend on the knowledge of the quantity and energy distribution of electrons impacting the upper atmosphere of Jupiter.

Joint investigations with the Photopolarimeter Radiometer experiment will help define the particulate properties of the Jupiter atmosphere. Ultraviolet brightness measurements at 1.3 μm resolution in the 200–320 μm wavelength region will complement visible and near-infrared polarimetric measurements. Measurements of brightness as a function of emission, incidence and phase angle will define the effective particle phase function over a broad wavelength range, providing constraints on cloud particle size, shape, and composition. Complementary ultraviolet spectrometer and photopolarimeter observations will provide information about the spatial extent and altitude distributions of these clouds.

Properties of the satellite surfaces will be measured in cooperation with the Near Infrared Mapping Spectrometer, the Photopolarimeter Radiometer, and the Solid State Imaging Instrument. The Ultraviolet and Infrared spectrometers provide extended wavelength coverage from 55 μm up to nearly 5 μm. Scattering properties as well as ultraviolet absorbers, e.g., sulfur dioxide, will be measured to add leverage to our understanding of the Galilean satellites.

3. UVS Instrument Description

The Galileo Ultraviolet Spectrometer (UVS) consists of a Cassegrain telescope and a Ebert–Fastie scanning spectrometer (see Figures 1 and 2 and Table I). Spectral scanning is accomplished using a fully programmable diffraction grating drive. Three separate photomultiplier detectors, located in the exit focal plane of the spectrometer, are used to cover the entire ultraviolet-near-visible spectrum from 113 to 432 nm. Spectral scanning, instrument command and control, data formatting, and spacecraft interface are all controlled using an RCA Sandia CDP 1802 microprocessor within the instrument.

3.1. Telescope

The optical design for the UVS telescope is a Dall–Kirkham configuration (elliptical primary mirror and spherical secondary mirror) with an effective focal length of 250 mm and a focal ratio of $f/5$. In order to measure accurate limb profiles, the telescope has been equipped with an external sunshade and an extensive baffle system for rejection of off-axis scattered light. The field of view is wavelength-dependent, being limited by

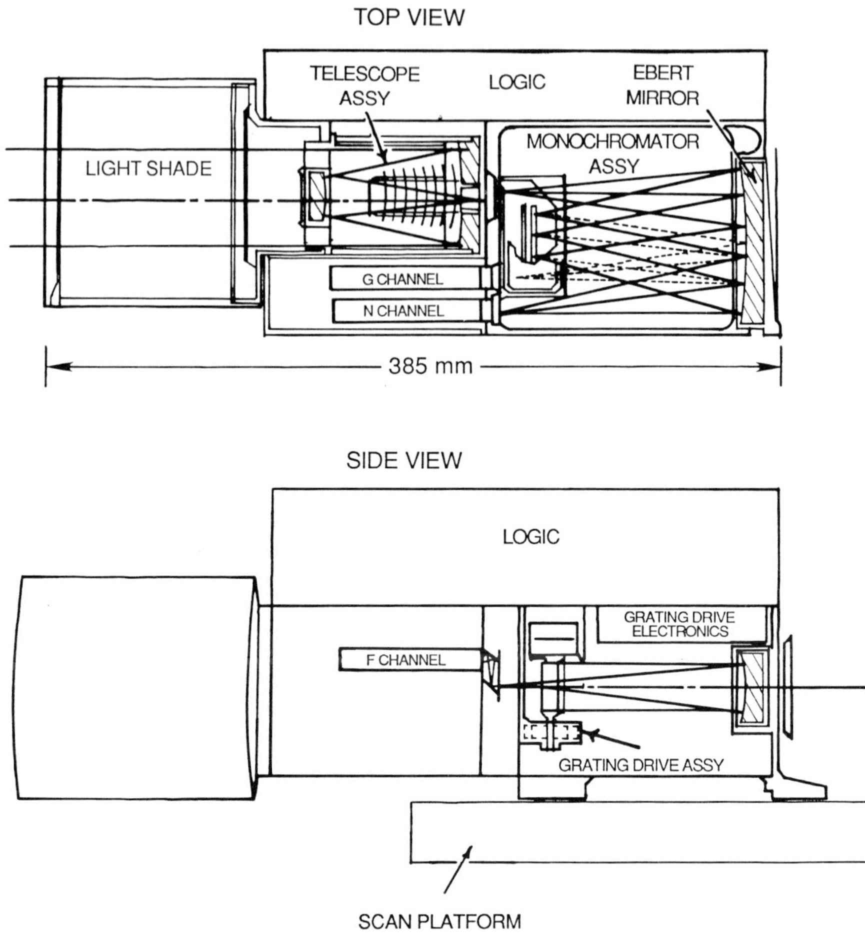

Fig. 1. UVS Mechanical Optical Configuration. The top panel is a view of the UVS looking down onto the spacecraft scan platform. Solid lines superimposed on the instrument figure show the light path through the instrument to the near-ultraviolet-visible detector (*N* channel). In order to reach the far-ultraviolet detector (*G* channel) light travels on a different path from the grating to the detector assembly as shown by the dotted line. The field of view of the telescope is 0.1° in this plane. The bottom panel shows a side view of the instrument mounted on the scan platform. The solid line represents the light path from the grating to the near-ultraviolet detector (*F* channel). In this plane the telescope fields of view are 1° for the *G* and *N* detectors and 0.4° for the *F* detector.

the spectrometer entrance slit to 1° × 0.1° for two of the detectors (113 to 192 nm) and (282 to 432 nm) and by one of the spectrometer exit slits to 0.4° × 0.1° for the third detector (162 to 323 nm).

A bright object sensor (limb sensor) with a 1.5° FWHM field of view located below the telescope sunshade structure is used to protect the long wavelength detector during atmospheric limb observations.

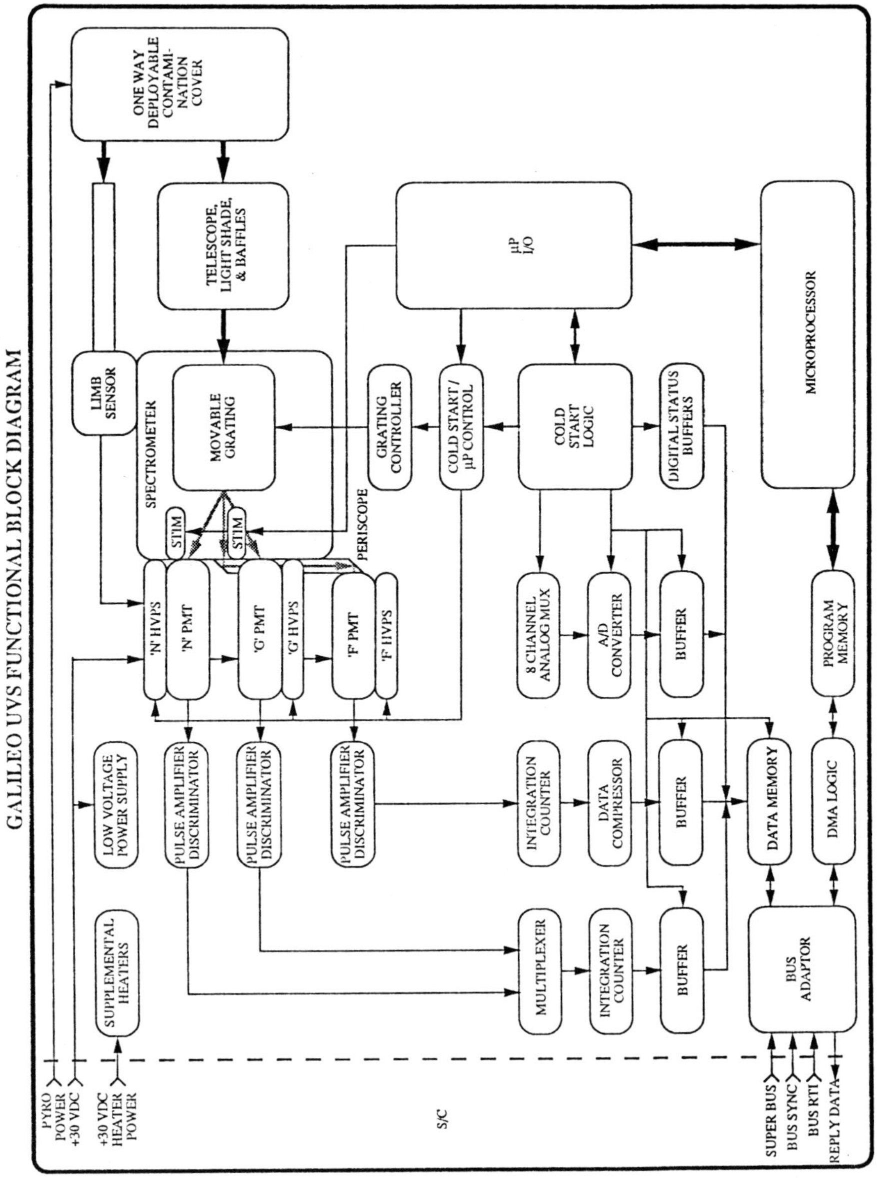

Fig. 2. UVS functional block diagram.

TABLE I
Summary of UVS characteristics

Telescope	
Focal length	250 mm
Focal ratio	*f*/5
Aperture	50.3 mm × 52.8 mm
Unobscured area	13.89 cm^2
Spectrometer	
Focal length	125 mm
Grating	
Ruling	2400 lines per mm
Blaze angle	16.75 deg
Detectors	
G channel	EMR 510G-09 CsI photocathode
F channel	EMR 510F-06 CsTe photocathode
N channel	EMR 510N-06 KCsSB photocathode
Nominal wavelength range	
G channel	113.3–192.1 nm 2nd order
F channel	162.0–323.1 nm 1st order
N channel	282.0–432.0 nm 1st order
Nominal resolution	
G channel	0.67 nm
F channel	1.36 nm
N channel	1.27 nm
Field of view	
G and *N* channels	0.1 × 1 deg
F channel	0.1 × 0.4 deg
Exit slit solid angle	
G and *N* channels	3.05×10^{-5} steradians
F channel	1.20×10^{-5} steradians
Instrument	
Mass	5.2 kg
Power	2.4 W

3.2. Spectrometer

The spectrometer is a standard, 125-mm-focal-length, Ebert–Fastie design which uses a single spherical mirror as both collimator and camera and a plane diffraction grating. A ruling density of 2400 grooves per mm provides a first-order dispersion of 2.9 nm per mm and an average spectral resolution of 200 for a 0.43-mm-wide entrance slit (0.1° telescope field of view).

Three EMR Photoelectric Corp. 510 photomultiplier tubes, located behind three separate exit slits in the focal plane of the spectrometer record the spectrum in three overlapping wavelength ranges. Each detector has its own high voltage power supply and pulse counting electronics, allowing for independent operation. Photocathodes and windows for the detectors were chosen to optimize measurements in narrow spectral ranges. The far-ultraviolet detector (*G* channel) is equipped with a MgF_2 window and a cesium iodide photocathode resulting in a solar blind detector with high sensitivity in the wavelength range 113 to 192 nm. The middle-ultraviolet detector (*F* channel) is equipped with a quartz window to block radiation below 160 nm and a cesium telluride photocathode to suppress its response to radiation from wavelengths longer than 350 nm. The near-ultraviolet-visible detector (*N* channel) is equipped with a quartz window and a bi-alkali photocathode and is sensitive to radiation in the wavelength range 160 to 450 nm.

All three detectors are mounted in a single mechanical housing along with their high voltage power supplies and pulse-amplifier-discriminators. The *G* and *N* photomultipliers are located directly behind their respective exit slits in the spectrometer housing. Volume constraints require that the *F* photomultiplier be mounted above the slit plane and light is directed to it by a small two mirror periscope located behind the *F* channel exit slit (see Figure 1).

Spectral scanning is accomplished by rotating the diffraction grating. The UVS grating drive uses a moire fringe pattern, generated by overlaying two radially etched transmission gratings, to control the angular position of the grating. One of the transmission gratings is fixed, and the other rotates with the diffraction grating housing. The fringe cycles are detected using two solid state detector-emitter pairs, located relative to the radial gratings so that one pair detects a sine modulation and the other detects a cosine modulation. The sine and cosine modulated signals are combined with a 10 kHz reference signal $\sin(\omega t)$ to produce a Control Phase Signal (CPS) $\sin(\omega t + a)$ (see Figure 3). In order to rotate the grating, the control electronics generates a second

GRATING DRIVE FUNCTIONAL BLOCK DIAGRAM

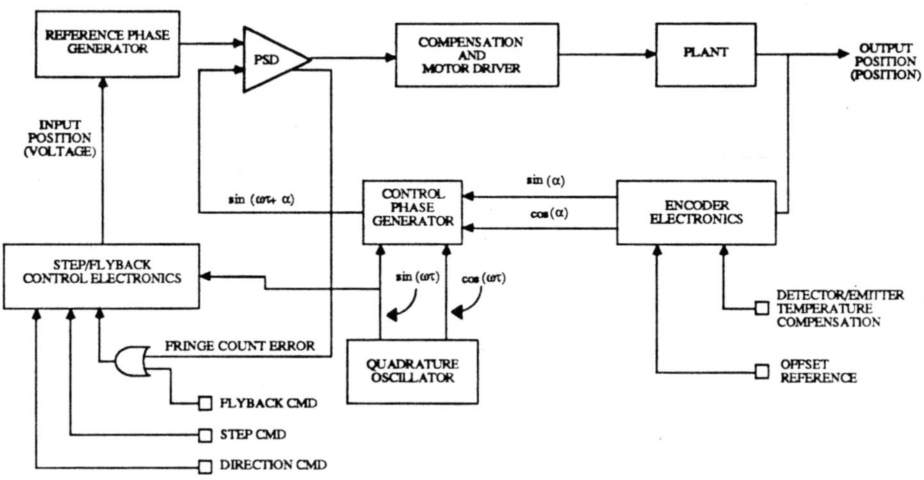

Fig. 3. UVS grating drive functional block diagram.

10 kHz Reference Phase Signal (RPS). The phase of the RPS is shifted in 64 equal steps through 2π and is used as a step input to grating drive control system. After the phase of the RPS is incremented, it is compared to the CPS using the Phase Sensitive Detector

(PSD) electronics. The PSD electronics produce an error signal which is proportional to the phase difference between the RPS and the CPS. This error signal is applied to the motor drive to null the difference in the two signals resulting in a step rotation of the diffraction grating. Since the grating drive is an incremental, rather than absolute, encoding system, the PSD electronics must also generate a fringe count error signal to insure that the system does not control on an adjacent fringe.

The transmission gratings have a ruling of 1500 lines per 360° rotation resulting in a single cycle of 0.024° and a single phase increment step size of 0.00375°. Each grating step for the UVS is a sum of six phase increment steps or 0.0225°. Thus a grating step results in a 0.1-mm displacement of the spectrum in the spectrometer focal plane so that the spectrum is sampled on the average of 4 times per spectral resolution element.

3.3. Microprocessor and Bus Adaptor

The UVS uses an RCA 1802 CMOS microprocessor for command parsing, spacecraft time recognition and synchronization, and instrument control. In addition, the UVS design incorporates additional electronics called the Cold Start Logic (CSL) that places it into a cyclical $F-G$ scan mode (see below) until microprocessor control is initiated by spacecraft command.

The microprocessor receives commands and spacecraft timing information via the Bus Adaptor and associated Direct Memory Access (DMA) logic.

The Bus Adaptor serves as the bi-directional interface between the Galileo spacecraft and the UVS. Its circuitry serves to isolate the UVS electrically from the spacecraft and to allow for 8-bit information flow to and from the UVS.

3.4. Modes of Operation

Modes of operation for the UVS are designed to be compatible with a data rate of 1 kbits per second and with the basic spacecraft major frame time period of $60\frac{2}{3}$ s. For the UVS operation each major frame is subdivided into 14 UVS 'science frames' each of $4\frac{1}{3}$ s duration.

The basic data frame for the spectrometer is a detector counting integration time (accumulation) followed by a single grating step. Each data frame is fixed in length to 7.6 ms with 1.6 ms allocated to step the grating. Output pulses from the G and N detectors are accumulated in an 8-bit counter and then routed to a circular output buffer. Output pulses from the F detector are accumulated in a 16-bit counter and then log compressed to 8 bits before routing to the circular output buffer. In order to minimize spectral smearing caused by step and settling time when the grating drive mechanism changes positions, only one step is allowed between successive data frames. Data from a single UVS science frame consists of 18 8-bit words of engineering data and 528 words of science data which are acquired in a 4-s period. The final $\frac{1}{3}$ s of each science frame is used for large motions of the grating drive; either fly-back to the zero-step position or step to a different wavelength to initiate a mini-scan (see below).

3.5. Scan modes

The UVS software supports two distinct operations modes: (1) 'Scan' mode, and (2) 'Fixed Wavelength Mini-Scan' mode.

When the UVS is commanded into any of the Scan modes, the instrument starts integrating at one end of the spectrum for a single detector, and steps the grating (one step per 7.6 ms) to the other end of the detector's spectral range, using 528 steps over a 4-s science frame.

In any given Scan mode, either one or two spectral regions can be covered. If only one region is desired then the microprocessor steps the grating up the wavelength scale during even-numbered Science Frames and down on odd-numbered Science Frames. If two regions are selected, for example the N and G channels, then the microprocessor scans up the N channel on even-numbered Science Frames and down the G channel on odd-numbered Science Frames. Any single or combination of two detectors can be used for a total of six scan modes.

3.6. Fixed-wavelength mini-scan modes

These modes were designed to perform detailed studies of various features in the spectrum. For this set of modes, only one detector may be selected for investigation per command. Within the spectral range for the selected detector, from one to four separate spectral regions can be scanned in a sawtooth pattern. The microprocessor will position the grating drive to the bottom of the first spectral region, and will then perform the mini-scan sawtooth motion for a programmable number of steps up and down for 4 s. During the $\frac{1}{3}$-s dead time the grating will be positioned at the bottom of the next feature. This feature will then be mini-scanned for 4 s. The process will continue in a 1–2–3–4–4–3–2–1 pattern until receipt of a new command (note that if only two features are commanded, the sequence is 1–2–2–1, etc.).

3.7. Calibration and laboratory performance

The instrument parameters which most directly influence the proper evaluation of science data returned by the UVS are: instrument absolute sensitivity, spectrometer wavelength scale and band pass, telescope scattered-light rejection, and instrument polarization. For an extended source the UVS count rate can be expressed as

$$C = I\omega A Q_e \tag{1}$$

where I is the source radiance, ω is the angular field of view of the spectrometer exit slit, A is the telescope collecting area, and Q_e is the instrument absolute efficiency.

If I is measured in units of 10^6 photons cm^{-2}/4π steradians s^{-1}, then C is expressed in counts per Rayleigh.

Instrument absolute sensitivity was measured using source standards which are traceable to the National Institute for Standards and Technology (NIST). For wavelengths greater than 200 nm an NIST deuterium lamp (200–350 nm) and an Optronics Laboratories Inc. FEL tungsten-halogen lamp (250–432 nm) were used to illuminate a barium sulfate scattering screen. Because the screen closely approximates a Lambertian radiator, its radiance (photons per second per unit area per unit solid angle) could be calculated directly from the lamp flux (photons per unit area per second) impinging on it. Observation of the scattering screen resulted in a direct measure of the instrument response to an extended source which fills the telescope field of view.

For wavelengths in the range 115–310 nm the instrument sensitivity was measured in the University of Colorado's Laboratory for Atmospheric and Space Physics (LASP) vacuum calibration facility (McClintock *et al.*, 1982) using detector standards traceable to the NIST. The instrument was illuminated with a nearly monochromatic collimated beam of photons from a laboratory spectrometer-telescope. The flux of photons impinging on the instrument aperture was measured by scanning the input beam with a pulse-counting photomultiplier tube which was calibrated using an NIST standard photodiode. The ratio of instrument counts per second to input beam flux measured the product of the UVS telescope area and the instrument absolute quantum efficiency, AQ_e. Because the input beam is highly collimated, it was necessary to scan the 'point' object along the 1° UVS entrance slit and fold that spatial response into the absolute sensitivity measurement made with the point source at slit center to measure the effective value of ω. Two sets of measurements were averaged: one taken with the UVS grating and the laboratory spectrometer grating grooves parallel, and one with the grooves perpendicular in order to avoid polarization-induced changes in efficiency.

Independently from the source and detector standards, relative sensitivity measurements for 115–220 nm were made using an electron lamp (Ajello *et al.*, 1988). Using both H_2 and N_2 as source gases allowed the relative shape of the lamp spectrum to be calculated to about 10%. Figures 4 and 5 show sample spectra of H_2 and N_2 taken with the UVS *G* and *F* channels, respectively.

Figure 6 shows results synthesized from all three calibration techniques. The ordinate is the sensitivity *C* in counts s^{-1} per Rayleigh for a monochromatic source. If the source radiance is expressed in Rayleighs per nanometer, then the results must be multiplied by 1.27, 1.36 and by 0.67 for the *N*, *F*, and *G* channels, respectively. Estimated accuracies in the absolute calibrations are 15% for the *F* and *N* channels and 20% for the *G* channel.

The pulse-counting electronics for the detectors have a dead time which inhibits counting of photo-events. Dead-time corrections were measured for each detector using a stable lamp and a set of precision pinholes. Within the accuracy of the measurements the number of true counts per second is related to the number of observed counts per second and can be calculated from the relation

$$Ct = \frac{C_0}{1 - C_0\tau}$$

where $\tau = 6.25 \times 10^{-7}$ s. The relation is valid within 10% for $C_0 \leq 5.2 \times 10^5$ counts s^{-1}.

Spectrometer wavelength scale and band pass were measured by observing a scattering screen illuminated by mercury and platinum emission line sources. Wavelengths determined by a best fit to the grating equation agree with published values to within

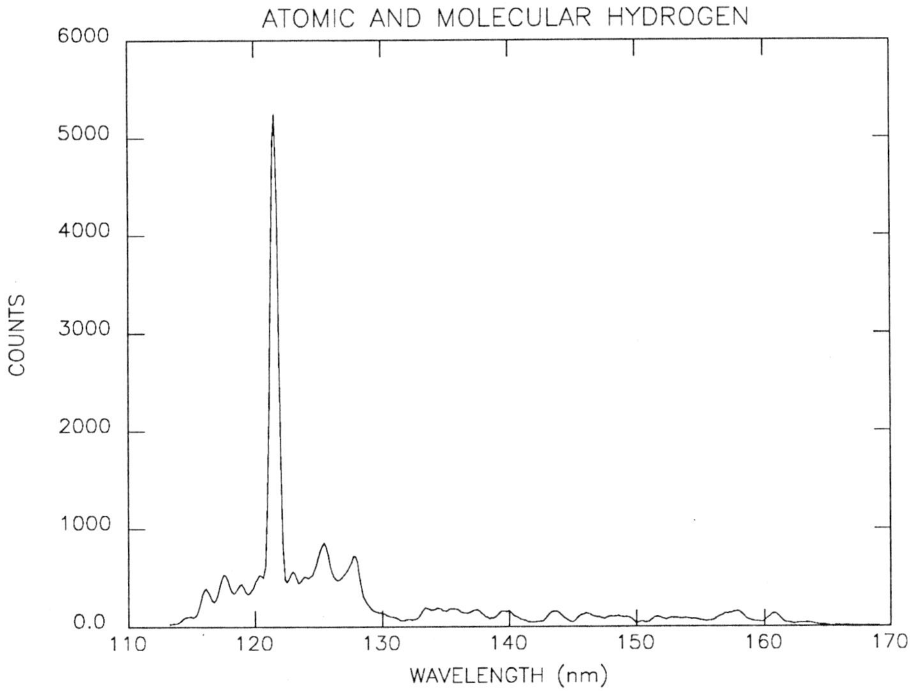

Fig. 4. An electron impact spectrum of hydrogen obtained with the UVS G channel. The prominent emission features are the Lα line of atomic hydrogen at 126 nm and the Lyman and Werner band systems of molecular hydrogen.

± 0.2 nm. The line shape is well fit by a Gaussian profile with an average FWHM of 4.5 grating steps. Measurements of grating scatter show no ghosts greater than 0.01% of the parent line within 50 nm of parent line center.

The amount of false signal caused by long wavelength radiation which scatters through the exit slits from surfaces within the spectrometer housing was measured for the F and N detectors using a tungsten-halogen lamp and a series of long wavelength pass spectral filters. These measurements were connected to an equivalent source radiance using the telescope collecting area and entrance slit field of view. For the F channel the scattered light false signal varies from 6.7×10^{-7} counts s^{-1} Rayleigh^{-1} at 350 nm to 6.7×10^{-9} counts s^{-1} Rayleigh^{-1} at 400 nm to 1.3×10^{-9} counts s^{-1} Rayleigh^{-1} at 450 nm. For the N channel the scattered light false signal varies from 3.5×10^{-5} counts s^{-1} Rayleigh^{-1} at 500 nm to 2.3×10^{-8} counts s^{-1} Rayleigh^{-1} at 750 nm.

The G channel operates in second order from 113 to 192 nm and second-order radiation for 226 to 394 nm which is imaged directly onto that exit slit and rejected by the very low response of the CsI photo cathode to those wavelengths. For 253.7 nm the

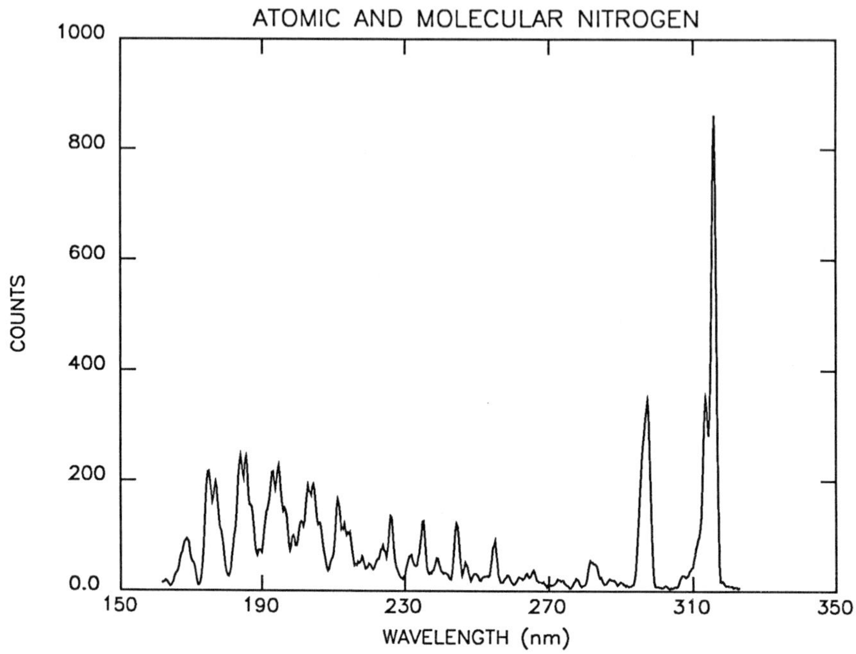

Fig. 5. An electron impact spectrum of nitrogen obtained with the UVS F channel. The prominent features are the Lyman–Birge–Hopfield bands and the second positive system of molecular nitrogen at 160–255 nm and 296–316 nm, respectively.

measured response is less than 8.5×10^{-7} counts Rayleigh^{-1} s^{-1}. For 365 nm the calculated response is less than 4.2×10^{-9} counts Rayleigh^{-1} s^{-1}.

Figure 7 is a plot of the normalized off-axis response of the telescope to a point source made as its image was scanned across the narrow dimension of the spectrometer entrance slit. Very close to the optic axis the shape of the scattering curve is dominated by the micro roughness of the telescope mirror surfaces. At larger angles, the curve is dominated by the telescope baffle system and sun shade.

Sensitivity to polarization in the UVS is due almost entirely to the diffraction grating. Figure 8 shows the polarization response of the UVS plotted as the ratio of $(I_{para} - I_{perp})/(I_{para} + I_{perp})$ where I_{para} (I_{perp}) are the instrument response to light polarized with its electric vector parallel (perpendicular) to the grating grooves.

4. EUVS Instrument Description

The Galileo EUVS channel is an objective grating spectrograph covering the wavelength range of 54–128 nm in 128 contiguous intervals of 0.59 nm. Figure 9 is an optical diagram. The Galileo EUVS channel consists of the flight spare Voyager Ultraviolet Spectrograph (Broadfoot *et al.*, 1977) and an electrical interface to adapt it to the Galileo command and data bus (Figure 10). We have modified the instrument from its Voyager

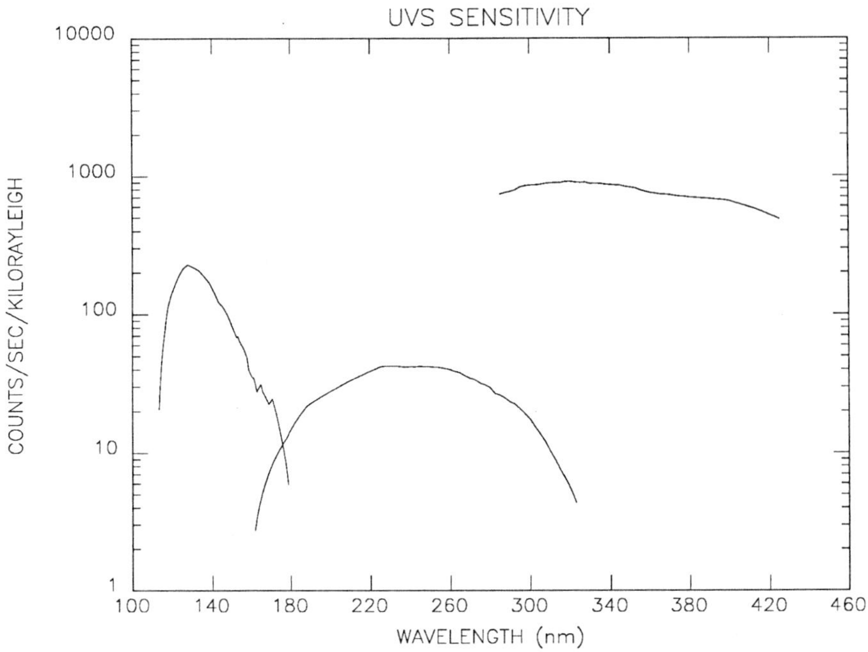

Fig. 6. Composite UVS sensitivity in counts per second per Rayleigh for a monochromatic emission line which fills the field of view. When the source radiance is expressed in Rayleighs per nanometer, the sensitivities must be multiplied by 1.3 for the *F* and *N* channels and by 0.67 for the *G* channel. Estimated accuracies in the absolute calibrations are 15% for the *F* and *N* channels and 20% for the *G* channel.

configuration to increase by 70% its field of view in the dispersion direction, and we have changed the grating to increase the dispersion by 40%. The characteristics of the EUVS channel are summarized in Table II.

4.1. Optical configuration

An objective grating spectrograph is commonly used for observations of point sources. An open, or reflective, optical system is required for wavelengths less than 104 nm, the shortest wavelength transmitted by a refractive material. The low reflective efficiencies at these wavelengths, except at grazing incidence, limit an instrument intended for the study of weak emissions to a single reflection. We have adapted the objective grating design to observe extended sources by restricting the field of view in the dispersion direction. The field is restricted by a series of mechanical stops that we call the 'collimator' because it controls the angle at which the grating is illuminated, as does a traditional optical collimator. The combination of grating and collimator has the properties described by Wadsworth in 1896 and Beutler in 1945 (Hale and Wadsworth, 1896; Beutler, 1945). The collimator was designed to induce negligible Fresnel diffraction, but the mechanical structure causes a loss of about $\frac{2}{3}$ in effective aperture, and scattering from the edges of the aperture plates complicates the analysis of the spectrum to some

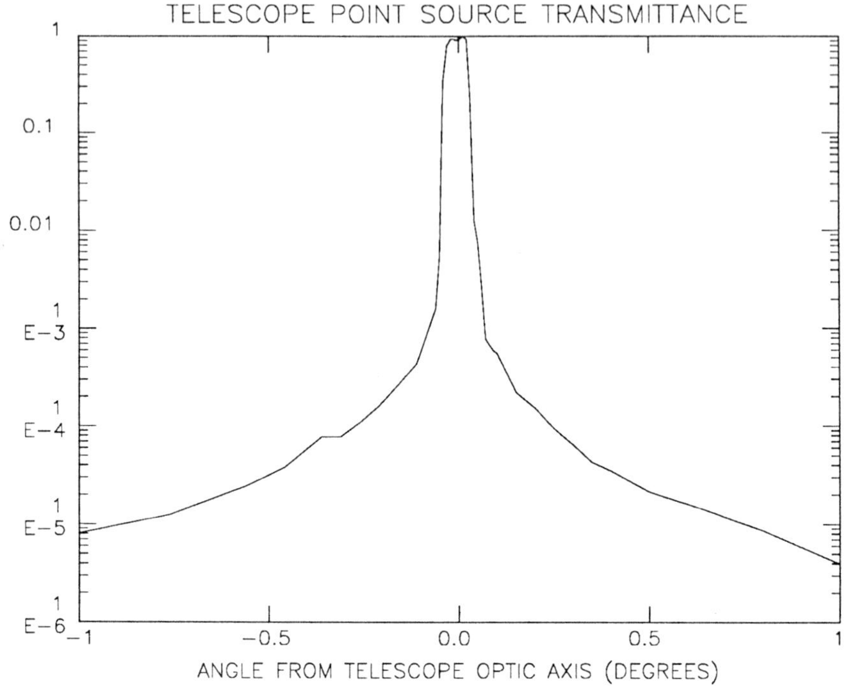

Fig. 7. Normalized off-axis response for the UVS telescope. Beyond 1° off axis the response continues to decrease by a factor of 2.1 per degree ($R = 10^{-6}$ at 4° off axis). No measurements were made beyond 4°. The rate of decline is expected to decrease beyond 15°, which is the largest angle at which light is directly incident on the telescope primary mirror.

TABLE II

Summary of EUVS characteristics

Location	Spinning section
Field of view	
Orientation	Perpendicular to spin axis
	long axis of slit parallel to spin axis
Size	0.17 deg (dispersion direction)
	0.86 deg (cross-dispersion direction)
Optical configuration	Objective grating spectrograph with mechanical grille collimator
Detector	Microchannel plates and self-scanned anode array
	128 channels scanned at 3125 Hz
Grating	Holographically fabricated
	842 lines/mm
	iridium coated
Dispersion	0.59 nm channel^{-1}
Wavelength range	54–128 nm
Focal length	200 mm
Aperture	40 × 60 mm
Spectral half-width	35 Å
Minimum integration time	20.8 ms
Best spatial resolution	0.36 deg

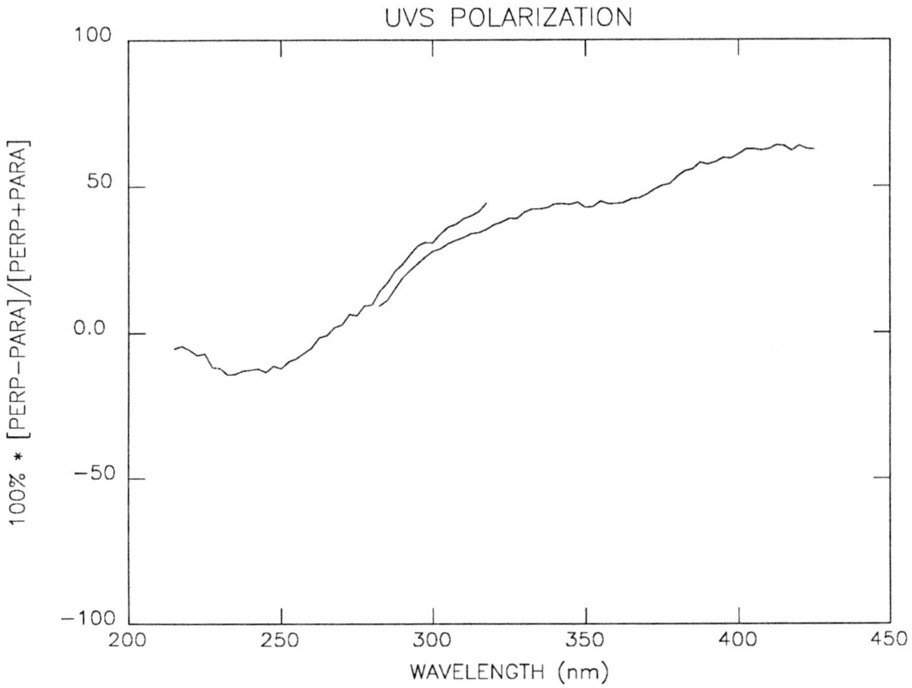

Fig. 8. The percent polarization for the UVS *F* and *N* channels. Perp and para are the response for the instrument to plane-polarized light which has the electric vector perpendicular and parallel to the UVS grating grooves, respectively. Differences in the measured polarization arise chiefly from the fact that the angles of incidence and diffraction for a single wavelength of light are different for the two detectors because their exit slits are located at different places in the spectrometer focal plane. Small differences may also arise from the additional optical elements in the *F* channel light path.

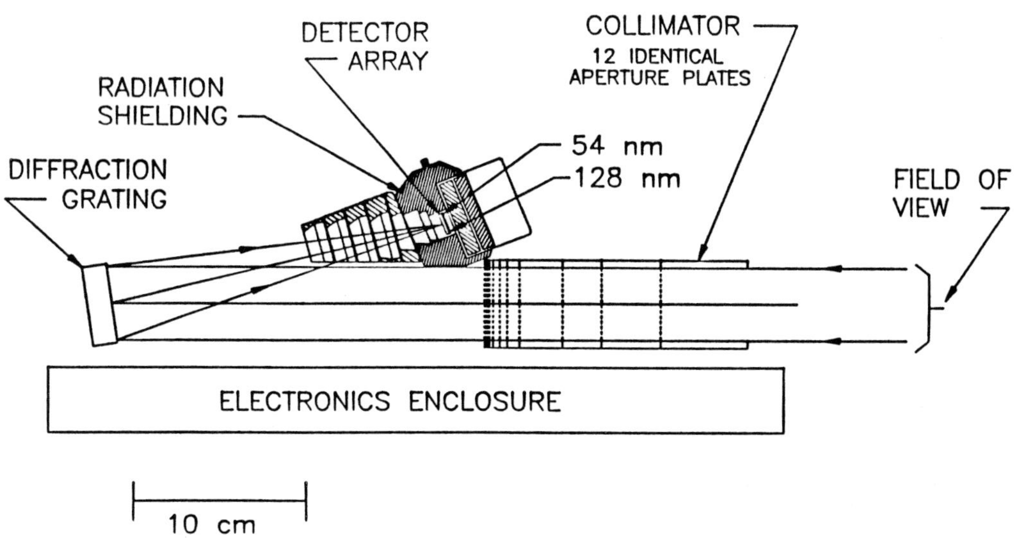

Fig. 9. EUVS mechanical-optical configuration.

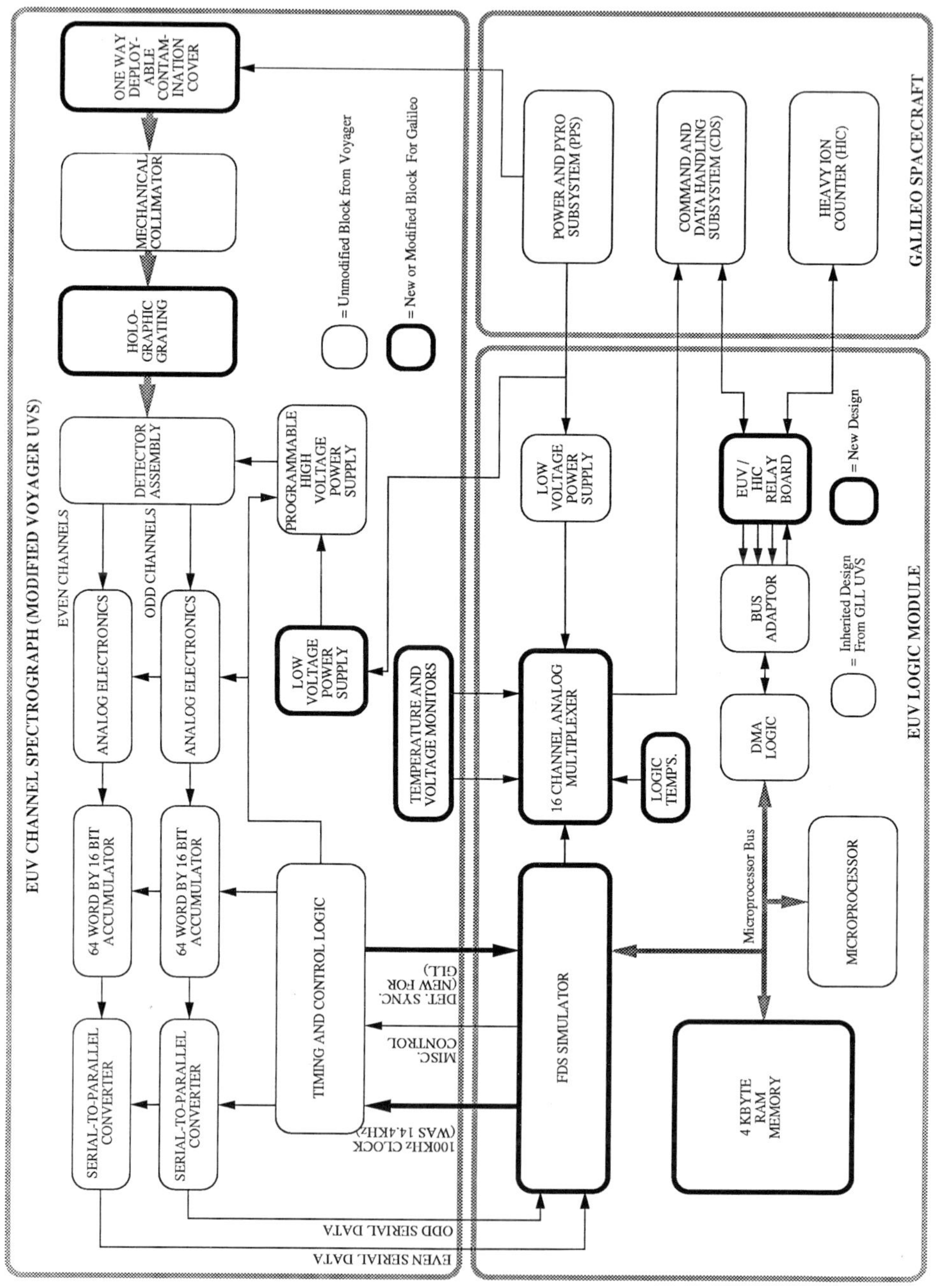

Fig. 10. EUVS functional block diagram.

extent. In practice, separation of spectral features from the scattering background is reasonably straightforward.

The collimator restricts the field of view in the dispersion direction to 0.17° FWHM, but is essentially open in the cross-dispersion direction; the field of view in this long direction is defined by the angular width of the detector in the image plane and is about 0.87°. The 40 × 60 mm diffraction grating has a spherical radius of curvature of 400.1 mm. The grating was fabricated holographically, and includes corrections to flatten the field at the image plane. The surface is coated with iridium to enhance the EUVS reflectivity. The dispersion required to match the desired wavelength range to the length of the detector is 5.9 nm mm^{-1}, which implies a ruling density of 851 lines mm^{-1}.

The instrument has two distinctly different spatial and spectral resolutions depending on the nature of the source. (1) An extended line emission source, which fills the field of view of the instrument, produces a triangular intensity distribution in the image plane 0.17° in half-width; this corresponds to a spectral half-width of 35 Å. (2) A point source is accurately imaged in the image plane and the width of a detector element, 0.029°, determines both the spectral and spatial resolutions. Both spatial and spectral resolution may be improved to some extent through spectral analysis.

4.2. Detector

The photon-counting detector uses a 128-element, linear, self-scanned anode array to collect the output of a dual microchannel plate (MCP) electron multiplier. This detector was developed for the Voyager instrument (Broadfoot and Sandel, 1977). The 128 narrow aluminium anodes, each 3 mm long, are deposited on 0.1-mm centers for a total collecting length of 13 mm in the dispersion direction. The specially designed 18-mm diameter MCPs have a rectangular active area corresponding to the collecting area of the anode array. Photons diffracted by the grating to the detector eject photoelectrons from the MCP. These photoelectrons undergo about 10^6 amplification through the cascaded microchannel plates. The pulse of electrons falls on the self-scanned anode array where the charge is stored. The anodes are accessed sequentially by a shift register and FET switches contained in the single integrated circuit. The scanning circuitry discharges each anode into a charge sensitive amplifier. If the charge pulse exceeds a fixed threshold, the memory location corresponding to the anode is incremented. The access time is 320 ms per anode; therefore, single random photo events can be recorded on any one of the 128 anodes with a rate of about 300 per second on each anode with a coincidence loss of about 10% of the events; such a loss can be corrected statistically. This is a satisfactorily high rate for the emissions from Jupiter and the plasma torus.

4.3. Calibration and sensitivity

The EUVS channel was calibrated in the EUVS Calibration Facility of the Lunar and Planetary Laboratory (LPL), and cross checked with the UVS at the Laboratory for Atmospheric and Space Physics (LASP).

LPL's EUVS Calibration Facility consists of a vacuum chamber with a gimballed support for the instrument. The gimbal system can rotate the instrument about two perpendicular axes and translate it in a plane perpendicular to the UV beam. Light from a DC glow discharge enters a modified Seya–Namioka monochromator that selects the wavelength and collimates the exit beam. The dispersed and collimated beam enters the test chamber. Here a portion is split off to a channeltron and a portion continues to the instrument under test. The channeltron monitors the beam intensity continuously during the time that the spectrum is integrated within the EUVS.

The source, monochromator, and test chamber are differentially pumped, because no window materials can be used to separate them. Pumping is by a combination of turbo-molecular, ion, and cryogenic pumps. All these are oil-free systems suitable for use with microchannel plates and EUVS optics.

A second channeltron whose response over the wavelength range of the EUVS channel is known by calibration traceable to the National Bureau of Standards can be substituted in the position of the test instrument. It is used to map the intensity of the beam, and to thereby determine the absolute photon flux into the entrance aperture of the instrument. A microcomputer controls most aspects of the operation of the calibration system and acquires data from the EUVS channel and system monitors.

An important goal of the calibration procedure was to determine the absolute response of the EUVS channel as a function of wavelength. Other important goals were to measure the shape and width of the nominally triangular response function of the collimator, and to characterize the response to scattered light. The first is needed to combine with the on-axis response to compute the sensitivity to a uniformly bright monochromatic source filling the field of view. The second is needed to compute the effects of scattering from a bright emission line when it is necessary to measure nearby weak emissions superimposed on the scattering background (Holberg, 1986). Because the response of a single channel 100 Å from the peak signal to scattered light is only 1 to 2% of the total signal, substantial integration time was required to adequately determine the scattering characteristics.

The counting rate summed over all channels for a 1-Rayleigh monochromatic source filling the field of view is given by Equation (1).

$$C = I\omega A Q_e \,. \tag{1}$$

For the EUVS instrument ω, A, and Q_e have the following definitions: ω = angular field of view, $0.87° \times 0.34°$, or 9.3×10^{-5} steradian; A = ruled area of the grating, 21.2 cm^2; Q_e = transmission of the collimator, $0.33 \times$ grating efficiency $\times$ detector quantum efficiency.

The sensitivity as a function of wavelength for this geometry (a uniform, extended monochromatic source filling the field of view) is shown by the solid line in Figure 11. Because of the triangular transmission function of the collimator, radiation from a monochromatic extended source incident on the detector will fall in a triangular intensity distribution over 12 anodes. The triangular half-width of six anodes will receive 75% of the signal. For an extended source emitting a continuum of wavelengths with

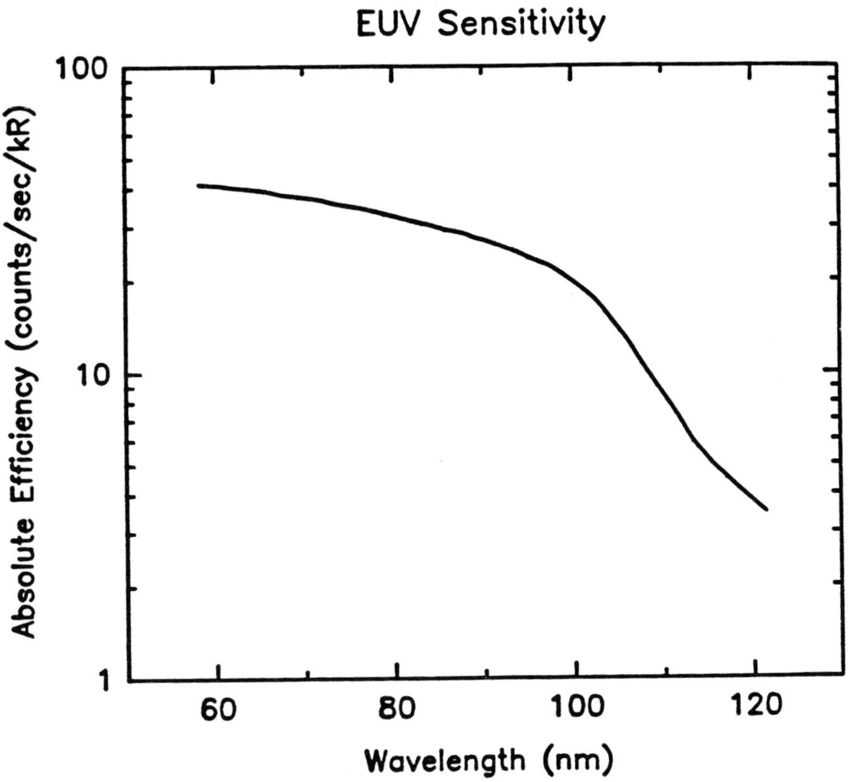

Fig. 11. EUVS sensitivity in counts per second per Rayleigh for a monochromatic emission line which fills the field of view. When the source radiance is expressed in Rayleighs per nanometer, the sensitivity must be multiplied by 3.5.

brightness 1 Rayleigh per nm, the count rate for a single anode is given by

$$C \times W, \tag{2}$$

where W is the spectral half-width, 3.5 nm.

The practical lower limit for the signal measurable by the EUVS channel is a function of both the instrument sensitivity and the dark count rate of the detector. An important advantage of the EUVS detector is its extremely low intrinsic dark count rate of less than 3×10^{-3} count anode^{-1} s^{-1}. In flight, this low dark count rate will not be realized because of noise induced by gamma rays from the RTGs. Our experience with Voyager has shown a dark count rate of 0.02 counts channel^{-1} s^{-1}, and on Galileo the dark count rate will probably be similar.

When the spacecraft is within about 15 R_J of Jupiter, noise induced by high-energy electrons trapped in the Jovian magnetic field will greatly exceed the intrinsic noise of the detector. Extensive shielding has been provided around the detector to reduce radiation interference.

4.4. Data handling and spacecraft interface

To ease the adaptation of the Voyager instrument to the Galileo spacecraft, changes to the spectrograph enclosure were minimized, and an external logic box to interface the EUVS spectrograph and electronics to the Galileo bus was constructed.

This design restricts the rate at which the spectrum can be read from the EUVS memory to 20.8 ms. During this time, the spacecraft will rotate by 0.37°, which defines the spatial resolution in the roll (dispersion) direction.

Because of its late addition to the spacecraft, the EUVS channel shares a telemetry channel with the Heavy Ion Counter (HIC). We have implemented a procedure in the logic box that time-tags detected photons as the information is read from the memory in the spectrograph enclosure. The procedure is based on the concept of a sector. A sector is a time corresponding to an integral number n of 20.8-ms intervals, or equivalently, a number n 0.37° sections of the band swept out on the sky by the EUVS field of view. The number n is programmable. For encounter operations, it will usually be 1 to give the best available spatial resolution, but for cruise observations it may be much larger. The crossing of sector boundaries is marked in the telemetry stream by a unique byte. Between sector boundaries, the wavelength position of each photo event recorded in the spectrograph memory is telemetered. Synchronizing the sector boundaries with spacecraft roll phase then permits reconstruction, on the ground, of the wavelength and origin on the sky of each detected photon.

To remain within the allocated telemetry rate, during encounter operations it will be necessary to restrict processing of data to that portion of the roll when Jupiter and the plasma torus are in the EUVS field of view.

Two additional data buffers will be used during encounter and orbit operations. The first of these, called the wavelength buffer, is a 128-element buffer that accumulates the spectrum over all sectors. The second, the sector buffer, has an element corresponding to each of the sectors and sums all photo events detected in a particular sector into the appropriate element. Thus data in the wavelength buffer represent the spectrum averaged over all sectors, and data in the sector buffer represent the spatial variation in the intensity summed over the passband of the EUVS channel.

Cruise observations will consist mainly of mapping He 58.4 nm and H Lα emissions from the local interstellar medium. For this investigation, it is desirable to record data over the full range of roll phase, but high spatial resolution is less important than for encounters.

4.5. EUVS logic module

The EUVS uses an RCA 1802 CMOS microprocessor for command parsing, spacecraft time and sector recognition and synchronization, and instrument control. The microprocessor and support logic, the bus adaptor, and the DMA logic are identical to that for the UVS. New designs were required for a 4 kbyte Random Access Memory (RAM) and a simulator for the Flight Data System (FDS) from the Voyager spacecraft. The simulator allowed us to transform the data from the EUVS channel to the Galileo spacecraft without modifying the original Voyager-style hardware.

The EUVS Microprocessor Logic differs from the UVS in two ways: (1) there is no Cold Start Mode, and (2) the microprocessor is responsible for data buffering, formatting, and transmission from the EUVS channel Spectrometer to the Galileo Command and Data System (CDS). It receives commands, and spacecraft timing and sector information via the Bus Adaptor and associated Direct Memory Access (DMA) logic. The DMA logic also handles the function of loading and verifying program memory, and of reading out telemetry data from the microprocessor telemetry data buffer.

The FDS Simulator performs all interface tasks to simulate the Voyager FDS including gated clocks, discrete I/O, and serial-to-parallel conversion of the two EUVS channel data lines. This logic also contains the handshaking circuitry necessary to achieve 20.8 ms integration periods.

The Bus Adaptor serves as the bi-directional interface between the Galileo spacecraft and the EUVS. The spacecraft bus is defined to contain four signals. These include a 806.4 kHz clock, the Real Time Interrupt (RTI) timing signal, and one line for serial communications in both directions. The bus adaptor circuitry serves to electrically isolate the EUVS from the spacecraft, and to allow for 8-bit information flow to and from the EUVS.

4.6. Operational modes and microprocessor software

The EUVS software supports two distinct operational modes. These modes are defined as: (1) encounter mode, and (2) cruise mode. Encounter mode is the primary science mode, and will be utilized when the spacecraft is in the Jovian environment. Cruise mode provides monitoring of EUVS sources while the spacecraft is traveling to Jupiter.

4.7. Encounter mode

The EUVS software includes two, 1 kbyte buffers (Encounter Mode only). The first, called the Pixel Data Buffer (PDB), is a circular buffer which holds data from the EUVS channel Spectrometer. Data is placed in this buffer whenever the FDS Simulator detects non-zero data. The second circular buffer, called the Spin Packet Buffer, is filled as the microprocessor removes data from the PDB in non-real time and transforms this data into the Galileo EUVS format. Twelve bytes are removed from this buffer every $\frac{2}{3}$rd s.

Because of the spinning motion, the microprocessor must monitor the spin position and gate the integration periods as the EUVS field of view sweeps past its target. This is accomplished by monitoring the spacecraft sector broadcasts. The minimum integration time corresponds to $0.3744°$.

The following four parameters for the EUVS can be set by spacecraft command: (1) High voltage level applied to the microchannel plates to select the gain of the detector. (2) The roll angle at which integration should start. This fixes the start angle granularity to approximately $1.4°$. (3) and (4) establish the time of integration, i.e., the angle over which the instrument will integrate. When the EUVS detects photons, it basically time tags the photons, so that when combined with the roll position, the photons can be linked to specific physical locations in the sky. The third parameter determines the minimum angular resolution over which target is subdivided, and defines the number of 20.8 ms (0.3744°) periods to integrate before incrementing the time tag. This parameter is called an EUVS Sector. The fourth parameter sets the total time the EUVS integrates each spin, by defining the number of EUVS Sectors that make up an integration period. Thus the total integration period is equal to 20.8 ms × parameter 3 × parameter 4.

The EUVS is body fixed on the spinning section of Galileo with its field of view pointing at right angles to the spin axis. Figure 12 is a schematic of the observing

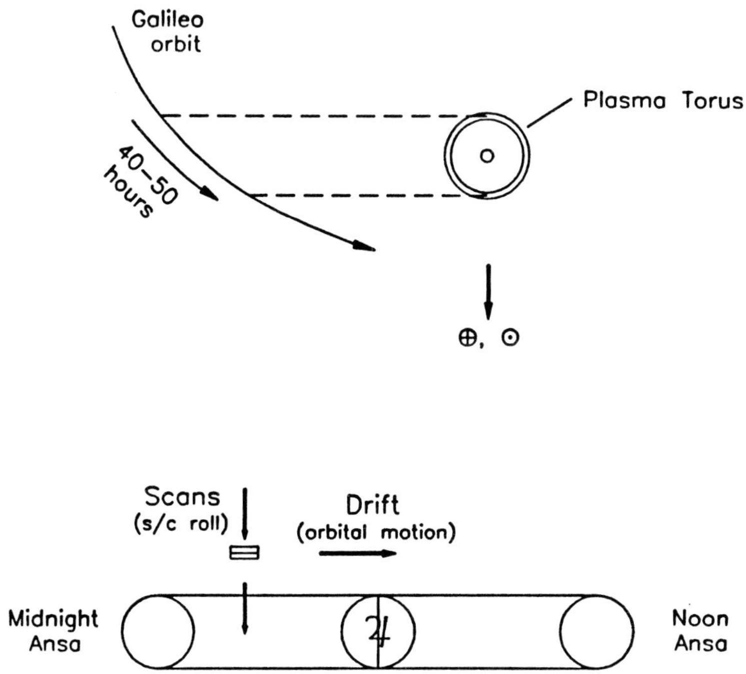

Fig. 12. *Top*: A schematic representation of a typical Galileo orbit, approximately to scale, as seen from above Jupiter's north rotation pole. Galileo's roll axis is pointed toward Earth, and the dashed lines show the positions at which the EUVS field crosses the ansas of the torus. *Bottom*: A schematic view of the plasma torus and Jupiter as seen from Galileo at an orbital position between the two dashed lines in the top panel. The EUVS instantaneous field is shown by the small lower rectangle; the spatial resolution implied by the minimum sector size is represented by the larger rectangle.

geometry. The orbital motion of Galileo carries the field of the EUVS across the plasma torus and Jupiter twice in each orbit. Because the spin axis points toward Earth, and therefore nearly toward the Sun, one of these sweeps occurs when Galileo is near Jupiter's dawn meridian, and the other when it is near dusk. Only the sweep near the dawn meridian is useful for the EUVS, because a typical orbit carries the spacecraft inside 25 R_J at the dusk meridian crossing where high radiation-induced dark counts will mask the data.

For a typical orbit, the EUVS field crosses the torus from ansa to ansa in about

45 hours, at a rate of about 0.3 R_J per hour. The spatial resolution element will be about 0.75 R_J parallel to the orbital plane of the satellites, and about 0.32 R_J perpendicular to the plane. Thus there will be about 6 resolution elements across the torus and from pole to pole on Jupiter. The spacecraft rolls some 450 times as the EUVS field moves by its own length parallel to the satellite plane. Thus, for some of the brighter emissions, it should be possible to detect spatial structure having scales smaller than the 0.75 R_J length of the slit.

4.8. Cruise mode

The mechanism for telemetering data in cruise is to use trickle memory readouts at about 15 bps. By using larger values in parameters 3 and 4, brightness measurements can be recorded at most values of the roll phase.

A single 1.5 kbyte telemetry buffer is read out. The contents of the buffer is a two-dimensional array, e.g., the first index defining an EUVS sector, and the second index defining a given pixel number (i.e., wavelength).

Whenever the FDS Simulator detects non-zero data, the appropriate register in the two-dimensional array is modified to include these data. Thus the value of each element in this array is the accumulation of photo events at a given wavelength in a particular section of the sky.

Acknowledgements

We would like to thank the many people who supported the addition of the Extreme Ultraviolet Spectrometer to the Galileo Orbiter science payload. Of the EUV's many fathers, we would like to single out John Casani for his support and confidence in our team.

References

Ajello, J. M., Shemansky, D. E., Franklin, B., Watkins, J., Srivastava, S., James, G. K., Simms, W. T., Hord, C. W., Pryor, W. R., McClintock, W. E., Argabright, V., and Hall, D.: 1988, *Appl. Optics* **27**.

Beebe, R. F., Orten, G. S., and West, R. A.: 1989, NASA Special Publication SP-494, p. 245.

Beutler, H. G.: 1945, *J. Opt. Soc. Am.* **35**, 311.

Broadfoot, A. L. and Sandel, B. R.: 1977, *Appl. Optics* **16**, 1533.

Broadfoot, A. L., Sandel, B. R., Shemansky, D. E., Atreya, S. K., Donahue, T. M., Moos, H. W., Bertaux, J. L., Blamont, J. E., Ajello, J. M., Strobel, D. F., McConnell, J. C., Dalgarno, A., Goody, R., McElroy, M. B., and Yung, Y. L.: 1977, *Space Sci. Rev.* **21**, 183, 205.

Caldwell, J., Cess, R. D., Carlson, B. E., Tokunaga, A. T., Gillet, F. C., and Nolt, I. G.: 1979, *Astrophys. J.* **234**, L155.

Clarke, J., Caldwell, J., Skinner, T., and Yelle, R.: 1989a, NASA Special Publications SP-494, p. 211.

Clarke, J. T., Trauger, J., and Waite, J. H., Jr.: 1989b, *Geophys. Res. Letters* **16**, 587.

Eshleman, V. R., Tyler, G. L., Wood, G. E., Lindal, G. F., Anderson, J. D., Levy, G. S., and Croft, T. A.: 1979, *Science* **204**, 976.

Fastie, W. G., Feldman, P. D., Henry, R. C., Moos, H. W., and Donahue, T. M.: 1973, *Science* **182**, 710.

Geirasch, P. J., Conrath, B. J., and Mahalhães, J. A.: 1986, *Icarus* **67**, 456.

Gladstone, G. R. and Skinner, T. E.: 1989, NASA Special Publication SP-494, p. 221.

Hale, G. E. and Wadsworth, F. L. O.: 1896, *Astrophys. J.* **4**, 54.

Hall, D. and Shemansky, D. E.: 1988, *Nature* **335**, 417.
Holberg, J. B.: 1986, *Astrophys. J.* **311**, 969.
Lawrence, G. M., Barth, C. A., and Argabright, V.: 1977, *Science* **195**, 573.
McClintock, W. E., Barth, C. A., Steele, R. E., Lawrence, G. M., and Timothy, J. G.: 1982, *Appl. Optics* **21**, 3071.
Metzger, A. E., Gilman, D. A., Luthey, J. L., and Hurley, K. C.: 1983, *J. Geophys. Res.* **88**, 7731.
Stewart, A. I. and Barth, C. A.: 1979, *Science* **205**, 59.
Strobel, D. F.: 1975, *Rev. Geophys. Space Phys.* **13**, 372.
Strobel, D. F.: 1989, NASA Special Publication SP-494, p. 183.
Thomas, G. E.: 1978, *Ann. Rev. Earth Planetary Sci.* **6**, 173.

GALILEO PHOTOPOLARIMETER/RADIOMETER EXPERIMENT

E. E. RUSSELL, F. G. BROWN, R. A. CHANDOS, W. C. FINCHER, L. F. KUBEL

Santa Barbara Research Center, Goleta, CA 93017, U.S.A.

A. A. LACIS and L. D. TRAVIS

Goddard Space Flight Center, Institute for Space Studies, New York, NY 10025, U.S.A.

Abstract. The Photopolarimeter/Radiometer (PPR) is a remote sensing instrument on the Galileo Orbiter designed to measure the degree of linear polarization and the intensity of reflected sunlight in ten spectral channels between 410 and 945 nm to determine the physical properties of Jovian clouds and aerosols, and to characterize the texture and microstructure of satellite surfaces. The PPR also measures thermal radiation in five spectral bands between 15 and 100 μm to sense the upper tropospheric temperature structure. Two additional channels which measure spectrally integrated solar and solar plus thermal radiation are used to determine the planetary radiation budget components. The PPR photopolarimetric measurements utilize previously flown technology for high-precision polarimetry using a calcite Wollaston prism and two silicon photodiodes to enable simultaneous detection of the two orthogonal polarization components. The PPR radiometry measurements are made with a lithium tantalate pyroelectric detector utilizing a unique arrangement of radiometric stops and a scene/space chopper blade to enable a warm instrument to sense accurately the much colder scene temperatures.

1. Introduction

The Galileo Photopolarimeter/Radiometer (PPR) is designed to measure the degree of linear polarization (and intensity) of reflected sunlight and the intensity of the thermally emitted infrared radiation from Jupiter and its major satellites. The PPR combines the high-precision polarimetry capability of predecessor instruments flown on the Pioneer 10 and 11and Pioneer Venus missions with the capability of obtaining thermal infrared measurements for temperature sounding and radiation budget studies.

Many of the basic optical, mechanical and electronic design features of the PPR are based upon proven techniques employed on precursor instruments designed and fabricated at the Santa Barbara Research Center. The Imaging Photopolarimeter (IPP) instrument (Pellicori *et al.*, 1973) was flown on both the Pioneer 10 and 11 spacecraft, successfully obtaining imaging, polarimetry and photometry of Jupiter and Saturn (flybys) and interplanetary observations of zodiacal light. These instruments are still functioning and are being used by mission control for spacecraft navigation purposes now that they have left the solar system. Using some of the features of the IPP, the Cloud Photopolarimeter (CPP) instrument (Watts *et al.*, 1977) was designed and built for the Pioneer Venus Orbiter spacecraft launched in 1978. With more than 12 years of flawless operation, the CPP continues to provide images and polarimetry maps of the Venus clouds, yielding the most extensive climatology for any planet other than the Earth.

Following a brief description of the scientific objectives of the PPR experiment, we describe the optical and electronic hardware of the instrument, its basic functions and

Space Science Reviews **60:** 531–563, 1992.

calibration. Finally, we present a brief summary of the expected instrument signal to noise ratio (SNR) performance characteristics.

2. Scientific Objectives

The primary science objectives and anticipated results of the PPR experiment are to: (1) determine the vertical and horizontal distribution of cloud and haze particles in the atmosphere of Jupiter, including their size, shape and refractive index; (2) determine the energy budget of Jupiter and the variation in amount and spatial distribution of reflected solar radiation and emitted thermal radiation for Jupiter and its satellites, including the thermal structure of the atmosphere and the vertical distribution of absorbed solar radiation in the atmosphere of Jupiter, and (3) measure and map the photometric, polarimetric, and radiometric properties of the major satellites of Jupiter.

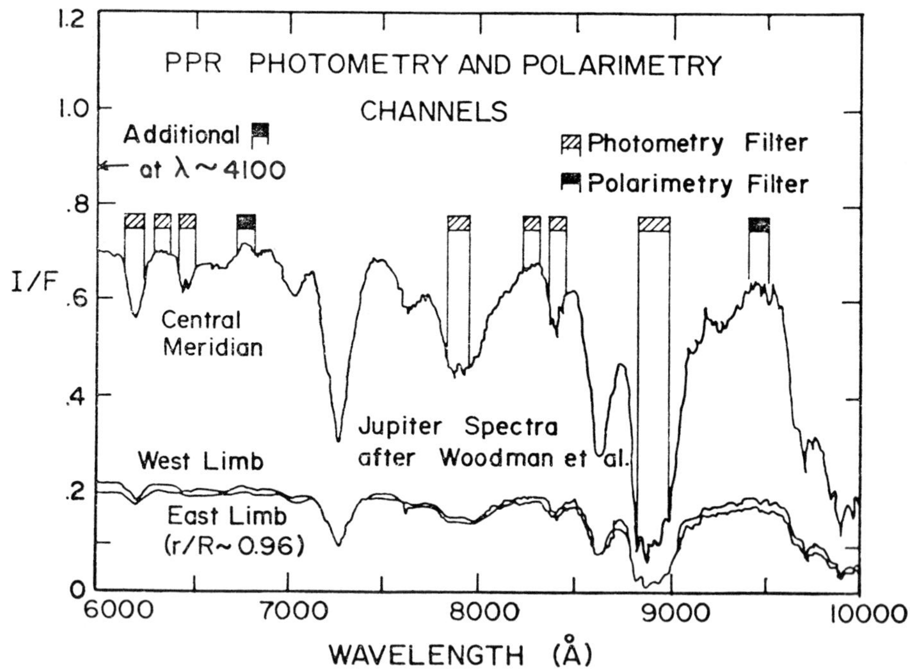

Fig. 1. Spectral locations of PPR photometry and polarimetry channels.

The polarimetric and photometric measurements made by the PPR will permit determination of the vertical cloud structure of Jupiter, including also the cloud particle size, shape, and refractive index. The spectral bands used for this purpose are illustrated in Figure 1 and their spectral characteristics are summarized in Table I. Basically, the polarimetric channels are situated in continuum regions at 410, 678.5, and 944.6 nm spaced as widely as possible to provide the broadest spectral baseline commensurate with SNR greater than 2000. The measurements yield the precision polarimetry necessary to separate cloud micro-structure information from Rayleigh scattering and thus enable deduction of cloud-top heights and cloud vertical distribution.

TABLE I

PPR filter spectral bandpasses

Measurement function	Center wavelength	Full width at half max	Filter/retarder wheel position
Polarimetry	410.0 nm	60.0 nm	13, 15, 17
	678.5 nm	8.7 nm	7, 9, 11
	944.6 nm	10.8 nm	1, 3, 5
Photometry	618.7 nm	7.0 nm	25
	633.3 nm	8.6 nm	26
	648.0 nm	7.4 nm	27
	788.7 nm	11.9 nm	28
	829.3 nm	11.9 nm	29
	840.3 nm	7.1 nm	30
	891.8 nm	11.1 nm	31
Radiometry	Solar + thermal	0.3–110 + μm	24
	Solar	0.3–4 μm	23
Channel: *A*	16.8 μm	4.2 μm	18
B	21.0 μm	3.0 μm	19
C	27.5 μm	7.2 μm	21
D	35.5 μm	6.9 μm	20
E	$\lambda > 45$ μm	45–110 μm	22

The photometric channels shown in Figure 1 also yield polarimetric information of comparable precision, provided the polarization direction is known. This follows because the photometric channels utilize the same detectors and optical elements used for polarimetry, except for the absence of repeated retarder position measurements used for polarimetry. The photometric channels at 633.3 and 829.3 nm provide additional spectral constraints in the continuum regions of the spectrum, advantageous in separating cloud, haze, and 'chromophore' contributions. The photometric channels at 618.7, 840.3, and 891.8 nm are located within medium, weak, and strong CH_4 bands, respectively, and will be instrumental in further constraining the cloud vertical structure. The 648.0 and 788.7 nm channels measure weak and medium NH_3 absorption, respectively, and will permit mapping of the NH_3 distribution at visible wavelengths.

A primary objective of the PPR radiometric measurements is to determine the radiative energy balance of Jupiter. This will be accomplished by differencing measurements obtained in the Solar + Thermal channel (0.3–110 μm) and measurements in the Solar Only channel (0.3–4 μm). Voyager results (Hanel *et al.*, 1981) indicate that the emitted thermal radiation from Jupiter tends to be globally uniform and exceeds the latitudinally dependent absorbed solar radiation by a factor of 1.7 (globally averaged).

The ratio of emitted-to-absorbed radiation and its latitudinal dependence, fundamental for understanding the internal structure and evolution of Jupiter, will be measured by the PPR with improved accuracy. The expected SNR for these measurements at Jupiter is ~500. With the more complete geometrical coverage available from an orbiting platform, the PPR measurements will significantly improve on our knowledge of the Pioneer-derived planetary phase integral (Tomasko *et al.*, 1978) and thus of the global Jovian radiative energy balance.

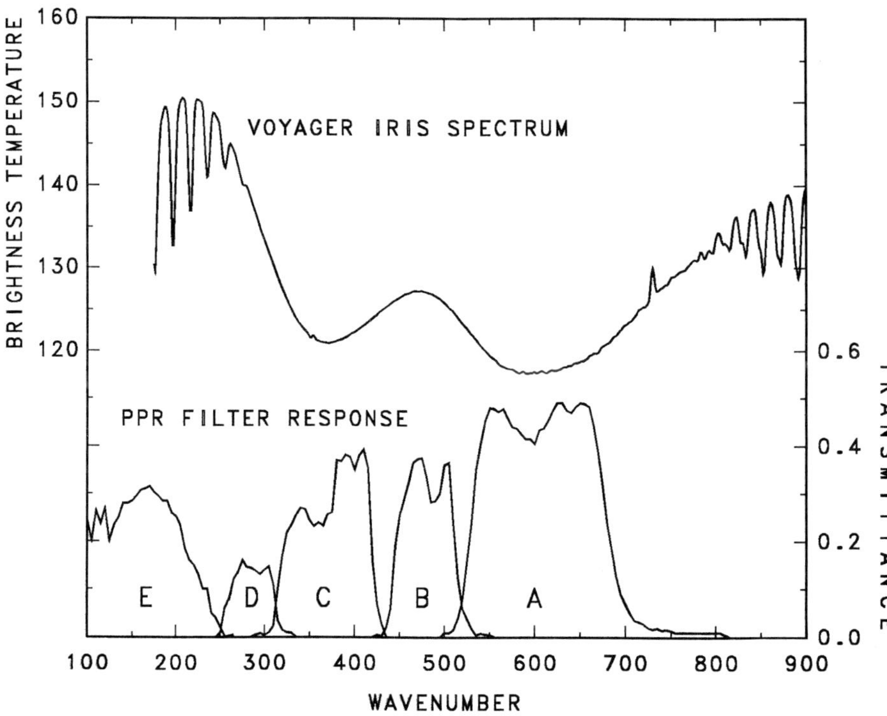

Fig. 2. Spectral locations of PPR thermal bands relative to the Voyager IRIS spectrum of Jupiter.

The PPR will also make radiometric measurements in five broad-band spectral channels spanning 15 to 110 μm. The placement of these channels relative to the Jovian thermal spectrum is shown in Figure 2, and their spectral bandpasses are given in Table I. These radiometric channels were incorporated into the PPR design to provide a measure of temperature sounding in the 60 to 600 mb region of the atmosphere. These measurements will help to relate the observed cloud and thermal structure with the atmospheric dynamics of Jupiter. Channels *A*, *B*, and *C* in Figure 2 will also measure the spatial variations in ortho-para distribution which is an important tracer of deep atmospheric circulation. Channels *D* and *E* will sample radiation from deeper atmospheric regions with channel *D* having some sensitivity to the ammonia cloud layer, while channel *E* will, in addition, have sensitivity to the vertical NH_3 distribution. The SNR performance of the temperature sounding spectral channels, however, is less than optimum (see Table VI), and will require measurement strategies to permit data averaging in order to build up the SNR.

The full complement of PPR photometric, polarimetric, and radiometric measurements will also be applied to the Jovian satellites and to other targets of opportunity. The photometric and polarimetric measurements over a wide range of phase angles will yield information on the nature of satellite surface material in regard to texture and composition. The radiometric measurements, taken under different conditions of solar illumination, will provide additional information on the texture and heat capacity of the satellite surface material. For typical satellite encounter altitudes of 1000 km, the spatial resolution of PPR measurements will be 2.5 km, comparable to much of the better Voyager imaging resolution.

The photometric, polarimetric, and radiometric measurements obtained by the PPR will provide phase angle and zenith angle coverage of Jupiter and of its satellites that is not possible for ground-based observations. These measurements will provide important new information that will permit more definitive determinations of the vertical cloud structure and of the cloud and haze micro-structure in the cloud-top region. They will also improve our knowledge of the global Jovian energy balance and will help characterize possible time and/or spatial variations in the solar and thermal components of the radiation budget of Jupiter.

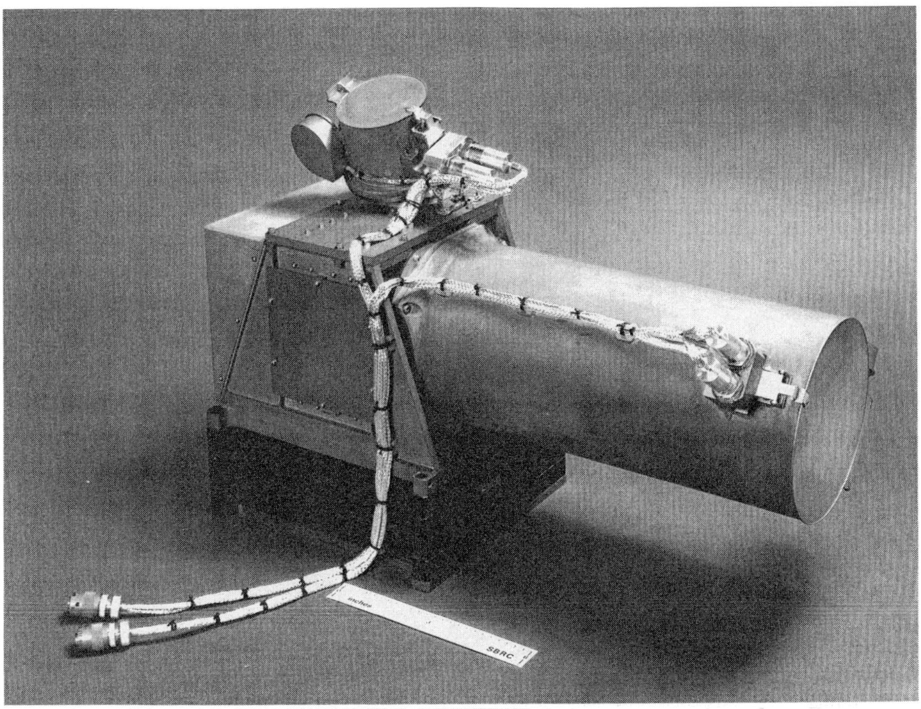

Fig. 3. The Galileo Photopolarimeter/Radiometer (PPR) with covers closed.

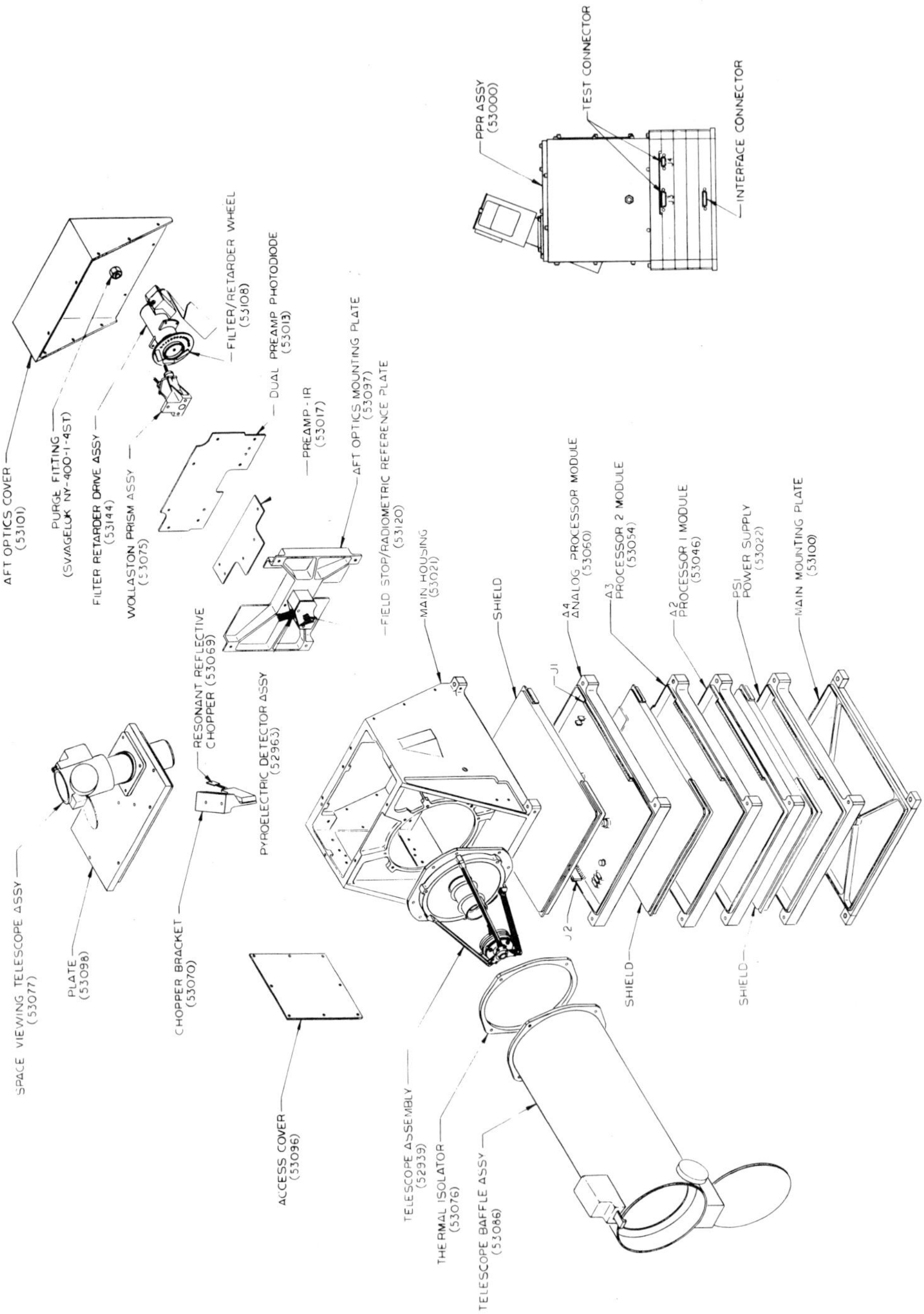

Fig. 4. Exploded view of the Galileo Photopolarimeter/Radiometer (PPR).

3. Instrument Description

The PPR is a multi-purpose and multi-function instrument designed for both solar wavelength photopolarimetric and thermal infrared measurements. The PPR is a 5.2 kg instrument with a 10 cm Cassegrainian Dall–Kirkham telescope. The overall dimensions of the PPR are 45 × 19 × 33 cm. A pictorial view of the PPR is shown in Figure 3.

Figure 4 is an exploded view of the PPR illustrating the main functional elements of the instrument. The main housing of the PPR mounts onto the top of the electronics, which consists of a set of four stacked modules which, in turn, are mounted onto the baseplate. The main housing supports the scene-view and space-view telescopes along with the ancillary optical elements, the filter/retarder motor, the chopper, and the detectors and associated pre-amplifiers. The key optical elements of the PPR contained within the main housing are shown pictorially in Figure 5.

For polarimetry and photometry measurements, the flux from the scene is collected by the scene-view telescope and focused onto a circular field stop subtending 2.5 mrad. Flux passed by the field stop then is modified by passage through optical elements located on the filter/retarder wheel. At the polarimetry positions, the flux passes through a halfwave retarder and a spectral filter, while at photometry positions it passes only through a filter. The relay lens directs the flux through the Wollaston prism which serves as a polarizing beam-splitter and produces two spatially-separated and orthogonally-polarized output beams. The detector lenses focus these beams onto the two silicon photodiodes. For polarimetry and photometry measurements the chopper is stationary and is positioned so as not to block the scene flux.

During radiometry measurements, the chopper is operated at 30 Hz and alternately directs the flux from the scene-view and space-view telescopes through the field stop. At each radiometry position the flux passed by a radiometry filter is reflected radially outward from the filter/retarder wheel by means of an ellipsoidal mirror (one mounted on the wheel at each of the seven radiometry positions). The flux reflected by the ellipsoidal mirror is collected by a condensor system consisting of a truncated conical reflector with a small diamond lens mounted onto the small end of the cone. The focused, modulated flux is detected by a lithium tantalate pyroelectric detector. The alternative views from the scene-view and space-view telescopes allow the scene radiance to be referenced to the space background (approximately to a 3K blackbody). Also indicated in Figure 5 are four optical elements designated as 'radiometric stops' which serve to restrict the modulated flux reaching the detector to that received from the scene or space, or from the internal instrument elements which are radiometrically 'balanced' to first order. These radiometric stops together with careful preflight calibration, temperature monitoring of key optical elements, and an accurately-monitored, high emittance inflight calibration target (the RCT-PPR), are key to performing radiometrically useful measurements on a 'cool' scene, such as Jupiter, with the 'warm' PPR instrument.

The primary characteristics and instrument description of the PPR are summarized in Table II. The upper portion of the table lists the general instrument properties, while

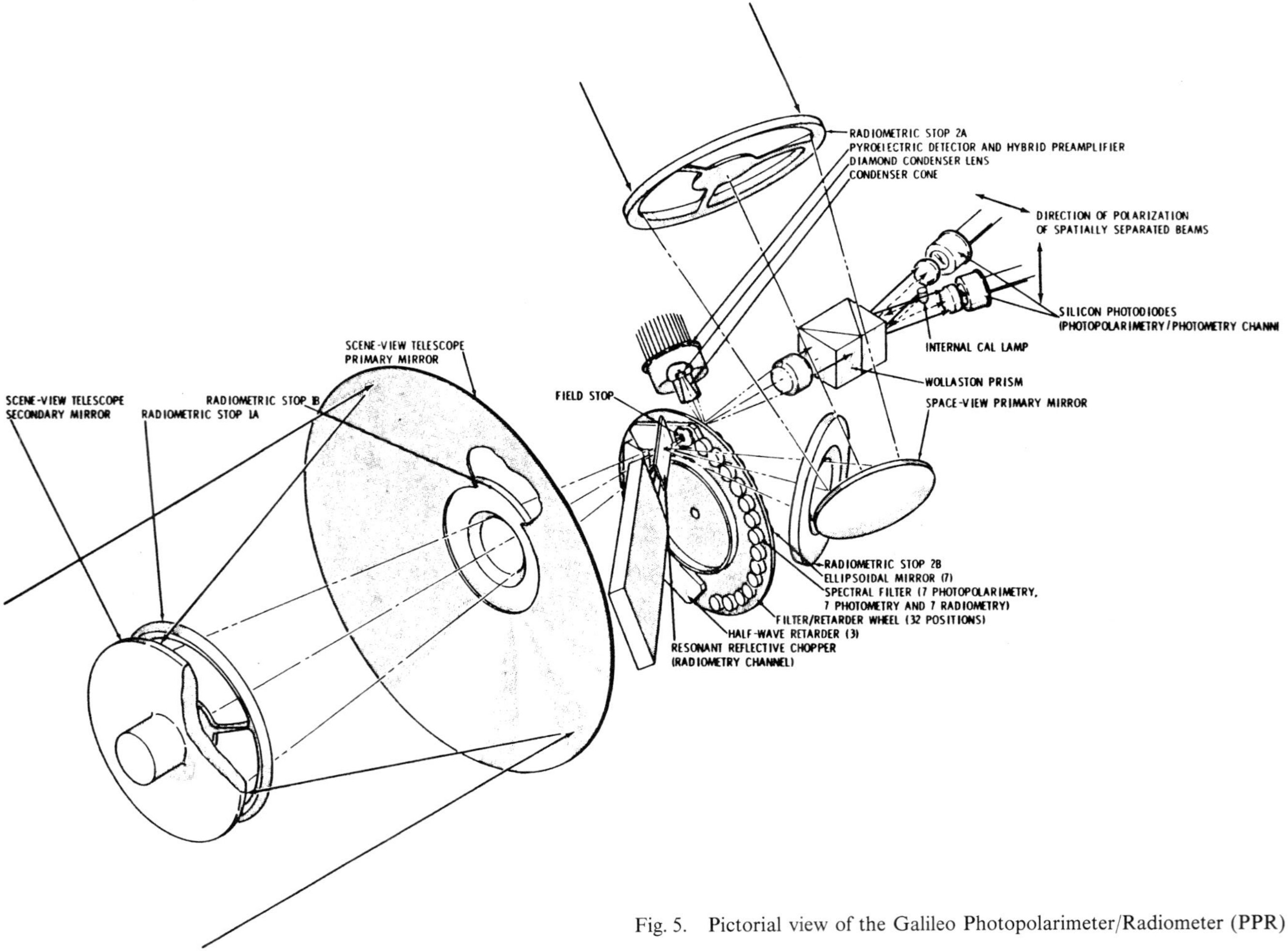

Fig. 5. Pictorial view of the Galileo Photopolarimeter/Radiometer (PPR) optical system.

TABLE II

Photopolarimeter/radiometer instrument parameter summary

Telescopes	Cassegrainian type (10-cm aperture diameter, 50-cm effective focal length) for scene view and plane mirror system for space view		
Command	One configuration control command (three 8-bit bytes)		
Interface signals	Interface bus signals (CDS sync., CDS RTI, CDS supervisory data and reply data); 30 VDC; replacement heater; radiometric calibration target temperature sensors and external cal lamp; and bellows actuator power		
Data format	48 bits status plus 96 bits science sample related data		
Data rate	144 bits (buffer fill) every minor frame ($\frac{2}{3}$ s) or 216 bps		
Location	Instrument located on scan platform		
Viewing and pointing requirements	Instrument optical axis aligned parallel to that of the SSI, NIMS, and UVS instruments; science photometric calibration target and radiometric calibration target for PPR to be viewable occasionally; direct viewing of the Sun to be avoided except on a transient basis		
Temperature limits	+50 °C to −40 °C design range; +40 °C to −30 °C protoflight test range		
Size	44.8 × 39.1 × 32.7 cm		
Mass	5.2 kg (11.4 lb) including radiation shielding		
Power	11.0 W (peak); 3.8 W (filter/retarder motor, chopper and cal lap off); replacement heater (actuates contamination covers) 4.5 W		
	Photopolarimetry	Photometry	Radiometry
Spectral bands	410 nm ($\Delta\lambda$ = 60 nm) 678 nm ($\Delta\lambda$ = 9 nm) 945 nm ($\Delta\lambda$ = 10 nm)	618 nm ($\Delta\lambda$ = 9 nm) 633 ($\Delta\lambda$ = 10 nm) 646 ($\Delta\lambda$ = 8 nm) 789 ($\Delta\lambda$ = 12 nm) 830 ($\Delta\lambda$ = 10 nm) 841 ($\Delta\lambda$ = 7 nm) 892 ($\Delta\lambda$ = 12 nm)	Solar ($\lambda < 4$ μm) Solar plus thermal (no filter) 17 μm ($\Delta\lambda$ = 4 μm) 21 ($\Delta\lambda$ = 4 μm) 27.5 ($\Delta\lambda$ = 9 μm) 37 ($\Delta\lambda$ = 10 μm) $\lambda > 42$ μm
Instantaneous geometric field of view	2.5 mrad circular	2.5 mrad circular	2.5 mrad circular
Detectors	Two enhanced silicon photodiodes (1.5-mm diameter) operated in photovoltaic (unbiased) mode	Same detectors as used for photopolarimetry	One lithium tantalate pyrolectric detector (0.57 mm circular)
Inflight radiometric calibration	Internal and external inflight calibrator lamps: science photometric calibration target	Internal and external inflight calibrator lamps: science photometric calibration target	External inflight calibrator lamp; PPR radiometric calibration target: space view reference
Polarization analyzer and calibration	Symmetrical Wollaston prism analyzer; achromatic half-wave retarders (optimized for each spectral band) positionable to produce $x + 0°$, $x + 45°$, and $x + 90°$ rotations	Symmetrical Wollaston prism analyzer	Not applicable

Table II (continued)

	Photopolarimetry	Photometry	Radiometry
Analog/digital conversion	12-bit A/D conversion of the two analog signal channels	12-bit A/D conversion of the analog signal channels (7- to 8-bit precision required with remainder of range utilized to assure adequate gain balance margin)	12-bit A/D conversion of the analog signal channel (7- to 8-bit precision required with remainder of range utilized to assure adequate gain balance margin)
Integration and sampling	Two channels simultaneously sampled (217 ms integration period) for 1, 4, 16, or 256 samples per position	Two channels simultaneously sampled (217 ms integration period) for 1, 4, 16, or 256 samples per position	One channel sampled (433 ms integration period) for 1, 4, 16, or 256 samples per position
Measurement accuracy	Polarimetric: $\pm 0.1\%$ Absolute radiometric: $\pm 3\%$ (precision better than 1%)	Absolute radiometric: $\pm 3\%$	Absolute radiometric for solar reflectance band: $\pm 5\%$ Absolute radiometric for thermal emission: ± 1 K (at 130 K)

the lower portion indicates features associated with the three primary measurement functions provided by the PPR.

3.1. Optical Design

3.1.1. *Telescopes*

The 10-cm aperture scene-view telescope is a Cassegrainian Dall–Kirkham design which gives excellent image quality for the 2.5 mrad (0.14°) instantaneous field of view, with the image quality being dominated by diffraction at the longer wavelengths. The view of space in radiometry mode is via reflection from the chopper mirror and the planar space-view telescope mirror. The telescope assemblies, including the mirrors themselves, are fabricated of beryllium which thermally match the beryllium main housing and together yield significant mass savings. The mirrored surfaces are nickel-plated and overcoated with aluminum. This choice of mirror finish yields a highly durable mirror with good spectral performance over the extended spectral range of interest.

3.1.2. *Radiometric Stops*

As indicated in Figure 5, two radiometric stops are used on the scene-view telecope assembly, one is mounted adjacent to the secondary mirror (between the primary and secondary mirrors) and the other on the back of the primary mirror (between the primary mirror and the field stop). These are designated as RS1A and RS1B, respectively, with the identifier '1' denoting use in the scene-view beam and the modifier 'A' or 'B' indicating the relative positioning along the optical axis. The optical surfaces of these radiometric stops are spherical with the radius of curvature of each being equal to the distance of the radiometric stop from the field stop (the aperture defining the instantaneous field of view). As viewed from the field stop, RS1A is configured to block the view of the secondary mirror obscuration, to block the three secondary mirror support arms, and to define the viewed outer perimeter of the primary mirror. In a similar manner, RS1B spans an annular solid angle which blocks the outer perimeter of RS1A and extends to beyond the angular response of the radiometry channel (pyroelectric detector plus lens/cone/relay mirror). This is shown schematically in Figure 6. The spherical surfaces of the radiometric stops are gold coated to provide a high reflectance (low emittance) in the spectral region where thermal emission of the instrument is significant.

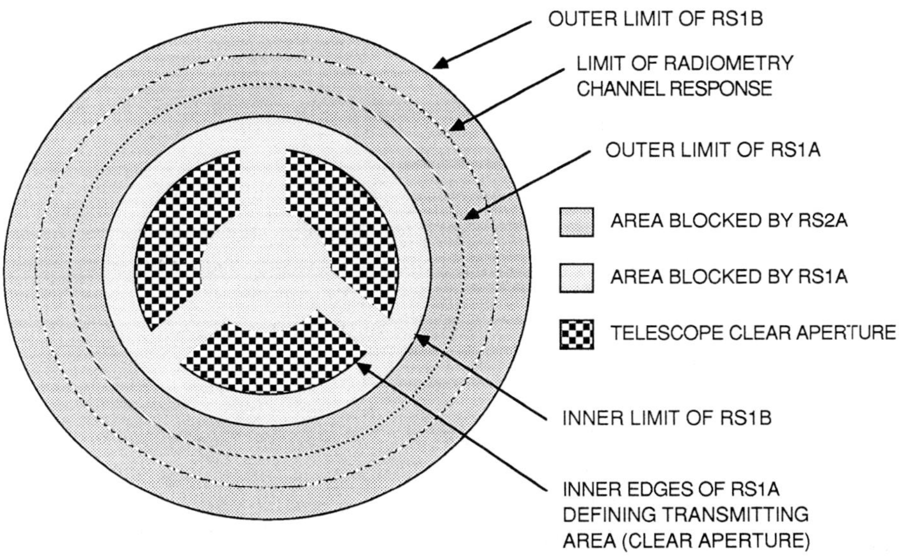

Fig. 6. Field-of-view angular blocking by radiometric stops.

During the installation of RS1A and RS1B, both elements are tilted slightly such that an area approximately 3 mm from the field stop is imaged by the two radiometric stops through the field stop. This area imaged has a machined, pyramidal array-shaped structure to increase the effective emittance of this reference target to near unity. (A more conventional, conical target could not be used since the depth dimension is limited to approximately 1.5 mm by the proximity of the chopper mirror and filter wheel.) Thus, the detector alternately views the scene radiance plus the radiance from the scene-view mirrors and radiometric stops and the reference target (via imaging by the radiometric stops) and then space radiance plus the near equivalent elements in the space-view optical path.

The rear surface of the chopper mirror is optically polished and coated in the same manner as the front surface. This feature together with a second high emittance target (mounted orthogonal to the plane of the field stop) prevents the detector from viewing ill-defined portions of the instrument during the portion of the chopping cycle when the scene-view is partially blocked, i.e., as the chopper mirror partially covers the field stop. This viewing situation results from the image inversion that occurs during reflection from the radiometric stops. During the transition period (between viewing only the scene or only space) a varying portion of the rear of the chopper mirror is viewed, and with the present design, a varying portion of the second target is viewed via reflections from the rear surface of the chopper and from the radiometric stops. Thus, this arrangement assures that no large transients associated with viewing ill-defined portions of the instrument occur during the chopper mirror transition period.

The space-view telescope assembly has two radiometric stops, RS2A and RS2B, that serve optically identical roles to the two radiometric stops in the scene-view telescope assembly. The radiometric stops RS2B and RS1B are identical parts, while stops RS2A and RS1A are optically identical, but the support structures are different as they mount on entirely different assemblies.

3.1.3. *Filter/Retarder Wheel Optical Elements*

Located on the filter/retarder wheel are spectral filters used to define the required spectral bandpass, the halfwave retarders used for the polarization analysis of the scene, and the ellipsoidal mirrors used to direct scene flux towards the radiometry detector. The sequential arrangement of the polarimetry, photometry, and radiometry filter positions are illustrated in Figure 7. The filter bandpasses and their spectral locations are summarized in Table I.

The photometry and polarimetry filters are of conventional multilayer design. The materials used for the blocking elements of the filters were prescribed based on the results of radiation testing of candidate optical materials. The difficult spectral requirements for the radiometry filters used designs based on the techniques developed by Seeley *et al.* (1981). A low transmittance problem for the E (> 45 μm) filter was solved by combining the long-wavepass multilayer element with a 0.7 mm Z-cut crystal quartz element AR-coated with parylene.

The halfwave retarders were similar in design to those used on the Cloud Photopolarimeter flown on the Pioneer Venus mission. Since the halfwave retarders are located in an $f/5$ beam, a key feature of the design was a large angular acceptance. This was achieved for the three polarimetry bands by combining crystalline elements of positive and negative birefringence (MgF_2/Al_2O_3 for the 410 nm band and SiO_2/Al_2O_3 for both the 678 and 945 nm bands). The crystalline elements were bonded together with a third element of fused silica to provide the same optical thickness for all three retarders. This configuration of each retarder has sapphire (Al_2O_3) and fused silica as the outer elements with the exposed surface of the sapphire AR-coated to minimize retardance variations due to internal reflections.

The seven small mirrors located on the filter/retarder wheel are off-axis sections of an ellipsoid. These direct scene flux passed by the radiometry filters radially outward toward the radiometry channel detector assembly.

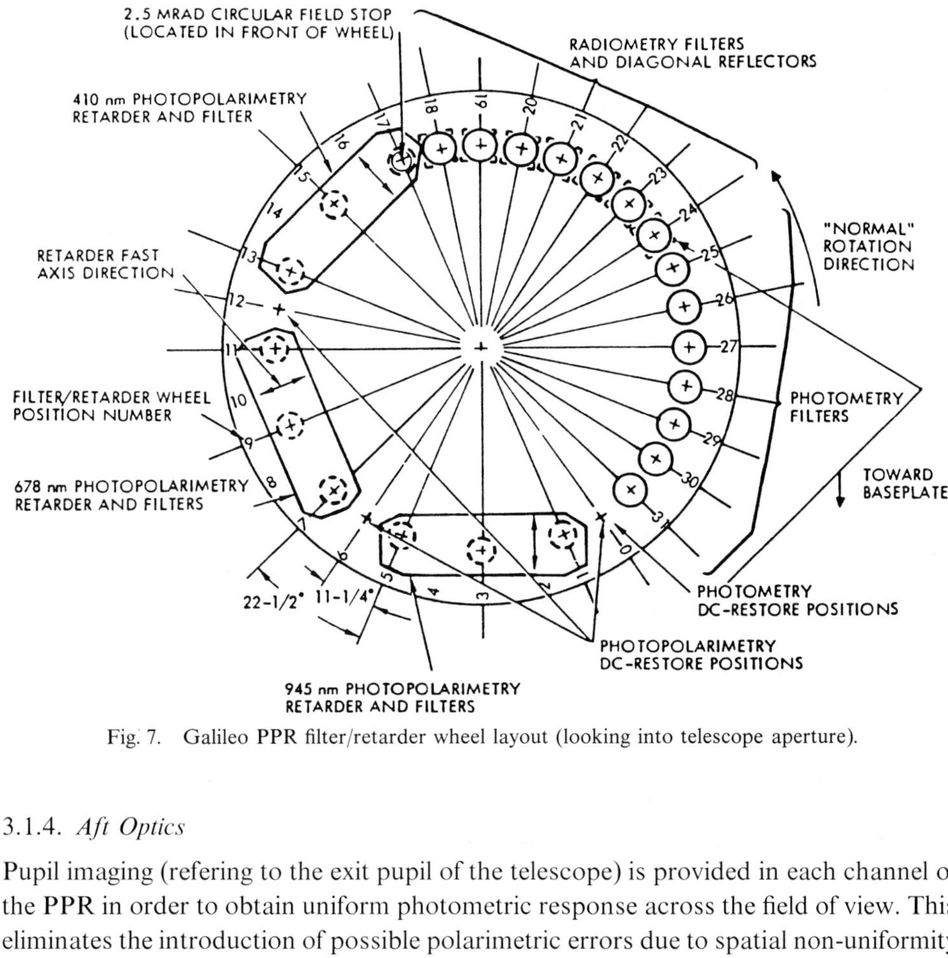

Fig. 7. Galileo PPR filter/retarder wheel layout (looking into telescope aperture).

3.1.4. *Aft Optics*

Pupil imaging (refering to the exit pupil of the telescope) is provided in each channel of the PPR in order to obtain uniform photometric response across the field of view. This eliminates the introduction of possible polarimetric errors due to spatial non-uniformity of the detectors and of the scene radiation sensed by the detector. In the case of radiometry, the ellipsoidal filter wheel mirrors together with the cone/lens of the radiometry assembly relays the scene flux onto the 0.57 mm lithium tantalate pyroelectric detector.

3.1.5. *Stray Light Baffles*

The baffle system for the scene-view and space-view telescopes are of conventional ring-baffle design. The ring baffles are sized and located to prevent any ray incident through the entrance aperture that reflects from the side wall of the baffle from reaching the primary mirror with less than one additional reflection. Due to the electrostatic discharge (ESD) requirements imposed on the Galileo instruments, it was necessary to passivate the beryllium baffle with a conductive chemical conversion coating. The baffle interior was painted, since the more usual black anodized finish results in a non-conductive surface. A small amount of carbon was added to the black paint (by ball milling to provide a uniform mix of the carbon), and this provided sufficient electrical conductivity to satisfy the ESD requirements while maintaining the low reflectance of the paint.

As part of the instrument calibration, preliminary measurements were made to characterize the off-axis light rejection of the PPR optical system. A more complete determination of stray light rejection as a function of angle from the PPR optical axis will be performed during inflight calibration opportunities (including also the boresighting of the PPR with respect to other Galileo instruments) by viewing such astronomical objects as stars and spatially unresolved planets.

3.2. Mechanical design

3.2.1. *Mechanical Structure*

Except for the electronic modules, the basic PPR structural elements are beryllium. This choice provides high stiffness with low mass and excellent stabilty. However, additional safety factors as a result of more stringent fracture analysis criteria applicable late in the program removed some of the relative advantages of beryllium versus aluminum. As noted in the Optics Section, the majority of the metal optics components were fabricated with nickel-plated beryllium. The electronics modules, for which substantial radiation shielding was required, were made of aluminum. Machinable tungsten was used to provide suitable shielding around the more radiation-sensitive elements such as the radiometry channel J-FETs and the analog-to-digital (A/D) converter.

3.2.2. *Mechanisms*

The principal mechanisms of the PPR are the filter/retarder wheel motor, the scene/space chopper, and the scene- and space-view telescope contamination cover actuators. The filter/retarder wheel motor is a 32-position (11.25°/step) direct-drive, stepper motor which is accurately indexed at each wheel position by means of frictionless permanent magnet detenting. Special low magnetic remanence materials and winding techniques maintain tightly coupled flux paths to minimize stray static and dynamic fields. A pin-disk mechanical encoder identical to that used on the Pioneer Venus CPP is used to indicate the filter/retarder position.

The resonant reflective chopper assembly is an electromagnetically driven tuning fork with a mirror and thermistor temperature sensor on one tine and a counterweight on the other tine. The basic design includes special modifications to withstand the launch environment.

Thermal motors, in the form of heated bi-metallic springs, are used to drive the contamination covers of the scene-view and space-view telescopes to their closed position. The actuator springs are wound in the shape of a watch spring with a heater wire attached to one side. The covers normally remain open, but when powered by the PPR replacement heater power, they can be closed during attitude control thruster firings. The mechanism is failsafe in the sense that if a heater were to fail, the cover would remain open. For launch, the covers were secured by latches that were released at the initial inflight checkout by a pair of redundant squibs (bellows actuators).

3.2.3. *Thermal Control*

The PPR is conductively isolated from the scan platform to allow lower temperature operation for improved radiometry channel performance. The instrument is blanketed with multi-layer insulation except for the telescope apertures, the telescope covers, and an exposed area on the side of the main housing. This area (140 cm^2) on the side of the housing is exposed to space to help maintain the PPR operating temperature near -15 °C. The instrument thermal analysis was used to establish the size of this area, with preflight verification being obtained, where possible, during spacecraft system thermal/vacuum testing.

3.3. Electronics

The low power electronics for the PPR instrument retains much of the radiation-proven circuitry from the predecessor instruments built for the Pioneer missions, but most of the discrete control logic has been replaced by a radiation-hard CMOS microprocessor system. The functional block diagram shown in Figure 8 indicates the principal elements of the PPR electronics. The analog circuitry consists of two silicon photodiode polarimetry/photometry channels, one pyroelectric radiometry channel, and an analog multiplexer which presents the three signal channels and the temperature telemetry channels to the 12-bit analog to digital converter. The digital circuitry decodes the serial spacecraft commands and formats the instrument signal and telemetry data for transfer to the spacecraft via the command and data bus. The digital system also provides timing signals for analog channel and mechanism controls. The power subsystem conditions the 30 vdc spacecraft input power and provides the necessary regulated and filtered voltages for instrument operation. This subsystem also contains the power driver circuitry for the actuator controls and calibration lamp.

3.3.1. *Analog Processing*

Pre-amplifiers for all the detector channels are mounted in the optical housing near the detectors. The silicon detectors are operated at zero bias by low-offset transimpedance pre-amplifiers with 100 megohm feedback resistors. Zero bias operation was chosen so that increased detector dark current caused by radiation damage would not saturate the sensitive pre-amplifiers. The slightly increased detector capacitance resulting from zero bias operation does not affect the pre-amplifier noise performance, which is limited by the thermal noise of the feedback resistor.

The pyroelectric detector, the 10^{11} ohm load resistor, and the low-noise junction field effect buffer transistor (J-FET) are mounted in a small custom detector package. The

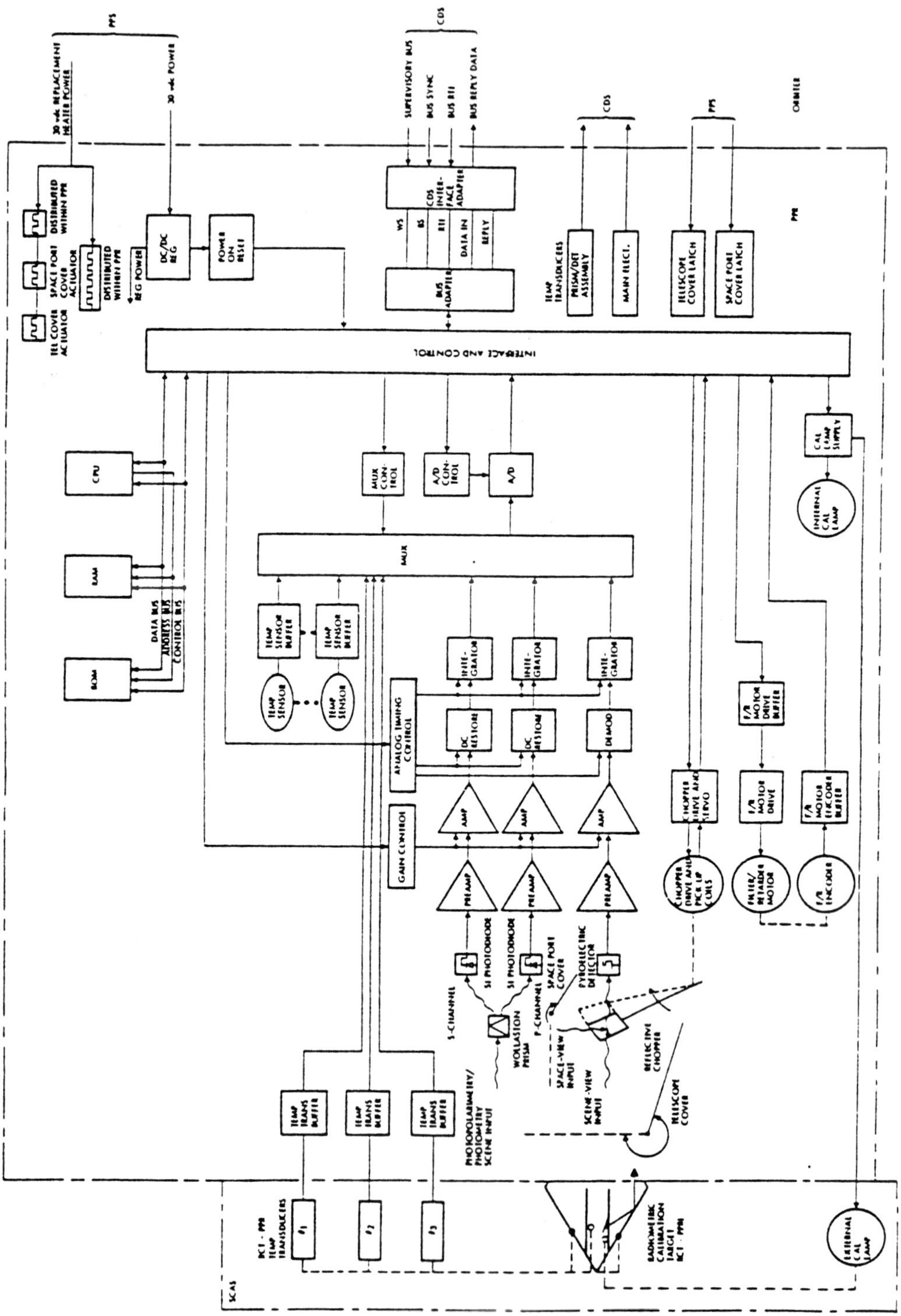

Fig. 8. Galileo PPR and RCT-PPR functional block diagram.

detector element alternately receives scene radiation and space reference radiation as relayed by the reflective 30 Hz optical chopper assembly, which operates only in radiometry filter wheel positions. The waveform is asymmetrical because the rest position of the chopper must be clear of the optical path to allow photopolarimetry operation. In spite of the extremely high impedance load resistor and J-FET circuitry, sample radiation testing has ensured operation at exposures exceeding 10^5 rads. The remaining pre-amplification stages include boost compensation for the electrical and thermal time constant of the detector element making the 30 Hz output waveform amplitude accurately represent the scene/space flux difference.

Variable-gain signal amplification is provided by electronics in the top section of the electronics module with the gains set in discrete steps in response to instrument commands. The polarimetry/photometry channels incorporate sixteen gain steps spaced a factor of 1.4 apart. These are mechanized by junction FETs switching feedback resistors in the three operational amplifier gain stages. Since it is possible for radiation induced offsets in these stages to degrade the data (particularly at high gains), a dc restore loop is included which acts as a sample and hold circuit, forcing the channel offset to zero when the filter wheel blocks the optical path at specified positions between active filter wheel positions.

The radiometry channel incorporates four FET-switched gain steps spaced a factor of 2 apart. DC restoration is not required because the stages are capacitively coupled to pass the 30 Hz signal waveform. Transient suppression clamp diodes are used at both ac coupling capacitors to rapidly settle the quiescent operating point with each filter position change. A gated full wave synchronous demodulator converts the 30 Hz waveform to a scene proportional dc level. Gating the demodulator serves both to eliminate the effects of spurious optical reflections as the chopper enters and leaves the optical path, and to balance the asymmetrical signal waveform. An intentional signal offset of about 30% of full scale is introduced in the demodulator to ensure that the channel will not saturate if the signal phase reverses. This can occur when viewing a low radiance scene if the optical elements in the space-view telescope are warmer than those in the scene-view telescope such that the net instrument thermal signal exceeded the scene signal.

Electrical low pass filtering is provided for all three signal channels by integration using a resetting integrator circuit. Integration times are synchronized to and include an integer number of radiometry chopper cycles. Integration times of 0.233 and 0.467 s are used for the polarimetry/photometry and radiometry channels, respectively. Four FET-switched resistors (controlled by the microprocessor) at each integrator input adjust the integrator gain in order to partially balance the signal levels for the various filter positions. When all four FETs are switched off, the integrator holds the final signal value, eliminating the need for a sample and hold circuit. Another FET switch then zeros the integrator for the next signal sample.

Signal multiplexing for the data channels and temperature telemetry circuits is accomplished by radiation-hard J-FET high level FET switches identical to those used in previous instruments. Higher density MOS devices were considered, but rejected

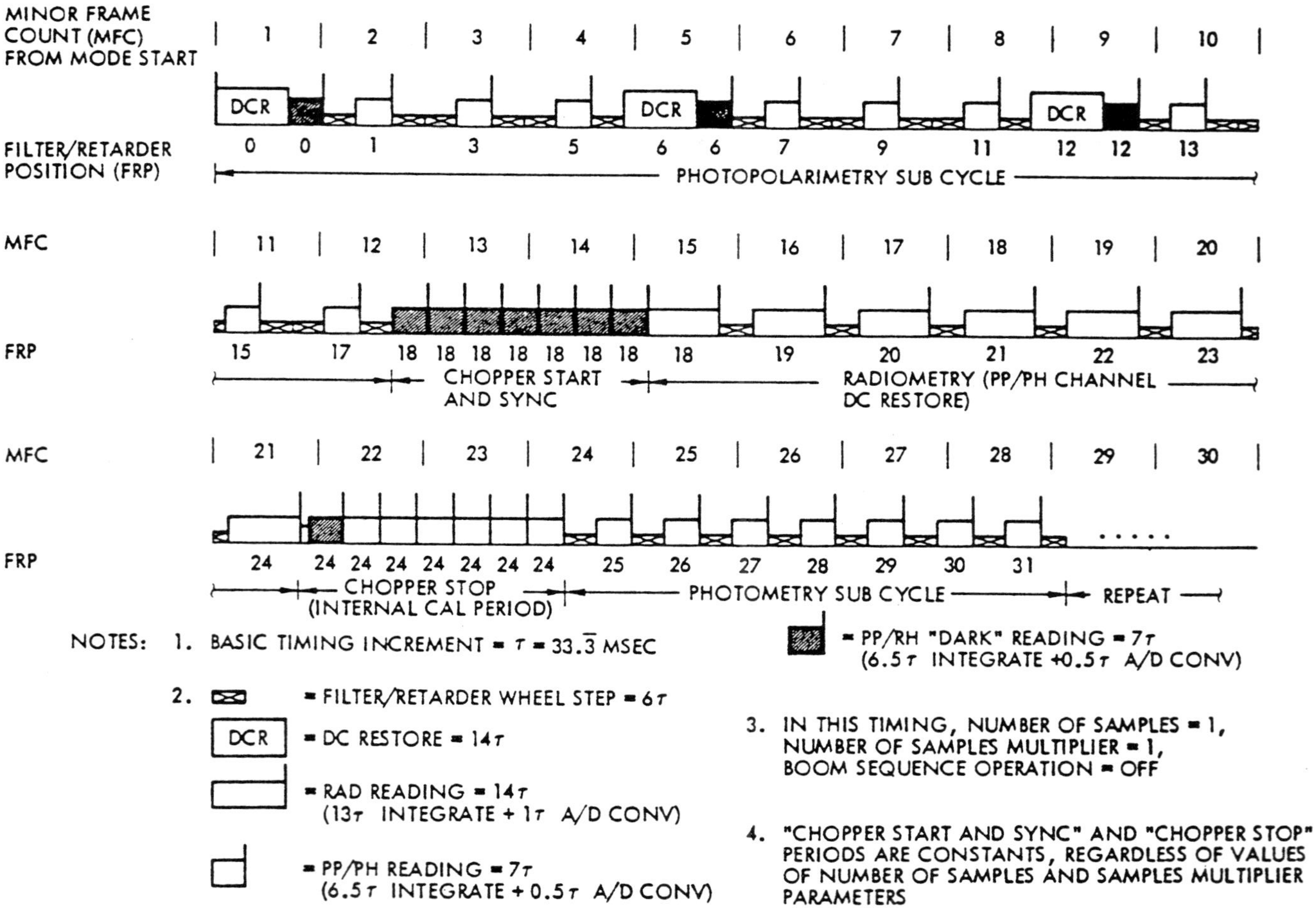

Fig. 9. Galileo PPR cycle mode sample timing diagram. Cycle shown requires 18.8 s (28.2 minor frames) for the parameter values indicated.

TABLE III

Commandable functions of Galileo photopolarimeter/radiometer (PPR)

Command	Command function
Power on	1. PPR is set to 'cycle' mode at power 'on' with gain step PP/PH = 5, gain step rad = 0, number of samples = 256, number of samples multiplier = 0, cal lamp = 1, dc-restore initiate = 1, boom sequence operation = 0, temperature range select = 0, and chopper heater = 0
Mode	1. Selects PPR mode from among cycle, photopolarimetry plus photometry (PP/PH), photometry (PH), radiometry (RAD), and position select modes 2. Initiates operation of selected mode after reaching starting filter/retarder position wherein: A. Filter/retarder wheel is stepped, and B. Sampling and dc-restoration occur as appropriate for selected mode
Programmed filter/ retarder position	1. Selects filter/retarder position to be used as starting position when position select mode in operation
Gain step PP/PH	1. Selects gain setting for photopolarimetry/photometry channels from among 16 levels with each being separated from adjacent levels by a factor of approximately 1.4
Gain step rad	1. Selects gain setting for radiometry channel from among four levels
Number of positions	1. Selects number of positions in addition to programmed filter/retarder position (starting position) from among 0, 1, 2, and 5 when position select mode is operating
Number of samples	1. Selects number of samples of data to be accumulated at each active filter/retarder wheel position from among 1, 4, 16, and 256 2. In position select mode, the commanded number of samples shall be used as the number of cycles of position select operation occurring prior to dc-restoration
Number of samples multiplier	1. Selects multiplier for number of samples as either 1 or 4, depending on whether bit is a 0 or 1, respectively 2. Increases the number of samples by the selected factor for the radiometry channel sampling only (256 maximum) 3. If external cal lamp is activated (see below) selects on period to be either six samples or the total number of samples to be taken during roll when bit is 0 or 1, respectively
Cal lamp	1. Enables or inhibits pre-programmed calibrator lamp operation depending on whether command bit is 1 or 0, respectively 2. Enables external cal lamp, if bit is a 1 and if boom sequence operation bit is also a 1 (with duration controlled as noted above). If either of these control bits is 0, the external cal lamp operation is inhibited
DC-restore initiate	1. Initiates special dc-restore cycle in position select mode or boom sequence operation wherein: A. Filter/parameter wheel is stepped to appropriate dc-restore position B. dc-restoration function is performed followed by a single dark sample, and C. Filter/retarder wheel is returned to next filter/retarder position of the sequence of the selected mode
Boom sequence operation	1. Normal stepping of filter/retarder wheel is inhibited (except in transition mode) 2. Stepping between adjacent functional filter/retarder positions is initiated once per roll based upon receipt of spacecraft spin angle and spin rate data 3. Affects external cal lamp operation as noted above under cal lamp
Temperature range select	1. Selects initial high or low range for thermistor temperature sensors depending on whether bit is a 1 or 0, respectively. Where thermistor output is out-of-range, software switches range as appropriate
Chopper heater	1. Initiates heating of chopper mirror when bit is 1 with function disabled when bit is 0

because of the cost of the extensive radiation testing required in this critical signal switching area. Thermistors for telemetering the optical element temperatures are sub-multiplexed using similar circuitry to that used in the main multiplexer. Each thermistor is unpowered when it is not being telemetered so that self-heating error is entirely eliminated. A ground command option, however, allows the thermistor on the tuning fork chopper to remain powered to slightly raise the temperature of the chopper mirror during one portion of the PPR radiometric calibration (optical element heater testing). Two ground commandable temperature ranges are provided in the thermistor scaling amplifier to ensure high readout resolution throughout the possible inflight temperature range. Platinum resistance thermometers are used to read out the temperature of the radiometric calibration target (RCT-PPR). These require more scaling circuitry to operate than the thermistor channels, but provide an accurate and stable, long-life temperature measurement capability for the RCT-PPR. All multiplexed channels are digitized to 12 bits by a low power CMOS A/D converter.

3.3.2. *Digital Circuitry*

The digital circuitry centers around the RCA 1802 microprocessor hardware which was supplied and radiation tested by the Galileo Project. Because of the low data rate and simplified conceptual design of the instrument, only 256 bytes of random access memory are required, and the entire instrument operating system is contained in less than 4K bytes of read-only memory. Clock reference for the microprocessor is formed by doubling the spacecraft supplied 866 kHz clock signal, and all instrument operations are synchronous with clock transitions. Bi-directional spacecraft communications of commands and data are performed using a transformer-coupled bus interface unit (also supplied by the Galileo Project). Microprocessor communication to the PPR electronics is accomplished using I/O port controllers. Most port signals are buffered to eliminate possible failure modes which could damage the microprocessor bus and cause total instrument failure.

In spite of the small ROM memory size, the instrument control program is extensive, controlling all of the commandable optional functions, the command processing and data formatting, the instrument analog and multiplexer timing, the closed-loop filter-retarder wheel and chopper operation, and the power system synchronization. A 3-byte PPR configuration control command is used to select the detailed operation of the PPR. Figure 9 illustrates typical PPR operation in the cycle mode as the instrument steps through the polarimetry, radiometry, and photometry measurement sequences, with the dc restore, chopper initiate, chopper stop and cal lamp activate functions indicated. The commandable functions of the PPR and their brief description are summarized in Table III.

3.3.3. *Actuator Controls*

The 32-position filter/retarder wheel is driven by three sets of bipolar transistor switches, which are in turn controlled from 6 bits of a microprocessor port. The binary brush-type encoder located in the filter/retarder motor is read into another port, thereby giving full closed-loop position control to the microprocessor. To ensure rapid motor damping in the final filter position, the processor momentarily turns on the appropriate motor switches to short the motor windings, thus allowing the motor to be stepped in less than 0.2 s. The brush-type encoder and filter motor were unchanged from the previous Pioneer Venus Orbiter CPP instrument, having demonstrated extreme reliability (more than 12-year operation in Venus orbit).

Amplitude and phase control are provided for the tuning fork chopper by a combination linear and switching servo system. The tine driver coil is energized by pulse-width-modulated bipolar transistor switches, and the signal from the tine position feedback coil is processed to provide a linear amplitude error signal. Zero crossings of the feedback signal are phase-compared with a chopper reference pulse from the microprocessor. The servo electronics then meter an appropriate width and timed drive pulse to correct both the amplitude and phase error during each chopper cycle. Servo control is complex because of the requirement to rapidly accelerate the chopper when entering the portion of the filter/retarder wheel containing the radiometry positions and then to damp the oscillations to unblock the optical path when starting the photometry filter positions. Rapid acceleration is obtained by applying a higher than normal power turn-on pulse, and active damping is achieved by reversing the phase of the pickoff signal, thereby providing negative feedback until the fork position feedback signal is small enough to ensure that the optical path is clear.

3.3.4. *Power Subsystem*

Power converter circuitry, located in the bottom section of the electronics module, operates from the 30 V balanced dc spacecraft bus. Bus protection is provided by redundant instrument fuses, EMI filters, and in-rush current limiting inductors. There are two separate transformer coupled power converters, one to supply the pulsed power for the actuators and calibration lamp, and the other to supply the 'cleaner' electronics power. A low-headroom, wide-bandwidth series regulator supplies the electronics converter to ensure that spacecraft bus voltage transients are not transferred to the electronics loads. Both converters are synchronized to a sub-multiple of the spacecraft clock signal. The power supply module also contains all of the power transistor switches for the actuator controls as well as the calibration lamp precision voltage regulator so that EMI generated by the current surges is confined inside the power supply. The 28 V motor voltage is used as input to the calibration lamp series regulator so that regulator failure will cause the 5 V lamp to burn out, though otherwise allowing normal instrument operation (but without calibration). A ground commandable relay switches the calibration lamp regulator from the internal lamp to another lamp located within the RCT-PPR mounted on the spacecraft.

Electronics supply voltages are LC-filtered and regulated using monolithic linear regulators operated with low headroom for efficiency. The logic supply is unregulated, relying on the pre-regulated input to the electronics converter. Nominal instrument power consumption is 3.8 W without actuator operation, increasing to approximately 11 W during the short duration filter/retarder motor steps.

4. Instrument Calibration

4.1. Preflight calibration

The preflight testing of the assembled PPR consisted of tests performed to verify the mechanical integrity and stability of the instrument (environmental tests such as vibration and temperature cycling) and those associated with the characterization of instrument performance and calibration. Consistent with the scope of this paper, only the latter type of testing will be discussed here. (The instrument thermal/vacuum testing often is performed in a manner to satisfy simultaneously the calibration and environmental test functions.) The performance and calibration testing can be broadly divided into those tests done at ambient temperature and pressure in a laboratory environment and those performed in a thermal/vacuum chamber. Some of the more important preflight calibrations are shown in Table IV. Several of these tests which are important

TABLE IV

Selected preflight calibration tests for the Galileo PPR

Test name	Measurement procedure
Radiometry (RAD channels except solar band)	View BCUs at various temperatures in T/V chamber while varying PPR optical element heaters with BCUs held at < 188 K
Radiometric calibration (POL/PH channels plus solar and S + T channels)	View halon reflectance standard illuminated by standard lamp (irradiance standard)
Gain/linearity (POL/PH channels plus S + T bands)	Measure PPR gain steps by viewing OCU with various selected OCU neutral density filters
Signal-to-noise performance	Collect applicable data during radiometric calibrations and gain and linearity testing
Polarimetric calibration	View OCU and/or standard lamp/reflectance standard with and without interposing rotatable polarizing elements for the PPR in its normal orientation and rotated by 90°
Relative spectral response	Measure spectrophotometer output with PPR 410 nm, solar and S + T bands with a calibrated silicon photodiode or electrically calibrated pyroelectric detector. Cascaded optical element measurements used for other bands
Temperature-dependent response	Performed as part of T/V calibration testing with BCUs or the OCU viewed with the PPR at various temperatures
Air-to-vacuum responsivity shift	Measured by viewing OCU through T/V chamber window with chamber at ambient pressure and evacuated
Integrated spectral out-of-band response	View known sources through colored glass and other materials with well-defined cut-on/cut-off characteristics
Field-of-view response	View source of small angular subtense scanned across PPR field of view
Scattered light (off-axis) response	View extended solar illuminated white target with a central circular portion of target removed and with target intact. Direct off-axis viewing of Sun for $> 18°$ off-axis response
Optical axis alignment	Compare alignment of field stop center and alignment mirror to PPR reference surfaces using theodolite

to the ultimate performance of the instrument or have novel or less well-known implementations will be described in the following.

4.1.1. *Ambient Calibrations*

Of particular importance here are the optical calibrations. These include the measurement of the radiometric calibration of the channels that measure the radiance resulting from reflection of the incident solar flux. Due to the spectral radiance differences between the sources used in the radiometric calibrations and the scene radiance, relative spectral response measurements were made on the wide PPR spectral bands, namely, the polarimetry band and the solar reflectance band. The relative spectral response for the other PPR bands was determined from individual component measurements. The majority of the detailed polarimetric calibrations also were performed in an ambient environment. Supplemental measurements to assess any air-to-vacuum and temperature-dependent instrument response were performed as a part of the thermal/vacuum environment testing.

4.1.2. *Radiometric Calibration* (*Photopolarimetry Channels*)

Halogen-cycle tungsten-filament lamps with spectral irradiance calibrations traceable to the National Institute of Standards and Technology (NIST) were used in a fixed geometry with a diffuse reflectance target of known characteristics. The lamp spectral irradiance uncertainty is better than 1.35% (3σ NIST uncertainty value) with an additional estimated transfer uncertainty of 0.7% over the photopolarimetry spectral range. The reflectance target consists of pressed Halon (polyfluoroethylene) fabricated by a well-controlled procedure to produce a target of known bi-directional reflectance (Weidner and Hsia, 1981). The target is illuminated at normal incidence at a fixed source-to-target distance (100 cm) and is viewed at a 20° angle of incidence. During each of these PPR radiometric calibrations (spanning a period of more than 6 years between the first and last 'final calibration'), measurements were taken with two different lamps to permit cross-checking between the resultant calibrations and to improve the likelihood that during any future calibration the same lamp would be available. The above considerations provide the basis for estimating the PPR radiometric calibration accuracy.

4.1.3. *Radiometric Calibration* (*Thermal Channels*)

The preflight radiometric calibration of the thermal channels was performed as part of the thermal/vacuum environmental testing of the instrument. For this purpose, blackbody targets were designed to overfill the apertures of the scene-view and space-view telescopes. These blackbody calibrator units, designated BCU-1 and BCU-2, are identical except that the aperture diameter of BCU-1 is larger than for BCU-2 in order to accommodate the larger scene-view telescope. The BCUs are conical blackbody sources with a 30° apex angle. The interior conical surfaces are painted with a gloss black paint applied with a thickness between 100 to 150 μm to provide a low reflectance coating such that the resulting BCU normal emittance is greater than 0.999. Temperature monitoring of the BCU is provided by two embedded 100-ohm platinum resistance thermometers (PRTs), with dynamic temperature control provided by a combination of liquid/gaseous nitrogen flow and servo control heaters contained in the rear of the BCU cone.

Two types of test sequences were used for the radiometric calibrations. In one sequence, the blackbody targets were stepped through a range of temperatures of interest while recording the thermal channel signal in response to target and PPR optical element temperatures. Similar tests were performed with the RCT-PPR replacing BCU-1. The second calibration sequence involved the sequential heating of individual optical elements (until stabilized with a temperature rate of change less than 0.02 K min^{-1}), followed by cooling to the original temperature. The above sequences were used to determine the PPR thermal responsivity to changes in the scene- and space-view radiances, and to changes in optical element temperature, and thus establish the radiometric calibration coefficients for individual optical elements.

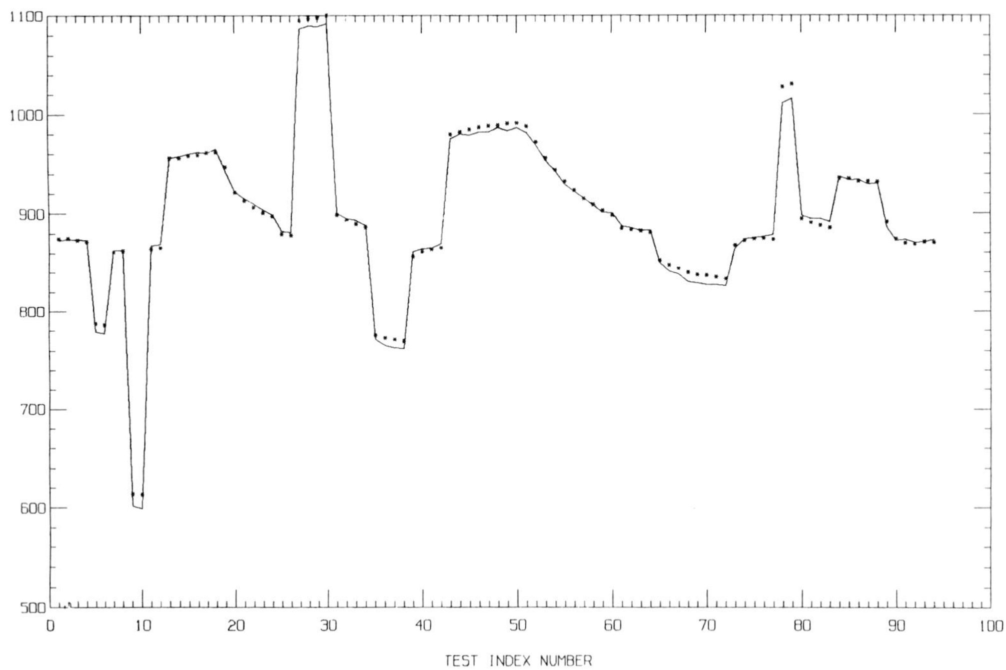

Fig. 10. Sample thermal/vacuum optical element heater test summary at +5 °C.

During the radiometric calibration tests, the instrument reference temperature was controlled to better than ± 3 K. Typically, these tests were 8 to 10 hours in duration for the BCU temperature sequences and 12 to 16 hours for the optical element sequences. For both types of tests, significant variations in instrument temperature (and temperature gradients) were found to occur the course of testing at each 'nominal' instrument temperature. As a result, it was necessary to perform a least-squares solution to the set of radiometric equations representing optical element, scene-view, and space-view contributions to model the instrument response signal.

Figure 10 summarizes the results of an optical element heater test for a nominal instrument temperature of 5 °C. Here, the time scale (which is discontinuous) is expressed in terms of an index number of the individual tests performed. The solid line is a piece-wise continuous readout of instrument response in data numbers (DN) as individual heater elements are turned on and off, while the plotted points are the expected instrument response using the least-squares fitted calibration coefficients and the recorded optical element temperature variations. Tests of this type were conducted for each radiometry band over the expected instrument operating temperature range.

While these radiometric calibration tests provide a good understanding of the instrument response to the scene/space target signal and to changes in instrument temperature gradients, the thermal vacuum test simulations could not be expected to reproduce the precise thermal environment that will exist for the Galileo spacecraft in Jupiter orbit. Thus, inflight calibration testing will be used to verify and refine the calibration coefficients by sequentially viewing (with the scene-view telescope) the RCT-PPR and then space.

4.1.4. *Polarimetric Calibration*

Converting the measured signals for each polarimetry band to polarization degree and azimuth can be done in a number of ways using the three pairs of measurements obtained as the halfwave retarder is positioned with the retarder fast axis at 0°, 22.5°, and 45° with respect to the instrument reference direction. It is convenient to denote the measurement pairs (with the instrument off-set subtracted) at these three positions by M_{1S} and M_{1P}, M_{2S} and M_{2P}, and M_{3S} and M_{3P}, respectively, where 'S' and 'P' represent channel designations sensing the perpendicular and parallel polarization azimuths, respectively. At these positions, the action of the halfwave retarder produces an optical rotation of the linearly polarized components of the scene flux by -2α, $-2\alpha+45°$, and $-2\alpha+90°$ where α is the azimuth of the linearly polarized component of the incident flux relative to the reference direction.

As indicated above, the third (45°) retarder position effectively produces an optical interchange of the roles of the S and P channels compared to measurements taken at the first (0°) retarder position. The optical interchange of channels provides a means to determine the responsivity ratio, K, relating the 'P' channel responsivity to the 'S' channel responsivity from

$$K = [(M_{1S}M_{3S})/(M_{1P}M_{3P})]^{1/2} .$$

Similarly, the effective band passfilter ratio, C_2 (to account for any small difference of filter transmittance between the filters used for positions 1 and 2), can be determined from

$$C_2 = (M_{1S} + KM_{1P})/(M_{2S} + KM_{2P}) ,$$

with an analogous relationship for position 1 and position 3 measurements.

The components of the Stokes' vector of the scene radiance can then be expressed in terms of the above quantities as

$$I = M_{1S} + KM_{1P} = C_2(M_{2S} + KM_{2P}) = C_3(M_{3S} + KM_{3P}),$$

$$Q = KM_{1P} - M_{1S} = C_3(M_{3S} - KM_{3P}),$$

$$U = C_2(KM_{2P} - M_{2S}).$$

The polarization degree, P (in %), and azimuth, α, of the linear polarization of the scene follows as

$$P = (100/I)(Q^2 + U^2)^{1/2}, \qquad \alpha = 0.5 \arctan(U/Q).$$

The photometry band measurements can also provide scene polarization information. These measurements are essentially equivalent to M_{1S} and M_{1P}. Thus, I and Q can be determined, and these are adequate to characterize the linear polarization of the scene in cases where the polarization azimuth is known (as can be established from measurements with the three polarimetry bands).

Polarization data reduction of cycle and PP/PH mode typically will use all six measurements obtained for each polarimetry band, i.e., outputs from both silicon photodiode detectors at each of the three successive retarder positions. While the desired radiance (when the appropriate instrument radiance responsivity factor is applied to I), linear polarization degree, and polarization azimuth constitute only three quantities to be determined, the redundancy of the six measurements can be exploited to monitor K, C_2, and C_3 or to improve the precision of the determination. In addition, since the intensity I is determined at each position, scene radiance changes are monitored throughout the measurements at the three retarder positions. Similarly, the equivalency of the position 1 and 3 measurements (see equation for Q) can be used to verify constancy of the scene polarization during the measurement period.

The approach of determining K, C_2, and C_3 from each set of the measurements has a disadvantage under special circumstances where K cannot be well-determined. Such a condition occurs when the radiance viewed has a linear polarization near 100% and the polarization azimuth is aligned nearly parallel or perpendicular to the PPR reference direction. In this situation, one term in the numerator and one in the denominator in the expression for K are near zero, and hence the equation is ill-conditioned. To avoid this problem, the data processing software was designed to allow the option to process the data using previously determined, read-in values of K, C_2, and C_3. The use of the appropriate read-in values has the additional advantage of better accuracy for calibration measurements since sufficient averaging of data can be used to eliminate any noise not associated with the quantization. (We note that K can be expected to be slightly dependent on the gain level being used since the gain step increments may be slightly different for the two channels.)

The calibration of the PPR polarimetry channels involved viewing various spatially uniform sources with predictable polarization characteristics. Scenes with linear polarizations approximating 100%, 8%, and 0% were used. Most of the measurements utilized either the standard lamp/reflectance standard or the optical calibrator unit (OCU). Since the xenon arc lamp in the OCU provides a near solar spectral distribution for the PPR 410 nm band, this yields a much better match to the expected spectral radiance of Jupiter, thereby improving the polarimetric accuracy for the intended application. For the narrow 678 nm band, either of the sources provides adequate source/scene spectral matching. The spectral line structure in the region of the 945 nm PPR band of the OCU xenon arc lamp together with the better stability of the tungsten filament lamp of the standard lamp/reflectance standard combination makes the latter source the preferred choice for the 945 nm band polarimetry calibrations.

TABLE V

Typical data to determine instrumental polarization
Measured degree of polarization in %, (azimuth in deg)

Source used	PPR in normal orientation			PPR rotated 90°		
	945 nm	678 nm	410 nm	945 nm	678 nm	410 cm
Optical calibrator unit	0.38 (18.0)	0.10 (143.8)	0.29 (165.8)	0.42 (15.4)	0.15 (155.7)	0.40 (167.6)
Standard lamp and reflective standard	0.46 (18.6)	0.13 (159.7)	0.34 (170.2)	0.35 (12.2)	0.18 (142.9)	0.36 (164.7)

To measure the instrumental polarization of the PPR, a low polarization source was viewed both with the PPR mounted in the usual orientation, i.e., with the instrument baseplate horizontal, and also with the PPR rotated about its optical axis by 90°. This was performed with the PPR mounted on the PPR handling fixture, since it is designed such that the PPR optical axis is equidistant from the two 'base' surfaces and also equidistant from the reference 'edge' for each of these surfaces. This allows the sources to be viewed sequentially with the PPR baseplate horizontal and then with the PPR baseplate vertical. Typical data taken to determine the PPR instrumental polarization are given in Table V. Since the small polarizations measured change only slightly in magnitude and the azimuth also is little changed, i.e., follows the rotation of the instrument, it is apparent that the small polarization values measured for each band are primarily instrumental polarization. From the data it can be concluded that the PPR instrumental polarization is approximately 0.35%, 0.15%, and 0.40% for the 410, 678, and 945 nm bands, respectively. Correspondingly, it is found that the linear polarization of both the standard lamp/reflectance standard combination and the OCU is less than 0.1% for all three spectral bands, thus indicating that these source configurations have achieved the desired low degree of source polarization.

With either of the source configurations, the approach for both the high (near 100%) and moderate (near 8%) polarization conditions was to locate the polarizing element between the uniform radiance source and the PPR. For the near 100% linearly polarized radiance, two parallel dichoic polarizers mounted with the high transmission directions oriented parallel to each other were used. An HN22 polarizer set was used for the 410 and 678 nm bands with a similar HR polarizer set employed for the 945 nm band. When in use, a polarizer set was mounted on a rotatable stage which permitted the polarizer azimuth to be set accurately to any angle through a full 360° range. Similarly, a plane parallel plate of Schott WG280 glass was mounted such that it could be installed on the rotatable stage so that it remained tilted at 45° to the rotation axis of the stage during rotation. This rotatable tilted plate allows a known (caculable using the known wavelength-dependent refractive index of the plate via the Fresnel equations) linear polarization to be viewed, i.e., for the 410, 678, and 945 nm bands, $P = -8.32\%$, -8.06%, and -7.97%, respectively, at 45° angle of incidence. (These polarization values for the transmitted beam include the effect of multiple reflections and are averaged over an integral number of cycles of interference – justified due to the width of the spectral bands and the 2.5 mrad field of view of the instrument.)

4.2. Inflight Calibration

Several means of inflight calibration will be utilized by the PPR to update the preflight calibration of the photopolarimetry channels and to provide the prime radiometric calibration of the radiometry channel. These include: (1) an internal calibration lamp within the PPR aft optics; (2) the third orientation position for each of the halfwave retarders which interchanges the roles of the two silicon detectors; (3) the radiometric calibration target (RCT-PPR) that is separately mounted on the spacecraft and can function either as a blackbody source or to provide a slightly polarized lamp output signal for the visible/near-infrared region; (4) a spacecraft supplied photometric calibration target (PCT) that provides a standard of spectral radiance that can be viewed by all instruments located on the spacecraft scan platform; and (5) viewing stars and spatially unresolved planets.

4.2.1. *Internal Cal Lamp*

The small tungsten filament lamp located within the aft optics of the PPR provides a means to track any changes with time of the silicon detector/amplifier channels. The spectral output of the lamp is modified with the use of a color glass filter (Schott BG-18). The stability of this long-life (50 000-hour) lamp is further enhanced by operating it only intermittently, at a derated power level, and with a controlled, slow turn-on characteristic. The lamp is energized intermittently only during cycle mode operation while the chopper is being driven to its rest position following radiometry mode sampling at the solar + thermal filter wheel position and if the cal lamp command bit is set to the ON state. (The estimated total on time is less than 200 hours during the 7-year pre-launch testing period plus the post-launch inflight period.)

4.2.2. *Internal Polarimetric Calibration*

As previously discussed, one key feature for achieving accurate polarimetry with the PPR is the ability to cross-calibrate the detectors by measuring simultaneously the orthogonal polarization components of the scene radiance. The third halfwave retarder position (with fast axis oriented at 45° to the plane of deviation of the Wollaston prism) effectively interchanges the scene polarization components incident on the two detectors. This permits maintaining the polarimetric accuracy even in the presence of slow relative changes of the two detector channels with time, temperature, radiation, etc.

4.2.3. *RCT-PPR Design*

The RCT-PPR will serve in a dual calibration role for the PPR. The primary role will be as a thermal calibration target which closely approximates a blackbody source when viewed along the RCT-PPR axisymmetric axis. Due to spacecraft space limitations, it was necessary to restrict the overall length of the target. To achieve the desired normal emittance ($\varepsilon > 0.998$) the geometry of the interior portion of the target has a truncated conical form with a center cylindrical section. This provides on-axis performance approximately equivalent to a cone with half the apex angle and twice the overall length. The end of the central cylindrical portion is not viewed by the PPR since this area is within the central obscuration of the PPR telescope. The interior of the target is a smooth (specularly reflecting) black-painted surface to achieve a higher on-axis emittance than would be possible with a rough (diffuse) surface in the same geometry. The calculated on-axis emittance of the RCT-PPR is greater than 0.998 based on the reflectance versus incident angle for the interior surfaces.

The RCT-PPR is designed and mounted such that it will be passively cooled at Jupiter to a temperature of 145 ± 15 K. The wall thickness is chosen to assure worst case temperature gradients of less than 0.5 K. The temperature of the RCT-PPR is monitored by two platinum resistance thermometers (PRTs) that are calibrated by the manufacturer (resistance versus temperature) to an accuracy 0.2 K. These PRTs are read out directly by the PPR along with a low temperature coefficient resistor also mounted on the RCT-PPR to allow a first-order correction for spacecraft cabling resistance. The annular aperture of the target is designed to accommodate the 3-σ worst case relative misalignments resulting from possible spacecraft environmental and mounting factors specified by the Galileo Project to assure that the PPR will view only the high emittance portion of the target during calibration. Through the use of the RCT-PPR and the preflight calibrations used to assess the influence of temperature changes of the PPR optical elements, it is expected that the overall radiometric calibration of the PPR thermal bands can be maintained within the desired ± 1 K over the duration of the Galileo Mission.

A small tungsten-filament lamp is mounted in one portion of the RCT-PPR interior surface. With the source commanded ON, flux from the source passes through an elliptically shaped, plane-parallel sapphire plate mounted such that the outer surface approximately conforms to the inner conical surface of the target. The flux transmitted to the PPR is partially polarized due to different S and P Fresnel reflectances of the inclined plate. Thus, this source will be useful in assessing possible photometric and polarimetric changes of the entire optical train of the PPR over the course of the mission.

4.2.4. *Spacecraft PCT*

The PCT is intended to serve as a standard of spectral radiance for the scan platform mounted instruments by reflecting sunlight from a diffusely reflecting surface with well-characterized reflectance properties. Since this target can be viewed by all scan platform mounted instruments, the PCT is expected to be particularly useful in the role of calibration intercomparison among instruments.

4.2.5. *Viewing Astronomical Objects*

Orienting the PPR to view such astronomical objects as stars or spatially unresolved planets will be used to provide both cross-check of the absolute photometric calibration of the PPR silicon photodiode channels and as an additional means to track any responsivity changes with time, temperature, radiation, etc. Sirius is a star which will provide an adequate signal-to-noise ratio for this purpose by aggregating a sufficient number of samples. Similarly, viewing unresolved (object subtending less than the PPR field of view) planets at phase angles accessible from Earth will allow intercomparisons to be made with ground-based photometric calibration.

4.3. Signal-to-noise performance

4.3.1. *Photopolarimetry Channels*

Four separate band gains are used for the photopolarimetry channels, with the value applied (as described in the electronics section) being dependent on the filter/retarder wheel position as determined by the encoder. This will provide signal outputs of similar magnitude for the three polarimetry and seven photometry bands for typical scene spectral radiances. The channel and band gains were set to provide signal levels at Gain Step 8 of approximately 2000 DN for the three polarimetry bands (each with a separate band gain) and of approximately 1500 DN for the 648 nm photometry band (a single band gain is applied to all seven photometry bands). For setting these levels, the Jovian albedo values of Woodman *et al.* (1979) were used.

The noise of the photopolarimetry channels is essentially independent of signal level, resulting primarily from the 100 megohm feedback resistors in the pre-amplifiers. As a result, the signal-to-noise ratio (SNR) varies only slowly with temperature over the PPR operating temperature range (since the Johnson noise varies as the square root of the absolute temperature). The measured SNR performance of the PPR photopolarimetry channels substantially exceeds the science-dictated, minimum SNR requirements of 1000 for the polarimetry bands and 200 for the photometry bands.

4.3.2. *Radiometry Channel*

Achieving the SNR performance desired for the PPR science investigations utilizing the radiometry channel produces far greater stress on instrument design that is the case for the photopolarimetry channels. The inevitable Galileo mission mass and size constraints on science instruments required substantial compromise on performance characteristics. To optimize the SNR performance of the pyroelectric detector required thinning the $LiTaO_4$ detector element to the maximum extent possible. For the PPR application, ion-beam milling was used to provide thicknesses in the 5 to 6 μm range. In order to provide good optical absorption with low mass, the detector was coated with an evaporated gold-black coating. The wide range of absorbing characteristics found in the literature for gold blacks increased the risk with this approach, but on balance offered the best overall choice for the PPR requirements.

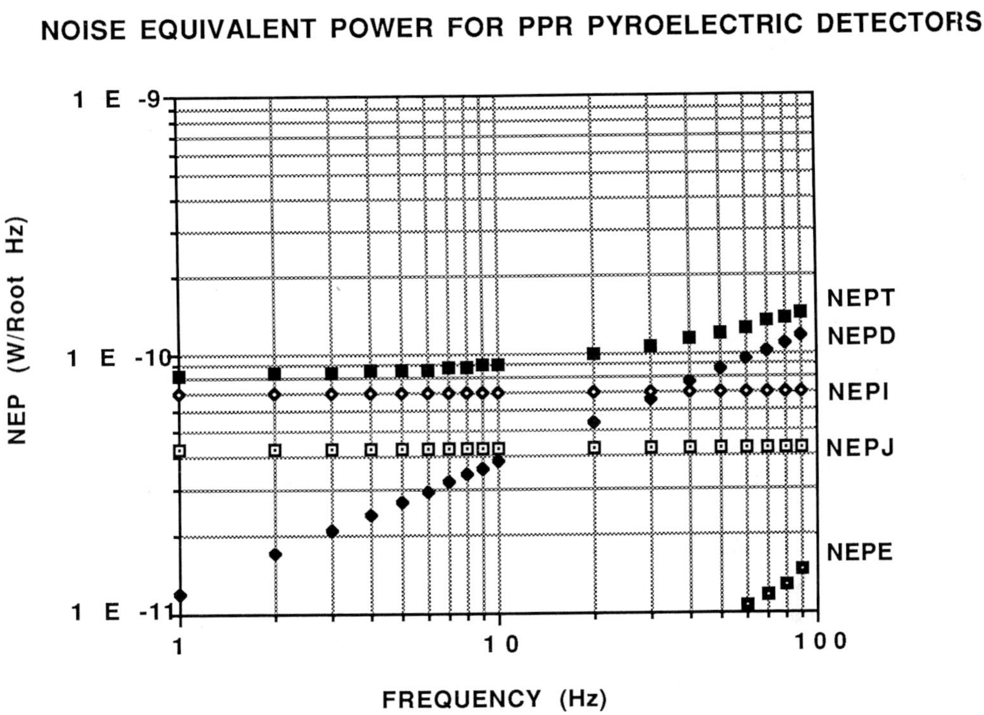

Fig. 11. Galileo PPR pyroelectric detector channel noise equivalent power (NEP). NEPT is total root-sum-squared combination of the NEP components due to: dielectric loss tangent noise (NEPD); pre-amplifier FET gate leakage current noise (NEPI); pre-amplifier voltage noise (NEPE); and Johnson noise of the detector load resistor (NEPJ).

Optimization curves for the noise components of the PPR pyroelectric detector is illustrated in Figure 11. Measured noise data for the PPR detector were near the levels predicted. However, relative spectral response measurements indicated levels substantially below specification. The lower than expected long wavelength responsivity, combined with the lower than specified filter transmittance for some of the filters (much lower for the 37 μm band) led to the inability to meet the instrument SNR performance specifications for four of the seven radiometry channel bands. The measured versus specified SNR performance is indicated in Table VI. The solar plus thermal band measurement tabulated includes only the thermal component; the solar band is not included in the table, but comfortably exceeds the specifications. Three of the four out-of-spec bands have SNR performances about 60% of specification, while the 37 μm band (*D* filter) is about one-third of the desired level.

TABLE VI

Measured vs specified signal-to-noise performance for PPR radiometry

Radiometry band	Signal-to-noise ratio (SNR)[a]	
	Measured[b]	Specification
17.0 μm	30	30
21.0 μm	20	35
27.5 μm	41	65
37.0 μm	12	40
$\lambda > 42 \pm 3$ μm	33	50
Solar + Thermal	460[c]	250

[a] SNR values are based on 4-sample averages with an instrument temperature of 0 °C.

[b] Data corrected to correspond to a 130 K blackbody scene (data measured with BCU-1 = 180 K and BCU-2 = 88 K.

[c] Measurement only includes thermal component for a 130 K blackbody scene.

Fortunately the mission profile and the flexibility designed into instrument operation allows for observational 'work-arounds' to achieve nearly all of the anticipated science goals. The obvious approach of increasing the number of samples to improve the SNR (by the square root of the increase factor) is the principal observation strategy to achieve the radiometry science goals.

Acknowledgements

We wish to acknowledge the significant contributions made by many others in the development of the PPR experiment. In particular, we want to thank J. Hansen, who as Principal Investigator has provided direction and leadership for the PPR investigation, and former Co-Investigator D. Coffeen for his important contributions and insights. Other members of the PPR science team are M. Allison, A. Del Genio, and W. Rossow, all at the Goddard Institute for Space Studies; T. Martin and G. Orton at Jet Propulsion Laboratory; Y.-L. Yung at the California Institute of Technology, and P. Stone at the Massachusetts Institute of Technology.

Key supporting effort at SBRC in the design, development, and fabrication of the PPR instrumentation was provided by E. Aasted (electronic test), J. Abid (digital electronics), J. Bauer, Jr. (thermal analysis), S. Benda (quality assurance), J. Brooks (dynamic analysis), R. Cline (optical design), C. Coles (digital electronics and test), D. Errett (optical design), G. Hughes (system test), G. Hyde (software programming), W. Jefferies (optical procurement), W. Johnson (electronics packaging), H. Kleeburg (servo control), P. Olson (mechanical design), S. Pellicori (optical filters), F. Pineda (mechanical assembly), C. A. Rodil (reliability), H. Sanchez (electronic assembly), J. Shields (optical assembly), G. Stark (manufacturing process), and L. Watts (program management).

We express our thanks to our technical officers, M. Davis and A. Peterson of the Goddard Space Flight Center for their dedicated effort. Other key personnel providing essential support were the instrument coordinators, D. Griffith, G. McSmith, R. Parrish, and L. White; the science coordinators, N. Devine and T. Martin; and the science instrument manager, W. Fawcett, all with the Jet Propulsion Laboratory.

References

Hanel, R. A., Conrath, B. J., Herath, L. W., Kunde, V. G., and Pirraglia, J. A.: 1981, *J. Geophys. Res.* **86**, 8705.

Hunten, D. M., Colin, L., and Hansen, J. E.: 1986, *Space Sci. Rev.* **44**, 191.

Pellicori, S. F., Russell, E. E., and Watts, L. A.: 1973, *Appl. Optics* **12**, 1246.

Seeley, S. J., Hunneman, R., and Whatley, A.: 1981, *Appl. Optics* **20**, 31.

Tomasko, M. G., West, R. A., and Castillo, N. D.: 1978, *Icarus* **33**, 558.

Watts, L. A., Russell, E. E., and Pellicori, S. F.: 1977, *SPIE* **112**, 28.

Watts, L. A., Russell, E. E., and Pellicori, S. F.: 1979, *SPIE* **172**, 326.

Weidner, V. and Hsia, J.: 1981, *J. Opt. Soc. Am.* **71**, 856.

Woodman, J. H., Cochran, W. D., and Slavsky, D. B.: 1979, *Icarus* **37**, 73.

GALILEO RADIO SCIENCE INVESTIGATIONS

H. T. HOWARD, V. R. ESHLEMAN, D. P. HINSON

Center for Radar Astronomy, Stanford University, Stanford, CA 94305, U.S.A.

A. J. KLIORE, G. F. LINDAL, R. WOO

Jet Propulsion Laboratory, U.S.A.

M. K. BIRD, H. VOLLAND

Radioastronomisches Institut, Universität Bonn, Germany

P. EDENHOFER

Institut für HF-Technik, Universität Bochum, Germany

M. PÄTZOLD and H. PORSCHE

Deutsche Forschungsanstalt für Luft- und Raumfahrt, Germany

Abstract. The radio science investigations planned for Galileo's 6-year flight to and 2-year orbit of Jupiter use as their instrument the dual-frequency radio system on the spacecraft operating in conjunction with various US and German tracking stations on Earth. The planned radio propagation experiments are based on measurements of absolute and differential propagation time delay, differential phase delay, Doppler shift, signal strength, and polarization. These measurements will be used to study: the atmospheric and ionospheric structure, constituents, and dynamics of Jupiter; the magnetic field of Jupiter; the diameter of Io, its ionospheric structure, and the distribution of plasma in the Io torus; the diameters of the other Galilean satellites, certain properties of their surfaces, and possibly their atmospheres and ionospheres; and the plasma dynamics and magnetic field of the solar corona. The spacecraft system used for these investigations is based on Voyager heritage but with several important additions and modifications that provide linear rather than circular polarization on the *S*-band downlink signal, the capability to receive *X*-band uplink signals, and a differential downlink ranging mode. Collaboration between the investigators and the spacecraft communications engineers has resulted in the first highly-stable, dual-frequency, spacecraft radio system suitable for simultaneous measurements of all the parameters normally attributed to radio waves.

1. Introduction

Galileo radio science comprises two main areas of investigation, each with its own experiment team. The radio propagation team, with members from Stanford, JPL, Bonn, Bochum, and DLR, has responsibility for investigations of the solar corona and a variety of aspects of the Jovian system. The celestial mechanics team, with members from JPL, will search for gravitational waves, conduct other experimental tests of general relativity, and investigate the gravitational fields of Jupiter and its satellites. The members of the propagation team, selected by NASA and BMFT/DLR*, and their associates are the authors of this article. The celestial mechanics investigations are described in a companion paper (Anderson *et al.*, 1992).

* Bundesministerium für Forschung und Technologie/Deutsche Forschungsanstalt für Luft- und Raumfahrt.

Space Science Reviews **60:** 565–590, 1992.

The radio science investigations employ telecommunications equipment onboard the Galileo spacecraft along with a network of tracking stations on Earth operated by the United States and Germany. One of the important preflight functions of the two radio science teams has been to use scientific experience gained from previous missions to influence the design of the relevant instrumentation. As will be seen in this paper, what has resulted from this collaboration with the Galileo project is a highly capable radio science system.

Figure 1 shows the trajectory of Galileo during its 6-year flight to Jupiter, which

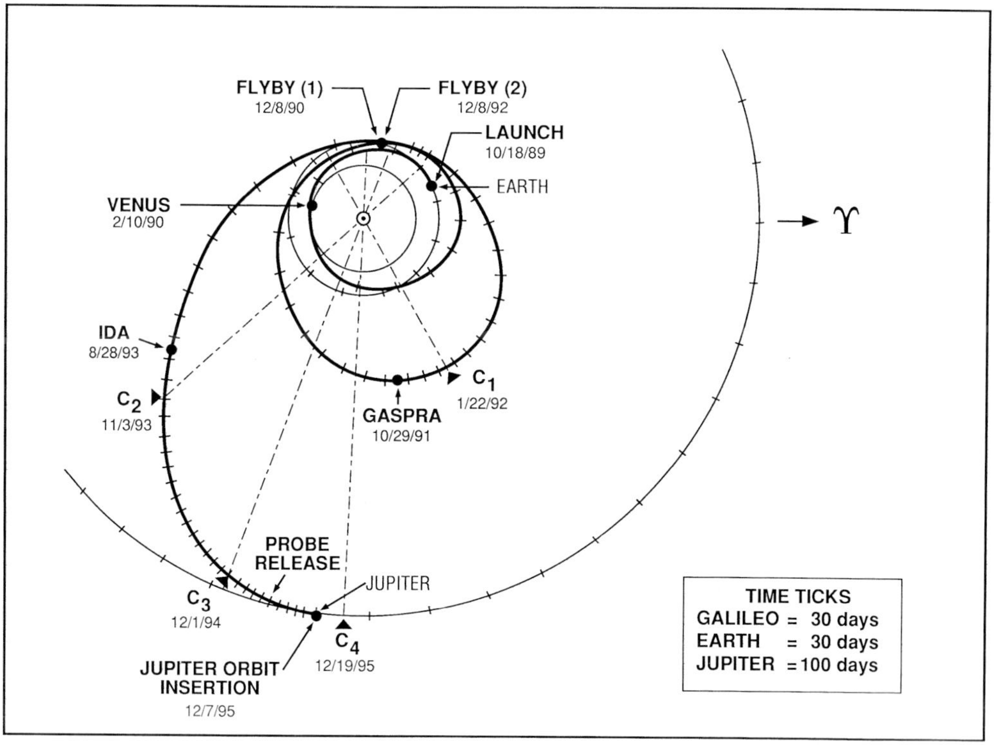

Fig. 1. Trajectory of the Galileo spacecraft (bold line) during its 6-year flight to Jupiter. The orbits of Venus, Earth, and Jupiter are also shown. Targets of opportunity for investigations during this 'cruise phase' of the mission include Venus, Earth, and the asteroids Gaspra and Ida. The trajectory also provides observing geometries favorable to a search for gravitational waves; the best opportunities occur near solar opposition where the influence of the solar wind on the radio link with the spacecraft is at a minimum. Conversely, remote sensing of the solar corona will be conducted during solar conjunctions when radio signals propagating between spacecraft and Earth pass near the Sun. Three conjunctions occur during cruise and one follows shortly after orbit insertion at Jupiter; the spacecraft positions for these events are labeled C_1, C_2, C_3, and C_4.

includes geometries favorable to several radio science investigations. Of interest here are the solar conjunctions during which radio signals propagating between spacecraft and Earth pass near the Sun, providing an opportunity for remote sensing of the structure and dynamics of the solar corona. After arriving at Jupiter, the spacecraft will execute a 2-year nominal mission comprising ten to eleven orbits of the planet. This tour of the Jovian system will include repeated occultations of the spacecraft by Jupiter as well as occultations by and close encounters with each of the Galilean satellites. Radio occultation and bi-static radar measurements during these events will extend our understanding in diverse areas ranging from atmospheric dynamics on Jupiter to the properties of the surfaces of the icy Galilean satellites.

This paper is organized as follows. The next section describes the instrumentation and some of the basic measurement techniques used in radio science experiments. Subsequent sections describe the various radio propagation investigations roughly in their order of occurrence in the mission: the solar corona, atmospheres and ionospheres in the Jovian system, and the surfaces of the icy Galilean moons. Readers who are interested primarily in anticipated results can skip directly to the later sections; we have attempted to write these so that each can be read and understood independent of all others.

2. Radio Science Instrumentation and Measurement Techniques

2.1. Instrumentation

The telecommunications equipment on Galileo closely resembles that of the Voyager spacecraft (Eshleman *et al.*, 1977; Tyler, 1987). The basic system includes the capability to transmit two radio signals at coherently-related frequencies of about 8400 MHz (X-band) and 2300 MHz (S-band). The exact frequencies of the downlink signals depend on the operating mode of the spacecraft equipment but are always in the ratio of 11/3. The two signals are radiated through the main spacecraft antenna, a 4.8-m paraboloid attached to the spinning portion of the spacecraft and aligned with the axis of rotation. Table I summarizes key characteristics of the Galileo spacecraft radio system.

Two operating modes, selectable by commands sent to the spacecraft, are used in radio science experiments. In the 2-way coherent mode, used, for example, in tracking the motion of the spacecraft, the downlink frequencies are controlled by the coherent turnaround of an uplink signal received at either S- or X-band. Alternatively, the downlink frequencies can be derived from an onboard, radiation-hardened, temperature-controlled, ultrastable oscillator (USO) identical to the ones flown on Voyager (Eshleman *et al.*, 1977; Tyler, 1987). This latter configuration is necessary for measurements of the atmosphere and ionosphere of Jupiter and for satellite occultations for two reasons. First, it is difficult to maintain a 2-way radio link with the spacecraft when signal strength and frequency exhibit dynamic variations as will occur in the Jupiter occultations. Second, the time delay in establishing a 2-way radio link after the downlink signal has been absent would result in a serious loss of data at emersion from the satellite occultations.

The Galileo radio system is enhanced relative to Voyager in three respects. First, the

TABLE I

Selected Galileo spacecraft radio system parameters (nominal)

	X-band	S-band
Transmitting parameters		
Frequency (MHz)	8415	2295
Wavelength (cm)	3.6	13.1
X/S coherency ratio	11/3	
Transmitting RF power (W)	12 or 21	9 or 27
4.8 m antenna gain (dBi)	50	38
Half-power beam width (deg)	0.6	1.5
Polarization	LCP or RCP	Linear
Axial ratio (dB)	2	32
Receiving parameters		
Frequency (MHz)	7167	2115
Wavelength (cm)	4.2	14.2
4.8 m antenna gain (dBi)	46	36
Polarization	LCP or RCP	Linear
Noise temperature (K)	270	1000
Ranging channel noise bandwidth (MHz)	1.5	

spacecraft can support ranging measurements not only in the conventional 2-way mode but also in a new differential downlink mode made possible by a tone generator on the spacecraft that can coherently modulate the S- and X-band downlink signals. Second, the spacecraft can receive uplink signals at both S- and X-band rather than S-band only. The ability to receive and coherently lock to an uplink signal at either frequency adds both redundancy and flexibility to the system and improves its scientific capabilities. As a result of a 20-year evolution in spacecraft radio equipment, Galileo is the first mission where the primary uplink and downlink communications will be conducted at X-band. This has made it possible to convert the S-band downlink from circular to linear polarization for scientific purposes (i.e., Faraday rotation measurements), the third enhancement relative to Voyager. With its strong magnetic field, Jupiter is a particularly good target for this type of measurement. Overall, these additions and modifications not only increase the speed and accuracy of command and data transmissions to and from the spacecraft but also advance the precision and scope of the science investigations.

The NASA Deep Space Network (DSN), operated by JPL, is responsible for tracking and commanding the spacecraft, receiving engineering and scientific telemetry, and for recording the basic measurements which become radio science data. It is a worldwide network of 70- and 34-m diameter antennas that are capable of simultaneously receiving and transmitting at S- and X-band in both left- and right-hand circular polarization (LCP and RCP, respectively). A hydrogen maser is used as a frequency reference for both the transmitted and received signals. Table II summarizes the expected performance of the DSN tracking stations. Figure 2 shows a partial block diagram of the relevant equipment. In addition, the 30-m antenna of the DLR located near Weilheim,

TABLE II

Selected DSN station parameters (nominal)

Antenna diameter (m)	70		34 Standard		34 High Efficiency	
Transmitting parameters	*S*		*S*		*X*	
Frequency (MHz)	2115		2115		7167	
Power (kW)	20/400		20		20	
Antenna gain (dBi)	63		55		67	
Polarization	LCP or RCP		LCP or RCP		LCP or RCP	
Receiving parameters	*X*	*S*	*X*	*S*	*X*	*S*
System noise temperature (K) (at zenith; typical)	21	17	25	21	20	38
Frequency (MHz)	8415	2295	8415	2295	8415	2295
Polarization	Simultaneous LCP and RCP	Simultaneous LCP and RCP	LCP or RCP	LCP or RCP	Simultaneous LCP and RCP	LCP or RCP

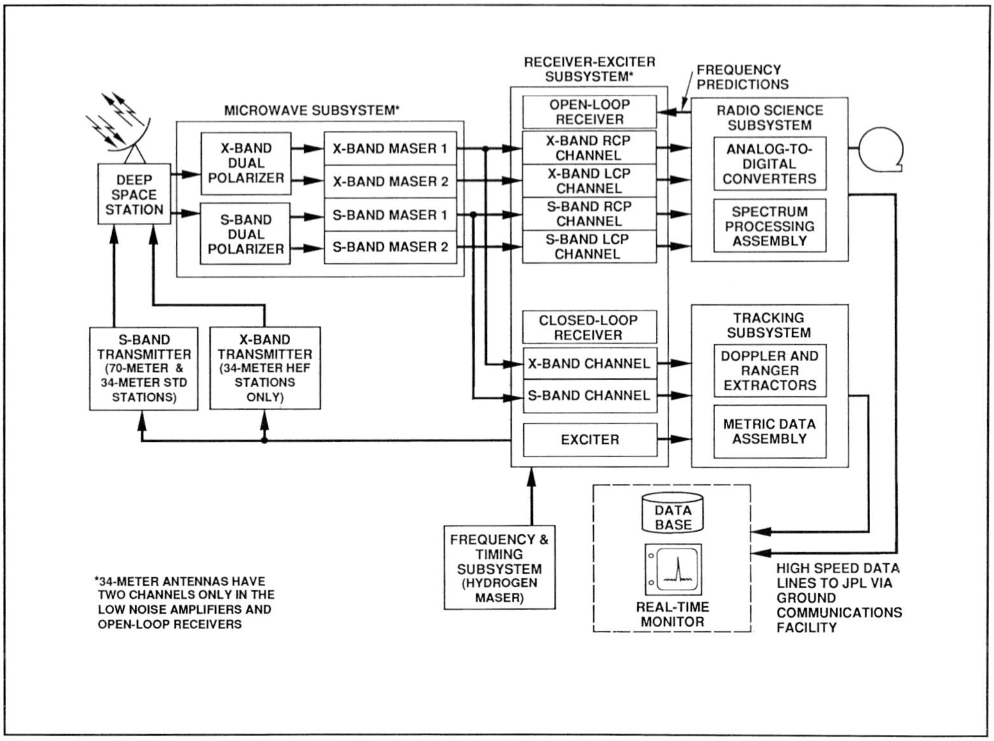

Fig. 2. DSN ground station functional block diagram for Galileo radio science.

Germany, has been subscribed for extensive tracking of Galileo during the interplanetary cruise phase.

The frequency stability of the radio link between spacecraft and Earth is a key parameter for many of the radio science experiments. The performance of the spacecraft USO and the hydrogen masers of the DSN can be characterized through use of the Allan variance (Barnes *et al.*, 1971); the square root of this quantity represents the size of a typical frequency fluctuation expressed as a fraction of the average frequency. Table III gives representative results on several time scales.

Procedures for recording radio science data at the DSN are mostly standard and will not be reviewed here. The reader should refer to the literature concerning radio science results from Voyager for further discussion.

2.2. Measurement techniques

Neutral gases and magnetized plasmas distributed along the propagation path between the spacecraft and a ground antenna can distinctively influence some or all of the following properties of radio waves: phase velocity, group velocity, direction of travel, amplitude, and polarization. Variations in these propagation parameters lead in turn to changes in the characteristics of the signals received on Earth. The Galileo radio

TABLE III

Performance characteristics of the USO and H_2 maser

	Integration time (s)	Frequency stability[a]
USO	1	3×10^{-11}
(spacecraft)	10	4×10^{-12}
	100	1×10^{-12}
	1000	1×10^{-12}
H_2 maser	1	2×10^{-13}
(ground stations)	10	3×10^{-14}
	100	4×10^{-15}
	1000	2×10^{-15}

[a] Square root of Allan variance (see text).

propagation experiments involve measurements at a ground antenna of the frequencies of the two signals, differential phase, absolute and differential propagation time delay, signal strength, and polarization. The relationships between each parameter and the properties of the media provide the basis for remote sensing in these experiments.

2.2.1. *Frequency and Differential Phase*

Any change in the number density of free electrons or of neutral gases along the propagation path between spacecraft and Earth will produce corresponding changes in the phase of a received signal. This effect produces an apparent Doppler shift in signal frequency. Moreover, the refractive index of a plasma depends strongly on frequency while that of a neutral gas generally does not for the frequencies considered here. By comparing phase measurements at two coherently-related frequencies, one can therefore distinguish between effects caused by neutral gases and those caused by plasmas. For example, the following phase difference is advantageous to studies of the solar corona as well as the ionospheres of Jupiter and its moons:

$$\delta\phi \equiv \Phi_S - \frac{3}{11}\,\phi_x \approx \frac{K_1}{f_s} \int_{\text{path}} N(S)\,\mathrm{d}s \quad \text{(radians)}\,,$$

where

$$K_1 = 7.82 \times 10^{-7} \quad \text{(mks units)}\,.$$

Here, ϕ is phase, f is frequency, and the 's' and 'x' subscripts refer to the S- and X-band signals, respectively. To a good approximation, the combination $\phi_s - \frac{3}{11}\phi_x$ removes from the data effects caused by neutral gases, relative motion of the spacecraft and ground antennas, and instability of the USO so that $\delta\phi$ is proportional to the number density of free electrons, N, integrated along the propagation path, s. We will refer to the path integral $I = \int N(s)\,\mathrm{d}s$ as the columnar electron content. While measurements of $\delta\phi$ can be used to monitor changes in I, an absolute measure of I cannot be obtained with this technique.

2.2.2. *Absolute and Differential Propagation Time Delay*

The time delay for signal propagation between spacecraft and Earth is obtained in the 2-way ranging mode by transmitting to the spacecraft a signal modulated by a ranging code. The coded signal received by the spacecraft is relayed back to Earth at the two downlink frequencies by the onboard transponders. The ground tracking station records the round trip propagation time as well as the difference between the arrival times of the range code at the two signal frequencies. Possible ambiguities in total range are resolved by transmitting the range code at several modulation frequencies. Alternatively, in the one-way ranging mode, ranging tones generated on the spacecraft are used to modulate the S- and X-band downlink signals. The tracking station records the difference in arrival times, yielding the differential but not the absolute range to the spacecraft.

Measurements of propagation time can be used for remote sensing because the propagation medium affects the group velocity of the radio signals. As with differential phase observations, the effect of a plasma on the group velocity is dispersive while the effect of a neutral gas is not. Consequently, the differential propagation time delay can be expressed as

$$\Delta\tau_{s/x} \equiv \tau_s - \tau_x \cong \frac{K_2}{f_s^2}\, I \,,$$

where

$$K_2 = 1.25 \times 10^{-7} \quad \text{(mks units)} \,.$$

Here, τ_s and τ_x are the propagation time delays at S- and X-band, respectively.

Differential time delay data differ from differential phase data in that an absolute measure of I can be obtained from the former but not the latter. However, either can be used for observing changes in I. For propagation experiments with Galileo, the choice of techniques is dictated by the experiment geometry. Solar conjunctions last several weeks and involve the use of multiple ground tracking stations which rise and set sequentially as the experiment progresses. In this situation, differential range data are essential for minimizing problems arising from gaps in station coverage and for establishing boundary conditions for integrating differential phase data. On the other hand, occultations by planets and satellites involve characteristic changes in I over shorter time intervals, typically a few hours or less. As the repetition rate for range measurements is constrained by practical considerations, observations of planetary or satellite occultations rely exclusively on measurements of differential phase.

2.2.3. *Signal Strength*

Variations in signal strength on both short and long time scales are important to several investigations. Radio-wave scattering in the solar corona as well as the atmosphere and ionosphere of Jupiter produces intensity fluctuations on short time scales, typically a few seconds or less. These observations are germane to studies of the small-scale structure and dynamics of these media. On somewhat longer time scales, ammonia – a trace constituent in the atmosphere of Jupiter – can cause a reduction in signal intensity through absorption. As this absorption is strongly dependent on frequency, the two-frequency capability is important for extending the dynamic range of the measurements.

2.2.4. *Polarization*

A linearly-polarized electromagnetic wave propagating through a magnetized plasma will experience a rotation in its angle of polarization (cf. Yeh and Liu, 1972). The equations governing this phenomena, which is known as Faraday rotation, depend on the frequency of the radio signal, f, relative to the characteristic frequencies of the plasma:

$$2\pi f_p = \left(\frac{Ne^2}{\varepsilon_0 m}\right)^{1/2} \quad \text{and} \quad 2\pi f_B = \frac{eB}{m} .$$

Here, e and m are the charge and mass of an electron, respectively, B is the magnitude of the magnetic field, ε_0 is the vacuum permittivity, N is the number density of free electrons, f_p is the plasma frequency, and f_B is the electron gyrofrequency. Under conditions where $f \gg f_p$ and f_B, the angle of rotation of the polarization vector can be expressed as

$$\Omega \approx \frac{K_3}{f^2} \int_{\text{path}} N(s)\mathbf{B}(s) \cdot \mathrm{d}\mathbf{s} \quad \text{(radians)} ,$$

where

$$K_3 = 2.36 \times 10^4 \quad \text{(mks units)} .$$

The path of integration coincides with the path followed by the radio signal propagating from spacecraft to Earth, s is the distance measured along the path, and $\mathbf{B}$ is the magnetic field vector. The Faraday data, when combined with the previously mentioned measurements of columnar electron content, can be used to infer a value for the magnetic field along the path through the plasma. Galileo is the first spacecraft to provide this capability.

The rotation of the Galileo spacecraft (and the high-gain antenna) at a rate of about 3 revolutions per minute causes an identical rotation with the appropriate time delay in the angle of polarization of the *S*-band signal received on Earth. Changes in Ω arising from Faraday rotation appear superimposed on this rotating polarization vector. In practice, the DSN does not directly measure the angle of polarization; instead, the *S*-band signal is received and recorded simultaneously in coherent RCP and LCP channels so that polarization information can be extracted. The data are then calibrated to remove the effect of spacecraft rotation through use of engineering telemetry received from the spacecraft.

3. Solar Corona

The solar corona experiments are performed for several weeks around Galileo's superior conjunctions, i.e., when the spacecraft is on the opposite side of the Sun from the Earth.

In this configuration the radio signals propagating between Earth and spacecraft traverse the solar atmosphere. The radio signals are scattered and refracted as they propagate through the hot, tenuous, turbulent plasma of the solar corona. These effects are responsible for degradation of command transmission on the uplink and high bit error rates in the data reception on the downlink. On the other hand, they intrinsically yield valuable information about the structure and dynamics of the solar corona. Coronal sounding investigations using the polarization and group delay characteristics of radio signals from spacecraft in solar occultation have been performed for many years (Bird, 1982; Bird and Edenhofer, 1990) and such an experiment will be carried out during the cruise phase of the Ulysses mission (Volland *et al.*, 1983). Similarly, scintillation and scattering techniques for observing the corona and interplanetary medium have become well developed and have produced results on bulk motion and accretion in the corona, the variation of electron density irregularities with solar cycle and distance from the Sun (Tyler *et al.*, 1981; Woo and Armstrong, 1979, 1981; Woo, 1988), the evolution of interplanetary disturbances, and coronal mass ejections (Woo *et al.*, 1982, 1985).

Although the Galileo mission is not primarily concerned with solar research, its interplanetary trajectory and radio system capabilities are well suited for coronal radio sounding investigations. Table IV gives the timing and geometrical parameters of the four Galileo superior conjunctions en route to Jupiter. The Galileo orbital geometry at each of the conjunctions is shown in Figure 3.

3.1. Scientific objectives

The Galileo Solar Corona Experiment (SCE) will be use a variety of radio sounding techniques for investigations of the otherwise practically inaccessible regions of the outer solar atmosphere. Whereas observations of the solar corona at optical wavelengths (e.g., with coronagraphs) are generally possible only out to a few $R_\odot$ ($R_\odot$ = solar radius), radio sounding experiments can be employed with good results out to 40 $R_\odot$ and beyond. In this way such methods can at least partially bridge the gap between the *in-situ* measurements of previous spacecraft in to about 60 $R_\odot$ and the optical observations in the immediate neighborhood of the Sun.

Specific scientific objectives of SCE include investigations of the following:

– The 3D-electron density distribution of the coronal plasma, and its relation to the photospheric magnetic field configuration as obtained from solar magnetograms.

– The structural differences between coronal holes ('open' magnetic configurations), active regions (presumably 'closed' magnetic field lines), and the 'quiet' Sun (open or closed?).

– Characteristics of the acceleration regions of the solar wind in coronal holes, streamers, and other parts of the corona.

– Coronal characteristics possibly related to the heating agent responsible for driving the temperature of the solar material from some 10^4 K in the chromosphere to over 10^6 K in the corona.

– The effects of resonant solar oscillations on the dynamical characteristics of the relatively tenuous solar atmosphere.

TABLE IV

Galileo solar conjunctions based on actual VEEGA orbit

Sup.[a] conj. No.	Solar proximate date	Proximate[b] solar offset ($R_\odot$)	Mean daily motion ($R_\odot$ day^{-1})	East limb ingress[c] $2° < R < 8°$ One 70-m pass day^{-1}	Center occultation $R < 2°$ 70-m continuous	West limb egress $2° < R < 8°$ One 70-m pass day^{-1}	S/C range (AU)
C_1	22 Jan., 1992	8.4 N	2.05	9–22 Jan. (14 days)	None	23 Jan.–6 Feb. (15 days)	3.25
C_2	3 Nov.,1993	2.5 S	2.57	22–30 Oct. (9 days)	31 Oct.–6 Nov. (7 days)	7 Nov.–14 Nov. (8 days)	4.38
C_3	1 Dec.,1994	0.5 S	2.97	21–27 Nov. (7 days)	28 Nov.–3 Dec. (6 days)	4 Dec.–11 Dec. (8 days)	5.96
C_4	19 Dec.,1995	0.9 N	3.03	9–15 Dec. (7 days)	16–21 Dec. (6 days)	22–28 Dec. (7 days)	6.26

[a] Values for C_4 are only approximate – Jupiter arrival data is 7 December, 1995.
[b] (N/S) = north/south pole crossing.
[c] Solar offset R: $2° \cong 7.5\ R_\odot$; $8° \cong 30\ R_\odot$.

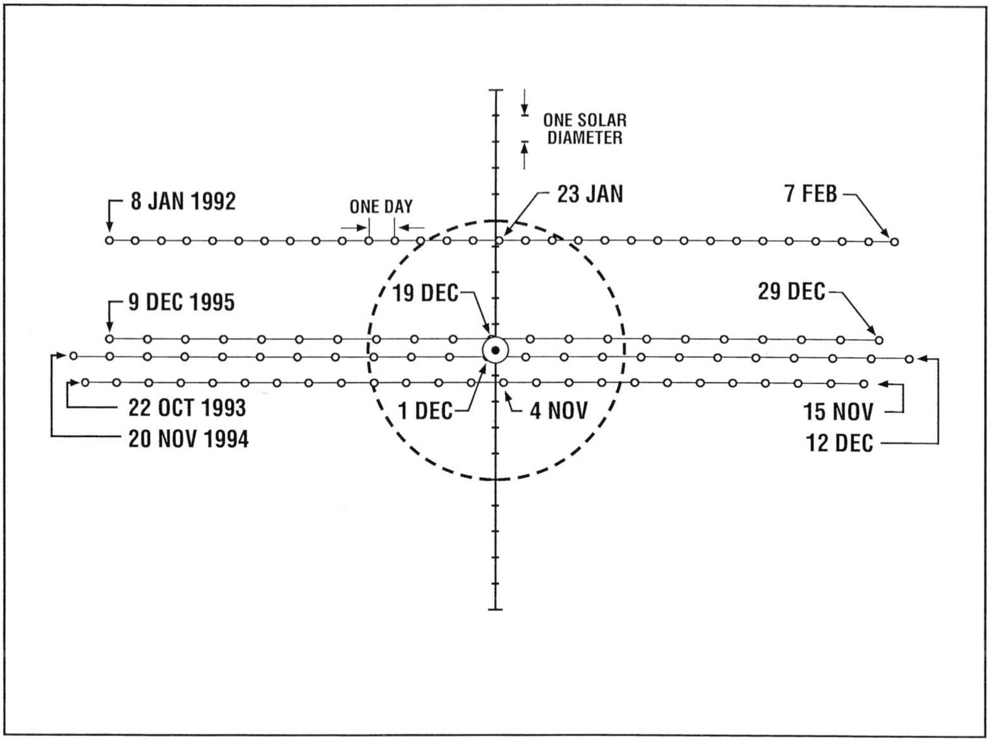

Fig. 3. View from Earth of the four solar conjunctions of the Galileo spacecraft (cf. Figure 1). The circle in the center is the solar disk; the dashed circle has a radius of 10 $R_{\odot}$. Spacecraft position at 00 : 00 UT each day is shown by small circles. Axis ticks are separated by 2 $R_{\odot}$.

– The excitation and propagation conditions for various types of plasma waves, e.g., Alfvén waves, magnetoacoustic waves, etc.

– The distribution of electron density irregularities over a wide range of spatial scale sizes and their variation with solar cycle, solar distance, and latitude.

– The topology and evolution of interplanetary disturbances near the Sun and their relationship to white-light coronal mass ejections.

The Galileo SCE, exploiting the Faraday capability as one specific example of meeting the above objectives, will place particular emphasis on achieving a better understanding of magnetic fields in the solar corona. It is anticipated that the magnetic field measurements obtained with Galileo will verify previous results for the background coronal field (Pätzold *et al.*, 1987) without recourse to independently derived coronal density models. An equally important aspect of these observations would be the unprecedented advantage of separating magnetic from electron density fluctuations in order to assess the spatial/spectral distribution of coronal MHD-wave activity. There exists strong evidence, based on previous Faraday/electron content data, that the magnetic fluctuations are dominant (Hollweg *et al.*, 1982). Finally, should we be fortunate enough to witness the transit of a coronal mass ejection through the spacecraft's signal ray path, the Galileo SCE measurements would be ideal for testing the alleged role of the magnetic field as a driving force for these poorly understood solar phenomena. An earlier analysis of the few rare and only partially observed examples was inconclusive (Bird *et al.*, 1985).

3.2. Experimental observations

The three main experiments to be conducted during solar superior conjunctions are:

Faraday, Doppler, and Range

The Galileo radio propagation experiment will be the first to utilize Faraday rotation plus dual-frequency range and range-rate measurements, thereby enabling a direct estimate of the coronal magnetic field. Faraday rotation is sensitive only to the mean line-of-sight component of the magnetic field $\langle B_L \rangle$, as weighted by the plasma density. The complex solar field can result in cancellation effects along the signal ray path so that very little rotation is sometimes observed even at small solar elongations. Nevertheless, the strength and orientation of the coronal magnetic field can often be approximated under appropriate and reasonable assumptions (Volland *et al.*, 1977; Pätzold *et al.*, 1987). A simple 'rule of thumb' formula for the mean longitudinal magnetic field in the corona (in gauss) can be written as

$$\langle B_L \rangle \cong \frac{\Omega}{\Delta\tau_{s/x}} ,$$

whee Ω is given in degrees at S-band, and $\Delta\tau_{s/x}$ is given in ns. Plasma differential delay times of up to a few tens of thousands of nanoseconds have been measured in previous solar conjunctions (1 ns $\cong 4 \times 10^{16}$ el m^{-2} for the S/X-band system).

Multiple Ground Station Observations

Whenever two or more ground stations are tracking simultaneously, a cross correlation of the signal fluctuations observed at the different ground stations can be used to compute a pattern velocity, which then can be related to the bulk velocity of coronal plasma or the phase velocity of a plasma wave. Any signal parameter can be used for this purpose. An example of the Faraday rotation fluctuations observed at two widely separated DSN ground stations during a solar occultation of Helios in 1983 (Bird *et al.*, 1985) is shown in Figure 4. Very large pattern velocities both away from and toward the Sun are often derived from these correlations, suggesting the presence of inward and outward propagating coronal Alfvén waves.

Scintillation at S- and X-Band

These investigations are based on radio scintillation and scattering techniques, with the observed scattering phenomena including amplitude scintillations, phase scintillations, and spectral broadening. Near-Sun observations will contribute to our understanding of electron density irregularities over 5 decades of spatial size scales (Woo and Armstrong, 1979).

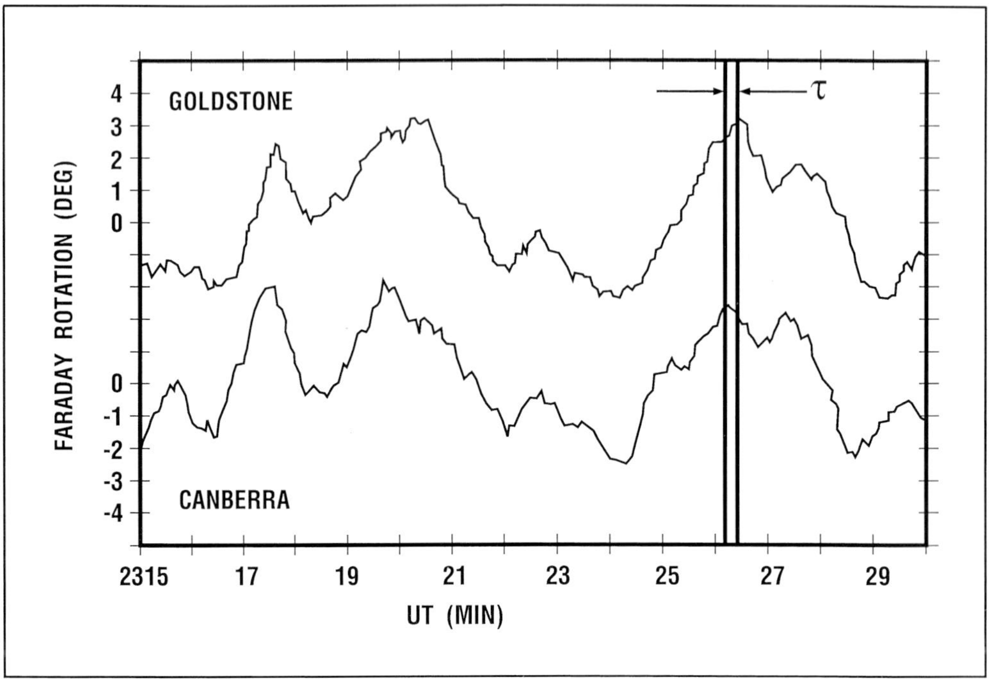

Fig. 4. Helios Faraday rotation measurements recorded at the widely separated Goldstone (California) and Canberra (Australia) ground stations on 9 January, 1983. The radial separation of the ray paths from Helios to the two ground stations in the corona was $\cong 2000$ km. In this case the time lag $\tau = 10.6$ s (correlation coefficient = 0.93) yielded a radially outward pattern velocity of ca. 200 km s^{-1}. Other examples could only be explained by pattern velocities moving toward the Sun – a possible indicator of MHD waves.

4. Jupiter and Satellite Occultations

Upon arrival at Jupiter in December of 1995, Galileo first has a close flyby of Io with no occultation and then spends about an hour collecting telemetry from the Probe as it enters the Jovian atmosphere. The Probe has an ultrastable oscillator which makes it possible to use the Doppler data from the Probe-to-Orbiter L-band link to characterize the entry path winds. A companion paper describes this experiment (Pollack *et al.*, 1992). Scintillations in the amplitude of the Probe-to-Orbiter radio signal will also be used to study small-scale irregularities and dynamics of Jupiter's atmosphere (Woo *et al.*, 1980).

The next event is a retropropulsion motor burn which injects Galileo into a 230-day orbit. The 1989 launch opportunity provided a particularly favorable geometry for obtaining Jupiter occultations which will occur on more than half of the planned orbits. At the time of writing this paper (1991) the tour selection process had not been started. Tour design begins in 1992, involves a number of operational and scientific trade offs and has the goal, from the radio science perspective, of producing multiple occultations of the spacecraft by Jupiter and Io along with close flybys of and occultations by Europa,

Ganymede, and Callisto. The orbit of the Galileo spacecraft is also expected to provide numerous opportunities for differential Doppler measurements concerning the spatial structure and temporal variations of the Io plasma torus.

4.1. Atmosphere and Ionosphere of Jupiter

The Galileo radio occultation experiments can provide the data needed to derive vertical profiles of refractivity and absorptivity in the neutral atmosphere at a quality comparable to that obtained from the corresponding experiments with Voyager (Eshleman *et al.*, 1977; Lindal *et al.*, 1981, 1985, 1987; Tyler, 1987). However, Galileo's multiple occultations will be at various latitudes and will occur over an extended period of time. This makes it possible to study atmospheric properties in a more comprehensive manner and over a wider range of conditions than could be accomplished with Voyager. In this way the Galileo radio occultations can extend the Voyager results and contribute to a better understanding of Jupiter in several key areas.

To be successful, the radio occultation experiments with Galileo will require precise control of the orientation of the spacecraft antenna. Each experiment must include a spacecraft maneuver designed so that the antenna beam tracks the virtual radio image of Earth as it moves along the limb of Jupiter (as seen from the spacecraft). More specifically, the pointing direction of the antenna will be carefully controlled so as to yield the strongest possible signal at the tracking antennas of the DSN after refraction in the atmosphere of Jupiter.

Multiple occultations and global coverage could be of particular importance for providing measures of atmospheric ammonia, as derived from absorptivity profiles. One of the four Voyager occultations produced a profile of ammonia vapor abundance with very good height resolution, from about the 400 mb level where the concentration is limited by saturation to a few parts per million, to the condensation level which occurred near 900 mb, and on downward to the one bar pressure level where the mixing ratio was 220 ± 80 ppm (Lindal *et al.*, 1981). That is, the measurement extended below the ammonia cloud base and thus reflects the composition of the lower well-mixed regions of the atmosphere. With Galileo it may be possible to extend such vertical profiles of the ammonia mixing ratio to global coverage. Other remote-sensing measures of the global distribution of ammonia, based on observations of thermal emission, do not provide fine height resolution and suffer from some ambiguity due to possible horizontal variations in temperature. Thus the radio occultation profiles and the global thermal emission measures are complementary in terms of coverage, resolution, and accuracy.

The scientific significance of these measures of ammonia can be summarized as follows. The abundance of ammonia on Jupiter is a consequence of the conditions in the primitive solar nebula, the process of planet formation, and the subsequent evolution of Jupiter over its lifetime. Several phenomena, acting at a variety of atmospheric pressures, determine the present spatial distribution of ammonia within the planet. These include photochemistry in the stratosphere, condensation in the upper troposphere, and at progressively deeper levels of the troposphere, chemical reactions with hydrogen sulfide and dissolution in water clouds. Vertical transport through convection and eddy mixing as well as large-scale atmospheric motions act to couple the effects of these various processes. Hence, global measurements of the abundance of ammonia and its horizontal and vertical variations are key to understanding not only a variety of atmospheric phenomena but also the origin and evolution of the planet.

Favorable occultation geometries will be required for Galileo to accomplish the desired ammonia measures. The single profile of ammonia abundance described above was obtained during egress of Voyager 1 and applies to the equatorial atmosphere. Absorptivity profiles could not be derived from the other three Voyager occultations due either to the geometrical conditions (grazing occultation and large limb-spacecraft distances) or to very weak and Doppler-spread signals (Lindal *et al.*, 1981). It is not clear what fraction of the Galileo occultations will allow determination of accurate absorptivity profiles, but careful attention to experiment design, orbital trajectories, spacecraft antenna orientation, and signal measuring and processing techniques will be required to optimize results. In this regard, it is inherent that the region we wish to measure, extending in particular to below the cloud base, is of course the same region where the signal is being strongly absorbed to the point of extinction. The margin between the pressure levels at the cloud base and at signal loss was very small for the successful Voyager example, and can never be large for Galileo. Nevertheless, the potential importance of the ammonia measurements merits special attention to ameliorate the difficulties.

Another area where a multiplicity of atmospheric profiles is needed is for the study and characterization of the spatial and/or temporal variability of the stratosphere. For example, the near-equatorial profiles of the stratosphere from Voyager 1 exhibit temperature oscillations with an amplitude of about 10 K and a vertical scale of about 3 pressure scale heights. The vertical variations in temperature are similar in appearance at ingress and egress but about 180° out of phase at the two locations. Such effects are absent from the mid-latitude profile of the stratosphere obtained with Voyager 2. Moreover, Voyager images of the equatorial atmosphere show a periodic pattern of cloud plumes encircling the planet at 8° N latitude with a zonal wave number of about 12. Allison (1990) showed that both sets of observations could be interpreted consistently as manifestations of a Rossby wave that propagates in the vertical and zonal directions but remains confined in latitude near the equator. His results have wide implications for atmospheric phenomena ranging from the eddy mixing of the middle atmosphere to the structure of the water cloud (and the abundance of water vapor) in the deep troposphere, as well as the dynamical coupling between the troposphere and stratosphere.

Against this background, the Galileo radio occultations can contribute to a more comprehensive investigation of planetary-scale equatorial waves on Jupiter. In distinction to the ammonia measurement, accurate stratospheric profiles between about 1 and 100 mb are expected to be obtained from every occultation of Galileo. Each occultation measurement at latitudes between about 15° N and 15° S, where wave amplitudes are expected to be appreciable, will yield a snapshot of the vertical structure of the wave, allowing meaurements of amplitude and vertical wavelength. Intercomparison of measurements at different locations, especially ingress and egress of the same occultation, will provide strong constraints on zonal wave number and meridional structure (cf. Hinson and Magalhães, 1991). Coordinated observations of zonally periodic cloud structure with the imaging instrument on Galileo will complement this aspect of the radio science investigation, providing independent estimates of zonal wave number and, possibly, zonal phase speed. These coordinated observations, when combined with *in-situ* measurements by the Galileo Probe of the vertical structure and composition of the equatorial atmosphere, can provide a far more complete picture of the behavior of equatorial waves on Jupiter, possibly allowing identification of the mechanism by which they are generated as well as assessment of their contribution to the energy and momentum budgets and eddy mixing of the middle atmosphere.

More generally, the Galileo refractivity profiles (extending from about 1 mb to 1 bar) will be used to search for latitudinal variations at all locations accessible to occultation measurements, and to refine the precision to which lapse rates can be determined. Accurate lapse rates in the troposphere could be used, for example, to help guide studies of the role of ortho-para conversion of molecular hydrogen in affecting atmospheric dynamics and stability (Gierasch and Conrath, 1985). Both tropospheric and stratospheric profiles as measured by radio occultation will also be used in conjunction with infra-red temperature sounding to refine the precision to which the atmospheric helium abundance can be deduced through intercomparisons (cf. Conrath *et al.*, 1984, 1987).

In all of the atmospheric studies described above, particular emphasis will be placed on comparison with results from the Galileo Probe experiments. The *in-situ* measurements will be used to calibrate and anchor the radio experiments for application to other regions of Jupiter, and in addition, to the other major planets where radio but no probe measurements have previously been made.

For the ionosphere of Jupiter, it is again the multiplicity and global coverage provided by the Galileo radio occultations that are of special significance. Profiles in height of the electron number density of the ionospheric topside have been obtained from both the Voyager and Pioneer encounters with Jupiter, but even though the Voyager radio instrumentation is generally more capable, the shorter limb-spacecraft distance in the Pioneer 10 and 11 encounters was fundamental in making possible a characterization of the very sharp, multiple, dense, low-lying (below 1000 km altitude as measured from the 1 bar pressure level) layers of the ionosphere at the occultation locations (Fjeldbo *et al.*, 1975, 1976a, b). This experience emphasizes the importance of designing the Galileo orbital tour so that at least some of the Jupiter radio occultations involve relatively short planet-spacecraft distances. The sharp low-lying layers of ionization are poorly understood and yet appear to be a general property of giant planet ionospheres, since they are also observed at Saturn, Uranus, and Neptune (Lindal *et al.*, 1985; 1987; Tyler *et al.*, 1989). From the greater areal and temporal coverage on Jupiter provided by Galileo, it may be possible to characterize the horizontal extent and temporal variations of the layers as clues to divining their source and other characteristics. The topside of the main ionization layer on Jupiter (which extends to at least 4000 km above the 1-bar pressure level) apparently consists of atomic hydrogen as the principal ion.

The plasma temperature of more than 1000 K may exceed the local temperature of the neutral species by about a factor of two. Lower down in the main layer, other ions and chemistry must play a role in determining the plasma density (e.g., see Horanyi *et al.*, 1988). The experimental results concerning the topside are now too sparse to investigate effects on the ionosphere of variations in solar activity, solar zenith angle, magnetospheric characteristics such as the Io torus, and position on Jupiter. Galileo radio occultations could fill at least some of the gaps in our knowledge.

For both the ionosphere and neutral atmosphere, the Galileo radio measurements will provide information not only about the profiles discussed above, but also about small-scale structures embedded within these regions. They manifest themselves in the form of scintillations of both the signal amplitudes and phases. The purpose of this investigation of the neutral atmosphere of Jupiter is to characterize the scintillations and deduce information on the global and spatial wave number distribution of temperature irregularities of scale sizes smaller than several kilometers. The goal is to interpret the results in terms of small-scale atmospheric dynamical processes, including but not restricted to turbulence and gravity waves (Woo *et al.*, 1980; Hinson and Tyler, 1983), and to infer estimates of vertical transport (Woo and Ishimaru, 1980; Hinson and Magalhães, 1991). For the ionosphere of Jupiter (and of satellites for which an ionosphere is detected), the investigation will be of the global morphology of scintillations and magnetic field orientation in the ionosphere, and the variation with altitude and latitude of the magnitude and spatial wave number spectrum of the electron density irregularities for scale sizes smaller than several kilometers (Woo and Yang, 1978; Woo and Armstrong, 1979; Hinson and Tyler, 1982; Hinson, 1984). These ionospheric measurements will lead to a better understanding of the nature and production mechanisms of the electron density irregularities, the interaction of the solar wind and magnetosphere with the ionosphere, and the configuration of the magnetic field near the planet.

4.2. Satellite Occultations

Another aspect of the Galileo radio propagation experiments concerns the study of Jupiter's satellites. As in the case of previous missions to the outer planets, radio tracking data acquired during the spacecraft's occultation by a satellite will be used to determine the satellite's diameter and, possibly, the properties of its atmosphere and ionosphere.

Measurements of satellite diameters involve the use of spacecraft and satellite ephemerides and the limb diffraction effects observed during ingress and egress. The accuracy of the measurements, which depends on the occultation geometry and the accuracy of orbit determination, should be about 1 km if the results from earlier missions can be used as a guide.

Integral inversion of the differential Doppler data acquired during ingress and egress may provide information on the vertical distribution of free electrons. The detection limit for plasma in a satellite ionosphere ranges from 10^8 to 10^{10} electrons per cubic meter depending on the extent to which the radio link is also affected by phase variations produced by plasma in the interplanetary medium and the terrestrial ionosphere. The neutral gas number density may in turn be inferred from the data on the vertical electron distribution – or, if the gas density is of the order of 10^{20} molecules per cubic meter or greater, directly from phase perturbations observed during ingress and egress.

The neutral gas detection limit for the phase data to be acquired with the Galileo radio system is comparable to that of Voyager 2, which was used to characterize the atmosphere of Triton whose surface gas density was determined to be 2.2 to 2.6×10^{21} molecules per cubic meter (Tyler *et al.*, 1989). The gases near the surfaces of the Jovian satellites may be too tenuous for direct measurements by radio occultation. However, a substantial ionosphere has been observed on Io by Pioneer 10 (Kliore *et al.*, 1975), and differential phase data permit inferences to be made about atmospheric regions where the gas density is much lower than 10^{20} molecules per cubic meter. For example, the peak electron concentration in Triton's ionosphere was detected with the Voyager radio link at an altitude where the gas density is only about 10^{16} m^{-3}, and dispersive phase measurements conducted during Mariner 10's occultation by Mercury (Howard *et al.*, 1974) have yielded an upper limit for the gas density near Mercury's surface of 10^{12} m^{-3} (Fjeldbo *et al.*, 1976b).

The vertical profiles of electron number density obtained with Pioneer 10 have helped to anchor theoretical models of the atmosphere of Io (e.g., Kumar, 1985; Ingersoll, 1989). Nevertheless, we assign a high priority to obtaining new radio occultation measurements of the ionosphere of Io for the following reasons. Given the substantial improvements in the instrumentation of Galileo relative to Pioneer 10, occultation experiments with Galileo can yield more accurate ionospheric profiles leading to improved constraints for the theoretical models. Moreover, repeated occultations at a variety of phases in the orbit of Io would provide valuable results concerning the spatial and temporal variability of Io's atmosphere and its interaction with the magnetosphere. These results would contribute to an improved understanding of the meteorology and aeronomy of this unique solar system satellite.

4.3. The Io Plasma Torus

Radio science experiments with Galileo can contribute to a better understanding of the Io plasma torus, a structure of considerable significance to the behavior of the Jovian magnetosphere (e.g., Connerney, 1987). When the radio link between spacecraft and Earth traverses the torus, free electrons will induce dispersive Doppler shifts on the dual-frequency radio signals. These data will provide a measure of the total number of free electrons integrated along the path of propagation with an accuracy comparable to previous results obtained with Voyager (Levy *et al.*, 1981). The integrated electron content in turn provides useful constraints on the spatial structure of the torus. With the frequent opportunities for such measurements afforded by an orbiting spacecraft, these radio science results will complement the *in-situ* measurements obtained during the flight of Galileo through the torus immediately prior to orbit insertion as well as remote observations of the torus with other instruments on Galileo.

4.4. Jupiter's magnetic field

The unique combination of radio equipment on the Galileo spacecraft makes possible for the first time Faraday rotation measurements concerning the strength and configuration of the internal magnetic field of Jupiter. We plan to exploit this capability to investigate the characteristics of the magnetic field in the ionosphere of Jupiter – a region that remains inaccessible to *in-situ* measurements.

The Faraday rotation experiment will be conducted during all occultations of the spacecraft by Jupiter that occur during the orbital tour. Each occultation presents two good opportunities for Faraday rotation measurements, one near immersion and the other near emersion. During these time intervals, the path followed by the radio signals propagating from spacecraft to Earth passes through the topside ionosphere of Jupiter but remains above the neutral atmosphere. For example, Figure 5 shows the view from

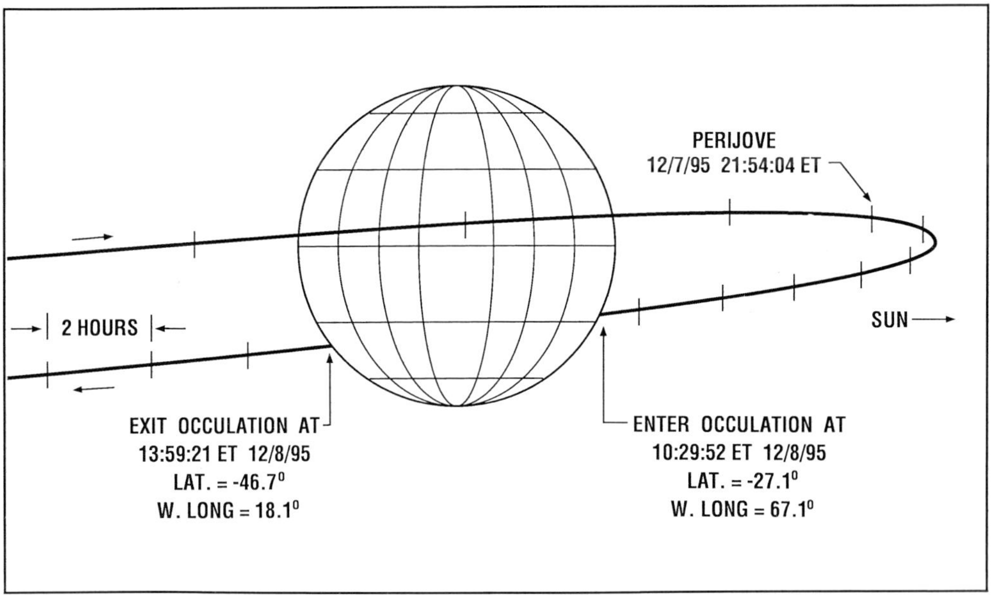

Fig. 5. View from Earth of the initial occultation of the Galileo spacecraft by Jupiter (provided by J. R. Johannesen of the Galileo Navigation Team). Distance from spacecraft to Jupiter is about 12 R_J at immersion and 14 R_J at emersion, where R_J is the radius of Jupiter.

Earth of the initial occultation by Jupiter that immediately follows orbit insertion. Faraday rotation measurements during this event will allow remote sensing of the ionosphere and magnetic field at two locations, both at mid-latitudes in the southern hemisphere. Thus the scope of this investigation of the magnetic field will be determined by the total number and the locations of the occultations by Jupiter, which are a direct consequence of the geometry of the orbital tour.

In an earlier section we gave a formula for the angle of rotation of the polarization vector, Ω, that results from Faraday rotation. (This approximate formula is valid in the ionosphere of Jupiter where $f_p \leq 4$ MHz and $f_B \leq 40$ MHz, so that $f_s \gg f_p$ and f_B.) Several conclusions follow from this expression for Ω. First, Ω depends only on the component of **B** that is parallel to the direction of propagation. For the geometry of the Galileo orbit and the inclination of the spin axis of Jupiter (cf. Figure 5) Ω thus responds most directly to B_ϕ, the longitudinal component of Jupiter's magnetic field. Second, B_ϕ cannot be inferred from Ω without independent measurements of N, the number density of the free electrons. We will obtain these through measurements of differential phase, $\delta\phi$, as described previously. The combined measurements of Ω and $\delta\phi$ will allow inferences of B_ϕ.

When properly normalized, the ratio $\Omega/\delta\phi$ gives the value of B_ϕ averaged along the path of integration with N acting as a weighting function. This is the basic result of the Faraday rotation measurements. The effective length of the path of integration is approximately the geometrical mean of the circumference of the planet and the scale height of the topside ionosphere:

$$L_{\text{eff}} = \sqrt{2\pi RH}\,.$$

For the ionosphere of Jupiter, $R \approx 74\,000$ km at the equator (Lindal *et al.*, 1981) and $H \approx 900$ km (Eshleman *et al.*, 1979a, b) so that $L_{\text{eff}} \approx 2 \times 10^4$ km, equivalent to about $16°$ in latitude at the equator. This represents the effective spatial resolution of the experiment.

We conducted numerical simulations to evaluate the potential of the Faraday rotation measurements. These involved a model for the ionosphere based on Voyager radio occultation measurements (Eshleman *et al.*, 1979a, b) along with a variety of models for the internal magnetic field which have been derived from Pioneer and Voyager magnetometer data (e.g., Acuna and Ness, 1976; Connerney *et al.*, 1982; Connerney, 1987). From these calculations we anticipate that Faraday rotation measurements with Galileo will yield results in the range $-\pi < \Omega < \pi$ (radians). The value of Ω depends strongly on the location of the measurements as a result of the complex structure of the magnetic field near the planet. Of greater significance, the differences among competing models of Jupiter's magnetic field result in variations in Ω of order $\pi/10$, which represents the accuracy in measurements of Ω required to obtain essentially new information about the magnetic field configuration. We have designed the experiment with the goal of exceeding this threshold of sensitivity by a factor of at least ten. At this level of performance, Faraday rotation measurements are highly complementary to magnetometer measurements for investigating the internal magnetic field of Jupiter.

5. Bi-Static Radar Studies of the Icy Galilean Moons

The Galileo mission provides the first opportunity to obtain critical angular data on the electromagnetic scattering properties of the surfaces of the icy Galilean moons of Jupiter. Ground-based radar measurements of Europa, Ganymede, and Callisto indicate that they may share surface and subsurface properties that are currently not understood, but which produce backscatter echoes that differ profoundly from those obtained from rocky planets and moons. Preliminary radar studies of Titan and the south polar cap of Mars suggest that these surfaces may be similar to the icy moons in this regard. More complete understanding in this area is important and possibly vital for the design and interpretation of radar mapping missions to icy planets and moons (particularly Titan), and for improved understanding regarding the formation, evolution, and current state of outer-solar-system surfaces. In a possibly related development, new insight on this subject could follow from recent experimental and theoretical studies of the three bodies of our solar system (Mars, Triton, and Pluto) that appear ot have sensible atmospheres whose main gaseous constituent is also present on the surface as ice.

Monostatic radar echoes from the icy Galilean moons are unusually strong with anomalous polarizations and Doppler spreads, as compared with echoes from the terrestrial planets and Earth's moon (Goldstein and Morris, 1975; Campbell *et al.*, 1978; Ostro *et al.*, 1980; Ostro, 1982). For example, the normalized backscatter radar cross section of our rocky moon is about 0.1, while, for icy Europa, it is 2.6 for echoes of all polarizations. For reference, the corresponding number for a perfectly reflecting sphere is 1.0.

For a circularly polarized incident wave, about 90% of the echo energy is in the expected circularly polarized sense for our moon, and the order of 10% is in the opposite or unexpected sense. For a perfectly reflecting sphere, the corresponding numbers are 100% and zero. For Europa, they are about 40% and 60%, respectively, so that most of the energy has the unexpected polarization. Linearly polarized incident waves are also anomalously depolarized by the icy Galilean moons.

For a homogeneous sphere, backscattered surface reflections come only from the center of the disk, since off-normal incident rays are scattered in other directions. Although rocky-planet echoes are mainly from this central zone, the icy-moon echoes are strong from nearly all areas of the disk. Thus, by comparison, the limbs appear to be remarkably bright when observed by radar so that the Doppler spreads of the echoes are unusually large.

The recent reports of the detection by Earth-based radar of echoes from the south polar cap of Mars (Muhleman *et al.*, 1989) and from Titan (Muhleman *et al.*, 1990) describe the echoes as being unusually strong with the suggestion that they may have characteristics similar to the returns from the icy Galilean moons of Jupiter. Thus the evidence is strong and growing that the icy surfaces that occur in the outer solar system may share common attributes that are not now understood, but which are fundamental to the way in which these surfaces interact with electromagnetic waves.

A number of theories have been proposed to account for these observations (Ostro and Pettingill, 1978; Goldstein and Green, 1980; Hagfors *et al.*, 1985; Eshleman, 1986a, b, 1987; Gurrola and Eshleman, 1990; Ostro and Shoemaker, 1990; Hapke, 1990). It is generally agreed that the scattering is not just from surface structures but must involve significant penetration of the radio photons to at least meters below the surface, and possibly to a few tens of meters. Cold ice has a low electromagnetic loss tangent so that if the near-surface regions are relatively pure ice, such penetration need not involve significant absorption. The strong backscatter echoes argue for a mechanism that can change photon directions over large angles with efficiency and with significant areas of coherence. Although not common to all theories, several invoke the phenomenon of total internal reflection at dielectric interfaces for such efficiency and coherence. Furthermore, even if every incident photon were returned to space, the observed globally-integrated echo strengths (at least for Europa and Ganymede) require that they display some preference for returning along the direction from which they arrived. They might simultaneously favor another direction, such as the specular one relative to the surface, but from conservation of energy it follows that the backscatter direction must involve stronger scattering than the averge of all directions when the normalized backscatter radar cross section exceeds unity. In several theories, subsurface refractive structures with approximate spherical or hemispherical symmetry are invoked for this purpose, where an axis of symmetry results in markedly increased backscatter coherence relative to other directions. The peculiar polarization properties of the observed echoes, while not understood in detail, might well involve the decoupling of two characteristic electromagnetic modes, such as would occur during total internal reflections (Eshleman, 1986a, b). In this mechanism, the complex reflection coefficients for the two modes both have a magnitude of unity, but their phase differs and this constitutes an efficient method of decoupling. The observed bright–limb character of the scattering, plus the lack of an observable backscattered specular component, is explained in most of the theories by invoking volumetric instead of surface scattering mechanisms. For example, Gurrola and Eshleman (1990) show a good match to the observed broad power spectra (and to the polarization and strength parameters) using a model with a uniform distribution of buried craters on the moons.

The principal requirement for progress in our understanding of electromagnetic interactions with icy outer-solar-system surfaces is obtaining data of a new type. This could be accomplished with the Galileo mission. Special emphasis should be placed on: (1) measuring the bi-static scattering as a function of time as the Galileo Orbiter passes between Earth and each of the three icy Galilean moons, as nearly as possible through the exact backscatter geometry; (2) measuring the specular and near-specular scattering from each of these surfaces for various geometries ranging up to the condition of near-grazing bi-static scattering; and (3) measuring scattering in the above geometries for different geological regions on the moons.

The first measurement would reveal the width and shape of the strong backscattered glory lobe, as well as its maximum strength. The width measurement would constitute a direct indication of a characteristic scale of the unknown scattering mechanism, such as the diameters of coherent scatterers or the mean path length of radar photons between subsurface scattering centers. The lobe shape and the polarization properties of the echoes near its peak could be diagnostic of significant properties of the surface and sub-surface material and could help guide theoretical modeling. For example, in one model the strongest scattering is in the form of a cone, with the exact backscatter direction being a local minimum on the axis of the cone, instead of being an isolated maximum. This local minimum, however, would be much stronger than the average scattering outside the conical region and would have different polarization properties. The second measurement would separate the 'usual' specular kinds of scattering from planetary surfaces from the strong backscatter glory produced by the icy moons, making possible an investigation into what is likely to be two fundamentally different kinds of interaction between electromagnetic waves and the near-surface material of these moons. The third proposed set of measurements would then provide the diagnostic results of the other two sets for characteristically different domains on the icy moons. Note that it is inherent that no radar investigation from Earth or from monostatic spacecraft systems could do the kinds of investigations described above, since they only measure the return at the exact center of the strong backscattered glory lobe.

The bi-static radar experiment would be based on dual-frequency downlink transmissions using the Galileo telecommunications system and the Earth tracking stations, although an investigation needs to be made of whether any useful information might also be obtainable from the uplink command links. The spacecraft high-gain antenna would need to be pointed and tracked for the required bi-static geometries. An investigation by R. A. Simpson of Stanford has been made of expected signal strengths for sample Galileo encounters with the outer Galilean moons. The results indicate that significant data could be obtained from the known strong scattering in and presumably near the backscatter geometry, and for the somewhat enhanced scattering that is expected for the oblique specular geometries (Simpson and Tyler, 1981). The detection of scattering in other directions is problematical, but the geometries discussed above appear to be the key ones for help in discriminating among existing theories and for guidance in the construction of new theories.

Europa, Ganymede, and Callisto must have atmospheres, which may consist only of the extremely tenuous gas that would be due to vapor pressure equilibrium (VPE) of atmospheric water vapor with the cold surface ice. The constant interchange of water between atmospheres and surfaces in VPE might play a role in producing surfaces that display the strange scattering behaviors discussed above. In this regard, it now appears that there are just three solar-system bodies with substantial (surface pressure greater than one-millionth that of Earth) atmospheres where VPE with surface ices is the controlling factor (Eshleman *et al.*, 1990). These are Mars, Triton, and Pluto, where the main ice-vapor involved in the VPE control is CO_2 for Mars, N_2 at Triton, and CH_4 or possibly CO or N_2 on Pluto. (For Titan, VPE of N_2 with nitrogen in solution in surface hydrocarbon liquids may be involved (Lunine *et al.*, 1983).) Recall that the south polar cap of Mars produces at least some of the anomalous radar characteristics of the icy Galilean moons (Muhleman *et al.*, 1989).

It is clear that there are major puzzles attending the interaction of electromagnetic waves with the icy surfaces that occur in the outer solar system. An increase in understanding of this problem is needed both for applications and for basic knowledge about these surfaces. The Galileo mission has the potential of providing new insights based on a bi-static radar study of Europa, Ganymede, and Callisto, as experimental targets of opportunity.

Acknowledgements

The authors gratefully acknowledge the assistance and support of the Galileo project staff at the Jet Propulsion Laboratory of the California Institute of Technology and at NASA Headquarters. Particular appreciation is due to members of the Radio Science Support team at JPL who represented the interests of the Radio Propagation team to the project from its inception in 1977 through launch in 1989 and are now conducting flight planning and operations. They are: Sami Asmar, Gerard Benenyan, Jay Breidenthal, Joseph Brenkle, Susan Borutzki, Peter Doms, Paula Eshe, Daniel Finnerty, Carole Hamilton, Randy Herrera, Dwight Holmes, Tony Horton, Rob Kursinski, David Morabito, and Yong Ho Son.

This work is supported by NASA, BMFT, and the DLR.

References

Acuna, M. H. and Ness, N. F.: 1976, *J. Geophys. Res.* **81**, 2917.

Allison, M.: 1990, *Icarus* **83**, 282.

Anderson, J. D., Armstrong, J. W., Campbell, J. K., Estabrook, F. B., Krishner T. P., and Lau, E. L.: 1992, 'Gravitational Celestial Mechanics Investigations with Galileo', *Space Sci. Rev.* **60**, 591 (this issue).

Barnes, J. A., Chi, A. R., Cutler, L. S., Healey, D. J., Leeson, D. B., McGunigal, T. E., Mullen, J. A., Smith, W. L., Sydnor, R. L., Vessot, R. F. C., and Winkler, G. M. R.: 1971, *IEEE Trans. Instr. Meas.* **IM-20**, 105.

Bird, M. K.: 1982, *Space Sci. Rev.* **33**, 99.

Bird, M. K. and Edenhofer, P.: 1990, in R. Schwenn and E. Marsch (eds.), 'Remote Sensing Observations of the Solar Corona', *Physics of the Inner Heliosphere*, Springer-Verlag, Heidelberg.

Bird, M. K., Volland, H., Efimov, A. I., Stelzried, C. T., and Seidel, B. L.: 1985, paper presented at the 5th General Assembly of IAGA, Prague, Czechoslovakia, August 10.

Bird, M. K., Volland, H., Howard, R. A., Koomen, M. J., Michels, D. J., Sheeley, N. F., Jr., Armstrong, J. W., Seidel, B. L., Stelzried, C. T., and Woo, R.: 1985, *Solar Phys.* **98**, 341.

Campbell, D. B., Chandler, J. F., Ostro, S. J., Pettengill, G. H., and Shapiro, I. I.: 1978, *Icarus* **34**, 254.

Connerey, J. E. P.: 1987, *Rev. Geophys.* **25**, 615.

Connerney, J. E. P., Acuna, M. H., and Ness, N. F.: 1982, *J. Geophys. Res.* **86**, 3623.

Conrath, B. J., Gautier, D., Hanel, R. A., and Hornstein, J. S.: 1984, *Astrophys. J.* **282**, 807.

Conrath, B., Gautier, D., Hanel, R., Lindal, G., and Marten, A.: 1987, *J. Geophys. Res.* **92**, 15003.

Eshleman, V. R.: 1986a, *Nature* **319**, 755.

Eshleman, V. R.: 1986b, *Science* **234**, 587.

Eshleman, V. R.: 1987, *Adv. Space Sci.* **7**(5), 133.

Eshleman, V. R., Gurrola, E. M., Hinson, D. P., Lindal, G. F., Marouf, E. A., Simpson, R. A., and Tyler, G. L.: 1990, paper presented at the XXVIII COSPAR Conference, The Hague, 4 July.

Eshleman, V. R., Tyler, G. L., Anderson, J. D., Fjeldbo, G., Levy, G. S., Wood, G. E., and Croft, T. A.: 1977, *Space Sci. Rev.* **21**, 207.

Eshleman, V. R., Tyler, G. L., Wood, G. E., Lindal, G. F., Anderson, J. D., Levy, G. S., and Croft, T. A.: 1979a, *Science* **204**, 976.

Eshleman, V. R., Tyler, G. L., Wood, G. E., Lindal, G. F., Anderson, J. D., Levy, G. S., and Croft, T. A.: 1979b, *Science* **206**, 959.

Fjeldbo, G., Kliore, A., Seidel, B., Sweetnam, D., and Cain, D.: 1975, *Astron. Astrophys.* **39**, 91.

Fjeldbo, G., Kliore, A., Seidel, B., Sweetnam, D., and Woiceshyn, P.: 1976a, in T. Gehrels (ed.), 'The Pioneer 11 Radio Occultation Measurements of the Jovian Ionosphere', *Jupiter*, University of Arizona Press, Tucson, p. 238.

Fjeldbo, G., Kliore, A., Sweetnam, D., Esposito, P., Seidel, B., and Howard, T.: 1976b, *Icarus* **29**, 439.

Gierash, P. J. and Conrath, B. J.: 1985, *Recent Advances in Planetary Meteorology*, Cambridge University Press, Cambridge, p. 121.

Goldstein, R. M. and Green, R. R.: 1980, *Science* **207**, 179.

Goldstein, R. M. and Morris, G. A.: 1975, *Science* **188**, 1211.

Gurrola, E. M. and Eshleman, V. R.: 1990, *Adv. Space Res.* **10**(1), 195.
Hagfors, T., Gold, T., and Ierkic, H. M.: 1985, *Nature* **315**, 637.
Hapke, B.: 1990, *Icarus* **88**, 407.
Hinson, D. P.: 1984, *J. Geophys. Res.* **89**, 65.
Hinson, D. P. and Magalhães, J. A.: 1991, *Icarus*, in press.
Hinson, D. P. and Tyler, G. L.: 1982, *J. Geophys. Res.* **87**, 5275.
Hinson, D. P. and Tyler, G. L.: 1983, *Icarus* **54**, 337.
Hollweg, J. V., Bird, M. K., Volland, H., Edenhofer, P., Stelzried, C. T., and Seidel, B. L.: 1982, *J. Geophys. Res.* **87**, 1.
Horanyi, M., Cravens, T. E., and Waite, J. H.: 1988, *J. Geophys. Res.* **93**, 7251.
Howard, H. T., Tyler, G. L., Esposito, P. B., Anderson, J. D., Reasenberg, R. D., Shapiro, I. I., Fjeldbo, G., Kliore, A. J., Levy, G. S., Brunn, D. L., Dickinson, R., Edelson, R. E., Martin, W. L., Postal, R. B., Seidel, B., Sesplaukis, T. T., Shirley, D. L., Stelzried, C. T., Sweetnam, D. N., Wood, G. E., and Zygielbaum, A. I.: 1974, *Science* **185**, 179.
Ingersoll, A. P.: 1989, *Icarus* **81**, 298.
Kliore, A. J., Fjeldbo, G., Seidel, B. L., Sweetnam, D. N., Sesplankis, T. T., Woiceshyn, P. M., and Rasool, S. I.: 1975, *Icarus* **24**, 407.
Kumar, S.: 1985, *Icarus* **61**, 101.
Levy, G. S., Green, D. W., Royden, H. N., Wood, G. E., and Tyler, G. L.: 1981, *J. Geophys. Res.* **86**, 8467.
Lindal, G. F., Lyons, J. R., Sweetnam, D. N., Eshleman, V. R., Hinson, D. P., and Tyler, G. L.: 1987, *J. Geophys. Res.* **92**, 14987.
Lindal, G. F., Sweetnam, D. N., and Eshleman, V. R.: 1985, *Astron. J.* **90**, 1136.
Lindal, G. F., Wood, G. E., Levy, G. S., Anderson, J. D., Sweetnam, D. N., Hotz, H. B., Buckles, B. J., Holmes, D. P., Doms, P. E., Eshleman, V. R., Tyler, G. L., and Croft, T. A.: 1981, *J. Geophys. Res.* **86**, 8721.
Lunine, J. I., Stevenson, D. J., and Yung, Y. L.: 1983, *Science* **222**, 1229.
Muhleman, D. O., Butler, B. J., Grossman, A. W., Slade, M. A., and Jurgens, R.: 1989, paper presented at the Fourth International Mars Conference, Tucson, AZ, 10 January.
Muhleman, D. O., Grossman, A. W., Butler, B. J., and Slade, M. A.: 1990, *Science* **248**, 975.
Ostro, S. J.: 1982, in P. Morrison (ed.), *Satellites of Jupiter*, University Arizona Press, Tucson, p. 213.
Ostro, S. J. and Pettengill, G. H.: 1978, *Icarus* **34**, 268.
Ostro, S. J. and Shoemaker, E. M.: 1990, *Icarus* **85**, 335.
Ostro, S. J., Campbell, D. B., Pettengill, G. H., and Shapiro, I. I.: 1980, *Icarus* **44**, 431.
Pätzold, M., Bird, M. K., Volland, H., Levy, G. S., Seidel, B. L., and Stelzried, C. T.: 1987, *Solar Phys.* **109**, 91.
Pollack, J. B., Atkinson, D. H., Seiff, A., and Anderson, J. D.: 1992, *Space Sci. Rev.* **60**, 143 (this issue).
Simpson, R. A. and Tyler, G. L.: 1981, *Icarus* **46**, 361.
Tyler, G. L.: 1987, *Proc. IEEE* **75**, 1404.
Tyler, G. L., Sweetnam, D. N., Anderson, J. D., Borutzki, S. E., Campbell, J. K., Eshleman, V. R., Gresh, D. L., Gurrola, E. M., Hinson, D. P., Kawashima, N., Kursinski, E. R., Levy, G. S., Lindal, G. F., Lyons, J. R., Marouf, E. A., Rosen, P. A., Simpson, R. A., and Wood, G. E.: 1989, *Science* **246**, 1466.
Tyler, G. L., Vesecky, J. F., Plume, M. A., Howard, H. T., and Barnes, A.: 1981, *Astrophys. J.* **249**, 318.
Volland, H., Bird, M. K., and Edenhofer, P.: 1983, in K.-P. Wenzel *et al.* (eds.), 'The ISPM Coronal Sounding Experiment', *The International Solar Polar Mission – Its Scientific Investigations*, ESA SP-1050, p. 243.
Volland, H., Bird, M. K., Levy, G. S., Stelzreid, C. T., and Seidel, B. L.: 1977, *J. Geophys. Res.* **42**, 659.
Woo, R.: 1988, *J. Geophys. Res.* **90**, 154.
Woo, R. and Armstrong, J. W.: 1979, *J. Geophys. Res.* **84**, 7288.
Woo, R. and Armstrong, J. W.: 1981, *Nature* **292**, 608.
Woo, R. and Ishimaru, A.: 1980, *Nature* **289**, 383.
Woo, R. and Yang, F.: 1978, *J. Geophys. Res.* **83**, 5245.
Woo, R., Armstrong, J. W., and Ishimaru, A.: 1980, *J. Geophys. Res.* **85**, 8031.
Woo, R., Armstrong, J. W., and Kendall, W. B.: 1980, *Science* **205**, 87.
Woo, R., Armstrong, J. W., Sheeley, N. R., Jr., Howard, R. A., Koomen, M. J., Michels, D. J., and Schwenn, R.: 1985, *J. Geophys. Res.* **90**, 154.
Woo, R., Armstrong, J. W., Sheeley, N. R., Jr., Howard, R. A., Michels, D. J., and Koomen, M. J.: 1982, *Nature* **300**, 157.
Yeh, K. C. and Liu, C. H.: 1972, *Theory of Ionospheric Waves*, Academic Press, New York.

GRAVITATION AND CELESTIAL MECHANICS INVESTIGATIONS WITH GALILEO

J. D. ANDERSON, J. W. ARMSTRONG, J. K. CAMPBELL,
F. B. ESTABROOK, T. P. KRISHER, and E. L. LAU

Jet Propulsion Laboratory, California Institute of Technology, Pasadena, CA 91109, U.S.A.

Abstract. The gravitation and celestial mechanics investigations during the cruise phase and Orbiter phase of the Galileo mission depend on Doppler and ranging measurements generated by the Deep Space Network (DSN) at its three spacecraft tracking sites in California, Australia, and Spain. Other investigations which also rely on DSN data, and which like ours fall under the general discipline of spacecraft radio science, are described in a companion paper by Howard *et al.* (1992). We group our investigations into four broad categories as follows: (1) the determination of the gravity fields of Jupiter and its four major satellites during the orbital tour, (2) a search for gravitational radiation as evidenced by perturbations to the coherent Doppler link between the spacecraft and Earth, (3) the mathematical modeling, and by implication tests, of general relativistic effects on the Doppler and ranging data during both cruise and orbiter phases, and (4) an improvement in the ephemeris of Jupiter by means of spacecraft ranging during the Orbiter phase. The gravity fields are accessible because of their effects on the spacecraft motion, determined primarily from the Doppler data. For the Galilean satellites we will determine second degree and order gravity harmonics that will yield new information on the central condensation and likely composition of material within these giant satellites (Hubbard and Anderson, 1978). The search for gravitational radiation is being conducted in cruise for periods of 40 days centered around solar opposition. During these times the radio link is least affected by scintillations introduced by solar plasma. Our sensitivity to the amplitude of sinusoidal signals approaches 10^{-15} in a band of gravitational frequencies between 10^{-4} and 10^{-3} Hz, by far the best sensitivity obtained in this band to date. In addition to the primary objectives of our investigations, we discuss two secondary objectives: the determination of a range fix on Venus during the flyby on 10 February, 1990, and the determination of the Earth's mass (GM) from the two Earth gravity assists, EGA1 in December 1990 and EGA2 in December 1992.

1. Introduction

The gravitational investigations discussed in this review are a subset of radio science investigations that use the Galileo telecommunication subsystem and Earth-station radio systems of the Deep Space Network (DSN). Unlike the Voyager mission to the outer planets, where there was one radio science team representing both gravitational science and propagation science, NASA has followed the examples of earlier missions and has selected for the Galileo mission two experiment teams representing respectively the two scientific disciplines (see the companion paper by Howard *et al.* (1992), for details and for a description of the overall radio-science system). Because the two teams have instrumentation in common, there is an inevitable overlap in the overall planning of radio science activities, both in the spacecraft sequencing and in the DSN scheduling. Nevertheless, we review here only the gravitational science which is organized as follows. Anderson is the Team Leader for Gravity and Celestial Mechanics; he proposed investigations in the area of celestial mechanics, including relativistic time delay and relativistic red shift. Estabrook and Armstrong are Team Members; they proposed a

Space Science Reviews **60:** 591–610, 1992.

search for gravitational radiation. Campbell is Science Coordinator for the team, and in addition is involved in the analysis of data for the celestial mechanics investigations. Krisher is participating in the relativistic celestial mechanics, including the relativistic time delay and red shift experiments. Lau is participating in applications of the JPL ephemeris system to the celestial mechanics experiments, in particular the tests of general relativity and the improvement of the Jupiter ephemeris.

At this early stage in the mission, we have received data from the Venus encounter, the first Earth Gravity Assist (EGA1), and the cruise phase. We are actively engaged in the analysis of these data and in planning our experiments for the cruise gravitational-wave opportunities and the next Earth flyby, EGA2. As with any planning activity on space missions, the optimization of spacecraft and DSN radio systems, spacecraft trajectories, and mission operations for our team must be accomplished in competition with other mission and science requirements, hence the ideal conditions for gravitational science will be achieved only rarely. Yet even under less than ideal conditions, we are satisfied with the current status of mission planning, not only for the cruise experiments prior to Jupiter arrival, but also for the high-priority measurements of the gravitational fields of the Galilean satellites during the satellite tour. The Galileo mission offers major new opportunities for gravitational science and celestial mechanics.

Radio data for our team consist of ranging and Doppler measurements generated by the DSN at its three sites at Goldstone California, near Madrid Spain, and near Canberra Australia. The DSN uses low-noise, highly phase-stable receivers, and a distribution system for frequency and timing based on hydrogen masers. These systems, continually under improvement, make use of the latest technologies in digital electronics and fiber optics, for example. Though these improvements are primarily motivated by the requirements for tracking and communicating with spacecraft in deep space, for example Pioneer 10/11 and Voyager 1/2 at distances beyond the orbit of Neptune, we recognize that new or improved radio science measurements are often enabled as well.

The importance of frequency stability in the end-to-end Doppler System will be discussed for each of the investigations in the following sections. The general idea is that gravitational fields, whether produced by asteroids, planets, satellites or even gravitational waves (GW), will affect the path of the Galileo spacecraft and hence the frequency of the radio link between DSN stations and the spacecraft. The lower the noise in the radio link over the frequency band of interest, the smaller the gravitational signal that can be detected and measured in that band. Our task is to identify and subsequently measure gravitational signals that yield information on masses, densities, and the internal structure of planets and satellites. In some cases the sensitivity is good enough to test the foundations of gravitational theory at the Einstein level, or to search for GW produced by extreme events at the galactic center or in external active galactic nuclei, or even in the early universe, including the Big Bang.

Until May 1991, all Doppler and ranging data are being generated with the spacecraft's low-gain antenna, thus the Earth-spacecraft uplink is limited to 2215 MHz (14.17 cm wavelength) in the radio S-band. With the unfurling of the spacecraft's high-gain antenna in May 1991, the uplink frequency can be either $\simeq 2115$ MHz or, by using selected 34-m radio antennas at each of the three DSN sites, the uplink can be transmitted at $\simeq 7167$ MHz in the X-band. Upon receiving either the S-band or X-band transmission from the ground, but not both simultaneously, the spacecraft's transponder and radio subsystem will generate the phase coherent, simultaneous downlink signals with an S/X carrier coherency ratio of 11/3. The transponded signals will be transmitted by the spacecraft with up to 20 W of power beamed to the Earth through a 5-m parabolic dish, the high-gain antenna. If the high-gain antenna is not fully deployed as planned, all our investigations will be carried out at S-band using the low-gain antenna.

The advantage of generating Doppler and ranging data with X-band on both the uplink and downlink lies in the reduction of plasma effects in radio-frequency measurements at higher frequencies. Because the refractive index of cold plasma is inversely proportional to the square of the frequency of the link, the perturbations of the radio phase are inversely proportional to the first power of the frequency. In terms of a figure of merit, defined by frequency fluctuations divided by the center frequency for the link, the improvement goes inversely as the radio frequency squared. The scintillation noise in the beamed radiation is reduced by a factor $(3/11)^2$ for X-band Doppler with respect to S-band. Similarly, the uncertainty in the group velocity of the ranging modulation (pseudo-random code) is reduced by a factor of $(3/11)^2$ for X-band ranging. Another advantage of the Galileo spacecraft is that the larger high-gain antenna, compared to the 3.66-m dish used on Voyager, will yield about a 38 dB antenna gain at S-band and about a 50 dB gain at X-band. Once the high-gain antenna is unfurled, the gravitational investigations will no longer be limited by a poor signal to noise ratio at the DSN receivers, whether located at 34 m or 70 m stations. Although in principle it is possible to integrate any coherent radio signal for a long enough time that the phase can be measured to a fraction of a cycle, in practice there are limits to what can be done with a weak signal buried in noise. Besides, for some of our measurements, in particular the GW search and the determination of mass signals rich in high-frequency harmonics, we are interested in relatively short Doppler integration times in the range of 1 to 10 s. For comparison, the GW search with Pioneer 10/11, limited by a weak signal from the spacecraft's 8 W transmitter and 1 m dish, requires Doppler integration times of 100 s or longer (Anderson *et al.*, 1990a). Of course at some point integration times can become so long that systematic effects from sources such as interplanetary plasma are important, not poor signal to noise, but for Galileo this occurs for periods longer than 1000 s at X-band. Fortunately, even at Jupiter distance, there will be plenty of signal at both X- and S-bands. Signal to noise is not a concern for any of our experiments.

A primary aim of the celestial mechanics experiments is to measure the shapes of the gravitational fields of Ganymede, Io, and Europa. The results will allow us to make a better selection of models for the interior of the satellites. The experiment provides data on the masses and moments of inertia, and these data constrain the central density, differentiation of materials within the satellites, and chemical composition and physical states of the interiors. This is possible because Galileo will approach the satellites much closer than did any earlier spacecraft, hence gravitational forces will be larger and easier to observe.

2. Gravitational Experiments During Cruise

Ongoing investigations and data analysis at the time of the writing of this review concern (1) an Earth–Venus range fix from the Venus encounter on 10 February, 1990, and (2) a test of the redshift of the onboard Ultra Stable Oscillator (USO) in the gravitational field of the Sun.

Galileo is only the second U.S. spacecraft to provide ranging data at Venus. The first transponded ranging at Venus was generated with Mariner 5, but an accurate range fix to the planet was not obtained. Instead, the emphasis was upon combining the spacecraft radio data with simultaneous radar ranging to Venus to determine the radius of the planet (Anderson *et al.*, 1968). Galileo's ranging data during encounter will provide data on the Venus ephemeris. Unfortunately, the two U.S. orbiters of Venus do not have a ranging capability: Pioneer 12 (PVO) because of a spacecraft radio system dating from the early days of the Pioneer missions (no ranging transponder), and Magellan because NASA rejected all relativity proposals that would have placed a requirement on the Magellan mission for ranging. With no requirement for ranging, the Magellan Project used the ranging port of their transponder for a telemetry channel, thus gaining a much-needed increased bit rate. But even in the absence of spacecraft ranging, data from radar altimeters on PVO and Magellan will result in improvements to the ephemeris. The limiting error in ground-based radar ranging to Venus is km-sized topographic variations on the planet. By calibrating the radar ranging for distances between the center of mass of Venus and the sub-radar points on the planet's surface, as determined by the altimeter data (Pettengill *et al.*, 1980), a much improved set of Venus radar ranging will result. The accuracy will not be as good as what could be obtained by ranging to an orbiter, but the r.m.s. radar ranging residuals will be improved from about 10 μs to about 1 μs, comparable to the accuracy of the Galileo range fix.

We are in the process of analyzing the Galileo ranging data generated during the Venis encounter. Ranging residuals referenced to a best-fit flyby orbit are shown in Figure 1. We estimate that the reduced range fix between the centers of mass of Earth and Venus will be accurate to about $\pm$ 1.0 μs ($\pm$ 150 m in distance). This information, in combination with radar ranging to Mercury and Venus dating back to 1966, as well as the Viking Lander ranging to Mars between 1976 and 1982 and the Mariner 10 range fixes to Mercury, will be used to improve the ephemerides of the inner planets and to check for agreement with current gravitational theory. For example the current determination of the excess relativistic precession of Mercury's perihelion, 42.96 arc sec per century in excess of the inertial 530 arc sec per century from planetary perturbations, is accurate to 0.2 arc sec per century and is in agreement with General Relativity (Shapiro *et al.*, 1976; Anderson *et al.*, 1987, 1990b). There is a potential for improving this result by about a factor of 2 by analyzing all available ranging data for the inner planets, including the Venus radar ranging calibrated for topography and the Galileo range fix. Improved accuracy for the general relativistic parameter α_1 may be obtained as well (see Will, 1981, for a definition of α_1 as well as other parameters that characterize general relativistic orbital corrections).

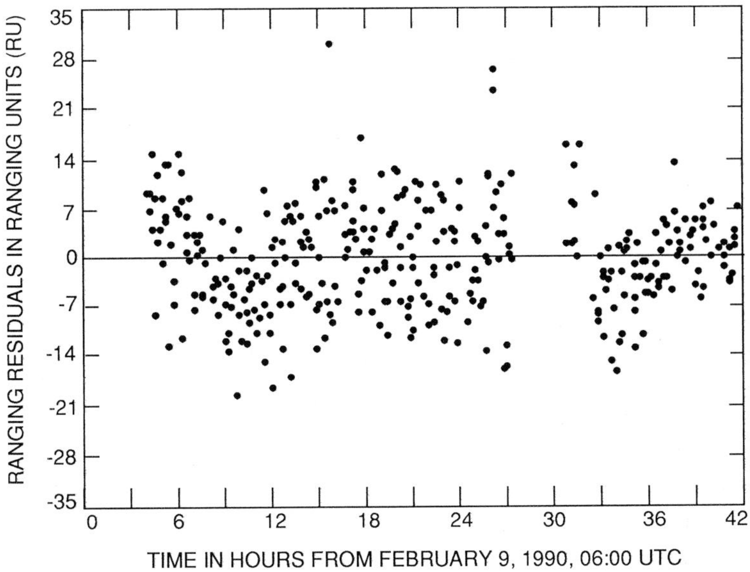

Fig. 1. Ranging residuals (observed range minus computed range) for the encounter of the Galileo spacecraft with Venus on 10 February, 1990. The residuals correspond to ranging data generated at three 70-m DSN stations (DSS14, DSS43, DSS63). The units for the residuals (RU) are expressed in terms of the output from the ranging hardware and are a function of the exciter frequency. For the Venus encounter, 1 RU = 1.056 ns of round-trip time delay, or 0.158 m in distance. The r.m.s. ranging residual is about 7 ns (1.1 m). Any unknown bias in the data is less than 80 ns.

Another application of more accurate ephemerides for the inner planets is to determine or limit a possible time variation in the gravitational constant G as measured in atomic units. The ephemerides are based on astronomical ephemeris time as independent variable, but the measurements of range are made in atomic-time units, hence it is operationally possible to detect a systematic difference between ephemeris time and atomic time that could be attributable to a time-varying G.

The expected effect is small. For a Hubble constant H_0 of 75 km s^{-1} Mpc^{-1}, the cosmic expansion rate is 7.67×10^{-11} yr^{-1}. The orbital motion of the binary pulser PSR 1913 + 16 yields a determination of $\dot{G}/G$ of $(1.0 \pm 2.3) \times 10^{-11}$ yr^{-1}, a result that is consistent with zero (Damour *et al.*, 1988). Lunar laser ranging to corner reflectors on the Moon, and Viking Lander DSN ranging, can be used to limit $|\dot{G}/G|$ (Williams *et al.*, 1978; Hellings *et al.*, 1983; Reasenberg, 1983), however both the Moon and Mars are affected by dynamical systematic noise (geophysical effects for the Moon and unknown asteroid masses for Mars), and it is generally agreed that a limit of $|\dot{G}/G| < 3 \times 10^{-11}$ yr^{-1} is the best that can be achieved with either of these bodies at present, although the work of Hellings *et al.* (1983) would suggest that a limit somewhat smaller than this could ultimately be achieved with confidence by means of a careful modelling of the asteroid belt between Mars and Jupiter. The motions of Mercury, Venus and Earth are not so affected by the asteroids, hence by turning our attention to ranging data for Mercury and Venus, and by using the Viking Lander data to determine the orbit of the Earth, it should be possible to determine $\dot{G}/G$ with a 1 σ accuracy of $\pm 0.2 \times 10^{-11}$ yr^{-1} (Anderson *et al.*, 1990b).

Another prediction of General Relativity that is being tested with the Galileo spacecraft during cruise is the gravitational redshift in the field of the Sun (Will, 1981; Krisher, 1990). Only one method has been used previously to test the solar redshift. This has involved determining the shift in the positions of spectral lines of elements in the Sun. This type of measurement is difficult to perform accurately, however, resulting so far in only a 5% test of the redshift (Snider, 1972, 1974). The measurement with Galileo depends on observing the gravitational shift in the frequency of an oscillator deep within the gravitational potential of the Sun. The location of the spacecraft in the Sun's field is determined by the phase-coherent Doppler data based on the transponded radio signal. Then the measurement of frequency at the spacecraft is accomplished by breaking the phase lock with the uplink. This noncoherent transmission is referenced to an oven-controlled crystal oscillator, the Ultra Stable Oscillator (USO), a spare Voyager USO with similar frequency stability to those flown on Voyager 1 and Voyager 2. The one-sided power spectral density of the Voyager 2 USO is shown in Figure 2 over a range of Fourier frequency from 2×10^{-5} Hz (13-hour period) to 0.5 Hz (2-s period). The spectral density represents the noise in fractional frequency $\Delta\nu/\nu$ for Voyager 2 noncoherent transmissions during its cruise. At low Fourier frequencies the noise is characterized by a combination of a flicker-frequency component (f^{-1}) and a random walk component (f^{-2}). At the Saturn flyby in 1980 we were able to determine the redshift in the Voyager 1 transmissions to an accuracy of 1% even in the presence of low-frequency noise characterized by Figure 2 (Krisher *et al.*, 1990). A 1% test in the solar field is a possibility with Galileo.

From 28 November, 1989 to the present, the command sequence for the spacecraft has included a switch from coherent to noncoherent (USO referenced) tracking on roughly a weekly schedule. Two hours of one-way USO Doppler data have been extracted and recorded by the DSN each week during these noncoherent periods. We intend to follow the frequency shift in the USO data as the spacecraft proceeds from Earth to Venus, back to the Earth at EGA1 and then to EGA2 (see Figure 3). The gravitational shift in the S-band transmission (2295 MHz) will be roughly 10 Hz (minimum to maximum frequency shift) over a period of 100 days, for a fractional frequency shift $\Delta\nu/\nu = 4.4 \times 10^{-9}$. By Figure 2, the worst-case estimate of the inherent USO random walk over 100 days is 0.5 Hz (one sigma) or 2×10^{-10} in fractional frequency, which would imply a potential measurement accuracy of 5%. However, by using properties of the predicted gravitational frequency shift, it might be possible to achieve at least another factor of 5 improvement in accuracy, provided that possible systematic errors do not corrupt the determination.

The USO stability characterized by Figure 2 is specified in terms of the frequency deviation from a straight-line fit to the oscillator drift. Hence, during the data analysis for the solar redshift, we will remove a bias and a linear trend in the USO frequency data. The residual frequency shift over the total interval of data will consist of the

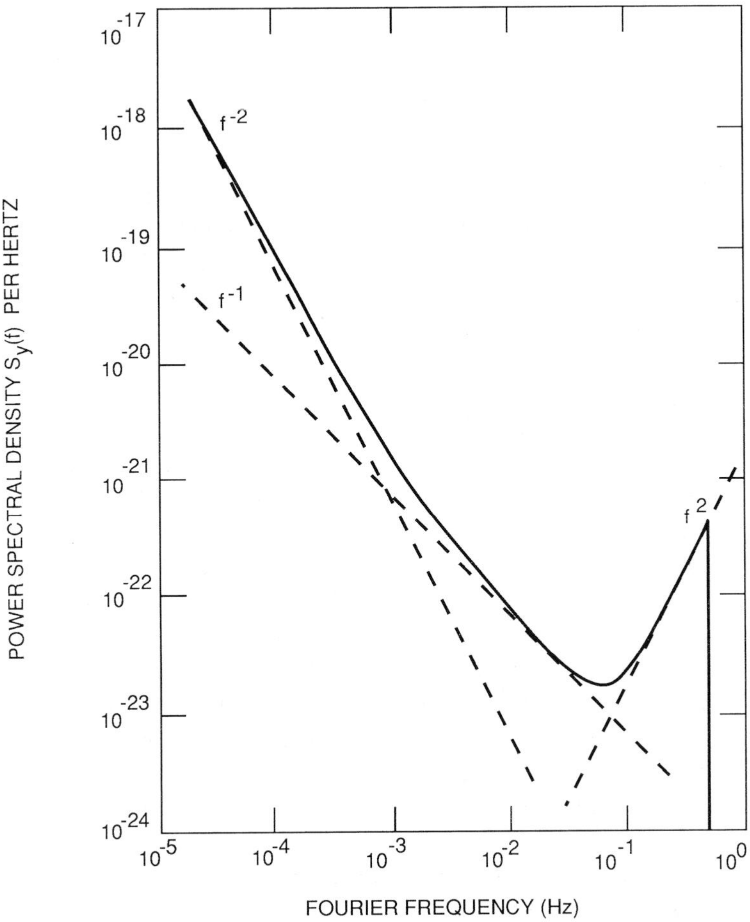

Fig. 2. One-sided power spectral density for measurements of fractional frequency for the Ultra Stable Oscillator (USO) flown on Voyager 2. The Doppler shifts in USO-referenced transmissions from Voyager 2 during cruise were referenced to ground-based hydrogen masers. The raw data were detrended by first removing effects of spacecraft-station relative motion, and then by fitting a bias and linear drift to the resulting Doppler residuals. The USO in the Galileo Orbiter's radio system is a Voyager spare, hence its noise properties should be similar. The spectral density can be represented by three power laws in Fourier frequency f as shown.

redshift signal and red noise. With a reasonable model for the noise at periods longer than the 13 hours of Figure 2, the signal can be extracted by standard statistical techniques (Wiener filter or minimum-variance regression analysis).

2.1. Determination of the Mass of the Earth

The masses of all the planets in the solar system, with the exception of Earth and Pluto, have been determined by radio tracking of spacecraft during flybys. Now with EGA1 and EGA2, the Earth will be added to the list of planetary flybys. The DSN will generate both transponded range and Doppler data with the Galileo spacecraft for two intervals,

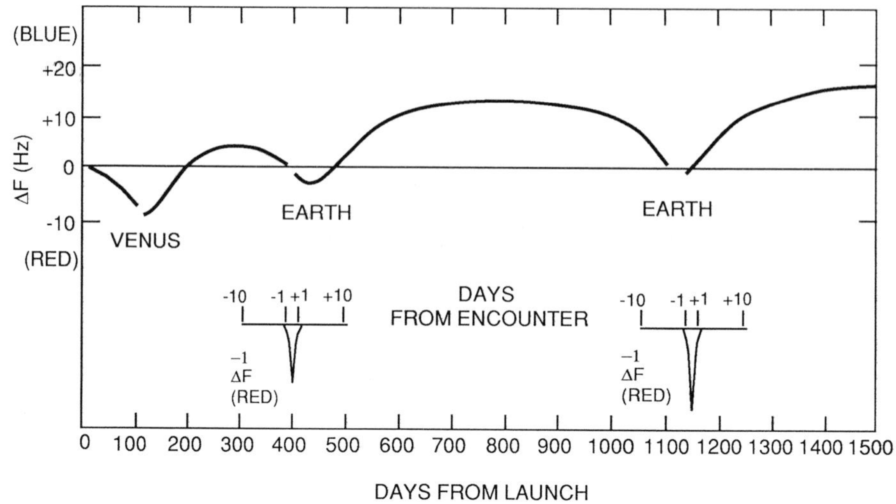

Fig. 3. Expected gravitational redshift caused by solar gravity as evidenced in radio transmissions from the Galileo spacecraft to Earth at 13 cm wavelength. The arrows indicate the times of the Venus flyby and the two Earth gravity assists, EGA1 and EGA2. The two insets in the figure show the expected redshifts caused by Earth gravity for each gravity assist. Figure courtesy of J. P. Brenkle.

defined by three days prior to encounter to three days after encounter at EGA1 and EGA2. Because the DSN's frequency and timing system is based on hydrogen-maser frequency standards, the instrumental error in measuring the Doppler shift will be better than 2×10^{-14} in fractional frequency $\Delta v/v$ for sufficiently long Doppler Integration times. The corresponding measurement of velocity will be accurate to about 3×10^{-6} m s^{-1}, with a seeming possibility of measuring *GM* of the Earth to an accuracy of about one part per billion. But in fact the actual error in the measurement could be as much as a factor of 100 larger than that. Several considerations contribute to this possibly larger error. Most importantly, the instrumental error, referenced as it is to the hydrogen masers, is a fraction of the total error budget for the determination of *GM*. Among other important error sources are (1) the correlations between *GM* and the determination of the spacecraft's trajectory, (2) noise introduced into the *S*-band ($\sim$2.3 GHz) radio signal by the Earth's troposphere and ionosphere, (3) uncertainties in the absolute location of the tracking stations at Goldstone California, near Canberra Australia, and near Madrid Spain, and (4) buffeting of the spacecraft by drag and by reaction forces generated by subsystems on the spacecraft itself. These error sources will be most important for the more distant flyby EGA1, where the error in *GM* could pessimistically be as high as 0.3 ppm. However they are less of a concern for EGA2, so by combining the Doppler data from the two flybys into a single determination of *GM*, along with a determination of the two assumed independent flyby trajectories, the realistic error (1 σ) in *GM* should be less than 0.07 ppm.

One would think that *GM* for our own planet would be so well known that another determination by the Galileo flybys would be at best superfluous. But in fact there are subtleties with previous determinations that make a relatively straightforward and independent determination by the flyby technique worthwhile. One recognizes that the external gravitational potential function ϕ for the Earth is best determined by observing secular and long-period changes in the orbits of as many artificial satellites as possible. Yet the complexity of ϕ, both on a local and on a global scale, necessitates a determination of a wide range of spherical harmonics, with the result that there is little room left for a good determination of the zero degree harmonic, the total mass GM.

In fact, it was not until 1978 or 1979 that good results from satellites first became available. Recently, Ries *et al.* (1989) have derived a value of GM from Earth-orbiting satellites, including LAGEOS. Their value is so accurate that in order to assure consistency with General Relativity, one must specify whether the result is given for geocentric coordinates or solar-system barycentric coordinates (Huang *et al.*, 1990). In geocentric coordinates, the natural coordinate system for Earth-satellite work, the result is $GM = 398600.4405 \pm 0.0010\ \mathrm{km^3\,s^{-2}}$, where $c = 299792.458\ \mathrm{km\,s^{-1}}$. In solar-system barycentric coordinates, the system used by JPL in orbit-determination software, the result would be about 0.015 ppm smaller, thus the GM result from the Galileo flybys should be consistent with a value of $GM = 398600.435 \pm 0.001\ \mathrm{km^3\,s^{-2}}$. We must wait until after EGA2 in December 1992 before we can verify that there is this agreement between the Doppler tracking of the Galileo spacecraft during Earth flyby and laser ranging to the LAGEOS satellite. A good agreement will provide increased confidence that flyby determinations for other planets are sound.

2.2. Search for Gravitational Radiation

There is strong theoretical support for the idea that matter undergoing asymmetrical motion will radiate gravitationally, and that the radiation will propagate at the characteristic speed c in the form of gravitational waves (GW) in the space-time metric, or tensor potential, of General Relativity (Thorne, 1987). Though GW have never been detected with certainty, there is observational evidence that the theory is correct; the predicted effect of gravitational radiation reaction is seen in the acceleration of the mean orbital motion of the binary pulsar PSR 1913 + 16 (Taylor and Weisberg, 1989).

For several selected periods during Galileo's cruise to Jupiter, its X-band microwave tracking link will be continuously monitored to detect any Doppler frequency fluctuations that could be caused by passing, cosmically generated, long period gravitational waves. These observations will be conducted when other spacecraft experimental activity is at a minimum and when the spacecraft is in the anti-solar direction (so that solar plasma interference with the microwave link is minimized), and will utilize the high precision H-maser time-keeping standards of the Deep Space Network. The resultant overall strain (Doppler) measurement precision will be $h \sim \Delta v/v \sim 5 \times 10^{-15}$, which is at the threshold theoretically calculated for pulses of millihertz frequency waves from several classes of extragalactic sources (for a comprehensive review see Thorne, 1987; Schutz, 1989). By obtaining up to 40 days of continuous X-band Doppler, as requested by our team, it may be possible to search for periodic waves to a threshold of $\sim 10^{-16}$.

Gravitational waves are propagating gravitational fields, 'ripples' in the curvature of space-time that carry energy and momentum and move at finite speed. All relativistic theories of gravity agree on the existence of these waves, although the theories may differ in number of polarization states, propagation speed, efficiency of wave generation, etc. In General Relativity, gravitational waves are transverse, have two polarization states, and propagate at the speed of light. As a wave passes through space, it changes the geometric curvature of the space-time, in directions transverse to the propagation direction. Thus the effect may be described as a strain in space, a dimensionless fractional change in the distances between any massive objects present, coupled with a similar fractional change in the rates at which separated clocks keep time. The amplitude of a gravitational wave is characterized by the dimensionless 'strain amplitude' $h = \Delta l/l$.

While gravitational waves can in principle be produced in the laboratory (e.g., by a steel bar spinning about an axis perpendicular to its length), the resulting signal would be far too weak to be detected with any foreseeable technology. Even the waves from the binary pulsar PSR 1913 + 16 are at far too low an amplitude to be detected directly. Extreme astrophysical events involving at least solar mass objects undergoing high acceleration are required.

Several experimental groups worldwide are building and operating resonant bar and laser-interferometer gravitational wave antennas, sensitive to kilohertz waves expected from various (solar-mass) sources in our Galaxy. For the millihertz frequencies expected from larger extra-galactic sources, for example from extreme events in the nuclei of active galaxies, massive binary black holes, or possibly the Big Bang, we require 'antennas' on a large scale of 1 AU or more such as the Earth–Galileo microwave link (for gravitational waves of period P, the antenna response falls rapidly if P is larger than the round-trip-light-time of the link, Estabrook and Wahlquist, 1975; Estabrook, 1985). Searches for millihertz gravitational waves can only be done in space. Previous searches have been performed with the Viking spacecraft, Voyager 1, Pioneer 10, Pioneer 11, and they have been scheduled with Ulysses (Armstrong *et al.*, 1979, 1987; Hellings *et al.*, 1981; Anderson *et al.*, 1984, 1990a; Anderson and Mashhoon, 1985; Estabrook, 1988; Armstrong, 1989).

Implementation of up-down coherent X-band microwave links on Galileo allows for the first time a spacecraft Doppler search at the lower strain thresholds for which plausible astrophysical sources may exist. The DSN has recently installed X-band transmitters, and implemented careful improvements to the phase stability of the timing and frequency stabilization electronics of selected 34-m radio antennas at all three sites in California, Australia, and Spain (Peng, 1988).

The Galileo high-gain antenna is scheduled to be unfurled in May 1991, after the first Earth encounter, and just before the first solar opposition (see Figure 1 in the companion paper by Howard *et al.* (1992) for a description of events important to Radio Science in general, and the GW search in particular). After necessary spacecraft control and pointing procedure have been tested, a week of continuous X-band Doppler data will be requested for purposes of categorizing thresholds for system phase noise and for demonstrating ground operations at the 34-m high precision tracking sites. This will also be the first significant gravitational wave search opportunity, inasmuch as the range on the Earth–Earth leg of the cruise trajectory will have opened rapidly to 73×10^6 km – or 487-s round-trip-light-time. We calculate the burst sensitivity for 10 mHz waves ($F = 100$ s) to be $\sim 8 \times 10^{-15}$. Similar searches will be done at the second opposition. November 1992 and third opposition, February, 1993.

Forty-day gravitational wave searches are scheduled for Opposition 4 and 5, enroute to Jupiter. The longer round-trip-light-times allow searches for millihertz waves ($P \sim 100$ s) at limits set by the X-band capability. At Opposition 4, May, 1994, Galileo will be 5×10^8 km from Earth, 3367 seconds round-trip-ligth-time; at Opposition 5, June, 1995, 6.3×10^8 km, or 4233 s round-trip-light-time. The expected sensitivities for these 40-day searches are: for bursts, 5×10^{-15}; for periodic (sine) waves 1 to 3×10^{-16}, depending on frequency in the millihertz band; and for a possible gravitational wave background, 2×10^{-15} (r.m.s. in 118 μHz band, limited by DSN station stability). If the high-gain antenna is not successfully deployed, these opportunities will still exist at S-band, but with decreased sensitivity.

Finally an opportunity may occur in April 1993 for a joint spacecraft Doppler tracking experiment. If present plans go through, the Mars Observer spacecraft will also be flying enroute to Mars. It also will have X-band capability, both receive and transmit. Both it and Galileo will be in roughly anti-solar directions, though not at strict opposition. In fact they will be sufficiently separated in the sky that they can be simultaneously tracked from DSN stations at different longitudes. A continuous 3-week joint tracking experiment is being proposed. Joint experiments would provide much more convincing evidence for any putative gravitational wave detection near the system sensitivity threshold.

2.3. Mass determinations for small bodies

There is the possibility that the Galileo spacecraft may fly close enough to one or more small bodies that a meaningful mass determination can be obtained. There are no plans to fly near small satellites of Jupiter or any known comets, but during two passes through the asteroid belt between Mars and Jupiter, once near aphelion in December 1991 at a distance of about 2.3 AU, and again in 1993 on the transfer trajectory between EGA2 and Jupiter arrival (see Figure 1 of Howard *et al.*, 1992), close encounters with one or two asteroids are likely. The accuracy in the determination of their masses depends on the distance of closest approach and on the relative spacecraft-asteroid velocity. The fractional error in mass is given by the expression,

$$\frac{\Delta m}{m} = kc\left(\frac{bV}{Gm}\right)\frac{\Delta v}{v},$$

where m is the mass of the asteroid, b is the impact parameter of the asteroid-spacecraft hyperbolic flyby, V is the flyby velocity on the hyperbolic asymptote, G and c are the usual physical constants, $\Delta v/v$ is the accuracy of fractional-frequency Doppler shift over a characteristic time for the flyby ($\sim 10\, b/V$), and k is a dimensionless factor that depends on both the geometry of the flyby as observed from Earth, and the sample interval for the Doppler data. If b can be determined from optical navigation data to an accuracy of a few pixels, rather than from a simultaneous Doppler solution for b and Gm, the value of k will lie in the range 0.4 to 1.0, thus it can be taken equal to unity for a conservative estimate of the mass error.

The mean density ρ of the small body is of fundamental interest, so both the mass and volume must be determined. For close approaches to asteroid-sized bodies, it is the mass that limits the determination of density, hence $\Delta\rho/\rho$ is given by the same expression as $\Delta m/m$. Volume is determined by imaging data to smaller fractional error. In order to minimize the error $\Delta\rho/\rho$, it is necessary to minimize the product $bV(\Delta v/v)$. We have no significant control over the flyby velocity, though it is small (of order 8 km s^{-1} for the first pass through the asteroid belt between EGA1 and EGA2), but the impact parameter is fairly easy to control by small midcourse corrections. The fractional Doppler error $\Delta v/v$ can be minimized for Galileo by tracking the spacecraft with the X-band uplink instead of S-band.

The Galileo Project has targeted the asteroid 951 Gaspra for a flyby on October 29, 1991. Unfortunately, the expected radius of the asteroid is only 8 km, hence the expected Gm is small, about 4×10^{-4} km^3 s^{-2}. The planned flyby distance is 1600 km, and with a flyby velocity of 8 km s^{-1} and a Doppler error of 5×10^{-14}, the formula for $\Delta m/m$ yields an uncertainty of 50%. Detailed numerical simulations with Navigation software confirm this result. Furthermore, we have shown that coherent X-band tracking during the flyby is an absolute requirement, even for a 50% determination. The only possibility for improving on this result is to fly closer. Because the error decreases linearly with the distance, a flyby at 800 km would yield a density determination to about 25%. With regard to distinguishing between a predominantly icy or predominantly rocky composition, 25% is better than 50%, but the improvement is probably not significant enough to change the Project's decision for a 1600 km flyby.

3. Experimental Celestial Mechanics During the Orbital Tour

For the first time we have an opportunity to conduct gravitational investigations with an orbiter of a giant planet. Not only that, but we expect a high level of performance from the Orbiter's radio subsystem, at least when compared to previous missions to the outer solar system, or even to the Viking Orbiter of Mars. Furthermore the DSN, with substantial improvements to its worldwide network of tracking stations, will be in a position to take full advantage of Galileo's improved capabilities. Our investigations will rely primarily on transponded Doppler and ranging data, supplemented by ground-based astrometric data on Jupiter and its satellites, and by star-satellite imaging data from the orbiter itself. All these various types of data will be available for analysis by means of archiving and software systems used by the Galileo Navigation Team. The Doppler and ranging data generated by the DSN specifically for the gravitational investigations will be transferred to Archival Tracking Data Files (ATDF) by the same procedures and in the same format as for the Navigation Team, thus navigational data and gravitational data can be merged for subsequent data reduction and analysis.

Whenever possible, Doppler data will be generated in the X-band ($\simeq$7167 MHz uplink and 8422 MHz downlink) for purposes of minimizing scintillation noise from the interplanetary medium. Similarly, for purposes of minimizing the effects of the interplanetary medium on the ranging data, modulation on the X-band uplink will be available on both the X-band downlink and the S-band downlink at 2296 MHz. The DSN's ranging assembly (SRA) at the 34 m stations, equipped with X-band uplink capability, will autocorrelate the pseudo-random code modulated on the uplink with the received modulation on the downlink, and simultaneously will determine the difference (S-X) in the group delay of the signals at the two downlink frequencies.

The end-to-end ranging system consists not only of DSN hardware but also software used by the Navigation Team. The final output of this system is S-band and X-band ranging residuals in units of microseconds, where the residuals are referenced to predicted ranges from a spacecraft trajectory computed by the Orbit Determination Program (ODP). For details regarding the computation of these range residuals at any given time of observation, as well as the computation of Doppler residuals, see Moyer (1971).

3.1. Expected accuracy of ranging data

The instrumental accuracy (1 σ) of the ranging hardware at a DSN station is about $\pm$7 ns ($\pm$1 m) for Galileo. There is also an unknown ranging bias, perhaps as large as 80 ns (12 m) because of group delays through the transmission media, through the spacecraft radio system, and through the transmission path at the station. This unknown bias can be minimized by calibration of the signal path. The delay through the spacecraft radio system, as measured prior to launch on a prototype system, is 713 ns at X-band and 715 ns at S-band, with unknown variations possibly as large as 30 ns because of environmental conditions on the spacecraft. The delay through the station signal path, on the order of 12 000 ns, is calibrated at least once for each ranging pass, with perhaps a few exceptions when there is no time in the schedule for station calibrations. However, it is usually a straightforward matter to interpolate the station calibrations when there are gaps.

A significant portion of the unknown group delay is caused by propagation of the radio signal through interplanetary plasma, but most of this contribution can be calibrated by means of the dual-band downlink. For an electrostatic plasma, the observed range residuals at X-band and S-band are. respectively,

$$\Delta l_x = \Delta l_g + P_u / \nu_u^2 + P_d / \nu_{xd}^2 \, ,$$

$$\Delta l_s = \Delta l_g + P_u / \nu_u^2 + P_d / \nu_{sd}^2 \, ,$$

where Δl_g is the non-dispersive group delay independent of the plasma, P is proportional to the columnar electron content of the beam, ν is the carrier frequency, and the subscripts u and d refer to the uplink and downlink transmissions. For Galileo the common uplink frequency can be either ν_u = 2115 MHz or ν_u = 7167 MHz. The dual-band downlink transmissions are at frequencies ν_{xd} = 8415 MHz and ν_{sd} = 2296 MHz.

To the first order, the downlink contribution to the X- and S-band residuals is

$$P_d = \frac{v_{sd}^2 v_{xd}^2}{(v_{xd}^2 - v_{sd}^2)} (\Delta l_s - \Delta l_x)$$

and with this measured value of P_d, an estimate of P_u can be obtained by interpolation (Muhleman and Anderson, 1981). We have made no attempts to predict the nature of the power spectral density for the calibrated ranging residuals during the orbital phase, but instead will compute periodograms empirically from the actual calibrated residuals after the unfurling of the high-gain antenna.

3.2. Expected accuracy of Doppler data

The characteristics of the Doppler noise were discussed in Section 2.2 for favorable geometries with Sun–Earth-Probe (SEP) or solar elongation angles greater than 160 arc deg. We expect an Allan variance with X-band uplink of $\sigma_y = 2 \times 10^{-14}$, and about $\sigma_y = 10^{-13}$ with S-band uplink, at periods typical of the celestial mechanics investigations (> 600 s). For SEP angles less than 90 arc deg, σ_y could be a factor of 10 to 100 times larger for the respective uplinks, and for SEP angles less than 10 arc deg, we have asked that no critical measurements be scheduled at all. The measurement of phase, or equivalently range change, is much more accurate than the measurement of absolute range, by a factor of about 10^3 for X-band, hence the calibration of Doppler by means of the dual-band downlink is far less effective than for range.

3.3. Non-gravitational accelerations

Another noise source of concern for the celestial mechanics investigations is buffeting of the spacecraft in the range of periods from 1000 to 100 000 s. During the design phase of the mission around 1980, we imposed two requirements on the spacecraft as follows:

(1) During a satellite encounter the orbiter shall not average non-mean-zero accelerations greater than 10^{-11} km s^{-2} on any axis when measured over a time period of 1000 s.

(2) The uncertainty in the unmodelled portion of any such non-gravitational accelerations shall not exceed 10^{-12} km s^{-2} when measured over any time period exceeding 1000 s.

It has yet to be demonstrated during the orbital phase that these requirements can be met, and until we gain experience with real data, the spectral distribution of the spacecraft-generated forces is unknown. During critical periods for the celestial-mechanics measurements, reaction forces from spacecraft subsystems could be minimized by keeping spacecraft activity to a minimum, but here again we need experience with real data during the Orbiter phase before we can be definitive on the level of acceptable spacecraft activity and before we can identify particularly troublesome spacecraft systems.

We anticipate that the spectral density of acceleration noise from spacecraft systems will increase with increasing frequency. However, because the corresponding Doppler spectrum (first integral of acceleration) is proportional to the acceleration spectrum divided by f^2, the Doppler spectrum will at best be white, and it could instead increase at lower frequencies. In either case we expect that it will be considerably larger than the buffeting noise expected from fluctuations in the solar wind, and it should be larger than the f^{-3} spectrum from fluctuating solar radiation pressure for periods shorter than 10^5 s (Woodward and Hudson, 1983). The Doppler error budget is probably dominated by the approximately f^{-1} plasma scintillation spectrum unless orbital fits over several days are required, in which case fluctuating solar radiation pressure could be a problem. It will probably be necessary to fit long arcs of Doppler data with a batch-sequential filter, where the duration of each batch is on the order of 12 to 24 hr. Of course our current evaluation of the spacecraft reaction forces could be too optimistic if it turns out that there are unexpected low-frequency components in the acceleration spectrum, as in the case of Voyager.

4. Gravitational Experiments During the Orbital Tour

During the orbital tour, the fundamental problem for a number of gravitational experiments is to determine the orbit of the Galileo spacecraft as accurately as possible with respect to the center of mass of the Jupiter system. With this fundamental problem solved, we can determine parameters of scientific interest by observing deviations in the barycentric orbit caused by various sources of gravitation, for example the rotational and tidal distortions of the Galilean satellites. Because the orbital tour will require a number of midcourse maneuvers in order to achieve the desired encounter conditions with the satellites, our best-determined orbit may in fact consist of segments between maneuvers, rather than a continuous spacecraft ephemeris for the duration of the tour. Even so, the orbit determination accuracy is expected to be consistently good over more than 99% of the orbital path. For the first time we are ranging to an orbiter of another planet with X-band uplink and dual-band downlink, also the Doppler accuracy is improved over previous missions to Jupiter, and we have the advantage of a spacecraft with reduced disturbances from the attitude control system, at least compared to Voyager.

Before discussing our primary scientific objective of the satellite gravitational fields, and our secondary objective of tests of general relativity, we should point out that an accurate orbit is needed by other investigation groups on Galileo. The required satellite-spacecraft geometries will be determined by the spacecraft ephemeris in combination with accurate ephemerides for the satellites. Although this information will be provided to the Galileo Project by the Navigation Team in agreed upon formats, it will be generated with the cooperation of our team and with merged files of navigation and radio science data. In practice the reconstructed trajectory determined by the Navigation Team for each satellite encounter probably will be identical to the reference trajectory used by our team to determine the values and 1 σ errors for the gravitational parameters. By making use of results generated by the Navigation Team, as well as relying on their facilities, we can proceed with our investigation in the most cost-effective manner. As a byproduct, Galileo investigators will receive the most accurate Jupiter-centered trajectory possible, for example for the analysis of wind profiles in Jupiter's atmosphere with the Probe–Orbiter Doppler link (Pollack *et al.*, 1992).

4.1. Gravitational fields for Jupiter and its satellite

Four spacecraft have visited Jupiter on flyby trajectories. The Pioneer 10 and 11 spacecraft came much closer to Jupiter than Voyager 1 and 2, and in fact the minimal improvement expected from the Voyager flybys caused the Voyager Project to scrub the Celestial Mechanics experiment at Jupiter. No radio science data were scheduled for celestial mechanics, although the Voyager Navigation Team scheduled enough data to allow a determination of the flyby orbits for the two Voyager encounters. Subsequently, Campbell and Synnott (1985) combined the Voyager navigation data with Pioneer 10/11 data archived for the Pioneer Celestial Mechanics Experiment, and with data from the four flybys they were able to improve substantially over the final results published by Null (1976) for the Pioneer 10/11 investigation. The accuracies for the masses of the Galilean satellites were improved by factors of 3 to 12, but no satellite was ever approached close enough to provide even a detection of second degree and order gravitational harmonics. A summary of our current knowledge of gravitational parameters for the Jovian system is given in Table I.

It is unlikely that Doppler data generated by the DSN with the Galileo Orbiter, even with X-band up and down, will yield improvements of any significance to the gravi-

TABLE I

Combined Pioneer 10/11 and Voyager 1/2 gravity results

Parameter[a]	Value	Realistic uncertainty[b]
GM (System)	126 712 767	100
GM (Io)	5961	10
GM (Europa)	3200	10
GM (Ganymede)	9887	3
GM (Callisto)	7181	3
J_2	14 736	1
J_3	1.4	5
J_4 $\times 10^6$	−587	5
J_6	31	20
C_{22}	−0.03	0.05
S_{22}	−0.007	0.05
α (pole)	268.001	0.005
δ (pole)	64.504	0.001

[a] GM units are km^3 s^{-2}, α and δ in arc deg are for the rotation pole in 1950.0 coordinates, the gravitational harmonics for Jupiter are referred to an equatorial radius of 71 398 km.

[b] 1-σ realistic errors as opposed to much smaller formal errors from a covariance analysis.

tational field of Jupiter. However, flybys of the Galilean satellites will definitely yield new results on second-degree gravity harmonics for these satellites. For purposes of mission planning of the orbital tour, we have imposed two requirements on the satellite flybys:

(1) At least one encounter with Io, Europa, and Callisto with uninterrupted tracking at less than 1400 km altitude and a solar elongation (SEP) angle greater than 10 arc deg.

(2) At least two encounters with Ganymede at varying latitudes with uninterrupted tracking at less than 1400 km altitude and a solar elongation (SEP) angle greater than 10 arc deg.

In addition we have requested 20 hours of Doppler data and 10 hours of ranging data, centered on the time of closest approach, for all encounters that satisfy one of the two mission requirements. The Io encounter will of necessity satisfy the conditions for a successful gravitational flyby, although the expected Doppler data will be of limited value because of the scheduled JOI maneuver. Nevertheless we expect to measure differences in the principal moments of inertia for Io to an accuracy of one percent or better. It should be possible to discriminate between plausible interior models for this satellite. Perhaps in combination with imaging data on its shape, a good model can be developed for the chemical composition and physical size of Io's core and envelope respectively (for a discussion of these measurements of Galilean satellite interior structure see Hubbard and Anderson, 1978).

In requirement (2) on the flyby conditions for Ganymede, we have recognized that the satellite is likely to be in hydrostatic equilibrium, and that an independent determination of the rotational and tidal response can be achieved by two flybys, one in a near equatorial orbit and the other in a near polar orbit. So far the mission planning for the orbital tour has been able to provide these two flybys. The constants that we will measure directly are $G[C - (A + B)/2]$ and $G(B - A)$, and with G known to four or five significant figures from laboratory Cavendish experiments, the differences in the moments will easily be accurate to ten percent. One percent or better is likely with X-band up and down. The two Ganymede flybys scheduled for gravity harmonics should give reasonably definitive information about interior structure for Ganymede.

Useful gravity data is also expected for Europa, although there is a possibility that departures from hydrostatic equilibrium may confuse the issue for this satellite (see Hubbard and Anderson, 1978). The experiment is most marginal for the outermost satellite Callisto with its relatively weak response to rotation and tides. Even so, the second-degree gravity harmonics will be determined for Callisto, and perhaps some useful information will be forthcoming.

4.2. Tests of General Relativity

Subsequent to some of the early radar and spacecraft ranging experiments in the mid to late 1960's, there has been a continuing interest in testing general relativity by means of planetary orbiters or landers (Will, 1981). The idea is to use the Doppler and ranging data to a spacecraft anchored to a planet for purposes of ranging to that planet to better accuracy than can be achieved with radar bounce. In the case of Jupiter, spacecraft ranging is the only technique presently available for measuring the distance between the center of the Earth and the center of Jupiter. Pioneer 10 provided a range fix on Jupiter for 4 December, 1973 to an accuracy of ± 6 km, and Pioneer 11 provided a similar measurement a year later on 3 December, 1974 to an accuracy of ± 1.5 km (Standish, 1990). The more sophisticated radio system on Voyager, including its ranging transponder and dual-band (S and X) downlink, provided even more accurate ephemeris data on 5 March, 1979 for Voyager 1 and 9 July, 1979 for Voyager 2, although the analysis of these Voyager data is still in progress at JPL.

Two or more years of ranging to the Galileo Orbiter will provide numerous range fixes on Jupiter to an accuracy on the order of 150 m, where the limiting accuracy is set by the orbit determination error along the Earth–Jupiter line, not by the instrumental error of the ranging system. In combination with the earlier radio measurements with Pioneer and Voyager, as well as radio and optical data from Earth-based observatories that provide angular positions on the sky, a much improved ephemeris for the planet will be available by early 1998. The most immediate application of an improved Jupiter ephemeris will be to improve our knowledge of the perturbations caused by Jupiter on the orbits of the inner planets, particularly Mars, with the result that existing data, for example ranging measurements to the Viking landers, will be more sensitive to small general relativistic orbital corrections, and to perturbations by asteroids.

It will be possible to measure the excess time delay in the Galileo ranging modulation caused by solar gravity. The detailed nature of this relativistic effect was first published by Shapiro (1964), and it has since been tested several times by radar and spacecraft ranging. The most accurate experiment with the Viking Orbiter and Landers agrees with the prediction of General Relativitiy to $\pm 0.1\%$ (Reasenberg *et al.*, 1979). For ray paths that pass near the Sun, the extra round-trip delay is given by (Will, 1981),

$$\Delta t = \frac{2R_G}{c} \ln\left(\frac{4r_e r_p}{r_i^2}\right),$$

where r_e is the Sun–Earth distance, r_p is the Sun-spacecraft distance, and r_i is the impact parameter, the closest approach distance of the ray to the center of the Sun. The gravitational radius is $R_G = 2GM/c^2$, where M is the solar mass and $R_G = 2953.25$ m.

The excess relativistic time delay is maximum for ray paths that graze the Sun, and at the distance of Jupiter it amounts to 271 μs. We expect to determine the Earth-spacecraft distance to 150 m, so a measurement of the excess time delay to about 0.5% seems feasible. Although this is about a factor of five less accurate than the Viking result, it is important to remember that only one solar conjunction for Viking yielded $\pm 0.1\%$ accuracy. Other published tests from radar and spacecraft ranging are accurate to at best $\pm 2\%$ (for a compilation of results see Will, 1981). With Galileo, the excess delay Δt can be measured annually at each solar conjunction for the duration of the mission. There is justification for determining Δt to the sub-1% level more than once, simply from the standpoint of good experimental practice, but in addition the predicted effect can be tested for a fairly wide range of directions in inertial space by means of the single Viking measurement plus two or more Galileo measurements.

Another experiment of possible interest is the measurement of the redshift in the spacecraft's oscillator (USO) caused by the gravitational field of Jupiter. After radiation hardening of the crystal by Jupiter's charged-particle environment, particularly during the close approach at the orbital radius of Io for JOI, the redshift will be measured for each orbital revolution to an accuracy of about $\pm 1\%$.

Finally, we should mention the possibility of measuring general relativistic effects on the orbits of the Galilean satellites and the spacecraft as discussed by Hiscock and Lindblom (1979). There is no doubt but that the ephemerides for the Galilean satellites will be improved by star-satellite imaging data and by spacecraft ranging data during a close satellite encounter. However, it is unknown whether the relativistic components of orbital precession for the satellites can be isolated from the far larger Newtonian precessions. All we can do is to perform enough data analysis to find out. Similarly with X-band Doppler up and down, the analysis of Hiscock and Lindblom suggest that the relativistic parameter β (see Will, 1981, for a definition) can be determined to better than $\pm 20\%$ from the spacecraft's motion, and given fractional frequency stability of $\Delta v/v = 5 \times 10^{-15}$ at solar opposition, the error on β could be as small as $\pm 3\%$.

Acknowledgements

The authors gratefully acknowledge the assistance and support of the Galileo Project and mission Analysis staff at JPL, the Galileo Radio Science Support Team, also personnel of the Deep Space Network and the Navigation Systems Section, and scientific colleagues at their respective institutions. We would like to thank J. P. Brenkle, D. J. Mudgway, T. K. Peng, C. T. Stelzried, and J. M. Rotenberry for extraordinary contributions to our team during the developmental phase of the investigations. Discussions with B. Bertotti and H. Wahlquist contributed to the development of the gravitational-wave investigation. The first draft of this paper was developed while one of us (JDA) was visiting in the Department of Mathematics, Monash University, Australia. We acknowledge helpful discussions and support by Monash staff, particularly A. J. R. Prentice, and we thank Mrs G. Nayak for her careful typing of the first draft. An anonymous referee contributed several comments that improved the text. This work was performed at the Jet Propulsion Laboratory, California Institute of Technology, under contract with NASA.

References

Anderson, J. D. and Mashhoon, B.: 1985, *Astrophys. J.* **290**, 445.

Anderson, J. D., Cain, D. L., Efron, L., Goldstein, R. M., Melbourne, W. G., O'Handley, D. A., Pease, G. E., and Tausworthe, R. C.: 1968, *J. Atmospheric Sci.* **25**, 1171.

Anderson, J. D., Armstrong, J. W., Estabrook, F. B., Hellings, R. W., Lau, E. K., and Wahlquist, H.: 1984, *Nature* **308**, 158.

Anderson, J. D., Colombo, G., Esposito, P. B., Lau, E. L., and Thager, G. B.: 1987, *Icarus* **71**, 337.

Anderson, J. D., Armstrong, J. W., and Lau, E. L.: 1990a, 'Application of Hydrogen-Maser Technology to the Search for Gravitational Radiation', in *Proceedings of the Twenty-First Annual Precise Time and Time*

Interval (PTTI) Applications and Planning Meeting, U.S. Naval Observatory, Time Service Department, Washington, DC.

Anderson, J. D., Slade, M. A., Jurgens, R. F., Lau, E. L., Newhall, X X, Standish, E. M. Jr.: 1990b, 'Radar and Spacecraft Ranging to Mercury Between 1966 and 1988', in *Proceedings Astronomical Society of Australia*, IAU Fifth Asian-Pacific Regional Astronomy Meeting, July 16–20, 1990, Sydney, Australia.

Armstrong, J. W., Woo, R. T., and Estabrook, F. B.: 1979, *Astrophys. J.* **230**, 570.

Armstrong, J. W., Estabrook, F. B., and Wahlquist, H. D.: 1987, *Astrophys. J.* **318**, 536.

Armstrong, J. W.: 1989, in B. F. Schutz (ed.), 'Spacecraft Gravitational Wave Experiments', *Gravitational Wave Data Analysis*, Kluwer Academic Publishers, Dordrecht, Holland, pp. 153–172.

Campbell, J. K. and Synnott, S. P.: 1985, *Astron. J.* **90**, 364.

Damour, T., Gibbons, G. W., and Taylor, J. H.: 1988, *Phys. Rev. Letters* **61**, 1151.

Estabrook, F. B.: 1985, *Gen. Rel. Grav.* **17**, 719.

Estabrook, F. B.: 1988, *Acta Astron.* **17**, 585.

Estabrook, F. B. and Wahlquist, H. D.: 1975, *Gen. Rel. Grav.* **6**, 437.

Hellings, R. W., Callahan, P. S., and Anderson, J. D.: 1981, *Phys. Rev.* **D23**, 844.

Hellings, R. W., Adams, P. J., Anderson, J. D., Keesey, M. S., Lau, E. L., Canuto, V. M., Goldman, I., and Standish, E. M.: 1983, *Phys. Rev. Letters* **51**, 1609.

Hiscock, W. A. and Lindblom, L.: 1979, *Astrophys. J.* **231**, 224.

Howard, H. T., Eshleman, V. R., Hinson, D. P. *et al.*: 1992, 'Galileo Radio Science Investigations', *Space Sci. Rev.* **60**, 565 (this issue).

Huang, C., Ries, J. C., Tapley, B. D., and Watkins, M. M.: 1990, *Celest. Mech.* **48**, 167.

Hubbard, W. B. and Anderson, J. D.: 1978, *Icarus* **33**, 336.

Krisher, T. P.: 1990, *Modern Physics Letters A* **5**, 1809.

Krisher, T. P., Anderson, J. D., and Campbell, J. K.: 1990, *Phys. Rev. Letters* **64**, 1322.

Moyer, T. D.: 1971, *Mathematical Formulation of the Double-Precision Orbit Determination Program (DPODP)*, TR 32-1527, Jet Propulsion Laboratory, Pasadena.

Muhleman, D. O. and Anderson, J. D.: 1981, *Astrophys. J.* **247**, 1093.

Null, G. W.: 1976, *Astron. J.* **81**, 1153.

Peng, T. K.: 1988, *Acta Astron.* **17**, 321.

Pettengill, G. H., Eliason, E., Ford, P. G., Loriot, G. B., Masursky, H., and McGill, G. E.: 1980, *J. Geophys. Res.* **85**, 8261.

Pollack, J., Atkinson, D., Seif, A. and Anderson, J. D.: 1992, 'Retrieval of a Wind Profile from the Galileo Probe Telemetry Signal', *Space Sci. Rev.* **60**, 143 (this issue).

Reasenberg, R. D., Shapiro, I. I., MacNeil, P. E., Goldstein, R. B., Breidenthal, J. C., Brenkle, J. P., Cain, D. L., Kaufman, T. M., Komarek, T. A., and Zygielbaum, A. I.: 1979, *Astrophys. J.* **234**, L219.

Reasenberg, R. D.: 1983, *Phil. Trans. Roy, Soc. London* **A310**, 227.

Ries, J. C., Eanes, R. J., Huang, C., Schutz, B. E., Shun, C. K., Tapley, B. D., Watkins, M. M., and Yuan, D. N.: 1989, *Geophys. Res. Letters* **16**, 271.

Schutz, B. F. (ed.): 1989, *Gravitational Wave Data Analysis*, Kluwer Academic Publishers, Dordrecht, Holland.

Shapiro, I. I.: 1964, *Phys. Rev. Letters* **13**, 789.

Shapiro, I. I., Counselman, C. C. III and King, R. W.: 1976, *Phys. Rev. Letters* **36**, 555.

Snider, J. L.: 1972, *Phys. Rev. Letters* **28**, 853.

Snider, J. L.: 1974, *Solar Phys.* **36**, 233.

Standish, E. M., Jr.: 1990, *Astron. Astrophys.* **233**, 252.

Taylor, J. H. and Weisberg, J. M.: 1989, *Astrophys. J.* **345**, 434.

Thorne, K.: 1987, in S. W. Hawking and W. Israel (eds.), 'Gravitational Radiation', *300 Years of Gravitation*, Cambridge University Press, Cambridge, 330–446.

Will, C. M.: 1981, *Theory and Experiment in Gravitational Physics*, Cambridge University Press, Cambridge.

Williams, J. G., Sinclair, W. S., and Yoder, C. F.: 1978, *Geophys. Res. Letters* **5**, 943.

Woodward, M. and Hudson, H. S.: 1983, *Nature* **305**, 589.

GUIDE AND EXAMPLES FOR USERS OF THE 'SPACEKAP' STYLE FILE

Basic Instructions

A. AUTHOR 1 and B. AUTHOR 2*
The Institute and its Address of authors 1 and 2.
and
C. AUTHOR 3
*Institute of Author 3***

Abstract. This document describes in some detail the use of the SPACEKAP style file for the preparation of articles to be submitted to journals or edited volumes published by Kluwer Academic Publishers. The 'spacekap' style file is very similar to the 'article' style sheet of LaTeX.

Key words: LaTeX – SPACEKAP – Guide

1. To Start

The 'SPACEKAP' style file is addressed by using the command: `\documentstyle{spacekap}`. The files SPACEKAP.STY and SPACE11.STY should be accessible from the directory you run LaTeX from.

2. The 'Opening' Environment

The title, subtitle, author names and institutes are handled by the *opening* environment in the preamble. Its structure is as follows:

```
\begin{opening}
\title{THE TITLE OF THE ARTICLE}
\subtitle{If There Is Any}
\author{Author's Name}
\institute{Institute and/or its Address}
\date{}
\end{opening}
```

2.1. Title and Subtitle

The title should be typed in CAPITALS. The subtitle in Initial Capitals and lowercase. The `\subtitle` command may be omitted. A missing `\title` command generates an error message.

* On leave from his original Institute
** affiliated to CMDA, Dordrecht

If your title or subtitle are too long for one line, you can split it by putting two backslashes at an appropriate place, e.g., the commands
`\title{GUIDE AND EXAMPLES FOR USERS OF \\ THE 'SPACEKAP' STYLE FILE}`
`\subtitle{Basic Instructions}`
have generated the title and subtitle of this Guide.

2.2. Author names and Institutes' Addresses

For each author's name the `\author` command should be used:
`\author{A. Author 1}`, `\author{B. Author 2}` etcetera. The names should be written in Initial Capitals and lowercase, but they are printed in Capitals.

The affiliation of authors is generated by the `\institute` command:
`\institute{The Institute and its Address}`
If there are more than one author at the same institute, you should first use the `\author` command for each author and then use the `\institute` command.

A second institute's name to indicate another affiliation can be handled through the `\thanks` command. For this Guide the following sequence of commands was used:

```
\author{A. Author 1}
\author{B. Author 2\thanks{On leave from his original institute}}
\institute{The Institute and its Address of \\authors 1 and 2.}
\author{C. Author 3}
\institute{Institute of Author 3\thanks {affiliated to CMDA,
Dordrecht}}
```

2.3. Date

The `\date` command gives you the possibility of entering the date (or a text) at the beginning of your article. If you do not wish a date, only the two brackets will do. In principle you can put any text between the brackets.

3. Running Heads

If the *title* of your article is longer than 60 characters, the running head will be omitted in the printout. You can make your own running title by using the `\runningtitle{}` command, where you put the shorter title in capitals between the curly brackets.

If the number of *authors* is more than two, please use the
`\runningauthor{}` command, and abbreviate the list of authors to `A. AUTHOR 1 ET AL.` which you put, in capitals, between the curly brackets (but ET AL. *not* in italics).

If you use one or both commands, please put these in the preamble, just before the `\begin{document}` command.

4. The Abstract Environment

This environment is standard LaTeX. You need not enter the word 'Abstract'. It prints the abstract in a smaller font than the body text.

4.1. Keywords

Key words are printed at the end of an abstract. The `\keywords` command should be used within the *abstarct* environment, but after the text of the abstract. For example, `\keywords \LaTeX -- SPACEKAP -- Guide` resulted in the key word line at the beginning of this Guide. Please, use `--` to separate the keywords.

5. Sections, Subsections and Subsubsections

A section starts with the command: `\section{The Title of This Section}`. If the (sub-)section title is too long, you can break it at an appropriate place by using `\protect\\`. (Only the two `\\`'s do not work here).

For the section, (sub)subsection titles you are kindly requested to use Initial Capitals and lowercase only, *not* entirely capitals.

6. Figures and Tables

6.1. Figures

The *figure* environment produces figure space and caption in the text. The numbering of figures is automatic.

```
\begin{figure}
\vspace{Space in mm or cm}
\caption{Caption Text.}
\end{figure}
```

The SPACEKAP style file puts figures wherever possible on the top of the page. This may not always work, in which case you need to be a little creative and consult the LaTeX manual.

6.1.1. Labeling of Figures

It may be convenient to label the figures in case you have to insert another figure later. The following can be done:

```
\begin{figure}
\vspace{}
\caption{}\label{short name}
\end{figure}
```

In the text you could write:
`... can be seen in Fig. \label{short name}.`

6.2. Tables

Tables can be generated in several environments for which we kindly like you to refer to the LaTeX manual.
An example of the convenient *tabular* environment for simple tables follows.

```
\begin{table}[h]
\caption{Caption text.}
\begin{center}
\begin{tabular}{lll}\hline
description 1 & description 2 & description 3\\
\hline\\
Entry1 & Entry 2 & Entry 3\\
Entry1a& Entry2a & Entry3a\\
\hline \end{tabular}
\end{center}
\end{table}
```

This produces:

TABLE I
Caption text.

description 1	description 2	description 3
Entry1	Entry 2	Entry 3
Entry1a	Entry2a	Entry3a

We wish to avoid both using vertical lines between columns and horizontal lines between rows.

6.2.1. Positioning of Tables

The position of a table in the text may not always be appropriate. As with figures, the style file tries to put it wherever possible on top of the page. In some occasions you may wish to put it at one particular spot in the text. The option `[h]` can then be used: `\begin{table}[h]`

It is acknowledged that positioning of figures and tables is by no means straightforward.

7. Acknowledgements

Acknowledgements can be given by using the `\acknowledgements` command. The text for this item can start on the same line as the command.

8. Appendices

An Appendix can be generated by using the `\appendix` command, followed by a `\section` command. This results in a Roman numbered Appendix section.

9. The References

References are inserted in the *thebibliography* environment:
`\begin{thebibliography}{}`. (The second pair of curly brackets is needed.)
For typing the references you may use `\bibitem{}`. In that case you must type the items as they should appear in the list:
`\bibitem{} Author, I.N.: year, 'Title of the Article', {\it journal}{\bf vol. no.}, page.`

9.1. (Author, Year) Referencing

For certain, rather standard, reference expressions there are a few commands available which generate the right output. Caution should be exercised in applying the correct number of items.

`\article{Author, A.N. and Secauth, N.}`
`{year}{journal abbrev.} {vol.no.}{page no.}` This command requires precisely *five* items. If an item is not known, the corresponding curly brackets should still be typed.

If the title of a reference is required or known, one can use the command:
`\reftitle {Author, A. and Secauth, N.}`
`{year} {title of article} {journal abbrev.} {vol.no.} {page no.}` This command requires precisely *six* items or {} combinations. If an item is not known, the corresponding curly brackets should still be typed.

`\book{Author, A.}{year}{title of book} {Publisher: city}{page no.}`
This command requires precisely *five* items. If an item is not known, the corresponding curly brackets should still be typed.

`\inproceedings{Author, A.} {year} {title of contribution} {title of proceedings} {editors} {Publisher:city} {page no.}`
This command is rather complex and needs some attention when typing the re-

quired *seven* entries. If an item is not known, the corresponding curly brackets should still be typed.

9.2. BIBTEX

You may use BIBTEX . Please be sure that you send all the necessary files with the original file, i.e. the `bbl` file.

10. Some Hints

A few typographical conventions should be adhered to in order to improve the readability of your text.

- As you will often be working in the mathmode, composing mathematical expressions, you will have to take action to prevent the subscripts from being printed in italics. For example, `$R_{max}$` produces: R_{max}. Convention is `$R_{\max}$` which gives: $R_{\max}$
- Do not use the mathmode to print italics, but use the `\it` command. $ITALICS$ in mathmode is different from *ITALICS* in italics.
- It is almost forbidden to use `\vspace` or `\hspace`. It often creates unexpected problems.
- Stacking of formulae must be avoided. The fontsizes often become too small to be read comfortably, or are even illegible.

11. Information

The '`spacekap`' style files are available from:

KLUWER ACADEMIC PUBLISHERS
P.O. Box 17, 3300 AA Dordrecht, The Netherlands
email: SCITECH@KAP.NL / fax: 31-78 334254

If you encounter problems, or when you have suggestions for improving the `spacekap` style file or this Guide, we would appreciate hearing from you.